water well technology

water well technology

Field Principles of Exploration Drilling and Development of Ground Water and Other Selected Minerals

Michael D. Campbell
Jay H. Lehr

McGRAW-HILL BOOK COMPANY
New York St. Louis San Francisco Düsseldorf London
Johannesburg Kuala Lumpur Mexico Montreal New Delhi
Panama Rio de Janeiro Singapore Sydney Toronto

Production Editor. . Jerrie H. Campbell
Design Artist . . . . Shaylah Fletcher
Illustrator. . . . . Larry Wood

The work on which this book is based was supported in part by funds provided by the U.S. Department of Interior, as authorized under the Water Resources Research Act of 1964, Public Law 88-379, as amended.

Library of Congress Catalog Card Number
TC 405.C2 622.'37 72-13735
ISBN 0-07-046097
Printed in the United States of America

1112131415161718l920 VBVB 09876

Preface

In this text, we have attempted to bring together all of the salient features of shallow well technology. Ground water is an economic mineral just as are uranium, petroleum, etc. All are commonly recovered by the well structure. The common denominator between the petroleum, mining and ground water technologies is also the well; hence by treating the needs of one industry (the ground water industry) the needs of the other two are also served.

We have attempted to reach a wide audience. The successful well contractor is hungry for detailed knowledge of his industry. The ground water geologist, engineer or hydrologist has never had a single comprehensive source of field information. To be useful to both groups in the ground water industry, we have stressed the basic principles of shallow drilling and development of a natural resource. No attempt has been made to describe the equipment presently in use. With a little assistance, manufacturers are more than ready and very capable of informing the field of the operational and maintenance aspects of their equipment.

We have endeavored to review the present state-of-the-art of the technology both in a practical field-oriented way and in a way that the research scientist will find appropriate for his research purposes. Coverage of certain topics has been limited but literature support has been included to serve as a base for further study.

We have incorporated both the English and metric units of measurement. With use, the metric system is without question far less cumbersome. Appropriate conversion tables have also been included.

During the preparation of this text, numerous organizations and industries contributed in a variety of ways. We have attempted to appropriately acknowledge credit for this and hope that no one has been overlooked as a result of the complicated process of writing, editing, rewriting and reediting the material treated. To a large extent, the numerous authors of the references cited have written this text; we have merely put it together into a useable form. We are sure oversights must exist; however, we also hope they are few and excusable.

We would like to acknowledge the complete cooperation, aid and understanding of H. Garland Hershey, Ph.D. and R. Schneider, U. S. Office of Water Resources Research and R. A. Jensen and J. T. Campbell of OWRR's Water Resources Scientific Information Center. We would also like to make a special mention of acknowledgement for the assistance, cooperation and dedication of the leading authorities in the petroleum, mining and ground water fields, who assisted the authors by reviewing the manuscript at various stages of the investigation, by offering suggestions for improvement and by supplying information on a variety of topics. Without their strong support this work could not have been completed. Of special mention are: J.T. Callahan, Chief, Ground Water Branch, Water Resources Division, U. S. Geological Survey, Washington, D. C.; R. K. Blankennagel, E. Shuter, W. E. Teasdale, W. S. Keys and L. M. MacCary, U. S. Geological Survey, Denver; G.R. Gray, Ph.D., and F. S. Young, Ph.D., N. L. Industries, Houston; Prof. J. C. Warman, Auburn University; Prof. R. D. Singer, and W. C. Walton, Ph.D., University of Minnesota; W. H. Walker and J. P. Gibbs, Illinois State Water Survey, Urbana; T. P. Ahrens and H. H. Ham, U. S. Bureau of Reclamation, Denver; W. C. Maurer, Ph.D., Esso Production Research Co., Houston; R. Alger, Schlumberger, Houston; J. S. Fryberger, Engineering Enterprises, Norman, Oklahoma; J. Mogg and R. L. Schreurs, Johnson Div., Union Oil Products, St. Paul; R. Moss, Los Angeles; L. Koenig, Ph.D., San Antonio; H. Guyod, Gerhardt-Owens, Houston; D. W. Emersen, Ph.D., University of Sydney; J. Johnson and V. Wilson, Australian Ground Water Consultants, Pty. Ltd., Sydney; E. H. Zander, United Nations Development Program, Sudan; R. Voight, Ph. D., Rheinsche Braunkohlenwerke, Aktiengesellschaft, Köln, W. Germany; R. B. Heater, NWWA Certification Committee; B. Cortell, NWWA Research Committee (1971).

A special mention of acknowledgement is extended to Mr. Garmon McCall, NWWA Past President, (1971) for his support during the conceptual stages of this effort; to Mr. Howard White, NWWA President, (1972) for his encouragement during the laborious period of production, and to Mr. Ira L. Goodwin, NWWA President-elect for his dedication to continuing and expanding the research and development efforts of the ground water industry.

This text is a reflection of the effort being made by the entire staff of the National Water Well Association Research Facility to develop a more thorough understanding of the world's ground water resources. The Research Facility, under grants from a variety of federal agencies is helping to expand our knowledge of ground water science. We hope the reader finds this text a valuable product of our research.

We consider that this edition only lays the foundation for up-dated future editions which will become necessary in the decades ahead. Serving as a base for future revisions, we welcome correspondence concerning errors of fact or interpretations, omissions, suggestions for improvement or criticisms of the text. Indeed, if we were to begin again, we would certainly make a great many changes. We are informed, however, that they who demand perfection never complete a text such as this. Perfection does not exist here for we have finished.

Michael D. Campbell
Jay H. Lehr

To our children, David and Julianne and Leslie and Tracy, in the hope their world will find the key to environmental protection with industrial development.

Foreword

"Water, water everywhere, nor any drop to drink" comes not only from the ancient mariner, but also from the contemporary one. For in looking down at the earth from the "bridge" of a spacecraft in orbit,one has the impression that nearly the whole globe is water — and definitely inaccessible! But there are, in reality, two vast oceans — the obvious universal ocean one can see draping the globe, and the unseen ground water resources filling the porous crust of mother earth. This bit of data is not new — but the academic nature of that knowledge has been transformed into a physical visceral reality for a few who have orbited the earth for days on end.

In a spacecraft circling the globe, when you're not busy with navigation, experiments, or housekeeping, it is great fun to drag out the map, set the orbit trace through your present position, and look ahead to see what is coming up. "Hey, we'll be going right over Cairo, troops," or "Acapulco coming up just south of track in 10 minutes." And then into the night, the earth's shadow, the lightning flashes of thunderstorms over equatorial Africa, the faint airglow in its nocturnal evolution defining the night horizon, and dawn! Rapid, majestic dawn; and below, almost invariably the western Pacific. The Pacific and the long wait. For even at 17,000 mph, it seems forever before that vast blue expanse of water surrenders to the West Coast of North America. The human eye can see the rivers of the continents pour back into the ocean from where they came. The space ship's infrared eyes can likewise photograph the flow of cool underground waters from their coastal aquifers welling up beneath the ocean depth.

Then the fun of skipping across the states. City to city, following the interstate highways, marveling at the white rockies disappearing into the northern horizon. All these different cloud patterns over the ocean — all apparently independent , sometimes arranged in groups of honeycomb cells with lines of ocean between them, like an onion skin through a microscope. And sometimes long parallel lines of clouds, streaking along wind lines from nowhere to nowhere. Why? All that sunlight falling on the scene, heating things at different rates. We know there are tremendous exchanges of energy going on down there — between the ocean and the air and, in turn, between the air and the land. But what specific processes are involved and what factors control the energy flow across the boundaries? Can we predict what sea state will result in four or five days as a result of our observations today? Can we measure the depth and thickness of our underground reservoirs destined to provide the earth's water supplies for centuries to come? Not yet! But with finer measurements, better "eyes," infrared sensitive ones, microwave sensitive ones — maybe.

And what about the land? If the ocean in its dance with the atmosphere is quietly mysterious about this relationship, the land puts out a veritable cacophony of intermingled data. The checkerboard of field and meadow, brown and green to the eye, releases a wealth of information to the trained camera or infrared detector: the presence of significant shallow ground water supplies made evident by patches of lush vegetation amidst only mediocre crop land; the stress of wheat with too little water, or a pine forest infested by bark beetles or citrus trees under attack by Brown Soft Scale — all broadcast observable symptoms. These are the eyes we need for diagnosis.

And water too — how much snow is trapped up there in those mountains? How deep is it, and what is its density? How fast is it melting? What significance does it have to us? Inaccessible by foot and shrouded by clouds, the secret is well protected; except that microwave radiation doesn't honor the overcast. It passes through, oblivious to the cloud's existence and bounces off the snow, coded now with valuable data. But we don't yet know how to fully interpret the data from this probe. One day we may, and combined with soil moisture information and meteorological data, we may be able to deduce the total input of water a mile or more into the earth where the availability of ground water is limited by the denseness of the earth. This same ground water may be drilled by either a rock bit or perhaps one day soon, by laser. And if we're smart enough to do that, we should be able to construct a model which would then allow us to follow that water through its cycle back to the ocean and atmosphere. But what a lot of work ahead! New eyes to see and larger and more flexible brains (computers) may yet let us better manage and control our actions to harmonize with the environment on which we depend.

Interdependence — a key word. As soon as one grapples with problems on a global scale (whether hydrology, agriculture, oceanography, demography, meteorology, etc.), one is struck by the interdependence of these disciplines. If the boundaries are narrow enough, one can isolate the subject of interest and approach a single academic, pure, variable problem. But we now circle the globe in ninety minutes, and we cross all boundaries, real and man-made. And as this descriptive fact becomes a repetitious experience, the old human tendencies toward isolation and separation become inadequate for the situation. This is true both technically and humanistically.

Not that specialization is outdated. Far from it. But specialization in isolation is. What the oceanographer is discovering on his frontier is of interest to the geologist. And the ground water geologist is finding that the precious resource he deals with can be better managed if he takes advantage of the capabilities of the latest drilling and development technology, techniques and knowledge being born the world over in related disciplines such as gas and oil development, mineral exploration, and mining engineering. Interestingly enough, these interdependencies have always existed; but we have preferred, for the most part, to ignore them. But the combined influences of expanding communications, the inexorable growth in the world population, and the realization of limits to world resources force us to think further and further ahead. To do this effectively, we must better understand the long term consequences of the actions we take today. This means improved understanding of the dynamics of our world and the technological relationship of one field of science to another. This text for the very first time attempts to bridge multidisciplinary gaps and bring to a common interface all known technology of well construction as it may be related to the recovery of precious underground water supplies as well as other subsurface minerals.

What is required, ultimately, is a new attitude. An attitude which reflects the perspective forced on man by facing his world from a quarter of a million miles away. A small brilliant blue and white jewel suspended in a very large, very black, very neutral universe. Of all the elements composing this cosmic scene, only man — the dominant life form on this small planet — cares. Now, slowly, the awareness dawning that in spite of the stupendous flows of energy and cataclysmic events in and among the stars, man is the predominant influence on his own future here on earth. The future is not only heavily committed to outer space but also to inner space below the earth's surface where latent resources must be captured and conserved. So much of the earth's surface has been desecrated, polluted, and misused that if we are to survive, greater care and increased knowledge must be brought to bear on our subsurface resources of which water heads the list. Failure in this last arena of the struggle to live with nature will surely reduce the quality of our life to an intolerable level; thus, the importance of this book which addresses itself squarely to the increased but orderly development of our underground water resources cannot be stressed too highly.

Some of the future centuries may well be written in the stars but most of the immediate future is, rest assured, written within our own planet. I am sure that the future will welcome us; I am also sure, alternatively, that we will not be missed. Perhaps Archibald MacLeish painted the most vivid picture of our situation in his essay for the *New York Times*, December 25, 1968. People were still reflecting on the Christmas Eve message from Apollo 8 as it circled the moon, when they read:

> "To see the earth as it truly is, small and blue and beautiful in that eternal silence where it floats, is to see ourselves as riders on the earth together brothers on that bright loveliness in the eternal cold — brothers who know now they are truly brothers."

Russell L. Schweickart
Astronaut, Apollo 9
National Aeronautics
& Space Administration
Houston, Texas

Contents

Chapter 1 Introduction 1

Purpose & Objectives 2
Approach 4
Definitions 7

Chapter 2 Ground Water Pollution 11

Pollution Potential 12
Well Structure 18
Oil Wells 23
Petroleum Exploration & Production 23
Natural Gas Exploration & Production 24
Oil Field Brines 25
Assimilation of Technology: A Finite Solution . 25

Chapter 3 Rock Drillability: A Review 29

Rock Properties 30
Laboratory Testing 32
Field Testing 32

Indirect Measurement: A New Approach . . . 37
Summary 39

Chapter 4 Cable - Tool Drilling System 41

Early History 42
Operations 42
Drilling Rate and Capacity 45
Samples 46
Field Use 46
Test Drilling 48

Chapter 5 Rotary Drilling System 49

Early History 50
Test Drilling 50
Drilling Fluids 51
Hole Problems 105
Research and Development 141

Chapter 6 Variations of Common Drilling Systems 121

Air-Rotary Drilling 121
Air-Percussion Rotary Drilling 132
Reverse-Circulation Rotary Drilling 136
Special Application Drilling 138

Chapter 7 New Drilling Systems 143

Turbine Drilling 143
Electro-Drilling 153

Chapter 8 Future Drilling Systems 155

Development 156
Research 157
Economic Potential 163

Chapter 9 Formation Identification and Evaluation 165

Borehole Geophysics 166
Novel Borehole Geophysics 209

Formation Samples (Fluid) 211
Formation Samples (Lithology) 216

Chapter 10 Well Hydraulics 223

Early Development 223
Well Loss 225
Optimization of Specific Capacity 226
Selected Design Criteria 228
Petroleum Technology: Well Hydraulics 237

Chapter 11 Well Design and Yield 239

Design Characteristics 239
Sedimentary Formations 242
Igneous and Metamorphic Formations 245
Alternative Well Design 251

Chapter 12 Well Construction Operations 257

Gravel Packing 257
Well Casing 265
Drill Pipe 281
Fishing 281
Cementing 288
Pitless Adapters 302
Plumbness and Alignment 302
Sterilization 303

Chapter 13 Well Efficiency and Maintenance 309

Efficiency 310
Physical Conditions 327
Corrosion and Incrustation 329
Stimulation 351
Abandonment 374

Chapter 14 Well Cost Analysis 377

Sedimentary Formations 377
Igneous and Metamorphic Formations 396

Chapter 15 Summation and Outlook 401

Drilling Techology 402
Rotary Drilling 402
Drillability 403
Drilling Economics 403
Other Drilling Methods 405
Novel Drilling Methods 405
Drilling Economics and Future Developments . . 406
Completion and Development Techology . . . 406
Formation Evaluation Methods 407
Well Casing 410
Well Cementing 410
Well Efficiency 411
Corrosion/Incrustation Control 412
Well Evaluation Methods 413
Well Stimulation 414
Well Abandoment 415
Ground Water Contamination 415
Conclusions: Research and Development . . . 416

Chapter 16 Annotated Bibliography 421

Chapter 17 Appendix: Miscellaneous Tables, Charts, Formulae and Technical Guides 573

Chapter 18 Selected Glossary of Inter-Industry Technology . . 627

Chapter 19 Subject Index 659

1
Introduction

As society demands larger and larger quantities of fresh, potable water, the utilization of one of its natural resources, ground water, will become vitally important to the individual, to the municipality, and to industry. Today, ground water resources, which constitute more than 95 percent of the world's total fresh water supply, are essentially uncontaminated in contrast to the growing polluted nature of many of its surface water sources. It is clear that greater utilization of ground water will be made in the future since at present only about 20 percent of the water being used in the United States comes from underground. Ground water, as an abundant natural resource, is a relatively inexpensive, drought-proof, fresh water supply.

Since ground water is part of an unseen system which is relatively difficult to study directly and, therefore, difficult for the general public to understand, early protection and conservation of ground water is a necessity if this fresh water source is to remain unpolluted. Likewise unseen is the apparatus which brings underground water to the surface — namely, water wells.[68,170] A detailed understanding of well construction technology is, therefore, of paramount importance to assure the protection of the ground water resource.[73] Furthermore, with the growing emphasis on ground water usage, more efficient and sanitary well construction methods must be sought if future generations are to have potable water in sufficient quantities at reasonable costs. Some ground water pollution has already begun in parts of the United States,[201, 202] but wide-spread contamination caused by faulty well construction can be abated if rationally and technologically sound well construction methods are applied.[345] This can only be accomplished by rigid application of the best available technology and by continuing research and development.

PURPOSE AND OBJECTIVES

Over the past five years, national concern has been focused on the problem of pollution of all our natural resources. A state-of-the-art review of water well construction technology is particularly timely now, during the early stages of this general awareness, while ground water pollution control is being sought by various state and federal agencies and by other workers in the ground water field within the United States. This new public awareness is felt in the areas of communication media, state and federal regulations, and the political arenas. Ground water, one of our most precious and irreplaceable resources, is naturally under scrutiny as well.

The primary purpose of this text is to review the current well construction methods and techniques presently in operation in the petroleum, mining, and ground water industries. The literature of these and related industries is voluminous, widely scattered, and contains various levels of applicability to water well construction technology.

A second objective of the investigation which brought this text into existence is to build an avenue of communication between the various industries and a central data collection organization in order to foster a more rapid transmission of technology in the future, which heretofore has been slow at best. Because the petroleum and mining industries have been highly efficient for many years in obtaining their goals of locating and developing oil and other high value minerals, their technology has reached elevated levels of sophistication. One of the first questions which might logically be raised here concerns the economic practicability of the many oil and mining field techniques treated in this text. It is obvious that the economic impetus of drilling for oil and of developing mining properties is different than that of drilling for water, but the difference in the order of magnitude is relatively small. However, sophistication of operations does not necessarily imply that these operations would be too expensive for the ground water industry because many oil and mining techniques have now been simplified, assimilated, modified, and put into field use by the ground water industry. The results have been technically successful, and the costs, unquestionably justified. As the quest for ground water continues further into the 1970's, the depths of water wells and their numbers will certainly increase.[70, 440] Water wells 1,500 to 4,000 feet in depth are not uncommon today and will become more widespread in the foreseeable future, especially in the irrigation regions of the western United States; the most advanced and efficient technology must be available in the years ahead to assure ground water protection and conservation.[490] State regulations and public health aspects of well construction practices will also come under increased pressure for appropriate revision.[152, 301]

The principles of operation and the concepts behind the petroleum and mining industries' techniques are examined in this text for possible reapplication. The ground water industry will, in the years to come, further modify these principles according to their needs if the techniques find practical use in the field. The industry is becoming well endowed with technical personnel to translate these new methods into practical field use, and with economic support, the rigid application of existing valid techniques can be promoted.

Another important purpose of this investigation is to provide computerized literature data for the U.S. Department of Interior's Office of Water Resources Research (OWRR), Water Resources Scientific Information Center (WRSIC) in Washington, D.C. The data-handling capabilities and the personnel who operate this agency are, without question, largely responsible for the growing technical awareness and the increased communication within the water resources field. Since the overt economic impetus was not inherent in the ground water industry, the U.S. Government has wisely provided the necessary economic and technical guidance.

This text is based on a review of a multitude of publications. After selected publications were abstracted, referenced, and key words chosen, the information was prepared as quick-retrieval, computer data input for inclusion in the data handling system of WRSIC and its twice-monthly publication: *Selected Water Resources Abstracts.*

APPROACH

During the initial literature search for this text, thousands of petroleum, mining, and water resource publications, reports, and technical papers were evaluated for possible applicability to the state-of-the-art of water well construction technology. Numerous trips to various centers of technology in the petroleum and mining fields were made by the senior author. Various academic and industrial institutions and organizations were visited during the early stages of the investigation in the hope of setting up avenues of communication for later information input into the National Water Well Association's Research Facility, the center for this investigation.

The response from the petroleum and mining industries was encouraging. Many of the industrial organizations were found to be most interested in the water well drilling market which is, at present, of relatively minor importance with respect to prime marketing objectives, i.e., oil and mineral exploration.[385]

The technical literature originates from publications in the numerous trade organizations, technical societies, and a few university centers which carry on petroleum and mining research. In order to reduce the time required for the literature search as much as possible, various petroleum information services were used for the initial review. Approximately 120,000 abstracts from the University of Tulsa's *Petroleum Abstracts* were reviewed for publications of applicability to this particular investigation. The initial search was conducted via microfilm cartridges and a reader-printer console. Each abstract selected was integrated into a card-file system for future, detailed review. After considerable discussions with numerous representatives of the major petroleum companies, related service organizations, and equipment manufacturers, it became clear that at least 60 percent of the technical data sought for this part of the investigation could be obtained by relying heavily on the University of Tulsa's Department of Information Services. This information service center, the largest of its kind in the world, began operations during 1961, financially supported by the major petroleum companies. In order to review the field's literature prior to 1961 as well as the assumed forty percent portion of the literature beyond the coverage of the primary information service, other library services and indexes were utilized.

The following major information services and libraries were employed during the course of this investigation:

(1) The University of Tulsa, Department of Information Services Tulsa, Oklahoma
(2) Rice University Houston, Texas
(3) Colorado School of Mines Golden, Colorado
(4) University of Minnesota Minneapolis, Minnesota
(5) Pennsylvania State University University Park, Pennsylvania
(6) Engineering Index
(7) Applied Science & Technology Index
(8) NWWA Research Facility Columbus, Ohio
(9) Ohio State University Columbus, Ohio
(10) Illinois State Water Survey Urbana, Illinois

The literature of the mining field is dispersed more than that of the petroleum field, primarily because a comprehensive information service is not available to the mining and mineral exploration industry. Therefore, technical information was sought directly from the various major academic centers of applied research in mining and mineral exploration within the United States and abroad. These extensive university libraries were searched for publications of possible applicability, and the investigation was given extensive assistance by numerous faculty members of these universities, without whose help the literature search would have been overwhelming. The references found to be of possible interest were cataloged in the master card-file system for later review.

The literature relating directly to water well construction technology is by far the most scattered of the three industries under consideration. Furthermore, a widely accepted central information service does not exist at present, as in the petroleum industry, nor are there academic centers where well construction techniques are treated in any continuing program of applied research. The literature on well construction technology is collected only by a few water well service and equipment companies, by numerous state and federal government agencies, as well as by the NWWA's Research Facility and other professional and industrial organizations. Usually, only selected aspects of well construction technology are of interest to the above companies, agencies, and organizations.

A variety of workers in the United States, Australia, and Europe were associated with this investigation and have contributed to the literature search of the ground water industry for information covering the broad interdisciplinary topic of well construction technology. An extensive literature search was conducted of the major publications found to generally offer pertinent papers or reports on the subject areas. The references and papers of interest were integrated into the master card-file system for later subdivision, review, and evaluation.

After the initial information input system for the petroleum, mining, and ground water industries was developed, a detailed review of approximately 2,000 potentially pertinent publications was undertaken. The master card-file was originally subdivided into general subject areas, within which additional topic subdivisions were later made to facilitate access to specific areas within the technology of well construction. Final selections were based on the degree of relevancy of each paper to the topic areas under consideration. The primary selections were reviewed in detail for the preparation of this text and were prepared for input to WRSIC data banks.

During the early stages of the preparation of this text, numerous decisions had to be made concerning the text's technical approach.[192] After considerable discussions with a host of technical personnel, industrial representatives, and the primary research staff of this investigation, the approach toward the preparation of this text was defined as follows:

(1) All state-of-the-art reviews of current technology should deal with the technology which in the forseeable future will be in general use in the specific industry of interest.

(2) Since the technology of the water well industry has, to an extent, resulted from modifications of principles originating in the petroleum and mining fields during the past 25 or more years, it was deemed advisable to define and then to explore the present technological principles and features of water well construction. Because the water well industry has never had the benefit of the historically close association of the driller and the well-site geologist found in the petroleum and mining industries, this loss of technical supervision must be replaced with a detailed explanation of technology that can be implemented in further advancement of water well construction techniques.[364, 579]

(3) It was deemed advisable to stress the principles of the current and new technology rather than to describe the physical appearance of current and new equipment incorporating such principles; therefore, the text's usefulness will extend beyond the equipment's obsolescence.[385]

(4) It was also deemed advisable to emphasize the results of the literature sources rather than to explore the merits of the numerous approaches without detailed field research which this text will hopefully stimulate.

DEFINITIONS

Water well construction technology has been defined for this investigation to include all technological features which relate to drilling, completing, developing, and maintaining water wells of various capacities for a variety of large and small domestic and industrial purposes in both consolidated and unconsolidated formations. It is assumed that the reader has a basic understanding of the technical field principles and concepts of well construction so as to reduce the extent of introductory material necessary for topic treatment. Furthermore, an annotated bibliography is provided should additional follow-up information be sought for the subjects treated.

To facilitate coverage of the various aspects which constitute well construction technology, the following broad topic subdivisions were incorporated in this text.

Research and Application of Well Drilling Technology
Research and Application of Well Completion/Development Technology
Research and Application of Well Maintenance Technology

Numerous factors must be considered in well construction, all of which bear heavily on the techniques of drilling. Rock characteristics play an important part, for example, in the selection of the most economical and efficient bit. In order to drill a well efficiently, a thorough knowledge of what happens on the bottom of the hole, as well as on the hole wall above is of vital importance. Rock drillability as explored in this text treats the laboratory and field achievements in technology as to the nature of the physical breakdown of rock in response to drilling variables, e.g., rock type, bit design, etc. The relationship of rock drillability and rock characteristics is explored with respect to penetration rate, rotary speed, bit wear, and other related variables.

The drilling fluid system is also examined in detail since it cools the bit, promotes bit longevity, conveys the cuttings away from the bit face to the surface for examination, plugs lost circulation zones, and promotes hole stability. These and other related factors are at present of top priority in drilling research and will, therefore, be treated in detail. In order to consider these new techniques and concepts, however, current drilling methods must also be explored to achieve an understanding of the relationship between the new concepts and the present stage of advancement of well drilling technology. The concepts in rock excavation, or drilling, will also be included to demonstrate new directions the industry may take in the not-too-distant future should a particular method prove to be of special merit.

Drilling either for oil, minerals, or ground water stands as one of the last remaining arts in an industry where virtually all other related operations have been translated into a science. The industry is now passing through a transition period which will elevate the art of drilling and herald the beginning of new drilling methods where drilling variables are more easily and reliably predictable.

The petroleum industry is well on its way toward making the drilling process a matter of controlled variables and predictable results. The mining industry has begun to look to this approach, although the magnitude of this industry's drilling parameters is more diverse and perhaps more difficult to define than those of the petroleum industry (within 2,000 feet or so of the surface, igneous, and metamorphic rock are drilled more frequently than common types of sedimentary rock). The mining industry drills under less severe hole conditions, e.g., less depth, less pressure, etc., but under relatively more severe economic pressure. However, the water well industry must drill under the strongest economic pressure; the industry at present cannot afford millions of dollars in research to control its own partially unique drilling variables. Although with increased technological transfer from the petroleum and mining industries, the water well construction industry will be able to control the variables which the petroleum industry and others have already defined. All variables lie within a system which has been defined for the water well industry but awaits further definition and industrial development.

Well completion or development is also composed of various systematic techniques. These techniques, such as cementing, aquifer stimulation, acidizing, blasting, well screen selection and placement, well logging, etc., are on the whole within the broad category defined herein as well completion. These techniques are interrelated, but should one system not be understood, well completion would not be fully acceptable nor would the well operate at designed efficiency. Any waste in these operations which is not anticipated is due to faulty techniques or to the lack of pertinent technology or perhaps, to a lack of a full consideration of the relationships between the technical systems involved.

A rapid advancement has occurred in recent years in the area of well completion and development. This text will treat in some detail the current techniques and the modifications or new concepts which have come from the petroleum and mining industries that are likely to find application in water well construction in the years ahead.

Water well maintenance is, in many instances, considered well rehabilitation. Maintenance also includes the various features which affect well life. Considerable work has been done by the petroleum industry and others on downhole maintenance procedures, cathodic protection, bacterial and chemical corrosion, etc. Aquifer stimulation is treated in conjunction with other "work-over" techniques.

The water well industry, as well as the petroleum industry, loses many thousands of dollars each year to "corrosion." It affects the well casings, the well screens, and the quality of the ground water produced by the well. Long term well performance depends not only on the design and construction techniques used but also on factors continuing after the well has been completed. The nature of the casing, the well screen, the pump, the aquifer characteristics, and the original ground water quality all play a vital role in determining the longevity of the well. The petroleum and water well industries have begun to deal with these problems. This text will explore the current practices and the new concepts which are in the process of adaptation by the water well industry. The oil industry's knowledge of maintenance problems has increased sharply in recent years. These problems differ considerably from those encountered in drilling operations. Not only have maintenance systems been defined, but many variables have been controlled in the well-maintenance field. Corrosion-incrustation systems, for example, have been examined in detail during the past 15 years, and the technology is available now to effectively reduce well failure due to "corrosion," if the economics merit such considerations.

In many cases, various topics are discussed briefly and then referenced as fully as possible for further study. The major factors of well construction technology are treated, however, in sufficient detail to delineate the techniques and problems as well as the benefits to be derived by the ground water and mineral exploration industries and the limitations involved from close contact and cooperation with the other extractive industries, i.e., petroleum and mining, etc.

The scientific aspects of ground water exploration are placed in proper perspective by considering ground water as a highly valuable mineral (to the public) as are other economic minerals (to company shareholders). All ground water geologists, hydrologists, mineral exploration geologists, oil geologists and engineers, government geologists and engineers, ground water planners, geology and engineering students, as well as drilling contractors and related vocational personnel, will more clearly understand the value of ground water when it is treated as an important natural resource consistent with other members of the family of economic geology.

The authors hope that after reading this text the oil geologist will have a more realistic respect for the upper 2000 feet of the earth's crust, the mineral geologist will be more clearly aware of his practical problems and of new tools available in his search for minerals, the government geologist and water resource planner will have a more realistic basis from which to educate and assist the public, the university researcher will expand his practical experience into the field and its problems, the drilling contractor will more clearly understand the problems of the professional workers and gain new respect for his responsibility in drilling and developing a very important natural resource, and finally, the university student who in the past has never known what to expect when he is asked to "sit on a well," interpret an electric log, or assist in any exploration program whether for ground water or for other minerals dealing with a drilling rig will more fully appreciate his duties.

2
Ground Water Pollution

Most recorded instances of ground water pollution have been fairly localized in extent and are of relatively low magnitude.[422] However, the increasing frequency of pollution occurrences now being recorded indicates that widespread contamination of aquifers may occur soon unless stringent precautionary measures are employed.[381, 382, 383]

In the past, dilution of liquid waste with large quantities of clean surface or ground water was commonly accepted as an adequate solution to pollution. This practice was successful as long as the quantity of pollutants was small in relation to the total volume of uncontaminated water available for dilution. The time is rapidly approaching when this solution will not be possible.[646] The total flow of some streams is already being used and reused several times, and with each use increased mineralization, heat or bacterial accumulation occurs. In many areas the clean water sources formerly used for dilution are now also polluted and in need of treatment before they can be classified as acceptable.

Ground water contamination caused by the disposal of solid waste in pits or land fills is already a very serious problem in some areas.[202] Chemical and thermal pollution caused by liquid waste disposal in pits or surface depressions also is increasing in some of the more heavily developed shallow aquifers. It is imperative that other methods of solid and liquid waste disposal be devised, perfected, and used. Until this is done, rigorous control of waste disposal site selection, operation, and maintenance must be employed to prevent serious contamination of ground water reservoirs.[434]

Domestic and farm water supplies are derived primarily from wells tapping shallow aquifers that have a high contamination potential. It is apparent that localized parts of these aquifers are already being seriously polluted, primarily with liquid waste derived from feedlots or poorly constructed or designed septic tank sewage disposal systems, and that many improperly constructed or poorly located wells in these areas act as conduits for vertical migration of contaminants into ground water aquifers.[340] Unfortunately, water samples from such wells are not analyzed on a regular basis, and the possibility of contamination often is not discovered until a high degree of contamination is present. A large percentage of the presently unaffected domestic and farm ground water supplies is in danger of serious contamination, unless proper disposal facilities for human and farm animal waste are provided and proper well construction practices are followed vigorously.

POLLUTION POTENTIAL

It is clear that a practical method of defining areas of potential ground water contamination is needed. In such areas various well design features can be considered on the basis of a well's possible role in contributing to local ground water pollution. Awareness of the pollution potential could serve to emphasize the necessity of technologically sound well-construction practices.

One such "early warning" method of potential ground water contamination has been explored by Walker.[645] He defines *pollution* as "an impairment of water quality by chemicals, heat, or bacteria to a degree that does not necessarily create an actual public health hazard, but that does adversely affect such waters for normal domestic, farm, municipal, or industrial use. The term *contamination* denotes impairment of water quality by chemical or bacterial pollution to a degree that creates an actual hazard to public health . . . ."

As an example of the above approach, Figures 1 and 2 roughly outline areas in Illinois where there may be danger of aquifer pollution now or in the future. The maps were prepared from the type of information that is typically available in most of the United States and elsewhere in the world. Using bedrock surface (outcrop) maps and surficial geological maps, the figures show areas where unconsolidated and bedrock aquifers *may* be subject to high,

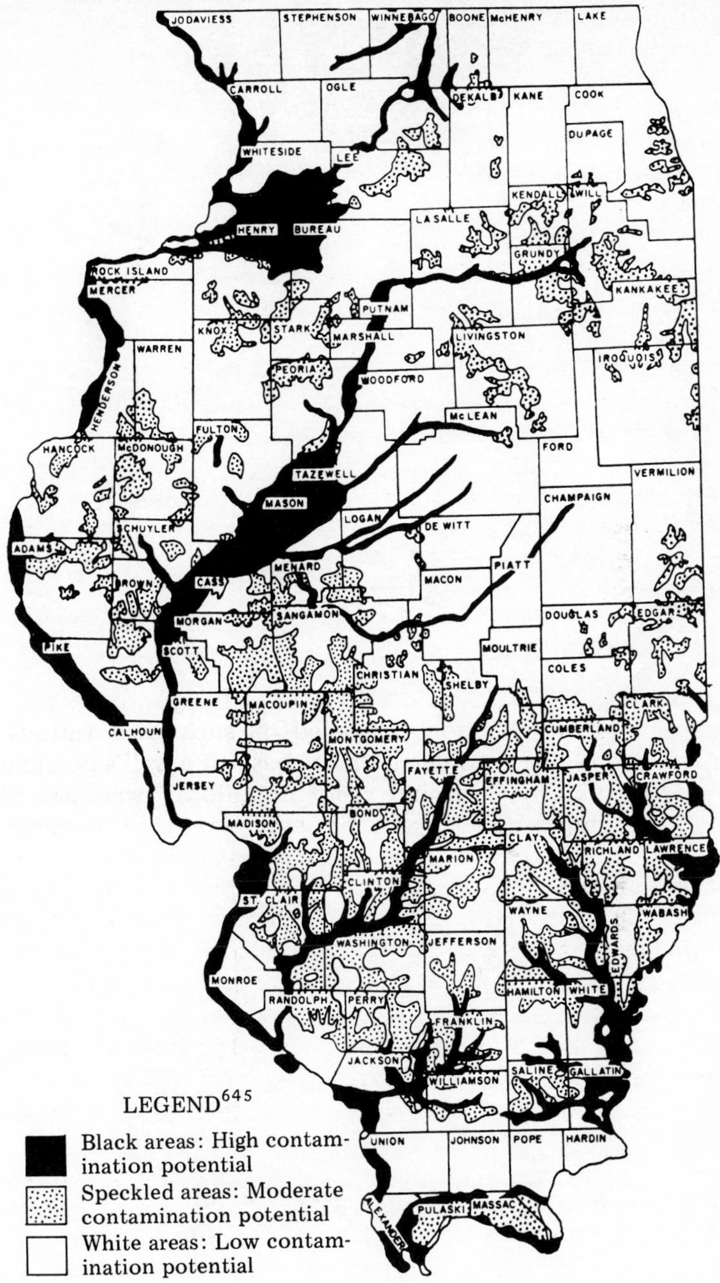

Figure 1[645] Unconsolidated Aquifer Contamination Potential. Surficial sand and gravel deposits are most likely to be contaminated from surface-derived sources.

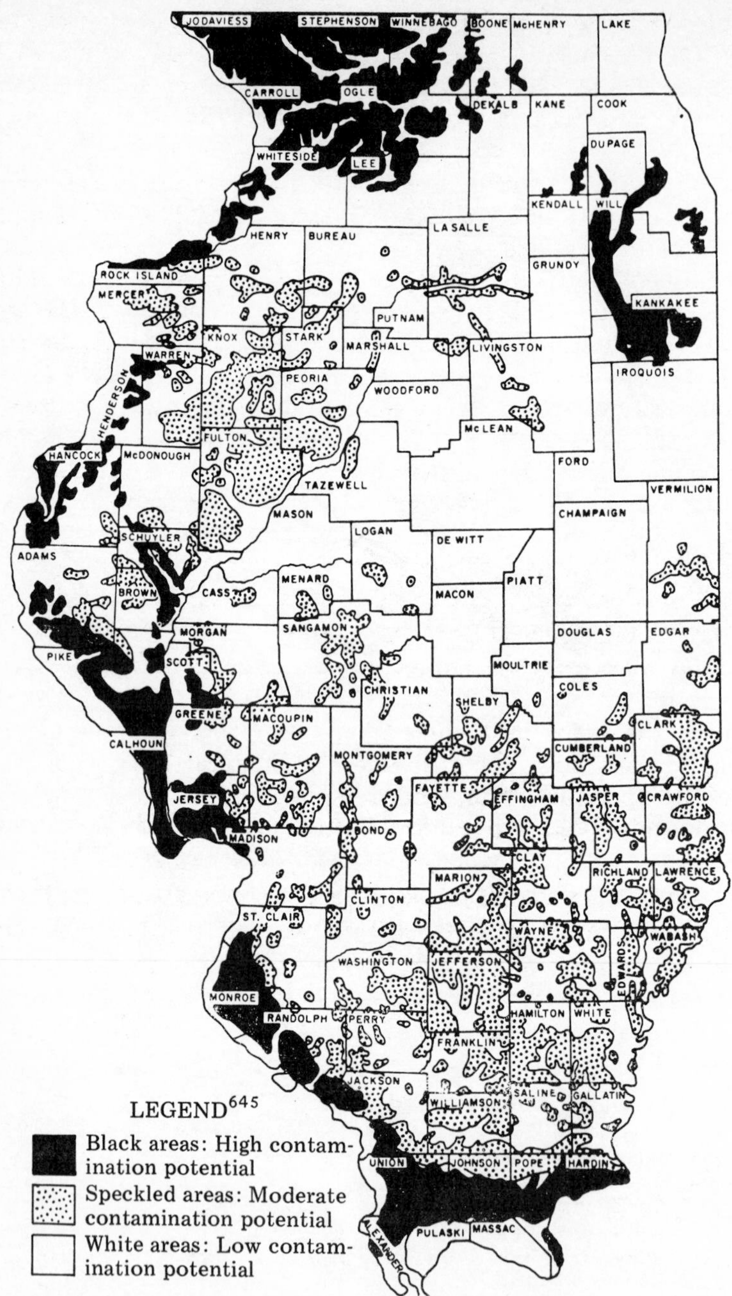

Figure 2 [645] Bedrock Aquifer Contamination Potential. Outcrop areas of creviced dolomite and limestone aquifers have a high contamination potential.

moderate, or low contamination potential. The maps serve as a guide in selecting reasonably safe areas for locating productive wells or waste disposal sites. Of course, only the general parameters can be established, and local field information should accompany any follow-up.

Shallow aquifers that intersect or lie near ground level readily receive, store, and transmit possible contaminants down gradient from points of entry to natural discharge areas or nearby pumping centers. Deeply buried water-bearing units overlain by relatively impermeable shale or clay beds are generally effectively sealed from the surface. These units must receive most of their recharge from outcrop areas of the aquifers.

Deeply buried parts of aquifers may be relatively free of contamination derived from surface sources, providing wells and test holes tapping these units are properly sealed. In the outcrop areas, however, waterborne contamination can directly enter exposed or thinly covered, weathered, or creviced limestone and dolomite beds, faults, etc., through interconnecting systems. Surface derived contaminants gain easy access to shallow sandstone or sand and gravel formations through interconnected uncemented pore spaces.

Outcrop areas of the aquifers are the primary controls used to delineate areas of high, moderate, and low contamination potential shown on the maps in Figures 1 and 2. Other information used include: drillers' logs, records of wells, chemical analyses of water samples, temperature data, and recorded case histories of heat, chemical, and bacterial pollution from the files of the local water survey, public health department, and geological survey.

Unconsolidated aquifers subjected to a high contamination potential (Figure 1) generally are contained within those areas covered with glacial drift or outwash deposits. The surfical aquifers usually consist of very permeable sand and gravel capable of producing from about 100 to more than 500 gpm for properly constructed wells.

The areas of moderate-contamination potential roughly coincide with those parts of Illinois, for example, covered only by glacial deposits. Water-bearing formations in these areas are generally thinner and less permeable than those of glacial deposits, and wells capable of producing in excess of 20 gpm are rare.

Areas having a low contamination potential are either underlain by deeply buried sand and gravel aquifers, isolated from the surface by relatively impermeable clayey deposits, or the unconsolidated materials present are thin and non-waterbearing. In parts of Illinois, for example, properly constructed and cased wells effectively seal deep aquifers from surface-derived contaminants.

Bedrock aquifers subject to a high contamination potential generally are creviced formations, shallow carbonate formations and are most likely to be contamined from surfaced-derived sources (see Figure 2). Areas of moderate contamination potential are underlain by shale, sandstone, and limestone formations. Also, highly mineralized water from deeper buried parts of the aquifers may be obtained from shallow wells. Low contamination potential bedrock aquifers normally are isolated from land surface by thick glacial till or bedrock shale formations.

The above approach, although developed for the state of Illinois, is directly applicable to other regions of the United States, in addition to areas outside the United States where the necessary hydrogeological data are available. The approach is especially applicable in areas of relatively high rural population density or in areas with a high projected agricultural or industrial expansion rate.

Ground water in urban and other areas of concentrated usage is subject to contamination from a number of sources.[294] Walker[645] and Deutsch[202] cite typical examples of "natural" contamination/pollution. Figure 3 illustrates a commonly encountered case of ground water quality deterioration caused by leachates from the city dump and the probable manner of entry and movement of chemical contamination from this source to the municipal well field. Interception and diversion of contamination by cones of influence of the nearby municipal production wells apparently began almost immediately after pumping started, as evidenced by the progressively increasing total hardness, sulfate, and chloride content of ground water withdrawn from this field.

Contaminants can enter the well as it is being drilled, during its operational life, or following its abandonment.[54] Because of this

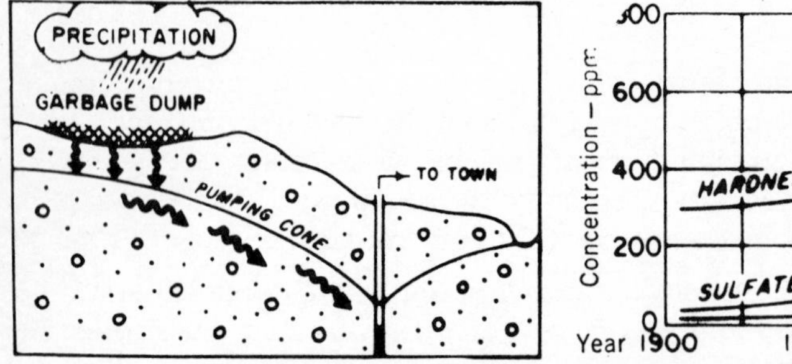

Figure 3[645] Typical ground water quality deterioration from surface waste disposal operation.

complex environment, such contamination is often difficult to rectify; however, Ham[300] summarizes the salient features of potential ground water contamination of water wells and delineates many of the problems encountered.

During the drilling of a well, a hole is opened (and sometimes cased) from the ground surface into the aquifer. Various substances, including water, drilling fluids, muds, chemical additives, hay, gasoline, etc., are, on occasion, introduced into the well to facilitate drilling and development. The volume of these materials is generally small compared to the volume of the aquifer. In essence, they are "one-shot" additions. If the well is promptly completed and test-pumped, most of the substances added during drilling are drawn back into the well and removed. However, under atypical conditions (including the presence of cavernous rock) highly permeable gravels or other conduit-like aquifers, such drilling materials may move too far from the well to be recovered and may contaminate nearby wells.

However, a special problem present during drilling is the danger of accidental entrance of flood waters or concentrated chemical/radioactive fluids into the well. Any uncompleted or unsealed well in proximity of a surface stream or stored toxic fluids is a potential avenue for contamination. A comprehensive review of the practical aspects of ground water pollution has been published which explores the numerous features of potential pollution of the ground water resource.[86]

Possibly of greater potential consequence is the introduction of biological agents into the aquifer via drilling muds or other materials. The presence of sulphate-reducing and other similar non-disease-causing bacteria in wells can be attributed to contamination of drilling fluids, muds, etc., introduced during the drilling process. Little information is available on the effects of this procedure; however, the method of bacterial introduction may well be an important contributor to later contamination of aquifers via the growth of specific types of corrosion/incrustation-causing bacteria in wells coupled with favorable local ground water chemistry, temperature, etc.[99,146,492]

The presence and severity of both chemical and biological contamination of the well during its operational life and post-abandonment period are closely related to well design, location, and construction. With respect to bacterial contamination causing human diseases, it has been established that contaminated surface water has entered a well either because of faulty well construction methods by transmission from some distant point through the aquifer (especially in creviced rock), or from sewage or other wastes that have been improperly disposed.[362,481,527,533] Proper well location based on a firm knowledge of the local geological condition can minimize the potential problem.[434]

WELL STRUCTURE

The well structure, however, offers another avenue for contamination. Any break or other opening in the casing or between the casing and the pump base or seal is a potential source of contamination (see Figure 4-A). Also, the discharge system offers opportunity for reversal of flow (see Figure 4-B).

Aside from the well structure, the disturbed zone immediately surrounding the casing frequently offers a passage for contaminants (see Figure 4-C). The presence of an improperly constructed gravel pack may also lead to contamination because of the necessity for maintaining a conduit from the surface into the well for installation

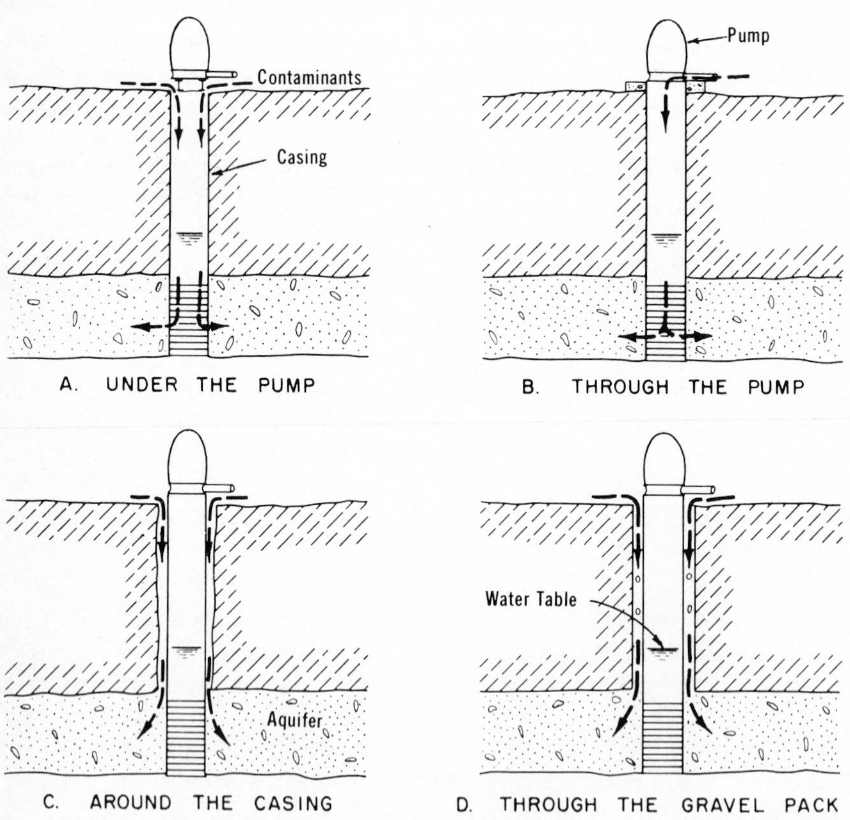

Figure 4 [300] Entrance of contaminants.

of the pack. In some examples of poor construction, the pack is continuous from the bottom to the surface without any method of sealing at the surface (see Figure 4-D).

The more advanced of the techniques presently used for preventing contaminants from entering around the casing and through the gravel pack are adequate under normal conditions. However, changing physical conditions of the well which often result from improper design, construction, operation, or lack of maintenance, in time tend to negate protective techniques no matter how sound in principle. These include:

(1) Subsidence, due to sand pumping, resulting in surface-grade reversals, destruction of surface protection, and reduction in the effectiveness of the cement-grout seal (see Figure 5).
(2) Desiccation or other factors causing shrinkage, cracking, or other alteration of grout or cementing material.
(3) Breaks or leaks in discharge pipes, leading to the failure of sanitary protection apparatus.

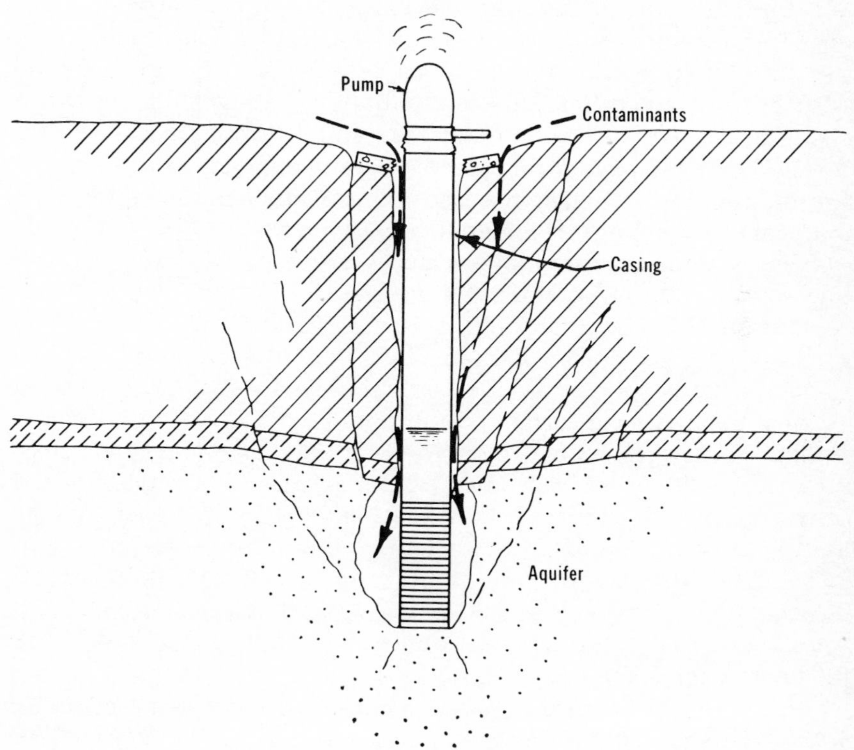

Figure 5[300] **Entrance of contaminants resulting from subsidence.**

Jones,[340] in a case study on well construction and water quality, reports that wells lacking adequate sanitary protection from improper well construction methods serve as uncontrolled ground water recharge points. He further states that, with the entrance of contaminated surface water into a well, the ground water quality deteriorates. This deteriorated ground water is often equated, by random water well samples, with ground water quality when, in fact, the contamination resulted from the entering surface water. Poorly constructed wells may ruin high quality ground water. This becomes critical when a well receiving dissolved solids from livestock wastes, agricultural chemicals, or fecal organisms from surface drainage becomes a source of ground water. The recent study by Jones [340] substantiates this possibility and suggests that improper sanitary protection of wells incorporating inadequate well-construction methods requires an early solution, especially with respect to locating such well defects *after* well construction has been completed.

When a well penetrates a single, relatively homogenous aquifer under water-table conditions, the possibility of contamination originating from anywhere but surface sources is remote. However, the penetration of two or more aquifers each under a different hydraulic head presents an entirely different situation. The well bore or gravel pack can act as a vertical conduit for natural flow from one aquifer to another. One aquifer may become contaiminated by water from another aquifer, although the two are separated by an appreciable thickness of impermeable material (see Figure 6).

When chemically unsuitable water is encountered during drilling, the contributing zone normally is sealed off with casing, liner, cement grout, or is totally plugged back. However, the presence of bacterial or other organic contaminants is not as readily detectable.

A potentially serious situation exists when a well is cased through a contaminated near-surface aquifer and completed in a deeper aquifer. Under favorable head differentials, contaminated near-surface ground water can enter any opening in the casing and be conveyed into the aquifer in use (see Figure 7-A). The opening may be a split seam, improper weld, joint failure, corrosion pitting, or inadequate seal below the casing shoe. The extension of an inadequately protected gravel pack upward into a near-surface aquifer can lead to downward drainage of contaminated water into the lower-lying aquifer (see Figure 7-B).

Disuse of a water well creates conditions that often lead to an increased potential for contamination and accelerates deterioration as in other metallic structures. Upon abandonment, the pump should

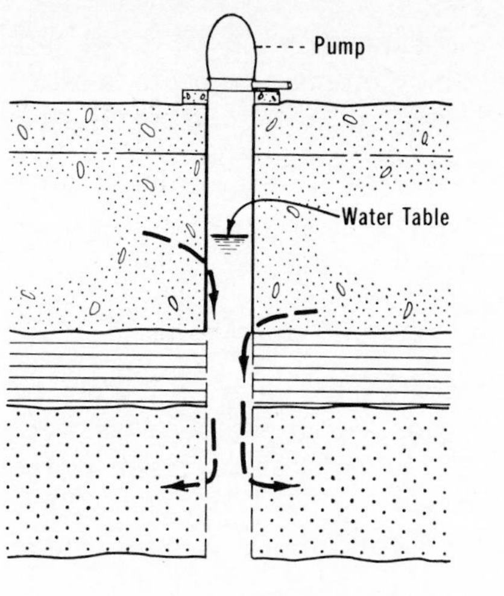

A. THROUGH THE CASING

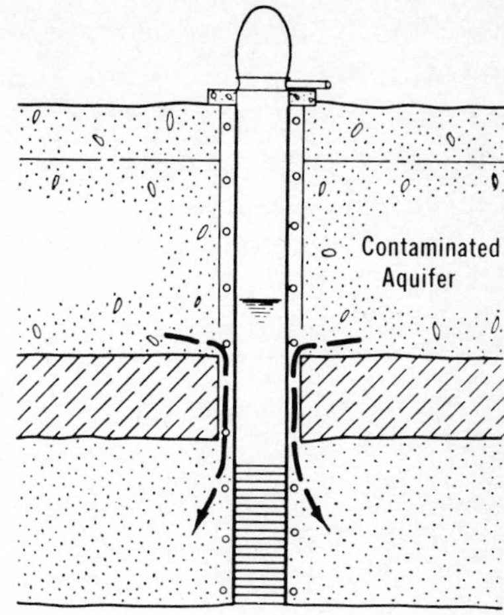

B. INTO THE GRAVEL PACK

Figure 6 [300] Entrance of contaminants.

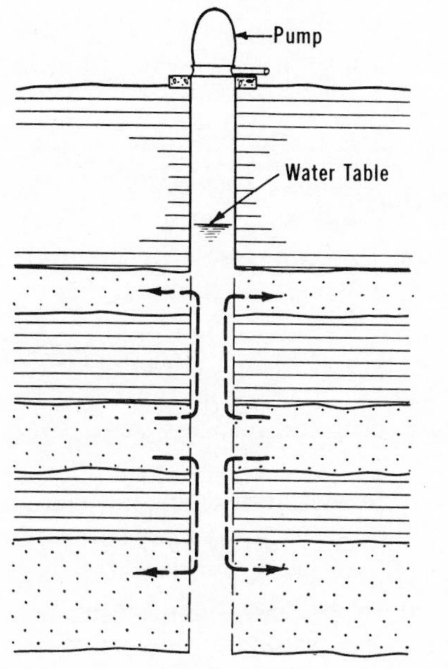

A. BY VERTICAL FLOW

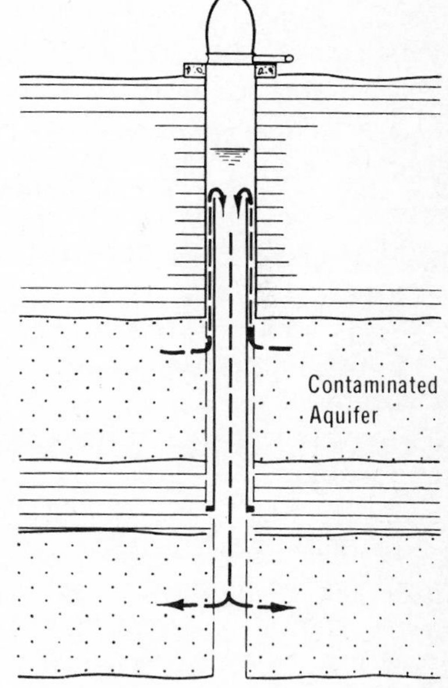

B. BY VERTICAL FLOW BY-PASSING A LINER

Figure 7 [300] Entrance of contaminants.

be removed and an attempt made to insure that the well does not constitute a public hazard. Common abandonment procedures often are less than ideal. The contamination present during the operational life of a well usually will continue after abandonment unless preventive measures are taken.

An abandoned well is a convenient receptacle for disposal of wastes, or it may become a haven for vermin. There are numerous cases of building activities uncovering old, unprotected, and unrecorded wells.

The presence of unrecorded wells abandoned in areas subject to natural flooding or in permanently flooded areas, such as new reservoirs, introduces unnoticed surface water into fresh water aquifers. The addition of reservoir operating head can cause high inflows with the attendant danger of continuous ground water contamination.

The surface contamination hazards associated with abandoned wells usually are obvious and can be rectified without great difficulty. However, subsurface contamination is more insidious, difficult to detect, and often expensive to control especially if the well has undergone deterioration. The corrosion of well casing and liners, originally installed to seal off contaminated aquifers, constitutes a major hazard if head differentials in the various aquifers are favorable for interflow between aquifers.

Of practical concern are wells that penetrate saline or other similar aquifers having high artesian heads. Normally, such aquifers are sealed off to permit the production of usable water from other aquifers. After abandonment, deterioration can result in the well becoming a subsurface hydraulic conduit which continuously adds contaminants to a fresh water aquifer.

Concerning abandonment of wells, it is generally accepted that restoration of the controlling geological conditions that existed before the well was drilled or constructed should be attempted as far as possible.

Ham[300] and Jones[340] state that the potential for serious water well-induced contamination of ground water is always present and will become intensified with increased use of ground water and concentration of facilities, especially in areas with high-population growth rates. Once recognized, however, the problem is alleviated by appropriate statutory and administrative measures and by the development of new and improved materials, equipment, and techniques. Appropriate monitoring tools and techniques are also needed to assure the sanitary protection of the water well and the valuable ground water resource.

OIL WELLS

All wells, regardless of their use, pose a potential threat to the ground water environment. The part that oil and gas wells play in local ground water contamination has been well documented, but obviously neither oil/gas wells nor water wells will be prohibited in the foreseeable future since contamination is unnecessary and a result of inadequate well construction techniques and practices. However, unusual or poorly understood local geology and hydrology also contribute heavily to many types of ground water contamination.

Studies of well blowouts and possible development of communication between a fresh water aquifer and an oil-bearing sand have been made as have studies of possible ground water contamination related to poor oil production practices. [554, 638] Brines produced with oil and gas are also known to contribute to ground water pollution, [184, 267] and a universally satisfactory method for their disposal has not been found to date.[611, 600] Some brines, however, contain valuable minerals that are economically recoverable, and treatment or disposal of such brines should be coordinated with mineral recovery processes whenever possible.[178]

Collins [179] cites several publications on oil field brine disposal by subsurface injection into permeable strata and describes gathering systems, pumps, treatment methods, and injection well construction techniques. Two recent symposiums, i.e., National Ground Water Quality Symposium [481] and the Underground Waste Management and Environmental Implications Symposium,[9] have dealt with major water quality problems and state-of-the-art of waste disposal techniques.

Disposal methods have been blamed as the possible cause of some earthquakes, and if a natural disaster such as an earthquake occurs, new faults or fractures in subsurface strata may provide an avenue between the strata containing the waste and the fresh water aquifer which obviously would be potentially disastrous to the potential usefulness of any ground water system involved.

PETROLEUM EXPLORATION AND PRODUCTION

When oil or gas wells are drilled into an abnormally high fluid-pressure environment, there is always the possibility of a blowout unless elaborate precautions are taken and proper drilling fluids are used. [179] This situation can develop, for example, if degradation or sloughing off around the casing in a high pressure zone occurs allowing the pressurized hydrocarbons to escape along the outside of the casing to an upper zone (see Figure 8).

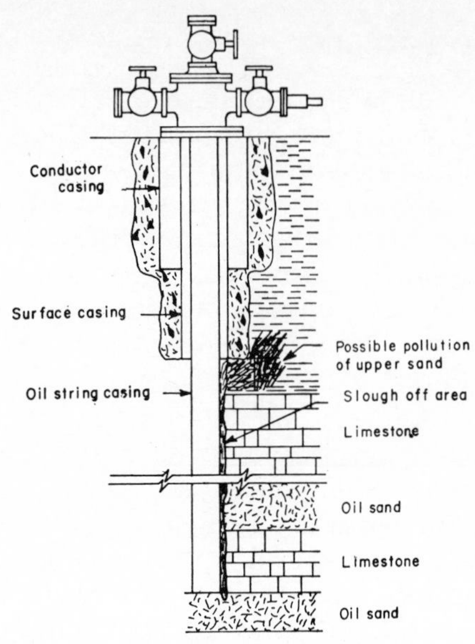

Figure 8 [179] Manner in which heaving shales or incompetent zones can slough off and allow a lower zone to communicate with an upper zone.

Collins[179] stresses that most states have laws requiring the setting of surface casing to protect the fresh water aquifers from invasion by brines and hydrocarbons from deeper horizons. A minimum of two strings of casing—the surface casing and the oil-string casing—are usually used and are generally adequate. Additional strings of casing may be employed if heaving shales are found while drilling progresses, if abnormal pressures are encountered, or if a zone of lost circulation is encountered. Each additional string of casing, however, requires more capital and increases the cost of the well.

If appropriate precautions are not taken in planning, drilling, and completing an oil or gas well, serious consequences can occur. For example, during drilling operations, a well may blow out if adequate mud pressure is not maintained. Such a situation may develop if the mud line is accidentally broken or if the well casing is not properly cemented to competent zones. Figure 9 illustrates what may well occur if fluid from a high pressure well escapes into an incompetent zone and develops communication between a lower hydrocarbon-bearing horizon and an upper aquifer.

NATURAL GAS EXPLORATION AND PRODUCTION

Blowouts of natural gas wells have also contributed to ground water pollution and are especially serious if the natural gas contains appreciable quantities of hydrogen sulphide. Many gas wells contain enough hydrogen sulphide to contaminate fresh water upon contact. Such contact may develop if well construction methods are faulty and communication between the gas zone and an upper freshwater

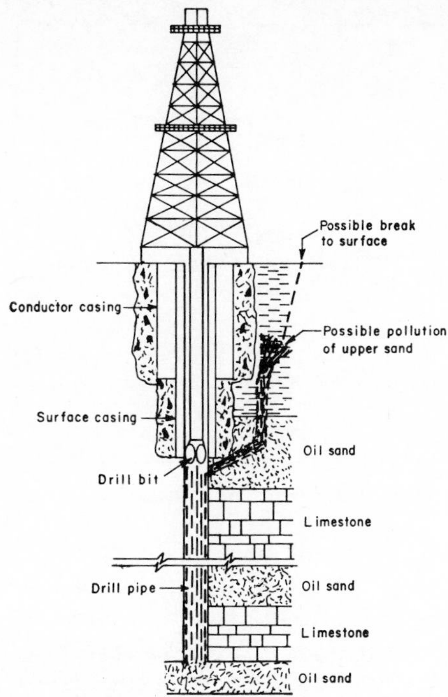

Figure 9[179] Probable manner whereby a well blowout can develop communication between an upper sand and a lower sand.

zone occurs. Brines associated with hydrogen-sulphide-bearing gas zones also contain appreciable quantities of the sulphide. Keech[345] cites a dramatic case history of widespread ground water contamination by a subsurface, natural gas blowout.

OIL FIELD BRINES

Waters associated with petroleum in subsurface formations usually contain many dissolved ions. Contamination of shallow ground water by the use of "evaporation pits" for the disposal of brines has been widespread. Such pits, although outlawed in some states, are still too often found in use because of the lack of energetic local enforcement. Salt water disposal through faulty disposal wells and increasing water flooding activities both pressurize lower formations and can result in the movement of brine up poorly plugged abandoned oil wells. Of course, this causes contamination of fresh water aquifers. Furthermore, sources of such contamination are difficult to trace.[660]

The stability of petroleum-associated brine is related to the constituents dissolved in it, to the chemical composition of the surrounding rocks and minerals, to the temperature, the pressure, and the composition of any gases in contact with the brine. Lehr[384] describes some of the consequences of local aquifer contamination by improper brine disposal methods. Commonly, evaporation pits facilitate both vertical and horizontal migration of oil field brines to the ground water reservoir, as a result of a surface-derived source of shallow aquifer contamination (see Figure 10).

ASSIMILATION OF TECHNOLOGY: A FINITE SOLUTION

The ground water industry must serve two masters or must seek two goals concurrently if it is to flourish in the years ahead. It must,

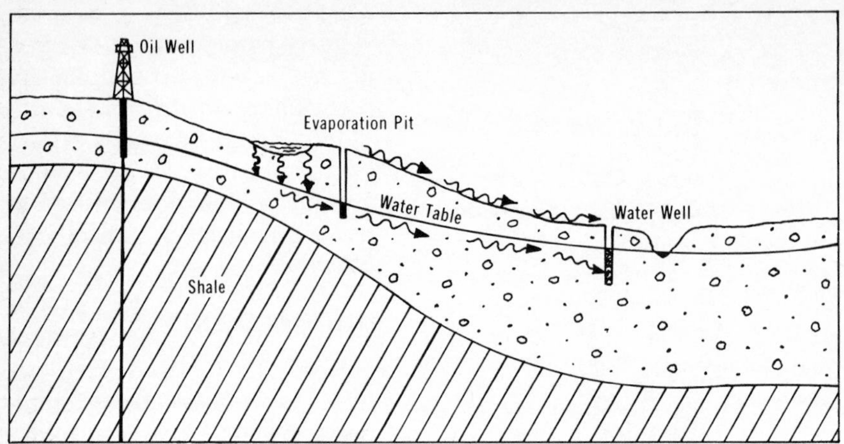

Figure 10[645] Contamination of ground water from brine disposal.

of course, make a reasonable profit, but it must also act as the day-to-day guardian of the ground water resource. This can be accomplished by incorporating the most up-to-date techniques of well construction, completion, and maintenance. The achievement of both goals, however, is not mutually exclusive, as any successful well contractor can testify. There is no tenable excuse for blatantly poor well construction or for inadequate profit in a healthy economy. Improved technology designed to increase profit also increases the quality of well construction and serves as the final answer to water well-induced contamination of the ground water resource.

Monitoring equipment is needed in the petroleum industry to continuously assure that all materials used in the construction of oil or gas wells remain as designed throughout the years of operation. Corrosion as a major cause of eventual ground water contamination is of serious concern to the petroleum industry as well as to the ground water industry. Furthermore, oil/gas well abandonment techniques bear heavily on potentially widespread ground water pollution. These techniques should be under continuous technological development, the most effective of which should be used consistently in the field as a matter for the common good, not as a matter of economics. Petroleum and potable water are not of equal importance to society's needs for there are energy-source substitutes for petroleum and related oil/gas products, e.g., uranium, geothermal[97, 605] and solar energy, etc.; but there is no known alternative for potable water. The development of petroleum and ground water is competitive only to the extent that the latter must be protected more jealously than the former. Both can be developed simultaneously if technology can serve to protect one from the other. Too often, strong industrial influences promote unfortunate compromises between the economic development of one resource over another.

Water well construction technology has developed considerably in the past 25 years. The impetus for such development has come from the inherently more wealthy oil and mining industries. New techniques and concepts have, for many years, been adapted by the water well construction industry after practical applications were realized by numerous service companies, equipment manufacturers, and technical personnel in the petroleum, mining, and ground water industries.[265]

The ground water industry, in recent years, has returned this assistance of the petroleum and mining industries by developing some of its own concepts and techniques which have found subsequent use in the petroleum and mining fields. Each industry is seeking more economical ways of developing the natural resources of the earth, e.g., petroleum, ground water (including geothermal resources), and other minerals of value which today's world must have in greater and greater quantities.

The U. S. Environment Protection Agency (EPA) in cooperation with the U. S. Office of Water Resources Research has recently published a three-part selected annotated bibliography on subsurface water pollution, i.e., Part I: Subsurface Waste Injection; Part II: Saline Water Intrusion; and Part III: Percolation from Subsurface Sources. These publications and others to be released soon represent the first major step in the wide-spread dissemination of research reports and studies of ground water pollution and contamination control efforts.

The assignment of responsibility for ground water pollution control is difficult because so many segments of society contribute to pollution. However, the primary responsibility can and should indeed be assigned to that segment of the industry contributing the greatest impact on pollution. The industries drilling for oil/gas, minerals, or ground water, who open the ground water reservoir to potential pollution, must design well construction technology and operations so that the risk of future pollution to the ground water reservoir is all but eliminated. To accomplish this, well construction should be based on pollution control factors rather than on economic factors which too often overshadow pollution control efforts.

The complaint is often heard that the well constructed in strict accord with the available technology is too expensive. What is too often not understood or not widely accepted is that if the efficiency of the various well construction operations were increased only slightly over present levels, the monetary savings derived would pay for well construction of the highest quality capable of meeting the strictest of well standards and specifications. So it behooves the drilling industry to look to technology for an answer to its very real economic problems. Strict standards and specifications are not designed to impoverish the drilling industry, only to assure the protection of one of the nation's most valuable natural resources.[674]

To accomplish the necessary increase in efficiency of many well construction operations, a more detailed appreciation of the principles of available technology must be realized if an economic edge on pollution control is to be accomplished in the near future. In the following chapters, various aspects of well construction are explored with an accent on increasing efficiency.

W.A. Pettyjohn has recently published another important contribution to surface and ground water quality control entitled: *Water Quality in a Stressed Environment*, (Burgess Publishing Company, Minneapolis, 1972). In addition, W.J. Powell, *et al.*, have also made an important contribution in the form: "Water Problems Associated with Oil Production in Alabama," 1963, Geological Survey of Alabama, Circular 22.

3
Rock Drillability: A Review

The early history of water well drilling relates the ingenious beginning of drilling technology.[135,157] And today, the technical literature abounds in papers on the flow characteristics of ground water to wells, the relation of well diameter to capacity, the percent of open hole, the drawdown, and other factors relating to well geometry, discharge, and aquifer response.[248,632] The practical field relationship of rock drillability, however, has remained obscure.

Recently, not only have the tools become available to the shallow drilling industry, but also advanced drilling techniques have begun to find application in water well construction, although the transition has been slow. The broad features of water well technology have had many years to become established, but the accent today is on efficiency, and an acceleration of the adaptation process is mandatory. "Optimization of operations," a method of maximizing efficiency, has been an important concern of the oil industry for a number of years,[232] because more efficient drilling practices promote higher profits.[505]

One of the most important aspects of an information transfer from the oil industry is the concern for detail. If the oil industry's attention for detail can be emphasized in the shallow drilling industry, one effect could be a substantially increased level of efficiency in water well and mineral exploration drilling.

More than 40 variables are generally assumed to influence drilling,[243] and many of the relationships of these variables must be established in the laboratory where conditions can be controlled. Mere simulation of field conditions in the laboratory is inadequate. Field conditions must be more closely duplicated to provide meaningful results. A parameter or criterion which describes both field and laboratory behavior is needed, and the results of any study must then be applied to drilling practices in the mining, petroleum, and ground water industries.

Frequently, drilling tests designed to convey meaningful data provide only conflicting results. The validity of any such tests cannot be established until general equations describing drilling behavior are fully delineated. Conversely, the applicability of any drilling equation is in doubt until valid results can be predicted.

The subject of drilling can be subdivided into (1) Rock Drillability and (2) Drilling Systems. Although the various systems of drilling are technologically interdependent, the next few chapters will explore these systems separately. In this manner, the review both of the current status of drilling technology and the potential for modification of water well and mineral exploration drilling practices can be facilitated.

ROCK PROPERTIES

In the broadest sense, the rate of rock drilling is directly proportional to rock drillability, *in situ.* If it were possible to isolate and express mathematically all the factors which affect drilling rate, some function of rock drilling strength would certainly be included. For example, it is known that while drilling an alternating sand/shale sequence under identical and controlled conditions, the observed drilling rate changes in response to the alternating lithology. These fluctuations in drilling rate are due to a difference in drilling resistance of each rock type, and this relationship is the inverse of rock drillability.

The method of testing for strength and elastic properties of rock has made rapid progress in the last 25 years.[17,252,485] An awakening interest in all branches of the mineral industry in the past ten years accompanied by standardization has resulted in the collection of a great amount of new physical data on the most commonly encountered rocks. The interpretation and correlation of these new data with technological properties, such as drillability, however, have not kept pace. One reason for this lag may be that test results of the rock properties were widely scattered in the technical literature of the mineral and petroleum industries.

In a paper by Wuerker,[666] annotated tables of strength and elastic properties of rocks are presented. Twenty-one rock properties including the importance of each are listed in tabulated form. Types listed include igneous, sedimentary, and metamorphic rocks classified according to their mineralogical content. Not all of the properties shown have been measured for any one rock type, although Wuerker has attempted to make the varied data as congruous as possible.

Of particular interest is the part of the tables that deals with compressive strength. Since percussive and rotary bits attack rock by impact, compressive strength intuitively would control the resistance to indentation by either mode of attack. Many reference books list *rock compressive strengths* at values no greater than 40,000-50,000 psi. This is the stress required for failure of a sample which is loaded at its ends and initially has a constant cross section. Wuerker lists a number of rocks with considerably higher values, for example:

Rock Type	*Compressive Strength, psi*
(1) Epidosite from Virginia	63,000
(2) Porphyry Syenite from Ontario	63,000
(3) Amphibolite from India	60-70,000
(4) Hematite from Minnesota	88,100
(5) Jaspilite from Minnesota	99,000
(6) Chert from Missouri	86,300
(7) Pink Quartzite	68,000

For comparison, it should be noted that some of the listed compressive strengths are equal to or greater than that of mild steel. A stress-strain diagram of the hematite sample (no. 4 above) shows a striking similarity to the elastic portion of the stress-strain diagram of mild steel under uniaxial compression. Since the reported values of strength in the above list are for commonly encountered rock types, some insight into the difficulty experienced in drilling these materials is essential.

Other rock properties listed by Wuerker are of interest to the water well and mineral exploration drilling industries, notably, confining pressure and porosity. Because all rocks at depth exist under a state of imposed stress, the effect of confining pressure on strength should be examined. Elastic properties of rocks are greatly influenced by the state of stress to which they are exposed. Shales, for example, become harder to drill as depth increases. This effect is due to the confining pressure on the cutting surface which controls the cuttings-removal efficiency. At extremely high pressure, weak rocks such as salt, anhydrite, and, to a lesser degree, shales become incompetent and will flow plastically. Studies of Payne and Chippendale[502] dramatically illustrate this transition. Porosity of a rock is known to influence drilling rate, although no quantitative data to illustrate this effect are available. Drillers and geologists have long used "drilling breaks" as an indication of porosity. The reduced compressive strength of porous zones probably accounts for this phenomenon.

LABORATORY TESTING

Simple indentation tests have been attempted under more complicated conditions of elevated pressure and temperature to evaluate the behavior of the rock upon impact. The concept of indexing, or spacing, of successive blows has been evaluated extensively by Gnirk,[250,251] who found that the optimum distance between successive blows to produce the maximum crater volume decreases substantially with increasing confining pressure. This relationship exists because of the transition from brittle to ductile behavior of the rock specimen. Cheatham[164] has studied the behavior of Cordova Cream Limestone to provide data for extending the plasticity theory to cover situations in which consolidation and strain hardening are present. Both consolidation and strain hardening are important considerations in describing the mechanical behavior of this limestone. Further, Cheatham's tests show that this rock possesses strength anisotropy, which indicates that the orientation of the sample is important when testing. This strength anisotropy has been studied for various laminated rock types,[331,648] and the role of strength anisotropy in natural hole deviation has been described. Based upon these studies, new concepts in bit design are being undertaken to control drilling deviations.[237,411]

FIELD TESTING

The U.S. Bureau of Mines has conducted drilling studies in an attempt to establish indices for predicting rock drillability which are based upon the application of drilling forces, the measurable physical properties of the rock and associated parameters.[147] Earlier work conducted by the Bureau of Mines on laboratory drilling tests of nine different rocks suggests that compressive strength of the rock could be a parameter which is functionally related to drillability. Field trials were performed to verify earlier laboratory findings in which 26 rocks were tested, ranging from soft limestone to hard, dense taconite. To obtain correlations between laboratory and field data, six of the nine rocks drilled in the laboratory were also among those drilled in the field. The field testing was performed with a commercially available diamond drill mounted on a two-wheel trailer. All the bits used were AX size, and the drilling experiments were conducted at all combinations of 600, 1,000, and 1,600 rpm with 1,000 and 2,000 pounds thrust.

The Bureau of Mines' research produced a theoretical equation for predicting core-drilling rates:

$$d = \frac{2\pi (T - \mu r F_v)}{S A - F_v} \qquad \text{EQUATION 1}$$

where: d = penetration per revolution (in./rev.)
T = torque on drill bit (in./lb.)
μ = coefficient of friction
F_v = thrust (lb.)
r = bit radius (in.)
S = rock drilling strength (psi)
A = kerf area of bit (in.^{2})

Note: Consult Appendix for conversion of English to metric units.

Although this theoretical equation satisfactorily describes the observed experimental data, the value of the coefficient of friction (μ) was difficult to measure and apply. Figure 11, reproduced from Bruce, [147] illustrates the relationship of the theoretical and the experimental curves for drillability.

In this same study, it is further observed that the relationship between penetration per revolution (d) and rock compressive strength (C) in the laboratory data was hyperbolic. Furthermore, the reciprocal of d, revolutions per penetration, plotted against rock compressive strength is found to be linear. A least-squares analysis of the data was performed to determine if the predicted equation might be expressed hyperbolically. On the basis of the similarity of the empirical expression produced and the theory (Equation 1), the validity of the general relationship is established.

By simplifying Equation 1, the following relationship results:

$$d = \frac{\pi T}{5 C A} \qquad \text{EQUATION 2}$$

where: d = penetration per revolution (in./ rev.)
$T = F_t r$
F_t = tangential force at bit face (lb.)
r = bit radius (in.)
C = compressive strength of rock (psi)
$A = (\pi/4)(D_o^2 - D_i^2)$
D_o = outside diameter of diamond core bit (in.)
D_i = inside diameter of diamond core bit (in.)

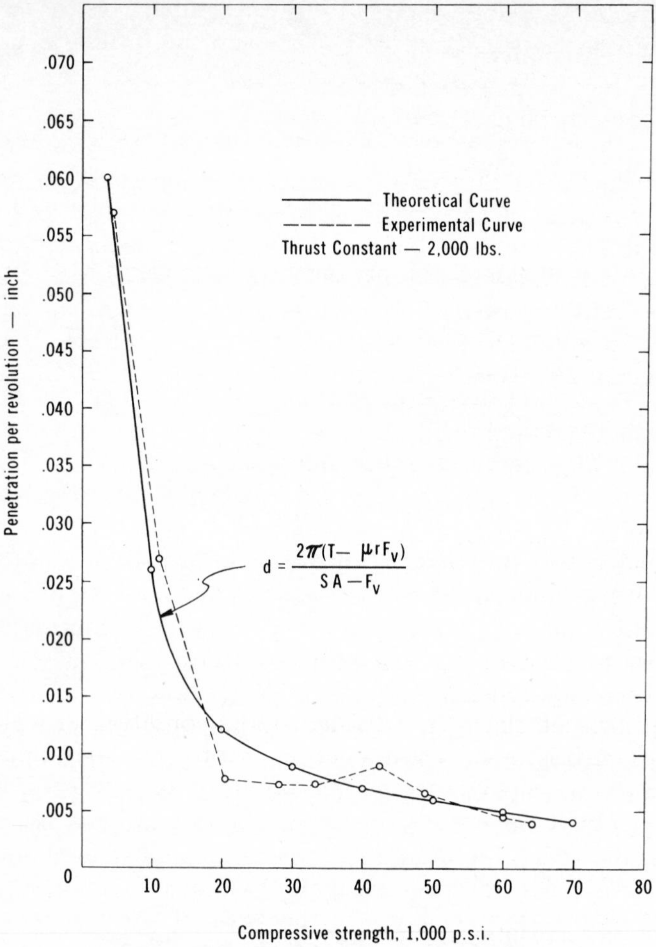

Theoretical Curve
Experimental Curve
Thrust Constant — 2,000 lbs.

$$d = \frac{2\pi(T - \mu r F_V)}{SA - F_V}$$

Penetration per revolution — inch

Compressive strength, 1,000 p.s.i.

Figure 11[147] Relationship of Theoretical and Experimental Curves for Drillability.

The Bruce study suggests that the rate of penetration of a diamond core bit is inversely related to the physical properties of the rock. Although the most significant single property seems to be compressive strength, other properties of the rock must also influence drillability. Uniaxial compressive strength is highly dependent upon test conditions such as confining and pore pressure and mode of rock failure.[302,303,529] It appears, however, that diamond core drilling rates may be predicted from drilling conditions and compressive strength of the rock material being drilled.

The approach of relating only compressive strength to a definition of rock drillability is not without criticism. Somerton,[583] for example, has conducted a laboratory investigation of rock breakage from rotary drilling. Although results of this investigation do not have direct field application, experimental verification of several concepts has been obtained. In this work, a method of comparing drilling strength of rock to measurable rock strength properties is presented. It was found that the ultimate compressive strength of a rock is not an adequate measure of its drillability. In addition, bit tooth condition was found to have a pronounced effect on penetration rate. Drilling chips become finer, and drilling energy requirements increase as bit tooth wear progresses. This is reportedly due to a decrease in the amount of rock chipping and an increase in regrinding of cuttings by the drill bit teeth. As the effective area of the tooth increases at constant loading, the stress on the rock at the contact surface decreases. Furthermore, it is determined in Somerton's investigation that in rocks containing two or more mineral constituents of different strengths, a greater amount of rock breakage will occur in the weaker constituent.

A drilling rate equation for roller-cone bits is derived from rock cratering mechanisms investigated by Maurer.[429] This equation expresses a relationship between drilling variables for a condition termed "perfect cleaning," defined as the condition in which all of the rock debris is removed between tooth impacts. Under these circumstances, usually encountered only during shallow drilling, the equation is as follows:

$$R = k \frac{NW^2}{D^2 S^2} \qquad \text{EQUATION 3}$$

where:
- R = drilling rate (ft./hr.)
- k = formation drillability (a constant)
- N = rotary speed (rpm)
- W = bit weight (lb.)
- D = bit diameter (in.)
- S = rock drilling strength (psi)

Upon impacting a rock surface, the bit tooth forms a crater at or near the bottom of the wedge which is produced beneath the tooth itself. The mechanism is shown in progressive stages in Figure 12. It is observed that the craters are formed at some distance beneath the bottom of the tooth. When craters are formed with a differential pressure across and into the rock, the shattered material will remain in the crater when the tooth is removed, held in place by the differential pressure. If the bit cones are skewed, the teeth will drag through these cuttings and gouge some of them from the crater. Because the broken material is held in the craters, and the craters are formed near the bottom of the wedge, Maurer contends that the increase in drilling rate produced by this dragging action is largely due to better cleaning and not to the actual breakage of virgin rock. If the dragging action is, in fact, a cleaning mechanism, it will have no effect on drilling rate under "perfect cleaning" conditions and, therefore, will have no effect on the validity of Equation 3.

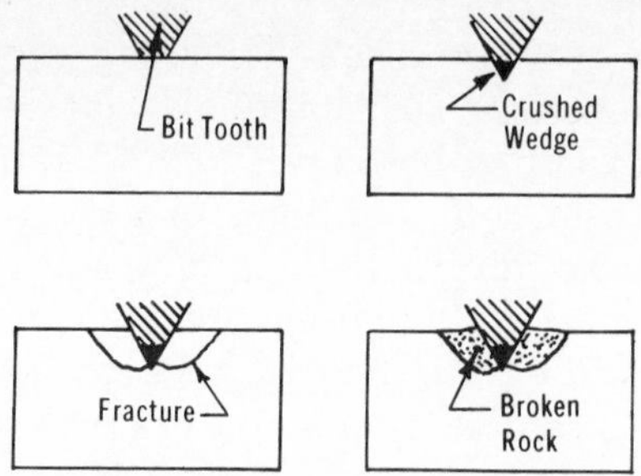

Figure 12[429] **Crater Formation Mechanism.**

Equation 3 relates four variables to drilling rate; namely, N, W, D, and S. With S defined as rock drilling strength, it is observed that drilling rate varies inversely with the *square* of rock drilling strength. No relationship between rock drilling strength and any physical rock property was presented. The drillability constant (k) in the equation is a mathematical manifestation that many other parameters *may* affect drilling rate, including other rock properties (other than "rock drilling strength"), drilling fluid properties, pressure conditions, and bit design.[470] The magnitude of k is constant for a given bit at a given depth, and all other variables are represented within this constant (k) if perfect cleaning is, in fact, obtained.

The "perfect cleaning" theory considers the roller-bit drilling process as two fundamental operations: (1) the formation of craters under each bit tooth and (2) the removal of the broken rock from the craters. To delineate the mechanisms of drilling, Maurer suggests that it is first necessary to understand the mechanisms involved in the formation of individual craters and then to relate those mechanisms to the overall drilling operation.

When a bit tooth impacts a rock, the rock is elastically deformed until the crushing strength of the rock is exceeded, at which time a wedge of crushed rock is formed below the tooth. As additional force is applied to the tooth, the crushed material is further compressed, and high lateral forces are exerted on the solid material surrounding the crushed wedge. When these forces become sufficiently high, fractures are initiated below the tooth and propagate to the face of the rock. The trajectories of the fractures intersect the principal stresses at a constant angle as predicted by the theories of both Mohr and Griffith.[173] This cratering mechanism and accompanying complications, including the effects of pore pressure phenomena and plastic rock behavior, have been studied in some detail,[163,239,477] and the basic mechanism is reasonably well understood.

A useful correlation has been found between the crater depth and the reciprocal of both the shear strength and the compressive strength of the rock.[123,569] These relationships broaden the scope of rock properties which must be considered in the evolution of a realistic drilling model by reducing the importance of compressive strength, *per se.* Several mechanical properties of rocks, including hardness and specific disintegration, have been related to rock drillability by Gstalder and Raynal.[286] In this work the hardness test, developed by Schreiner, was used to measure the breaking strength of rocks. Rock breaking strength was determined by application of an increasing load on the rock face through a flat-faced cylindrical punch until rupture occurred.

INDIRECT MEASUREMENT: A NEW APPROACH

Relationships among the hardness value so determined and rock specific disintegration (volume of rock broken per unit of work input) and Young's modulus and sonic velocity have been demonstrated. The hardness-drillability postulate was verified by laboratory drilling tests under constant drilling condition in which the pressures closely simulated downhole magnitudes. The results of this work suggest that rock drillability might be determined from sonic, wire-line logs or *vice versa.* An extension of this idea, considering variations in drilling parameters, such as bit weight and rotary speed, has been explored by Somerton and El Hadidi.[584] Using a drilling equation relating drill rate to bit weight, to rotary speed, to bit diameter, to a tooth-dulling parameter, and to rock drilling strength, sonic velocity was found to correlate with drilling strength if the general mineralogical condition of the rock specimen did not change. A plot of drilling strength and interval transit time is shown in Figure 13. *Transit time*, in sonic logging, is defined as the time required for a sound wave to travel through a specific thickness of rock. The speed of the wave depends on the elastic properties of the rock, the porosity of the formation, and the fluid content and confining pressure.

Many other studies on rock drillability and bit efficiency have been undertaken, and these expand the knowledge of rock properties.[108, 276, 309, 431, 501, 585] It is clear that while the

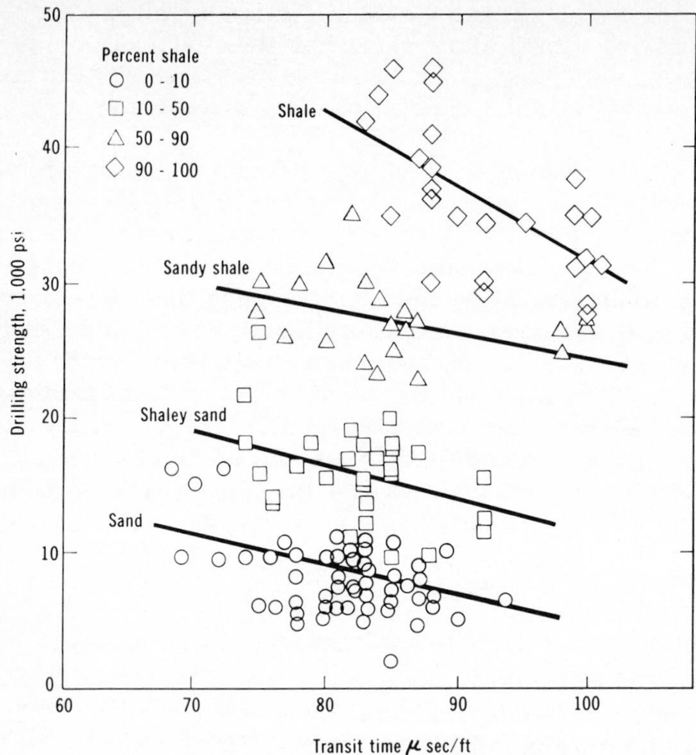

Figure 13[584] Rock Drilling Strength 1000 psi vs. Transit Time.

understanding of rock properties is growing, the overall drilling mechanisms in operation are, at present, not clearly understood. Although the variables are being defined, agreement on the basic theories which control rock failure is not unanimous.[583] Many approaches have been proposed previously for determining the relative drillability of rock. These approaches generally considered only one drilling method, and many include only a limited range of rock hardness. A comprehensive rock drillability index proposed by White[655] incorporates the three major drilling systems — rotary, percussive, and rotary-percussive — and considers a wide range of rock types from the very soft, highly altered rocks to the extremely hard taconites. This index approach, however, is based on the penetration rate achieved by a 3/4-inch diameter bit drilling a hole four inches deep. Triaxial tests[571] were conducted on two rock types, using specimens ranging from 2 to 32 inches in diameter to determine the influence of specimen size on strength and mechanical properties. While the effect of specimen size on mechanical properties was found to be small, the effect on strength and character of failure was significant. Additional support data will be required to negate the value of the many drillability indices which have been constructed over the past few years using small laboratory rock specimens.

SUMMARY

In the meantime, rotary-drilling technology has evolved to the point where the limitations of the method are becoming obvious, and totally new approaches to the problems of rock excavation are being seriously investigated. The need to relate drilling parameters to the properties of the rock is compelling. Only after a correlation has been established between some measurable rock properties and drillability, can an estimation be made of the ideal conditions. Actual performance then can be expressed in terms of the ideal, and the relative merits of various approaches toward improvement can be evaluated. It might appear that the true fracturing ability of each type of rock bit, for example, would be needed to establish a criteria of rock drillability; however, perhaps consideration of a wider range of parameters would serve the same purpose.[123] Because the shallow drilling industry operates under relatively severe economic pressure, the need for such a definition is urgent. Improvement in shallow drilling economics is a prerequisite for healthy industrial growth.

In reviewing the present knowledge of the rock drillability system, the development of any drilling system must first give consideration to the rock because its composition, properties, and behavior control the drilling process. All else is secondary in rock destruction. A relatively new science, rock mechanics, has arisen. This field is vitally important in rock drilling because the theories and laws of this science describe the material being destroyed. *Rock drillability* is defined here as the inverse of the resistance of a rock to be drilled. This resistance is related not only to the material composition of the rock but to its environment. Among many physical properties which have been described, basic indentation tests have shown that rock compressive strength, shear strength, and impact resistance control drillability. Rocks can be hard, strong materials, in some cases exceeding mild steel in strength. Under perfect cleaning conditions, the behavior of rock bits can be mathematically described, but rock drillability does not appear to depend only on uniaxial compressive strength. The relationship of rock properties to drillability controls the economics of the rock drilling process.[238,332,410,484,564,589]

Although some variation of the rotary drilling method will eventually be replaced by a totally new, more efficient method, a comprehensive understanding of the present cable-tool and rotary systems is necessary to hasten the transfer of one method to another. To this end, the degree of complexity of the presently available knowledge on rotary drilling, for example, is relatively great. This complexity demands simplification by the engineer if it is to reach the drilling contractor and which must assist in his day-to-day problems of controlling econmics and increasing production. Academic research must be partly realigned from its present course of examining the frontiers of science toward a course where the man in the field benefits today, not tomorrow.

4
Cable-Tool Drilling System

Having surveyed the general features of rock drillability, the technological developments applicable to the two commonly used drilling systems will be explored in the following chapters: the cable-tool system and the rotary system. In the past, the field of shallow well drilling has included only a description of equipment design and the manipulation of this equipment. Only minor attention has been focused on an understanding of the variables involved in the drilling process and of the interrelationship of these factors to the direct costs of well construction.

The staggering cost of deep drilling, especially offshore, has concentrated the drilling research effort of the petroleum industry on means of reducing the time spent on deep wells. Such efforts have been directed toward faster penetration rates and have been concerned with the minimization of a variety of hole problems.[274,540] All of the developments that have been made do not offer the same economic rewards in shallow hole drilling, but many of the principles and practices can be beneficially adapted to shallow drilling conditions.

Certainly the shallow hole drilling industry must progress according to its economic means, but knowledge of the developing technology is a necessary prerequisite to growth in any industry. In the following survey of drilling systems, little attention will be given to equipment design; but developing concepts and basic principles will be emphasized.

EARLY HISTORY

The cable-tool method of drilling, often referred to as the standard, percussion, or "yo-yo" method is the oldest and most versatile drilling procedure. The cable-tool system has a long history of use in drilling for ground water and, under special circumstances, for other minerals located at relatively shallow depths.

The first recorded use of percussion boring tools was in China about 600 B.C. for drilling brine wells a few hundred feet in depth. By 1500 A.D., holes were being drilled to a depth of 2,000 feet. Spring poles and similar variations were used to drill water and brine wells in the early 19th Century in the United States.[157] Following the completion of the Drake oil well in 1859, great advances were made in the mechanical equipment and technical skills involved in cable-tool drilling. Colonel Drake drilled the first commercial oil well at Titusville, Pennsylvania, in the 1860's, using a steam-powered cable tool which was the mainstay of the drilling industry until early in the 20th Century.[137] The early drilling and fishing tools were locally made by blacksmiths, and although modern tools are stronger and heavier, they show little or no improvement in principle or basic design over the early tools.

CABLE-TOOL OPERATIONS

The cable tool drills by lifting and dropping a string of tools suspended on a cable. A bit at the bottom of the tool string strikes the bottom of the hole, crushing and breaking the formation material. A string of tools in ascending order consists of a bit, a drill stem, jars, and a swivel socket which is attached to the cable. The joints by which the parts of the drill stem and other tools are connected are now well standardized in the United States according to American Petroleum Institute specifications.

Drilling is accomplished with a tight line so that the bit strikes the bottom of the hole when the cable is stretched. Because of the elasticity and lay of the steel cable, stretching causes it to unwind and recover when tension is released. This imparts a turning action to the tool string through the swivel socket. The bit is then turned a few degrees between each stroke, so that the cutting face of the bit strikes a different area of the hole bottom after each stroke. Thus, the hole remains circular and vertical. On most rigs, the cable is led from the swivel socket over a sheave at the top of the tower then down through a sheave on the *spudding beam*. The spudding beam, driven by a *Pitman arm/arms* attached to the *spudding gear*, imparts the reciprocal motion to the cable and tool string. From the spudding beam, the cable passes to the *bull reel* on which it is wound. Today, most rigs have shock-absorbing, spring steel, or rubber cushions for the sheave at the top of the tower. These cushions prevent must of the drilling shock from reaching the machine and improve the drilling effectiveness of the tool string. The spudding beam usually has two or more wrist pins to which the Pitman arm or connecting rod can be attached, so that the length of the stroke may be extended from 16 to as much as 48 inches, depending upon the size and make of machine.

In consolidated rock, open hole can be drilled; but in unconsolidated or raveling formations, casing is driven down the hole during the drilling. In some unconsolidated materials, the casing is driven, and the resultant plug is drilled and bailed out. In other cases five to ten feet are drilled below the casing, the casing is driven to the bottom of the hole, and the hole is bailed clean. Above the water table or in otherwise dry formations, water is added to the hole to form a slurry of the cutting so that the cuttings can be readily removed by the bail. Casing is driven by attaching steel-drive clamps to the wrench squares of the drill stem and by a steel-drive head with an opening for the drill stem. The regular drilling action of the rig is used to drive the casing with the full weight of the tool string.

The bottom of the casing is fitted with a heavy-walled, hardened, steel drive shoe with a slightly larger O.D. than the casing. Usually the inside, lower edge of the shoe is beveled to form a cutting edge. The casing sits on a shoulder inside the drive shoe. The drive shoe prevents collapse of the casing, shaves off irregularities from the side of the hole during driving, and forms a tight seal to exclude undesirable water.

As casing is driven in unconsolidated formations, the vibration causes the sides of the hole to collapse against the casing. Frictional forces increase until the casing cannot be driven. When this occurs, a smaller diameter casing is telescoped inside of the casing already in place, and drilling continues using a smaller diameter bit. On deep holes, as many as four or five reductions may be required.

One approach to driving casing and assuring a tight seal around the casing has been developed by Church.[169] The essence of this development consists of maintaining hydraulic pressure on an envelope of drilling mud surrounding the well casing being driven (see Figure 14).

The mud lubricates the outer surface of the casing reducing the skin friction and also seals the annular space as the casing is driven through clay or other similar materials.

The drive shoe having a larger O.D. than the casing generally opens an oversized hole. The Church method was designed to prevent the occurrence of an upward artesian flow of water around the casing.

As drilling proceeds, a heavy bentonite mud is used, and its pressure maintains a seal around the outside of the casing. One disadvantage of this method is that mud remains in place if the "pull-back" method of screen emplacement is employed. However, backwashing with water (see: "Well Hydraulics"), use of the horizontal-jetting technique to breakup the mud cake, and use of a polyphosphate dispersing agent should remove any remaining drilling mud.

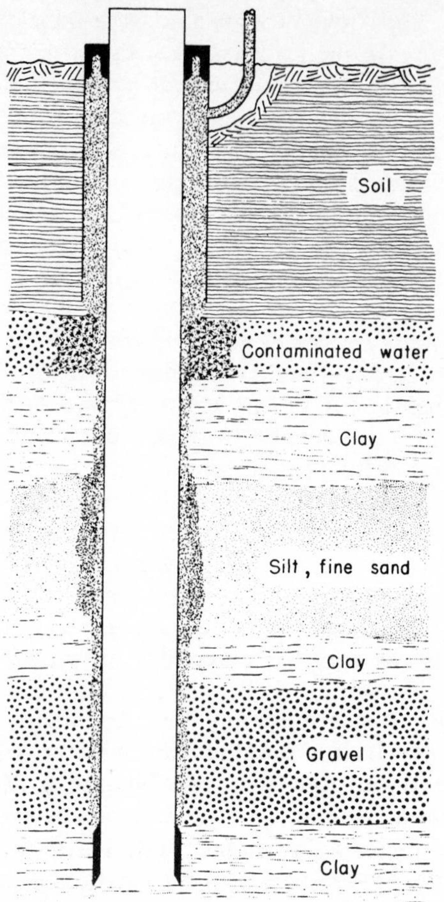

Figure 14[169] Church method of driving casing.

To further seal the well, cement grout is pumped into the annular space, and the casing is perforated from the inside at the desired level to allow the bentonite mud into the casing as the cement grout displaces it outside the casing. Water and mud are pumped or bailed from the well as grouting proceeds, and a water-soluble coloring agent is added to the grout mixture. The appearance of the color in the water being removed signals that the grout has reached the level of the perforations in the casing. Church suggests, however, that it is difficult to assess whether the grout has completely circled the casing. The heavy bentonite mud is an excellent sealing material at depths where the possibility of shrinkage is remote.

The Church method offers the following advantages:

(1) A single string of casing may be driven to greater depths.

(2) Migration of water along the outside of the casing from one aquifer to another is prevented.

(3) Flow around the casing from artesian aquifers under high pressure is prevented.

CABLE-TOOL DRILLING RATE AND CAPACITY

The speed of drilling and rate of progress with a cable-tool rig depends upon a number of basic factors: (1) the hardness of the rock, (2) the diameter and depth of the hole, (3) the dressing of the bit, (4) the weight of the tool string, and (5) the stroke rate and length. Cable-tool drilling more than any other drilling method is to a large extent an art rather than a science; and the capable driller continually increases his skill and knowledge.

The capacity of a cable-tool rig depends on the weight of tools which it can handle safely. Tool weight is determined largely by the diameter of the bit and the diameter and length of the drill stem. However, as the hole is deepened, the rig must support more and more of the cable weight in addition to that of the tools. There is a limit to the depth at which a cable-tool rig can be safely used to drill a given size hole. On the other hand, the more shallow the hole, the larger is the feasible bit diameter. Smaller, commercially available cable-tool rigs, capable of handling 300 pounds of tools, can usually drill a five-inch hole to a depth of about 300 feet, while the largest rigs, capable of handling three tons of tools, can drill a similar size hole to 5,000 feet. With medium and high capacity rigs, 18- to 24-inch holes are commonly drilled to depths of several hundred feet. With increased depth, more time is required in removing the tools, bailing, and reinserting the tools. Increasing cable length requires a reduction in tool weight and rate and length of the stroke. These and other factors decrease the rate of drilling, increase costs, and necessitate smaller hole diameters with increasing depths. However, it may be economical to drill to depths of more than 2,000 feet, although water wells are rarely drilled to such depths using the cable-tool drilling method.

The cable-tool rig in the hands of a skilled operator is probably the most versatile of all rigs because this rig can drill satisfactorily over a wide range of conditions. The method's major drawback, compared to other rig types, is its slower rate of progress and depth limitations.

Although cable-tool drilling offers certain advantages under special circumstances, the method, as mentioned previously, has not significantly developed since its inception. It can, under certain conditions, compare favorably with the general economic efficiencies of rotary-type methods.

CABLE-TOOL SAMPLES

Cable-tool sampling and formation logging are simpler and more accurate than with most other rig types. The cuttings bailed from each drilled interval usually represent about a five-foot zone. When casing is used, there is little chance of contamination of the sample, thus making the method particularly useful in certain types of mineral exploration. A change in lithology can usually be recognized by the response of the rig to the changed drilling condition, at which time samples can be taken at a shorter interval. In some unconsolidated formations, casing can be sunk by merely bailing and driving,[52,7] so that samples are relatively unbroken and representative. The samples are not contaminated by drilling mud and clay, and shale and silt fractions are not likely to be lost by dispersion in the drilling fluid. Cuttings of unconsolidated formations are usually not finely crushed, and some are of sufficient size to permit geological identification and description. When a potential aquifer is encountered, tests for yield and quality of water are made by bailing or, if of sufficient importance, by pumping.

Samples of unconsolidated formations are highly representative if taken with a "sand-pump" type bailer; whereas, a "dart-valve" bailer, used primarily for consolidated rock drilling, usually requires additional drilling and breaking of the gravels, hence the samples are less useful, especially for sieve analyses purposes.

CABLE-TOOL FIELD USE

The obvious advantages of the cable-tool rig are its suitability for use in rugged terrains and in developing countries where wages are low. The initial cost of a cable-tool rig complete with tools is generally one-half to two-thirds that of a rotary rig of equivalent capacity. The rigs are usually compact requiring less accessory equipment than other types and are more readily moved in rugged terrain where roads are poor. The simplicity and ruggedness of design and ease of repair of these rigs and tools are particularly advantageous in isolated areas. In general, the method requires less skilled operators and smaller crews than other drilling rigs of similar capacity. The low horsepower requirements are reflected in lower fuel consumption, an important consideration when fuel costs are high or sources of fuel are remote. While slower than other rigs in drilling some formations, cable-tool rigs can usually drill through boulders and fractured, fissured, broken, or cavernous rocks. The method requires much less water for drilling than most other commonly used rigs, an important consideration in remote arid and semiarid locations. The cable-tool method is also readily adaptable for use in hollow-rod jetting and mud-scow drilling. (See Chapter 6.)

One of the most important advantages of the cable-tool method is the ability to acquire qualitative data on the water-bearing characteristics and static heads of various formations as casing is being driven. Water quality data can be obtained by bailer samples as each formation in turn is opened to the bottom of the casing, and upper formations are cased off.

The disadvantages — a relatively slow rate of progress and economical and physical limitation on depth and diameter — have been mentioned previously. A further disadvantage of the cable-tool rig is the necessity of casing while drilling in unconsolidated materials. The necessity of driving casing requires a heavier-walled pipe than would otherwise be required in some installations. Screens often must be set by "pull-back" or "bail-down" methods. (See Chapter 10.) The former method is sometimes extremely difficult or impossible in deep or large-diameter wells, and the latter may give rise to problems in alignment.

The advantages and disadvantages of the cable-tool system in comparison with the rotary system have been summarized.[132] The major advantages of the cable-tool system as opposed to other drilling systems (rotary, etc.) are listed below:

(1) Economics:
  (a) Lower initial equipment cost, hence lower depreciation.
  (b) Lower daily operating cost, including maintenance, personnel, and water requirements.
  (c) Lower transportation costs.
  (d) Lower rig-up time and expense.
  (e) Drilling rates comparable to rotary in hard rocks at shallow depths.

(2) Better cutting samples.
(3) Easy identification of water-bearing strata.
(4) No circulating system.
(5) Minimum contamination of producing zones.

The major disadvantages of the cable-tool method as opposed to other drilling systems are summarized below:

(1) Limitation on penetration rate.
(2) Limitation on depth (the depth record for the cable-tool method is 11,145 feet, drilled in New York, 1953.)[326]

(3) Lack of control over fluid flow from penetrated formations.
(4) Lack of control over borehole stability.
(5) Frequent drill-line failures.
(6) Lack of experienced personnel.

The more important general aspects of cable-tool drilling have been treated in a number of publications.[43,45,199,259,260,480]

The mechanics of cable-tool drilling have been subjected to analytical treatment. Sprengling and Stephenson[588] and Mills[447] have presented formulas for evaluating the natural vibration frequency of a cable-tool drill string as a function of the modulus of elasticity, weight, length, and cross-sectional area of the drill line and tool weight. According to this theory, drilling speed should be equated with the natural frequency of the drill string or some even multiple thereof. A practical speed in the range of 20-40 strokes per minute which satisfies this criterion should be used. According to Griffith,[282] the greatest stroke amplitude does not necessarily result in the most economical drilling conditions.

Bonham[132] has analyzed the impact of a string of cable tools to evaluate the optimum tool weight, minimum stroke length, horsepower input to the formation, and drilling speed. The published equations can be used to design a drilling program for depths greater than 1,000 feet. At more shallow depths, greater tool weights and drilling speed are practical in special applications, as previously mentioned.

CABLE-TOOL TEST DRILLING

One major use of the cable-tool drilling system is for the drilling of test holes. These holes are drilled at or near the site of a proposed production well for the purpose of obtaining data for sound well design and to permit evaluation of the potential aquifer.

For large, deep installations where high capacities are involved, a test hole represents about ten percent of the total cost of the well and usually can be justified economically.[7] Cable tools frequently are recommended for test holes in loosely consolidated formations. Their use permits ready measurement of static-water levels and test pumping of individual aquifers. Since cable-tool holes are usually cased, only the gamma, neutron, and related logs can be run, although special logging techniques of cased holes have been partially developed (see section on "Borehole Geophysics").

5
Rotary Drilling System

The conventional rotary drilling method is becoming the most effective and widely practiced procedure of all available drilling systems, having been introduced by the mining industry and later adopted by the petroleum industry as the standard drilling system.

The distinguishing characteristic of drilling by the conventional rotary method may be generally described as the process of forcing drilling fluid, by means of a suitable pump, down through the inside of the drill pipe and out through the bit openings. The drilling fluid, being under pressure from above, flows back to the surface by way of the annulus formed between the outside of the drill pipe and the hole wall or casing. With the bit on the bottom, the drill stem is rotated and excavation begins.

More technical knowledge is available on the rotary system than on any other system of drilling.[138] This results from a wide acceptance of the method by the oil industry[405] whose extensive research toward more efficient and therefore less costly systems has promoted the natural evolution within the water well industry to the presently employed rotary drilling system.

The rotary-drilling process involves boring a hole by using a rotating bit to which a downward force is applied. The bit is supported and rotated by a hollow stem composed of high-quality steel, through which a drilling fluid is circulated. The fluid leaves the stem at the bit, thereby cooling and lubricating the cutting structure. By flowing across the cutting surface, the drilling fluid drags cuttings from the hole bottom and transports them to the surface by traveling up the hole-stem annulus.

EARLY HISTORY

Rotary drills were used by the early Egyptians in quarrying stone for the pyramids. Much later, in 1823, water wells were drilled in Louisiana with boring tools, but the cuttings were removed by bailing.[136] The first reference found of a drilling machine using a rotating tool, hollow drill rods, and circulating fluid to remove cuttings is in an English patent issued to Robert Beart in 1844. United States patents, incorporating the basic devices and principles in use today, were issued between 1860 and 1900. Early applications of the rotary method in the United States were in water well drilling. In 1900, Baker[404] moved equipment from South Dakota to Corsicana, Texas, and rotary drilling was used to drill the soft rocks of that area. The rotary method became commonplace after 1901, when Captain Lucas drilled the Spindletop discovery well near Beaumont, Texas, with rotary tools. Today, the rotary method is used to drill approximately 90 percent of all wells drilled for petroleum, gas, and other minerals, and the ground water industry is slowly making the transition to some form of rotary drilling.

TEST DRILLING

Direct-circulation rotary rigs are generally less desirable than cable tool for test-hole drilling. Drill cutting samples tend to be mixed from different depths and contaminated by the drilling fluid unless special care is exercised. Sample lag time can also become troublesome in constructing a reliable driller's log, and single-point electric logs are usually an important supplement for adequate interpretation. Despite the many advantages of rotary drilling, the above problems make its use questionable for drilling test holes when the completeness and reliability of the data obtained are considered.

Measuring static water levels, taking representative water samples, or acquiring pump tests of individual aquifers are not generally practical with the conventional rotary drilling system. Furthermore, in large diameter holes, i.e., those drilled by the reverse-circulation rotary method, a special apparatus is required to hold the logging tool or sonde near the side of the hole in order to obtain useable resistivity, gamma, or other types of logs.

The direct-circulation rotary rig is generally considered to be faster, more convenient, and less expensive to operate than the cable-tool rig, although the depth and the size of the hole govern the cost.

Because of the extent to which the rotary method is used today, most of the recent developments in drilling equipment and technique relate to it. Therefore, rotary drilling will be discussed in detail here, beginning with the drilling fluid, followed by factors which affect drilling rate, economic optimization of the process, and a discussion of the salient features of the more common hole problems which are encountered. Adaptations of the conventional rotary system, including air-rotary drilling, air-percussion-rotary drilling, and reverse-circulation-rotary drilling, etc., will be briefly surveyed in Chapter 6.

DRILLING FLUIDS

Water was the common circulating medium employed in early rotary drilling, although Chapman in a patent application filed in 1887 utilized a stream of water along with a quantity of plastic material. Mud-laden fluids resulting from the clayey formations drilled or from the addition of surface clays were employed later. Little attention was given to the distinction between weight, or density, and thickness, or viscosity. Studies of the control of blowouts in Louisiana oil fields led Ben K. Stroud in 1921 to recommend the addition of finely ground hematite and barite. Bentonite, as a suspending agent for these heavy minerals, was recommended in patent applications filed in 1929 by Cross and Harth. Thus, from barite and bentonite, the drilling fluids industry was developed in the United States.[269]

Rogers[530] traces the growth of the technology and discusses the manufacture, use, and properties of drilling fluids. The American Petroleum Institute has recommended procedures for testing and has set specifications for certain products.[71,98] From the publications available, this investigation integrates those directly applicable to the drilling and completion of water wells[206] and general mineral exploration.

Functions of Drilling Fluids

The rotary-drilling method is heavily dependent upon the circulating or hydraulic system.[130] The drilling fluid performs several primary tasks, namely:

(1) The relief of vertical axis stress in the formation immediately below the bit.
(2) The conveyance of energy
 (a) to remove the cuttings as they are formed beneath the bit.
 (b) to transport the cuttings to the surface.
(3) The separation of the cuttings at the surface.
(4) The maintenance of hole stability.
(5) The cooling of the bit.

A number of other functions varying in importance with local conditions must occasionally be served by the drilling fluid:

(1) The prevention of fluid entry from the porous rocks penetrated.
(2) The reduction of drilling fluid losses into permeable and loosely cemented formations.
(3) The lubrication of the mud pump, bit bearings, and the drill string.
(4) The reduction of wear and corrosion of the drilling equipment.
(5) The assistance in the collection and interpretation of information from cuttings, cores, and borehole geophysical surveys.

Classification

The drilling fluid is classified according to the principal-fluid phase as: (1) gas, (2) water, or (3) oil.[273] A gas-base drilling fluid may be dry air; air containing droplets of water or mud carried as a mist; foam, i.e., bubbles of air surrounded by water containing a foam-stabilizing substance; or stiff foam containing the film-strengthening materials, such as organic polymers and bentonite. Water may contain several dissolved substances, such as alkalies, salts, and surfactants in addition to droplets of emulsified oil and various insoluble solids carried in suspension. Oil-base drilling fluids may contain oil-soluble substances, emulsified water, and oil-insoluble materials in suspension. The term *mud* is applied to a suspension of solids in liquids and *water muds* are common drilling fluids. Oil-base muds have numerous applications[272] in the oil industry but are not used in drilling water wells.[565]

Value of Mud Programs

The composition of the drilling fluid frequently affects the economic success of the entire drilling operation. Consequently, the composition should be selected on the basis of the total cost of drilling, not on the price of the mud additives alone. In planning the drilling-fluids program, consideration should be given to the primary purpose for drilling the well; the nature of the formations to be penetrated; the site, i.e., accessibility to supplies; the layout of the rig; the disposal of wastes; the limitations as well as the capabilities of the drilling equipment; and, in particular, to the skill and experience of the operator.[271] The drilling fluid and the necessary mud-mixing equipment should be considered as an important part of a well-planned drilling program.[51,60]

Characteristics of Make-up Water

The major constituent of most drilling fluids is water; therefore, the quantity, quality, and on-site cost of the water used for drilling may affect the selection and the performance of the mud. The properties of bentonite in water are seriously impaired by dissolved acidic and salty compounds. When water is below a pH of 7, it may be unsatisfactory for use in mud without preliminary treatment. Highly mineralized water, with dissolved calcium and magnesium salts, impairs the suspending and sealing qualities of bentonite. A few simple tests will establish the suitability of the water — field measurements of pH and hardness are usually sufficient.

If the water is acidic, it should be treated with soda ash to raise the pH to 8.0 or 9.0 prior to the introduction of any mud additives. Most of the mineralized constituents of the water are removed by soda ash. Usually between 0.5 and 3.0 pounds of soda ash per 100 gallons of water is common; however, the treated water should then be tested for pH and calcium. If sulfides are relatively high in the make-up water, pH should be maintained above 10.0 to reduce possible premature casing and drill-pipe corrosion.

Knowledge of the source of the water usually indicates the possibility of contamination by salts, such as sodium chloride. Practical treatment to remove sodium and potassium salts does not exist; and, consequently, if salty water is to be used, the mud program must be planned according to the source water available. For instance, organic polymers are used to supplement (or to completely replace) bentonite in saline source water.

Mud Additives

Mud additives supplement water and water muds to make them:

(1) Thicker.

(2) Better wall-builders.

(3) Thinner.
(4) Heavier.
(5) Adaptable to special conditions.

Mud treatment is rarely simple. Often a substance is added to bring about a change in one property and simultaneously affects other properties either favorably or unfavorably. Bentonite and the organic polymers, for example, not only thicken but improve wall-building properties when added to fresh-water muds. Most "thinners" also reduce filtration. The practical approach to drilling-mud control is to adjust the composition to provide those properties which perform adequately at the lowest cost.[95]

Many of the products developed for oil-well drilling fluids are not adaptable for use in water wells. Other additives that generally are undesirable in water well drilling include: heavy metal compounds that are toxic, i.e., chromium; substances that decompose to produce objectionable odors or flavors or require the addition of toxic materials to prevent decomposition, such as starch; and substances that impart color to the mud and the filtrate, such as most tannins.

The technology of water-base muds is largely involved with the colloid chemistry of clays or *bentonite.* The economic importance of clays in agriculture and industry has led to extensive research on their behavior. Van Olphen[637] has prepared an authoritative survey on the subject of clay-colloid chemistry. Clay is broadly defined as a very fine-grained, unbedded rock; and if bedded or laminated, this rock is called shale. The clay minerals are alumino-silicates of varying compositions; and X-ray diffraction, chemical, and other methods are used to identify the various clay minerals. The minerals of major interest may be grouped as kaolinite, illite, chlorite, montmorillonite, and attapulgite.

Early field experience revealed that clay-water mixtures made viscous, slippery muds that promoted favorable hole conditions with little trouble in drilling. Extensive scientific studies have established that these desirable results are due to the active colloidal fraction of the clay. Because actual colloidal content is difficult to determine, performance tests have been developed.[71] The "yield" test is a determination of the relation between the solid content and the viscosity of clay-water mixtures. Thus, the highest yield is represented by the clay that will impart a specific viscosity to the largest volume of water. Yield is expressed as the number of barrels of mud having an apparent viscosity of 15 centipoises that can be prepared from one ton of clay (1 bbl. = 42 gallons). Typical yield curves are shown in Figure 15.

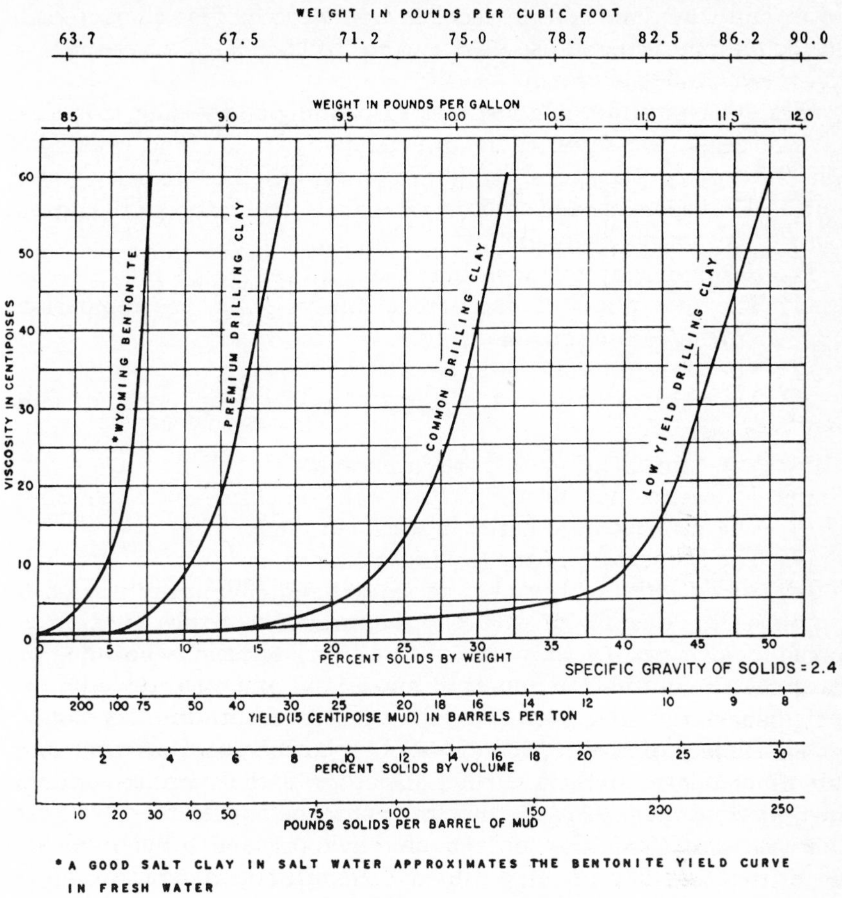

* A GOOD SALT CLAY IN SALT WATER APPROXIMATES THE BENTONITE YIELD CURVE IN FRESH WATER

Figure 15[76] Typical Clay Yield Curve.

Studies of clay mineralogy show that western bentonite is predominately a sodium montmorillonite, and southern bentonite is predominately a calcium montmorillonite. This basic difference in composition accounts for the observed differences in behavior of different types of bentonites. Structurally, bentonite is composed of thin, flat plates of the clay mineral, montmorillonite. When placed in water, the plates are separated by the entry of water molecules, and the particle "swells." Western bentonite may show at least a tenfold volume increase when placed in water.

Another property of bentonite is that of "gelling." A clay suspension that appears to be fluid while being mixed stiffens after standing. All of the liquid is suspended by the internal framework developed by the particles. If the gel is stirred again, the suspension returns to a fluid state, and the process can be repeated. This characteristic has led to the general use of the word "gel" in such expressions as "gel mud" and "gel cement."

Western bentonite demonstrates suspending and sealing properties in fresh water as a result of the dispersion, hydration, and swelling of the thin, platy particles of sodium montmorillonite. In salty water, much less dispersion and swelling take place, and many of the advantages of gel-forming clay are lost.

Practical applications for western bentonite are based on:

(1) The utilization of its colloidal properties in the removal of cuttings from the hole.
(2) The prevention of minor circulation losses.
(3) The suspension of cuttings in the hole during periods of shutdown.
(4) The consolidation of loose formations.
(5) The deposition of a thin filter cake on permeable formations.
(6) The suspension of barite in weighted muds.
(7) The lubrication of the drilling tools.

Peptized bentonite has been processed and chemically treated to improve the quality of standard bentonite. By the addition of organic polymers to the naturally occuring sodium bentonite, the quantity of bentonite required to make mud of certain viscosity can be reduced by half; therefore, the added cost of the "extra-high yield" bentonite is more than offset by reduced transportation costs for isolated drill sites. A further advantage is that peptized sodium bentonite mixes more easily than the natural product.

Because bentonite is not an effective thickening agent in salt water, another type of clay is used. *Attapulgite* has a fibrous structure and suspension properties depend on the mechanical dispersion of needle-like units which entangle to resist flow. This structural characteristic makes attapulgite useful as a thickener regardless of the salinity of the water. The needles do not form a tight filter cake; therefore, the filtration rate is high. The property, having a high filtration rate, is used to advantage in filling fractures to prevent loss of circulation. Standards of performance have been set for attapulgite.[71]

Native clay (also called drilling clay) is applied to clays obtained from local outcrops or mixed into the mud from clays penetrated by the bit while drilling. These clays consist mainly of illite and kaolinite, although calcium bentonite and other clay minerals may be present. Usually, fine sand and silt are associated with the clay. Such clays are lacking in colloidal properties, contribute to a rise in density, and are objectionable by increasing the possibilities for stuck pipe, loss of circulation, and plugging of the aquifer due to the lack of a suitable filter cake.

When hole conditions are such that water is adequate as a drilling fluid in all respects except in the removal of cuttings, the carrying capacity of water can be increased by the addition of chrysotile asbestos.[351,581] Cuttings of sufficient size to provide geological information can be recovered by adding as little as five pounds of asbestos per 100 gallons of water.

Organic colloidal materials are used in drilling fluids to increase viscosity, reduce filtration, stabilize clay, flocculate drilled solids, and serve as emulsifiers and lubricants. Several improvements in mud performance often result from a single additive. The *organic polymers* used in drilling fluids have a strong affinity for water (i.e., they are hydrophilic colloids). The polymers develop highly swollen gels and are adsorbed by clay particles, thereby protecting the clay from the flocculating effect of salts. Although these polymers do not swell as much in salt water as in fresh water, slimy particles are provided offering great resistance to the flow of water through a filter cake.

Some polymers are produced by the modification of naturally occurring starches and gums. Others are synthesized by building the polymer chain through the interaction of selected units or monomers.

Pregelatinized starch is an inexpensive agent for reduction of filtration in salt-saturated muds.[275] If the water is fresh and the mud is to be used for several days, fermentation can be prevented by the addition of a biocide, or the pH can be maintained at approximately 12.0 by the addition of caustic soda.[268] A *modified polysaccharide* product having a fermentation preventive is particularly effective in retarding disintegration of calcium bentonite.[456]

Sodium carboxymethylcellulose is produced in several grades which differ in purity and viscosity. The pure grade, high viscosity CMC allows the greatest increase in viscosity and does not ferment. Both suspension and sealing properties are present at low concentrations. Various other *cellulosic polymers* can be made by substitution of other groups in the cellulose chain. By selectively blending certain polymers, the outstanding features of each can be combined advantageously. Exceptional properties, such as a protective colloid for the prevention of the swelling and disintegration of bentonitic clays, have been reported for such polymers.[200,270,361]

Modified guar gum products provide suspension and sealing properties in either fresh or salty waters.[177,420] In very low concentrations (two to five pounds per 100 gallons — if used alone), guar gum flocculates small cuttings if the solids content of the mud is less than ten percent by weight. Furthermore, the gel strength of a guar gum drilling fluid is much lower than other common fluids. The fluid is also not thixotropic. Therefore, a relatively small mud pit would be required for the cuttings to settle as compared with the use of a bentonite-based fluid. A modified guar gum provides viscosity and filtration control in a drilling fluid that would normally become thin and watery after a certain time as a result of enzymatic action. An indicator in the additive changes color as the degradation progresses.

In recent years, new drilling-fluid additives have found use in the water well industry. One such additive is a food-grade, biodegradable material which when used with the drilling fluid converts to the viscosity of water through enzymatic action after a certain period of time.[672] (See Figures 16 and 17) Being a low solids, self-destroying fluid, this new additive does not contaminate water-bearing sands with clay particles and can be totally removed from the well during development. Carrying no extraneous clay particles, this particular additive does not hydrate natural clays and allows rapid mud-pit settling of cuttings (more representative samples can be obtained when test drilling). However, this material has limited gel strength.

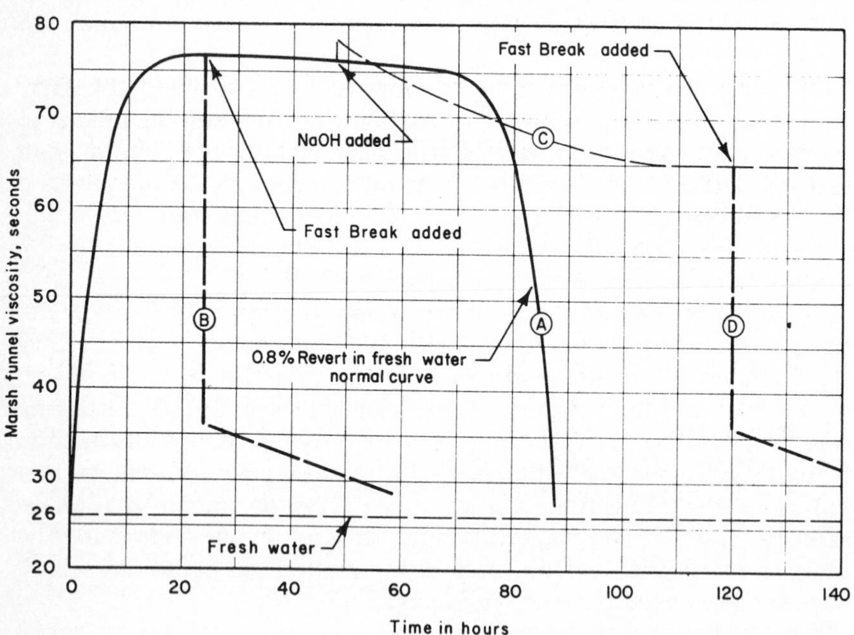

Figure 16[672] Viscosity of biodegradable drilling fluid can be extended or broken down chemically as necessary to meet job conditions. These curves show behavior of fluid with and without additives at a water temperature of 70°F.

Several *acrylic polymers* find application in drilling muds. Polymers based on acrylamide flocculate suspend drilling solids at concentrations of 0.05 to 0.20 pound per 100 gallons of low solids mud (density below nine pounds per gallon).[233] A large settling area or cone-type centrifugal separator is necessary to complete the removal of the flocculated solids. Sodium polyacrylate reduces filtration while causing only a small rise in viscosity.[549] Certain acrylic polymers in low-solids muds (less than five percent by volume solids) reduce friction losses as shown by pumping tests.[400] Copolymers of vinyl acetate and maleic anhydride selectively flocculate cuttings and leave bentonite in suspension.[396,495] Such polymers greatly increase the viscosity of bentonite muds and are constituents of extra-high-yield bentonite products. Highly mineralized water limits the effectiveness of the acrylic polymers; therefore, soda ash should be added as needed to precipitate unwanted minerals in water.

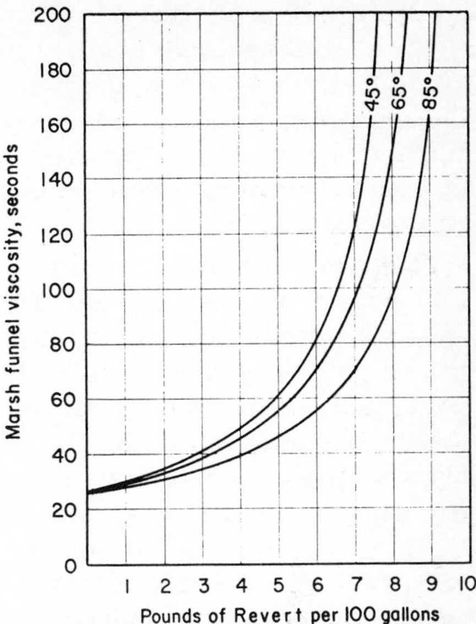

Figure 17[672] Viscosity of drilling fluid containing a given quantity of biodegradable agent will vary somewhat with water temperature.

The term *thinner* is applied to a substance that reduces the apparent viscosity and gel development of mud without lowering the density.[530] The addition of thinner affects the colloidal clay fraction of the mud. In terms of flow properties, as measured with the two-speed viscometer, the effect is a reduction in the yield point. The attractive forces between the clay particles are diminished, and the structural rigidity of the system is lowered.[437]

Water, excluded as a thinner in the above definition, is, however, the only economical viscosity-reducing agent for muds containing a high concentration of native clay or pulverized-rock solids. The plastic viscosity component of the apparent viscosity is reduced by water.

Thinners are classified in four groups:

(1) Complex phosphates.

(2) Tannins, such as quebracho extract.

(3) Lignitic materials.
(4) Chromelignosulfonates.
The effectiveness of the thinner depends upon the composition of the mud, the pH, and the amount (and chemical nature) of the contaminants present. More than one type of thinner may be added to insure the highest possible improvement in properties at the lowest possible cost.[566]

Complex phosphates, such as sodium tetraphosphate and sodium acid pyrophosphate, are very effective degelling agents when added to a fresh-water suspension of bentonite. As little as 0.2 to one pound per 100 gallons of mud usually will reduce the apparent viscosity and allow sand to settle. A solution of the polyphosphate should be added slowly to the mud. At shallow depths, if gummy clays tend to stick to the bit or pipe, approximately a pint of the dry polyphosphate should be added to the mud at the pump intake. Because the complex phosphates counteract the effect of calcium ions on clays, moderate contamination by anhydrite and cement can be eliminated by this additive. Solutions of sodium tetraphosphate are nearly neutral, and solutions of sodium acid pyrophosphate are acidic; consequently, sodium tetraphosphate is more effective against anhydrite contamination, and sodium acid pyrophosphate is more effective against cement.

Tannins are extracts produced from certain wood. Quebracho extract, made from a South American hardwood, is a widely used tannin, although extracts of domestic fir, redwood and hemlock, are important. Tannins are dissolved in a solution with a nearly equal amount of caustic soda, and this solution is slowly added to the mud.

Lignitic material, selectively mined from deposits in North Dakota, is used as a thinner to reduce filtration. Lignite is less acidic than quebracho and, therefore, less caustic soda is required to make lignite soluble in water. A readily soluble, dry product is made by reacting lignite and caustic soda in the proportion of six parts lignite to one part caustic soda.

Chromelignosulfonates are made from the spent sulfite-liquor by-product of the manufacture of cellulose pulp from wood. Ferrochromelignosulfonate is an all-purpose thinner and is effective in muds contaminated by cement, anhydrite, or salt. In sufficiently high concentration, ferrochromelignosulfonate exerts a protective action on bentonite; and, in this way, stable muds can be made which contain calcium salts in amounts sufficient to inhibit disintegration and dispersion of certain clays.[567] By adding chromelignosulfonates before cement or anhydrite is drilled, mud properties remain stable in spite of contamination. The caustic solutions of the tannins, lignite, and chromelignosulfonates are deeply colored and, therefore, should be used sparingly, if at all, in drilling the water-bearing section.

Barite is used to make muds heavier. The mineral, barite, is barium sulfate, but the commercial product contains small amounts of iron oxide, silica, and other minerals. The specific gravity of commercial barite is 4.2. Barite is ground to such fineness that not over three percent is coarser than 200 mesh, and not less than five percent is retained on 325 mesh.[71]

In most shallow drilling, a column of water provides more pressure than fluids present in a porous formation that is penetrated. A column of fresh water exerts a pressure of 43.3 pounds per square inch for every 100 feet of depth. Under ordinary conditions, drilling mud weighs between 8.5 and 9.5 pounds per gallon and, thus, affords a pressure of from 44.2 to 49.4 pounds per square inch per 100 feet of depth. When gas or water under greater pressure is encountered, the weight of the mud must be increased to maintain control.

Clays are 2.5 to 2.6 times as heavy as water. Native clays can be used for only small increases in density. The rise in viscosity, gel strength, and cake thickness with the increased volume content of clay solids retards drilling progress to such an extent that a maximum mud weight of about ten pounds per gallon (52 psi per 100 feet of depth) is the practical limit, even with an optimum addition of thinners.

Increasing the mud weight with barite, which has a specific gravity 1-2/3 times that of clay, causes less increase in volume of solids. For example, a clay-water mud weighing 10.5 pounds per gallon contains 17.4 percent by volume of solids as compared to 9.0 percent by volume of solids in a bentonite-barite-water mud of the same density. Furthermore, the shape and size of the particles of barite cause a smaller increase in viscosity than an equal weight of clay particles. Bentonite or other suspending agents, however, must be used with barite to prevent settling.[18]

Minor Additives

Caustic soda (flake caustic, lye) is used: to make tannins, lignite, and chromelignosulfonates soluble and effective as thinners; to raise the pH of acidic makeup water; and to reduce corrosion of iron by maintaining the pH above ten. Caustic soda is a hazardous chemical and should be handled carefully to avoid burns.

Lime (hydrated lime) is sometimes added to bentonite mud to rapidly develop a stiff gel in order to "wall off" loose sands and gravels. The "clabbering" destroys the filtration properties of the mud; and therefore, such mud should be discarded as soon as the problem section has been drilled. Lime is much less injurious to the skin than caustic soda but should be washed off to avoid irritation.

Soda ash (washing soda, sodium carbonate) is used: to remove hardness from water by precipitating calcium and magnesium salts; to raise the pH of acidic waters; and to treat anhydrite-contaminated mud. *Baking soda* (sodium bicarbonate) is used to counteract cement contamination of mud.

Surfactants concentrate in liquid interfaces, adsorb on the surfaces of solids and, thereby alter interfacial tension and wettability. Surfactants, combining detergent and water-wetting properties, increase the rate of penetration by causing water to spread along the freshly exposed surface of the rock as the bit penetrates. The cleaning action of water in flushing the cuttings from the bit is enhanced. Effective detergents are beneficial in concentrations as low as one part to five-hundred parts of water.

Lubricants reduce frictional drag or torque. Although lubrication of the tools is one of the functions the drilling fluid is expected to perform, lubricant measurements are not made until a modified *Timken-lubricant tester* is employed in a demonstration of the effectiveness of sulfurized oil derivatives.[537,610] With a modification of the testing device and an emphasis on the avoidance of oil in the mud, a composition containing triglycerides and higher alcohols is introduced. This composition is effective in fresh-water muds without the need for an oily vehicle.[456, 458]

Mud Performance as Related to Properties

Satisfactory performance of the drilling fluid depends not only on the composition selected but also on the drilling equipment employed and the rock type drilled. Frequently, alterations must be made in the properties of the mud to compensate for deficiencies in the drilling equipment.

Several physical properties that affect the performance of the mud are: the density, viscosity, gelation, lubricity, and filtration qualities of the slurry; the size, shape, number, hardness, and abrasiveness of the suspended solids; and the chemical and interfacial forces that exist within the suspension in the borehole. However, as a practical means of control, only a few of these properties can be measured.

Experience has shown that certain measurements of properties can be related to performance under field conditions. A committee of the American Petroleum Institute has surveyed new methods and instruments and published recommendations.[98]

Mud Density

A significant but easy measurement to be made by the drilling operator is mud weight or density. A reliable visual estimate cannot be made — density must be measured.

The pressure exerted by the drilling fluid is directly related to the density. To prevent subsurface flow of fluids into the hole, the drilling fluid must furnish a pressure greater than that of the fluids in the porous rocks that are penetrated by the bit. Loss of circulation may result from excessive pressure exerted by an overly dense drilling fluid. As density increases, the buoyant effect on cuttings and the carrying capacity of the mud increases, but the settling of cuttings at the surface is retarded. With simple waterbase muds, density can be regarded as a measure of suspended solids.

Solids that do not contain useful properties are definitely objectionable. Solids interfere with the transfer of heat from the bit to the circulating liquid. Abrasive solids cause excessive wear on pumps, drill string, and bit body. When unnecessary solids are recirculated after accumulating in the mud, the rate of drilling is retarded, a thick filter cake is deposited on permeable formations, and unnecessary work is performed by the pump. Under conditions of normal formation pressure, the effects of excess mud weight increase as the hole deepens because of the increase in differential pressures.

Density or mud weight is measured with a mud balance. Density usually is expressed as pounds per gallon; although other units are also used (specific gravity as pounds per cubic foot, hydrostatic gradient as psi/1,000 ft). The *mud balance* consists essentially of a cup attached to one end of a graduated arm along which a rider counterbalances the cup filled with mud.

Flow Properties

The removal of rock chips from the cutting face of the bit and the transport of these cuttings to the surface depend on the flow properties and the velocity of the drilling fluid. *Viscosity* is defined as the resistance to flow offered by a fluid (liquid or gas). More precisely, viscosity is the frictional drag (of one platelet of liquid sliding over another) divided by the relative velocity of the platelets. The component factors of this relationship are: the viscosity of the liquid phase; the size, shape, and number of suspended particles; and the forces existing among the particles and between the particles and the liquid. The carrying capacity of the drilling fluid depends upon the relationship between the shear stress and the shear rate that exists under the conditions of flow. Even if known, actual flow conditions throughout the path of the mud could not be reproduced in a simple instrument acceptable for field use. For convenience, the flow properties are measured under strictly arbitrary conditions.

The *Marsh funnel* is a means for making viscosity measurements that are useful only when used by an experienced operator. Such measurements cannot be used to calculate pressure drop or carrying capacity and may be misleading unless the limitations of the method are recognized. The "funnel viscosity" is the time in seconds required for one quart of mud to flow from the filled funnel. Time of outflow of water at 70°F is 26 ± 0.5 second. Funnel viscosity of a mud is affected by the density and by the development of the structure resulting from the attractive forces between the particles.

The property of *gel development* is associated closely with the flow properties of most water-base muds. If the mud stops moving (or gels), the forces between the suspended particles may create a structure that will require exerting a measurable shearing stress to initiate movement. The force necessary to break the gel is called the "gel strength." High gel strength may require a high pump pressure to start circulation after a period of shutdown, thereby causing mud to be lost to the formation.

Rapid gel strength has the disadvantage of retarding the settling of cuttings in the mud pit, but the same property is an advantage in retarding the settling of cuttings in the borehole when circulation is stopped. Cuttings can settle down around the bit in the borehole during a breakdown or when circulation is stopped, and this can create problems in regaining circulation and even in binding the bit. This problem, although not eliminated by high gel strength fluids, is reduced.

As a means of more effectively evaluating the flow properties of drilling muds, a multispeed rotational viscometer was employed in conjunction with the Bingham model or plastic flow.[171] By suitable adjustment of speed settings and instrument constants, a viscometer was designed that permitted the calculation of plastic viscosity and yield point from torque readings made at two fixed-rotor speeds.[548] The shear stress/shear rate relations for three liquid models are illustrated in Figure 18. Based on the Bingham model, measurements made at rotor speeds of 300 rpm and 600 rpm permit direct calculation of plastic viscosity and yield point as follows:

Plastic viscosity (centipoises) equals 600 rpm reading minus 300 rpm reading.

Yield point (lb. per 100 sq. ft.) equals 300 rpm reading minus plastic viscosity.

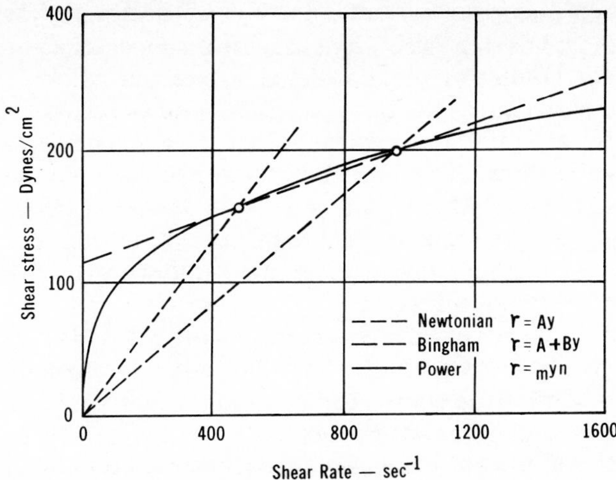

Figure 18[643] Stress-rate relation for three liquid models.

Apparent viscosity (centipoises) 600 rpm reading divided by two. These relations are shown in Figure 19.

The measurements are applied in hydraulics calculations and in the interpretation of the factors that regulate flow properties of mud.

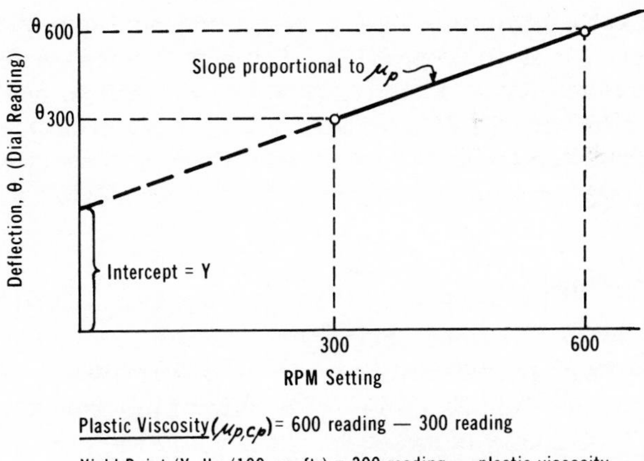

Figure 19[530] Interpretation of Fann viscometer readings in terms of plastic viscosity and yield point.

Apparent viscosity can be regarded as an "averaging" of several factors affecting the flow of mud. *Plastic viscosity* can be related to the mechanical friction among the solids, friction between the solids and the liquid, and the resistance to flow of the liquid. Not only the quantity of solids but also their size and shape affect plastic viscosity. *Yield point* can be considered a measure of the forces between the particles while the mud is in motion. These forces are electrostatic in nature and involve both attraction and repulsion. *Gel strength* is a measure of the effect of the forces between the particles while the mud is at rest. Gel strength is read from the viscometer dial in pounds per 100 square feet after the mud has been stirred thoroughly and then allowed to remain undisturbed, first for ten seconds and then, ten minutes.

Practical applications of the measurements made with the direct-indicating, two-speed viscometer have led to improvements in drilling practices in addition to more efficient and economical control of the flow properties of the mud. For example, if apparent (funnel) viscosity is high, addition of dispersants (thinners) lowers the yield point component while dilution by water is necessary to lower the plastic viscosity. Thus, dispersants need not be wasted to lower the apparent viscosity if the yield point is low, and plastic viscosity is high. These measurements can be used to calculate the pressure drop in the various segments of the flow system and the equivalent circulating density of the mud.[182,438]

Walker[643] points out that the Bingham model does not apply to solutions of non-ionic polymers and that shear stress/shear rate relations, even for a bentonite mud, are more closely represented by the Power model (refer to Figure 18). The significance of the velocity profile in the annulus as affecting hole cleaning was emphasized, and the relation of the Power constant (N) to the readings of the viscometer at 300 and 600 rpm was demonstrated. This approach has been very useful in problems of hole cleaning. (See "Hole Cleaning.")

Mud Filtration Properties

The ability of the solid components of the mud to form a thin filter cake of low permeability on a porous formation is a property intimately related to hole stability, freedom of movement of the drill string, and satisfactory completion of the water well. Clear water can be pumped into a permeable formation, such as sandstone, under a given pressure at a rate dependent on the size and configuration of the openings in the rock. When water carrying suspended solids comes into contact with a porous, permeable formation, solid particles immediately enter the openings. As the individual pores become bridged by the larger particles, successively smaller particles are filtered out until only the liquid passes through the openings.

Thus, the mud solids are deposited as a filter cake, proportional in volume to the relative volumes of the liquid and solid fractions in the mud. As soon as the bridging of the openings has occurred, the sealing property is dependent upon the amount and physical state of the colloidal material in the mud and not on the permeability of the formation.[677]

While the mud is being circulated, part of the cake is continually washed away. The amount of liquid (filtrate) entering the porous rock depends on the sealing qualities of the thin sheath at the bore wall. When circulation stops, the filter cake begins to build up on the wall. Several problems (often attributed to other causes) may then arise if the mud has a high solids content and a high-filtration rate. If, when rotation is stopped, the drill pipe is in direct contact with the filter cake on permeable, porous rock, the pipe may be held firmly in place by the differential pressure.[311] (See "Stuck-Pipe," this chapter.) The pipe becomes wall-stuck and cannot be rotated even though there is free circulation of the mud. Even if the pipe is not stuck, severe "swabbing" may occur as the pipe is pulled. When re-entering the hole, a semi-solid cake may be encountered and interpreted as "tight spots" or "bridges." The texture as well as the thickness of the filter cake is significant. A gritty, sticky texture creates more frictional drag on the pipe than that of a smooth, slick cake.

Filtration properties of mud are measured by means of an API standard filter press. A sheet of hardened filter paper having an area of seven square inches is supported on a screen in the base of the metal cell. The cell is filled with mud, and a pressure of 100 psi is applied. The liquid, passing through the paper, is collected and measured (cubic centimeters) after a period of thirty minutes. The pressure is released, the mud is poured from the cell, and the thickness of the filter cake is measured after removing any adhering mud. The test is arbitrary and can be related to downhole conditions only through experience.

Sand Content

The sand content of the mud is abrasive, makes a thick filter cake, and causes difficulties associated with settling in the hole. The sand-content test measures the amount of solids coarser than 200-mesh. The test set consists of a glass tube having a tapered, graduated lower section, a plastic funnel, and a 200-mesh sieve. The tube is filled with the appropriate amount of mud, water is added, and the diluted mud is washed onto the screen until only the sand remains, the funnel is then fitted on top of the screen, the screen is inverted, and the sand is washed into the tube. After settling, the percent of sand is indicated in the graduated section of the tube. Sand content should be measured at the pump intake to determine how much is being recirculated and should be kept to a minimum.[207]

Routine Testing Program

Time and money can be saved by recording mud properties. Measurements of mud weight and funnel viscosity in many cases furnish sufficient information for adequate control of mud characteristics. Mud weight should be measured at the mud pit and at the pump intake to determine how effectively the cuttings are separating. Emphasis must be given to the measurement of density. An increase in solids from cuttings reduces drilling penetration rates, increases bit wear, and increases filter cake and pressure downhole with the possibility of sticking the drill pipe and losing circulation. Funnel viscosity should be just adequate to carry the cuttings and provide a stable hole. Based on experience with standard mud, limits can be set for weight and funnel viscosity which will assure satisfactory filtration properties. For example, if a fresh-water bentonite mud has a funnel viscosity of 32 to 38 seconds and weighs less than nine pounds per gallon, satisfactory performance usually can be expected for average drilling.

Drilling Fluid Surface System

Attention should be directed to the importance of the surface mud system in its relationship to efficient operations. In brief, the surface mud system involves:

(1) Separation of unwanted materials, principally cuttings.
(2) Storage of active, reserve, and waste mud.
(3) Addition and mixing of ingredients.
(4) Agitation to maintain uniformity.

The components of the surface mud system have been examined in detail by McGhee[408] and Bobo[129] and these publications should be consulted for additional information on rig layout and equipment involved.[176]

Recirculation of sand and silt is recognized as an objectionable practice, but little attention is given to the removal of sand from the drilling fluid. Ample evidence has been collected to prove that removal of sand and silt results in a cost reduction of rig maintenance, in an improved penetration rate, and in reduced hole problems.[213,426,487,595,667]

FACTORS THAT AFFECT DRILLING RATE

A number of factors which affect rotary-drilling rate have been listed by both Gatlin[243] and Cunningham and Goins,[191] and a combined list should include the following:

(1) Formation Characteristics.
  - (a) Strength.
  - (b) Abrasiveness.
  - (c) Drillability.
  - (d) Physical state of existence.
    1. Permeability.
    2. Porosity.
    3. Fluid content.
    4. Interstitial pore pressure.
    5. Confining pressure.
    6. Borehole pressure, including drilling-fluid pressure plus annular, circulation-pressure losses.
    7. Temperature.
  - (e) Stress-strain relationship.
  - (f) Stickiness or balling tendency.

(2) Mechanical Factors.
  - (a) Weight on bit.
  - (b) Rotary speed.
  - (c) Condition of bit.
  - (d) Bit diameter.
  - (e) Bit type.
    1. Intended mode of deformation — scraping, crushing, or combination of both.
    2. Geometry of cutting structure.
    3. Mode of drilling fluid introduction — conventional water course or jets.
    4. Location and aiming of fluid-entry ports.

(3) Hydraulic Factors.
  - (a) Circulation rate.
  - (b) Friction losses.

(4) Drilling Fluid Properties.
  - (a) Density.
  - (b) Filtration (static and dynamic).
  - (c) Solid content.
  - (d) Viscosity.
  - (e) Yield point.
  - (f) Oil content.
  - (g) Wetting characteristics.

(5) Intangible Factors.
  (a) Personnel efficiency.
    1. Competence.
    2. Psychological factors.
  (b) Rig efficiency.
    1. State of repair.
    2. Proper physical size for work intended.
    3. Ease of operation.

These factors will be discussed in order. A rotary-drilling rate equation can be developed to relate mathematically those factors mentioned where such relationships have been defined. This equation will be utilized in the following section on economic optimization to discuss the control of rotary drilling costs.

Formation Characteristics

Rock mineralogy, pressure, and fluid environment in place determine the drilling strength of rock. Rock-crater volume varies with the depth of wedge indentation, squared, as follows:[429]

EQUATION 4

$$V_C \sim X^2$$

Where: V_C = Crater volume

X = Depth of penetration

Note: $\sim$ = Is proportional to

This relationship holds for craters produced in rock by the impact of spheres[429] and wedge-shaped chisels.[504] Furthermore, good correlations have been reported between crater depth and the reciprocal of both rock-shear strength and compressive strength.[235,428] Therefore, with constant force on the tooth:

EQUATION 5

$$X \sim \frac{1}{S}$$

Where: X = Depth of penetration

S = Rock-drilling strength

Combining relationships, Equations 4 and 5, show that crater volume is proportional to the inverse of rock-drilling strength, squared, or:

EQUATION 6

$$V_c \sim \frac{1}{S^2}$$

Where: V_c = Crater volume

S = Rock-drilling strength

Since rock-drilling rate will be directly proportional to the rate at which craters can be formed, the equation below applies:

EQUATION 7

$$R \sim \frac{1}{S^2}$$

Where: R = Drilling rate

S = Rock-drilling strength

Thus, by defining rock-drilling strength — rock compressive strength and shear strength are not considered in developing a drilling relationship. The physical state of the rock will have the effect of changing drilling strength, and drilling rate will vary in accordance with Equation 7. Thus rock-drilling strength will control the entire drilling process and is the most important factor which affects drilling rate.

Mechanical Factors

Mechanical factors, principally *weight on bit*, *rotary speed*, and *bit type* are quite easily adjusted, and the influence on drilling rate, at least in a qualitative sense, is well understood.[231] In an analysis of field data, Gatlin[242] concludes that penetration rate is a linear function of weight on bit over a limited range of bit weights if other drilling conditions remain the same. This work was later qualified by Moore and Gatlin.[465]

> Penetration rate in "soft" (easily drillable) formations is directly proportional to the applied bit weight, i.e., by doubling the weight, the drilling rate doubles, as long as the drilling fluid removes the cuttings or until the bit "flounders." In "hard" rocks, penetration rate increases with bit weight at an increasing rate, finally becoming essentially linear above a "critical" bit load.

Woods and Galle,[664] in conjunction with an American Association of Oil Well Drilling Contractors' Committee, studied several examples of field tests and concluded that penetration rate increased linearly with weight on bit, but that analysis of such field data was greatly hampered by formation changes.

In theory, drilling rate varies with the square of bit weight when "perfect cleaning" of the hole bottom occurs.[429] In reality, "good"

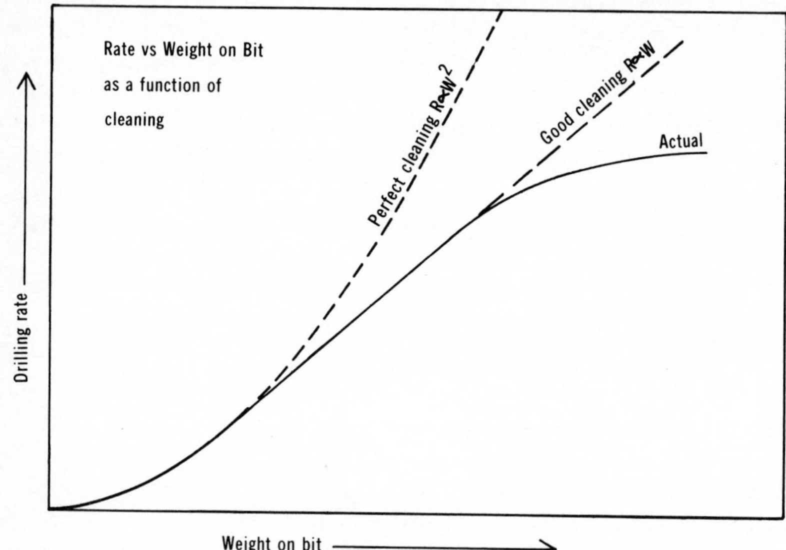

Figure 20[468] Rate vs. Weight on bit as a function of cleaning.

cleaning is obtainable, and in this case, penetration rate varies with weight in a linear manner. When bottom-hole cleaning is inadequate, increases in bit weight do not give incrementally proportional increases in penetration rate, and the point at which this condition arises (on an R vs. W curve) is defined as the "flounder point." These three responses are shown in Figure 20. Figure 21 shows the location of the flounder point for various levels of hole cleaning.[464] Laboratory data illustrating the effect of bit weight on penetration rate for various rocks is shown in Figure 20.

To mathematically describe the effect of bit weight on penetration, Figure 22 demonstrates that all the curves are linear beyond a minimum weight, and no flounder is present. Therefore, if only weights greater than a threshold and less than the flounder are considered, the following applies:

EQUATION 8

$$R \sim (W - W_o)\ ;\ W_o < W < W_f$$

Where: R = Drilling rate

W = Bit weight

W_o = Threshold weight

W_f = Flounder Weight

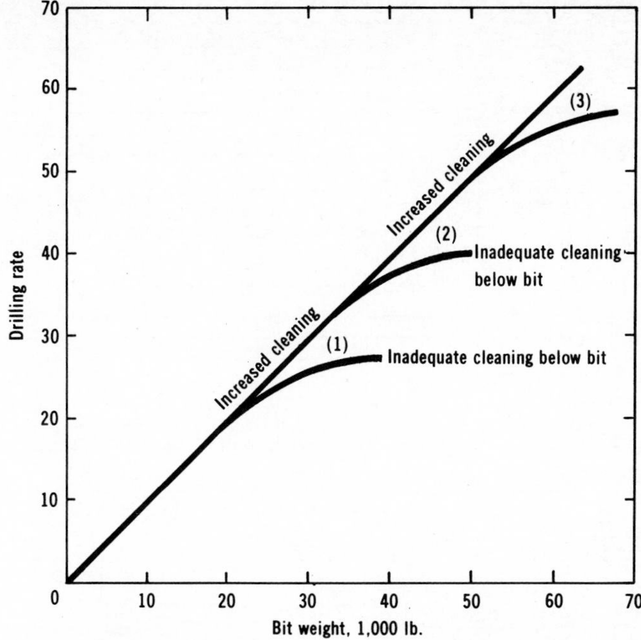

Figure 21[237] Drilling Rate vs. Bit Weight at various hydraulic levels.

Note: The value of W_o is obtained by extrapolating the straight-line portion of the curve to the zero on the ordinate, or zero penetration rate. In the case of the pink quartzite shown in Figure 22, W_o = 26,500 pounds.

Curves, as shown in Figure 22, may be obtained in the field by performing a five-spot test as suggested by Young[669] or by a drill-off method as proposed by Lubinski.[394]

The benefits of using additional collar weight to increase drilling efficiency has long been recognized; and, of the mechanical factors affecting drilling rate, bit weight is significant.[465,466]

The effect of *rotary speed* on drilling rate is not well established because of the interrelationship of this variable with such complicating factors as hole cleaning, rate of rock loading, and drill-string vibrations. In general penetration rate increases with rotary speed. At high rotary speeds, however, the increase in drilling rate is progressively less than the increase in speed. Figure 23 shows this relationship, and field observations verify the general shape of the curve. Figures 24 and 25 show laboratory and field data for various levels of bit weight.

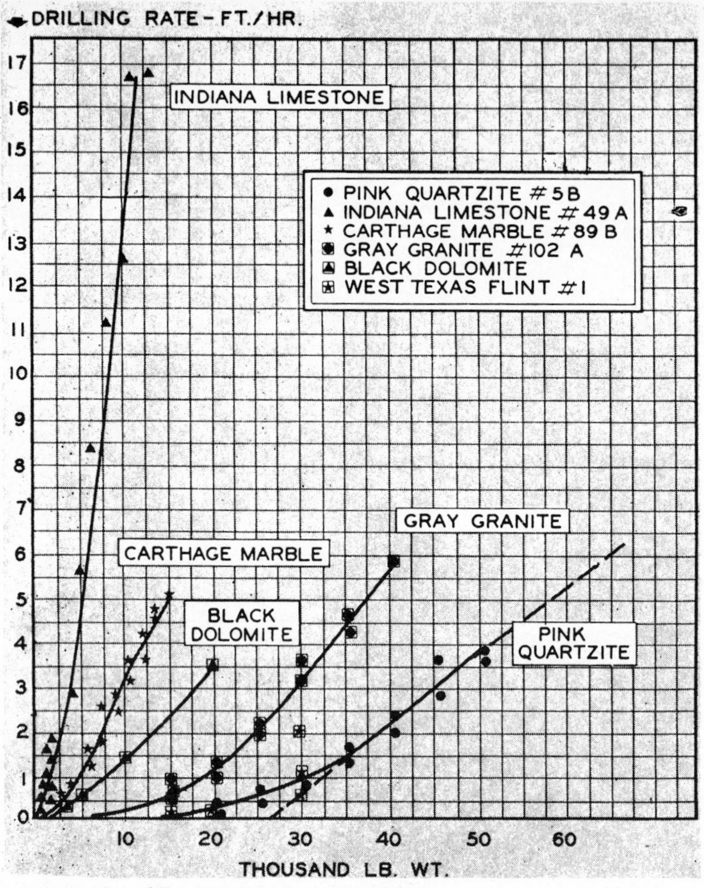

Figure 22[465] Drilling Rate vs. Bit Weight — laboratory data.

In general mathematical form, penetration rate is related to rotary speed according to the following equation:

EQUATION 9

$$R = f\,(N)^{\lambda}$$

Where: R = Drilling rate
f = A constant
N = Rotary speed
λ = Exponent

The exponent λ is an empirically derived value and normally varies between 0.5 and 1.0. "Perfect cleaning" theory suggests λ = 1.0[429]

Field data, however, show that λ varies with rock-drilling strength.

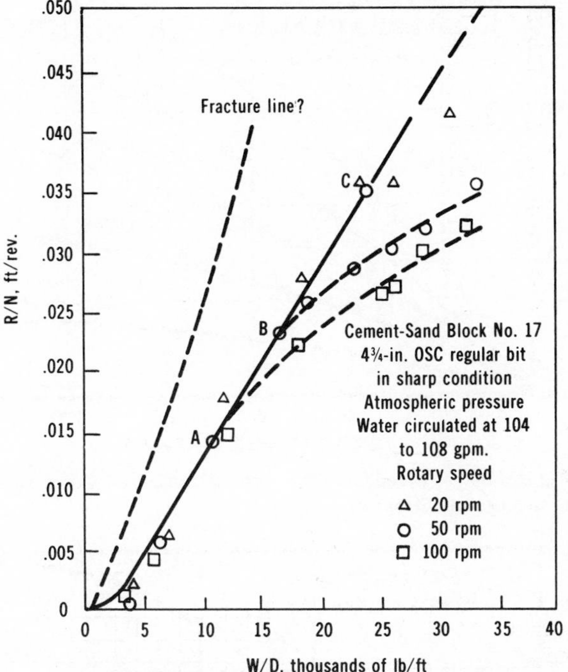

Figure 23[466] Plotting against penetration per revolution.

The dependence of λ on rock strength is shown in Figure 26. Because wear of drill pipe, drill collar, and bit teeth is primarily a function of rotary speed, excessive rotary speed (above 150 rpm) is usually avoided in all but the softest rock.

The value of the exponent λ can be determined in the field by plotting penetration rate vs. rotary speed on log-log graph paper with all other variables constant. The slope of the resulting curve is equal to the exponent, λ.

In an extensive analysis of field data, Eckel and Rowley[213] conclude that the penetration-rate vs. rotary-speed curve flattened even at high circulation rates. The analysis also indicates great dissimilarities among formations penetrated and shows a 50 percent difference in penetration rate over a 40-foot interval using constant bit weight, rotary speed, and circulation rate. While the effect of rotary speed on drilling rate is secondary to the effect of bit

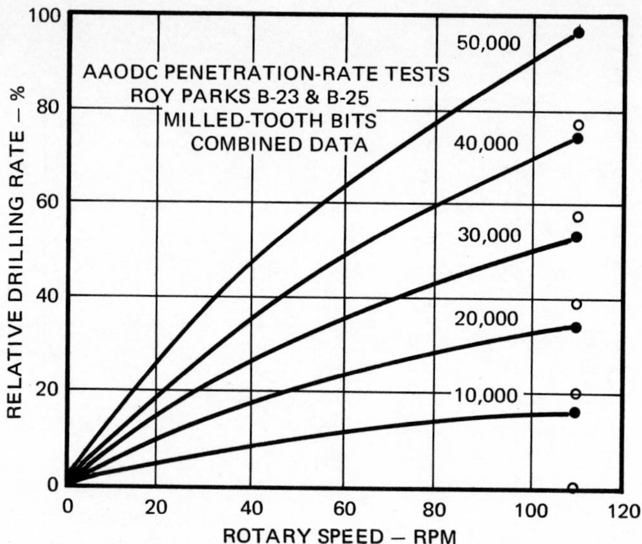

Figure 24[467] Effect of rotary speed on penetration rate at various bit weights.

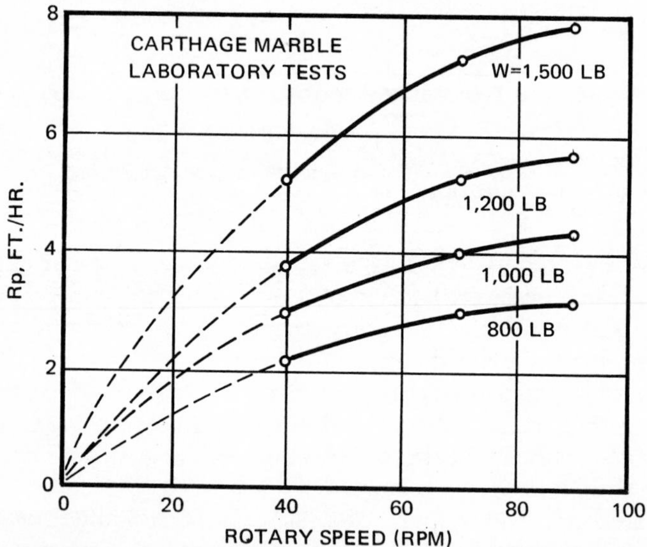

Figure 25[467] Effect of Rotary Speed on Penetration rate at various bit weights.

weight, rotary speed deserves consideration in view of the complications described above.

As any bit dulls, efficiency is reduced because the stress concentrated on the cutting surface decreases as the area of contact increases. Gray, *et al.*,[211] found that in fixed-blade studies, the rake angle of the cutting edge significantly influenced the forces on the blade tip. The rake angle is defined as the angle between the upper-blade surface and the vertical angle. A reduced-rake angle simulates *bit wear*, since the force required to initiate rock failure is concentrated at the blade tip. To maintain a constant cut depth (constant drilling rate) with a decreasing rake angle (bit wear) greatly increased resultant forces on the tip were required.

A dull bit normally shows a slower drilling rate than the rate measured for a sharp bit. The degree to which this effect is observed in the field is related to the formation being drilled, the bit type, the pressures at the cutting surface, and the type of drilling fluid.[191] The relationship may be determined by noting drilling-rate behavior as the bit is dulled at constant bit weight and rotary speed, as is shown in Figure 27.

Mathematically, the curve in Figure 27 may be expressed as follows:[274]

EQUATION 10

$$R \sim \frac{1}{1 + C_2 H}$$

Where: R = Drilling rate

C_2 = A constant

H = Normalized tooth height

The American Association of Oil Well Drilling Contractors adopted a "Code of Bit Condition" as a method of grading bit wear.[40] Tooth wear is graded in eights; thus, a new bit is 0/8, a bit with one-half the cutting structure height removed is 4/8, etc.

The constant, C_2, in Equation 10 is empirically determined. The drilling rate of a worn bit, R_1, is measured, and after replacing the worn bit with a new one, the new drilling rate, R_2, is measured:

EQUATION 11

$$\frac{R_1}{R_2} = \frac{1 + C_2 H_2}{1 + C_2 H_1}$$

Where: R_1 = Drilling rate (worn bit)

R_2 = Drilling rate (new bit)

C_2 = A constant

H_1 = Normalized tooth height (worn bit)

H_2 = Normalized tooth height (new bit)

Because H_2 = 0 for a new bit, Equation 11 may be simplified to:

EQUATION 12

$$C_2 = \left(\frac{R_2}{R_1} - 1.0\right)\frac{1}{H_1}$$

For example, if R_2 = 100 ft/hr (new bit), and R_1 is 50 ft/hr, and the dull bit is graded a 4/8, then Equation 12 would read:

$$C_2 = \left(\frac{100}{50} - 1\right)\frac{1}{0.5} = 2.0$$

Correction of penetration rate values for condition of the bit teeth is important when logging wells during drilling because observed

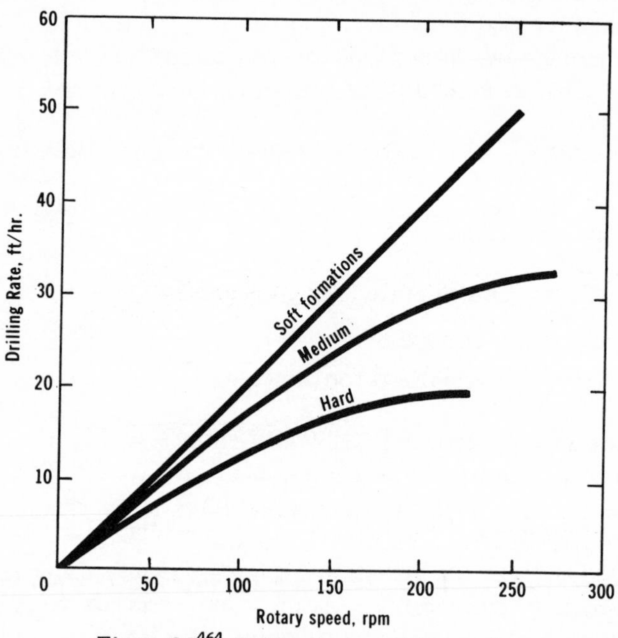

Figure 26[464] Drilling Rate vs. Rotary Speed.

trends can be used for correlation purposes, and increases in bit weight and rotary speed may be justified.

Drilling rate is independent of *bit diameter* where the ratio of bit weight to diameter (lb/in) is held constant. With constant bit weight and rotary speed, however, drilling rate varies inversely with the diameter, squared:[429]

EQUATION 13

$$R \sim \frac{1}{D_2}$$

Where: R = Drilling rate

D = Bit diameter

Equation 13 describes the correlation of drilling rate to area of the hole drilled by the cutting structure. Figure 28 illustrates the effect of bit diameter on drilling rate.

The *bit type* selected for drilling plays an important role in the balance of performance with capability. The intended mode of deformation, cutting structure geometry, and method and location of fluid introduction to the cutting area are equally important considerations. Experimentation and research have resulted in bit designs that provide a specific bit type for a wide range of drilling conditions. These bits are divided into three broad classifications: the drag bits, the rolling-cutter bits, and diamond bits. Drag bits drill by scraping; roller bits drill by a combination of scraping and crushing; and diamond bits drill by ploughing.

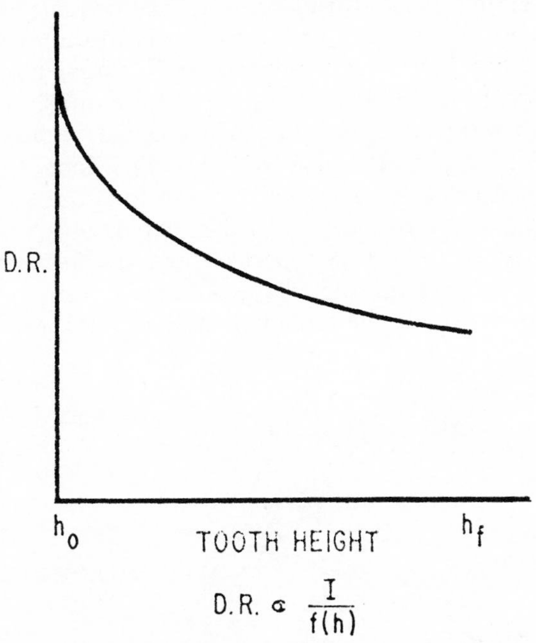

Figure 27[215] Effect of tooth wear on drilling rate.

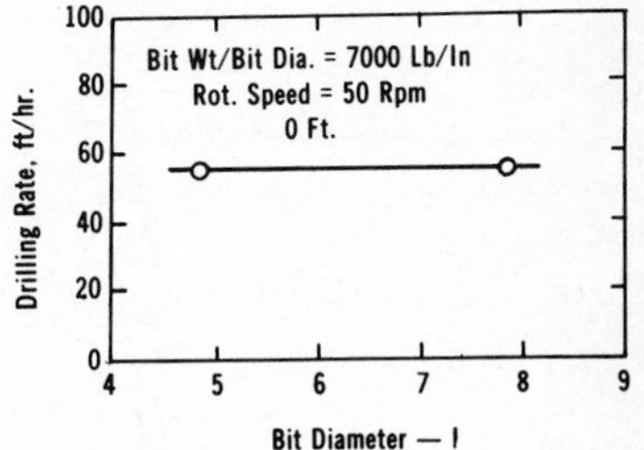

Figure 28[429] Drilling Rate vs. Bit Diameter.

The *drag bit* generally is equipped with short blades and a body fitted with water courses that will direct the drilling fluid stream to keep the blades clean. This arrangement assists penetration by means of a jetting action against the bottom of the hole. These bits are best adapted to the drilling of gumbo, sticky shale, and near-surface, poorly consolidated formations.[568]

The *rolling-cutter bits* (cone bits) provide a wide range of capability for drilling harder formations. These bits may be classified according to the number of cones, the method of fluid introduction, the cutting structure, and the bearing type.[324] Roller bits are available in both double-cone and triple-cone configurations. Drilling fluid may be introduced either through a single water course or through jets. Regular water-course bits contain a large "throat" or opening at the center; whereas, jet bits introduce fluid through one, two, or three openings above the cones. The choice of nozzle number depends upon desired nozzle area, and many combinations are available. Nozzles of varying internal diameter are chosen to achieve a desired pressure drop at the selected drilling-fluid flow rate. A

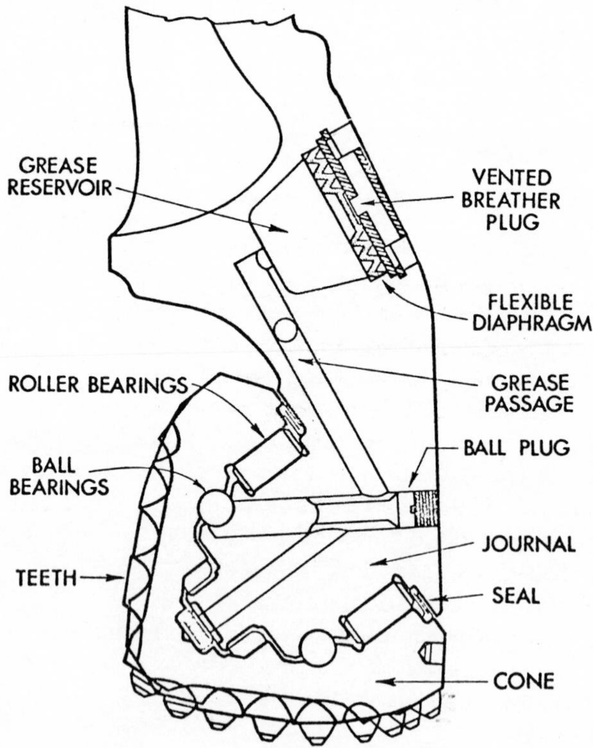

Figure 29[189] Sealed Bearing Bit.

variety of tooth configurations are available to drill rocks of varying composition.[47] The teeth may be either milled from steel or constructed of carbide inserts of varying shapes. The individual cones roll on bearing surfaces, and three bearing types are available: (1) nonsealed roller, (2) sealed roller, and (3) sealed friction.[219] Sealed-bearing bits have a closed, grease-lubrication system activated by mud pressure; whereas, unsealed-bearing bits are lubricated by the circulating fluid. Roller-bearing bits are constructed to provide cone support through a combination of roller and ball bearings. The friction bearing was developed after the sealed, lubricating system was designed, and there are three types: segment, solid bushing, and journal.[253] Figure 29 illustrates the basic design principle and construction of the seal and lubricating system in a sealed-bearing bit. The most important part of this design is the grease seal between the cone and the shank. In recent years, design improvements in allowable tolerance and metallurgy have led to increased use of the sealed-bearing bit. Figure 30 shows the friction-bearing design principle. By taking the cross section (A-A') through the cutaway view of the sealed-roller-bearing bit, the roller cross section is indicated. The load, imposed upon the journal, is transmitted from the roller cone to the journal through the roller bearings and represents the reaction of the formation to applied bit weight. In the segmented- and solid-bushing-type bearing, the segment or bushing simply replaces the roller bearings. In the journal-type bearing, the inside surface of the cone shell directly contacts the journal.[253]

Friction-bearing bits have a decided advantage over roller-bearing types because in the friction bearing, the load is more distributed over the journal. This distribution reduces spalling of the bottom of the journal race, thereby increasing bearing life.[59]

The combination of shaped inserts and friction bearings has a strong impact on drilling economics in certain areas.[328] Although the cost of these new-style bits is relatively high, the results are impressive even in shallow drilling. Tables 1 and 2 illustrate this point. The economic applications in shallow drilling require careful analysis. A method of analysis will be discussed in a later section on optimization of the rotary-drilling process.

Whereas the inserts in the recently developed bits are all composed of tungsten carbide, different grades of this carbide are being tested. Various types of sintered tungsten carbide are being evaluated to develop inserts suitable for drilling specific formation types. The longer inserts are made of softer tungsten carbide which is less resistant to abrasion but also less subject to brittle breakage. Thus,

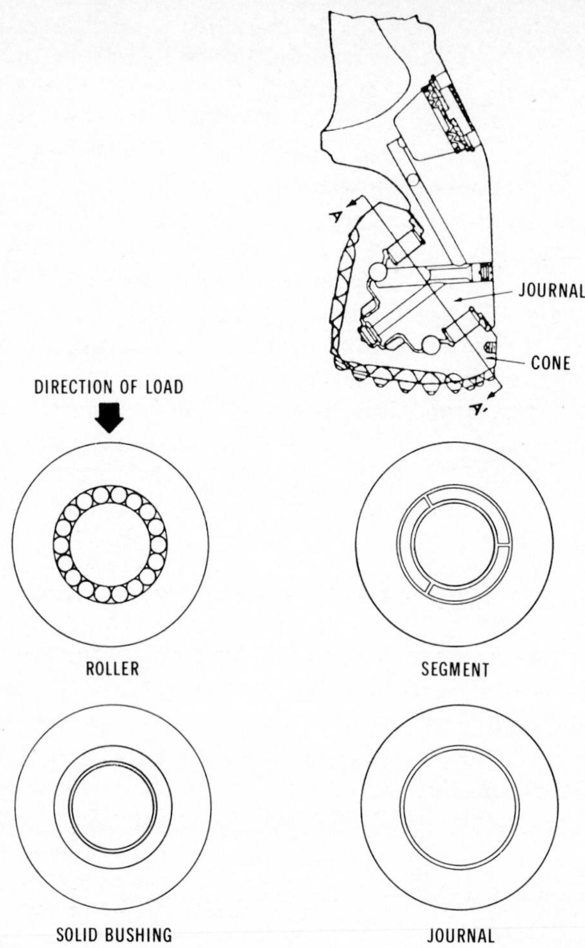

Figure 30[189] Cross Section A-A' Illustrates Friction Bearing Design Principle.

tungsten-carbide insert bits exhibit bit wear prior to bearing failure. The effect of dulling on drilling rate, as described earlier, can also be observed. The drilling characteristics of these bits are now being established, but a better understanding of existing limitations is needed.[373,542]

A classification of various milled-tooth, tricone roller bits is shown in Table 3. This classification, a modification of Bentson,[117] is based upon bit type and formation drillability. On a horizontal row, bit type designations A-K correspond to the formation types as described in bit manufacturer's catalogs.

TABLE 1[253]

Relative "Cost Number" Comparison Chart for 7-7/8" Bit

BIT TYPE	JET BITS	JET BITS WITH GAGE INSERTS "G"	SEAL BEARING BITS	SEAL GEARING BITS WITH "G"	FRICTION BEARING BITS		
					SEGMENT	SOLID BUSHING	JOURNAL
Mill Cutter	1.0	1.2	1.3	1.4			†1.4
Shaped Inserts	4.1		4.6		4.8	6.1	
Round Top Insert	4.3		4.7		4.9	* 6.2	

*Estimated
†Tentative Cost Number

TABLE 2[253]

Use of Shaped Insert Bits in Shallow Drilling

FIELD NAME, HOLE SIZE, AND TOTAL DEPTH	WELL NUMBER	INTERVAL DRILLED	FOOTAGE DRILLED	SHAPED INSERT	MILL TOOTH	TOTAL BITS	ROTATING HOURS	TRIP HOURS	COST/FT.
(A)									
HONDA FIELD	1	2002 – 4855	2853	–	13	13	229	34.5	6.70
IRION CO., TEX.	2	1980 – 4845	2865	–	13	13	264	35.4	7.42
HOLE SIZE: 7 7/8"	3	2175 – 4840	2665	4	–	4	213	11.1	6.60
TD 4950′	4	1855 – 4935	3080	2	2	4	219	12.0	5.41
(B)									
TIPPETT FIELD	1	950 – 4760	3810	–	12	12	215	29.3	4.64
CROCKETT CO., TEX.	2	1000 – 4811	3811	–	10	10	206	25.0	4.38
HOLE SIZE: 8 3/4"	3	987 – 4768	3781	2	3	5	203	12.2	4.36
TD 5500′	4	955 – 5193	4238	2	2	4	233	9.8	4.21

Rock bits are usually classified by water well contractors on the basis of the formation for which they are designed to drill. Four major classifications are in use and are described as follows.[84]

(1) *Soft formation rock bits* have widely set teeth and cut deeply so that large cuttings may be dislodged at a fast rate. This tooth arrangement prevents tracking and promotes cleaning of the cutting surfaces. The cones are offset.

(2) *Medium-to-hard formation rock bits* have a greater number of shorter teeth per cone than no. (1). As the formation becomes harder, this bit tends to chip more readily; therefore, the added number of teeth, more closely spaced, promote greater footage per bit. This bit has less cone offset and greater bearing capacity than is found in the soft formation bits.

(3) *Hard formation rock bits* require a cutting structure designed to withstand the high weights demanded to penetrate these dense, high-strength formations. The teeth are shorter and are milled with a large included angle to resist breakage and battering. The cones are offset to a lesser degree than in bits for softer rocks.

(4) *Very hard formation bits* have short teeth, closely spaced. To prolong the life of the gage and the cutting structure, the cones are not offset. The rows of teeth intermesh to clear the grooves.[319]

Cross-roller rock bits constitute a broad class of bits with the main cutting elements arranged at a right-angle cross. All four cutters

TABLE 3[117]

Three-Cone Rock Bit-Formation Classification

Affixed Type Designation	ROCK BIT MAKE, DESIGN, AND TYPE DESIGNATION						FORMATION TYPES (Combined from information contained in each of the manufacturers catalogs)
	H. C. SMITH	HUGHES	CHICAGO PNEUM.	SECURITY	GLOBE	MURPHY	
	3 Cone	Tri-Cone	3 Cone	3 Cutter	3 Cone	3 Cone	
A	3C-DT	OSC-3	ES-1C	S3	SS3C	YT-3	Soft formations having low compressive strength and high drillability (sticky shales, clays, red beds, salt, soft limestone, unconsolidated formations, etc.)
B	3C-DT2	OSC-1 OSC-1G	ES-1	S4	S3C	YT-1	Soft to medium formations or soft interspersed with harder streaks (firm, unconsolidated, or sandy shales, red beds, salt, anhydrite, soft limestone, etc.)
C	3C-K2P	OSC	ES-2 ES-3	S6	M3C	YT	Soft to medium formations interspersed with hard streaks (medium hard and unconsolidated shales, red beds, chalk, salt, anhydrite, gypsum, medium hard limestones, unconsolidated sands, etc.)
D	3C-SV2	OWV	EM-1V	M4N		YS-1	Medium to medium hard formations (harder shales, sandy shales, shales alternating with streaks of sand and limestone, etc.)
E	3C-2	OW		M4	MH3C	YS	Medium hard formations (hard tough shale, sandy shale, hard limestone, anhydrite, dolomite, hard rock interbedded with tough shale, etc.)
F	3C-2	OWS	EM-3	M5	MH3C		Medium abrasive to hard nonabrasive (hard shale, hard lime, hard anhydrite, dolomite, chalk, slate, hard rock interbedded with tough shale, etc.)
G	3C-T2	OWC	EH-1	M4L	MHT 3C	YM	Medium hard abrasive to hard formation (high compressive strength rock, dolomite, hard limestone, hard slaty shale, chert streaks, etc.)
H	3C-4	W7	EH-1	H7		YH	Hard semi-abrasive formations (hard sandy or chert bearing limestone, dolomite, granite, chert, abrasive sand, etc.)
J	3C-4W	W7R	EH-2 EH-3	H7W		YH W	Hard abrasive formations (chert, quartzite, pyrite, granite, hard sand rock, etc.)
K		R-1		H9			Extremely hard, abrasive formations (chert, quartzite, granite, flint, novaculite, taconite, basalt, quartzitic sand, etc.)

CLASSIFICATION of types of bits in relation to drillability of formations using several manufacturers' terminology. This does not include types of bits of all manufacturers.

engage the bottom of the hole, but only two cut the hole to gauge. The tooth structure varies according to the formation to be drilled.

Jet-type roller cone bits have found widespread use in the petroleum industry because of an inherently faster drilling rate due to improved bottom-hole scouring. The bit is designed so that flushing nozzles allow a high-velocity jet stream of drilling fluid to be aimed directly at the bottom of the hole which aids penetration by the jetting action.[88,464] While this bit type warrants considerations by the shallow-well drilling industry, the overall advantage must be questioned because of the investment required in high volume, high pressure pumping equipment.

A classification chart showing manufacturers, milled-tooth bits, sealed and unsealed, and insert and friction-bearing bits is given in Table 4. Bits grouped on the same line of this chart are generally designed to withstand the same operating conditions. A description of the rock-bit classes is shown in Table 5. Table 6 shows the correlation of mill-tooth and insert classes of bits with types of formations. The three tables are from a review by Estes[219] that includes an excellent discussion of bit classification, class differences, and recommended bit-operating conditions.

TABLE 4[219]

Rock Bit Classification Guide, Sept., 1971.

	DRESSER – SECURITY				HUGHES						G. W. MURPHY					SMITH				
MILL-TOOTH CLASS	STD	"T" GAGE	"G" GAGE	"S" SEAL	STD	"T" GAGE	"G" GAGE	"S" SEAL	"SG" SEAL	"J" BEAR	STD	"T" GAGE	"G" GAGE	"S" SEAL	"SG" SEAL	STD	"T" GAGE	"G" GAGE	"S" SEAL	"SG" SEAL
1 - 1	S3S			S33S	0SC3A			X3A			YT3A			ST3A		DS			SDS	
2	S3	S3T		S33	0SC3			X3			YT3	YT3T		ST3		DT	DTT		SDT	
3	S4	S4T	S4TG	S44	0SC1G	C1C	ODG	X1G	XDG		YT1A	YT1T	YT1AG		ST1AG	DG	DGT	DGH	SDG	SDGH
4	S6		S6G		0SC											K2		K2H		
2 - 1	M4N		M4NG	M44N	0W4		ODV		XDV		YS1		YS1G		SS1G	V2		V2H	SV	SVH
2					0WV			XV												
3	M4L		M4LG	M44L	0WC			XC			YM		YMG		SMG	T2		T2H		
3 - 1	H7	H71	H7TG	H77	W7		WD7	X7	XD7		YH		YHG		FHG(J) SHG	L4		L4H	SL4	SL4H
2	H7U		H7UG	H77U	W7R2			X7R			YHW		YHWG	SB7		W4		W4H		
3			H7SG	H77S																
4 - 1	HC		HCG	H77C	WR		WDR	XWR	XDR	J8 JD8	YBV		YBVG	FV(J) SBV	SBVG	WC		WCH	SWS	SWCH

	DRESSER – SECURITY		HUGHES			G. W. MURPHY				SMITH		
INSERT CLASS	REG.	"S" SEAL	REG.	"S" SEAL	"J" JOURNAL	REG.	"S" SEAL	"J" FC	"J" FBC	REG.	"S" SEAL	"J" SEGMENT
5 - 1												
2		S84						FCT	FBCT			
3		S86			J33		SCS5	FCS5	FBCS5		3JS	SS3
6 - 1		S88		X44R	J44		SCM5	FCM5	FBCM5		4JS	SS4
2					J55C		SCM	FCM	FBCM		4-7JS	SS4-7
7 - 1		M88	55R	X55R	J55R	YC5G	SCH5	FCH5	FBCH5	5	5JS	SS5
2							SCH	FCH	FBCH	7	7JS	SS7
8 - 1	H8	H88	RG7	RG7X		YC4G	SC4G	FCH4	FBCH4			
2				RG1X	J88		SCG			8	8JS	
9 - 1	H10	H100	RG28	RG2BX		YC2G	SC2G	FCH2	FBCH2	9	9JS	SS9

The above types are generally available in popular sizes. Obsolete types not listed may also be available in some sizes.

Deviation Control Bits:

Class:	0 – 1	1 – 2	1 – 5	2 – 4
Smith	BHDJ	DJ		
Dresser		S3SJ	DS	DM

NOTES: (J) Indicates a journal bearing mill-tooth bit.
Class 0 bits are two-cone.

TABLE 5[219]

Description of Rock-Bit Classes

Type	Class	Formation Type	Tooth Description	Offset
Steel Cutter Mill-Tooth Bits	1-1, 1-2	Very Soft	Hard-Faced Tip	3° - 4°
	1-3, 1-4	Soft	Hard-Faced Side	2° - 3°
	2-1, 2-2	Medium	Hard-Faced Side	1° - 2°
	2-3	Medium Hard	Case Hardened	1° - 2°
	3	Hard	Case Hardened	0
	4	Very Hard	Case Hardened, Circumferential	0
Tungsten-Carbide Insert Bits	5 - 2	Soft	64° Long Blunt Chisel	2° - 3°
	5 - 3	Medium Soft	65 - 80° Long Sharp Chisel	2° - 3°
	6 - 1	Medium Shales	65 - 80° Medium Chisel	1° - 2°
	6 - 2	Medium Limes	60 - 70° Medium Projectile	1° - 2°
	7 - 1	Medium Hard	80 - 90° Short Chisel	0
	7 - 2	Medium	60 - 70° Short Projectile	0
	8	Hard Chert	90° Conical or Hemispherical	0
	9	Very Hard	120° Conical or Hemispherical	0

TABLE 6[219]

Correlation of Mill-Tooth and
Insert Classes with Types of Formation

FORMATION DESCRIPTION	MILL-TOOTH CLASS	INSERT CLASS
1. Very Soft Shales	1 - 1, 1 - 2	
2. Soft Shale/Sand	1 - 3	5 - 2
3. Medium Soft Shale/Lime	1 - 4	5 - 3
4. Medium Lime/Shale	2 - 1, 2 - 2	6 - 1
5. Medium Hard Lime/Sand/Slate	2 - 3	6 - 2
6. Hard Lime/Dolomite	3 - 1, 3 - 2, 3 - 3	7 - 1
7. Hard Sand/Dolomite	4 - 1	7 - 2
8. Very Hard Chert		8 - 1, 8 - 2
9. Very Hard Granite		9 - 1

When considering penetration rate, the soft-formation bits generally drill faster than the harder formation counterparts — possibly due to the deeper penetrating ability and offset action of these bits. There is approximately a 15 to 20 percent difference in penetration rate between different classes of mill-tooth bits operating at identical weight and speed in the same formation. The lowest drilling penetration rate is observed with the hardest formation-bit type. As a general rule when a selection is available, the softest formation bit type should be used for the formation being drilled. Roller-bit type will not be mathematically related to penetration rate, but the importance of properly engineering each bit run should not be overlooked.[313]

Diamond bits, including core bits, have found a broad but limited application in deep drilling in the petroleum industry. In deep wells where round trips are costly and time consuming, the extended life of the diamond bit offsets the high cost so that overall cost per foot of hole drilled is reduced. Cutting action of a single diamond,[102] fluid dynamics in a diamond-drill bit,[476] and the construction of diamond bits have been studied extensively.[85] As in the case of roller bits, proper engineering design of a diamond bit and usage can result in reduced operating costs; however, diamond bits find a very limited application to shallow well drilling.[87,241,541]

Hydraulic Factors

The hydraulic factors relating to penetration rate affect the drilling-fluid, pressure drop, and velocity at the bit. Hydraulic power must be supplied by the circulating system in sufficient quantity to insure the cleaning and removal of rock debris from the cutting surface. Thus, analysis of the entire circulating system is necessary.

Extensive analytical work has defined the various criteria which can be used to optimize circulating-system hydraulics.[128,134,350,462,559]

These works, however, relate to jet-bit drilling. Only by using jet-bit nozzles can the flow area, open at the bit, be adjusted to regulate pressure drop across the bit. Although jet bits are not widely used in shallow drilling, application is found in areas where low drillability limits drilling rate. The available surface hydraulic horsepower should be distributed for maximum cleaning at the bit, even with small pumps;[39] therefore, a hydraulic program should be designed for jet-bit drilling since the importance of a hydraulic analysis even for a "jet-action" drag bit has been demonstrated.[74]

A fluid-jet system produces two mechanisms to clean the hole — an impact-pressure wave beneath the jet and crossflow at the bottom of the hole.[412] Laboratory measurements have shown that the maximum velocity of the crossflow is directly proportional to the square root of the product of the rate of flow and the jet-nozzle velocity:

$$V_c \sim \sqrt{Qv} \qquad \text{EQUATION 14}$$

Where: V_c = Crossflow velocity

Q = Flow rate

v = Nozzle velocity

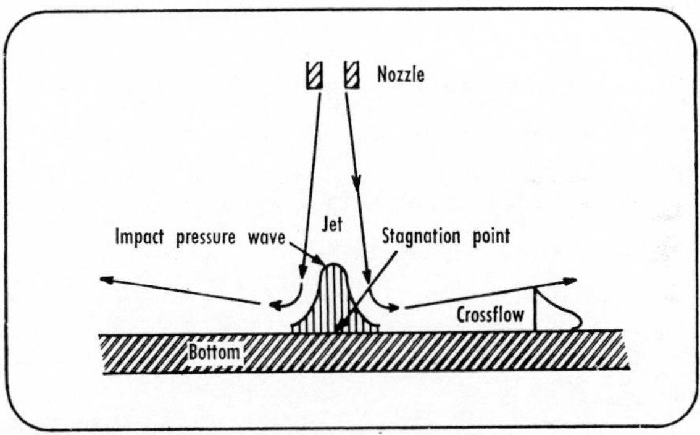

Figure 31[412] Impact Pressure Wave and Crossflow.

Whereas crossflow nearly covers the bottom, the impact pressure wave reaches only a limited portion of the area (see Figure 31). Velocity, kinetic energy, and shear in crossflow beneath a jet bit may influence cleaning of the bottom of the borehole and the bit teeth, and each is related to hole diameter, as follows:[413]

Van Lingen[635] found that drilling rate increased if the bit nozzles were extended to near the bottom of the hole. Using a full-scale rock bit in laboratory studies, Feenstra and Van Leeuwen[222] found that

$$V_c \sim \frac{(Qv)^{1/2}}{D}$$ EQUATION 15

$$E_k \sim \frac{(Qv)^{3/2}}{D^2}$$ EQUATION 16

$$S_s \sim \frac{(Qv)^{7/8}}{D^2}$$ EQUATION 17

Where: V_c = Crossflow velocity; D = Hole diameter
Q = Flow rate; E_k = Kinetic energy flux
v = Nozzle velocity; S_s = Bottom shear stress

for rocks of very low permeability, increased jet velocity influenced drilling rate more than an increase in flow rate. Using a microbit drilling apparatus, Eckel[212] has shown that drilling rate is related to a Reynolds number involving flow rate, nozzle diameter, fluid density, and viscosity. The viscosity is measured at the shear rate experienced by the fluid exiting the nozzle. Mathematically:

$$R \sim \left(\frac{Q\varphi}{d\mu}\right)^c$$ EQUATION 18

where: R = Drilling rate
Q = Flow rate
φ = Fluid density
d = Nozzle diameter
μ = Fluid viscosity (at nozzle shear rate)
c = A constant

A recent study by Sutko and Myers[609] examined the effect of nozzle size, number, and extension on the pressure distribution under a tricone bit. This study showed that hole cleaning and drilling rate could be improved by reducing the number of nozzles and extending the nozzle length.

Because of studies such as these, hydraulics programs are normally based on the bit using, as a design criterion, the maximization of bit hydraulic horsepower, jet velocity, or jet impact. The most efficient of these criteria is difficult to assess because the role of hydraulics in improving bottom-hole cleaning is not known. However, the response of penetration rate to changes in bit weight and rotary speed at

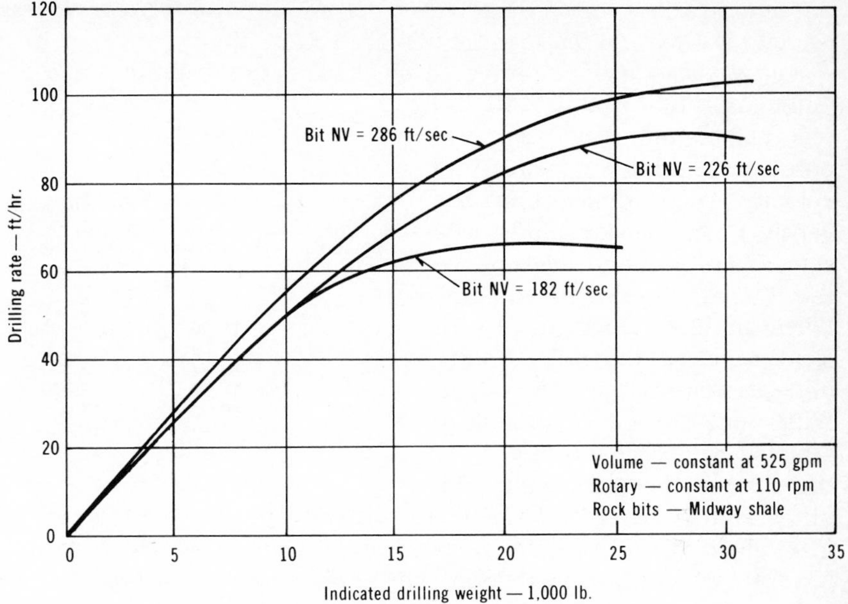

Figure 32[213] Drilling Rate vs. Bit Weight-Effect of nozzle velocity.

various hydraulic levels has been observed (see Figure 32). Therefore, if increases in bit weight give corresponding increases in penetration rate, bottom-hole cleaning (or hydraulics) is sufficient.[464]

A large pumping capability will provide better drilling performance than a small capability. This observation is true in all drilling areas. A reduction in hole size, however, will achieve similar results. A 4-3/4-inch hole, for example, produces 25 percent more cuttings than a 4-1/4 inch hole; therefore with a fixed-pump capacity, a higher percentage of cuttings per unit volume is produced in the larger hole (see Table 7). Annular velocity of return cuttings with a 2-3/8-inch drill stem in a 4-3/4-inch hole is reputed to be 36 percent slower than the velocity in a 4-1/4-inch hole. Therefore, a small change in the hole size can greatly affect drilling conditions. A 1/4-inch change in hole size shows an annulus area change of 30 percent for small diameter holes and 15 percent change for large holes, indicating a similar change in annular velocity. An up-hole velocity of 200 feet per minute has been demonstrated as being very efficient since hole cleaning is related to the velocity at which cuttings will fall in the annulus. Piggott[513] has shown that the slip velocity of round, thin particles (0.5 inch in diameter) is 54 feet per minute in water, using 2-3/8 inch O.D. "N"-type drill stem. Thus, for the anticipated velocity of 200 feet per minute, the relative cutting velocity is only 146 feet per minute (see Figure 33).

Annular velocity is determined by the area of the annulus and the pump output. About 150-200 feet per minute has been proposed as the ideal upward mud velocity (depending upon the density). In heavier muds, less velocity is required because of the buoyancy of the cuttings. When annular velocity is insufficient to clean the hole, "booting" or wall-packing may occur. Wall-packing is a condition affected to varying degrees by the size of the mud pump. All other factors being the same, a constant upward velocity can be maintained by a constant annulus area. That is, as the hole size increases, the drill stem theoretically should be increased in order to hold constant the difference in the squares of the diameters of the hole and the pipe. Such an arrangement usually is impractical; thus, a change in flow rate is normally required.

TABLE 7[39]

Annular Areas

Hole Diameter (in.)	Hole-Area (sq. in.)	Annulus with 2-3/8" O.D. Drill Rods (sq. in.)
3-1/2	9.62	5.19
3-3/4	11.04	6.61
3-7/8	11.79	7.36
4	12.57	8.14
4-1/4	14.19	9.76
4-1/2	15.90	11.47
4-3/4	17.72	13.29
5	19.64	15.21

In shallow drilling of soft formations, the high slip velocity of large cuttings can cause bit plugging. Cuttings easily settle and start compacting (plugging) in a very short time if the return velocity of mud or water is too low.

Figure 34 illustrates the relationship between annular velocity and hole size. The up-hole velocities require 70 strokes per minute and 80 percent volumetric efficiency of the pump. Note that the 3-1/2 inch and 3-7/8 inch holes show good return velocities for a 4 x 5-inch pump. Figure 34 is a typical chart for 2-7/8 inch drill pipe.

In the engineering design of a hydraulic program for any hole, the objective is to select nozzle sizes and flow rates for the hole geometry in use. The design criterion may be either (1) the maximization of bit-hydraulic horsepower, (2) jet velocity, or (3) impact. Primarily, a suitable, minimum flow rate which will effectively clean the hole must be chosen. This flow rate, then, may be combined with the available surface horsepower in order to analyze the pressure drops in the circulation system. These pressure drops may occur in the surface equipment, in the drill stem, at the bit, or in the annulus. Figure 35 represents a typical chart for determining friction losses through the inside of a drill stem at various flow rates. Figures 36 and 37 illustrate the tremendous pressure drops which can

Up Hole Velocity in Feet Per Minute

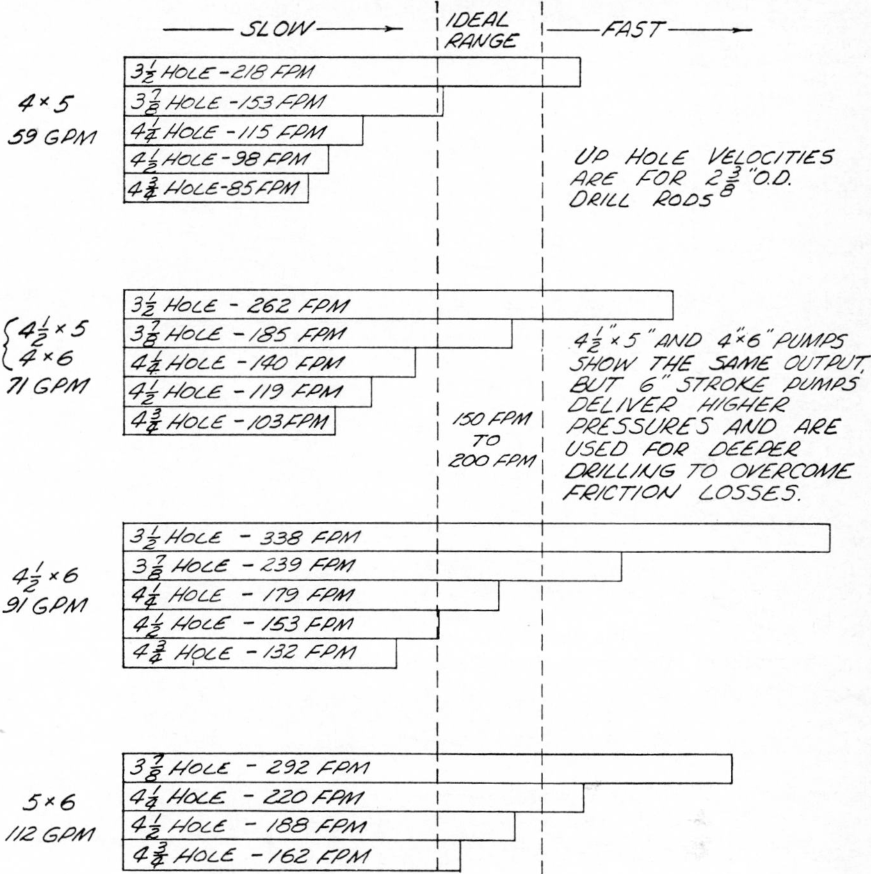

Figure 33[39] Cuttings up-hole velocity.

occur at the bit if the nozzle area is decreased in order to increase the fluid velocity through the bit. A convenient method of analyzing the circulation-system pressure drops is by the use of a log-log plot of pressure vs. flow rate, as shown in Figure 38. For various depth levels, the pressure drop at a practical flow rate is determined. A line with a positive slope of 1.83 is then drawn through the point, representing the pressure drop at the assumed flow rate. Thus, the pressure drop at all flow rates can be observed graphically.

A nozzle-pressure drop can be determined according to the chosen optimizing criterion. Two design programs for optimizing either bit hydraulic horsepower or hydraulic impact are described in Table 8. A

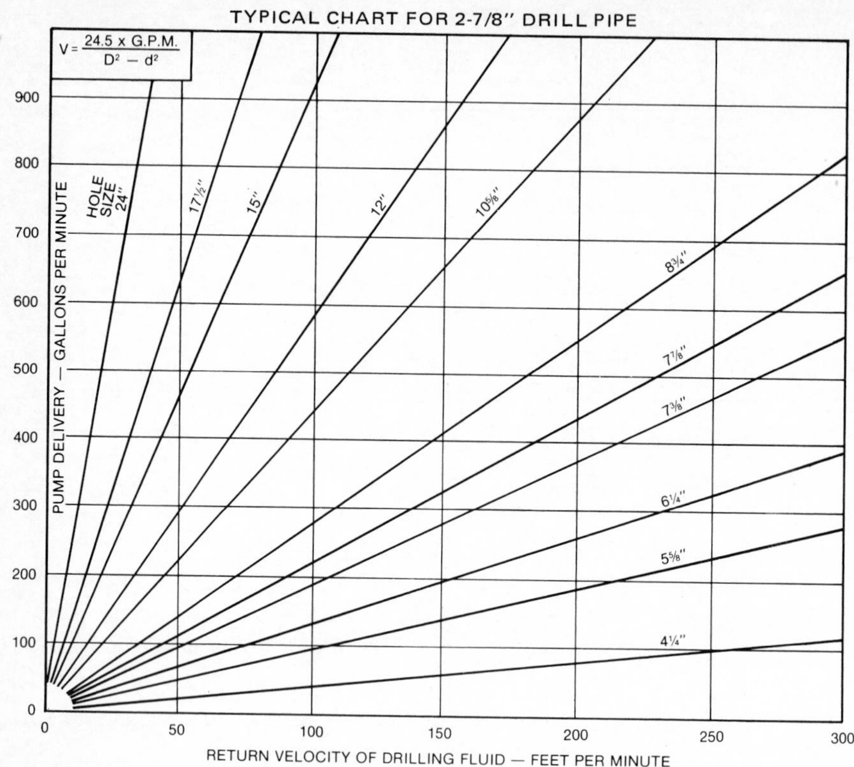

Figure 34[261] Pump Delivery vs. Annular Velocity.

nozzle area is selected which will provide the appropriate pressure drop at the bit for the flow rate to be used. By knowing the area required, a combination of nozzle diameters may be selected and placed into the bit receptacle. In general, to achieve optimum hydraulics, 50 percent or more of the available surface pressure should be lost by the drilling fluid in passing through the bit nozzles.[462]

TABLE 8[464]

Two design programs

Bit hydraulic horsepower	Hydraulic impact
1. Determine minimum annular velocity. 2. Select smallest liner that will give the required annular velocity. 3. Use a surface pressure limitation equal to the liner rating. 4. When the ratio of bit horsepower to surface horsepower is reduced to 65%, keep this ratio constant by reducing circulation rate. This has the effect of reducing the surface horsepower.	1. Using the maximum surface horsepower, select a circulation rate that will allow the use of 75% of the surface horsepower at the bit. 2. Allow this distribution of power at the bit to be reduced at a constant circulation rate until ratio of bit horsepower to surface horsepower is 48%. Keep this ratio constant by reducing circulation rate.

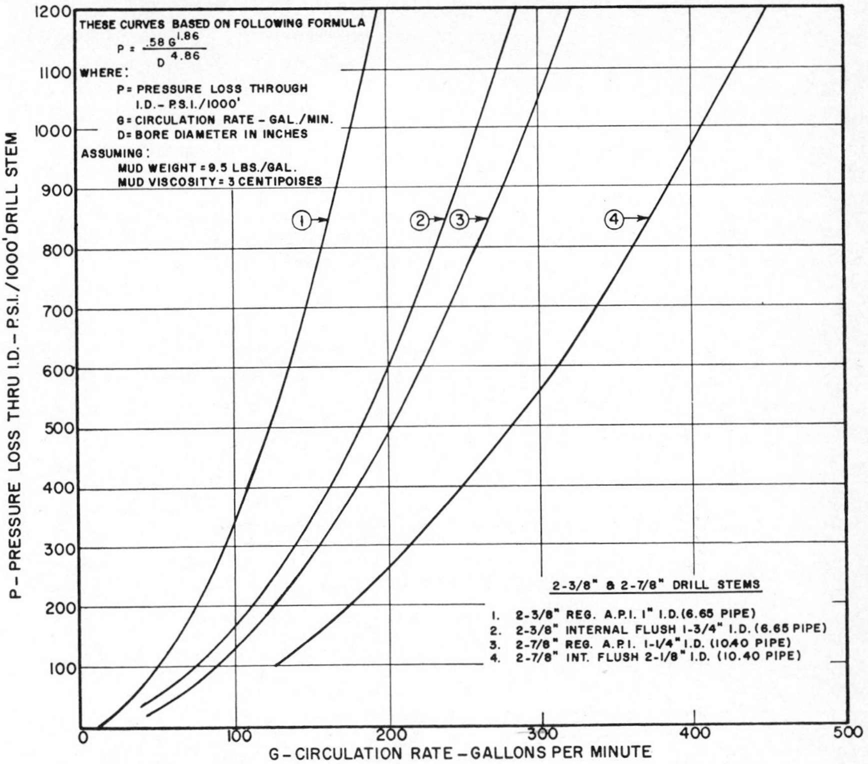

Figure 35[261] Typical chart for friction losses through inside of Drill Stem vs. Circulation Rate.

Drilling Fluid Properties

At depths greater than a few hundred feet, the destruction and removal of rock by rotary drilling is limited by the presence of a viscous drilling fluid. Rock drillability decreases with increased depth, and samples of rock originally tested under bottom-hole conditions increase in drillability when brought to the surface and tested.[266] Two factors, then, are important in altering drilling behavior at depth: pressure environment and drilling-fluid properties.[21]

The rock pressure environment is controlled by: (1) the weight of the overburden, (2) the pressure of the fluid in the rock pores, and (3) the pressure that the drilling-fluid column exerts on the hole bottom. The most important pressure relationship affecting drilling rate is the difference between the pore pressure and the drilling-fluid column pressure. For shallow wells, the formation pore pressure is normally equivalent to a fresh water gradient if the water table is

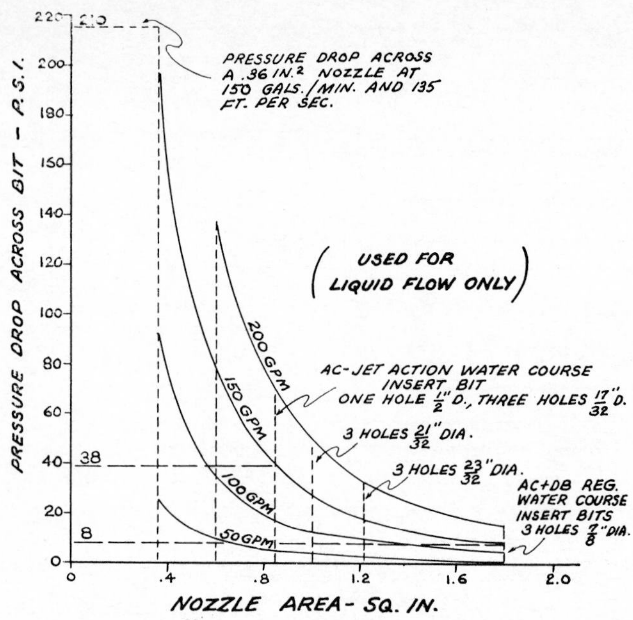

Figure 36[39] Bit Pressure Drop vs. Nozzle Area.

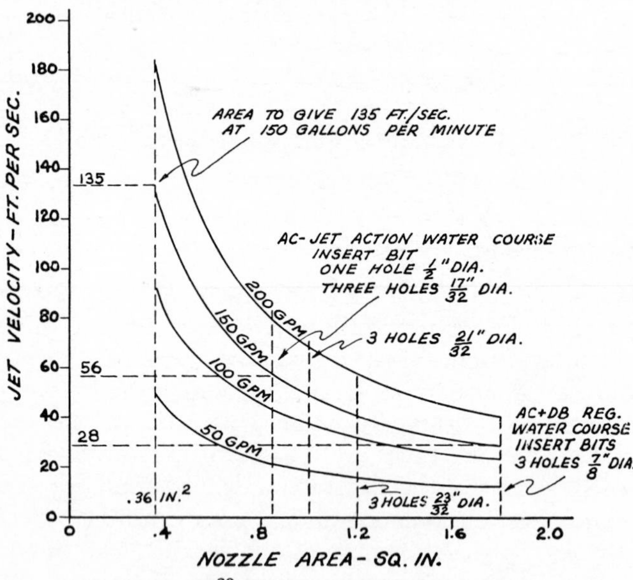

Figure 37[39] Jet Velocity vs. Nozzle Area.

near the surface. Since fresh water weighs approximately 8.33 pounds per gallon, rock pore pressure with water table at the surface, is related to well depth as follows:

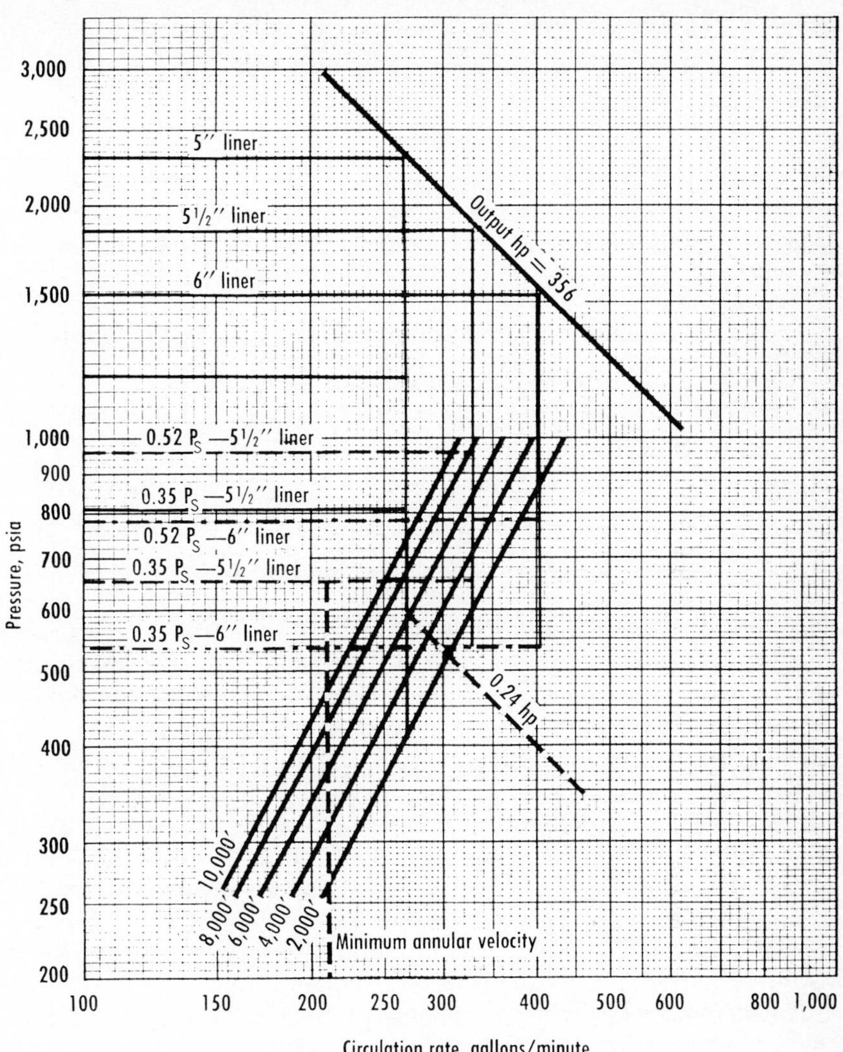

Figure 38[464] Design of Hydraulic Programs.

EQUATION 19

$$P_p = 0.052\ (D_W)\ (8.33\ \text{lb./gal.})$$

where: P_p = Rock pore pressure (psi)

D_W = Well depth (ft.)

or, with water table at some depth below the surface, rock pore pressure is:

EQUATION 20

$$P_p = 0.052\,(D_w - D_t)\,(8.33 \text{ lb./gal.})$$

where: P_p = Rock pore pressure (psi)

D_w = Well depth (ft.)

D_t = Water table depth (ft.)

Thus, for a 200-foot well, the rock pore pressure at bottom is 86.6 psi if the water table is at the surface, and is 43.4 psi if the water table is at the 100-foot level.

The static drilling-fluid column pressure is related to the density of the circulated fluid as follows:

EQUATION 21

$$P_m = 0.052\,(D_w)\,(\varphi m)$$

where: P_m = Drilling fluid column pressure (psi)

D_w = Well depth (ft.)

φm = Density of the drilling fluid (lb./gal.)

When the drilling fluid is being circulated, an actual mud column pressure in excess of that calculated by Equation 21 will result due to the impact of the fluid on the hole bottom and the annular friction loss. These pressures are in addition to the static value.

From the example above, if the circulating fluid density is 9.5 pounds per gallon (an average for a "native," uncontrolled mud), the bottom-hole static pressure will be 99 pounds per square inch. Thus, the pressure differential across the hole bottom in this case (for *no* circulation, water table at 100 feet) is 56 pounds per square inch (99 pounds minus 43 psi). Excess pressure not only retards drilling rate but could possibly cause a loss of mud into more permeable sections of the formation, particularly during circulation.

The difference between the drilling-fluid column pressure and rock pore pressure is given by the following:

EQUATION 22

$$P_m - P_p = 0.052\,\{D_w[\varphi m - (D_w - D_t)\,(8.33)]\}$$

where: P_m = Drilling-fluid column pressure

P_p = Rock pore pressure

D_w = Well depth (feet)

φm = Density of the drilling fluid (lb./gal.)

D_t = Water table depth (feet)

The pressure difference is directly related to the density of the circulated fluid and is most important in controlling drilling rate. Therefore, air drilling operations in many areas have been successful when the density of the drilling fluid is extremely low.[3, 210, 211, 239, 420, 473, 670] The effect of differential pressure on drilling rate has been examined in the laboratory by employing both single bit tooth and microbit.[190] Figure 39 illustrates the effect of overburden and differential pressure on drilling rate. Increases in overburden pressure do not reduce crater volume; whereas, increases in the differential pressure reduce crater volume significantly. This differential pressure effect on microbits can be seen in Figure 40 which extends the crater volume conclusions to an actual drill test.

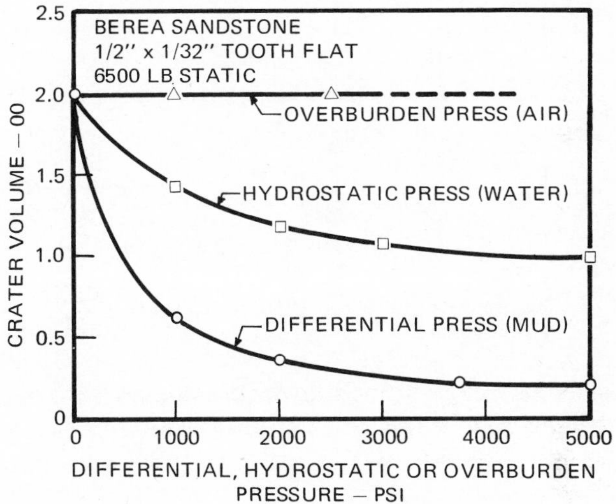

Figure 39[430] Effect of Bottom-Hole Pressure on Crater Volume.

The limiting effect of differential pressure on drilling rate has been verified by field tests, where a 70 percent reduction in drilling rate was noted as the differential pressure increased from 0 to 1,000 psi.[639]

While density (modified by differential pressure) is the predominant mud property which affects drilling rate, other properties have been identified, e.g., viscosity, filtration, oil content, and wettability effects have been reported.[210,464,467]

To mathematically relate the effect of fluid properties on drilling rate, only the two more important properties — density and viscosity

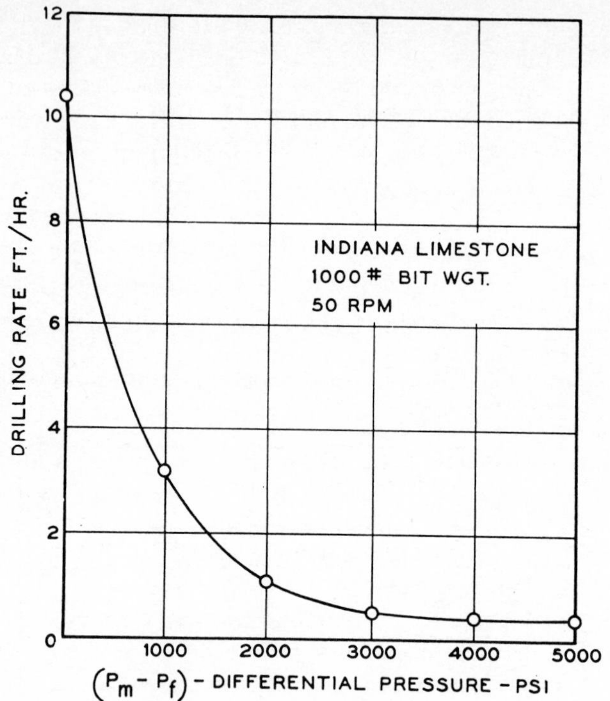

Figure 40[190] Effect of Differential Pressure on Drilling Rate.

— are considered. Eckel[212] has shown that mud density and viscosity relate to drilling rate in microbit studies. This relationship (shown earlier as Equation 18) is:

$$R = \left(\frac{Q\varphi}{d\mu}\right)^c \qquad \text{EQUATION 23}$$

where:

R = Drilling rate

Q = Flow rate

φ = Fluid density

d = Nozzle diameter

μ = Fluid viscosity (at nozzle shear rate)

c = A constant

For low values of differential pressure, Gray and Young[274] proposed the linear relationship:

EQUATION 24

$$R = \frac{(Ro - \alpha \triangle P)}{Ro} \text{ for } R \leqslant Ro$$

where:
- R = Drilling rate
- Ro = Drill rate at zero differential pressure
- α = A constant
- P = Differential pressure

Equation 25 shows that increases in rock drilling strength, differential pressure, tooth dullness, and mud viscosity will reduce the drilling rate, while increases in rock drillability, rotary speed, bit weight, and flow rate tend to increase the drilling rate.

From a combination of Equations 7, 8, 9, 10, 15, 18, 24, a final drilling rate equation is obtained as follows:[274]

EQUATION 25

$$R = \left(\frac{KN^{\lambda}\ (W - Wo)\ (Ro - \alpha \triangle P)}{R_O DS^2\ \ (1 + C_2 H)}\right) \left(\frac{Q\varphi}{d\mu}\right)^c$$

where:
- R = Drilling rate
- K = A constant
- N = Rotary speed
- λ = An exponent
- W = Bit weight
- Wo = Threshold weight
- R_O = Drill rate at zero differential pressure
- α = A constant
- P = Differential pressure
- Q = Flow rate
- φ = Fluid density
- D = Bit diameter
- S = Rock drilling strength
- C_2 = A constant
- H = Normalized tooth height
- d = Nozzle diameter
- μ = Fluid viscosity
- c = A constant

Intangible Factors

Other important factors which are often overlooked but which may significantly affect drilling rate, include the efficiency of the drilling equipment and the operating personnel. Not only aptitude, training, and experience but also such ill-defined characteristics as pride of accomplishment, appreciation of recognition, and reward for achievements determine the efficiency of the rig crew. Rig efficiency involves the type of operation, the skill of the crew, and consistent maintenance. Maximum drilling rate can be achieved only when these intangible factors are considered, some of which are not subject to mathematical evaluation.

Drilling Rate Equation

The preceding review has briefly discussed the present state of the art of the factors which affect drilling rate. The observed laboratory and field performance results have been related to mathematical terms. A combination of these effects, resulting in a drilling rate equation, can be utilized to study the present operations and predict future performance. Such predictability has significant economic incentives, even in shallow drilling.[466, 467, 639]

Drilling Optimization

The primary objective of the water well contractor is to produce water; consequently, any attempt to reduce cost of drilling that may interfere with the efficient completion of the well must be critically examined. Improper drilling practices contribute significantly to the cost of water well construction. In addition, drilling efficiencies play a vital part in the overall economics of mineral exploration drilling programs and mining operations.

Although this section of the chapter deals with optimization of drilling, the drilling aspects of water well construction should not obscure the real objective — the efficient development of a water supply. Nevertheless, the importance of total well planning should be highly emphasized as a means of reducing total expenditures. Accordingly, when surveying the rotary drilling system, attention should be given to optimization of drilling.

Rotary drilling optimization is defined as minimization of well cost per foot. Total drilling cost is composed of tangible and intangible costs. Tangible costs are those capital costs such as pipe, casing, and well equipment. Intangible costs are all other costs including rig operation, choice of bit types, mud control, and supervision.

Well productivity normally dictates pumping requirements, and the pump dimensions, in turn, establish the size of the hole and of the tubular goods. Well completion methods and equipment are discussed in later chapters. The discussion of optimization will be limited to the intangible costs of drilling.

Drilling cost may be defined as dollars expended for a bit run divided by the footage drilled, or:[461,586]

EQUATION 26

$$C_f = \frac{\text{Dollars}}{\text{Footage}}$$

where: C_f = Hole cost

Two operations are normally involved in a single bit run — hoisting and drilling. The efficiency of hoisting operations is determined by the equipment employed — the size and condition. Drilling operations, on the other hand, are efficient to the extent that the controllable factors which affect penetration rate are successfully employed. Certain alterable and unalterable drilling variables are shown in Table 9. Rotary-drilling optimization requires a knowledge of each variable

TABLE 9[399]

Drilling Variables

ALTERABLE	UNALTERABLE
Mud	Weather
Type	Location
Solids Content	Rig Conditions
Viscosity	Rig Flexibility
Fluid Loss	Corrisive Borehole Gases
Density	Bottom-Hole Temperature
Hydraulics	
Pump Pressure	Round-Trip Time
Jet Velocity	Rock Properties
Circulating Rate	Characteristic Hole Problems
Annular Velocity	Water Available
Bit Type	Formation to be Drilled
Weight-on-bit	Crew Efficiency
Rotary Speed	Depth

in order to achieve a proper balance. For example, mud density should not increase to the extent that the bit and rotary speed become ineffective. Likewise, poor hole cleaning can result in drilling problems which will adversely affect costs so that other factors become subordinated. Hole economy, then, requires the application of knowledge regarding all the alterable drilling variables.

To expand the relationship shown in Equation 26:

EQUATION 27

$$C_f = \frac{C_B + C_R\,(T_R + T_T + T_C)}{F}$$

where:
C_f = Hole cost ($/ft.)
C_B = Bit cost ($)
C_R = Rig cost ($/hour)
T_R = Rotating time (hours)
T_T = Trip time (hours)
T_C = Connection time (hours)
F = Feet drilled

Equation 27 relates all the costs associated with a single bit run to footage drilled. Rig cost is defined as the cost of operating the rig for one hour, including expenses, salaries, depreciation, and overhead. Since tripping and connection time per foot of hole are fairly constant for a bit run, rotating time and footage drilled will control the cost of a single bit run. Rotating time may be altered by the use of variations in bit weight and rotary speed (a high weight and speed will wear out the bit more quickly than lower levels of weight and speed).[461]

The relationship shown in Equation 27 is useful in evaluating different drilling methods. For example, for a rig costing $50 per hour, drilling at 500 feet, the sequential bit selection should be based on the least cost per foot as follows:

	Bit A	**Bit B**
Bit cost ($)	50	150
Rotating time (hours)	4	8
Tripping and connecting time (hours)	5	6
Footage drilled	40	120

In solving Equation 27, a run with bit A would cost $12.50 per foot, a run with a bit B would cost only $7.08 per foot, regardless of the fact that bit B cost three times as much. A savings of nearly $5.50 per foot over a 120-foot interval could be realized in this example.

The minimization of drilling cost has long been a goal of the petroleum industry.[266,409,461,586] Extensive work on the relationship of operating variables to bit life has defined the manner in which bit cutting structure and bearings fail.[154,360] This analysis allows consideration of the optimum bit weight and rotary speed during the life of the bit. Examples of constant vs. varied weight and speed during the life of the bit have been studied.[266,549,230] Even the use of constant rotary speed with variable weight has been considered.[122] These techniques and methods of analysis have been responsible for the savings in drilling cost and have emphasized the importance of proper engineering design of a drilling operation. Figure 41 shows how drilling cost is affected by changes in bit weight and rotary speed.[669]

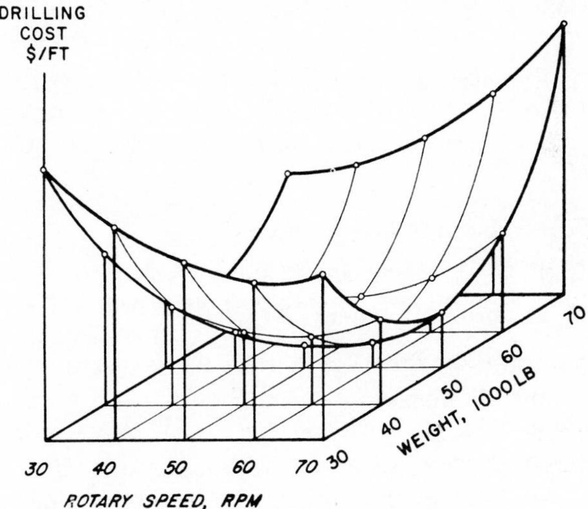

Figure 41[669] Effect of Bit Weight on rotary speed on drilling cost.

The drilling process may best be mathematically expressed in the form of a drilling "model." This model would relate drilling rate to the independent variables and consider those factors which affect bit life. A model of drilling, then, consists of equations for:

(1) Penetration rate.
(2) Bearing wear rate.
(3) Tooth wear rate.
(4) Cost per foot.

Equations for penetration rate and cost per foot have been presented previously. Typical bearing life and tooth wear equations are:

Bearing wear

EQUATION 28

$$T_f = \frac{b}{N\sqrt{W}}$$

where: T_f = Time to failure (hours)

b = Bit bearing constant

N = Rotary speed (rpm)

W = Bit weight (thousand lbs.)

The equations which constitute this model can, however, become tedious to solve; consequently, a computer is often used in the solution.[263,344,353,669]

Tooth wear

EQUATION 29

$$T_f = \left(\frac{(-D_1 W + D_2)}{A_f (PN + QN^3)}\right)\left(\frac{C_1}{2} + 1\right)$$

where: T_f = Time to failure (hours)

C_1, P, Q, D_1, D_2 = Bit constants

A_f = Formation abrasiveness factor

N = Rotary speed

W = Bit weight

The successful application of optimized drilling techniques requires not only a sound analytical procedure, but the careful examination of dull bits and proper instrumentation of the rig.[215,377,435] As a minimum requirement, bit weight (or hook load), rotary speed, and penetration rate should be measured and recorded. Proper records of bit performance through different formations are necessary. But primarily, a sound fundamental knowledge of the process is the best way to benefit from the previous work.

Although water-well drilling involves considerable expense for well completion, penetration costs can also be sizable. Total drilling optimization, including the mud system, circulating system, bits, and other mechanical factors that affect drilling rate, is a concept recently expanded by the oil industry for reducing the costs and enhancing operational efficiency. An awareness and adaptation of the approaches used are of significant potential to drilling for ground water and other economic minerals.

HOLE PROBLEMS

Hole Cleaning

A basic function of the drilling fluid is the removal of cuttings from the hole. Emphasis on fast drilling with fluids of minimum density and viscosity has at times created problems in hole cleaning. Failure to clean the hole adequately can result in debris on the bottom and in stuck pipe, two costly and time-consuming situations. Often, symptoms of inadequate hole cleaning are misinterpreted as hole instability, and mud weight is increased as a remedy. While increasing mud weight will provide added buoyancy to the drilled cuttings and assist in removal, penetration normally will be impaired, and other problems such as loss of circulation may occur. Thus, a knowledge of the flow properties of drilling fluids under conditions of use must be coupled with an understanding of slip velocity of cuttings in order to assess and correct the hole-cleaning problem.

The ability of mud to lift cuttings has been extensively studied by means of empirical and experimental techniques.[297,513,661] These studies have attempted to describe how fast particles will fall in the annulus under varying conditions of mud flow rate and fluid properties. This rate of fall is defined as *slip velocity*. Of importance in the hole cleaning is the relative velocity of the cuttings with respect to the hole wall, or:

EQUATION 30

$$Vr = Va - Vs$$

where:
- Vr = Relative velocity of particle (ft./min.)
- Va = Annular velocity of fluid (ft./min.)
- Vs = Slip velocity of particle (ft./min.)

A positive relative velocity is required to remove cuttings, and the measurement, or prediction, of this relative velocity is important in hole cleaning as well as in sample logging where cuttings must be collected at the surface and correlated to original depth of occurrence.[555] The slip velocity of a particle depends upon a number of factors, such as: the density and flow properties of the drilling fluid; the volume, specific gravity, shape, and roughness of the particle; and the shape and area of the projected face of the particle at right angles to the direction of the relative movement of solid to fluid. Because densities of cuttings are normally somewhat less than 22 pounds per gallon, depending upon porosity, slip velocity in very heavy fluids is low due to buoyancy effects. In light, thin fluids, however, particle slip velocity can be appreciable and in many cases, is a problem in shallow drilling.

In a series of laboratory tests, Hopkin[319] experimentally measured the slip velocity of glass particles of varying shapes in an eight-foot, vertical column. Sandstone discs and shale cuttings also were used. Thirteen drilling fluids, with viscosities ranging from 26 to 100 sec./qt. (Marsh funnel), were circulated at four flow rates in the test apparatus. The results of these tests are shown in Figure 42. The slip velocity was essentially constant up to a funnel viscosity of approximately 90 sec./qt., then decreased to lower values at mud viscosities in excess of 200 sec./qt. For the muds, the yield point, as determined with the Fann Viscometer, appeared to be a more reliable measure of the effect of viscosity on slip velocity, shown in Figure 43.

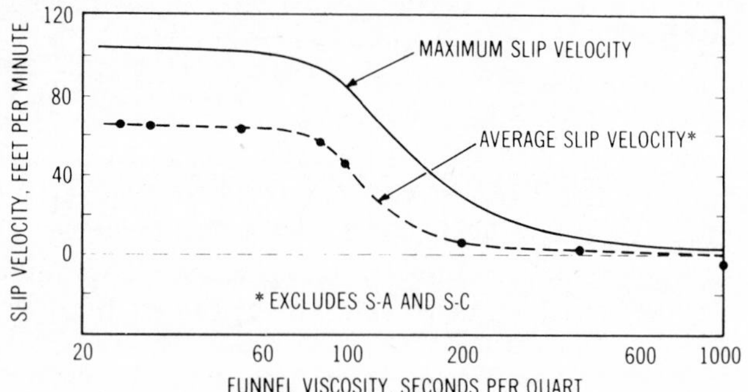

Figure 42[319] **Particle Slip Velocity vs. Funnel Viscosity.**

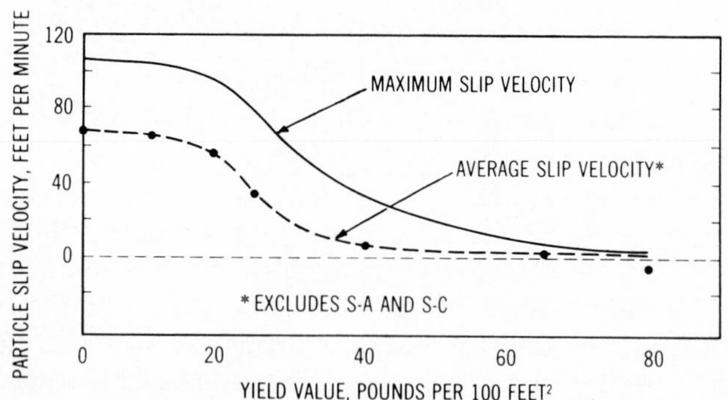

Figure 43[319] **Particle Slip Velocity vs. Yield.**

Moore[464] has presented an approximate relationship for slip velocity in laminar flow in which the mud viscosity can be related to rotational viscometer measurements. The relationship is based on Stokes' Law for a freely falling particle in a viscous medium:

EQUATION 31

$$Vs = \frac{8310\ d_c^2\ (P_s\text{-}P_m)}{\mu}$$

where: Vs = Slip velocity
μ = Equivalent mud viscosity
d_c = Particle diameter (in.)
P_s = Cuttings density (lb./gal.)
P_m = Mud density (lb./gal.)

and:

EQUATION 32

$$\mu = \mu_p + \frac{399\ YP}{V_f}\ (D_h\text{-}D_p)$$

where: μ_p = Plastic viscosity (cp. from rotational viscometer)
YP = Yield point (lb./100 ft.^{2} from rotational viscometer)
V_f = Average fluid velocity (FPM)
399 = Units conversion constant
D_h = Hole diameter (in.)
D_p = Pipe diameter (in.)

Equations 31 and 32 relate slip velocity to flow properties as measured at 300 and 600 rpm. As an example of how these equations can be used, consider the following:[464]

D_h = 7 7/8 in.
D_p = 4 1/2 in.
P_m = 10 lb./gal.
P_s = 21 lb./gal.
μ_p = 20 cp.
YP = 5 lb./100 ft.^{2}
V_f = 120 ft./min.
d_c = 0.20 in. (from screen analysis)

Slip velocity could be computed using Equations 31 and 32:

$$\mu_p = \frac{20\ +\ (399)\ (5)\ (3.375)}{120} = 76 \text{ cp.}$$

$$\therefore Vs = \frac{(8310)\ (0.20)^2\ (21\text{-}10)}{76} = 48 \text{ ft./min.}$$

The net upward velocity in this example would be (120-48) or 72 ft./min. Note that if the yield point in this example were increased to ten, the slip velocity would be reduced to 28 ft./min. If the plastic viscosity is doubled, the equivalent velocity is reduced only to 38 ft./min. Thus for lifting capacity, yield point contributes more substantially than plastic viscosity.

The relationships discussed above are strictly empirical but clearly illustrate the general relationship of the parameters which affect slip velocity. Limitations, however, do exist in the effectiveness of these relationships. For example, by applying Stokes' Law to the computation of slip velocity, an interparticle interference is not apparent, which is not true in normal drilling operations. Also, the effects of pipe rotation have been ignored.

Brown[95] and Williams and Bruce[661] have developed equations for slip velocity in turbulent flow, but these relationships are not discussed here because of the low annular shear rates normally experienced in "shallow" drilling.

From the work of Hopkin,[319] the following conclusions were stated:

(1) Maximum slip velocity of normal size bit cuttings, as measured in the laboratory, was found to be approximately 100 to 110 ft./min. in low-viscosity, low-density drilling muds.

(2) Slip velocity of bit cuttings and cavings was found to be a function of the mud viscosity. The yield point component of viscosity appeared to be most important in affecting the slip velocity.

(3) Slip velocity was approximately constant up to a funnel viscosity of 90 sec. per qt. or to a yield point of 15 lb. per 100 sq. ft. Above these values, increases significantly reduced particle slip velocity.

(4) Slip velocity is reduced by increasing mud density.

(5) Field experience indicates that annular mud velocities of 20 to 30 ft./min. in excess of the maximum slip velocity will keep the hole clean during slow, hard rock drilling.

(6) Field experience indicates that annular mud velocities of up to 200 ft./min. in excess of the maximum slip velocity of the cuttings may be required to clean the hole and to prevent sticking of the drilling string during very fast upper-hole drilling.

Velocity Profile

At the rates of shear, normally encountered in annular flow, mud is traveling in a streamline, or viscous, flow pattern. The layers of flow are parallel, and fluid velocity at the boundaries (hole wall and drill pipe) is zero. In the case of low-viscosity, low-solids mud systems, the shear stress/shear rate curve is nearly Newtonian. Such a fluid, however, can be rheologically characterized by a Bingham model by using the rotational viscometer to determine the plastic viscosity and yield point. (See "Drilling Fluids," this chapter). The shear stress measurements at 300 and 600 rpm can be related to a power law approximation according to the equation:[669]

EQUATION 33

$$n = 3.31 \log_{10} (\tau 600/\tau 300)$$

where:
n = Power law exponent
$\tau 600$ = Shear stress at 600 rpm
$\tau 300$ = Shear stress at 300 rpm

The magnitude of the power law exponent (n) which is proportional to the logarithm of the ratio of the plastic viscosity to yield point affects the velocity profile in the annulus. Figure 44 shows the liquid velocity plotted as a function of distance from the drill pipe at various values of the "n" factor. As "n" decreases, the velocity profile flattens, i.e., more of the annular fluid is moving at a uniform velocity. Conversely, for high values of "n," the profile becomes pointed with the inner layers of fluid moving at a high velocity, relative to the outer layers. In this case, cuttings would have a tendency to tumble and fall to the outer boundaries, possibly becoming trapped at a low upward velocity.

To clean cuttings more efficiently from the borehole, three alternatives are possible:[669]

(1) Reduce the slip velocity of the particles.
(2) Increase the fluid viscosity.
(3) Flatten the velocity profile.

The slip velocity may be reduced by increasing the equivalent mud viscosity, by increasing the mud weight, or by reducing the particle size (demonstrated in Equations 31 and 32). An increase in fluid velocity will not reduce the slip velocity but will increase the fluid's relative velocity. By increasing the pumping rate, adverse settling rates may be overcome. In many cases, the velocity profile may be flattened by adding bentonite or organic polymers in order to increase the yield point and reduce the "n" factor so that most particles will be removed from the annulus. Addition of bentonite will increase the gel development, thus reducing slip velocity during periods of quiescence. While ease of lifting is enhanced by an increased yield point, the problem of separation of cuttings at the surface is exaggerated, and this factor must be considered. Consequently, the addition of shear-thinning organic polymers is often the preferred method of altering the velocity profile.

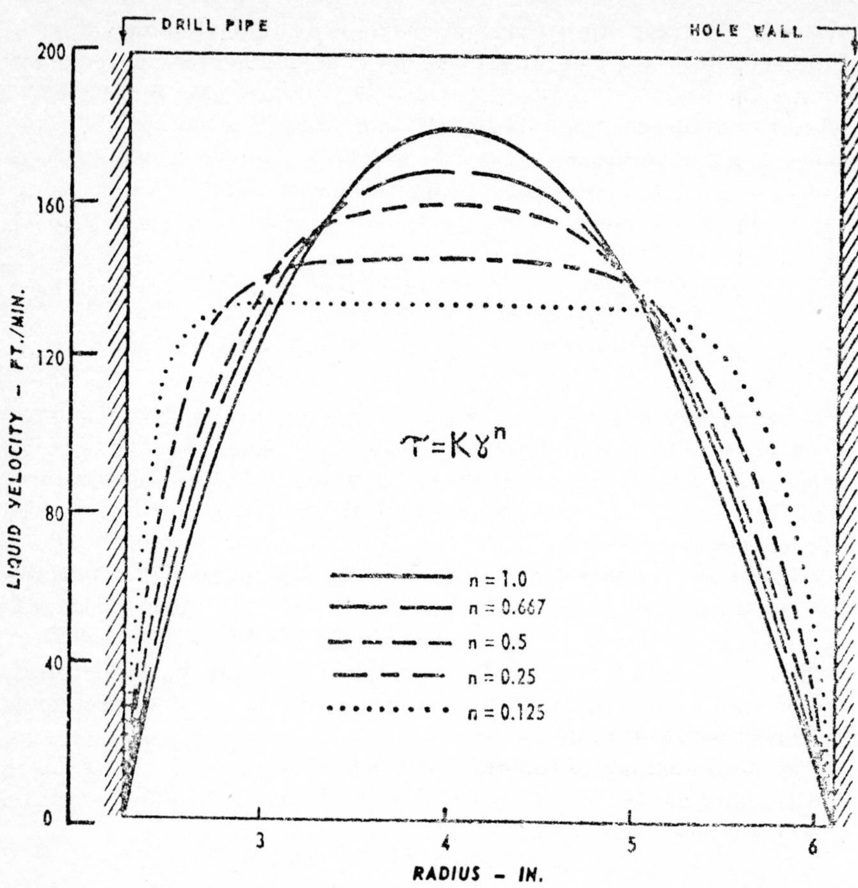

Figure 44[669] Calculated Velocity Profiles in the Annulus of 4½" Drill Pipe and 12¼" Hole, Average Mud Velocity is 120 Ft./Min.

Borehole Stability

Drilling problems referred to as heaving, sloughing, caving shale, tight hole, and "washouts" are included in the general topic of borehole stability. Stability of the borehole depends upon many factors; principally: rock strength; natural or tectonic stresses; stresses arising from the creation of the hole; pore pressure; and hydration effects. Instability of the hole in shallow drilling is usually caused by reduction of the strength of clays and shales because of absorption of water from the drilling fluid.[194] Shales are quartzitic-feldspathic rocks that contain clay. The drilling engineer applies the term "shale" to rocks, ranging from montmorillonitic clays that are highly reactive to water, to inert claystones, and to slates.

Darley[193] states that a test of dispersibility of shales in water serves as a means of characterization. Shales containing a high proportion of montmorillonite (especially the sodium type) tend to more readily disperse. Increases in geologic age and in degree of consolidation decrease ease of dispersion.

Based on laboratory tests made by circulating various drilling fluids through reconstituted shale cores, Darley[194] concludes that in young, dispersible montmorillonitic shales, collapse of the borehole wall occurs with fresh water because of swelling and plastic deformation. Older, less dispersible shale with fresh water in the hole fails by the heaving of firm fragments. Hole enlargement in brittle shales (i.e., shale that breaks into fragments when placed in water do not soften or swell) results from the entry of filtrate from the mud along fracture planes causing the weakening of the cohesive forces and subsequent erosion by the flowing mud stream. Instability of soft shale, caused by high earth stresses, results in plastic flow with water mud in the borehole in contrast to spalling with oil in the hole. Spalling was suppressed by an oil-base mud, presumably because of the plugging of incipient fractures.

Darley[194] discusses some practical implications from the laboratory studies. Borehole stability in soft, swelling shales is least affected by:

(1) Low-filtration mud.
(2) Minimum use of thinners.
(3) Salty mud (although heaving of hard fragments may occur).
(4) Oil-base mud.

Fractured, brittle shale undergoes less hole enlargement with: (1) minimum filtration, (2) avoidance of erosion, and (3) control of flow rate and flow properties.

Chenevert[165] measured swelling, not only of shales containing montmorillonite, but of hard illitic shale. Adsorption of fresh water by confined shale produced internal stresses that led to hydrational spalling, verticle fracturing, and reduction in compressive strength. Measurement of the water-adsorptive properties of the shale under controlled humidity conditions affords a means of calculating hydrational stress. The use of an oil-base mud (containing emulsified salt solution) with controlled salinity of the water phase provides a means of preventing hole instability caused by adsorption of water in shale.[166,457,459] This approach is rarely applicable in shallow drilling although possibly of interest in deep waste-disposal wells.

Control of the chemical and physical properties of the drilling fluid cannot in all cases prevent caving. The problem of hole instability frequently arises from mechanical factors. As with the loss of circulation, drilling practices may be responsible for difficulties in weak formations. As a supplement to maintenance of optimum mud properties, good drilling practices can often alleviate many hole problems.[76,270] Specifically:

(1) Avoid high annular velocities that cause hole enlargement by erosion.
(2) Minimize whipping action of pipe by keeping the drill string in tension.
(3) Avoid pressure surges resulting from rapidly raising or lowering the drill string.
(4) Keep the hole full of mud.
(5) Plan the drilling program to minimize the time of exposure to troublesome sections.

Loss of Circulation

"Lost circulation," or "lost returns," means the partial or complete loss of the drilling fluid into voids in the formation. Loss of water into any permeable or unsaturated porous section may occur while drilling and should be distinguished from "water loss" or filtration of water through the mud-cake solids formed on the borehole face of a permeable formation. Loss of water frequently can be stopped by the addition of finely divided materials, such as, bentonite, finely ground nut hulls, or mica, although damage to the permeability of the aquifer should be avoided.[354]

Howard and Scott[321] emphasized the importance of induced fractures in loss of circulation:

> The loss of drilling fluids and their costly constituents is only one of the detrimental effects experienced when circulation is lost. The loss of drilling time, plugging of potentially productive formations, stuck drill pipe, blow-outs, excessive inflow of water, and excessive caving of formations are other effects which contribute to make the control and prevention of loss of circulation one of the most challenging problems of the industry.

For loss of circulation to occur, openings in the formation must be large enough to accept the mud, and sufficient pressure must be exerted to force the mud into the openings.[149] Conversely, to stop loss of the whole mud, the voids must be plugged so that a filter cake can be formed on the porous section. The plugging material must have the consistency and particle size to prevent entry of the drilling fluid into the voids so that the movement is upward through the annulus. Not infrequently with small rigs, limitations in capability of the pump prevent this solution to the problem.

Types of formations that cause, or lead to, loss of circulation can be classed broadly as in Figure 45:

(1) Natural or intrinsic fractures.

(2) Induced or created fractures.

(3) Cavernous formations, crevices, and channels.

(4) Unconsolidated or highly permeable formations, loose gravels.

The identifying features of these types are given in Table 10:

No other problem as loss of circulation is so dependent upon the practices of the well contractors. "If good preventive measures are used, more than half of the lost circulation problems can be

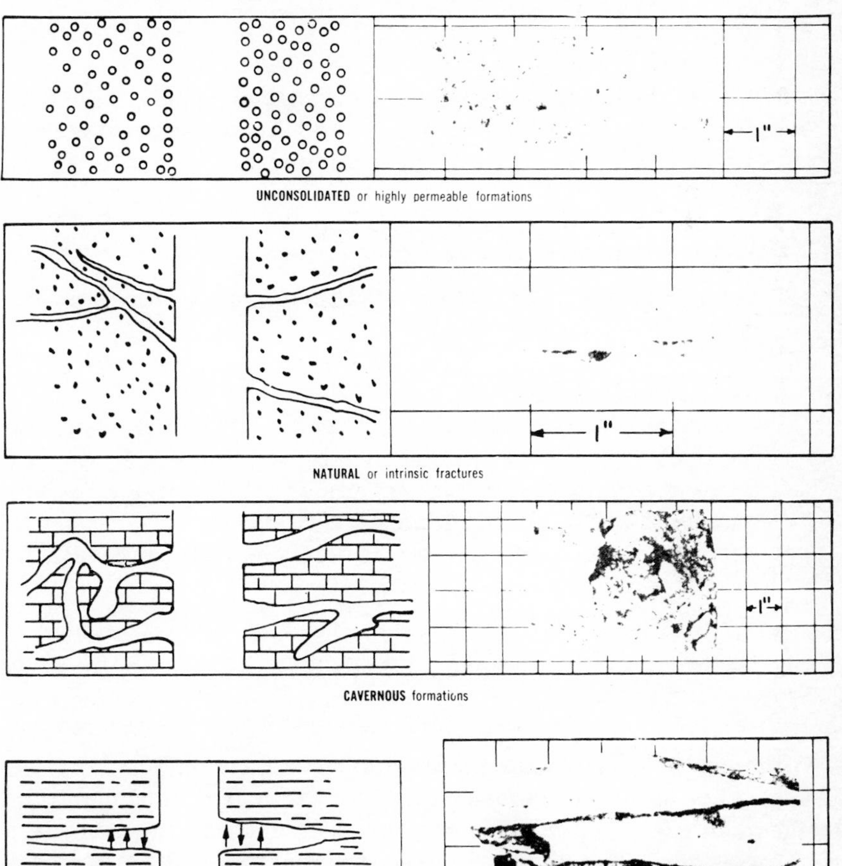

Figure 45[321] Subsurface Conditions that Cause Lost Circulation.

TABLE 10[321]

Identifying features of lost circulation zones

Unconsolidated formations:	Natural fractures:	Induced fractures:	Cavernous zones:
1. Gradual lowering of mud level in pits. 2. Loss may become complete if drilling is continued. 3. Since it is known that the rock permeability must exceed about 10 darcys before mud can penetrate and that oil or gas sand permeability seldom exceeds about 3.5 darcys, it is improbable that loose sands are the cause of mud loss to an oil or gas sand, unless the loss can be attributed to the ease with which this type of formation fractures.	1. May occur in any type rock. 2. Loss is evidenced by gradual lowering of mud in pits. If drilling is continued and more fractures are exposed, complete loss of returns may be experienced.	1. May occur in any type tock, but would be expected in formations with characteristically weak planes. 2. Loss is usually sudden and accompanied by complete loss of returns, Conditions are conducive to the forming of induced fractures when mud weight exceeds 10.5 lb./gal. 3. Loss may follow any sudden surge in pressure. 4. When loss of circulation occurs and adjacent wells have not experienced lost circulation, induced fractures should be suspected. 5. Can be in competent or incompetent formations.	1. Normally confined in limestone. 2. Loss of returns may be sudden and complete. 3. Bit may drop from a few inches to several feet just preceding loss. 4. Drilling may be "rough" before loss.

avoided."[280] Pressure surges resulting from pipe movement have been proven to be a cause of loss of circulation.[171] Circulation losses can often be avoided by minimizing downhole pressures, according to the following:[595]

(1) Raise or lower the drill string slowly.
(2) Do not spud; drill through any tight sections.
(3) Begin rotation of the string, start the pump at a low rate, and gradually increase the rate.
(4) Operate the pump at the lowest rate that will assure adequate removal of cuttings.
(5) If seepage losses occur while drilling with water, change to mud before the losses become severe.
(6) Prevent balling of the bit, do not drill soft formations fast enough to overload the annulus.
(7) Make frequent measurements of mud properties and treat the mud to maintain:
  (a) Minimum weight (hydrostatic head and solids content).
  (b) Minimum viscosity (pressure in the annulus while circulating).
  (c) Minimum filtration (filter cake restriction of the annulus).

Although the only "sure cure" for loss of circulation is to set casing, other measures frequently are successful. The first step to be taken when loss of circulation occurs is to make sure the drill string is in a safe zone. Operating practices and mud properties at the time the loss occurred should be examined and any deficiencies corrected. Sometimes the waiting period and the needed modifications are sufficient to restore circulation.[280]

Messenger,[439] in a systematic survey of the problem, recommends matching the remedial techniques to the severity of the loss (see

Table 11). The nine techniques proposed are listed below (parts 2 and 3 of his paper should be consulted for details):

(1) *Pull up and wait several hours:* stop drilling at the first indication of loss of circulation, pull the bit to safety and, while waiting, prepare for technique 2.

(2) *Plug of bridging agent:* prepare a batch of mud containing a mixture of coarse, medium, and fine bridging agents of granular, fibrous, and flake types and place the plug through an open-ended drill pipe opposite the suspected zone of loss. If the loss persists after two patches have been placed, proceed to technique 3.

(3) *High filter-loss slurry squeeze:* add diatomaceous earth (or proprietary mixtures) to a slurry of attapulgite in water or bentonite slurry treated with lime. Then add finely ground bridging agents. Depending on the severity of loss, increase the size of the bridging agents in the mixture. Displace the slurry to the bottom of the drill pipe, then gradually squeeze the slurry into the zone of loss, and maintain the squeeze for one-half hour.

TABLE 11[439]

Matching remedial techniques to zone severity

	Lost-circulation remedial techniques to be applied
Seeping loss (1-10 bbl/hr)	Pull up and wait 4-8 hr (See Technique 1,) squeeze with mud or high-filter-loss slurry containing fine mica, fine walnut or almond hulls, fine cellophane flake, fiber and shredded leather. (See Techniques No. 2 and 3.)
Partial loss (10-500 bbl/hr)	Pull up and wait 4-8 hr (See Technique 1), squeeze with mud or high-filter-loss slurry containing regular size granular lost-circulation materials (sawdust, ground walnut or almond hulls), cellophane flake and fiber. (See Techniques No. 2 and 3.)
Complete loss (Hole full to mud level at 200-500 ft)	Squeeze with mud or high-filter-loss slurry containing granular lost circulation material (up to ¼ in. in size), large cellophane flake and fiber. (See Techniques 2 and 3.) Use neat or bentonite cement or gilsonite cement. (See Technique 4.) Use DOB-cement slurry squeeze. (See Technique 5.)
Partial or complete loss to deep induced fractures	Pull up and wait 4-8 hr. Apply soft plug squeeze. (See Technique 6 or 7.)
Severe complete loss (Mud level 500 ft-1000 ft + down; long-loss intervals in honeycomb, fractures, or caverns. Water moving within or into loss zone.)	Squeeze either with high-filter-loss slurry containing granular lost circulation materials (up to ¼ in. to ½ in. in size) large cellophane flake and fiber (Technique 3) or large amounts of DOB-cement slurry (Technique 5). Use ac-set cement tool (Technique 8); drill blind or with aerated mud and set pipe (Technique 9).

(4) *Cement:* prepare a neat cement slurry, or lower-density slurries containing bentonite or gilsonite. Place the slurry by the "casing" or "plug" method to minimize contamination of the slurry by mud (see "Well Cementing").

(5) *Diesel oil-bentonite-cement squeeze:* mix equal amounts of bentonite and cement into diesel oil (200 lb. of each per bbl.). Follow the slurry with a slug of diesel oil and displace from the drill pipe while pumping into the annulus at the same time.

(6) *Surface-mixed soft plug clay cement:* recommended for sealing-induced fractures.

(7) *Down-hole-mixed soft plug, "gunk" squeeze:* slurry of bentonite in diesel oil (400 lb. per bbl.) placed similar to technique 5.

(8) *A-C set cement tool:* cement slurry is placed through special tool into a nylon bag placed in the zone of loss.

(9) Drill blind or with aerated mud and set casing.

Materials that have been used to stop loss of circulation include almost anything that is bulky.[241] Testing has employed several[279,656,665] methods under varying conditions, because all factors which affect loss in a particular instance usually are not known.[98] A mixture of granular, flake, and fibrous materials with a definite size distribution and adequate bridging strength is likely to be applicable and has been applied in the control of loss of circulation in both fractured and highly permeable formations.[397,398]

In small-diameter holes, drilling blind (i.e., pumping just enough drilling fluid to cool the bit and lift the cuttings into the zone of loss) in some instances is less expensive than the cost of regaining circulation.[595] When mud is lost into a potentially water-productive formation, efforts to regain circulation may result in plugging the aquifer. Drilling with foam or stiff foam (to be discussed later) may afford a solution to the problem.

Stuck Pipe

In shallow drilling, "caving hole" is usually blamed for stuck pipe; other causes are: "junk" in the hole; balled-up bit; hole collapse from loss of circulation; settling and packing of cuttings; and thick filter cake. Differential pressure as a major cause of stuck pipe in oil-well drilling is discussed by Helmick and Longley.[311] A theoretical analysis of the problem by Outmans,[491] defined the primary cause for differential-pressure sticking as cessation of pipe movement, and the severity of sticking as determined by the differential pressure and the standing time.

Characteristics of differential-pressure sticking of drill pipe are illustrated in Figure 46. When movement of the pipe stops (in contacting permeable formation), the part of the filter cake between the pipe and the hole wall is isolated. With the pump off, filtration continues because of the difference in pressure between the mud and the fluid in the porous rock. The pressure, initially carried by the isolated filter cake, is transmitted to an equivalent area of the drill pipe. As filtration proceeds, the isolated area of the pipe increases.

The ratio of pipe/hole diameter affects the area of pipe isolated from the hydrostatic pressure and the relative increase in isolated area as filtration continues. The force required to free the stuck pipe depends on: (1) the area of contact, (2) the magnitude of the pressure differential, and (3) the friction between the pipe and the mud cake.

In the mechanism of wall-sticking, certain practices in mud control inhibit stuck pipe:

(1) Minimum mud weight to assure least pressure differential and low solids content for a thin-wall cake.
(2) Low filtration rate for slowest build-up of cake when circulation is stopped.
(3) Minimum friction between the wall cake and the pipe.[590]

The least friction can be assured by keeping the mud free of sand, by adding a lubricating agent, and by emulsifying with a surfactant that promotes wetting of the pipe and the mud solids.[14,295,625]

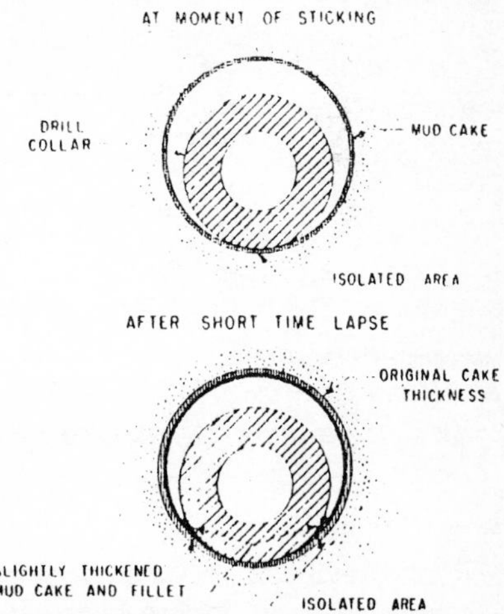

Figure 46[311] Stuck Drill Pipe results when it becomes motionless against a permeable bed. After cake buildup,-hole pressure forces pipe against wall.

By reducing the area of contact between the pipe and the wall of the hole, the possibility of sticking is lessened. The area of contact can be reduced by the use of stabilizers in the collar string and by the use of grooved pipe. Spiral grooves in drill collars and drill pipe repeatedly break the contact surface at short intervals. The spiral grooves are so designed that in the plane of rotation a cross section of the pipe does not have any concave surface.[227]

If pipe becomes stuck, the usual treatment is to place oil in the hole, although total removal of the oil is difficult and, therefore, limits the applicability of this method. Because oil is lighter than water or mud, the differential pressure is reduced. Displacement of mud by oil also stops build-up of filter cake. The seepage of the oil between the pipe and the filter cake is enhanced by adding a surfactant to promote oilwetting of the steel. Any means of lowering the pressure in the borehole may be effective—for example, a drill-stem tester has found application.[546] With the packer seated above the stuck pipe, the tester is opened, and the pressure drop inside the borehole causes the release of the drill pipe.[41]

Fishing

The great majority of downhole tool loss is due to either mechanical failures or carelessness of the personnel. Equipment that is overloaded causes the majority of fishing operations.

Fishing technology of the petroleum industry has developed considerably over the past five years. Because of the recent trend toward smaller-diameter oil wells and completions, "slim-hole" tools have been developed for many purposes. The newly developed fishing tools have been listed according to type, purpose, and company names.[55] Tool handling problems have been reduced by the development of such equipment as hydraulic vices and portable tool-servicing units which can be taken into the field.

Of particular use in fishing is a grappling tool, similar to a pair of ice tongs, described by Bader,[107] effectively used to remove "junk" or debris from wells. New methods of pump recovery have also been developed by petroleum industry personnel. For example, when a water well pump was dropped down a 280-foot hole, a special fishing tool was designed to slip over and grip the pump with four-inch lengths of drill pipe. In this case, 22 lengths of four-inch drill pipe were added until the pump was recovered; the operations required two days of effort.[31]

Basic fishing techniques are explored in a later chapter.

Hole Deviation

Hole deviation is caused by mechanical problems that result from an imbalance of forces between the hole bottom and the bit or the hole wall and drill string. This imbalance causes the bit direction to change; therefore, the objective in control of the hole deviation is to minimize the rate of change of the resultant angle.[395,411]

The most effective method of combating hole deviation is the proper design of the bottom-hole assembly, including the use of drill collars and placement of stabilizers.[519,659] "Packed hole" assemblies provide lateral bit support by stabilizers that are placed directly above the bit and at other strategic locations in the collar string.[531,532] Weight on the bit should be limited by a balance of the buoyed collar weight, so that all of the drill pipe remains in tension. Stabilizers near the top of the collar string will prevent the bottom-hole assembly from leaning to the low side of the hole (the "pendulum effect" that builds hole angle in deviated wells).[663] Hole deviation can be minimized, if not eliminated, by sufficient attention to these factors. (For further information, see "Well Plumbness")

In many areas it is not practical either mechanically or economically to drill a truly vertical hole; in fact, few, if any, vertical holes have been drilled which have remained within one or two degrees of vertical every point; many exceed five degrees in inclination.

Little was known of hole deviation until 1929 when a new, more accurate means of measuring hole inclination was introduced. The instrument consisted of a steel cylinder immersed in a copper sulfate solution. The solution would deposit copper with a top "level line" on the cylinder, thereby allowing the angle of inclination to be measured, after correcting for the lower side pull of the meniscus. Numerous instruments have been developed and marketed which measure (1) hole inclination (2) hole inclination and compass direction.

With the development of the instruments and standards for maximum hole inclination, it became necessary to examine and develop drilling techniques which would permit economic drilling of nearly vertical holes through so-called "crooked-hole" zones, and to determine as far as possible the basic causes of crooked holes. In recent years there has been fairly common agreement on these causes, e.g. (1) Formation Effects and (2) Mechanical Equipment.

In drilling moderately inclined stratified formations, it has been shown by survey that the tendency of the bit is to drill "into the hill." This process has been demonstrated in laboratory drilling tests. In severely inclined stratified formations, the bit will tend to drill parallel to the bedding planes, since the formation fractures more easily along each bedding plane than in a direction perpendicular to bedding. In local variations in drillability, such as chert modules, pyrite, boulders, caverns, the bit may be deflected from its normal course by the harder (or softer) rock or it may drift to follow the direction of fissures, etc.

The use of the "stiff" bottom assembly, large drill collars and stabilizers, has now become widespread in the petroleum industry, but not in the mining and ground water industries.

In order to reduce weight (and therefore stress) bit life and strain on other equipment, air drilling has gained popularity in the shallow drilling industry. The need for drill collars and other bottom-hole assemblies is not a prerequisite in controlling hole deviation.

The rotary drilling system has evolved into a number of variations, all basically rotary drilling but with specific differences of approach, air rotary, reverse-circulation rotary, etc.

6
Variations of Common Drilling Systems

AIR ROTARY DRILLING

The term "reduced pressure drilling" is applied to drilling with a circulating medium having a density less than water. This drilling method employs drilling fluids of dry air, mist, foam, "stiff foam," "gel foam," and aerated mud. Chambers[162] has summarized in tabular form the five forms of air drilling (Table 12).

The primary objective when using air and aerated drilling fluids is to increase the penetration rate by lowering the differential pressure (see "Factors that Affect Drilling Rate").[3,210] As the ratio of air to liquid decreases, the drilling rate is reduced by the increase in borehole pressure. Consequently, after a water-producing formation is penetrated, the amount of water produced will control the drilling rate. Reduced-pressure fluids can be used to circumvent severe loss of circulation. Foam and gel foam make possible the removal of cuttings from the hole at substantially lower velocities than is possible with dry air and at less annular pressure than is exerted by water. Air is the transport vehicle in dusting and misting; the liquid film carries the cuttings in foams, and mud carries the cuttings in aerated mud.[318]

An annular velocity of 3,000 feet per minute is often considered adequate to clean the hole of cuttings when drilling with dry air, although rates ranging from 2,000 to 5,000 feet per minute have

TABLE 12[162]

FIVE BASIC METHODS FOR USING AIR FOR DRILLING

	DUSTING	WATER MISTING	MUD MISTING	STIFF FOAM	AERATED MUD
LIFT MEDIUM	Air lifts cuttings.	Air lifts cuttings; water and foamer only remove water from the hole. Water wets hole.	Air lifts cuttings; fluid protects the hole and helps remove cuttings and produced fluids.	Foam lifts cuttings; air only adds volume	Mud lifts cuttings; air only lightens column.
ASSUMED FLUID VOLUMES**	None	10 bbl per hour*	10 bbl per hour*	10 bbl per hour*	10 bbl per minute*
ASSUMED AIR VOLUMES**	2000 to 3000 scfm	2000 to 3000 scfm	2000 to 3000 scfm	100 to 300 scfm	500 to 2000 scfm
AIR: FLUID RATIO (cu ft/ct ft)**	None	2000:1 (at 2000 scfm) 3000:1 (at 3000 scfm)	2000:1 (at 2000 scfm)	100:1 (at 100 scfm) 200:1 (at 200 scfm) 300:1 (at 300 scfm)	10:1 (at 560 scfm) 20:1 (at 1120 scfm) 30:1 (at 1680 scfm)
BASIC REASON FOR USING SYSTEM	Fastest penetration rates with dry air	Fast penetration rates with air. Foamer helps remove produced fluids and wets formation	Fast penetration rates with air. Stabilizes unstable formations. Superior removal of cuttings and fluids	Good cleaning ability in large holes. Stabilizes unstable formations	To combat lost circulation by lightening the hydrostatic head
TYPE SYSTEM	High energy type system	High energy system. High volume air with just enough air to clear the hole of produced fluids and cuttings.	High energy system. High volume air with just enough mud to give good hole protection.	Very low energy type system. Low volume of air and low volume of fluid	Very high energy type system. High volume of air and high volume of fluid

* All figures are approximations, based on drilling a 10-in. hole that is producing some fluid

** All numbers given above are for illustration only and must not be construed as recommendations

been recommended. Angel[12,13] using 3,000 fpm as the required air velocity, assumes that the air and the cuttings form a homogeneous mixture with the flow properties of a perfect gas. Drilling rate is included as a parameter, and a correction for downhole temperature is introduced. An equation is developed, and a computer used to calculate air requirements for typical hole and pipe sizes.[13] A table was prepared from which the approximate circulation rate can be calculated,[12] and this table is in common use.

Moore[463] states that "air drilling has been a trial and error technique primarily." The point of most difficult lift is at the top of the drill collars; therefore, if the air (gas) volume is too low to lift the larger cuttings through the relatively short zone at the top of the collars, the accumulation of additional cuttings in the zone will result in solids slugging and eventually in the breakdown of lift.

Lack of sufficient air velocity to clean the hole may be due to insufficient capacity of the compressors or to hole enlargement. On the other hand, excessive air can cause hole enlargement in certain soft formations. Relatively small back pressures on surface equipment can significantly reduce the annular velocity near the surface.

When formations that contain enough water to produce wet cuttings are penetrated, a "mud ring" may form that will be indicated by a rise in standpipe pressure. If there is not an actual flow of water, the introduction of drying agents or the placement of plugging substances[602] may serve to permit dust returns. If the flow of water entering the hole is too high to be handled by drying agents, the transition is made to foam drilling.

Foam drilling offers these advantages:[474] (1) wet cuttings can be removed from the hole with less pressure; (2) sloughing is reduced because of stabilization of pressure; and (3) penetration rate and bit life are improved because the pressure is stabilized at a lower level than if wet cuttings were allowed to build up. Studies have shown that flow in the annulus is probably "slug flow" as illustrated in Figure 47.

Water is injected with the foam to disperse the cuttings if insufficient quantities are entering the well bore from the formation. For example, water influx, as high as 350 gpm at 4,000 feet, can be handled with foam.[474]

The "stiff foam" technique is a method of drilling unconsolidated formations with a low density drilling fluid.[162] When using stiff foam, the required annular velocity is between 100 and 200 feet per

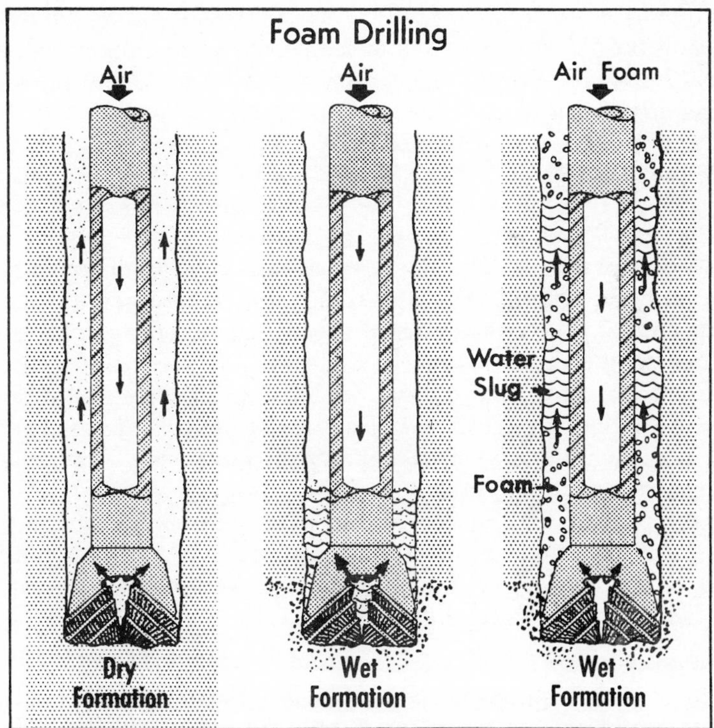

Figure 47[474] Foam drilling demonstrating "slug flow."

minute. Consequently, compressor requirements are much less than in other forms of air drilling. The slurry to be introduced into the air stream commonly consists of: 20 to 25 pounds of bentonite; one to three pounds of soda ash; and 0.5 to 2 pounds of organic polymer thoroughly mixed in 100 gallons of water. To this slurry 0.5 to 1 percent by volume of foaming agent is carefully added to avoid entrapment of air. The mixture is injected at such rate as to furnish a return flow having the consistency of shaving cream. A controllable, low volume injection pump is necessary for foam drilling. By varying the amount and composition of the film-stablizing organic polymer in the slurry, drilling in unconsolidated and water-sensitive formations can be successful. The stiff foam technique is widely adaptable but, at present, can be described only in terms of specific experience.[3,299,306,463,474]

Air-rotary drilling is especially advantageous in highly fractured and cavernous rocks such as carbonates, etc., with high secondary permeability. The loss of drilling fluids using conventional rotary methods is not only very costly to correct but results in plugging the aquifer. Air rotary drilling, often using foam, has proven very successful even in drilling wells up to a 10,000 gpm capacity by increasing the volume of air, by adding compressors in parallel, and if necessary, by increasing the pressure with the addition of compressors in a series.

Field Operations

Drill Bits

Jet-type nozzles in the bits are less effective than low pressure nozzles for air drilling. The air stream should be directed onto the cutters, and jet type nozzles direct the air stream between the cutters and directly upon the bottom. In some abrasive formations, directing the air stream onto the cutters will have an adverse sand-blasting effect on the cutting teeth. For drilling formations of this type, the jet bit can be used. In general, standard nozzle shapes in small-size openings are usually effective.

Rotary speeds are usually about the same as for conventional rotary. Any bit that performs successfully with drilling mud will usually be satisfactory with air. Weight carried on the bits is usually about half that used in mud drilling.

Crooked Holes

Control of hole deviation is a more serious problem with air as the circulating system than with mud or water. Too much bit weight is the primary cause of deviation.

Water Probelms

Small amounts of water in the formation may be detected by a significant reduction in the amount of dust returns and the presence of mud pellets in the dust exhaust. This is considered a precarious condition since the pellets tend to agglomerate in the hole near the bit and cause the drill pipe to stick. Partial remedies are to (1) stop drilling and blow the formation in an attempt to dry it; (2) reduce the drilling rate; and (3) inject water to increase the water-to-rock chip ratio.

Almost complete elimination of dust and replacement with a substantial water spray imposes no immediate drilling problem if the compressors have sufficient power to lift the water, and there are no caving or sloughing formations above the water producing zones. If such exist, the hole should be drilled with drilling mud until the casing is set. If sloughing is not expected, air drilling can continue until there is a tendency for the bit to stick.

In a "weeping" zone, calcium stearate or silica gel dry powder, added to the air stream by a solids pump, is effective. An injection rate of one to four percent of the cuttings weight is required. The finely ground powder coats the cuttings and effectively repels any water.

Foaming agents used to overcome water problems are organic materials added approximately one part to 1200 parts of water. In general, a flow of about 2000 gallons per hour marks the end of economically drilling with air.

Air-Lift Capacity

To determine the required air velocity enter Figure 48 at annulus air pressure (barometric pressure plus friction head) and intersect temperature line and read horizontal intersection on density coordinate. Continue horizontally and intersect with particle size and read required air velocity. Then, enter Figure 49 at determined air velocity and intersect given annulus area. Project horizontally to determine volume at hole conditions of pressure and temperature and continue to intersect with determined density. Read required air capacity.

Air-Drilling Efficiency

As mentioned previously, the injection of compressed air as a circulating medium is not new to the drilling industry. It has been in common use since the late 1800's. The 1950's, and the first big uranium boom provided the greatest impetus to the use of air. Prior

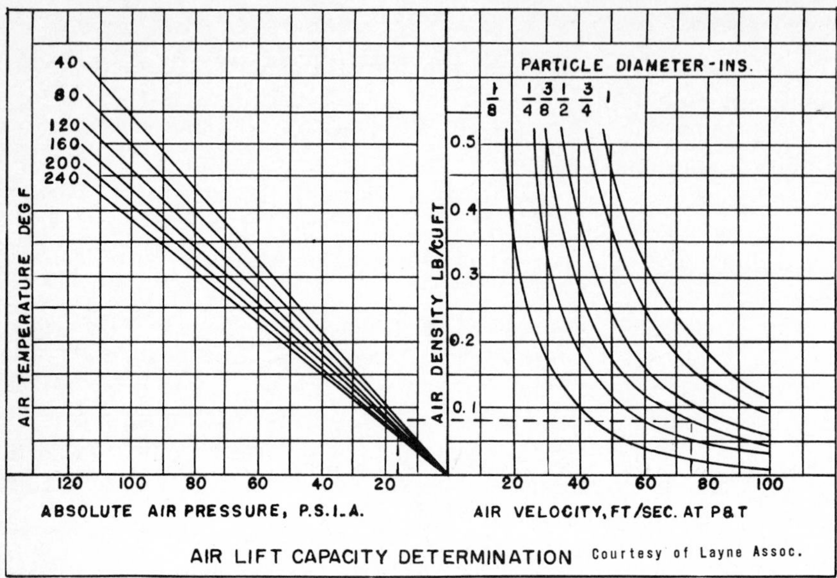

Figure 48

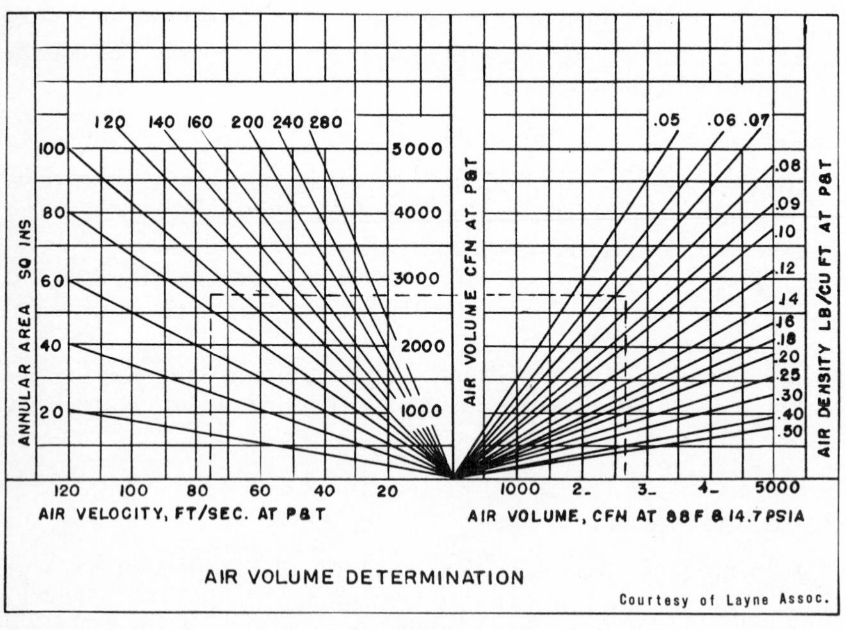

Figure 49

to this, most air uses were for operation of mining and construction percussion equipment, and the prime object was to provide the percussive impact. During the 1950's, many steps were taken in the use of air in exploratory drilling work, but all operations were not successful; the common problem was thought to be "not enough volume — not enough pressure." It was not unusual to find one or two large construction-type compressors (costing $25,000 to $50,000) providing air for a drilling machine costing $10,000. The extremely high cost of the operations forced many of the operations out of business. More importantly, some individuals were forced to reexamine the principles involved in air circulation. This reexamination and much subsequent research has given the industry a basis for a logical, economical approach to the use of air as a circulation fluid.

H. E. Davis of Mobile Drilling (*Water Well Journal*, in press) reports that air, as used with conventional rotary drilling equipment (where no impact force is involved) need not be delivered at high pressure. This provides for significant economy. Most exploratory holes used for civil engineering work are less than 250 feet in depth. Air pressures of 40-50 psi are more than adequate for recovery at such depth. A blower delivering approximately 375 CFM @ 50 psi can be purchased for as little as $4,000. A compressor delivering approximately 375 CFM @ 105 psi will cost approximately $16,000. Thus, the dollar value saved in the use of an adequate volume of low pressure air versus superfluous high pressure at the same volume can be as much as $12,000. The terms blower and compressor are not well defined; however, there does seem to be a tendency to refer to air blowers when considering pressures below 100 psi and air compressors when referring to 100 or more psi.

Most blowers and compressors use oil in the air stream for cooling and/or lubricating internal parts such as pistons or vanes. A minimum pressure of 30 to 50 psi must be maintained internally for the presently available machines to insure adequate lubrication of the mechanism. The minimum pressure of 30 to 50 psi is a machine requirement rather than a circulation requirement. Recent experiments, reported by Davis, with a radically new blower, delivering 400 CFM @ 20 psi, demonstrate that the pressure is adequate for circulation. This unit utilizes a special hydraulic circuit for internal lubrication. While it is apparent that even pressures as low as 20 psi will provide cuttings removal, pressure also plays an important part in bit cooling. Too low a pressure could cause inadequate bit cooling.

Davis further reports that all circulation fluids have two primary functions: (1) cooling of the bit and (2) removal of cuttings from the hole. Liquid-mud fluids can be altered by using barites and other weighting additives to stabilize the hole. Air cannot meet this function; however, air is seldom guilty of causing hole caving.

Bit cooling is accomplished by refrigeration. Air which is pulled into the blower and compressed is carried under pressure through the mast standpipe, swivel, hose, and tool string to the bit. The sudden release of pressure as the air leaves the bit causes a considerable drop in temperature of the air. It is not uncommon, under certain atmospheric conditions, to find ice on the bit. The actual temperature drop will be dependent upon, among other factors, the pressure released. This cooling principle is essentially the same as that used in modern refrigerating units.

Air Requirements

Chip or cuttings removal is accomplished by the flotation effect or buoyancy of the released air on its way up from the bit to the surface. The velocity of this upward-moving air will be the most important factor in determining how effective air drilling will be. Davis believes that this velocity is effected by three factors: (1) hole diameter, (2) drill stem O.D., and (3) blower output volume. The first two of these factors concern the annular area. The *annular area* is defined as the area of the hole minus the area of the drill stem. As an example, the annulus area of a four-inch diameter hole drilled with 2 7/8-inch O.D. drill stem is:

EQUATION 34

$$(4^2 \times .785) - (2.875^2 \times .785) = \text{Annular Area}$$

$$12.57 \quad - \quad 6.49 \quad = 6.08 \text{ sq. inches}$$

The annular area must be kept full of upward moving air in order to assure good chip removal. Logically, the smaller the annulus area, the less the air volume requirement to assure good recovery. The annular area can be reduced in two ways: (1) the hole diameter should be the smallest possible considering the use intended and the need for about 5/16-inch minimum radial chip clearance between the drill stem and the hole wall. The cost to adjust hole diameter is usually limited to the bit cost. (2) stem size (O.D.) should be the largest possible consistent with the need for 5/16-inch radial chip clearance as mentioned above. The cost to adjust stem size is limited to the purchase price of a new larger string which, even though expensive, will be much less than the only remaining alternative which is to use a blower with greater output.

The minimum recommended up-hole velocity for good chip removal has been determined by field work to be about 4,000 feet per minute. When drilling in soils, as in the case of most engineering investigations, the recommended minimum is 4,500 ft/min.

The conversion of blower volume to up-hole velocity, or vice versa, is based on the equation:

EQUATION 35

$$Q = AV$$

where: Q = Volume (cu. ft.)

A = Area (sq. ft.)

V = Velocity (ft/min)

For example, a hole of any depth must be filled with air moving at 4,500 ft/min. and the volume must be the same that would be required to fill a 4,500 ft. deep hole *once.*

The 4,500 FPM needed for good recovery is obtained as follows:

EQUATION 36

$$\text{CFM} = \frac{\text{A x (4500 feet x 12)}}{1728}$$

Referring to Equation 35, the four-inch hole with 2 7/8-inch stem had an annular area of 6.08 square inches. Using the above, the volume needed to remove chips in the example hole is:

EQUATION 37

$$\frac{6.08 \text{ x } 4500 \text{ x } 12}{1728} = 190 \text{ cu. ft/min}$$

As an example of what can be done to minimize air requirement, assume the bit size in the above example could be changed from four inches to 3 1/2 inches. In this case the air demand would be:

EQUATION 38

$$\frac{(3.5^2 \text{ x } .785) - (2.875^2 \text{ x } .785) \text{ x } 4500 \text{ x } 12}{1728} = 96.5 \text{ CFM}$$

The bit diameter at 3 1/2 inches was the minimum tolerable since it allowed only 5/16 inch radial (5/8 inch on the diameter) chip clearance. To demonstrate that the best advantage is gained by reducing bit size, the following example shows the volume required if the hole remains at four-inch diameter, and the stem is increased to maximum which will provide chip clearance. Use: 4 inches — 5/8 inches equals 3 3/8 inches O.D. Stem.

EQUATION 39

$$\frac{(4^2 \text{ x } .785) - (3.375^2 \text{ x } .785) \text{ x } 4500 \text{ x } 12}{1728} = 113 \text{ CFM}$$

Even though the 5/16-inch radial chip clearance is maintained the air demand has increased 16.5 CFM to accomplish the same up-hole velocity.

Air Supply Equipment and Selection

Having determined the actual volume required, it is necessary to assure that the blower purchased or used is actually capable of delivering the volume needed. Davis contributes three factors to the problem of selecting the correct blower: (1) Many blowers have dual ratings — one will be the actual delivery, and the other is displacement. There can be in excess of 40 percent difference between the ratings. Selection must be made on the basis of actual delivery.(2) The internal combustion engine which powers the blower will lose power at the rate of 3.5 percent per each thousand feet above sea level. All engines are rated at sea level, so at 5,000 feet elevation only 82 1/2 hp can be expected from a 100-hp engine. (3) All engines lose power at a rate of about one percent per 10° over 60°F. Engine ratings are usually based on 60° F air. Operation at 100° F means the 100-hp engine only produces 96 hp.

All these losses and other minor losses can combine in such a way as to make one big loss. It is, therefore, essential that the required air and the power to provide that air be based on the location of the drill site.

Davis suggests that various other factors can be of considerable importance in conducting an efficient air-drilling program:

(1) *Selection of Drill Stem:* Outside flush pipe or D.C.D.M.A. drill stem should be used whenever possible in order to minimize obstructions in the path of the returning air. Tool joints, when used, cause the air to deflect outward against the hole wall and continually "sandblast" the wall, thus enlarging the hole. Stems should be selected which will provide a minimum coupling or joint bore of one inch for volumes less than 300 CFM and 1 1/2 to two inches for delivery in excess of 300 CFM. Very large deliveries — 800 CFM or more — should utilize even larger bores. The downhole air as it passes through each joint is alternately compressed and expanded which causes heating and cooling, respectively, of the pipe. This phenomena is generally observed only in relatively deep holes. It does occur, however, to a limited extent in shallow holes.

(2) *Standpipe and Swivel:* No portion of the plumbing from blower to swivel should be smaller in cross section than the pipe itself. This is true with any circulation medium but is especially true with air. Swivels which are used exclusively with air utilize special asbestos-base packing which is resistant to the drying and cracking effect of air. Mud swivels should be used when both air and water will be used and swivel fluid courses should be equal to or larger in diameter than the pipe.

(3) *Liquid Injection:* Air drilling is most efficient in dry formations and nearly as efficient in saturated formations. Intermediate formations, especially those which tend to "boot" or ball the bit, are the most difficult. Difficult formations can be handled by increasing the moisture content of the formation drilled by several different methods ranging from dumping water down the hole, or injecting water into the downhole air, to metering detergents into the downhole air. The most effective method, as determined by an experimental contract let by the Atomic Energy Commission, appears to be with sodium-based detergents. These detergents are superior to water because they lubricate the stem, thereby eliminating some booting and wall packing — and because the sodium base minimizes the formation of ice at the bit. Both water injection and detergent mists will increase the chip-carrying capacity of the air and will assist in minimizing the dust from the air drilling operation.

(4) *Dust and Chip Catchers* have three purposes:

(a) Proper geological sampling.

(b) Prevent dust and grit from entering the precision-machined surfaces of the drill and associated machinery.

(c) Minimize the health and sanitation dangers inherent in a dusty and gritty environment.

Sample catchers may be as simple as a "rod wiper" running on the stem a foot or two above ground, or as complex as a complete cyclone-type dust collector. The complexity and efficiency of the collector or catcher is largely dictated by specific requirements. All present collector systems have one basic weakness. The packing glands used to seal the rotating stem to the deflector box are of the contact type and wear rapidly due to friction and sandblast abrasion.

(5) The recommended uphole velocity of 4,500 FPM advocated by Davis is reportedly valid for formation materials up to 3.0 specific gravity and, therefore, is adequate for more than 95 percent of all formations. In exploratory drilling in materials having a specific gravity in excess of 3.0, uphole velocities must be increased in proportion to the specific gravity. For instance, 6.0 Sp. Gr. requires 9,000 FPM, 9.0 Sp. Gr. requires 13,500 FPM, etc.

(6) Table 13 can be read directly and provides data for most common drill rod sizes and holes ranging from 1 3/16 inches to **ten** inches diameter. The 6,000 FPM velocity is utilized for preliminary and/or tentative recommendations. A similar table based on water as a circulation medium can be found in Catalog 960, published by Mobile Drilling Co., Inc. The proper volume for 4,500 FPM may be obtained from Table 13 by multiplying the appropriate 6,000 FPM figure by 0.75, etc.

TABLE 13

DRILL ROD or DRILL STEM

	HOLE DIAM	RW	E	EW	A	AW	B	BW	N	NW	KWY	3¼	HWY	3-7/8	4¼	5¼
RW	1-3/16	7														
	1-1/4	12														
	1-3/8	23														
EW	1-1/2	35	16	12												
	1-5/8	47	30	25												
	1-3/4	61	44	40	12											
AW	1-29/32	80	63	57	33											
	2	92	76	68	44	32	12									
	2-1/4	127	108	104	80	64	48	20								
BWR	2-3/8	145	128	124	96	84	64	32								
	2-1/2	165	148	144	120	104	84	36	20							
R	2-5/8	186	169	164	139	129	107	78	41							
	2-3/4	208	192	188	160	148	128	100	64	24						
R	2-15/16	243	226	221	196	186	164	135	98	57						
NW	3	255	236	232	208	192	176	144	102	68	24					
R	3-1/8	280	264	256	232	220	200	172	136	96	48					
R	3-1/4		288	284	260	244	228	200	160	120	76					
R	3-1/2			340	312	300	280	252	216	176	132	55				
R	3-3/4				372	360	340	312	276	216	188	115	60			
LSR	3-7/8					388	373	344	308	264	220	145	90			
	4						408	376	340	300	252	178	123	32		
R	4-1/4							444	408	364	320	245	190	100		
R	4-1/2								476	436	392	317	262	172	71	
R	4-3/4									552	468	393	338	248	147	
	5										548	473	418	328	227	
	5-1/4											556	460	408	311	
LS	5-1/2											644	589	496	399	88
	5-3/4											736	681	588	491	180
	6											832	777	684	587	316
	6-1/2												981	888	791	480
	7															700

Compressor or Blower volume (CFM) to provide 6,000 FPM Up-Hole velocity is found by intersecting hole size HORIZONTALLY with rod size VERTICALLY.

i.e. 3" hole with BW rod requires 144 CFM to provide 6,000 FPM Up-Hole velocity.

The 6,000 FPM velocity shown here provides a theoretical safety factor of 25%. This safety factor may or may not actually exist depending on many factors-altitude, efficiency, losses etc. Values obtained from the table may be multiplied by 0.75 in order to eliminate the safety factor and arrive at an up-hole velocity of 4,500 FPM — Air Drilling is rarely successful at less than 4,500 FPM. (Courtesy of Mobile Drilling, Inc.)

Davis has thoroughly examined the basic aspects of air circulation. The economics are extremely variable and dependent on many diverse factors. Air drilling is claimed to be economical in over 75 percent of all cases. The advantages of air drilling are many and include: (1) instant cuttings recovery, (2) minimum setup time, (3) no water hauls, (4) "as-is" moisture samples, (5) no recirculation tanks, (6) no suction hoses, (7) no water trucks, and (8) no washed cores.

AIR-PERCUSSION ROTARY DRILLING

Another form of air drilling is percussion or downhole-hammer drilling. Percussion drilling tools and techniques are gaining acceptance and are no longer considered a specialty application.[112] Percussion rotary drilling is the rotary technique in which the main source of energy for fracturing rock is obtained from a percussion machine connected directly to the bit. The circulating system, drill stem, etc., are unchanged from that used in the conventional air rotary method. The percussion method incorporates a single-cylinder, reciprocating, air engine driven by the circulating air. The extremely rapid hammer blows — delivered with relatively light weight on the bit — increase bit life and help to control deviation while maintaining relatively high penetration rates.

Until the last few years, bottom-hole percussion tools were made to operate on 100 psi air pressure primarily because of the availability of 100 psi compressors. Since reliable higher pressure compressors (200-250 psi) are now available, the trend is toward the use of higher pressure tools, and turbine compressors will probably be available in the not-too-distant future. The reason for commercial development of 100 psi percussion tools is that the limited piston area restricts the air that can be ported because the piston diameter has to be approximately two inches less than hole diameter. Consequently, the amount of air required to operate the piston is less than that needed to clean the hole.[118]

High-pressure tools (200 psi) operate with approximately twice the hammer-blow frequency of 100 psi tools. Thus, all of the energy in the compressed air can be ported to the piston, resulting in penetration rates double that obtained with 100 psi tools. Most carbide bit wear is the result of the carbide scrubbing against the rock face during rotation; therefore, bit life is generally longer with high-pressure tools because, with faster penetration, less carbide is worn per foot of hole drilled.

One of the latest developments in downhole percussion air drilling is a button bit with replaceable, tungsten-carbide buttons. The bit, as originally designed, has buttons set from the face side to a predetermined height above the bit face. Buttons are removed with a special punch through access holes at the back of the drill bit.

Within the last two years, the button bits have been modified, so that either button setting or removal can be accomplished within a few seconds from the face side of the bit. The new system utilizes a gas-pellet "gun" — the first gas pellet sets the buttons to a specified height above a special base and sleeve arrangement, and the second loosens the sleeve for easy removal of the entire three-piece button assembly.

The proper size drill pipe is critically important. In the typical case previously discussed, if five-inch drill pipe were used, only 350 cfm would be required to maintain the recommended 4,500 feet per minute annular velocity. Thus, the same hole-cleaning ability can be obtained with less air (110 cfm) by selecting a larger size drill pipe. In many cases, hole-cleaning problems can be solved more economically by increasing drill pipe size rather than by increasing compressor capacity.

In the typical case of a 5 1/4-inch diameter tool (used to drill a 6 1/4-inch hole with a 4 1/2-inch drill pipe) only 200-225 cubic feet per minute of air can be ported to the piston, yet 310 cubic feet per minute is required to maintain an annular velocity of 3,000 feet per minute, the minimum to clean the hole in this type of drilling (see Figure 50). A recommended 4,500 feet per minute annular velocity requires 460 cfm of air.[65]

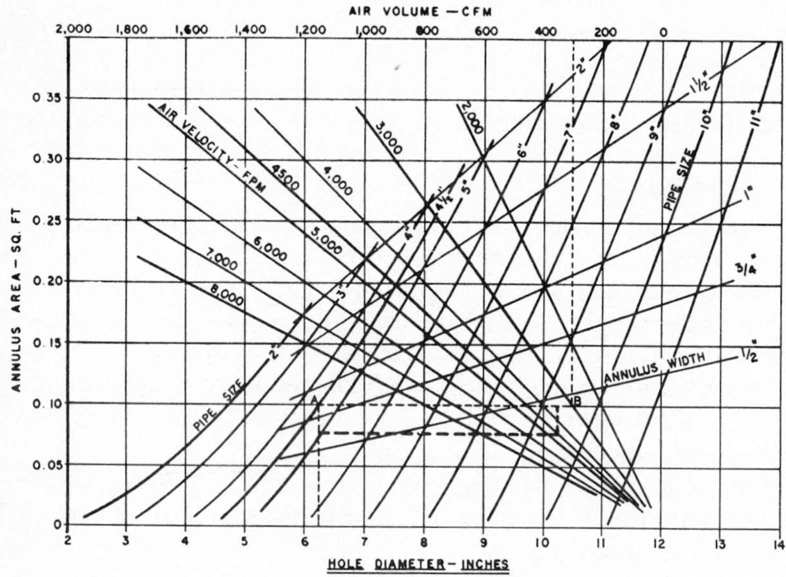

Figure 50[236] Air Velocity Determination Chart.

Thus in a normal operation, from 110 to 235 cfm of air, compressed to 100 psi, is used solely to provide enough return velocity to assure that the hole is cleaned properly. If an adequate supply of air is not ported to the bit, the energy of the available air is utilized in power for the bit, and adequate cleaning is not accomplished. Some of the more recent 100 psi tools have multiple pistons to increase piston area and, thus, to utilize energy that is wasted with the single-piston tools. These tools have penetration rates from 10 to 40 percent faster than single-piston tools, depending on the size of the compressor used.

Bottom-hole pneumatic tools generally require only sufficient "pull down" (2,000 to 4,000 pounds) to keep the bit closed in the tool. Additional pull down (rig weight) does not materially increase penetration rate but does result in accelerated bit wear.

Water injection into the air stream is recommended as an economical method to control dust, to prevent cuttings from "balling up" around the drill pipe which could bridge the hole and prevent circulation, and to seal any air leakage inside the tool between closely matched parts. Water injection cools the drilling air to assure that the air temperature supplied will not be excessive.

The downhole percussion technique can be modified to combat excessive ground water influx by incorporating foam (see previous section). Drilling for ground water in crystalline, metamorphic, or other rocks with high compressive strength is drawing particular interest and no doubt will expand as this method of drilling improves.[119, 175, 197]

Field Operations

This method of drilling is most efficient in consolidated rock formations which do not require casing. It is not usually satisfactory for use in boulders or in unconsolidated formations although the method has found some success in such material. Continued drilling in wet clay may plug the air holes in the bit and stop the hammer operation. Hard and abrasive cuttings which are wet should not plug the bit and hammer and can be blown from the drill hole. However, some cuttings do get into the hammer cylinder which requires cleaning after 500 to 1,000 feet of drilling.

The depth of drilling is limited by the diameter of the hole and the volume and pressure of the compressor in use. The depth at which cuttings can be effectively removed from the hole is also governed by the weight of the rock and the volume of water in the hole.

Often pull-down pressure with the rig at ground surface is not recommended to increase the penetration rate. As with any other type of rotary drilling, such procedure places the drill pipe in compression and causes crooked holes, key-seating, etc. The use of drill collars for weight with the drill pipe rotated in tension is to be preferred. Depending upon the type of rock formation, the air percussion method may penetrate 50 to 100 percent faster than the standard tricone rotary bit.

Recommended pull-down is in the range of 3500-4000 pounds. This can be obtained advantageously with drill collars. The hammer percussive blows at operating air pressures of 80-200 psi range from 600 to 1,000 per minute, depending upon the pressure and hammer design. Rotary speeds at an optimum bit index angle of 11 degrees vary according to the bit diameter, and type of formation, and hammer blows per minute. For the optimum index angle, rotary speed would vary from 18 to 30 rpm. The most economical rotating speed is that which gives the highest rate of penetration without excessive bit wear and, therefore, depends on specific field conditions.

REVERSE-CIRCULATION ROTARY DRILLING

Reverse-circulation rotary drilling, with little application in the petroleum industry, has found considerable success in the mineral exploration and ground water industries.[46, 547] In mineral exploration, reverse-circulation drilling is applicable when the quality of the samples is critically important. Reverse-circulation coring has also found widespread acceptance in recent years, e.g., "Con-Cor," etc.

Reverse-circulation rotary drilling has a reversed flow of drilling fluid when compared to the system used in the conventional rotary method. The drilling fluid flows through the suction end of the rig pump — rather than the discharge end — and through the swivel, the kelly, and drill pipe. The drilling fluid and cuttings move upward inside the drill pipe and are discharged by the pump into the mud pit. The fluid returns to the borehole by gravity flow, moving down the annular space around the drill pipe to the bottom of the hole, picking up cuttings, and re-entering the drill pipe through ports in the drill bit.

The drilling fluid used with this method can be described as muddy water rather than as drilling mud. Suspended clay and silt which recirculate with the water are mostly fine materials that are picked up from the subsurface formations as drilling proceeds. Bentonite or other drilling fluid additives are seldom added to the water to make a viscous fluid.

To prevent caving of the hole, the fluid level is kept at ground level at all times. The hydrostatic pressure of the water column, plus the inertia of the body of water moving downwards outside the drill stem, support the borehole wall. Erosion of the wall is not a problem because velocity in the annular space is low.

Water is lost from the hole into all permeable formations that are penetrated. Some of the suspended fine particles in the fluid are filtered out on the wall of the hole, resulting in a thin mud deposit that partially clogs the pores and reduces the loss of water. A considerable quantity of make-up water is required, however, which must be readily available at all times when drilling in highly permeable formations.

Water loss can increase suddenly, and if the fluid level in the hole drops below the ground surface, caving can result. Water loss can be reduced by adding clay to the water (usually avoided unless necessary). One of the major problems in reverse circulation rotary drilling is preventing caving of clays and shales. Caustic soda in the fluid to raise the pH to about 10.5 may be successful, provided the clays are wetted in their native state. Dry porous clays and shales do not stabilize with increased pH.

Sodium silicate, in the ratio of four percent to ten percent, may be effective with clays and shales of this type. If the clay interval is thin, the treatment may be made directly at the hole. If the interval is relatively thick, then all of the fluid should be treated. If the treatment is still unsuccessful, the section should be cased, or high viscosity, low weight, low water-loss bentonitic muds prepared for use as the drilling fluid.

From 20 to 500 gpm of make-up water, depending on the hole diameter, may be needed at times when drilling through highly permeable sediments. Drilling in coarse, dry gravel poses the greatest difficulty because of the high water loss potential. Much of the high water losses commonly occur above the water table. This loss can be reduced by drilling the hole with a large auger or similar rig to the water table or into a relatively tight formation near the water table. The surface casing is then installed and grouted prior to deepening the hole with the reverse-circulation rig.

The mud pit or water supply pit should have a volume of at least three times the volume of material to be removed during the drilling operation. Circulation rate for the water used in drilling is commonly on the order of 500 gpm or more.

A centrifugal pump, with large passageways, is employed in order to handle large cuttings. One type of rig uses an ejector, operated like a large jet pump, which prevents the cuttings from passing through the centrifugal pump. The limitations of some rig pumps necessitate the use of drill pipe in lengths of ten feet. However, many other rigs are equipped with a pump arrangement which enables the drill pipe to be filled with water before starting circulation with the regular centrifugal or jet pumps. Using this system, 20-foot and longer lengths of drill stem can be used. Six-inch O.D. drill pipe is commonly used so that rock material up to a five-inch diameter can be brought up through the pipe.

The drill pipe commonly has flanged joints of approximately 11 inches in diameter. The smallest borehole that can be drilled by this method is about 18 inches in diameter which provides sufficient annular space at each flanged joint.

Wells with diameters to 60 inches can be drilled; the diameter of the hole must be large in relation to the drill pipe, so that the velocity of the descending water will be slow. Descending velocity of one foot per second or less is the general rule. The bit and drill pipe are rotated at speeds varying from ten to 40 revolutions per minute.

Reverse circulation offers an inexpensive method of drilling large-diameter holes in soft, unconsolidated formations. Where geologic conditions are favorable, the cost per foot of borehole increases little with increase in diameter. Drilling cost for a 36-inch or 40-inch hole is only moderately higher than for a 24-inch hole.

Most wells drilled by this method, therefore, have a diameter of 24 inches or larger — completion of the well by gravel packing is dictated by factors peculiar to this drilling method.

Conditions that favor the use of the reverse circulation method of drilling are sand, silt, or soft clay formations; absence of clay or boulders; or static water level of ten feet or more below surface. However, if the static water level is within ten feet of the surface, it is possible to construct a high retaining wall around the pit for rig placement in order to obtain the desired head differential, thereby, maintaining hole stability. Conditions that may limit the use of this method are:

(1) A static water level which is too high.
(2) The lack of an adequate water supply to supplement drilling fluid.
(3) Stiff clay or shale formations.
(4) A considerable number of cobbles or boulders.

Any cobbles or boulders, larger than the drill pipe or the openings in the drill bit, cannot be brought up in the drilling operation. The bits used cannot break cobbles, as soon as a few collect in the bottom of the hole, no further progress can be made. The drill pipe and bit must be pulled periodically, and the stones fished out by means of an orange-peel bucket so that drilling may continue.

The low content of suspended solids in the water of the annulus forms a relatively thin filter cake preventing excessive fluid loss. This thin filter cake is cleaned out more readily when developing the well than is the less pervious mud cake that is common to the mud rotary method.

Reverse-circulation rotary rigs are sometimes used to drill 16-inch diameter test holes in areas where the formation characteristics are fairly well known. When suitable equipment and procedures are used, this method provides reliable samples. However, in areas where the aquifer consists of interbedded thin sands and clays, samples should be taken through the drill pipe using a "split-tube drive sampler" or similar device.

SPECIAL APPLICATION DRILLING SYSTEMS

Other types of drilling systems are in use, e.g., jet drilling, chilled-shot core drilling, rotary-shot drilling, hydraulic percussion drilling, earth-auger drilling, percussion reverse-circulation drilling, etc.[145,427] These systems are generally designed for specialized drilling applications and will not be treated in any detail here, except for a brief description of selected methods.

Auger-Bucket Drilling

The auger-bucket system of drilling has been used primarily for surface water development, or water-table wells. Other economical and practical applications are: gravel testing, foundation holes, pier holes, seep holes, soil testing, etc.

The hole is bored without casing until the water-bearing formations are reached or hole caving begins. Concrete casing with a steel shoe is then lowered into the hole. The steel shoe fits over the bottom of the concrete casing to keep it from breaking and also to provide a cutting edge.

A suction is created when the auger bucket is removed for emptying. The removal of the formation materials and the weight of the casing force the casing to settle and shear the walls of the hole. If caving occurs at shallow depths or the well is relatively deep, temporary metal casings 1/4-inch to 1/2-inch thickness are used because the concrete casing generally has a wall thickness of at least two inches. The steel casings are of telescoping diameters like the ones used in standard caisson drilling. When satisfactory water capacity has been obtained, the screen and well casing are centrally located within the drilled hole, the gravel placed, and the caissons removed in the usual manner.

In gravel testing, after the topsoil has been penetrated, "sand flaps" are installed in the bottom of the auger bucket to trap the gravel. Steel casing is used to retain the walls of the hole and aid in penetration. A clamshell or trapping bucket, or plunger-type bucket, is used if sand in the gravel tends to seep past the sand flaps. Clamshell buckets, stone hooks and tongs, dynamite, and rams' horns are used for rock and boulders in large diameter shallow holes.

In difficult formations, chopping buckets and a pilot hole are often employed. A smaller auger bucket is used to bore a "lead hole" which enables the cutting blades of the larger bucket to later ream the hole.

Hollow-Rod Drilling

The hollow-rod drilling method differs from the cable tool drilling method in that the cable does not enter the hole, and the jetting drilling method is a reversal of the hollow rod drilling method.

The drill bit is connected to a string of hollow rods and drilling is accomplished with a reciprocating motion similar to cable tool drilling but with more rapid and shorter strokes. The bit is equipped with a small-ball check valve, and water is kept in the hole continuously. Downward motion of the tool string in the hole forces water and cuttings through the holes in the bit into the hollow rods. At ground level, the cuttings discharge through a hose into a tank where the cuttings settle, and the water flows back into the hole.

The drilling tool string consists of the drill bit, hollow drill rods, water swivel, and driving weights. The type of bit selected is based on the characteristics of the formation to be penetrated. In starting to drill, a short piece of pipe used as casing may be necessary if the topsoil caves, and the usual precautions are taken for maintaining a straight hole.

The formations penetrated and depth of the hole determine the type of operation method. For depths in excess of approximately 60 feet, the starting stroke should average 16 to 18 inches. Tools should be run slowly at about 35 to 40 strokes per minute until a depth of six to ten feet is reached— when the speed may be increased. The system is adapted for small holes up to about four inches in diameter because the tools cannot be made to pump the water when the diameter of the hole greatly exceeds the O.D. of the stems.

Jet Drilling

This method of drilling is similar to hollow-rod drilling except for the method of removing the cuttings from the hole. Water is pumped down the rods and out the bit into the drill hole. The water and cuttings return in the annulus to discharge at ground level. The cuttings settle to the bottom of a discharge tank, and the water is returned to the pump for reuse.

Jet drilling is advantageous in larger diameter holes of four to six inches, particularly where large amounts of sand or clay are to be penetrated.

Because of residual soil and grit in the circulating water, the pump should be of the displacement type for ordinary conditions, or the diaphragm type for extremely gritty water, and designed to provide a steady pressure on the rod line.

Usually the first few feet are drilled without water and with a blank valve over the bit openings. The valve is then removed before circulation is started. A thin short bit is necessary for drilling clays with a stiff tension maintained on the drilling cable. A short thin bit is most effective in sands with heavy water pressure. For "quick" sand, a check valve should be placed just above the bit to allow water to be forced down the rods, preventing sand from being forced up the rods. A thick bit is most effective for drilling through sand with boulders or heavy gravel. A regular rock bit with sharp cutting edge is most satisfactory.

Care is required in rock drilling to prevent sealing of the rock pores with the mud particles in the drilling fluid (water). After the rock has been penetrated to the required depth, the hole should be flushed with clean water.

Driving Pipe

Pipe driving methods are similar to hollow rod and jetting drilling methods. The driving tool is a block of steel weighing about 150 pounds. A series of blocks are clamped on a swivel stem with an eye at one end to provide a driving weight of 300-400 pounds. A driving plug at the bottom of the stem is screwed into the pipe coupling, and the stem works up and down inside the plug and pipe allowing the driving weights to strike the drive plug. The same tools are used to pump the pipe out of the hole.

References on other field-developed methods of practical value for special applications are listed for purposes of further study.[46, 64, 158, 287, 288, 315, 379, 425, 427, 599]

Horizontal Well Drilling

Horizontal wells are constructed by drilling a horizontal or slightly inclined hole into a steep slope, installing casing and screen, and grouting as required.[606] The principle difference between horizontal and vertical wells aside from their attitude are (1) All horizontal wells drain by gravity, (2) they generally require specifically built drilling rigs, and (3) their location depends upon the presence of steep slopes.

The U.S. Bureau of Mine's Information Circular 8392, *Horizontal Boring Technology: A State of the Art Study* describes a wide variety of horizontal drilling equipment and techniques available as of 1968. Although almost any hydraulic rotary drilling machine can be modified to drill horizontal holes, such a rig can not compete with one designed for the purpose.

RESEARCH AND DEVELOPMENT

The stage has been set for a dramatic appearance of one or more entirely new drilling systems for shallow drilling applications. Cable-tool drilling will only slowly disappear from the field because of its economic advantages. Rotary drilling and variations of it have nearly reached the limits of the method's capabilities. With economic efficiencies becoming of paramount importance, any new drilling method which offers a strong economic advantage will be received by industry far more quickly than previous modifications of the common methods now in use. If the "ultimate cost concept" continues to flourish, a new drilling system will be adopted even if the initial cost is somewhat higher than present rig costs. Incremental cost advantages spread over a year's operation will make the difference.

The operation of any new drilling system will not be simple. Well drilling contractors will be faced with more and more engineering problems, the solution of which will assure optimization of drilling operations. This will certainly require strong engineering support from within the contractors own organizational structure. A drilling engineer will eventually become commonplace, and in the not too distant future. Engineering control over much of the well contractor's operations is already exercised by some clients in certain types of drilling operations.

Engineers and geologists are now commonly on the staff of the larger well contractors. This will allow internal research and development of new drilling systems stimulating a research environment where economics will become the guiding principle not tradition as is so apparent even today in the field. Only an environmentally conscious, technically strong, economically based drilling operation will survive in the strongly competitive market of the water well and mineral exploration industries.

7
New Drilling Systems

Turbodrilling has more prospects today of becoming a main branch of world-drilling technology than the rotary method had 50 years ago when the cable-tool method was popular. Turbodrilling, in some cases, is ten times faster than the common rotary drilling; however, turbodrilling is not necessarily more economical. Since the method or a modification thereof offers considerable possibilities in shallow-drilling operations, turbodrilling will be examined in some detail. The present status of turbodrilling in the petroleum industry is on the narrow demarcation line between a difficult past and a future rich in possibilities.[15,616] The turbodrill concept has a long history of development in France and Germany, and since the late 1940's, the Soviet Union has contributed greatly to the present design.

TURBINE DRILLING

Of the characteristics of the various drilling methods, the main factor put forth by Tiraspolsky[616] is the specific power that is transmitted to the bit. *Specific power* is the power per unit section of the hole drilled. The number of horsepower per square inch is the measurement suitable for normal borehole dimensions. The equation for specific horsepower is:

EQUATION 40

$$P_{sp} = \frac{P}{\frac{\pi D^2}{4}} = 1.27 \frac{P}{D^2}$$

where: P_{sp} = Specific power (hp./sq. in.)
P = Useful turbine power (horsepower)
D = Bit diameter (in.)

In Figure 51 the abscissas represent the bit diameter in inches, and the ordinates represent the specific powers of Russian and French turbines. The series of curves shows the course of the specific power for different diameters and turbine outputs, ranging from 10 to 500 horsepower. The lower zone (I) corresponds to the specific power/horsepower for cable-tool drilling. Rotary drilling extends roughly between 0.5 and one horsepower per square inch (zone II). Above that, between one and four horsepower per square inch is within the capability of the turbine system. The power which can be transmitted to the bottom of the hole in rotary drilling has been discussed

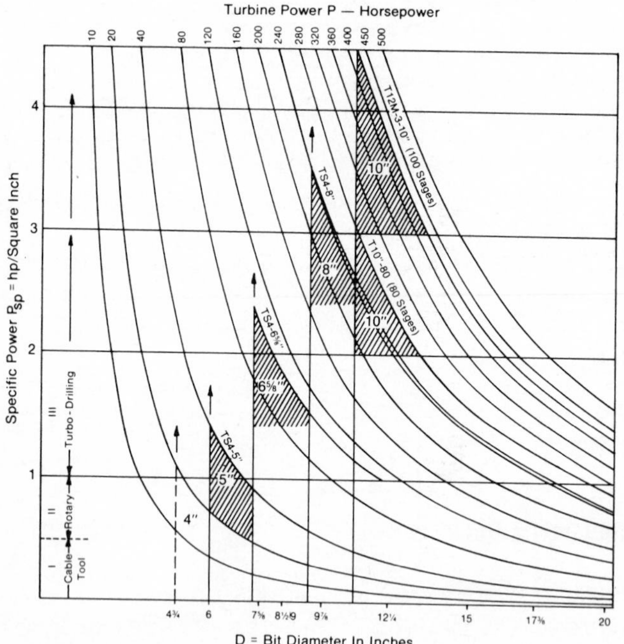

Figure 51[616] Comparison of specific power for basic drilling systems and equivalent turbine power according to bit diameter.

previously. Except for the special field of jet-bit drilling, the American oil industry has achieved rates of penetration many times higher than the French. Rotary rig weights, as high as 80,000 pounds or more on an 8 3/4-inch bit, are common, and the rate of revolution attained is rarely higher than 60 rpm. In hard formations, added weight can be regarded as one of the best performance features of American drilling.

Turbine Development

In Figure 51, the zones of application of the various Russian turbines (from ten inch to five inch) and of the French rotary turbine (ten inch) are cross-hatched. The four-inch turbine, still in the experimental stage, is shown by a dashed line. This figure illustrates that the specific power of the ten-inch and eight-inch turbines exceeds the potential of rotary drilling but decreases as the bit diameter decreases. Structural reasons which are difficult to overcome cause this decrease. The wall thickness, the tolerances, the blade thickness, and shaft diameter cannot be proportionally reduced as the external diameter decreases. At present, the drilling turbines have to be used with bit sizes of 8 to 18 inches. In Figure 51, the possibilities of further development of the turbine are indicated by arrows.

Russian technology has made significant advances with the introduction of double-sectional turbines.[22] Based on an analysis of the data collected on Russian turbodrilling and French ten-inch, 80-stage turbines, Tiraspolsky[616] states that:

> In every case where an appreciably higher power was transmitted to the bit by the turbine than was possible with the rotary, drilling proceeded more rapidly. In some cases the turbine drilled much faster because no load. could be placed on the rotary bit due to the risk of borehole deviation. In In some cases the turbine drilled much faster, because the drill stem was so overfatigued that the rotary table could not be made to rotate fast enough. These cases do not prove that in other circumstances or with better equipment, drilling would not have been better by the rotary method. The cases have proved that the turbine was more economical in such cases. Theoretical investigation has shown that turbodrilling is cheaper than rotary drilling, in cases where time saved by faster drilling is more valuable than the extra cost of turbodrilling.

During late 1956, the drilling industry recognized the possibilities of producing a downhole source of power for driving the drill bit. In the late 1950's, with the reported success of Russian turbodrilling, there was a widespread interest in this method in the United States. Forty Russian turbines were imported for study during that period. In excess of 13,000 feet were drilled with the Soviet imports to establish operating procedures and to verify turbodrill characteristics under American conditions. Hole sizes ranged from 7 7/8 to 17 1/2 inches. Drilling rates were from 2 1/2 to three times those for comparable rotary drilling under the same conditions, as reported by the Soviets.[449]

As an example, industry entered the field of turbodrilling with the acquisition of the EDCO turbodrill in the 1950's.[15] This drill consists of a multi-stage, direct-drive, impulse turbine, similar to an early Russian model. Development work began, but when completed the power output and efficiency of this turbine were too low for economical drilling. Other turbodrills were tested unsuccessfully, but the U.S.S.R. was the first to develop a turbine that seemed to work economically and efficiently.

The first Russian turbodrill, developed by M. A. Kapelyushinikov in 1925, was a single-stage, high-speed turbine, driving the bit through a planetary-reduction gear. The use was discontinued in 1934 because this turbodrill was considered less efficient than conventional rotary drilling.

In 1935, a new type of multi-stage, direct-drive turbodrill was designed by Russian engineers but not fully perfected until after World War II. However, for improvements on the turbodrill, R. A. Ioannesyan and M. T. Gusman were awarded the Stalin Prize. The first direct-drive, multi-stage turbodrills were equipped with sealed roller-bearings which failed frequently because drilling mud contaminated the lubricating system. The Russians decided that the turbodrill must be "de-refined" into a rough and rugged tool to meet oil field conditions. The result was the development of a rubber thrust bearing which is lubricated by the drilling fluid and need not be sealed against abrasives.

Table 14 shows how quickly the development advanced in Russia between the years 1948 and 1953, and the percentage of the drilling that was performed by turbines.[284]

Another approach incorporates a turbine with a vertical axis which is attached to the lower end of a stationary drill stem. High pressure pumps force the drilling fluid through the stem into the turbine situated immediately above the bit; and this fluid rotates the turbine shaft. The bit is joined to the shaft of the turbine, rotating approximately 600 revolutions per minute. As in the rotary method, the mud flows to the surface between the drill stem and the well bore. By regulating the pump pressure, the speed of the bit can be adjusted for various rock formations. The drill stem does not rotate inside the hole but merely serves to conduct the fluid into the turbine. The

TABLE 14[284]

Percent of Total Wells Drilled By Turbodrilling Method

Year	Percent
1948	31.1
1949	44.4
1950	48.6
1951	57.1
1952	61.1
1953	75.8
1954	------

entire drilling process takes place at the bottom of the hole.

The U.S.S.R. has developed different types of turbines with an overall length of 26 feet, differing mainly in the number of stages and diameters. The technical data of the types of turbines used at present are given in the literature.[284] The pump output should range between 635 and 930 gallons per minute using standard drill stems. For example, the 100-stage turbine of type T-12M transmits a torque of 2,170 foot pounds to the bit with a pump output of 875 gpm. The pump pressure in this case is between 1,175 and 1,470 pounds per square inch at a depth of 6,560 feet. The pressure drop for the turbine model mentioned is proportional to the torque, or about 800 psi.

The number of revolutions of the bit while drilling is about half the revolutions while idling. The drilling is regulated by a drillometer, a pump manometer, and a pump flow meter. The use of a bit revolution counter is being developed. The efficiency of the turbine drill is about 60 percent, significantly higher than any rotary drilling system now in operation.

An increase in pump output results in an increase in revolutions, torque, and drilling rate. However, mechanical factors limit the increase of the pump output. Therefore, the diameters of the turbochannels must be enlarged, and the pump capacity as well as the torque increased without increasing pressure in the turbine. With these turbines, the efficiency is boosted by 15 to 30 percent to a peak efficiency of approximately 85 to 90 percent.

In other experiments, the turbine stages were augmented, without changing the pump output or the pressure inside the turbine. The experimental length made the construction of machines of this kind very difficult; therefore, initial experiments were made by joining two turbines of 100 stages each.

Turbine Field Operations

A higher working capacity at the bottom of the hole results in drilling that is more mechanically efficient. Even when bits are employed larger than those customarily used with rotary drilling, this advantage applies. Although the bit speed in turbodrilling is higher than in rotary drilling, the bit weight is less, and the neutral point of the drill stem is lower down the drill collars than when more weight is used. Since the drill stem does not joggle, and the drilling is straight, a hole is made without bit walking. This is an advantage when running and cementing casing.

The wear on the outside of the pipe and the tool joints is greatly reduced because the string rotates in some cases with only 10 to 15 rpm or in other turbine models, rotation is not required.

In turbodrilling, most of the causes of parted pipe are eliminated. The pipe is not exposed to sudden dangerous momenta, such as special loads or bending pressure which may occur at any time while rotary drilling. The pipe does not drag on the wall of the hole except at very low speeds; twisting off does not often occur; and fishing for bit cones and other parts of bit debris does not present a serious problem in turbodrilling.

The danger of twisting off is always present while rotary drilling because of high rotational pipe speeds. Furthermore, the formation of keyseats and bridges frequently necessitates fishing operations. With turbodrilling, these problems occur less often because the hole is straighter, and the stems turn slowly or not at all. A straight, tension-loaded drill stem and very low rotation speed significantly reduce the danger of a parted drill string. Turbodrilling does not create any new drilling difficulties but may eliminate or reduce many of the existing problems in rotary drilling.

The strain on the swivel and on the rotary table is considerably reduced because the base of the rig mast vibrates only slightly, resulting in a reduced strain on the rig in general. The lifting capacity of the drilling mud is improved by the higher annular velocity; the drilled particles are of finer grain because of the higher speed of the bit and can, therefore, can be brought up more efficiently. If difficult drilling conditions prevail requiring a heavy and viscous mud, the turbodrill cannot be used.

The speed of the turbodrill bit varies between 200 and 600 rpm; therefore, certain rock types may require rotary drilling with a bit rotational speed of 50 to 70 rpm. The final answer, however, will be found only through extensive drilling tests and field experience.

The strain on the mud pumps will be significantly higher in turbodrilling. For this reason, the sand content of the fluid must be kept as low as possible. For practical purposes, a sand content of less than three percent by volume is considered acceptable.

Thiery comments on the development of a slow rotating turbodrill which incorporates a modified rock bit.[613]

A gap existed between optimum turbodrill and roller bit speeds. The continued development of turbodrilling in Europe and the United States was held back because turbodrills performed at speeds beyond rock bit capabilities.

Economics of rock bit performance are poor at high rotating speeds beyond 400-500 rpm, and conventional turbodrills operate in the 600-650 rpm range. Turbodrill rotation also is not stable below 400 rpm. For this reason, diamond bits which withstand high rotation rates have long been used with turbodrills.

In most cases, roller bits are pulled because of: (1) Tooth wear from conventional rotary drive. (2) Bearing wear from high speed turbodrilling. These factors indicate that standard rotary speed is too low, while turbodrill rates are too high for roller-type bits.

Soviet experience indicates that roller bits perform best at 200-275 rpm in soft formations; 300-400 in medium-hard formations; and 600-700 rpm in very hard formations. Researchers in the United States have found that optimum roller-bit speed is between 300-400 rpm, or above rotary and below turbodrill speeds.

For each type roller bit, there is a critical rotating speed above which contact between each tooth and the formation is too short for effective action.

Contact period should be:

8×10^{-3} second for brittle and elastic rocks.

7×10^{-3} second for elastic and plastic rocks.

6×10^{-3} second for non-brittle rocks.

French technicians observed during 13 years of turbodrilling that in hard rock a turbodrill-driven roller bit may dig a hole faster and still drill more footage than when rotated with a standard rotary. This held true when circulation rate was reduced, when rotation rate was held below 400 rpm, and when weight on bit was increased to near stalling point. That the operation of roller bits at speeds between 200 and 400 rpm — a range beyond standard table capability — is advantageous was confirmed.

Figure 52 shows torque and power plotted against conventional turbodrill speed at a given flow rate. Rotation is unstable below 400 rpm. Conventional turbodrills normally operate at one-half idle speed which corresponds to maximum available power. For normal drilling, circulation rates and conventional blade types, speed ranges from 600 to 650 rpm. At this speed, enough torque reserve and turbine rotor inertia are available to prevent stalling, considering possible fluctuations of resisting torque.

To improve turbodrill stability at slow rotating rates, several approaches have been evaluated. One method consists of using other types of downhole hydraulic motors. Only one type, the helicoidal pump has proven successful.[528] Speed of rotation depends on the drilling fluid circulation rate. Torque is proportional to the input pressure, and torque variations require variations of pump output.

Circulation rate should be adjusted in response to pressure variations in order to provide the desired torque-curve characteristics. Two methods have been used: one involves attaching a gaged valve at the turbodrill inlet, and the second uses a pumping and mud system with variable volume capacity.[432] Considerable development work has been done on the basic turbodrill design, generating a number of models differing only slightly in design.[15] One model incorporates a single-stage turbine which rotates a diamond-faced cutter wheel at 5,000 to 10,000 rpm at the bottom of a drill pipe. The drill pipe is rotated at 30-75 rpm to produce a hemispherical, bottom-hole contour. The load applied to the turbine drill is transmitted partly to the rock through diamond reamers at the bottom of the drill and partly to the spring-loaded cutter wheel. Sufficient load is applied to the drill to ensure that the reamers contact the rock. Operating characteristics of a field model turbine drill are given in Table 15.

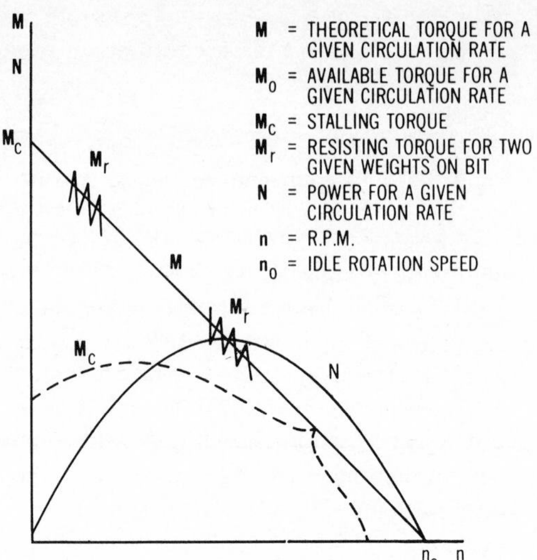

Figure 52[148] Torque and power curves versus conventional turbodrill speed at a given circulation rate. Revolutions per minute are roughly proportional to circulation rate. Below 300 rpm, torque reserve and rotor inertia reach a point where turbine stalling is frequent. Sliding thrust bearings reduce torque below a theoretical value by the difference between the straight torque line and the dotted practical torque curve.

The rotary-power output (P) of the turbine drill equals:[432]

$$P = 2\pi NT \text{ (kg-m/min)} \qquad \text{EQUATION 41}$$

where: P = Power output
N = Turbine rotary speed, rpm
T = Torque on the cutter wheel, kg-m

Note: See Appendix for Conversions of metric to English units.

TABLE 15[156]

Turbine Drill Operating Characteristics

Turbine Speed	5,000-10,000 rpm
Drill Pipe Speed	30-75 rpm
Hole Diameter	20 cm.
Total Load	1,500-2,500 kg.
Spring Load (Wheel)	500-1,500 kg.
Hydraulic Power	200-300 hp.
Turbine Power	20-60 hp.
Turbine Efficiency	10-20 %

Turbine speed decreases as the torque increases; the cutter wheel "runs away" at low torques and stalls at high torques. Maximum power output occurs between 5,000 and 10,000 rpm. Since the turbine's power output is controlled by the cutter-wheel torque, the thrust on the spring-loaded cutter wheel should be optimized. Some power is transmitted to the rock through the reamers, but this power is usually small compared to the turbine's power output (because of the low rotary speed of the drill pipe).

Several turbine drills were tested in oil field drilling from 1953 to 1956. Initially, failure of the cutting elements and fluid erosion of the turbine buckets were problems, but these difficulties were overcome. Results of field tests comparing rotary and turbine drills are given in Table 16.

TABLE 16[156]

Turbine Drilling Data.

Well number	Drill	Hole dia. (cm)	Drilling distance (m)	Drilling time (hr)	Drilling rate (m/hr)	Specific[b] energy (joules/cm^3)
1	Turbine	20	33	7.3	4.6	360
	Rotary	20	67	10.9	6.1	–
2	Turbine	20	16	12.1	1.3	1300
	Rotary	20	18	6.1	3.0	–
3	Turbine	20	21	9.9	2.2	750
	Rotary	20	67	10.7	6.3	–
4	Turbine	20	16	10.9	1.5	1100
	Rotary[c]	20	10	7.9	1.3	–

Exact turbine power output unknown, specific energy based on 20 h.p. output. (b)
Bit load reduced 50 percent by a hole deviation problem. (c)

This table indicates that rotary drills penetrate about twice as fast as the turbine drills, except in Well 4 where the rotary-bit load was reduced 50 percent because of hole deviation. In this case, the turbine drilled 15 percent faster than the rotary drill.[284]

Turbine Economic Potential

Application of the American turbine drill, however, appears to be limited; the single-stage turbine is only 10 to 20 percent efficient and has a power output of only 20 to 60 hp. This power output is comparable to that of rotary drills. Since diamond bits require about twice as much specific energy as roller bits, turbine drills are limited to drilling rates that are about half those of rotary drills. Significant improvements in the efficiency and power output of the turbine drill will require the modification to either a different type of single-stage turbine or a multi-stage turbine, being developed in the U.S.S.R. and other European countries. The present application of the American turbodrill is generally in directional drilling, turbocoring, and offshore drilling.[148,557]

The turbodrilling method promises an important contribution towards increased capabilities in relatively deep drilling, either as a general successor or as a supplement to the rotary method.

In summary, the most extensive development in drilling over the past ten years is the turbodrill. The basic concept of the turbodrill is to transmit relatively large amounts of power to the bit through the drilling fluid, thereby overcoming the limitations of transmitting power by means of a long, rotating drill string. With currently available turbodrills, the increased power at the bit is in the form of much higher rotary speeds than can be used safely with conventional rotary practice.

In turbodrilling, an axial-flow mud turbine is placed at the bottom of the drill string, and the bit is attached to the turbine output shaft. The early Russian designs incorporated a high-speed turbine with a gear reduction between the turbine shaft and the bit. A more recent design, however, does not employ gear reduction. The major difference in recent turbodrills has been in blade and bearing design. One American turbodrill has used machine ball bearings for thrust bearings and rubber radial bearings at each turbine stage, plus radial ball bearings near the bottom of the tool. Another tool uses ball thrust bearings and a rubber lower radial bearing. Additional tools have made wide use of rubber in both the thrust and radial bearings. All three of the turbodrills have been extensively field tested. The Russians have been working with turbodrill developments continuously since the late 1920's or early 1930's.

Field testing of an air turbine, suspended in late 1957,[148] resulted in an improved version. However, further work has been discouraged by the high cost of the turbine, by low bit-life, and by inadequate pumping equipment.[449]

Ionnesyan[287] reports on the development of new turbodrill designs in the U.S.S.R. The earlier Soviet turbodrill's flow rate was artifically increased, thus making the tool less competitive than rotary drilling in many cases. A major redesign feature eliminates rubber bearings that inherently prohibit low rpm. At present, Soviet turbodrills with ball bearings and packing glands operable at pressure drops up to 60 atmospheres are under production in the U.S.S.R. A new design of a turbodrill with rotating casing for diamond drilling is also under development by the Soviets. Field testing has shown that the American rotary tools and practices give lower costs per foot of hole than the turbodrill, except in very special cases such as large-diameter surface holes or certain directional drilling problems of the oil industry. The turbodrill is commercial in Europe for two reasons: (1) European contractors have not had drill pipe or collars of sufficient quality to permit rotary drilling without excessive pipe failures and (2) a premium is often placed upon maximum holes per year in Europe not on minimum dollars per foot. In the United States, further turbodrill test and development work has been suspended, and the effort in this area is being directed toward the development of suitable bits and pumps for use with the turbodrill. Summary work[327, 616] indicates that the turbodrill-bit problem is related more to cutting structure configuration and basic rock-failure principles than to bearing failures.

ELECTRO-DRILLING

In addition to the turbodrill, a bottom-hole rotation device, called the "electrodrill," is being developed. The basic concept of the electrodrill is the application of mechanical power to the bit without rotation of the drill pipe and, in some cases, the elimination of the drill pipe. The object also is to increase the amount of power available at the bit as compared with the rotary drilling method.

The final development, dealing with bit rotation, is the downhole-fluid motor. The function of such motors is to supply bottom-hole rotation power at greater revolutions per minute than the standard method, rotary, but without the complexity and cost of downhole turbines.

One downhole motor is operated by pumping mud along a helical rotor inside a specially designed housing (a Moyno pump uses this principle in reverse). This motor is designed to rotate the bit at speeds ranging from 200 to 400 rpm and to deliver a maximum of about 40 rotary horsepower to the bit, for 7 7/8- to 9-inch bit sizes. The drilling tool is still under development and has found significant field application, particularly in directional drilling.

Another similar drilling approach has two air motors connected by a gear reduction to the rotary bit. This tool is eight inches in diameter and is designed to give a maximum of about 40 hp at the bit with about 350 rpm. Of course, the tool is applicable only in drilling that uses air as the drilling fluid which, as previously stated, is a very small percentage of the total footage drilled in the petroleum industry. In the water well industry, air drilling has particular applicability and has found considerable use.

The drills have been field tested and are finding field application in the petroleum and mining industries as a tool to continuously deviate hole direction in all types of formations under many severe subsurface environments of relatively high temperatures, pressure, etc. The Dyna-Drill's penetration rate, for example, reportedly not only equals that of rotary drilling in medium to hard formations but surpasses rotary in softer and unusually hard formations.[240]

8
Future Drilling Systems

During the past three decades, the oil industry has expended increased energy to improve drilling tools or systems and to reduce drilling costs. The extent of these efforts is unknown but certainly amounts to tens of millions of dollars in expenditure. Many of the concepts of the "new" systems that are being investigated because of economic pressure in actuality were developed as much as 100 years ago.

In seeking to implement new drilling systems, investigations have evaluated the following:[432]

(1) Impact at frequencies ranging from 6 to 300 cycles per second.
(2) Electrical, mechanical, and hydraulic means of actuating percussors.
(3) Bit rotary speeds to 2,000 rpm.
(4) Electric and hydraulic bottom-hole means of rotating bits.
(5) Bottom-hole machines with power outputs up to 400 hp.
(6) Shock waves.
(7) Explosives.
(8) High-velocity pellets.
(9) Flame.
(10) Arc.

(11) Grinding wheels.
(12) Abrasive jets.
(13) Erosion by high-velocity gases.
(14) Chemical attack.
(15) Electric current.
(16) Magnetic waves.
(17) Retractable rock bits.
(18) Reelable drill pipe.
(19) Continuous coring with reverse circulation.
(20) Automation of drilling rigs.

In spite of these efforts to discover new and improved drilling systems, rotary drilling maintains its economic leadership. Undoubtedly, rotary drilling costs will continue to be reduced by rigid application of the best available technology and by development of new rotary technology. In view of the extensive past development programs, however, significant long-range improvements appear to be a research, not a development, problem. Research must postulate and prove theories and principles governing various subsurface rock-failure processes pertinent to both rotary and new drilling systems. Also, research must produce physical and engineering data relative to these processes. When such information is available, major improvements in rotary drilling can be expected, and the systematic evolution of an improved drilling method can be initiated — with a strong probability for success.

DRILLING SYSTEM DEVELOPMENT

The oil industry is developing drilling methods other than rotary drilling as a potential means for reducing drilling costs in the future. Major cost-reduction efforts, however, have been centered on engineering development work that is aimed at incremental improvements in rotary drilling, e.g., equipment, bits, high velocity jet drilling, etc. A considerable effort has been expended to establish basic physical principles and engineering data pertaining to the earth-boring process which can serve as a foundation for the development of cost-cutting drilling hardware.

Current oil industry economic trends have added impetus to the need for effective programs to reduce drilling costs. However, areas for expanding future efforts should be selected only after careful study of past investigations.

This section of the text will review past and present industry efforts in developing novel drilling tools or systems and will define areas where additional efforts are needed. Past reviews of drilling methods have been limited to a few processes prominent at the time the reviews were made.[380]

More engineering and development work has been focused on "new" systems than is generally realized. An understanding of the scope, nature, and status of the industrial research will give new insight into the urgent need for accelerating basic studies of the drilling process.

The numerous concepts that have been evaluated by the petroleum industry have cost from at least $13 million to $25 million in development and test work. Maximum expenditures on a single tool or system have ranged from $200 thousand to over $1 million per year. This work has been conducted by both oil-industry and commercial research organizations. In at least one case, a sizable research and development organization was created and staffed solely to develop a particular drilling tool.[505] Although industrial research and development is a costly venture, the financial return from any one success can be very high.

To date, this new system development work has not produced a drilling tool or system that can appreciably cut general drilling costs. In some cases, of course, the tools developed are more economical than rotary drilling under specific operating conditions not directly applicable to general field conditions.

Because the drilling rates continue to increase, the rotary method is likely to retain a place of leadership in the oil industry for the predictable future, possibly evolving from the conventional "jet-bit rotary-drilling system" into a "rotary-erosion drilling system."

In addition, some U.S. $600 million are being spent for the construction of new offshore drilling ships, most equipped with automated but conventional rotary equipment. Improvements in rotary performance often inhibit the development of novel drilling concepts. Ledgerwood, in his comprehensive review and analysis of novel drilling methods developments, stated in 1960 that "in spite of industry's efforts to develop a superior rock-drilling method, rotary drilling maintains its place of leadership as the most versatile and generally economical earth-boring method."[380]

DRILLING RESEARCH

Maurer's[432] excellent evaluations of novel-drilling techniques extend Ledgerwood's original list of possible rock-drilling methods and provide means for calculating the power consumption and the potential drilling rates for conventional rotary and novel drills.

Maurer and Ledgerwood report on forced-flame drills, plasma drills, laser drills, turbine drills, spark drills, and numerous others including rotary. Considerable development work is underway in the United States and other countries on at least several of the new drilling systems having the high potential drilling rates — primarily, erosion (with and without abrasives), modified turbine, and explosive.

In recent years, publications relating to oil field drilling emphasize progress in the study and development of novel drills while continued and steady improvements in rotary-system performance have been less spectacularly treated.[540]

Before the potential of a new drilling system can be evaluated, the potential drilling rate must be established.[432] The maximum drilling rate is the amount of power that can be delivered to the rock limited by the power transmitted down the borehole, by the power output of the drill, or by the power that the rock will accept from the drill.

The following drilling systems are primarily limited by the amount of power that can be delivered down the borehole:

(1) Chemical.
(2) Erosion.
(3) Explosive.[489]
(4) Implosion.
(5) Pellet.
(6) Turbine.

These drills utilize all the power that can be delivered to the bit, so the drilling rates are proportional to the power delivered down the drill pipe. The drilling rate of the *chemical drilling system* is limited by the rate at which chemicals can be delivered to the hole bottom. Research on the *erosion*, *explosive*, and *implosion drilling systems* is aimed primarily at delivering more power to the hole bottom and, therefore, reducing the specific energy requirement for rock removal.[433] Research on the *pellet* and *modified turbine drilling systems* is directed toward increasing the efficiency of these tools, since only 4 to 20 percent of the hydraulic power is converted to mechanical power and transmitted to the rock. Maurer[432] presents an excellent review of the principles involved in the novel methods presented here. His work should be consulted for details.

The following drilling systems are limited primarily by their power outputs:

(1) Conventional percussive.
(2) Electric arc.
(3) Electron beam.
(4) Forced-flame.
(5) Jet-piercing.[482,570]
(6) Laser.[159]
(7) Plasma.
(8) Spark.
(9) Spark-percussive.
(10) Ultrasonic.
(11) Electrohydraulic.[374]

Drilling rates of these systems are proportional to the power output, so research is aimed primarily at increasing power outputs and at reducing the specific energy required for rock removal.

The following drilling systems are limited by the amount of power that the rocks will accept from the systems:

(1) Conventional rotary.
(2) Electric disintegration.
(3) Terra jetter.
(4) High frequency electric.
(5) Induction.
(6) Nuclear.

Research on this group of drills is aimed primarily at increasing the rate at which rock will accept energy from the drill and at reducing the specific energy required to remove the rock. In highly conductive rocks, such as magnetite, for example, the rock will accept all of the power which can be delivered by the *electric disintegration drilling system*, in which case the drilling rate will be limited by power transmission down the borehole.

The maximum drilling rate for each of the novel drilling methods is estimated in Table 17 for a 20 cm.-diameter hole in "medium-strength" rock. The specific energy and power output are estimated for each of the drills based on available data presented by Maurer[432] and on the operating characteristics of commercially available equipment. Considerations used in making these estimates of specific energies and power outputs are discussed in more detail by Maurer. As new equipment and techniques are developed, the reader can easily revise the values of specific energy and power output in Table 17 and use Maurer's equation to recalculate the maximum-potential drilling rate.

For a 20 cm.-diameter hole, Maurer's equation reduces to:

EQUATION 42

$$R = 141 \frac{P}{E}$$

Where: R = Drilling rate (cm./min.)
P = Power transmitted to rock (hp.)
E = Specific energy (joules/cm^3)

Note: Consult Appendix for conversion of metric to English units.

Figure 53 is a graph relating drilling rate, power output, specific energy, and hole diameter. For the example shown, a drilling system capable of generating a 100 hp. drill and operating at a specific energy of 10,000 joules/cm^3 would drill a 10 cm.-diameter hole at 5.6 cm./min.

Table 17 shows that the *spark*, *erosion*, and *explosive drilling systems* can drill faster than conventional drills in some rock types.

TABLE 17[432]

Estimates of Maximum Drilling Rates for 20 cm Diameter Novel Drills in Medium-strength Rock

Drill	Status	Rock Removal Mechanism	Specific energy (joules/cm^3)	Maximum power to rock (h.p.)	Maximum potential drilling rate (cm/min)
Rotary[a]	Field Drill	Mechanical	200–500	20–30	14-85
Spark[a]	Laboratory drill	Mechanical	200–400	100–200	35–140
Erosion[a]	Laboratory drill	Mechanical	2000–4000	1000–2000	35–140
Explosive[a]	Field drill	Mechanical	200–400	75–100	26–70
Forced-flame	Field drill	Spalling[b]	1500	300–600	28–56
Jet-piercing	Field drill	Spalling[b]	1500	100–200	9–18
Electric disintegration	Laboratory drill	Spalling[c]	1500	100–150	9–14
Pellet[a]	Laboratory drill	Mechanical[a]	200–400	10–20	4–14
Turbine[a]	Field drill	Mechanical[a]	400–1300	30–40	3–14
Plasma	Laboratory tests	Spalling[b]	1500	80–120	8–11
Electric arc	Laboratory tests	Spalling[b]	1500	45–90	4–8
High-frequency	Shatters rocks	Spalling[b]	1500	30–60	3–6
Plasma	Laboratory tests	Fusion	5000	80–120	2–3
Electric heater	Laboratory drill	Fusion	5000	50–100	1–3
Electric arc	Laboratory tests	Fusion	5000	45–90	1–3
Nuclear	Conceptual	Fusion	5000	1250–2500[e]	1–3
Laser	Small holes	Spalling	1500	12–24	1–2
Electron beam	Small holes	Spalling[b]	1500	10–20	1–2
Microwave	Shatters rock	Spalling[b]	1500	10–20	1–2
Induction	Shatters rock	Spalling[d]	1500	5–10	0.5–1.0
Laser	Small holes	Fusion	5000	10–20	0.3–0.6
Electron beam	Small holes	Fusion	5000	10–20	0.3–0.6
Electron beam	Small holes	Vaporization	12,000	10–20	0.1–0.2
Laser	Small holes	Vaporization	12,000	7–14	0.1–0.2
Ultrasonic[a]	Laboratory drill	Mechanical	20,000	5–10	0.04–0.07

[a] Water-filled borehole.
[b] Limited to highly spallable rock such as taconite.
[c] Limited to highly spallable rock with high electrical conductivity.
[d] Limited to highly spallable rock with high magnetic susceptibility.
[e] 100 cm diameter drill.

These three systems are versatile by the fact that they can remove rock mechanically and drill any type of rock. *Forced-flame* and *jet-piercing systems* can also drill faster than conventional systems in hard, spallable rocks such as taconite, but since rock is removed by a spalling mechanism, and many rocks will not spall, the systems are not as versatile as the conventional types.

Electrode consumption and capacitor life are important factors to consider when evaluating the economic potential of spark drilling systems. The explosive-capsule cost is an important item in explosive systems, since up to 720 capsules per hour are used.[489] Erosion drilling systems can drill at very high rates, but nozzle erosion may be a problem, limiting the time which this drill can remain at the hole bottom. Erosion systems require high power outputs, necessitating the development of large capacity, high pressure pumps. Because of high potential drilling rates, the economic potential of the spark, erosion, explosive, and forced-flame drilling systems should be carefully evaluated. The jet-piercing system is already widely used for drilling highly spallable rocks.[433,482,570]

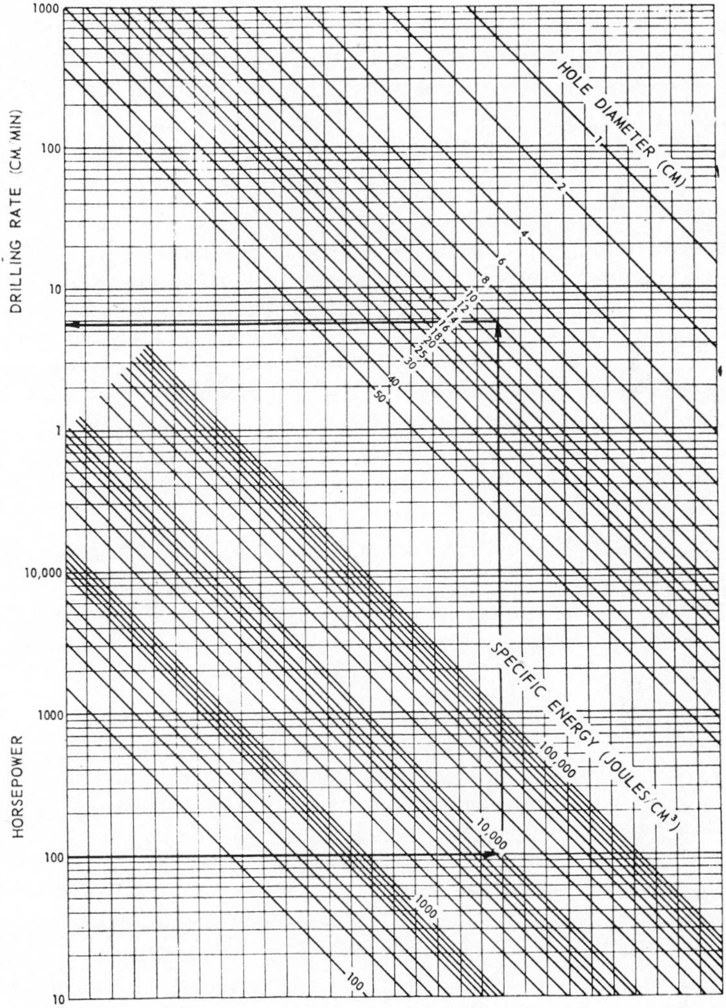

Figure 53[432] Drilling rate nomograph.

The *electric disintegration drilling system* can produce high drilling rates in highly spallable rocks which have high electrical conductivity, but only a few rocks have this property so the system has only limited potential application. The *pellet system* can drill at rates comparable to conventional systems, limited by low power output and low efficiency. The *electric arc* and *high frequency electric drilling systems* can possibly drill hard, spallable rocks faster than the rotary system with little advantage over the jet-piercing system.

Plasma systems consist of high heat-transfer devices and should be considered for any application where rock or other materials must be rapidly heated. With the current state-of-the-art, plasmas appear to have little advantage over jet-piercing systems, except possibly in small diameter holes or in problems from combustion products in the oxygen-fuel oil flame.

Because of the low drilling rates, the remainder of the novel systems in Table 17 appear to have limited potential for use in shallow drilling.[380]

> These low drilling rates result primarily from low power transmission to the rock and from the high specific energy required to fuse and vaporize rock. For example, a 10 kw. laser, transmitting 70 percent of its output power to the rock, would fuse a 20 cm.-diameter hole at only 0.3-0.6 cm./min. Even this is optimistic, because the largest current commercial laser has a power output of only one kw. A laser output of 10-20 kw. was used as the example [in Table 17] because of the rapid advance being made in laser technology. For example, the power output of lasers has increased over fifty-fold during the last three years; their efficiency has increased by 1,500 percent.

As the power output of lasers and other drilling systems increases, drilling rates and potential applications will increase.[159] For this reason the development of these systems must be continually re-evaluated as new equipment and better techniques are developed.[432]

> The power density and drilling rate for most of the drills [in Table 17] are nearly independent of the diameter of the drill. A few of the drills such as laser, electron beam, and erosion can produce very high power concentrations in application where the drill can be placed outside of the hole. For example, electron beams and lasers can produce power concentrations of 10^9 and 10^{12} w/cm^2 respectively. Assuming that 50 percent of the output power is utilized, these power concentrations could fuse small-diameter holes in rock at rates of 10^6 [one million] and 10^9 [one billion] cm./sec. A drilling rate of 100 cm./min. could be produced by a 20 hp. electron beam drill, in a 0.75 cm.-diameter hole, and by a 2 hp. laser drill, in a 0.24 cm.-diameter hole, assuming 50 percent of the output power is utilized. This is faster than conventional drills can drill most medium and hard rocks, which indicates that these devices have potential application for drilling small diameter holes. Erosion drills can also produce high power concentrations and high drilling rates in small holes. For example, a 690 hp. erosion water jet has drilled a 1.31 cm.-diameter hole in sandstone at an instantaneous rate of 200 cm./sec.

ECONOMIC POTENTIAL

Many factors must be considered when evaluating the economic potential of these new drilling methods. Oil field drilling is currently accomplished with very large, expensive rigs requiring crews of four or more men. New drilling methods that would use smaller rigs, smaller crews, or flexible drill pipe to increase trip speed could be economical although penetration rates would be lower. In oil field drilling, the cost of the power transmitted to the rock by the bit is relatively insignificant compared to the other costs, so novel-drilling systems that require considerably more power could be economical if the drilling rate increased or the operating costs decreased.

Consideration should be given to combining two or more of these novel drilling methods. For example, the laser system could be used to cut slots in rocks and, if combined with the spark or erosion jet systems, could be used to crush and remove the unsupported rock.[159] Novel systems could also be combined with conventional systems, e.g., lasers, plasmas, or electron beams could be used to heat and degrade rock, and roller or drag bits could be used to remove the weakened rock. Obviously, some of these new drilling methods can drill large diameter holes faster than conventional methods in some rock types.

In conclusion, a considerable amount of research on new drilling techniques is in progress throughout the world, and some of these drilling systems may find initial field application in the next few years. These systems will probably be tested for special applications where drilling rate is more important than unit drilling cost (military). Once tested and developed, efficiencies, power outputs, and drilling rates of these systems will be increased until more widespread use in industry is merited which would mean reduced costs and possible application to the shallow drilling operations of the water well and shallow mineral exploration industry.

Any advancement in a particular novel drilling approach may require many years before the approach can become sufficiently simplified and, therefore, economically feasible for industry to eventually adapt, however, the development of a novel or an entirely new drilling concept may be more applicable to shallow drilling than to the petroleum industry particularly in the early stages of research. Therefore, the petroleum industry's drilling research should be monitored closely for a potential breakthrough.

Many of the new techniques discussed in this chapter have been demonstrated in laboratory tests to drill rock effectively, but few of them have been tested on a large scale. One problem in developing these new drilling techniques is that they are competing with conventional tools that have undergone many years of improvement. For example, Maurer[432] calculates that the penetration rates of oil-field rotary drilling systems are ten-100 times faster then when this system was first introduced during the late 1800's. Similar improvements can be expected with new drilling methods once they are put into routine field use, which is in the not too distant future.

9
Formation Identification and Evaluation

The salient features of formation identification and evaluation research and technology will be treated in this chapter, with special emphasis on the techniques and knowledge developed by the petroleum and mining industries. Current well completion and development techniques of the water well industry will also be integrated.[422,522] Some topics will be given more attention than others, not to imply their relative importance but to emphasize selected topics which have not been given adequate attention in the literature or in the field. Well known topics are fully referenced for further study, while lesser known topics are expanded according to applicability and pertinency.[287,288,544]

Although the petroleum industry has supplied much of the present general knowledge of formation evaluation techniques which have found applicability in the water well industry, aspects such as detailed well design features, aquifer productivity evaluations, pump test methods, etc., have nearly all evolved from the water well industry and its field-developed efforts over the past 25 years.[223,288,551] There is, however, a close similarity between the technology of crude oil production and the technology of ground water production, both of which deal with the flow of fluids in porous, underground formations (see Figure 54). The physical laws governing the flows are

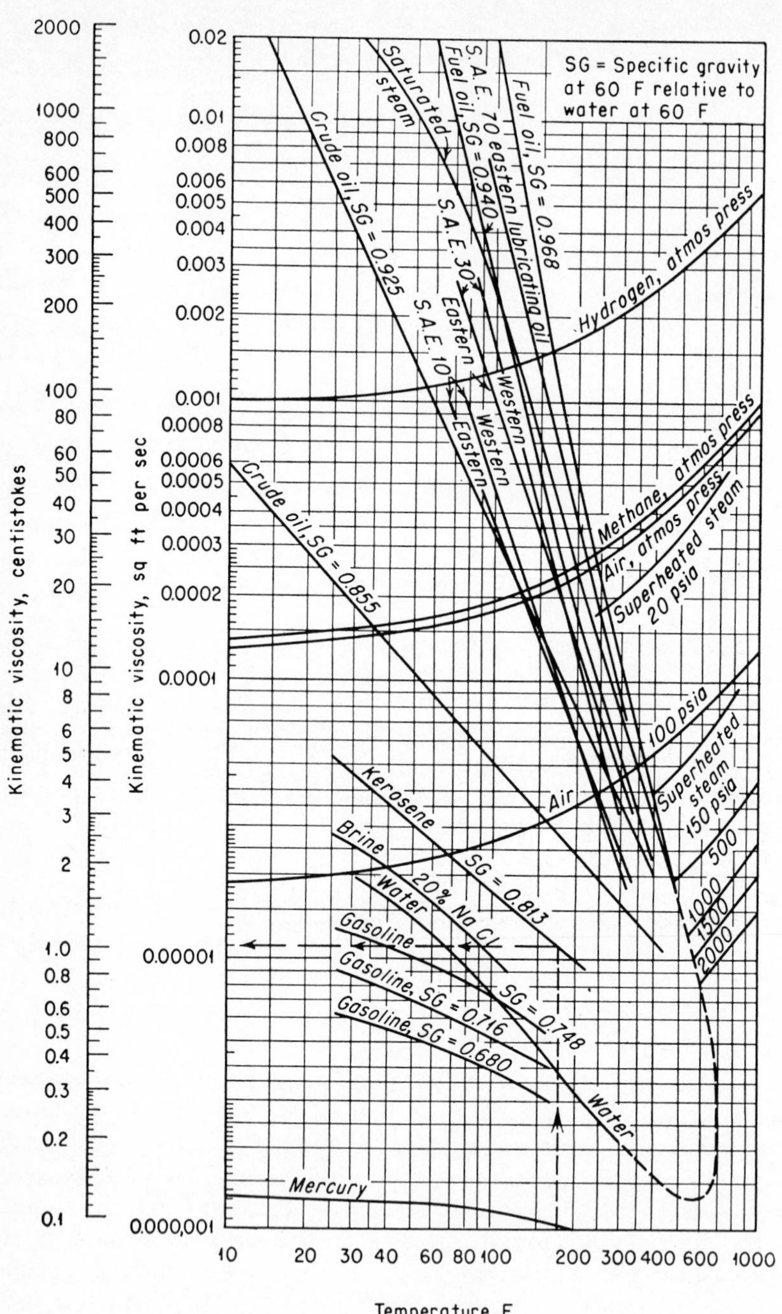

Figure 54[312] Viscosities of Common Liquids

identical (although the mathematical expressions for the flow of crude oil are more complicated), and the wells used are nearly the same for both technologies and are constructed by similar methods. With the increase of water-flooding oil recovery operations,[161,490,558,578,604,620] both for primary and secondary completions, the petroleum industry can draw from the available technology of the ground water industry. Both industries, in fact, can prosper by working together closely and by overtly supporting the interdisciplinary transfer of technology.

The most important technological feature which is under considerable research and development in the petroleum and mining industries is in the area of formation evaluation, e.g., borehole geophysics, drill-stem testing, etc. Other features, such as cementing techniques, casing designs, formation stimulation techniques, etc., have also made strong advances in the past 15 years.

The inability to sample formation fluids and to test production as drilling proceeds has delayed acceptance of the rotary-drilling method for a long time. It was recognized that the rotary method could drill quickly, but it was also evident that when using this method the hole might penetrate productive zones without the evidence of oil, gas, or water being noticed. The introduction of coring tools in 1921, drill-stem testing (DST) in 1926, and electric logging in 1933 changed the picture completely for rotary drilling. These innovations and others that followed made it possible to obtain detailed information of oil and gas shows when they were encountered.[62] Although conventional coring was followed by wire-line coring, both were almost completely supplanted by well logging techniques. Drill-stem testing and well log analysis have progressed to the point in the petroleum industry that formation evaluation can be accomplished without setting casing.[289]

BOREHOLE GEOPHYSICS

Exploratory drill holes or wells are the only means of direct access to the subsurface environment for information on fluid production or disposal. Drill holes are an expensive means of access to the lithosphere, and techniques such as well logging are economically justified because they produce sufficient data from each hole to extrapolate this information laterally. Borehole geophysics, or well logging, includes all techniques of lowering a sensing device into a well to make a record which can be interpreted in terms of the rock characteristics, of the contained fluids, and of the construction of the well.[217] Although the main objective of this investigation is to explore the use of borehole geophysics in ground water investigations, these methods are applicable to other fields of mineral development. More holes are drilled for water than for all other purposes combined. However, a significant total increase in the number of mineral exploration holes drilled is noted in the western United States, particularly in the relatively small sedimentary basins of New Mexico, Arizona, Wyoming, Colorado, etc., as a result of the uranium exploration activity of the late 1960's, which is still in progress and which may further expand as the uranium market improves during the mid-1970's. Valuable fresh water aquifers, although located in presently remote areas, can be indirectly damaged as a result of this drilling activity and can cause serious water supply problems in any future development of these areas. Each hole, however, is a valuable sample of the geological environment, and this information can aid in the proper development of water supplies; of sand/gravel deposits; of other nonmetallic and metallic mineral deposits; of petroleum; of underground waste disposal and storage sites for liquids and gases, etc. Engineering data is also provided for future building construction.

The petroleum industry is responsible for the development of almost all of the logging equipment and interpretive techniques used today. It is improbable that borehole geophysics would have reached its present state of development without the economic impetus provided by the search for oil. The worldwide search for ground water may never reach this economic level, but the industry can benefit from the existing techniques financed by the oil companies. A thorough knowledge of developments in petroleum well logging is essential to avoid duplicating the basic research in petroleum logging techniques that can be applied to other fields of mineral exploration including ground water; these modifications function to bring equipment and service costs in line with the value of the product.[499,500]

The Water Resources Division of the U.S. Geological Survey is conducting research in borehole geophysics as applied to geohydrology. A one- to two-man research project has been under way for eight years (since 1964) with headquarters at the Denver Federal Center, Denver, Colorado, under the leadership of W. Scott Keys. The project's objectives are to investigate and appraise the current and potential applications of qualitative and quantitative borehole geophysics in ground water hydrology and to report on the value of these techniques in the various ground water environments throughout the United States. Other objectives are to develop and to modify geophysical logging equipment as an aid in the solution of specific ground water problems, to improve methods of quantitative and qualitative log interpretation, and to adapt these methods to ground water studies. A comprehensive discussion of all logging methods either with present application or with potential application in ground water hydrology is beyond the scope of this text. Keys[356,357,358] and others,[96,388] however, have published various reviews. Additional papers offer emphasis on specific ground water applications and interpretations, e.g., basaltic aquifers, unconsolidated formations, etc.[8,168,292,294,401,500,508,615,668]

The application of geophysical well logging is as restricted in ground water hydrology today as in petroleum exploration in the 1930's. In 1964, approximately 440,000 new water wells were drilled in the United States; it is doubtful if more than one percent of these wells were logged with geophysical equipment. The conventional driller's log of cuttings is made for most water wells, but these are not of consistent value. Properly made geophysical logs are unbiased and, thus, may be utilized over wide geographic areas and for long periods of time without loss of meaning.

Cost is the single most important factor in the present minor use of geophysical logging by the water well industry. In 1966, the estimated annual investment in water wells in the United States was one-half to three-quarter billion U.S. dollars, not including pumps and plumbing.[356] Although the volume of drilling and total money invested in ground water development is very high, the unit value of water is rarely equal to the unit value of oil. The cost of logging equipment and commercial logging services must be compatible with well costs and the value of the product. In 1961, the estimated average cost for each United States water well was $1,500 (see Chapter 14); however, this reflected the preponderance of shallow, small-diameter domestic wells in the East.[356] In the western United States, municipal, irrigation, and disposal wells were comparable in cost to some oil wells. The general trend was toward deeper, more expensive wells which have required and have economically justified more advanced knowledge of the ground water environment and better well construction and completion practices. As a result of recent developments in electronics, small and relatively inexpensive logging units, which can be operated by one man, are now widely available, with the costs involved unquestionably justified in many instances.

A second basic deterrent to the widespread use of geophysical well logs in ground water is that few geologists, hydrologists, or engineers are working on water well drilling projects and many do not have the knowledge and experience necessary for detailed interpretation of geophysical logs. Therefore, interpretations are usually provided along with logs by commercial well logging companies. If logging is to have practical application to the solution of hydrologic problems, not only must the usefulness and concept of logging be accepted by the well contractor, but more detailed knowledge of the interpretation of the logs must be attained by the professional geologist, hydrologist, etc.

The many water wells being drilled throughout the world provide an invaluable fund of geological knowledge if the proper data are obtained. By interpreting geophysical logs with a single objective, such as discovering the presence of water or oil, valuable mineral deposits may be overlooked. For example, in southeastern United States and in Australia, gamma logging of small-diameter, domestic water wells has yielded information valuable in phosphate explorations. Uranium exploration, as well as exploration for other minerals such as phosphate, sulphur, etc. which are "strata bound," relies heavily upon electric logging. The logs shown in Figure 55, for example, were made for a hydrologic study. However, a consideration of clay content of a core sample in relationship to the neutron log provides a means for measuring the degree of alteration of granite — a guide to ore deposits in some areas. In a few areas, water wells have been logged for stratigraphic or structural information as a guide in oil exploration. Significant additions to the geological knowledge of the subsurface, which can guide the development of all types of mineral resources, will be derived from logging of water wells more frequently.[515] Table 18 is a review of the presently available logging tools and their uses.

Most geophysical equipment used by the ground water industry is generally much smaller, lighter, and less expensive than oil-well logging equipment. This equipment may be permanently mounted in a lightweight field vehicle or a passenger car. Simple logging equipment utilizes single-conductor cable, a pen-and-ink recorder, and is usually operated by one man. The same types of logging techniques may be used in ground water and petroleum development; however, the equipment, the purpose of logging, and the log interpretation are generally different. Most of these differences are dictated by the need for economy and permitted by the lower temperatures and pressures encountered at shallow depths.

Differences in log interpretation techniques are a result of the different borehole environments, e.g., temperature, pressure, etc. The interpretation methods developed by the petroleum industry are not always directly applicable to the water well industry. Furthermore, some of the lithologic and other physical information required for interpretation is unique to ground water hydrology.[293,341] It is not

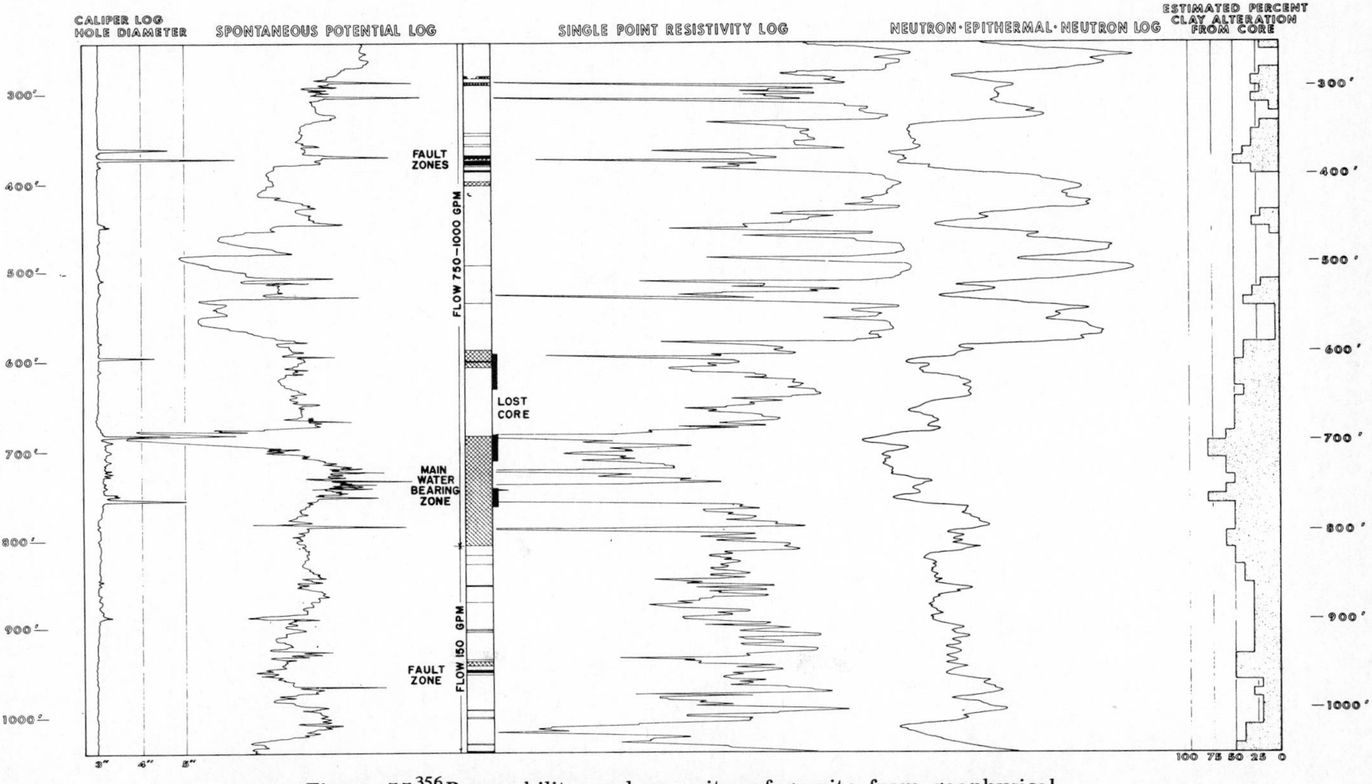

Figure 55[356] Permeability and porosity of granite from geophysical drill hole CX111, Clear Creek Co., Colorado

TABLE 18[356]

Suggested Logging Techniques for Groundwater Investigations

Information Needed on Properties of Rocks, Fluids, Wells or Groundwater System.	Conventional Logs which Might be Utilized.
Lithology of aquifers and associated rocks.	Electric, sonic, or caliper logs in open holes. Radiation logs in open or cased holes.
Stratigraphic correlation of aquifers and associated rocks.	Electric, sonic, or caliper logs in open holes. Radiation logs in open or cased holes.
Total porosity or bulk density.	Calibrated sonic or gamma-gamma logs in open holes. Calibrated neutron logs in open or cased holes.
Effective porosity or true resistivity.	Calibrated resistivity logs.
Clay or shale content.	Natural-gamma logs.
Permeability.	No direct measurement by logging. May be related to porosity, injectivity, sonic amplitude.
Secondary permeability – location of fractures and solution openings.	Single-point resistivity, or caliper logs, sonic amplitude, borehole television.
Specific yield of unconfirmed acquifers	Neutron logs calibrated in percent moisture.
Grain size.	Possible relationship to formation factor derived from electric logs. Clay content from gamma logs.
Location of water level or perched water outside of casing.	Electric, fluid resistivity, gamma logs in open hole or inside casing. Neutron of gamma logs outside casing.
Moisture content above water table.	Neutron logs calibrated in percent moisture.
Rate of moisture infiltration.	Time interval neutron logs or radioactive tracers. Temperature.
Direction, velocity and path of groundwater flow.	Single well tracer techniques – point dilution and single well pulse. Multiwell tracer techniques.
Dispersion, dilution and movement of waste.	Fluid resistivity and temperature logs, gamma logs for radioactive wastes, sampler.
Source and movement of water in a well.	Injectivity profile, flowmeter, or tracer during pumping or injection. Differential temperature logs. Time interval, neutron, or gamma-logs.
Chemical and physical characteristics of water – including salinity, temperature, density and viscosity.	Calibrated fluid resistivity and temperature logs in hole, neutron chloride logging outside casing.
Determining construction of existing wells, diameter, and position of casing, perforations, screens.	Gamma and caliper logs, collar and perforation locator, borehole television.
Determining optimum length and setting for screen.	All logs providing data on lithology, water-bearing characteristics and correlation and thickness of aquifers.
Guide to cementing procedure and determining position of cement.	Caliper, temperature, or gamma-gamma logs.
Locating corroded casing.	Under some conditions caliper, or collar locator.
Locating casing leaks or plugged screen.	Tracers and flowmeter.

the purpose of this section of the text to describe detailed interpretation procedures but rather to summarize some of the more important borehole logging techniques of the petroleum industry which have found application in ground water development.[160,346,347]

In general, effects due to higher temperatures, pressures, etc., on interstitial (pore) water require a change in present oil logging interpretation in order to be useful to the ground water industry, i.e., the change depends upon the relative difference in the fresh water's ionic makeup and upon the fact that its resistivity is about two orders of magnitude higher than that of water commonly encountered in formations associated with the occurrence of oil and gas.

Furthermore, the water well industry is interested in the evaluation of parameters that are different from those generally sought in the oil industry. The ground water geologist seeks to evaluate the quantity and quality of the water, and the well contractor desires a practical knowledge on the suitability of the various sands, etc., for completion, i.e., permeability and grain size, in addition to water quality. A comprehensive review of the practical features of geophysics and its application to ground water development has been published.[96]

For convenience, the information that may be obtained from logging can be applied to three subject areas: (1) aquifer characteristics, (2) fluid characteristics, and (3) well construction, although there is overlap among these areas.

Aquifer Characteristics and Well Logging

The lithologic character and stratigraphic correlation of aquifers and associated rocks are important in ground water geology and water well construction.[168,668] This information is usually obtained from gamma logs and singlepoint electric logs.[293] More recently, caliper logs and to a very limited extent, neutron and gamma-gamma logs have been used for aquifer identification and evaluation. Radiation logs (neutron, gamma, etc.) have found use in cased wells in unconsolidated sediments. Such logs provide the only reliable means of obtaining detailed lithologic information. Total porosity can be determined from neutron logs or calculated from gamma-gamma logs since fluid density and grain density are generally known or are assumed within fairly close limits.[183] The gamma log is used to distinguish units having a high-clay content, which causes a low effective porosity relative to the total porosity. In some aquifers, a linear relationship has been established between porosity and the hydraulic conductivity or permeability.[139] *Hydraulic conductivity* is the hydrologic term now used in place of the *coefficient of permeability* and is defined as the unit volume of ground water that will flow through a unit cross-section area of rock per unit time under a hydraulic gradient of unit change in head through unit length of flow. The hydraulic conductivity multiplied by the thickness of an aquifer is the *transmissibility*. Permeability is related to grain size as well as porosity; and a recent paper which demonstrates that the formation factor in fresh-water sands increases as the grain size increases is an important contribution to ground water hydrology.[8] Although recently developed techniques of multiple or composite interpretation utilize neutron, gamma-gamma, and sonic logs, they have not been integrated into standard procedures in the ground water industry to date.[618] However, research on the application of these techniques is planned and should produce new logging methods capable of superior aquifer resolution.

No one type of log gives a complete picture of rock-aquifer characteristics. However, several common logging tools can be combined to determine the local stratigraphy which affects ground water production, quality, etc.

The spontaneous potential (SP) log on the left-hand side of Figure 56 is one of the most common logs used in all down-hole logging today. The straight line portions of the log represent zones in which fine-grained rocks predominate, e.g., shales.

When the curve moves toward the left, away from this "shale line," sandy zones are indicated. Normally the curve will shift to the left (negative side), but if the water in the sand is very fresh, the curve may reverse and shift to the right, as in the most shallow sand intervals shown on the figure.

Resistivity values (Figure 56, second column) increase to the right. Dense rocks, such as granite and some limestones, are indicated by high resistivities. Medium high values in combination with negative or extremely positive SP logs indicate water-saturated sands.

The dotted line log (called a "long normal" log) indicates resistivity values at some distance from the borehole wall. Long normal logs are rarely run today. The curve illustrated in Figure 56 is also similar to an Induction Resistivity Log now common in the petroleum industry. This log, however, requires a more complex instrumentation than is normally used in water well logging. The "short normal" resistivity log (shown as a solid curve in Figure 56) can also be of particular value in determining the extent of mud cake invasion of the aquifer as well as in determining lithologic characteristics.

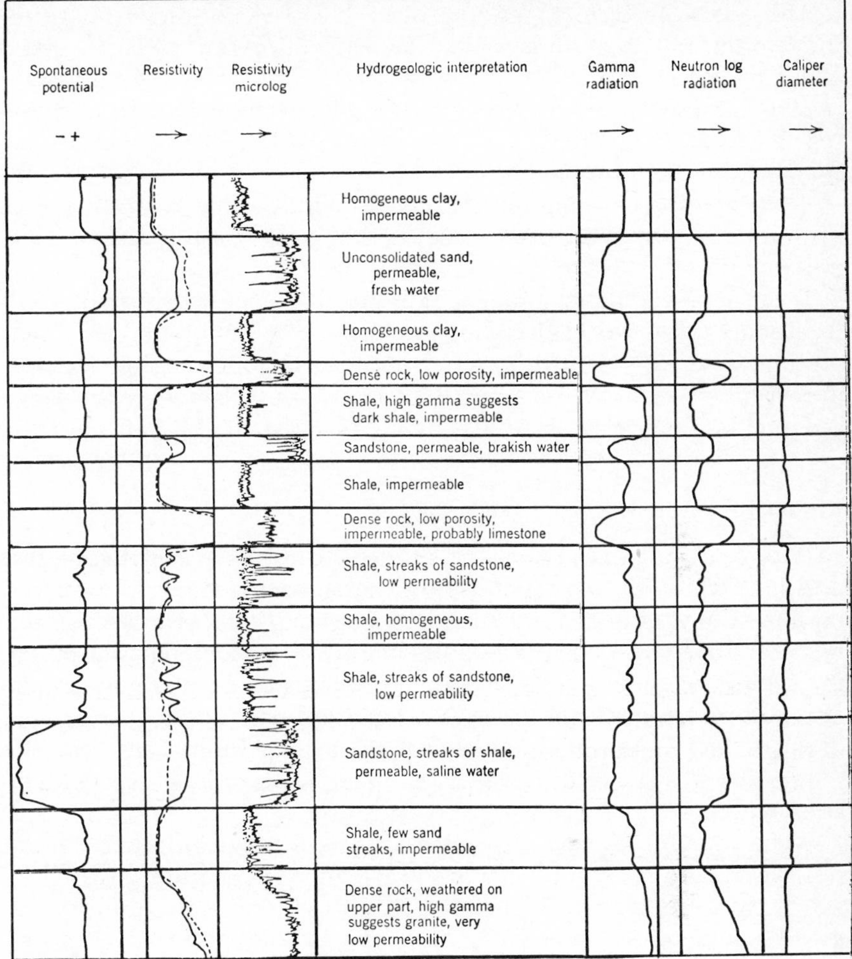

Figure 56[198] A combination of six common logs of a hypothetical test hole showing the hydrogeologic interpretations.

Gamma (natural) radiation under normal conditions is high in clays and shales because such rocks are composed of a very small but significant percentage of radioactive minerals. Sands or sandstones, containing quartz grains and other minor constituents, produce less natural radiation than clays, although exceptions do exist which will be examined later. Limestones typically show low radiation values; whereas, granites or granite washes show moderate radiation.

The neutron log in Figure 56, often looks like a mirror image of the gamma ray log and is useful for delineating general rock types. Neutron logs are obtained by bombarding the formation with artificial radioactivity and recording the results. The configuration of the log is affected by the fluctuation of the hydrogen content of the fluids within the pore spaces.

The microlog is primarily used in oil-field logging. Its main use is for locating very thin beds, such as stringers of sand within shale horizons.[262]

It must be emphasized that geophysical logs have a variety of uses. In ground water work, these logs are most commonly used for interpreting an already drilled borehole and can be indispensable for predicting what would be encountered in future wells within the same geographical area assuming the geology does not change radically.

Fluid Characteristics from Well Logging

Important information can be derived from well logs including the position, amount, movement, and character of water in the formation. Conventional fluid resistivity logs and temperature logs are used for this purpose if the well has had time to attain equilibrium in the ground water reservoir, thereby producing an adequate hydraulic connection between the hole and the rock penetrated. Formation damage due to inadequate removal of drilling fluids can seriously reduce the efficiency of a well.[205,442] This topic will be expanded in later chapters.

Water quality as well as temperature is a very important factor relating to the useability of ground water. Temperature, resistivity, and gamma logs have all been used to determine the dilution of dispersion and the movement of chemical and radioactive wastes underground. Continuous logs are more economical than numerous point samples and can be used to determine the location of check samples for laboratory analysis. Fluid-temperature and resistivity logs are used to calculate specific conductance which can then be used to calculate the quantity of various ions in solution if representative chemical analyses are available as a guide.[115,626,628] Temperature has an important bearing on the movement of water because of its relationship to viscosity, and water quality affects fluid density. Anomalies in temperature logs and fluid-resistivity logs are used to locate the source or sources of water in either a pumped or injected well; however, the recently developed differential-temperature logging tool finds application as well since it is much more sensitive than the gradient-temperature log. Of course, the position of the water level within the well is readily identified by fluid resistivity, electric, gamma-gamma, and neutron logs.

The location of the water table (or perched water table) outside the casing can be accomplished with neutron logs. The location of a fresh water/brine interface through casing, for example, has been a problem.[357] The use of neutron-epithermal neutron and neutron-gamma logs have proved successful in some cases. This type of logging is a potentially useful method for detecting changes in salt content.

An important application of well logging is in the unsaturated zone of aeration between the land surface and the zone of saturation. This is the zone through which much ground water recharge takes place. The measurement and movement of the moisture content in this zone is very important to many ground water evaluations. Various types of neutron logs are used for this purpose. The most common is the neutron moisture log. This equipment utilizes a short spacing between the detector and the neutron source so that count rate increases with higher moisture content. Most of this equipment does not provide a continuous log and is not suitable for large-diameter or deep wells. Recently, conventional neutron logging equipment, similar to that developed in the petroleum industry, has been used. In Figure 57, for example, Keys cites a comparison of

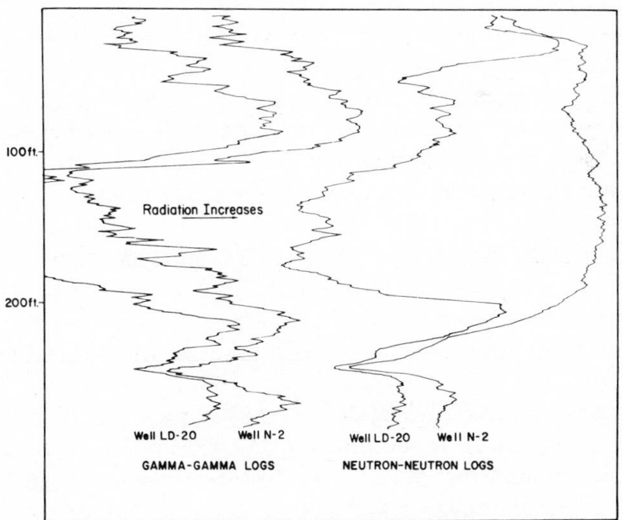

Figure 57[356] Measurement of porosity and moisture above the water table - Los Alamos, New Mexico.

neutron-epithermal neuton logs and gamma-gamma logs for two adjacent wells in a volcanic tuff near Los Alamos, New Mexico.

Core samples show that the gamma-gamma logs of these two wells correctly reflect the porosity, which averages 43 percent in the first 100 feet and 34 percent from 100 to 200 feet. The neutron log of well LD-20, however, reflects an increase in moisture from 50 to 200 feet due to water which was artificially introduced near the well. The neutron log at well N-2 shows that these same tuffs are normally dry except for a moist zone at a depth of about 250 feet. Time-interval neutron logging has been successfully used to follow the downward infiltration of water which begins when the frost goes out in northern climates or which follows heavy rains.[356]

An important new use of neutron logs is the determination of the specific yield of an unconfined aquifer as described by Meyer.[441] *Specific yield* is defined as the ratio of the volume of water that saturated rock or soil will yield by gravity to its own volume. This technique utilizes a neutron probe, calibrated in percent moisture, to measure the difference in moisture between the saturated material and the same materials after it has been drained. The method provides results quite similar to those derived from conventional aquifer tests.

The source of water in a well can be determined by time-interval neutron or gamma-gamma logging.[359] During development, testing, and cleanup of a well, large quantities of fine-grained sediment may be dislodged. Most of this material is derived from those parts of the aquifer where the ground water velocity is highest. Therefore, since the porosity of the intervals is significantly altered, these zones can be located by comparing neutron or gamma-gamma logs made before and after development, although subsequent hole enlargement during the development process may complicate the comparison.

The movement of water in a well responding to natural differences in head between aquifers, pumping, or injection is measured with an impeller flowmeter or by a radioactive or salt tracer. Conventional "injectivity profiles" may be constructed in order to locate the zones of highest relative permeability. Radioactive tracers are also used to locate permeable zones, producing or taking water outside the casing.

Most single-well and multiple-well techniques, using radioactive tracers for determining the direction and velocity of ground water movement in an aquifer, do not involve conventional logging tools. Single-well tracer techniques are largely experimental at this time. In the point-dilution method, a small quantity of radioactive tracer is injected into the center of the well, and its horizontal velocity and direction of movement are measured by a series of detectors placed around the circumference of the well. In the single-well pulse technique, a slug of tracer is induced into a stream of water being injected into the well. After a period of injection, the well is pumped, and the time required to recover the slug can be related to the ground water velocity. If a radioactive tracer is used, the rock section accepting and returning the slug of tracer can be located. Multiple-well tracer tests usually involve physical sampling rather than logging since tritium is generally used, and it cannot be detected with conventional logging equipment. Some gamma-emitting radioisotopes, however, have been located and traced, utilizing conventional gamma logging equipment. The identification of radioisotopes in waste disposal is possible with sophisticated logging equipment.

The quality of the water produced by the well is of major importance in the evaluation of an aquifer.[607] The U.S. Public Health Service has placed limits on the maximum dissolved salts in water for public use. A maximum of total dissolved solids (TDS) of 500 ppm, and a maximum chloride (Cl) content of 250 ppm is usually stated in addition to other recommendations. Water for irrigation or industrial use is not required to meet such rigorous standards. These standards will be revised in the future to meet more stringent water quality regulations. The responsibility for nationally uniform water quality standards was recently shifted to the Environmental Protection Agency (EPA). The EPA will be publishing the much needed revised standards in the near future.

SP Log

The SP log (Spontaneous Potential) in conjunction with its associated log — the Resistivity log — generally provides the best logging approach for determining water quality. The relationships between SP and resistivity, ionic concentration, etc., have been well established for oil-field brines. These relationships are based on research studies and are substantiated by extensive oil industry applications.[256,668] However, when these oil-field relationships are applied to fresh sands, the results can be misleading. The following relationship is used as the standard oil industry approach:

$$SP = -K \log \frac{Rmf}{Rw} \qquad \text{EQUATION 43}$$

where: SP = Spontaneous potential (in ± millivolts)
K = A constant as a function of the formation temperature
Rmf = Resistivity of mud filtrate
Rw = Resistivity of formation water

In a fresh water solution, the dissolved salts are not generally dominated by sodium chloride as in most types of oil-field formation waters. Therefore, the petroleum industry relationship based on sodium chloride solutions does not apply directly. It becomes necessary to re-examine the relationships between ionic concentration and resistivity, etc., for significant quantities of other ions.

Divalent cations (Ca^{++}, Mg^{++}, etc.) in ground water have a much stronger effect on the SP than does Na^{+}, with the effect producing an SP curve which suggests a higher "salt" content than indicated by the corresponding resistivity curve. Gondouin, *et al.*,[256] show that for such cases, the magnitude of the electro-chemical SP is as follows:

EQUATION 44

$$SP = -K \log \frac{\left({}^{a}Na^{+}\sqrt{{}^{a}Ca^{++} + {}^{a}Mg^{++}}\right)_{w}}{\left({}^{a}Na\right)_{mf}}$$

where: $({}^{a}Na^{+}, {}^{a}Ca^{++}$ and ${}^{a}Mg^{++})_{w}$ = solution activities of sodium, calcium, and magnesium ions of the water sample

K = f (x) of formation temperature

mf = mud filtrate

In the latter equation, the mud filtrate (mf) is considered to act as a sodium chloride solution, although there are cases where divalent ions are present in significant concentrations in the mud filtrate. Base exchange in clay additives or shales tends to reduce the divalent concentration of the mud makeup water. It is possible that the anion concentration has a small influence on the SP and some effect on the solution's resistivity.

For sodium chloride solutions, the relationship between "activity" and resistivity is shown in Figure 58. The resistivity of the mud filtrate (Rmf) used in the denominator of the latter equation can be determined when Rmf is known. The relationship shown in Figure 58 is correct only for a solution temperature of 77°F (25°C). For Rmf values at other temperatures, Figure 59 presents resistivity-temperature concentration data for dilute sodium chloride solutions which can be used for a conversion to standard temperature conditions.

The relationships between ion concentration and resistivity for other ions differ from that of sodium chloride. Alger[8] suggests that conversion factors (multipliers) are required to convert concentrations of common ions to equivalent sodium chloride concentrations for formation-resistivity determinations, as follows:

$Na^+ = 1.0$
$Ca^{++} = 0.95$
$Mg^{++} = 2.0$
$Cl^- = 1.0$
$SO_4^= = 0.5$
$CO_3^= = 1.26$
$HCO_3^- = 0.27$

By computation using Equation 44 and Figure 59 conversions, an SP log value can be confirmed from a specific well by properly accounting for the types and concentrations of ions present.

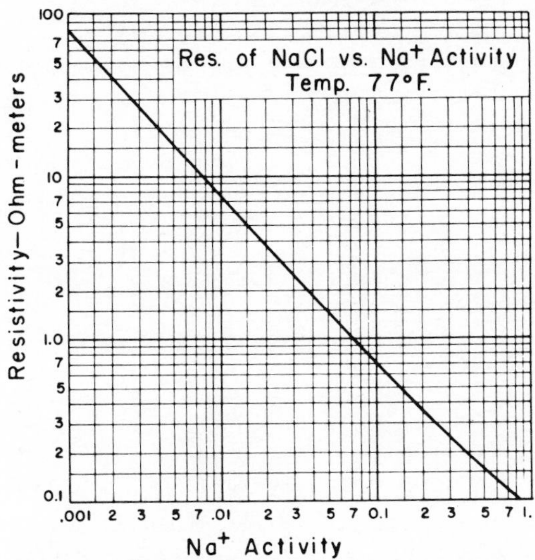

Figure 58[8] Relationship between sodium ion activity and resistivity for a sodium chloride solution at 77%.

Alger[8] shows that the relative ion assemblage, although variable from one geographical region to another, is predictable on a local basis given a standard from which to work.

In addition, fresh water often contains bicarbonate (HCO_3^-) as the predominant anion. If Na^+ is the accompanying cation, the effect on an SP curve is usually similar to that for a sodium chloride solution having the same Na^+ concentration. However, the resistivity of sodium chloride (NaCl) and sodium bicarbonate ($NaHCO_3$) solutions (with the same Na^+ concentrations) is different. Alger reports that the bicarbonate ion (HCO_3^-) contributes only 27 percent as much conductivity (reciprocal of resistivity) as an equal weight of chloride ion (Cl^-), i.e., the resistivity of a sodium bicarbonate solution is 1.75 times the resistivity of a sodium chloride solution having the same Na^+ concentration. Thus, empirical relations between SP, R_W, and TDS can be used to determine the quality from log data.

Because the relationship between resistivity of the formation water (R_W) and SP varies in accordance with the type of ions present, it is convenient to consider the water resistivity determined

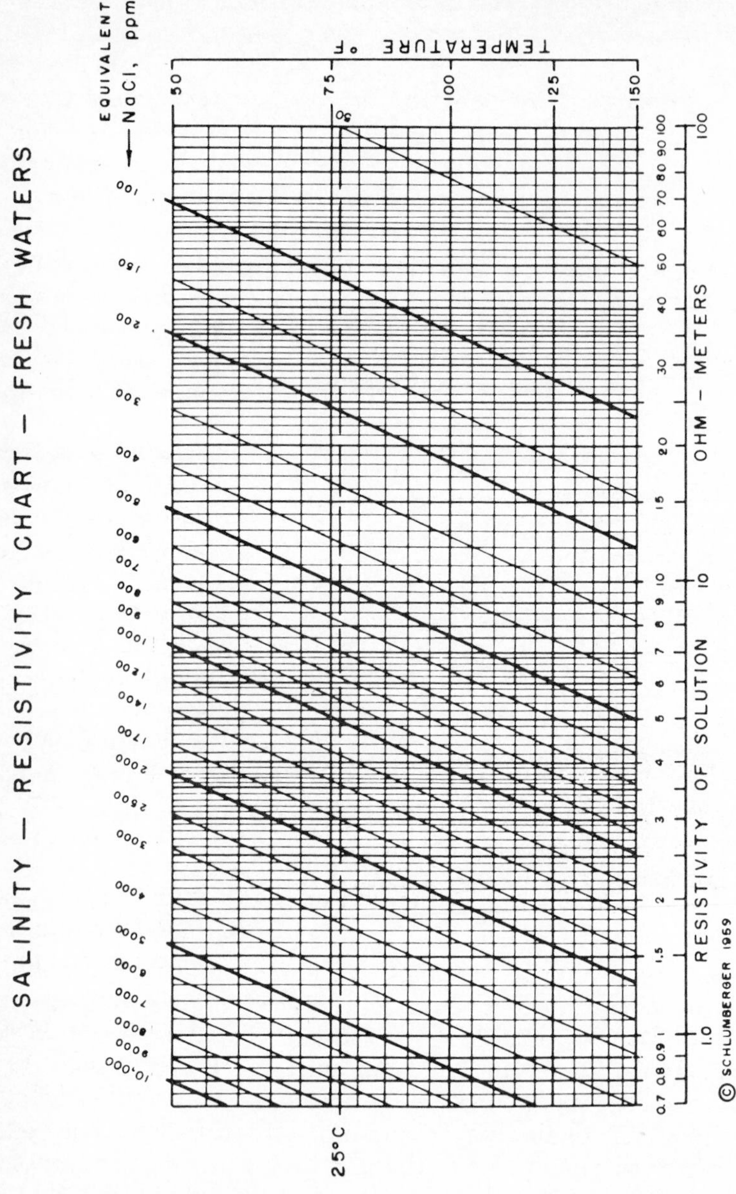

Figure 59[8] Relationship between resistivities, temperatures, and concentrations for dilute sodium chloride solutions.

from Equation 43 as an equivalent water resistivity, (R_{we}). Furthermore, it is also convenient to consider the mud filtrate as an NaCl solution. Then, depending on the ion assemblage for the locale, an appropriate relation may be used to obtain R_w from the SP-derived value of R_{we}. This is illustrated in Figure 60.

For water with only NaCl in solution, the $R_W - R_{We}$ relationship is given by the diagonal line between the lower left and upper right corners of the figure. The line for $NaHCO_3$ water parallels the NaCl line. It is displaced upward, so that for the same value of R_{We}, the R_W value is 1.75 times greater than for an NaCl water. The leftward line is for water with only $CaCl_2$ in solution. This $CaCl_2$ line represents the upper limit most likely to be found in a fresh water evaluation from SP data.

The points plotted on Figure 60 were taken from data on wells drilled for fresh water. In each case, the value of R_{We} was obtained from the SP using Equation 44. Also in each case, the mud filtrate was considered as an NaCl solution. The corresponding value of R_W for each point was determined from resistivity measurement of a sample of the formation water.

Experience, for example, shows that most Gulf Coast water wells drilled with fresh mud plot near the $NaHCO_3$ line. This is usually true even when the formation water contains significant quantities of divalent cations. The effect of the divalent cations is offset, to a large extent, by similar concentrations in the mud filtrate.

A different relationship between R_W and R_{We} is noted for fresh water wells drilled with salty muds. In such wells the $R_W - R_{We}$ plot is displaced toward the $CaCl_2$ line on Figure 60. For example, the log in Figure 61 was run in a Gulf Coast water well in which the drilling mud was deliberately salted to develop large (but positive) SP deflections opposite sands. The resulting plot is approximately midway between the $NaHCO_3$ and $CaCl_2$ lines. Salting the mud reduced the importance of the divalent cations in the mud filtrate and, thus, displaced the $R_W - R_{We}$ plot from its usual position for the Gulf Coast (near the $NaHCO_3$ line) and moved it toward the $CaCl_2$ line.

A diminishing effect of divalent cations in the mud filtrate is also noted opposite fresh water sands in wells drilled to deeper oil zones. This is due both to increased concentrations of NaCl in the mud filtrate and to reduced concentrations of divalent cations through base exchange. The dashed line on Figure 60 (taken from Schlumberger's Chart A-12) was empirically derived from brackish water zones encountered in wells drilled for oil and gas.

Determination of Formation Resistivity

Alger[8] states that an alternate method of determining R_W is based on measurement of formation resistivity. In his application, R_W is obtained by dividing the formation resistivity (R_O) by a formation resistivity factor (F). This approach has become well known and widely used in oil-field interpretations.

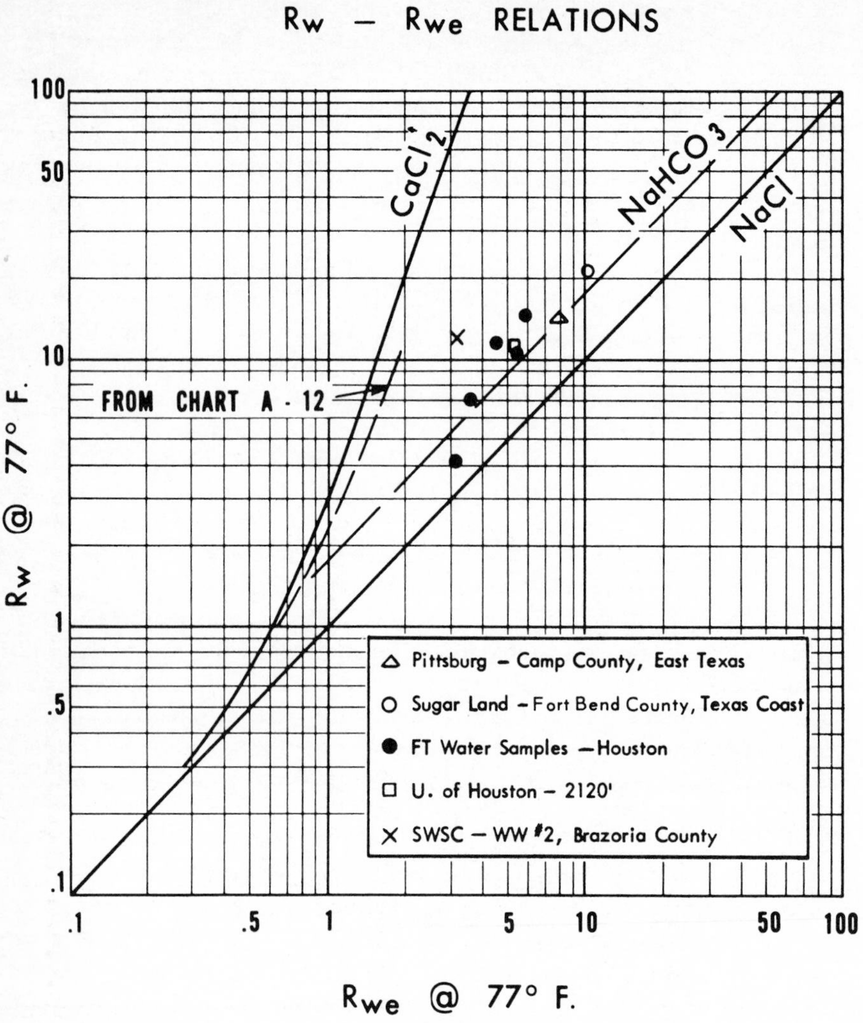

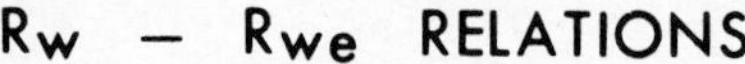

Figure 60[8] R_w vs R_{we} for fresh formation waters. Plots from Gulf Coast wells usually are close to the line for $NaHCO_3$ water.

However, the commonly known relationships between F and porosity used in oil-field interpretations usually do not apply to fresh water sands. F varies in fresh water sands not only with porosity, but also with R_W and grain size. Thus, F must be defined by other methods.

The best results to determine R_W from R_O measurements were obtained by Alger[8] using F values based on local empirical studies

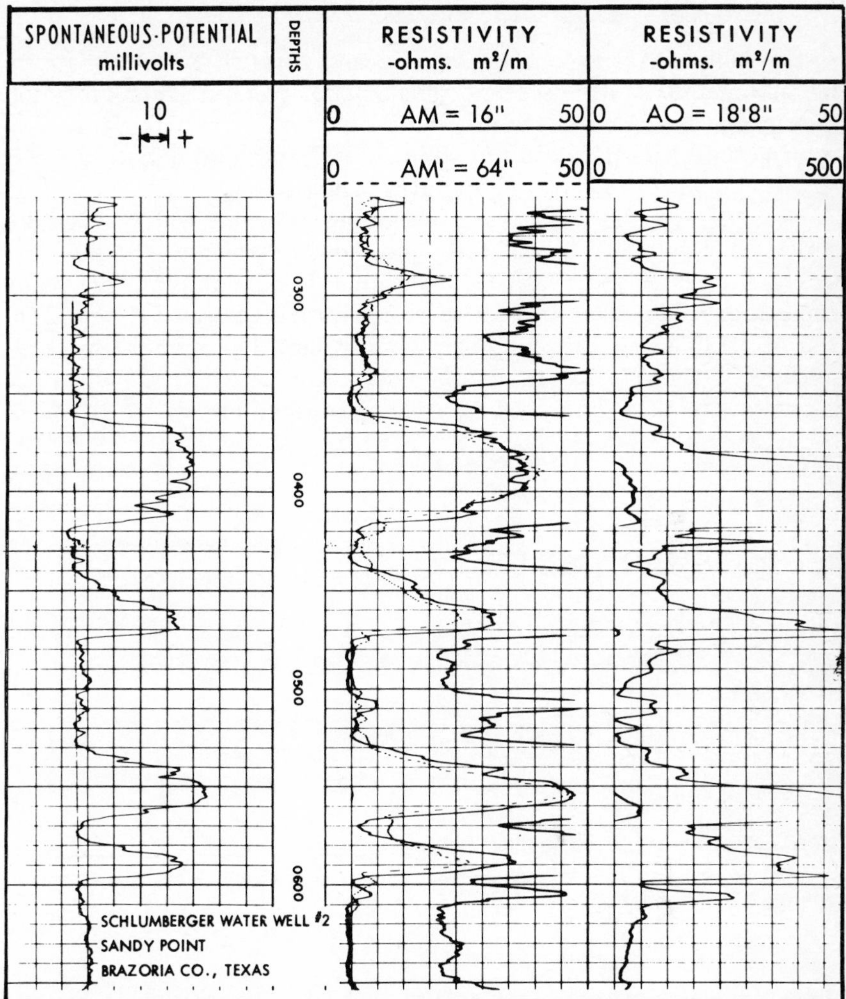

Figure 61[8] Electric Log recorded in a Gulf Coast water well drilled with salted mud (R_m = 1.0 at 77°F).

($F = R_o / R_w$). For example, Turcan[626] applies "field formation resistivity factors" (F_f) in shallow, clean sand formations in Louisiana. For sands in the Wilcox group, values of F_f ranging from 1.7 to 3.0 were reported. The use of these values of F_f necessarily demands that the sands are clean and relatively constant in grain size — and that R_w does not vary too widely. In conclusion, similar studies in other local areas will provide a simple and useful means of determining R_w from measurements of formation resistivity.

Determination of Total Dissolved Solids and Chloride Content

The value of R_W, whether determined through SP or resistivity measurements, is also used to evaluate water quality. The relative ion assemblage is reasonably predictable on a local basis[626]. Thus, empirical studies permit determination of both the total dissolved solids (TDS) and Cl content from computed values of R_W.

Plots of R_W vs. TDS from a number of fresh water sands are shown on Figure 62. All of the points plot between the lines appropriate for NaCl and $NaHCO_3$ solutions. If similar plots are made using local data, a more precise relationship between R_W and TDS can be established. For example, work in east Texas has shown that reliable values of TDS are obtained using the $NaHCO_3$ line when R_W is greater than 7.0 ohm-m, and using the dashed line when R_W is less than 7.0 ohm-m (all R_w values at 77° F).

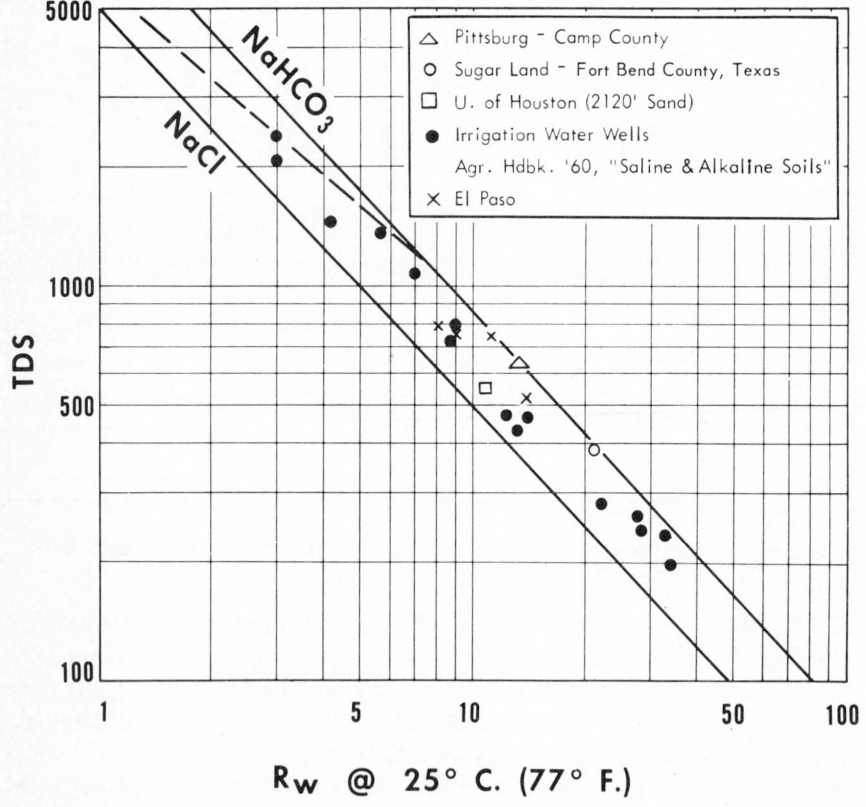

Figure 62[8] Empirical studies relate Total Dissolved Solids (TDS) to R_W.

Alger[8] stresses that care should be exercised, however, in extending such empirical methods from one area to another. This is illustrated in the work done by Turcan[626] for fresh water zones in the Mississippi embayment area.

Figure 63 is a re-plot of his relationships between R_W, TDS, and chlorides for the Wilcox sands in northwest Louisiana. He found that the TDS concentration was close to that for NaCl solutions for R_W values of approximately 10.0 ohm-m (at 77°F). However, for both lower and higher values of R_W, the TDS concentration departed from that for NaCl solutions.

It will be noted that two abscissa scales are shown on Figure 63; one is an R_W scale, the other, specific conductance (customarily used by water well contractors and engineers in reporting electrical measurements of water samples). *Specific conductance* is the electrical conductivity of a water sample at 25°C (77°F), expressed in micro-ohms per centimeter. It is related to resistivity, expressed in ohm-meters, by the following expression:

EQUATION 45

$$R_W = \frac{10{,}000}{\text{Specific Conductance}}$$

Turcan's data in Figure 63 shows an increase of chloride concentration as the TDS concentration increases. Observations in most areas show that the chloride ion becomes proportionally more abundant as the TDS concentration increases. This provides a basis for determining chloride ion concentration. However, local data is necessary to provide an appropriate relationship. For example, ion concentrations reported by Jones and Buford in Central Louisiana suggest the following relationship:[341]

EQUATION 46

$$Cl_{ppm} = 0.6\ (\text{TDS-}400)$$

This form of empirical relationship is useful in other areas, although the constants may vary. Below are examples of equations that seem appropriate for several other areas.

EQUATION 47

$$Cl_{ppm} = 0.5\ (\text{TDS-}400) \text{ in East Texas}$$

EQUATION 48

$$Cl_{ppm} = 0.53\ (\text{TDS-}200) \text{ near El Paso, Texas}$$

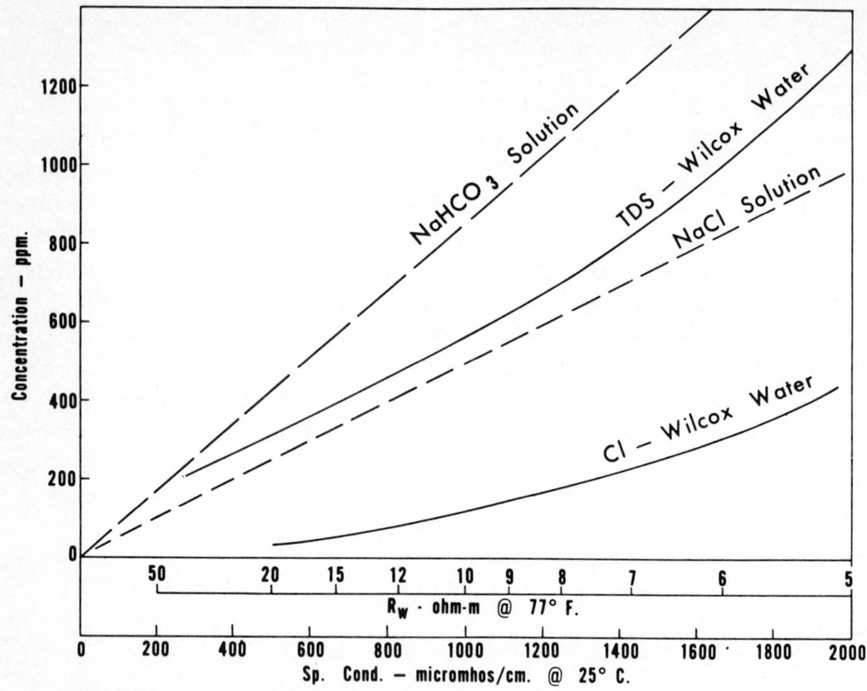

Figure 63[8] Relationships between R_w, TDS, and Chloride concentration for Wilcox sands in Northwest Louisiana.

EQUATION 49

Cl_{ppm} = 0.40 (TDS-250) Wilcox sands in Northwest Louisiana (from Turcan's data)

EQUATION 50

Cl_{ppm} = 0.25 (TDS-200) Houston, Texas

Alger[8] concludes that well logs can give reliable information which is required in the evaluation of water quality. However, for this information to be most effective, empirical relationships should be established for local areas. Much of the data is already available since many water wells have been logged and sampled and subsequently analyzed. This information can provide valuable assistance in defining many of the semi-quantitative aspects of water quality.[256, 518, 628, 640, 642]

SP Log Applications

The only possibility of using the SP data for quantitative salinity calculations in fresh water bearing formations would be to establish empirical data for the waters of an area and then use an appropriate chart or formula, as mentioned previously. Guyod[293] cites some significant difficulties with the SP:

(1) The numerical value of the K factor in Equation 43 can be accurately evaluated only when the clay formation [bound by] the aquifer is a perfect cationic permeable membrane. Actual K values can be determined only from the laboratory measurements.

(2) The chemical composition of the borehole fluid must be taken into account.

(3) A streaming potential is usually superimposed on the electrochemical potential. Although the former is very small at shallow depths, its relative value may not be negligible in deep water wells where the SP amplitude is low, and it is difficult to ascertain this fact. [Gondouin and Scala[255] suggest that a streaming potential (E_k) can occur in deep shales also, but the SP deflection from the shale-line is, however, unaffected by E_k]

(4) Even if there is no streaming potential, the measured SP is only a portion of the total electrochemical potential, $(SP)_e$, developed in the ground. The reduction, $SP/(SP)_e$, is a function of several factors, in particular: the aquifer resistivity and thickness. [A correction can be made, but it is difficult to accomplish in practice.]

The unpredictable behavior of the SP is well demonstrated. While the SP curve should not be applied to quantitative determinations of the salinity of fresh waters except with proper restraint, Guyod[293] suggests that it is permissible to use it qualitatively, according to the following rules of thumb:

(1) Aquifers that exhibit a positive SP almost invariably carry waters of low salinity provided the borehole fluid has a resistivity greater than about five ohm-m.

(2) In the intervals where the SP amplitude [measured from the shale line] in . . . thick aquifers remains nearly constant with depth, all the formation waters have about the same salinity.

(3) If the SP of the aquifers penetrated by a borehole becomes more and more negative with depth, the salinity of the aquifers probably increases with depth. If, simultaneously, the aquifer resistivities decrease with depth, the evidence is considerably stronger.

(4) Aquifers that exhibit a fairly large negative SP generally carry waters that are much more saline than where the SP has a low amplitude or is positive.

(5) Erratic changes in SP polarity, provided that the SP amplitude remains small (less than 25 mv), may or may not correspond to significant changes in water salinity.

Actual SP curves do not always conform to the suggested patterns. The most notable exceptions are described by Guyod[293] and summarized below:

(1) Drift in the clay base line. — Commonly the clay base line is essentially straight and vertical, especially below a few hundred feet. But in certain wells at shallow depths the SP curve gradually wanders, either as a whole or only in clay intervals, and generally to the left as the depth decreases. No satisfactory explanation has been offered for this phenomenon which appears to be more prevalent in arid areas.

(2) Shift in the clay base line. — This is frequently observed when there is a rather fast change in the salinity of formation waters (in a specific aquifer). A shift can also be caused by a change in the nature of the clay.

(3) Unstable SP. — This is observed in the upper part of holes in which there is an appreciable movement of water, as in artesian wells or above thieving zones; the signal changes constantly even if the logging electrode is kept stationary. This condition is due to an unstable electrode potential caused by the water flow. The instability disappears below the zone of water movement.

(4) Polarity reversals. — Numerous polarity reversals in the aquifers of a given well are sometimes noted even though the waters have salinities of the same order. These reversals are usually due to changes in the type of ions or in the quantities of some of the ions. (However, reversals may also be due to a logging equipment malfunction.)

Although the presence of permeable rocks having intergranular porosity and situated between clay beds can generally be inferred from the shape of the SP curve, neither the curve shape nor the amplitude provides a basis for direct calculations of water quality, porosity, or permeability. MacCary[402] suggests that resistivity and neutron logs can be used in a semi-quantitative, empirical manner to estimate water quality and formation porosity in carbonate zones. His study, however, did not reveal any significant relationship between the formation factor and permeability of the particular carbonate aquifer examined.

When the changes in the permeability of a rock are caused by the presence of some clay material within the pore space, they can be quantitatively estimated from the resulting changes in SP amplitude by using empirical data. Obviously, the method is applicable only if there are no changes in the water's ionic composition within the formation of interest.

It is possible to estimate permeability from streaming potential measurements made under several well-head pressures where the face of the specific formation is free of mud cake. Intake wells in water-flooding operations have been used to test this possibility. Results indicate only whether a formation is relatively permeable or impervious. This concept has been updated in a recent paper by Traugott.[622]

Although the electrochemical potential $(SP)_e$ is not influenced by porosity, the amplitude of the SP curve is indirectly affected by porosity changes. In fact, a decrease in porosity increases the rock resistivity, and this in turn reduces the SP amplitude in thin zones. However, if the zone is thick enough, a static SP is reached even when resistivity is relatively high. In particular, dense beds, located within thick clay intervals, exhibit no measurable SP deflections.

The SP curve is most reliable in formations comprising clay and granular aquifers, especially below a few hundred feet from the surface. For interpretation purposes, however, the SP is always analyzed simultaneously with the resistivity curve and all other available data.

Where formation waters are much more saline than the drilling mud, the SP is generally more negative in aquifers than in the adjacent clays; this permits using the SP curve for identifying formations, correlating, and determining the depth and thickness of certain beds. Supplemented by a resistivity curve, the SP indicates where the formation water changes from fresh to brackish.

A repeatable SP curve can be obtained in uncased empty holes if the measuring electrode is nonmetallic and makes a rolling contact against the bore wall. The SP curve is generally different from that obtained from a hole containing water or mud. An example is given in Figure 64.

An SP curve recorded in a steel casing is related primarily to the corrosion at the time the measurements are made.[349] In theory, the data can be used to obtain information on the condition of the casing, but interpretation is difficult and usually problematical. This approach is, however, amenable to future research and development.

The SP curve in a plastic casing is generally a straight vertical line and therefore of no value.

Resistivity Log

Alger[8] states that formation resistivity measurements are an integral part of most logging programs for oil and gas exploration. The applications of resistivity logs in determination of water saturation, detection of movable oil, correlation, etc., are widely understood in the petroleum industry.

Resistivity logs are also of considerable value in the exploration and production of ground water. However, since the petroleum and ground water industries encounter different down-hole conditions and different basic relationships between formation resistivity and other formation properties, these differences necessitate modifications.[104,304]

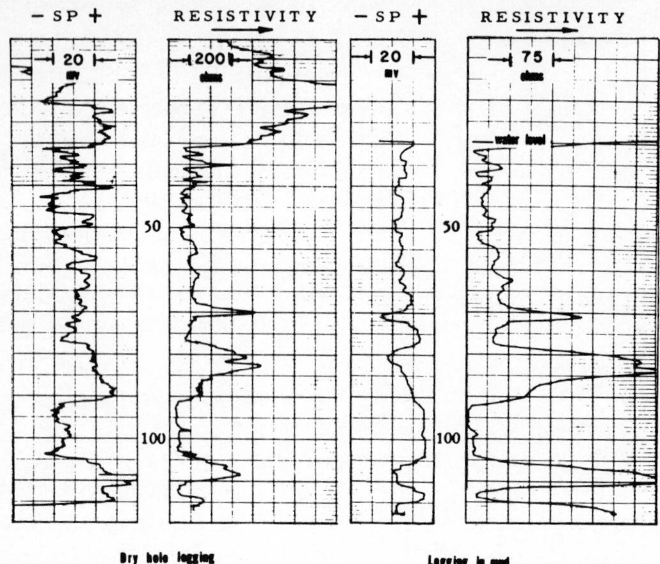

Figure 64[293] Comparison of electric logs made in an empty hole and in the same hole containing mud.

The ground water industry is concerned primarily with determining the quality and quantity of water that can be produced. Hydrocarbons are seldom encountered in significant quantities in the formations drilled for ground water. Thus, the ground water industry is not concerned with determining the fractional saturations of hydrocarbons and water.

In the shallow, unconsolidated sands drilled for ground water, porosities are relatively high. Furthermore, variations in porosity are generally small between sands encountered both in a single well and other wells in the same area. Thus, determination of porosity, so important in interpreting logs for hydrocarbon saturation, has less use in evaluation of ground water aquifers.

Determination of formation resistivity from logs is, for the most part, simpler for water wells than for oil wells. In most cases the resistivity value from a 16-inch "short normal" curve is similar to the true formation resistivity. This is illustrated in Figure 65 where an induction-electrical survey is compared with an electrical survey. Both of these logs were recorded in a shallow experimental well at the University of Houston.[536]

The usefulness of the "short normal" curve is due to several factors. First, most of these wells are drilled with fresh muds; the contrast between mud and formation resistivities is, therefore, low.

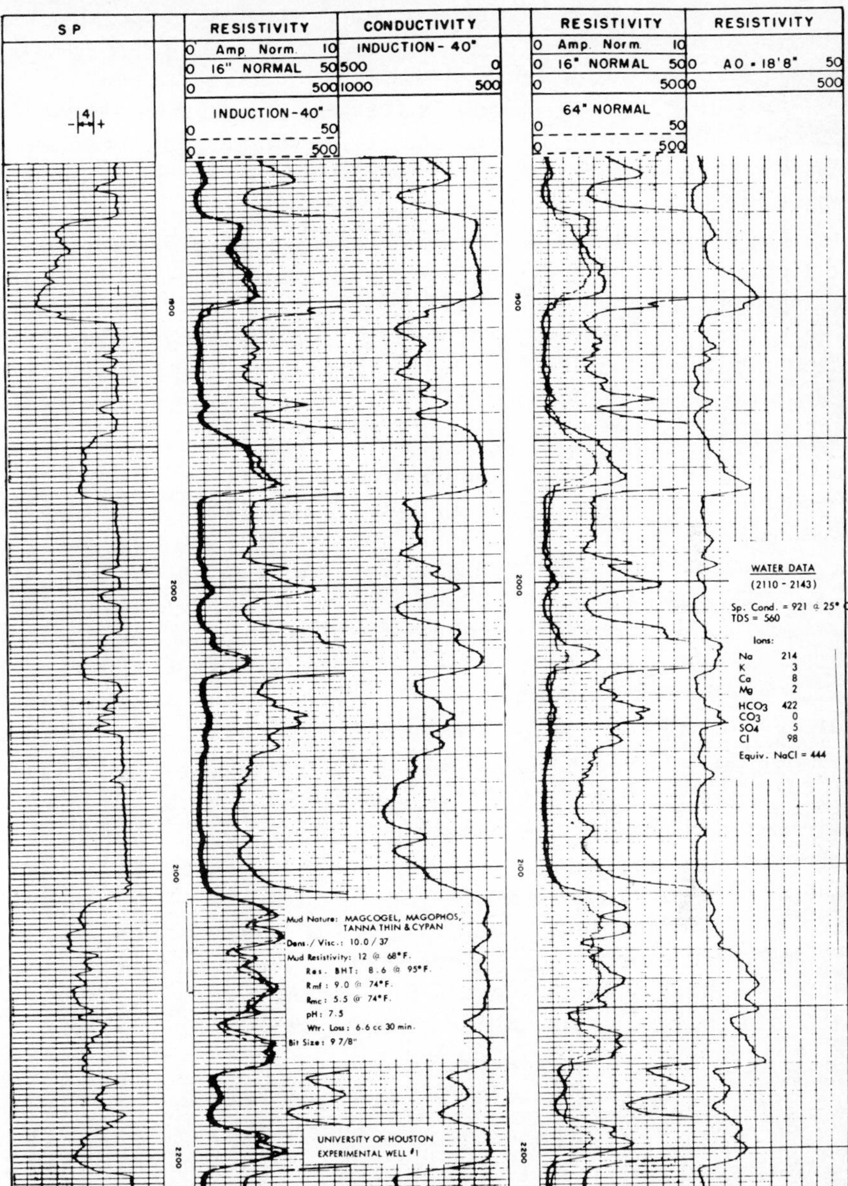

Figure 65[8] Comparison of Induction-Electrical Survey and Electrical Survey recorded in a shallow well at University of Houston.

Second, mud invasion is usually limited because the pressure differential between borehole and formation is relatively low. Furthermore, because the resistivity of the mud filtrate is similar to that of the formation water, the invaded zone's resistivity approximates the true formation resistivity. Finally, to be of interest for ground water production, sands must be relatively thick. The low formation pressures in ground water sands necessitate much higher values of permeability-per-foot than are required for economic oil production. Therefore, in the thicker zones of interest, there is little effect from surrounding beds on the "short normal" response.

Thus, a "short normal" curve is usually adequate for determining formation resistivity. Only for the deeper wells or for those drilled to test for oil or gas, are other tools necessary. In these cases, an induction log is used for a more precise measurement of formation resistivity. The induction tool overcomes the effects of the borehole and surrounding beds and produces less distortion of formation resistivity.

In view of the foregoing discussions, it would seem simple to determine R_W from a "short normal" measurement of formation resistivity. Saturation and porosity variables have been eliminated — as have problems in determining R_O. The R_W of an aquifer with low clay content can be obtained from solution of the customary equation, $F = R_O/R_W$, (with F determined from a porosity value appropriate for the area). However, in fresh-water sands the usual relationship between F and porosity, $F = a/_{\theta}\,m$, is not constant, since the value of F is a function of R_W.

Variation of F with respect to R_W is illustrated by Sarma and Rao,[545] shown in Figure 66. In these investigations, the R_O of washed and graded river-sand samples are measured for water saturants of various values of the R_W. For each of the three samples the formation resistivity factor (determined as R_O/R_W) decreased as R_W increased. The data shows that variations of F are most pronounced for high values of R_W. For fresh water in unconsolidated sands, computed formation-resistivity factors are appreciably lower than described by the F vs. porosity relationships commonly used in petroleum industry log interpretations.

Similar variations of F with R_W have been noted in shaly sands. For example, in investigations of shaly sands, Hill, *et al.*[314] found it necessary to modify the F by including the resistivity of the saturant used in the determination. They found the higher values of R_W tended to lower the values of F. They used this modified F to indicate the maximum value of formation factor for a given shaly sand.

Thus, it appears that the variations of F in shaly sands and in clean fresh-water sands are similar. Alger[8] suggests that surface conductance is the primary factor in variations of F in fresh-water sands.

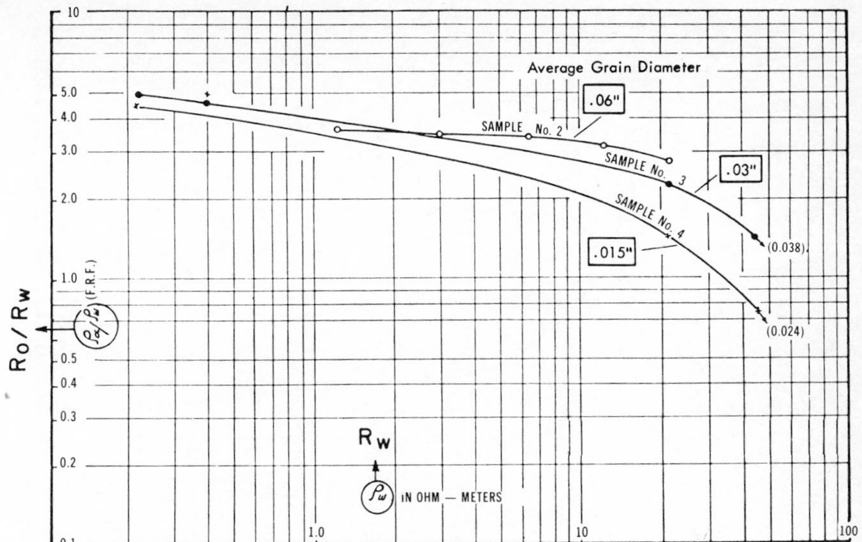

Figure 66[8] Relationships between Formation Resistivity Factor, R_w, and grain size.

The magnitude of surface conductance is related to the ion concentration of the saturant solution. As the concentration decreases, the magnitude of surface conductance also decreases. But, in the low conductivity environment of a fresh-water sand, even this reduced surface conductance is significant.

Another factor that affects the magnitude of surface conductance put forth by Alger is the surface area exposed to the saturant solution. If more surface (per unit volume) is exposed to the electrolyte, the total surface conductance increases. This is of particular importance since internal surface area of sands is related to both grain size and permeability, parameters frequently used in the water well industry. Sands in which the d_e (effective grain size) is less than 0.25 mm are difficult to screen and gravel pack and are, therefore, generally avoided. Furthermore, because permeability is related to grain size, the finer grain sands usually do not afford sufficient capacity. (See Figure 67.)

Determination of Grain Size

The Sarma and Rao data in Figure 66 indicate a relationship between grain size and R_O/R_W for fresh waters — the smaller the grain size, the smaller the value of formation factor (R_O/R_W). This is a reverse of the relationship normally encountered in oil-field interpretations. However, surface conductance, although important in

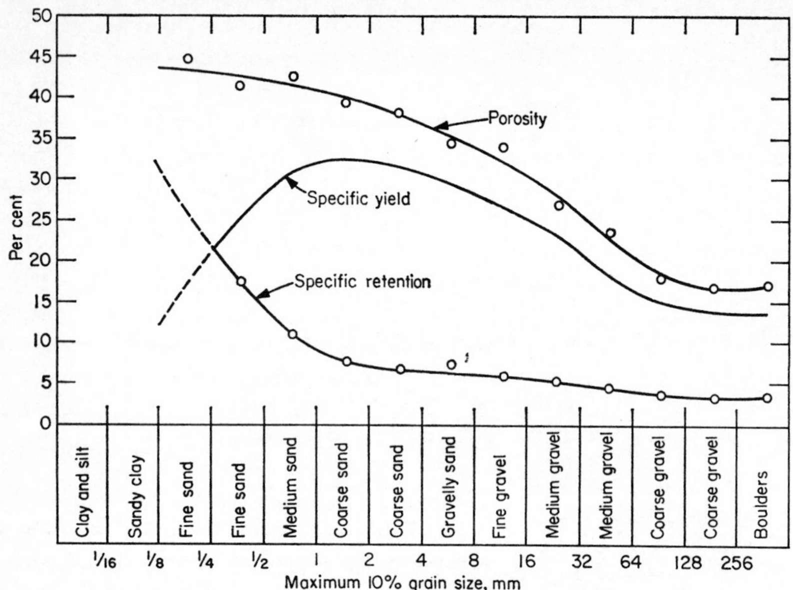

Figure 67 Porosity, specific-yield, and specific-retention variations with grain size, South Coastal Basin, California. The grain size in which the cumulative total, beginning with the coarsest material, reaches 10 per cent of the total sample. (After Eckis, R.P., 1934, Calif. Bull. 45, Sacramento)

fresh-water zones, is relatively unimportant in salty formation waters. Therefore, this relationship between grain size and formation factor in fresh-water sands is attributed to effects of surface conductance. The variation of F with grain size in fresh-water sands was confirmed by laboratory measurements performed by Alger[8].

Determination of Permeability

The relationship between permeability and grain size is well understood and has been reported by other authors.[520] Alger[8] has also reported on resistivity cell measurements of formation factor for a number of samples. A very fresh saturant solution (R_W = 35 ohm-m,) was used for this work. Both permeability and the formation factor demonstrated semi-orderly progression with an increase in grain size.

Alger[8] concludes that the relationship between permeability and formation factor is very important in evaluating water wells from log data since this relationship is the opposite to that commonly expected in oil-field work. It was demonstrated that as the permeability increases for brine-filled cores of Gulf Coast sandstones, the formation factor decreases. Alger supports these results. However, because the importance of surface conductance increases with decreased water salinity, the concept must be modified for sands containing fresh water.

Another feature is that electric logs (Sp and Resistivity) are useful in the evaluation of fresh-water sands. The water quality (TDS) is related to R_W: as the value of R_W increases, the TDS content in the water increases; as the grain size increases, screening or gravel packing is more effective; and as the sand becomes more permeable, the flow per unit of sand thickness increases. Because the above characteristics lead to increased formation resistivity, the best zones for water production are readily identified by the resistivity log. The most prospective water-bearing zones are indicated by the highest resistivities as a general rule, except where significant cementation of a water-bearing sand is suspected, i.e., based on other information sources such as a lithologic log, rate of penetration log, etc.

Resistivity Log Applications

Tangible benefits can be derived from an inspection of a log. The resistivity curve, even that recorded by inexpensive equipment, is beneficial since the depth and thickness of many lithologic units can be determined. In brackish-water zones, resistivity generally decreases. Furthermore, if the bit penetration rate remains constant, it can be assumed that the porosities of these zones are of the same order of magnitude. Therefore, a decrease in the apparent resistivity is an indication of a salinity increase.

When the quality of the water does not vary appreciably in the aquifers penetrated, changes in resistivity can generally be interpreted as changes in porosity, grain-size, or clay content. The simulataneous use of the SP or gamma curve will generally suggest which of these three situations exists. In practice, log interpretations are not made with the resistivity curve alone — the SP and all other available data must be analyzed simultaneously.

Good resistivity curves can be obtained in uncased empty holes providing the tool makes contact with the bore wall. Since the borehole effects are greatly minimized, there is usually less departure between true and apparent resistivities. On the left of Figure 64 is an example of an electric log made in an empty hole; on the right is a log made after the hole was filled with a drilling mud.

Resistivity (and SP) measurements made in cased holes are not related to formation characteristics. The resistivity obtained in a steel casing that is not too old is extremely low; its primary usefulness is to determine the casing depth. Since old steel casings are often corroded and the resulting iron oxide is non-conductive, the resistivity changes reflect primarily the degree of corrosion. The log can occasionally be used to obtain some information on the condition of the casing, but the interpretation is difficult and usually problematical.

Resistivity measurements in undamaged plastic casings are a result of the resistivity of the fluid contained in the casing. Large holes in casing cause local resistivity decreases; and therefore, damaged plastic casing can be located.

Gamma Log

The only borehole geophysical methods that give dependable data on the formations situated behind casing are those based on radiation measurements. Gamma logs can also be used in uncased holes, with the added advantage that the measurements are not adversely affected by the nature of the borehole fluid.[520]

There are two basic radiation logging methods — gamma ray and neutron. Gamma ray logging equipment is fairly simple to operate, inexpensive, and valuable in ground water investigations.[53]

The atoms of a few naturally occurring elements spontaneously disintegrate. This disintegration is slow but continuous, and it is accompanied by the production of radiation: alpha rays, beta rays, and gamma rays. Although alpha rays and beta rays are stopped after traveling less than one inch through matter, gamma rays can go through two feet of water and more than six inches of most formation types. All formations contain radioactive isotopes in varying amounts in the following elements: potassium, thorium, and/or uranium; thus, gamma ray measurements are valuable for formation logging.

It is convenient in gamma logging to classify formations into two groups, clay and non-clay. Although there is no specific amount of radioactivity that a given formation may produce, non-clay intervals generally are less radioactive than clay intervals regardless of porosity and fluids contained, but a few non-clay formations have a radioactivity equal to or exceeding that of clay or clayey formations. This exception will be explored later.

The gamma intensity of clay also varies from area to area. In Tertiary deposits and more recent formations, such as those found in the Gulf Coast and California, the radiation level averages five microroentgens per hour and generally doubles in older clays. Some organic marine clays produce much higher gamma radiation than the other clays of the same area; thus, these marine clays act as an excellent geologic marker on the gamma log.

The non-clay formations that generally have a very low radioactivity when they are totally free of clay material are: sands and sandstone; limestone; dolomite; anhydrite; gypsum; salt; most lignites; and most coals. Those formations having a high radioactivity (although generally less than that of clay) include the arkose and feldspathic sands and sandstones; a few volcanic and igneous rocks; potash; some phosphate; and all uranium ores. A short but interesting discussion of the gamma activity of common sediments has been given by Patten and Bennet.[500]

Gamma Log Applications

The interpretation of a gamma curve is based on the following observations:[293]

(1) In a given area, only the relative intensity measured for the various formations is of significance.

(2) Formations exhibiting a low gamma intensity are clean sand, gravel, sandstone, limestone, dolomite, anhydrite, salt, lignite, or coal. A low gamma reading may indicate a very porous and permeable aquifer, or it may indicate an impervious rock. Geological information is needed to remove the ambiguity.

(3) If it is known that the [sediments] in the area of interest have only a very low radioactivity, all the intervals of the log exhibiting a high gamma intensity are probably clay. The intervals of intermediate intensity correspond to [intervals] — generally aquifers — containing some clay material; the clay content can be assumed to increase nearly in proportion with the gamma intensity.

(4) If nothing is known on the radioactivity of the rocks of the area, it is not possible to interpret the intervals of the log that exhibit a high or intermediate gamma ray intensity. Some of the resulting ambiguity can be removed if an electric log or local geologic knowledge is available.

(5) The gamma ray curve should always be correlated with a lithologic log and all other data available.

A few exceptions to the above rules are listed by Guyod:[293]

(1) When water, instead of a properly conditioned mud, is used for drilling, clay and other cuttings may settle and may increase the gamma curve's amplitude in the bottom five to ten feet of the hole.
(2) If thick drilling mud remains behind the casing or if clay material clings to the face of some non-radioactive intervals, the increase in gamma radiation intensity indicates erroneously that these intervals are sandy clays or clayey sands.

(3) In gravel-packed wells, the gravel absorbs an important amount of the gamma radiation that would normally reach the detector, thus reducing the gamma amplitude.
(4) If the material selected for gravel packing is radioactive (volcanic or granitic rocks), the gamma log deflection indicates the presence and thickness of this material.

All the potential difficulties listed above highlight the importance of obtaining as much information as possible on the well condition and on the formation traversed.

The principal uses of gamma curves are:

(1) The gamma curve produces reliable data from cased wells. Wells that do not produce enough water or that produce water that is unfit for its intended use may be reperforated after the gamma log identifies the presence of other prospective aquifers through the casing.[359,512]
(2) The gamma curve is also useful for the logging of open holes in cases where an electric log would not be reliable, i.e., the presence of brackish or salty borehole fluid or a large hole size or because of an inappropriate probe.
(3) The depths and thicknesses of clay and non-clay beds can be obtained from the gamma curve, but the accuracy in measuring thicknesses of less than two feet is generally poor.
(4) The gamma data are valuable as a supplement to the electric log, particularly in identifying clay beds and porous zones in dense rocks.
(5) The permeability reduction in a rock can be estimated from the gamma data. This estimation is based on relative clay content in the pore spaces.

Mineral Exploration Logging

The gamma log as well as the SP and resistivity logs has become a valuable exploration tool in the mineral industry, e.g., uranium, coal, phosphate, etc. In 1969, millions of feet were drilled in the search for economic uranium ore bodies in Wyoming, Texas, New Mexico, Colorado, Utah, etc.[131,619] Many mineral exploration holes were either completed as water wells or abandoned; some have been sealed with either natural or commercial drilling mud, while many others have been left to cave with time.

These holes are generally less than 300 feet in depth and are in sandy intervals which could produce either ground water or commercial uranium. The ground water geologist should be aware of the "ground rules" of "cell-type" uranium exploration, especially since the same geophysical logs needed for superior water well construction can also be used to evaluate the uranium potential of prospective aquifers (in geologically favorable regions). Figure 68 summarizes the salient features of some common uranium exploration techniques.[521] Figure 69 shows an example of a typical log of an encouraging uranium lead in Wyoming, indicating that a potentially economic uranium ore body may not be too distant up-dip from the hole logged.[155,296,521]

Well Construction and Logging

Geophysical logs are often used in conjunction with drilling rate information to aid in identifying formation characteristics. During drilling, a record of bit penetration rate can be made. This data is either quantitative (in feet per minute) or qualitative, using such terms as "fast, slow, or very slow."

After the common suite of logs is run, the penetration rate is matched with the logs as shown in Figure 70 and 71. The combination of data helps to interpret more rock properties than can be obtained from the individual logs or the drilling rate alone.

For instance, in Figure 71 the geophysical logs distinctly separate the clays from the sand intervals. But the fast drilling rate for the entire section indicates that the sands are loose (uncemented) and will probably require a screen for proper well development.

The use of logs as a guide to well construction is similar to that employed in oil wells. A collar locator is used to determine the position of casing, collars, and perforations in old wells and to locate depth markers where necessary.[229,424] The gamma-gamma log also has proven applicable for locating one string of casing outside another. The caliper is employed to measure hole diameter for cementing, gravel packing, and casing operations and also to check casing size and to locate collars and badly corroded sections of pipe. Both gamma-gamma and temperature logs locate cement tops outside the casing, but the cement-bond log, which is undergoing further research, has not yet been widely used in water well cementing. This log will be of considerable use in water well construction. Casing leaks and plugged screens may be located with tracers or with the flowmeter. Another important application of logs, previously described under " Aquifer and Fluid Characteristics" is the accurate measurement of depth and thickness of potential water-producing zones as a guide to screen length and setting.[234,538]

In much of the present water well logging, SP, single point resistivity, and often gamma logs are coordinated. They form a package that produces the most information at minimal cost.

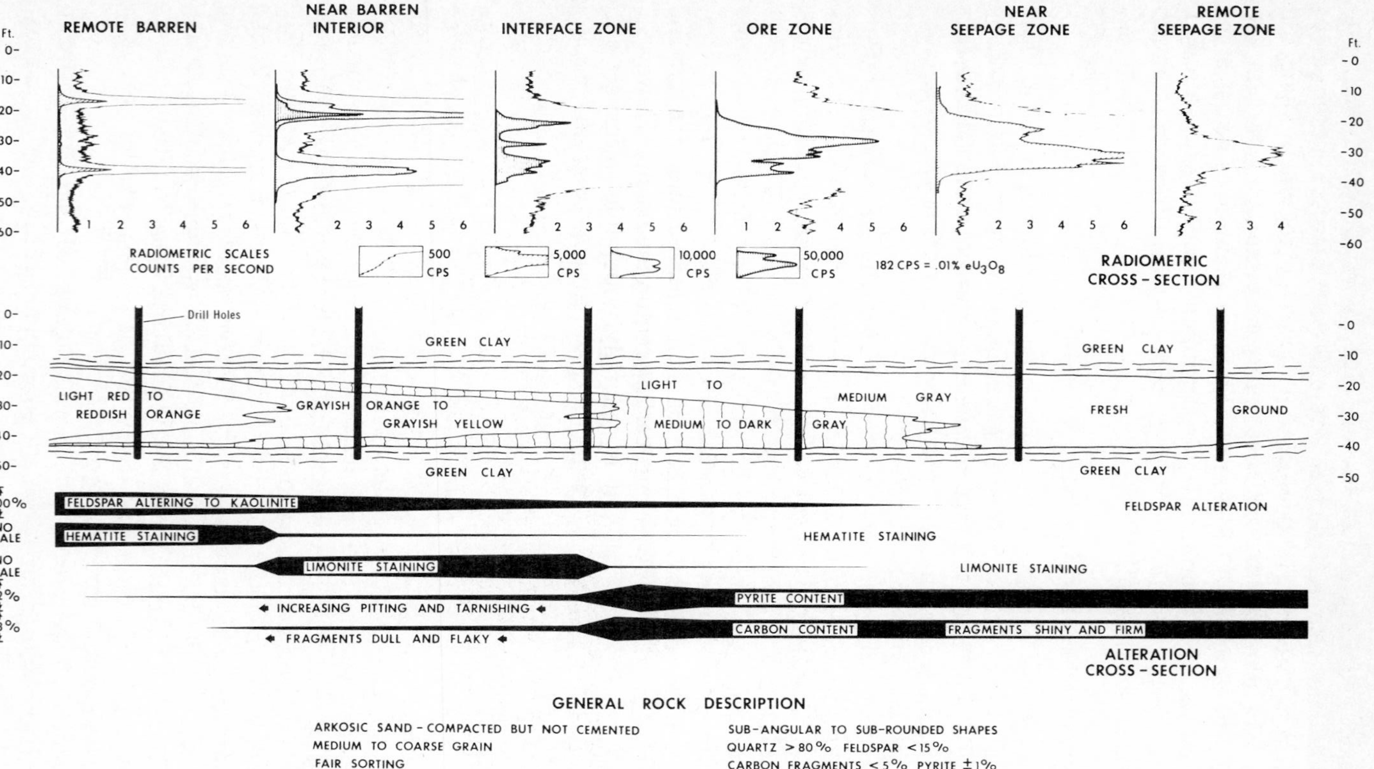

Figure 68[521] "Type" geochemical cell, Powder River Basin, Wyoming.

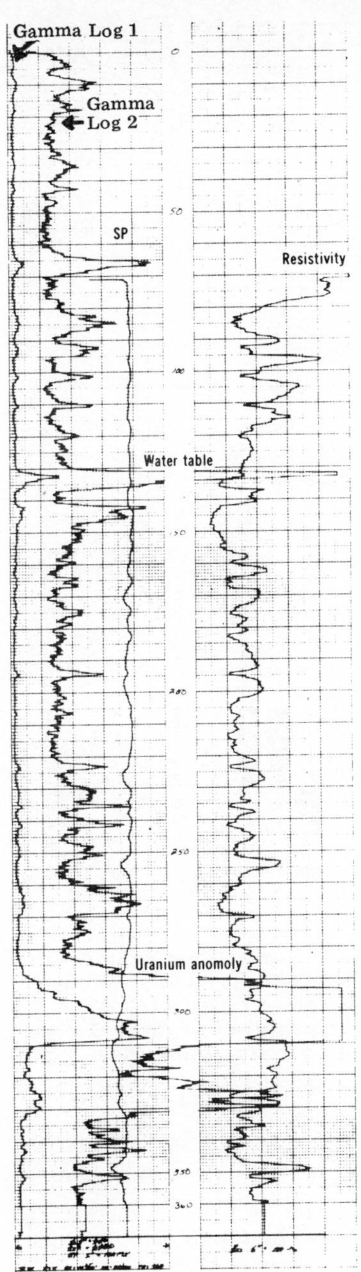

Figure 69[155] Electrical (SP and Resistivity) and Gamma Log Showing Encouraging Uranium Lead.

Most logging work is done on a contractual basis. Logging cost is determined by the expense of mileage, set-up, and the actual logging, either per foot or a total sum for the well. In some parts of the United States, wells 500 feet deep or less can be logged (including the above three variables) for a maximum cost of several hundred dollars.

A few state geological surveys and some federal projects also log water wells for their own information. Copies of these logs generally are available to the geologist or well contractor. Available oil-well logs can occasionally supply lithologic information for the upper 200-300 feet of stratigraphy.

Most states now have statutes designed to regulate well drilling and construction. Certain mandatory construction criteria are vitally important, e.g., grouting, casting requirements, etc. However, it is generally impossible to inspect all wells during the construction phase. Therefore, tools and techniques have been developed which allow regulatory agencies to inspect wells some time after the well has been completed and in operation for many years.

Such well logging tools and techniques can determine:

1. Whether the well has or has not been grouted.
   a. how effective the cement is bonded to the casing.
   b. location of grout.
2. Size and depth of casing.
   a. integrity of the casing.
   b. location of advanced corrosion
3. Well screen setting.
   a. screen position
   b. screen condition

Tools which some states and many well contractors are using as a regular procedure in certain types of wells include: the gamma log, the gamma-gamma log, the sonic log (cement bond log), the caliper log, the neutron and resistivity-SP logs.

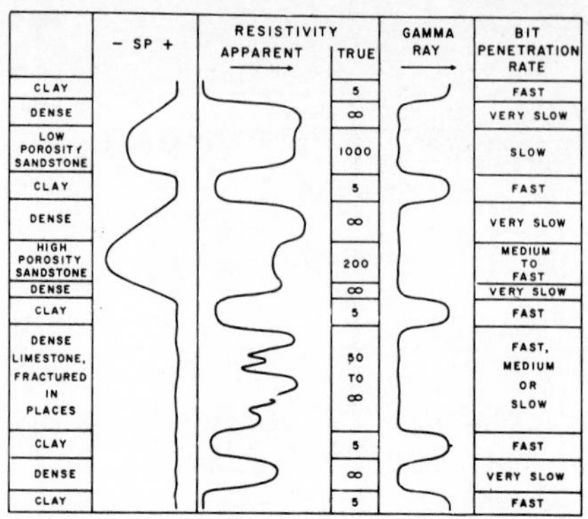

Figure 70[293] Artificial geophysical logs of consolidated rock layers separated by clay beds.

Caliper Log

The caliper log on the far right side of Figure 56, delineates various formations on the basis of the diameter of the borehole. Soft unconsolidated intervals have a tendency to cave and enlarge the hole, whereas consolidated, hard intervals do not cave and any hole penetrating such intervals will remain approximately the size of the drill bit. Calipers are highly sensitive borehole diameter-measuring devices. The caliper tool has independently operating measuring arms which ride the wall of a borehole and detect variations as small as 1/4 inch in the hold diameter. Because of the independent action of each arm, the diameter recorded is that of the circle described by the tips of the arms.

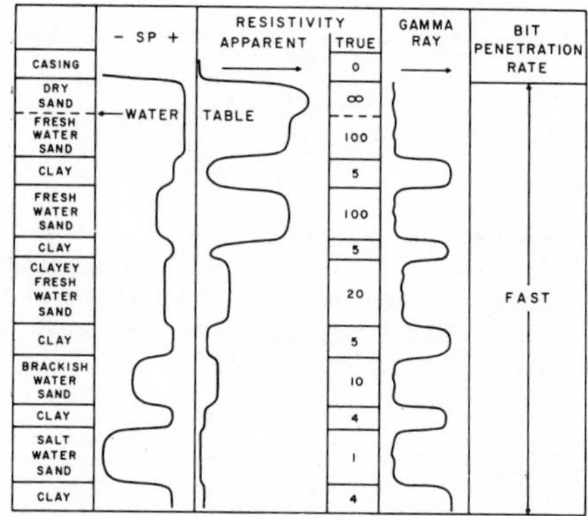

Figure 71[293] Artificial geophysical logs of a sand and clay sequence. Mineralized water in lower sands reduces their apparent resistivities. to that of clay, but gamma ray log distinguishes clay from sand layers.

Hole volumes can also be determined by a caliper log and casing variations can be measured.[216] The caliper is used to locate packer seats, washed out areas, and casing shoes. Calipers have a varying number of caliper arms, usually three to six. The reading, an average of all the arms, is recorded at the surface as a single curve.

Fluid Velocity and Conductivity Logs

Figure 72 shows that a combination of fluid velocity and electrical conductivity can be used to identify separate aquifers in an artesian

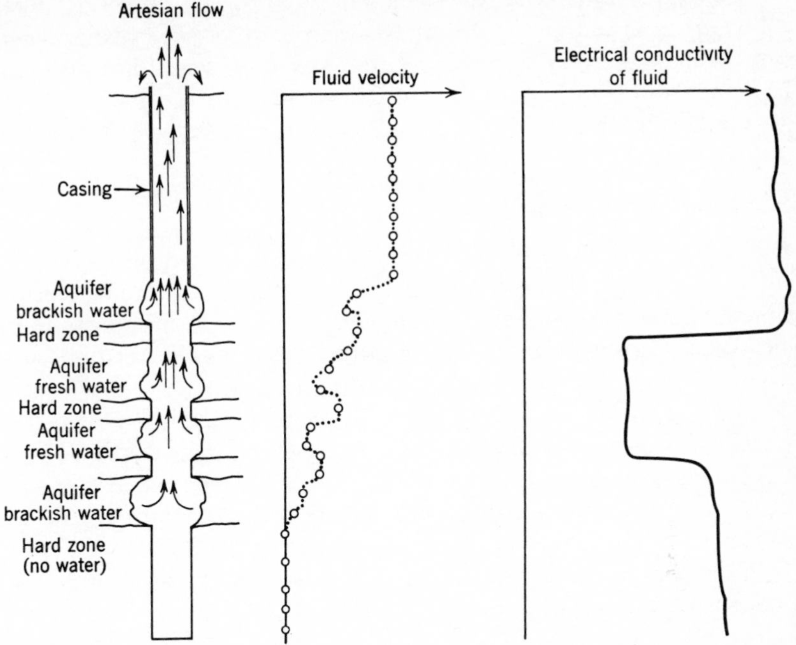

Figure 72[198] Fluid velocity and electrical conductivity logs of a hypothetical artesian well. Restrictions of the well diameter in the uncased portion cause local zones of maximum water velocity. Electrical conductivity log indicates that fresh water enters the well only through the two middle aquifers.

well. The velocity measurements indicate that four aquifers are present.

The small diameter of the borehole in the hard rock zones increases the flow rate in these zones, but each aquifer is still distinct. Only the middle two aquifers contain fresh water as shown by the low electrical conductivity. High mineral content of water usually gives a high conductivity reading.

With information obtained from this combination of fluid velocity and conductivity, the well can be completed so that it yields only fresh water. The upper and lower artesian zones must be cemented-off to prevent brackish water from mixing with the fresh water.

The impeller flow-meter is also a useful tool to determine the percentage of flow from various depths in an uncased, screened, or multi-aquifer well. The flow-meter is installed first just below the depth of the pump setting. The test pump is installed taking care to keep the flow-meter cable from being pinched. As the well is pumped at a constant rate, the flow meter is lowered, and the meter revolutions counted at regular intervals in an open well just above or below each screen in a multi-aquifer screened well. The resulting profile of water inflow to the well can be used to determine the lowest safe pumping level, hence the maximum safe yield of the well.

Temperature Log

One of the more useful and relatively inexpensive surveys is the temperature log.[110] If made within a short time after the well casing has been cemented, the temperature log can be used to determine the location of the cemented zone. Figure 73 shows that cement seals the annular space between the outer and inner casing.

Recording thermometers for very deep holes cost several thousand dollars, but a non-recording temperature probe, with up to 500 feet of cable, costs only a few hundred dollars. This type of probe is lowered to the desired depth, allowed to come to equilibrium, and then the temperature is read on the meter.

Bird[124] and Blankennagel[126] give details, other uses, and interpretations of the temperature log.[343,503]

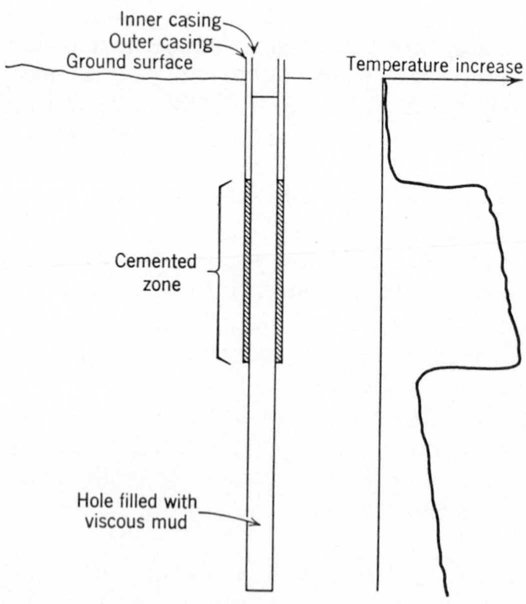

Figure 73[198] Schematic diagram showing heat given off by cement as it hardens.

Temperature logs are in common use in both drilling and production efforts. Because of the basically empirical development and semi-quantitative nature, the oil industry was slow to accept this instrument. Specific applications of temperature logs provide comparable information to other investigative logging tools. In each case, however, qualitative results depend strongly on equipment capability to well condition (before and during logging) and to down-hole logging methods. Six accepted temperature survey uses are:

(1) Cement tops.
(2) Gas-entry channels.
(3) Water-production logs.
(4) Fluid-injection profile (shut-in temperature profiles, water injection wells).
(5) Fracture-evaluation logs.
(6) Acid-evaluation logs.

These six topics provide sound information for most normal or routine temperature applications.[343]

In 1950, a temperature survey was routinely used to locate cement tops in the oil industry. At that time, the first concentrated effort to improve temperature logging tools, techniques, and applications was undertaken by men such as H. Guyod.[290] Electronic advancement, complemented by vivid imagination, resulted in the development of accurate and sensitive gradient and differential temperature surveys. This beginning led to many accepted logging techniques, such as temperature profiles in water-injection wells (FIL), fracture-evaluation logs (FEL), and acid-evaluation logs (AEL). These have been used as water production logs (for locating old cement tops, gas entries, and water channeling) and to determine proper cement-retarding agents in deep oil wells.

An accurate differential temperature survey technique has been used successfully to precisely locate a down-hole water loss in a waterflood injection well which was previously undetectable by any other means. The differential temperature log accurately measures fractional-temperature anomalies associated with fluid movements down hole. Besides being used to detect fluid communication down hole in water injection wells, the technique is being used to find tubing/casing leaks, gas communication, productive zones, lost circulation zones, oil and water production profiles, and gas/oil/water contacts in addition to other water well applications.[526]

Acoustic Log

Of the many logging tools available today, the acoustic log offers considerable promise in determining the density of cement behind casing, in analyzing fluid saturation, in locating cement tops, and in determining the presence and extent of formation fractures.[127,630,673]

The recent development of a superior acoustic log instrument has made possible the collection of porosity and lithology data from cased wells.[460] At the same time, a valuable evaluation of the cement condition can be made from the data recorded. Recent advancements in electronic technology as well as improvements in acoustic transducers made the design of this acoustic equipment feasible.[106]

In most situations, 40 to 50 percent cement bond is necessary for the accurate recording of formation travel times. Under certain fast velocity formation conditions, accurate interval transit-time logs have been obtained with an even smaller percentage of bonding.

The versatility of the acoustic log and the nuclear logging equipment makes it possible to simultaneously record the acoustic, gamma ray, neutron, caliper, and collar logs. These data are obtained routinely in both open and cased wells.

Shortly after the first acoustic log was introduced, a good acoustic signal was occasionally received which had traveled through the formation, even though the well was cased. It was soon realized that these results were associated with the zones where the cement was completely bonded to the casing and to the formation. In the early days of acoustic logging, runs were occasionally made over the section of the well that had already been cased. From such logs the formation signal could be tracked; for this signal correlated quite well with the open-hole log. These logs were of the single-receiver type and could not be used to determine porosity. As the dual-receiver log was being developed, many attempts were made to obtain a continuous and accurate acoustic log in the cased hole. Most of the attempts failed for one of the following reasons: first, the cement bonding to the casing and the sidewall was usually not continuous; second, the acoustic instruments did not have sufficient output signal nor did they receive signal amplification. Thus, the first compressional signals from the formation could not be logged successfully.

As the electro-acoustic transducers were improved, much knowledge was being gained through constant efforts to improve the acoustic tools. In a Soviet report, Titkov, *et al.*[617] suggest that the velocity of sound pulses indicates both the growth of strength and the deterioration of the cement structure. The report stresses that a coefficient of sound attenuation is a specific, sensitive index of microfracture development in cement. The use of this knowledge along with improved electronic technology resulted in the development of a superior acoustic tool — one capable of transmitting and receiving a good acoustic signal through the formation around the casing and cementing material. These tools have been used quite successfully in West Texas. Accurate acoustic-gamma-neutron logs have been obtained in new oil wells with high gas pressures or critical hole conditions which were cased. The log has also been used with very good results in old oil wells that were completed before the development of the acoustic porosity tool.[673]

In summary, acoustic amplitude logs are a record of changes in the amplitude of the received sonic signals, instead of the travel time of a pulse as in sonic velocity logging. Attenuation of sonic amplitude is used to interpret the quality of cement bonding in the annulus between the casing and the rock and to locate zones of secondary porosity, such as solution openings or fractures. The relationship of sonic attenuation and the lithological characteristics of aquifers which affect permeability has considerable potential in the ground water industry.[511]

NOVEL BOREHOLE GEOPHYSICS

Several logging and interpretive techniques, used experimentally in petroleum logging, have considerable potential for ground water investigations.[356] These sophisticated tools and others that will be developed in the future will be widely used in hydrology only when the equipment is perfected for routine use, and the size and the cost are reduced. Currently, the techniques with the greatest potential appear to be: (1) neutron life-time logging, (2) several types of spectral logging, (3) nuclear magnetic logging, and (4) computer analysis of logs.

Neutron life-time, or thermal neutron decay-time, logging employs a pulsed-neutron generator and a synchronously gated gamma detector which measures gamma and, under special conditions, neutron radiation as well. With this system the neutron or gamma field is measured at two or more preselected times after a pulse of neutrons is emitted by the generator. A consistent relationship has been demonstrated by the pulsed-neutron-log response between formation porosity and fluid content for a given formation. The system has several advantages over the continuous neutron logs. This system is much less affected by borehole parameters and produces more accurate quantitative results in cased holes than continuous neutron logging equipment; and the generator is occasionally capable of higher neutron fluxes than a radioactive source, yet requires no shielding when not in operation.

A similar type of portable neutron generator makes possible in-hole neutron activation analysis. In activation logging, a delay of a few seconds between neutron bombardment and recording gamma rays is common. Thus, the gamma rays result from radioactive decay of elements with short half-lives; these elements were created by absorption of a thermal neutron. Gamma radiation, with an energy characteristic of the element, is emitted when neutrons are thermalized and captured. A multichannel analyzer provides a record of the energy distribution of the gamma photons that reach the scintillation detector. Although borehole effects introduce many complications, this technique has been used for qualitative activation analysis in drill holes. Neutron generators, now available, are too large, expensive, and require excessive electrical power for widespread use in hydrology.

Multichannel-spectrum analyzers and energy-discriminating logging equipment also provide means for enhancing the quality of the information produced by neutron and gamma logs. Potassium 40 and the daughters of uranium and thorium, present in varying quantities in different rocks, and their relative abundances can generally be used to identify rock units. Spectral equipment has also been used to identify and track radioisotopes present in deep waste disposal materials and radioactive tracers in the ground water system.[355] Spectral analysis and energy discrimination improve the quality of neutron and gamma-gamma logs. Threshold and upper level discriminators can be adjusted so that most radiation below or above these energy levels is rejected.

Nuclear-magnetic logging is a relatively new technique that has not yet been applied in ground water investigations, and the logs are available only from a commercial logging company. A relationship between parameters derived from nuclear magnetic measurements and the permeability of a number of sandstone samples has recently been demonstrated. The logging device measures the nuclear magnetic properties of hydrogen in formation fluids and provides two values: the thermal-relaxation time and the free-fluid index which is related to effective porosity and to the amount of moveable fluid. Both the theory and equipment of nuclear magnetic logging are too complex to be adequately described in detail here. Briefly, the down-hole equipment consists of a coil, powered from the surface, which creates a strong magnetic field. The magnetic field causes spinning nuclei of hydrogen atoms to realign themselves within the artificial field. Procession of the spinning protons when the field is cut off induces a weak AC voltage in the coils of the inactive electromagnet. The most important factor in this logging system is that the system is almost uniquely responsive to hydrogen contained in low viscosity fluids. Furthermore, a detectable response from fluids mechanically bound in fine-grained detrital sediments are not included in the system. In contrast, the conventional neutron log responds to hydrogen, regardless of the chemical or physical form in which it occurs.

Computer interpretation and collation of geophysical logs are now available from commercial well logging companies, and computer interpretation of logs is also used in uranium ore reserve calculations by the U.S. Atomic Energy Commission. Some factors such as corrected total porosity, effective porosity, grain density, shaliness, and water resistivity may be calculated from logs digitally recorded on magnetic tape. These parameters may be played back as a continuous record, and the mode of graphical presentation can be selected to fit the problem. Existing geophysical logs can also be digitized and recorded on magnetic tape; thus, they are amenable to computer analysis. To justify programming and computer time for the interpretation and collation of gound water logs, a large volume of standardized logs must be available. The few experimental logs now available do not warrant the high expense, except in rare applications where costs are of secondary importance. However, one use of computers that may be justified now is the calculation of correlation coefficients as a means of testing the depth fit of core or sample values to geophysical logs.

The petroleum and mining industries will, in the future, continue to refine the various features of borehole geophysics. An increasingly aware ground water industry will take advantage of these developments and perhaps make special modifications which the oil industry can also use.[203, 291, 310, 352, 514, 515]

REPRESENTATIVE FORMATION SAMPLES (FLUID)

In the field, water quality data is generally obtained by bailing when drilling by the cable-tool method. Qualitative data can be acquired by this method on the aquifer characteristics, static heads of the various formations encountered, etc. Water quality data can be obtained by bailer samples as each formation, in turn, is opened to the bottom of the casing, and upper formations are cased-off.

Interpretations of the various geophysical logs can be verified by isolating specific zones in the drill holes with inflatable packers and spacers using drill stem testing methods.[126] Drill-stem tests provide the petroleum industry with information on various properties of subsurface formations.[140] As it is increasingly necessary to study the hydraulic and chemical properties of deep-lying rocks in order to understand the behavior of ground water, data on drill-stem tests made by the petroleum industry will become an important source of information which otherwise is unobtainable because of the high cost. Data from these tests made by methods currently in use are highly useful in ground water studies.

As utilization of ground water and ground water reservoirs increases, the importance of obtaining knowledge on regional ground water systems also increases. Understanding these systems often requires hydrologic data from depths far in excess of the present drilling limits of ground water development. In some areas, such information may be available from pumping tests made during the course of petroleum exploration.

The usual hydrologic test performed by the petroleum industry is the drill-stem test. The oil industry developed the drill-stem test as a method of sampling the fluid in a subsurface formation during the course of drilling operations. Most modern drill-stem tests, however, yield three types of hydrologic data: (1) a sample of the formation fluid, (2) the undisturbed formation pressure, and (3) a coefficient of permeability for the stratigraphic interval tested.

During the drill-stem test, the stratigraphic interval of interest is isolated in the hole by the use of inflatable packers attached to the drill string and is allowed to yield fluid into the drill pipe under the influence of the formation head (see Figure 74).

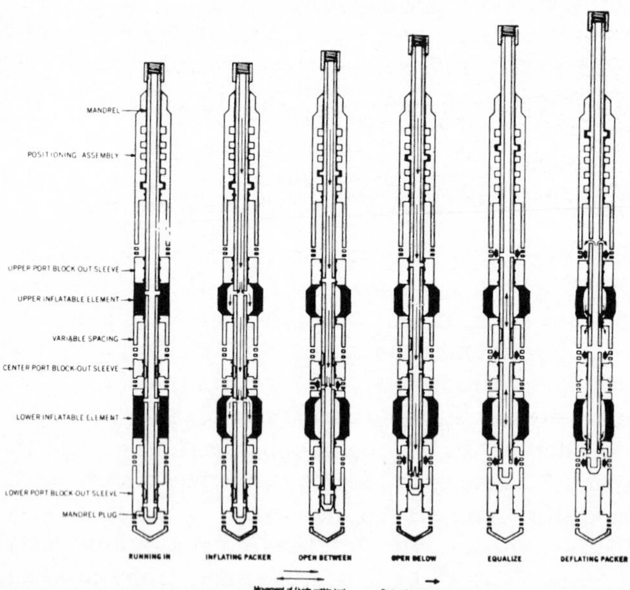

Figure 74[126] Operation of inflatable straddle packer used in open hole or casing.

In the usual relatively shallow test, the drill pipe initially is completely empty and open to atmospheric pressure. By opening the tester valve in the drill string, the operator allows the formation to yield fluid into the drill column for a short period of time. The period usually ranges from 30 minutes to two hours. Following the production period, the tester valve is closed, shutting in the formation and causing the formation pressure to recover. Pressures are recorded throughout the test by a Bourdon-tube, pressure-recording gage contained within the drill string near the bottom. Bredehoeft[140] subdivides the test into five phases:

(1) The initial phase of the test involves lowering the drill string into place in the hole. The pressure guage records the increasing pressure of the drilling fluid as the pipe is lowered into the hole.

(2) Once the perforated section of pipe is in the desired position, the packer is expanded or "set" isolating the interval of interest as well as the gauge from the column of mud above the packer. The tester valve momentarily allows the formation to produce for a few minutes, reducing the pressure in the isolated zone to very nearly atmospheric, then closes. The pressure builds up again to some point which asymptotically approaches the undisturbed formation pressure. The initial shut-in period commonly ranges from 15 to 45 minutes.

(3) After the initial production and shut-in period, the tester valve is again opened, and the formation fluid flows into the drill pipe. Within the isolated interval, the pressure reduces to essentially atmospheric, but, as the column of produced fluid builds up in the drill pipe, the pressure increases. Commonly this production period lasts from 30 minutes to two hours.

(4) After the production period, the tester valve is again closed isolating the formation and allowing the pressure to recover. During this final shut-in (recovery) period, the pressure guage records the pressure build-up in the isolated section of the hole. The final shut-in time is often equal to the time of production.

(5) After the final shut-in period, the packer is unseated, and the pressure returns to that produced by the column of drilling mud in the hole, the so-called hydrostatic mud pressure. The drill pipe is then pulled out of the hole. The pressure recording gage records the decrease in pressure as the pipe is removed from the hole. The column of produced formation fluid remains in the drill pipe until the pipe reaches the surface. This affords a sample of fluid produced from the formation tested. From the measurement of the length of drill pipe filled with fluid, the volume produced can be calculated.

A wide variation is possible in the combination of operations as well as in the components that make up the test string. Tests in recent years record both an initial shut-in pressure and a final shut-in pressure as earlier described; whereas, most older tests were run without recording the initial shut-in pressure.

Drilling Fluids and Drill Stem Testing

Drilling mud is designed to form an impervious layer along the wall of the hole. One of the principal functions is to reduce the loss of fluids from the hole. The filter cake built up on the well-bore face may penetrate the formation for short distances, thereby reducing the formation permeability around the borehole. This zone of reduced permeability affects the rate of pressure build-up in the drill-stem test in the same manner as a partly plugged screen in a water well. As the time of pressure recovery increases, the effect of the reduced permeability zone near the well bore diminishes which is largely a function of the formation permeability beyond the mud-affected zone.[10]

The reduction of permeability by the mud, if large, may greatly reduce the quantity of fluid produced during a short-term production test. The concept of "skin effect" (damage produced by the mud) was introduced by Van Everdingen[634] in order to evaluate the effect of the drilling mud in reducing production.

Dolan, *et al.*[204] introduced another expression, the "damage ratio" which relates the quantity of fluid that might have been produced if no mud damage had occurred to the quantity of fluid actually produced. Both the skin effect and the damage ratio are discussed theoretically by Dolan.

The problem of mud damage to the producing formation is very similar to the problem of well losses in inefficient water wells.[556] The productivity index, defined later in Chapter 13 relates quantity of flow to pressure drop at the well and is analogous to the specific capacity (yield per unit drawdown) used by ground water investigators.

The usual units used in drill-stem tests are, of course, biased for the petroleum industry. Table 19 lists these units along with the corresponding units generally used in ground water studies. Petroleum production rates are usually expressed in barrels per day; permeability, in millidarcies, and depth and formation thickness, in feet.

Bredehoeft[140] also effectively summarizes some of the known limitations of the drill-stem test.

> The most obvious difficulty in the drill-stem test is that of accurately recording bottom-hole pressures over the range involved. However, with

the present Bourdon tube recording device, calibrated frequently and micrometer chart reader, the guage error can be reduced to approximately ±1 to ±2 psi at pressures as high as 4,000 to 5,000 psi.

Difficulties involved arise from the fact that actual field conditions only approximately satisfy the assumptions on which the mathematical model is based. Factors such as: (1) the increased pressure produced in a formation by the column of mud in the hole during the course of drilling, (2) an imperfect seat of one of the packers, and (3) uncertainties in interpretation due to the short duration of the test, partial penetration of the formation tested, etc., make quantitative interpretation increasingly difficult.

Experience with tests in the Big Horn Basin suggests that a reasonable estimate of the error involved in obtaining the true undisturbed formation pressure is often on the order of ± 25 psi under the best of conditions.

TABLE 19[556]

COMPARISON OF UNITS USED IN PETROLEUM INDUSTRY WITH THOSE NORMALLY USED IN GROUND WATER STUDIES

1 barrel	42 gallon 9,702 cubic inches 5.615 cubic feet
1 darcy	18.24 gallons per day per square foot (60°F)
1 millidarcy (1×10^{-3} darcy)	0.01824 gallons per day per square foot

Packers have been used for many years in testing and completing oil wells but have not been used extensively in testing water wells until recently. The use of inflatable straddle packers are now feasible for testing various types of aquifers.[126,370,608]

Another method for ground water sampling has been developed by McMillion and Keeley.[415] This consists of portable pumping equipment, capable of sampling to depths of 300 feet, with pumping rates ranging between seven and 14 gpm, with rate variation depending upon sampling depth. The equipment size is convenient and easy to operate since only one line has to be handled during its operation. A recent modification of this method incorporates an inflatable packer. The packers are inflated by the pressure produced by the pump and deflated when pumping ceases. This allows the sampling of isolated intervals. Other tools, recently developed by the USGS for use in measuring down-hole pressures, fluids, and gases are described by Fournier and Truesdell[225] and Cherry.[167]

REPRESENTATIVE FORMATION SAMPLES (LITHOLOGY)

Much of the success of a completed well depends upon the degree of care exercised in obtaining formation samples. Generally, samples are taken as soon as a water-bearing formation is encountered in the drilling operation. However, in areas where the geology is not well known, samples should be taken and stored at every lithologic change. Records on drill samples above potential aquifers have become very valuable in many instances in later exploration for ground water and even other strata-bound economic minerals, e.g., sedimentary uranium, silver, etc.[56,155,318,417,469,594]

Cable-tool samples are usually representative of the interval drilled between bailing operations, although some contamination of the cuttings from holes in consolidated formations are knocked from above the uncased parts of the hole by the cable. The measurement of depth on the drilling cable is relatively easy, but correction at intervals by steel-line measurements is necessary because of the stretching of the cable. The cuttings are relatively fine as a result of the crushing action of the bit; and, in some wells, the cuttings are ground to powder by the continued use of a dull bit or by drilling in a hole full of water. The fineness of the cuttings is a distinct disadvantage in determining both the lithology and fossil content. Another disadvantage in logging cable-tool samples is that electric-logs cannot be run to check and supplement the sample interpretation because most holes drilled in unconsolidated sediments by cable tools are cased, although radioactivity and acoustic logs may be run.

Cable tool samples from granular unconsolidated formations are usually excellent. In some instances, casing may be inserted by bailing the loose formation without having to drill. The samples are obtained in a relatively unchanged state. Samples of unconsolidated formations are highly representative if taken with a "sand-pump" type bailer; whereas, a "dart-valve" bailer, used primarily for consolidated rock drilling, usually requires additional drilling and breaking of gravels, hence the samples are less useful especially for sieve analysis purposes. However, when fine-grained, saturated sands and silts are involved, care must be exercised in bailing to avoid heaving of formation into the hole.

Direct-circulation-rotary-tool samples are usually collected at regular intervals of five to ten feet. This uniform system facilitates the plotting of the sample log and aids in the detection of omissions.

The rock fragments in the samples normally range in maximum dimension from 1/16 to 1/2 inch, with most fragments larger than 1/4 inch, although excessive bit weight and drilling speed can pulverize the samples to dust in some cases which makes them of little value for interpretive purposes. Because of their relatively large size, rotary-tool cuttings usually can be examined rapidly under low magnification and generally can provide some whole specimens of microfossils or microstructures in the rocks. Ordinarily, numerous cores are taken with rotary tools; these may supply important lithologic details and even some megafossils.

Rotary samples usually contain some cavings and fragments recirculated by the mud pump. The proportion of cavings in the samples may be large when the viscosity and circulation of the drilling mud have not been properly controlled. Differential settling rates of the heavy and light fragments in the mud fluid also mix cuttings from different beds. Because the collection of samples at the surface lags behind the actual cutting of the given bed at depth, the samples usually represent a depth somewhat less than that recorded on the sample sack.[319] This lag may amount to 20 or more feet in a 300-foot hole; and can be eliminated by taking the samples after circulation without preceeding with drilling so that a sufficient time interval permits the latest cuttings to reach the surface. Sample depths can be corrected to some extent by timing a round trip of some marker material in the hole (like rice) and by applying a correction factor to the samples. The most accurate bed-for-bed correction can be made with an electric or radioactivity log while the samples are logged in the laboratory. An easily identifiable shale marker in the log can be matched with the sample, thereby defining the lag in the interval. In general, as the penetration rate increases, the lag increases unless circulation is continued while adding drill pipe which reduces the lag.

Direct circulation rotary rig samples from unconsolidated formations may be poor, even when the hole is carefully logged and drilled with clear water or self-degradable, organic-based fluid, particulary when relatively thin beds of sand, gravel, and clay are being drilled. Soft clays are mixed and become dispersed in the drilling fluid. At times, it is impossible to differentiate from the cuttings between clayey sand, sandy clay, and intercollated beds of relatively pure clay and sand. Drive sampling or coring is recommended in such formations.

Samples obtained during air rotary drilling are generally superior to samples collected during much drilling. Air drilling for mineral exploration is now widespread. However, formations containing soft, finely divided minerals of potential economic value can be difficult to evaluate since dilution of the sample can occur as it travels up the hole. Casing is usually required to reduce such dilution and mixing of the sample with soft, fine grained material up the hole.

Collection of reverse circulation rotary rig samples requires some experience, practice, and special equipment, but a skilled operator can obtain excellent samples. The high turbulent fluid velocities up the drill stem, usually in excess of 400 feet per minute, result in little or no separation of the fines with a minimum lag time. The bit tends to loosen materials rather than grind them up, and the cuttings are immediately drawn into the bit and delivered at the surface. This method is particularly suited to shallow mineral exploration where sample quality is critical. Because of the high velocity and large volume of water involved, equipment such as a core sampler and a 55-gallon drum sample catcher is generally required. Samples caught in buckets or screens are generally useless.

Cores

Coring, of course, is a method of sampling as well as a method of drilling. This can be done with either cable or rotary equipment but is more commonly done with rotary drilling. When coring is undertaken as an accurate sampling procedure rather than a drilling method, the individual cores generally are less than four inches in diameter and ten feet in length, although short cores as large as 30 inches in diameter have been used in some detailed engineering studies. Wells have been drilled by coring with the rotary method as the most economical means in areas underlain by especially hard rocks. Usually diamond bits are used, and the core may be four to eight inches in diameter and as long as 50 feet. When complete cores are recovered and several hundred feet of cores are laid out in sequence, the opportunities for detailed study are generally superior to those with drill cuttings or scattered outcrops. The outcrop characteristics, except weathering, are apparent, and megafossils may be preserved intact; the location, sequence, and thickness are accurately known. The large cores are sometimes split with a diamond saw and preserved intact. More often, however, the core is described at the well site and then broken into sets of representative chip samples which can be re-examined in the laboratory at any time if a particular future operation merits the effort.

Double-tube core barrels are the type in general use today. The core-catcher assembly consists of an upper catcher with spring fingers and a lower catcher with spring-actuated pivoted attachment held in the correct position in the bit by a hardened adapter sleeve. The same assembly is used in both the hard and soft formation cutter heads and is placed as close to the bottom as possible.

The inner barrel is vented to the outside of the drill stem through the strainer cap, ball seat, stuffing box, and a vent "spider" passage. This arrangement allows free entrance of the core, independent of pump pressure, by allowing fluid entrapped above the core to exhaust into a low-pressure area.

The core-barrel plug is a refinement to prevent the catcher and the inner barrel from being fouled with cavings or debris encountered in the hole. During normal operation, the plug strikes bottom first, and as the weight of the drill stem is applied, the rivets shear, and the plug floats freely above the core, up into the core tube.

Special conditions in certain localities demand equipment somewhat different from the types previously described. In some instances, where holes cave and good mud is not available, a large portion of the coring time is spent getting to the bottom. In these cases, some provision must be made to pump through the inner barrel to enable the operator to wash the bottom more forcibly. Several devices are employed for this purpose, the most common is the "drop ball" type of inner barrel vent. Mud is pumped through the inner barrel as the tools are washed through the cuttings and cavings to the bottom. The ball is then allowed to drop into a seat diverting the mud to the bit head through the regular fluid passages. One disadvantage is that the fluid inside the barrel must be forced out against the pump pressure as the core enters the barrel. Another objection to this method is the damage done to core catchers, through which formation material and mud are forced while reaching bottom.

Diamond core heads have been developed in the past few years and are being used to core some of the very hard and tightly cemented formations, when the cost of coring with roller-cutter bits is high. Although the initial cost of the diamond bit is roughly 20 times that of the roller-cutter, hard-formation cone bit, the longer life, faster drilling speed, and reduction in the number of round-trips required in drilling these types of formations result in lower costs per foot of hole cored. The diamond core head consists of a threaded steel body to which is fastened a matrix having a series of diamonds set in the surface to form the cutting elements. The diamond pattern extends across the face and covers a portion of the outside and inside diameter of the matrix to provide cutting elements for maintaining the size of the hole and the core. A temperature of approximately 2,500°F. will cause permanent damage to the diamonds.[138,405]

Since the diamond core bits are capable of drilling (under ideal coring conditions) several hundred feet before the bit head is dulled beyond usefulness, they are usually run on longer core barrels than are used for the conventional soft- or hard-formation head. The longer barrels will accommodate longer cores, thus reducing the number of round trips required for completion. For shallow drilling operations, the conventional core barrel is usually ten to 20-feet long. Since the size of the cuttings are very small or powdery, less fluid circulation is required with a diamond bit core head than with conventional types.

The most important part of a core barrel is the core catcher. The first core barrels were used without a core catcher, but the results were very poor. The oldest of the three popular core catchers is the spring-type with spring fingers that bear against the core and start a bridge to hold the core when the bit is lifted. This type has its most valuable application in very soft, unconsolidated materials. The second type of core catcher is the spring-actuated dog type. A core catcher of this design has greater strength and is adaptable to a wider range of formations. A third catcher is a combination of the second type, with the spring-actuated dogs and a series of slips adopted to a wedge in a tapered bowl around the core when the bit is lifted. The slips tighten when the core resists movement, and the core is broken off at the slips or below them. There are many other types of core catchers on the market that utilize mechanical means to operate the core-retaining device. Very few offer advantages great enough to warrant the increased cost and additional care required.

There are several designs of retractable core bits from which the inner barrel and core can be retrieved at intervals through the drill stem. Roundtrips are necessary only when the bit is dull.[337]

There are numerous formations which can be cored successfully with any core bit, but soft sand, fractured limestone and dolomite, conglomerate and interbedded chert, and siltstone formations are the most difficult to core.

The most important phase of core drilling is the operation of the tool. For soft to medium-hard sands and shales and for formations ordinarily drilled with fishtail or drag bits, the use of a blade-type cutter head is preferred. For formations requiring a rock bit, a roller-type cutter head is used. There are, however, some formations, such as conglomerate and some hard shale, ordinarily drilled with rock bits, which can be cut more successfully with drag-type core heads than with hard-formation core heads. If the formation is limestone, dolomite, anhydrite, or a harder formation, blade-type heads will not cut enough core to warrant use. Diamond core bits are used in sandstone, siltstone and even shale and marl.[363]

A rotary speed of 20 rpm to 40 rpm with a relatively light but increasing weight is generally applied in coring. Pump speed is adjusted according to the formation being cored, i.e., hard or firm formations and sticky strata need almost as much circulating fluid as that required by a drilling bit of the same size. Soft, unconsolidated formations are cored successfully when the pump is only run enough to prevent sticking. It is very important to increase the weight gradually and continuously.[138,405] The mining industry as well as the oil industry has pressing needs for maximum efficiencies in coring operations.[337]

Poor core recovery may be due to a number of causes, the more common of which are summarized by Brantly:[138]

(1) Part of the bit will be plugged if there is formation trash in the mud, causing off-center action of the bit head, which will make a small-diameter core or cause all of the core to be lost.

(2) If too much weight is applied, the core will be burned or plugged, if it is soft; broken up, if it is brittle or hard.

(3) A crooked inner tube will prevent easy entrance of the core, and a crooked outer barrel will cause eccentric action on bottom. Crooked drill stem near the bit will have the same effect.

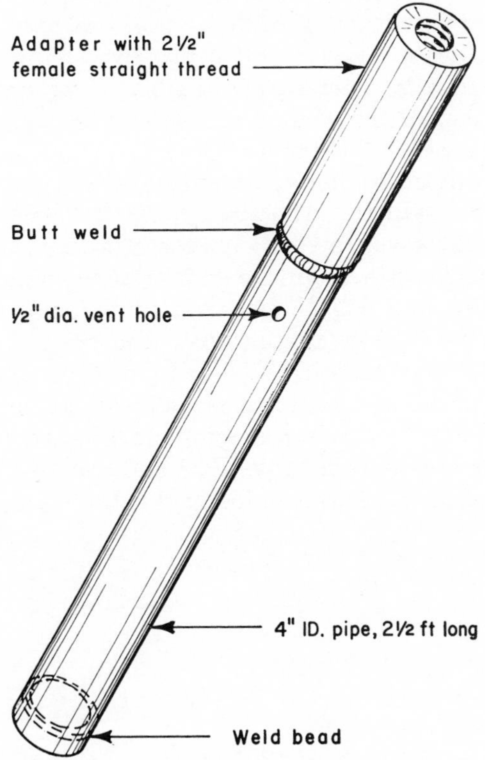

Figure 75[52] This sketch shows a small diameter core barrel suitable for coring inside 6-in. pipe.

(4) Poorly dressed, soft-formation cutter heads: blades of different lengths; core teeth dressed to make over-sized core water courses not rebuilt properly; lack of proper hard-facing to make vital parts of the bit last.

(5) Insufficient weight or intermittent feeding of weight will cause the bit to drill off center at times, thus allowing soft cores to be washed away or brittle cores to be broken up.

(6) Junk in the hole usually prevents good core recovery. Even a small amount can become lodged in the cutter head and destroy the core.

(7) Length of core cut may be excessive for the size of the bit used and the formation being cut. In broken or brittle

formations, as much core will often be obtained from a cut of three feet as can be recovered from a 20-foot cut. Cores cut from this type of formation shatter and wedge themselves so that no further core enters the barrel.

(8) Formation changes in the course of cutting a core. Hard core taken above soft strata will frequently offer sufficient resistance to prevent additional core from entering the core tube.

(9) Connections made in the course of cutting the core are sometimes harmful. Each time the core bit is raised from bottom, the core catcher should act. Thus, when returning to bottom, the catcher has caused a bridge condition which may prevent additional core entering. Also, the catcher becomes "second-hand" and its condition is doubtful.

(10) Rapid rotation of a core bit is usually harmful. The bit operates more roughly, tending to break up the core. Hard-formation heads have small-diameter cutters which turn several revolutions for each rotation of the drill stem. Rapid rotation causes bearing wear and short cutter life.

Interbedded, unconsolidated formations of soft clays, sand, and gravel are often difficult to core, although some success has been reported.[390,573] A split-core drive sampler will often furnish excellent samples from such materials. There are a number of coring devices available, ranging in length from 1½ feet to five feet, and in diameter from 1½ inches to six inches.[653] Such coring devices may be driven or forced into the formation by hydraulic pressure from the rotary rig or by "drive-coring" with the cable-tool rig.[52,390,573,653] (See Figure 75.) These tools, however, are not always effective in coarse sediments. This equipment has been successfully used with reverse circulation rotary rigs by installing the sampler and drive pipe through the column pipe to the bottom of the hole. This type of sampler was developed for soil investigations, such as foundation or borrow pit exploration, but it has a definite capability for use in ground water and mineral drilling.[348,372] Side-wall sampling, however, is quickly replacing conventional coring in the oilfield and will find general application in the ground water industry in the near future.[338]

10
Well Hydraulics

Well pump equipment will not be reviewed since commercial literature is readily available from the numerous pump manufacturers. Any advancement in pump design will attain national attention via sales promotion. Furthermore, a review of available pump types has been published by the Water Systems Council[50] — an industry-sponsored organization composed of the major pump manufacturers and related service companies. Furthermore, Hicks and Edwards[312] have published an excellent text on practical pump application engineering. (See Table 20 for general pump characteristics.) In addition, the AWWA has published a "standard" for vertical turbine pumps.[30]

EARLY DEVELOPMENT

Water well completion and development technology relating to well hydraulics in general is relatively advanced.[16,20,23,25,26,34,36,42,46,105,141,150,334,365,393,451,452,507,603,651,657] Most of the literature on water well hydraulics considers the relationship of the well to an ideal clastic aquifer and to laminar flow. Relatively little attention has been given to the well structure itself. Lohman,[393] Ahrens,[5,6,7] Reinke, et al.,[525] give excellent general reviews of well hydraulics and related construction features. The ground water literature is replete with statements to the following:[5,507]

TABLE 20[50]

CHARACTERISTICS OF PUMPS USED IN WATER SUPPLY SYSTEMS.

Source: **Manual of Individual Water Supply Systems,** USDHEW.

Type of pump	Practical suction lift[1]	Usual well-pumping depth	Usual pressure heads	Advantages	Disadvantages	Remarks
Reciprocating:						
1. Shallow well	22–25 ft.	22–25 ft.	100–200 ft.	1. Positive action. 2. Discharge against variable heads. 3. Pumps water containing sand and silt. 4. Especially adapted to low capacity and high lifts.	1. Pulsating discharge. 2. Subject to vibration and noise. 3. Maintenance cost may be high. 4. May cause destructive pressure if operated against closed valve.	1. Best suited for capacities of 5–25 gpm against moderate to high heads. 2. Adaptable to hand operation. 3. Can be installed in very small diameter wells (2" casing). 4. Pump must be set directly over well (deep well only).
2. Deep well	22–25 ft.	Up to 600 ft.	Up to 600 ft. above cylinder.			
Centrifugal:						
1. Shallow well						
a. straight centrifugal (single stage)	20 ft. max.	10–20 ft.	100–150 ft.	1. Smooth, even flow. 2. Pumps water containing sand and silt. 3. Pressure on system is even and free from shock. 4. Low-starting torque. 5. Usually reliable and good service life.	1. Loses prime easily. 2. Efficiency depends on operating under design heads and speed.	1. Very efficient pump for capacities above 50 gpm and heads up to about 150 ft.
b. Regenerative vane turbine type (single impeller)	28 ft. max.	28 ft.	100–200 ft.	1. Same as straight centrifugal except not suitable for pumping water containing sand or silt. 2. They are self-priming.	1. Same as straight centrifugal except maintains priming easily.	1. Reduction in pressure with increased capacity not as severe as straight centrifugal.
2. Deep well						
a. Vertical line shaft turbine (multistage)	Impellers submerged.	50–300 ft.	100–800 ft.	1. Same as shallow well turbine.	1. Efficiency depends on operating under design head and speed. 2. Requires straight well large enough for turbine bowls and housing. 3. Lubrication and alignment of shaft critical. 4. Abrasion from sand.	
b. Submersible turbine (multistage)	Pump and motor submerged.	50–400 ft.	50–900 ft.	1. Same as shallow well turbine. 2. Easy to frost-proof installation. 3. Short pump shaft to motor.	1. Repair to motor or pump requires pulling from well. 2. Sealing of electrical equipment from water vapor critical. 3. Abrasion from sand.	1. Difficulty with sealing has caused uncertainty as to service life to date.
Jet:						
1. Shallow well	15–20 ft. below ejector.	Up to 15–20 ft. below ejector.	80–150 ft.	1. High capacity at low heads. 2. Simple in operation. 3. Does not have to be installed over the well. 4. No moving parts in the well.	1. Capacity reduces as lift increases. 2. Air in suction or return line will stop pumping.	
2. Deep well	15–20 ft. below ejector.	25–120 ft. 200 ft. max.	80–150 ft.	1. Same as shallow well jet.	1. Same as shallow well jet.	1. The amount of water returned to ejector increases with increased lift—50% of total water pumped at 50 ft. lift and 75% at 100 ft. lift.
Rotary:						
1. Shallow well (gear type)	22 ft.	22 ft.	50–250 ft.	1. Positive action. 2. Discharge constant under varible heads. 3. Efficient operation.	1. Subject to rapid wear if water contains sand or silt. 2. Wear of gears reduces efficiency.	
2. Deep well (helical rotary type).	Usually submerged.	50–500 ft.	100–500 ft.	1. Same as shallow well rotary. 2. Only one moving pump device in well.	1. Same as shallow well rotary except no gear wear.	1. A cutless rubber stator increases life of pump. Flexible drive coupling has been weak point in pump. Best adapted for low capacity and high heads.

[1] Practical suction lift at sea level. Reduce lift 1 foot for each 1,000 ft. above sea level.

For maximum efficiency, a well should fully penetrate and be screened throughout the saturated thickness of the aquifer; specific capacity of a well in an ideal artesian aquifer is proportional to the percentage of open hole; in a free aquifer the specific capacity increases to a drawdown of about 65 percent of the saturated thickness; additional drawdown to possibly 90 to 95 percent increases total yield, but specific capacity decreases rapidly as the drawdown increases above 65 percent; and doubling the diameter of a well theoretically will increase the specific capacity between seven and 11 percent in the same ideal aquifer under conditions of laminar flow.

While all these statements are true for an ideal aquifer, they are inadequate as a basis for well design in commonly encountered aquifers.

WELL LOSS

Zangar,[671] when with the Bureau of Reclamation, found by electric-analogy investigations that the effect of a screen on a partially penetrating recharge well under laminar flow conditions would reduce the effective diameter of the well according to this equation:

EQUATION 51

$$r_e = r \frac{Ap}{Ac}$$

Where: r_e = Effective radius
r = True radius of the screen
Ap = Area of perforations
Ac = Cylinder wall area

Corey[180] demonstrates that there is a critical percentage of screen open area (about 60 percent), beyond which well loss is no longer a function of the open area. Loss decreases rapidly with an increase in open area (about 15 percent). The curve becomes progressively flatter above 15 percent.

Peterson, *et al.*,[509] developed an equation based on laboratory tests to estimate the length of screen beyond which no additional significant well-head loss would occur and concluded that the major flow through the screen occurred within this length, regardless of the total length.

Vaadia and Scott[631] took the above approach on different types of screens. The conclusion was that for a given length and diameter of screen, within limits, the open area is the major factor in well-head loss; and for gravel-packed wells, the head loss decreased with increased uniformity, roundness, and grain size of the pack material relative to a given slot size. A procedure was devised for estimating head loss based on screen diameter, slot type, and patterns for screens commonly used in California.

Wen[654] and Soliman[582] investigated the flow into a well in an ideal aquifer and concluded that the flow increases progressively toward the discharge point of a screen. These conclusions suggest the possibility that in small-diameter wells, savings are possible by decreasing the diameter of the screen without seriously decreasing well efficiency. Wen also stressed that there is a concentrated flow near the bottom of a partially penetrating well in the ideal aquifer. However, in a highly anisotropic aquifer, Ahrens[5] suggests that the bottom concentration of flow is of minor importance.

Rorabaugh[534] concluded that head loss through a screen is only a small percentage of the total loss, most of which occurs within the aquifer. The apparently large increase in efficiency results from an increase in well screen diameters, and additional study of this factor is needed.

Perhaps the most critical factor (and one difficult to duplicate by laboratory tests) is the ability to develop the formation surrounding the screen. According to Rorabaugh, head loss through the screen is not generally considered a major factor (pers. com., J. S. Fryberger), but if the permeability of the formation can be improved appreciably for a significant distance into the formation, the well efficiency can also be improved. Maximizing the open area of the screen may be more important from the point of view of being able to develop the formation more fully than from the position expressed by the conventional entrance velocity argument.

A major screen manufacturer[24,46,67,69,75,78] discusses the advantages of interspersing short lengths of screen with blank casing on the screen assembly in a thick aquifer in order to reduce formation loss, among other features of well development. Others[78,483] treat related features of the shallow-well point systems and their use.

The investigations and analyses on which much of the above summary of well hydraulics is based, in almost all cases, have considered the ideal isotropic, homogeneous aquifer and laminar flow. Tests, when made, used steady flows in a limited range with apparatus having fixed geometry which, in most cases, differed considerably from that of production wells.

OPTIMIZATION OF SPECIFIC CAPACITIES

Ideally, 100-percent open hole is desired for detailed evaluation of production wells in thick aquifers. This is, however, often not feasible from the standpoint of cost and desired yield. Under some circumstances, tests to determine the gross transmissibility and storage coefficient of an aquifer are usually omitted for economic reasons. In either case, a compromise should be made.

Stoner[596,597] considered the performance of a number of partially penetrating wells in a relatively thick and uniform aquifer in Pakistan and derived an equation generally applicable to the local conditions. However, the opportunity and conditions for such study and application are seldom encountered.

In an ideal aquifer, a test using partially penetrating observation holes and a pumping well will give a low value of transmissibility and a weakly determined, but possibly high value, of storitivity if the observation wells are too close to the pumping well. Tests using partially penetrating observation holes and a pumping well ina strongly anisotropic aquifer will give an approximate transmissibility value applicable to the portion of the aquifer tested.[596]

Ahrens[5] suggests that the theoretical increase in specific capacity with increased well diameter, in many artesian wells in consolidated sediments, appears to be in the correct range; but, the increased capacity in a screened well may amount to as much as 15 to 25 percent, with respective increases in well diameter, a result also noted by Rorabaugh.[534] The increase may be due to the effects of the open area as determined by (1) premises of Zangar[671] and Corey,[180] (2) reduced turbulence in the well, and (3) classical laminar flow considerations.

The DL/D>6 ratio of Petersen, *et al.*[509] (where C is a constant determined by screen slot geometry, L is the length of the screen, and D is the diameter of the screen) often cannot be realized in a thin aquifer because of yield and entrance velocity limitations. But if the screen diameter is increased to give a suitable open area for the desired entrance velocity, an efficient well will still be obtained, although the ratio may be less than six.

Desirable entrance velocity is usually considered to be 0.1 to 0.25 foot per second, based on the open area of the screen. This is an average velocity and varies from quite low at the bottom of a screen to many times as great at the discharge end or point in an ideal aquifer or opposite more permeable zones in a naturally developed aquifer.

The percentage of open area in slotted pipe and screens varies considerably. In slotted pipe, it ranges from about one percent for 0.030-inch slots to about 12 percent for 0.250-inch slots in slotting patterns which do not seriously weaken the pipe. Brand name, manufactured punched and slotted screens have open areas between four and 18 percent, depending upon slot size and pattern. Louvered-screen open areas range from about three percent for 0.020-inch slots to about 33 percent per 0.200-inch slots. Cage-type, wire-wound screens' open areas range from about two percent for 0.006-inch slot sizes to as much as 62 percent of 0.150-inch slot sizes. [See Appendix.]

From the standpoint of strength and feasibility of manufacture, Corey's 60 percent open area is obviously impractical except in cage-type, wire-wound screens with large slot openings. Experience has indicated that for a suitable ratio of screen diameter to well yield there is practically no measurable increase in performance for an eight-inch diameter and larger screens when open area is about 50 percent. This percentage of open area can usually be obtained with a maximum 0.125-inch slot size with cage-type, wire-wound screens.[509]

SELECTED DESIGN CRITERIA

Screen Size, Length and Gravel Packs

Most well design equations and methods presently used for selecting the proper screen size, gravel pack or length of screen required to produce a specific Q volume of sand-free water, ignore the permeability of the formation and of the gravel pack material. One method commonly used by the Illinois State Water Survey includes formation and gravel pack permeability factors as well as factors such as critical entrance velocity and effective open area of screen. Such a method of approach is as follows:[650]

EQUATION 52

$$L = \frac{Q}{A_e V_c (7.48)}$$

where:

L = Length of screen (feet)

Q = Discharge (gpm)

A_e = Effective open area/foot of screen (sq. ft./ft.) - e.g., approximately one-half actual open area.

V_c = Critical velocity (fpm) - e.g., velocity above which sand particle is transported.

Necessary support data for the above calculations are shown in Table 21, and Figure 76. A_e is obtained from effective open areas listed in Table 22, if a typical continuous slot well screen is used. The A_e factor for other screen types can be obtained from the Appendix or screen manufacturers. (See Appendix for comparisons of percent open area for typical screen types.)

TABLE 21[650]

Permeability (gpd/ft^2)	Velocity_c (fpm)
5000	10 (max)
4000	9
3000	8
2500	7
2000	6
1500	5
1000	4
500	3
0-500	2 (min)

The grain sizes of the formation must be determined before a successful gravel pack-screen size is chosen. See Chapter 12 for operational aspects of gravel packing. The standard method for determining these sizes is to sieve a formation sample using a standard set of sieves. Many of the screen manufacturers have this service available, although a few consulting firms and well contractors have their own sieve sets.

All weight-graphical mechanical analysis equipment and methodology now used to obtain and evaluate data required for accurate well screen slot size selection utilize the equipment and procedures to be described.

TABLE 22 [650]

Nominal Screen Size	Effective open screen area per foot of screen (sq. ft.)					
	Slot 10	Slot 20	Slot 40	Slot 60	Slot 80	Slot 100
8" I.D.	0.098	0.178	0.302	0.392	0.463	0.517
10" I.D.	0.124	0.226	0.384	0.498	0.587	0.654
12" I.D.	0.147	0.268	0.453	0.590	0.603	0.687
15" O.D.	0.173	0.315	0.533	0.595	0.710	0.804
16" O.D.	0.187	0.340	0.575	0.640	0.766	0.870
18" O.D.	0.168	0.312	0.541	0.720	0.860	0.980
20" O.D.	0.191	0.355	0.617	0.820	0.890	1.017
24" O.D.	0.230	0.427	0.744	0.886	1.070	1.225
26" O.D.	0.226	0.423	0.750	1.010	1.222	1.397
30" O.D.	0.246	0.462	0.819	1.101	1.222	1.520
36" O.D.	0.284	0.530	0.936	1.262	1.527	1.745

Four to six shallow metal pans with screen bottoms are stacked with the coarsest screen pan on the top and the finest on the bottom of the series. A known weight of dry water-bearing material, usually from 300 to 600 grams is poured on the top sieve and then the interlocking stack of pan-sieves is shaken either by hand or by a mechanical vibrator until the material is well sorted. Once this has occurred, the material retained on each screen is weighed and this weight expressed as a percent of the total weight of the original sample. The cumulative percent retained on each pan sieve is then plotted as a point on graph paper against the sieve opening in thousandths of an inch and these points connected to form a smooth curve.

The sand analysis curve is then interpreted to determine how much of the total sample is smaller or larger than a given particle size. Using this interpreted value, an optimum well-screen size is

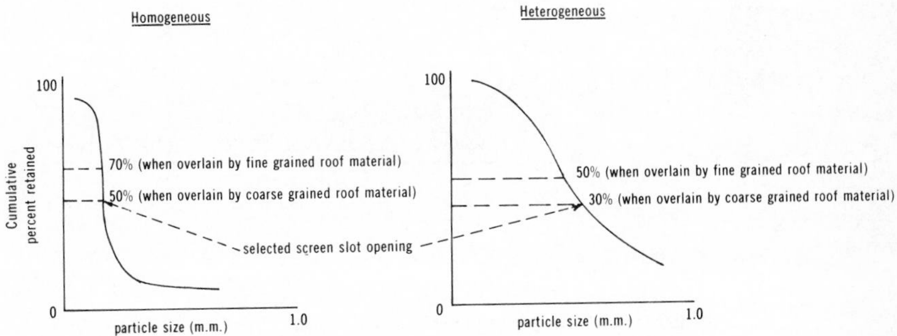

Figure 76[650] **Selected screen slot opening**

selected which, when installed opposite the formation material, will permit development of a final well that will retain the aquifer material yet permit free entry of water from it into the well.

The equipment and procedure required for this method of well design tends to be elaborate and time consuming. Also, it requires a relatively high degree of laboratory competence and expertise not normally at the disposal of field personnel. Instead, field personnel generally must depend on specialists and facilities in state and federal laboratories or well-screen company laboratories to make and interpret such analyses. Several days usually are required before results of these analyses are available for ordering and installing the proper-sized well screen for a particular well.

However, a well-screen selector field kit has been developed by W. H. Walker, of the Illinois Water Survey (see Figure 77). This device assists in the selection of the proper screen slot size. A small, representative sample of the aquifer to be screened is placed in the upper compartment of the kit and is thoroughly shaken.

The field kit well-screen selector permits immediate visual or graphical selection of the proper well-screen slot open size to install in a formation to obtain sand-free water. This determination can be made by the driller, engineer, geologist, or hydrologist on the well site within only minutes after the sample has been recovered. A series of removable and interchangeable screens in the kit readily separates the aquifer's grain sizes into standard well-screen slot-size percentages.

◆Well-screen Selector Field Kit

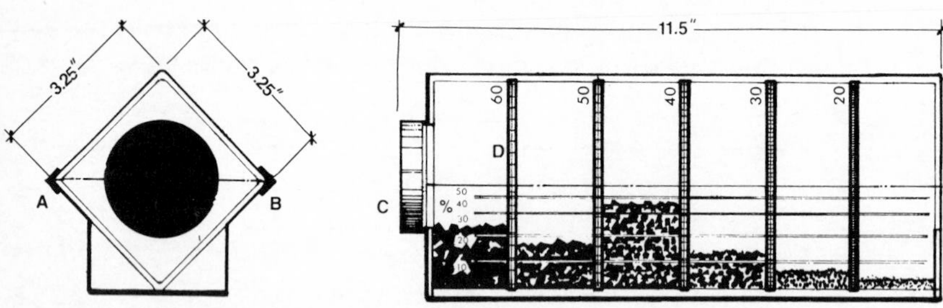

A. Latch
B. Hinge
C. Cap
D. Removable Screens

Figure 77 (W. H. Walker)

The principle involved is volumetric-visual rather than weight-graphical as is the standard method generally employed in well design work.

After loading, and after a brief interval of shaking, selection of the proper well screen slot size is apparent from a visual inspection of the relative volumes of material retained on each of the numbered screens. If a sand-analysis curve is desired for detailed design work, this can be determined volumetrically from a network of grids permanently molded into the outer surface of the polypropylene container.

Screen Installation

After the proper screen size and length have been determined by one or more of the methods previously described, placement of the well screen must be carefully considered.[16,24,46,69,78,580] Well screens are normally installed in wells constructed by the cable-tool method, by "jacking back" the outer casing exposing the screen and then sealing the two by a lead swedge.

In rotary-drilled wells, the screen may be attached by welding or by screw threads to the casing and lowered into the hole as one integrated unit, or as in the above cable-tool method, a telescoped screen may be lowered inside the casing. The casing is then pulled back, and the well screen sealed by a lead or other type of packer. (See Figure 78.)

In place of the "pull-back" method, a well screen may be set in the open hole drilled below casing by rotary method after casing is cemented in position. (See Figure 79.) The "bail-down" and "wash-down" methods are illustrated in Figures 80, 81 and 82.

A relatively new innovation is the self-sealing slip packer (see Figure 83). Before 1964, the standard method of sealing a telescoped well screen inside the casing was to expand a lead packer attached to the top of the screen or to an extension pipe above the screen. In that year, however, a self-sealing packer was introduced after several years of development and field testing.[91]

The packer consists of a flexible, Neoprene sealing ring, securely mounted on a heavy metal fitting, which is screwed or welded to the top of the screen or to an extension pipe above the screen.

This kind of packer is called self-sealing, because, unlike a lead packer, no swedging or expanding operation is necessary. The design is such that the packer seals the annular space between two cylinders. It resists being displaced to one side or the other even under considerable lateral pressure.

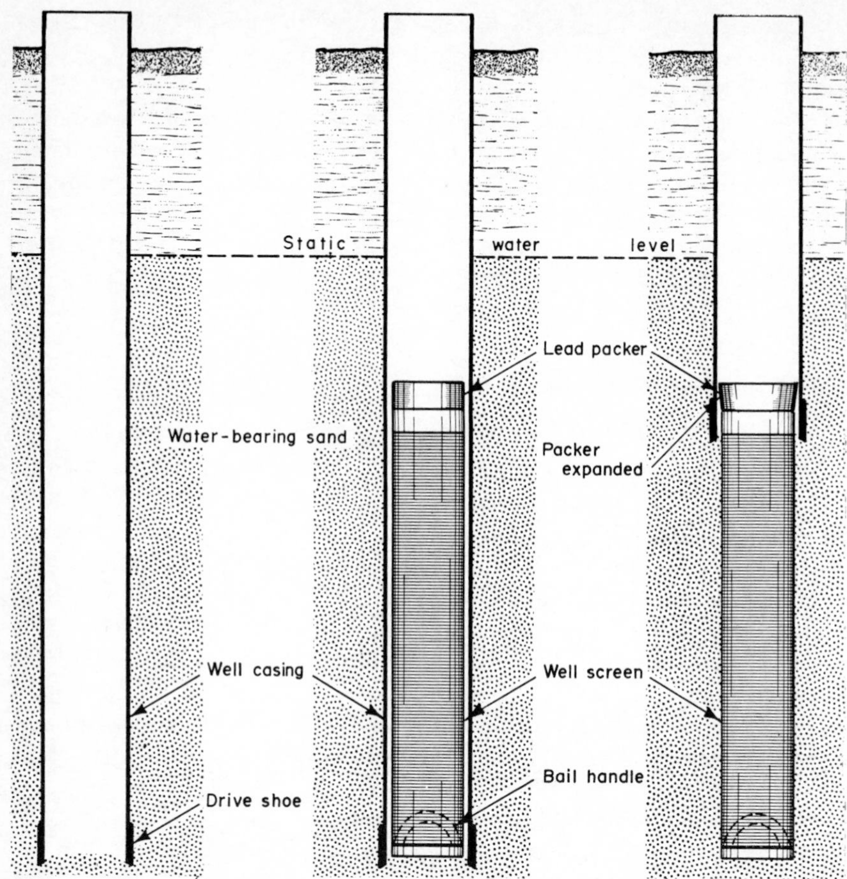

Figure 78[69] Basic operations in setting a well screen by the pull-back method. Casing is sunk to full depth of the well, well screen is lowered inside the casing, and casing is pulled back to expose the screen in the water-bearing sand.

A well screen, fitted with a self-sealing packer, is installed through the casing or, in some cases, through a liner by simply lowering the screen assembly with the packer attached. If the weight of the screen assembly is not sufficient to overcome the friction of the packer inside the casing, the added weight of a bailer or other tool may be used to help push the screen assembly to the bottom of the well.

Several obstacles had to be overcome to make this packer functional. A Neoprene rubber is used which has the hardness, stiffness, and toughness needed for the wide range of ground water conditions found in different regions. Good resistance to corrosion is essential, especially when such chemicals as chlorine or strong acid are used in periodic treatments for disinfection or removal of incrustation.

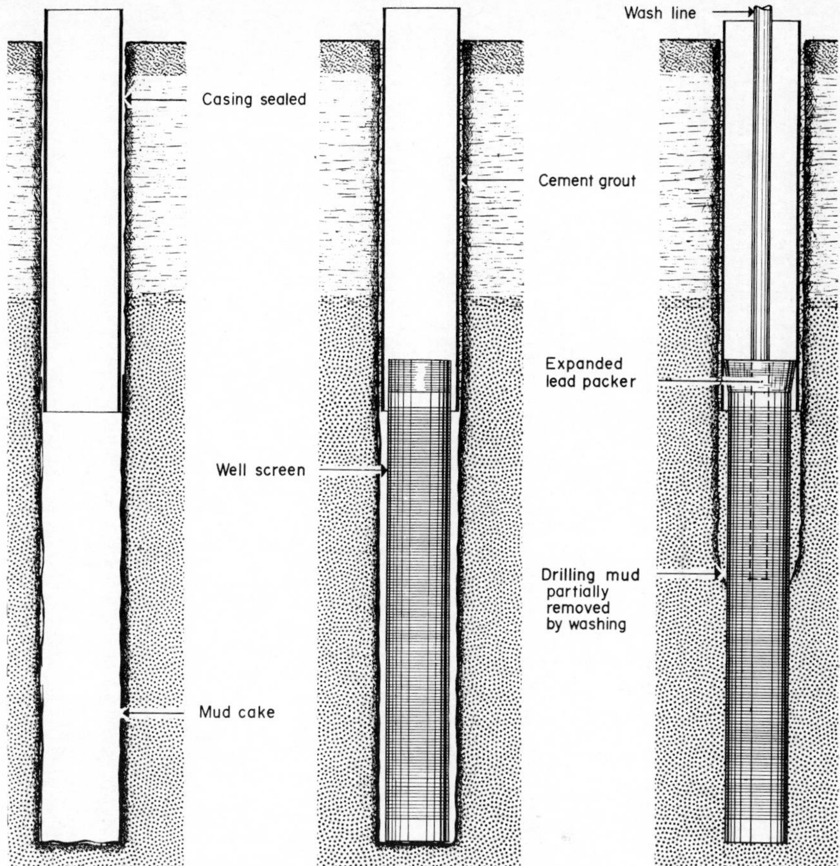

Figure 79[69] In place of pull-back method, well screen may be set in open hole drilled below casing by rotary method after casing is cemented in its permanent position.

There are, however, limitations to the use of self-sealing packers. The principal limitations are:

(1) The self-sealing packer is not intended for use when the screen installation requires that water recirculate on the outside of the screen, because the packer blocks circulation at the top of the screen.

(2) Casings which are in poor condition or inadequate in size are likely to prevent proper setting/sealing of the packer.

(3) This packer is not designed for sealing a liner or casing in an open borehole in a consolidated rock formation.

Since the self-sealing packer is not designed as an open-hole packer, as noted above, it has been used with much success to seal a liner

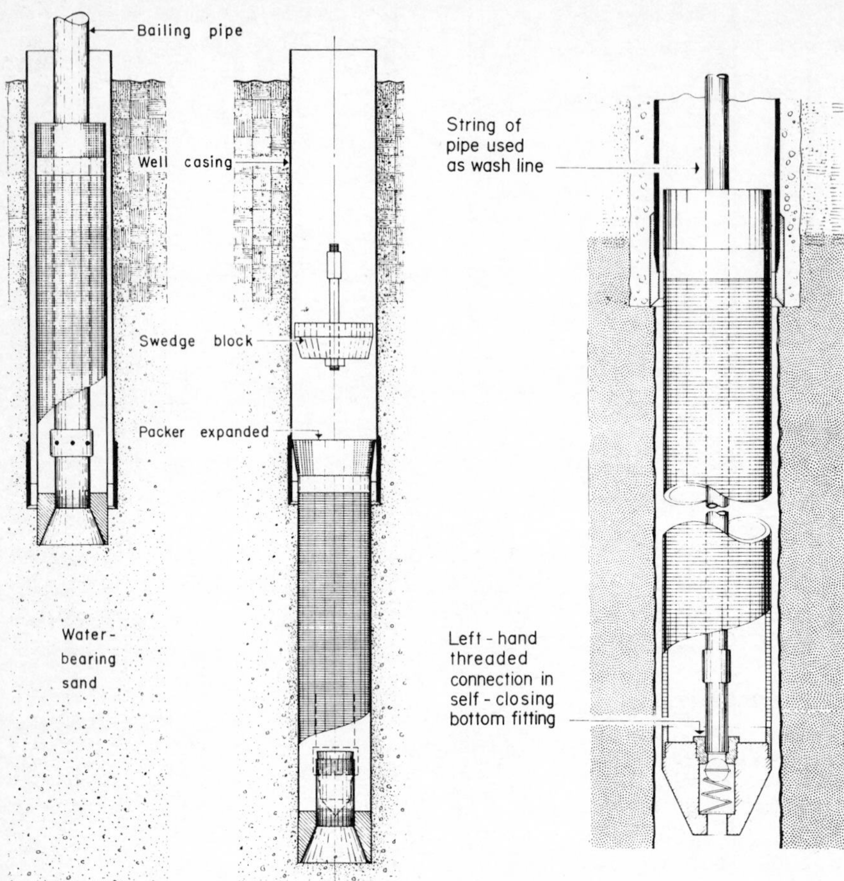

Figure 80[69] Assembly of well screen and bail-down fittings at start of operation (left), and final step in completing the screen installation by the bail-down method (right).

Figure 81[69] Wash-down bottom with spring-loaded valve permits washing screen into place. Space around the lead packer allows return flow outside the well screen.

inside a large-diameter casing. This is usually done where the large-diameter casing has developed a break or leak. Since the self-sealing packer can be adapted to changes in design and fabrication, it is advantageous for special installations.

PUMP TESTS

The literature of the ground water industry on pump test methods and variations thereon is voluminous and well known; hence, the topic will not be treated in any detail here, save for a few general remarks.

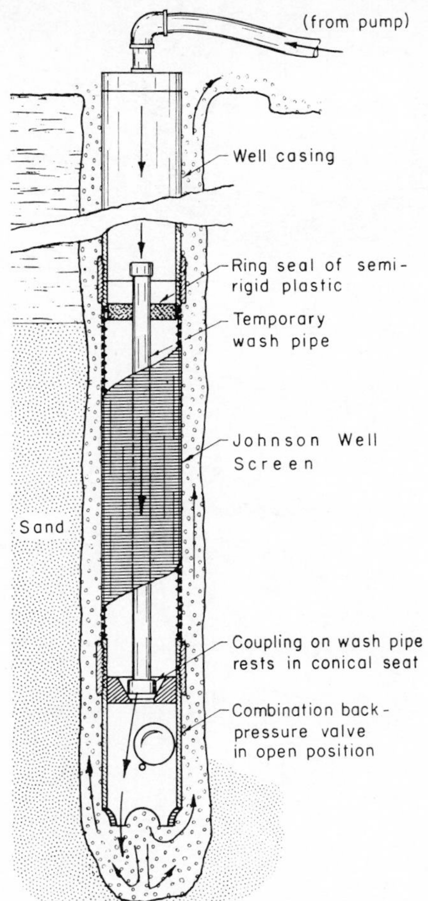

Figure 82[24] Small-diameter screens can be washed into place by jetting through temporary wash pipe and wash-down bottom with floating-ball valve.

Procedures and tables for field testing of well-flow rates can be obtained in the available literature.[46, 116, 249, 339, 418]

A paper by Trescott and Pinder[624] considers a truck engine air pump apparatus for pumping small diameter piezometer wells using the air-lift method of testing.

Water well deterioration is seldom sudden and castastrophic unless a casing or screen collapses. Decrease in yield and increase in draw-down progresses slowly and unnoticeably. However, a point is reached where the deterioration accelerates and trouble develops rapidly. When this occurs, corrosion, incrustation, and related conditions have become so aggravated that the possibility of successful rehabilitation is drastically reduced. (See Chapter 13.) Water wells should be step tested when initially completed and at least annually thereafter. A comparison of test results in conjunction with static water level changes will permit an evaluation of well condition, including the desirability of photographic or similar surveys and the need of rehabilitation.

The petroleum industry literature has little to offer in the way of directly pertinent technology on pump tests and well hydraulics, although the original work done on fluid flow and permeability came from the petroleum industry, e.g., Muskat[475] and Hubbert.[323]

Jacob,[329] Boulton,[133] Ferris,[223] Theis,[612] Walton,[649] Mogg,[454] Siddiqui,[562] Hantush[305] and others.[5, 6, 7, 33, 35, 57, 103, 185, 325, 330, 342, 386, 389, 392, 443, 497, 498, 563, 591, 597, 627, 650, 651] form the literature foundation for determining specific capacity or well yield. Of practical field value, however, is a simple method for establishing the order of magnitude of a well's production, both before and after well development.[451]

This simple "inflow test" provides an illustration of the effectiveness of development work in relatively low capacity wells. The test is performed by filling the well casing with water, then noting the rate at which the water level drops as water flows out of the bottom of the well and into the water-bearing formation. During the water-input test, the depth to water is measured every minute or every half minute after filling casing full. The data are plotted on chart paper with the depth to water as the vertical scale, and time after filling in minutes as the horizontal scale. The resultant curve shows the rate of fall of the water level. Tests performed after development work on the well indicate the degree of effectiveness of the well development. The transmissibility of the water-bearing formation can also be roughly calculated from the results of this test.

A modification of the above field test is to inject a specific volume of water into the well measuring the declining water level two or three times at specified time intervals. Using the same sized slug and the same time intervals for successive tests, the results can be

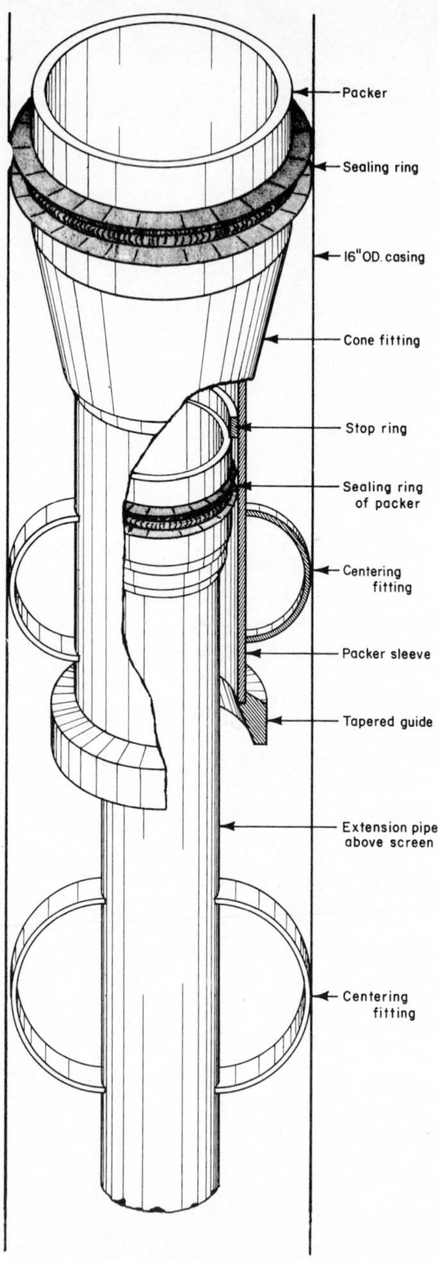

Figure 83[91] Self-sealing slip packer designed to seal annular space between casing and riser pipe after gravel pack is placed.

compared, which would eliminate the need for plotting curves. This test would be useful in determining when further well development ceases to increase the well yield.

Scheidegger[552] in his publication, "The Physics of Flow Through Porous Media," gives a good review of many of the factors which relate to permeability studies.

PETROLEUM TECHNOLOGY: WELL HYDRAULICS

One area where the petroleum industry does offer considerable technological assistance is the general area of permeability damage repair.[371] It is interesting to note that most, if not all, of the work done by the oil industry on formation investigations for secondary recovery purposes incorporates little of the technology available in the ground water industry. With the increase of water-injection practices, the petroleum industry relies to a large extent on their oil-biased completion practices. This is analogous to the water well industry employing its completion techniques on shallow oil wells. Technological efficiency is lost in both of the above cases. The water well industry, however, completes few oil wells. It is clear that inter-industry interchanges must be made to assure maximum efficiencies in well completions (an overt aim of petroleum industry in other aspects of their operation) and to guarantee maximum ground water pollution control, which should be the goal of both industries. With increased technological communication between the petroleum, mining and ground water industies, improved and more flexible interchange of ideas and practices can be achieved. Involvement of the respective technical personnel in professional societies, meetings, etc., other than their own, can assist, promote and widen the engineers' or geologists' understanding of new methods. Another way is for the technician in one industry to contribute to the literature of the other, eventually resulting in a uniformly steady flow of ideas between all these industries, not to mention the subsequent flow between all supporting industries, e.g. the chemical, steel, electronic and other related industries.

Before inter-industry communication can occur to any organized extent, the ground water industry must become fully cognizant of its own technological developments which have been achieved over the years by men and women both in the laboratory and in the field. Field application of the most effective methods of establishing specific capacity and well yield is of immediate concern for the rational development of the ground water resource.

The U.S. Geological Survey has spearheaded the development of aquifer testing methods. *U.S.G.S. Water Supply Papers* form the supporting literature base, i.e., Nos. 1536-A, B, E, F, G, I; 1544-C; 1545-A.

More recent contributions to the field have been made by H.B. Eagon and D.E. Johe, i.e., "Practical Solutions for Pumping Tests in Carbonate Rock Aquifers," *Ground Water*, Vol. 10, No. 4, 1972, and W.K. Summers, i.e., "Specific Capacities of Wells in Crystalline Rocks," *Ground Water*, Vol. 10, No. 6, 1972.

The implementation of many of these methods still require considerable development for use in the field under non-ideal conditions. A strong need exists for intra-industry communication of the many solutions achieved in the field which remain out of the main stream of technological transfer because of limitations of time, funding, etc. Commercial organizations should be economically practical but for every solution made, a significant return for the organization could be achieved by publishing any results of recognized merit.

11
Well Design and Yield

There are many variations in well design which are necessitated by local geological conditions and intended use of the well. However, those variations generally employed for sedimentary formations can be summarized in eleven basic configurations.[44] These are shown in Figure 84-85. Gibson and Singer,[246] Ahmad[4] and others [46,453] have written about the various features of water well design construction and maintenance.

The well design criteria presented here represent a compromise between the various criteria found in the literature and those adapted for the various conditions found throughout the United States, with applicability to other parts of the world. The well designs apply to domestic, industrial, municipal, and irrigation installations.

DESIGN CHARACTERISTICS

Water supplies from bedrock aquifers will need to be developed and utilized as demand for water increases with population growth and industrial expansion.[658] Adequate ground water supplies from bedrock wells for domestic and industrial use could be developed in many geographical areas, which are distant from shallow sand and gravel deposits commonly capable of producing major ground water supplies. Of course, the development of bedrock aquifers depends

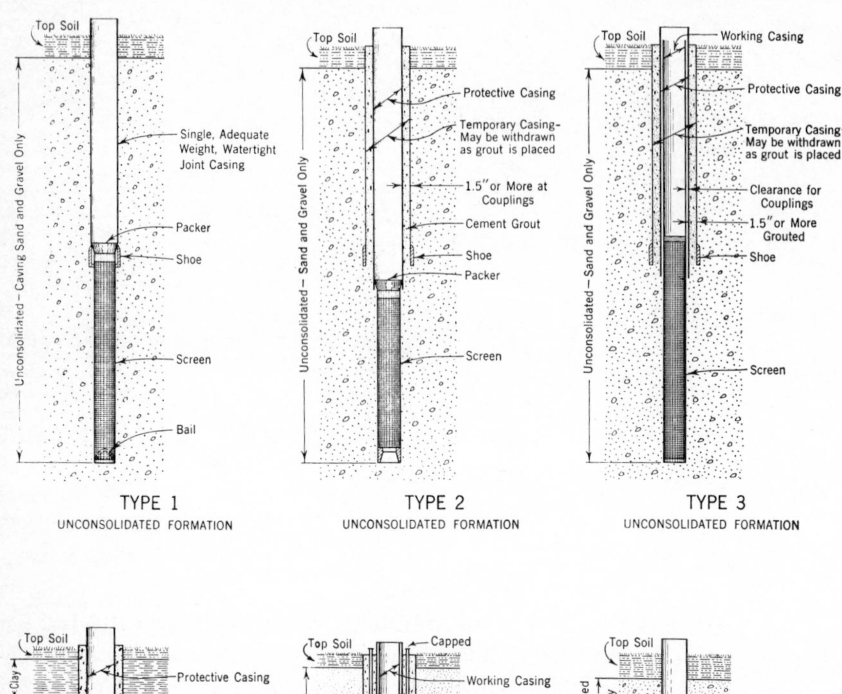

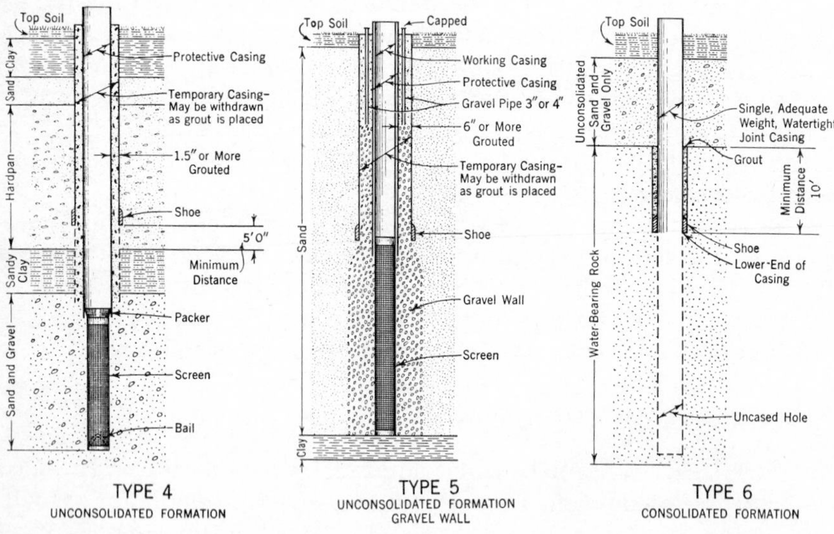

Figure 84[44] Typical well designs (type 1 through 6) for unconsolidated and consolidated formations. *Note:* all designs showing "uncased" hole intervals imply use of liner.

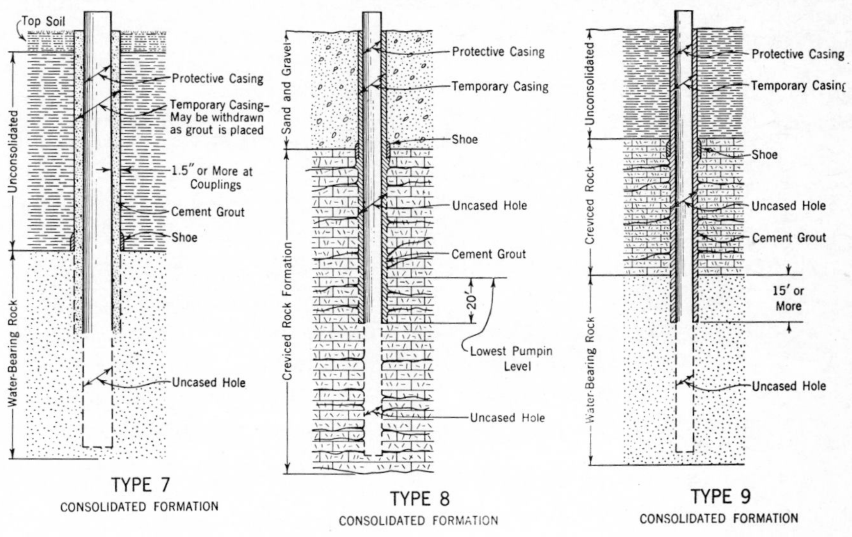

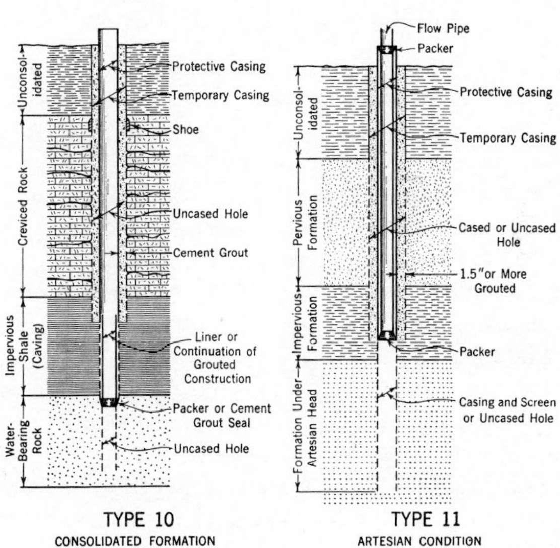

Figure 85[44] Typical well designs (type 7 through 11) for unconsolidated and consolidated formations. (See Note in Figure 84.)

heavily on an adequate knowledge of the local hydrogeological environment and on a knowledge of current well construction practices and design technology applicable to the specific formations encountered.[486,606,674]

WELLS IN SEDIMENTARY FORMATIONS

The yields of shallow, limestone/dolomite wells,[375,494] deep sandstone wells,[186,652] sand and gravel wells,[35,57] etc., are affected by many factors, e.g., leaking casings, partial clogging of the well bore, partial penetration of aquifers, improper well design (with respect to local hydrogeology), well losses and effects of acid treatment and shooting.[658] All of these factors must be carefully considered in appraising the relative local and regional requirements of well design and yield. The deeper the well the more complicated the above factors become. Allowance, for example, must be made for relatively deep sandstone wells since the yield from the deeper units can diminish because of caving or bridging.

The Illinois Water Survey has investigated many aspects of well yield and the inherent design features of individual bedrock aquifers.[188,652] Their studies emphasize that "the specific capacity of a multiunit (multi-aquifer) well is the numeric sum of the specific capacities of the individual aquifer units."

As an example of their approach. Csallany and Walton[188] prepared specific-capacity frequency graphs (See Figures 86) to determine the relations between well yields and the following geologic controls: 1) the glacial deposits penetrated by the well immediately above dolomite bedrock are predominately till (clayey materials); 2) the glacial deposits penetrated by the well immediately above dolomite bedrock are predominately sand and gravel; 3) sand and gravel deposits immediately above a dolomite bedrock are less than 10 feet thick; 4) sand and gravel deposits immediately above bedrock exceed 10 feet in thickness; 5) the well is in a bedrock valley; and 6) the well is in a bedrock upland. Figure 86 suggests that the productivity of the dolomite aquifer is greater in areas where sand and gravel directly overlie and are in hydraulic connection with the dolomite than in areas where relatively impermeable till directly overlies the dolomite. Figure 87 indicates that the productivity of the dolomite aquifer increases as the thickness of sand and gravel directly overlying the dolomite increases. Figure 88 indicate that the productivity of the specific dolomite is greater in bedrock uplands than it is in bedrock valleys.

Graphs in Figures 86 and 87 suggest that there is a good hydraulic connection between the carbonate rocks and overlying glacial drift

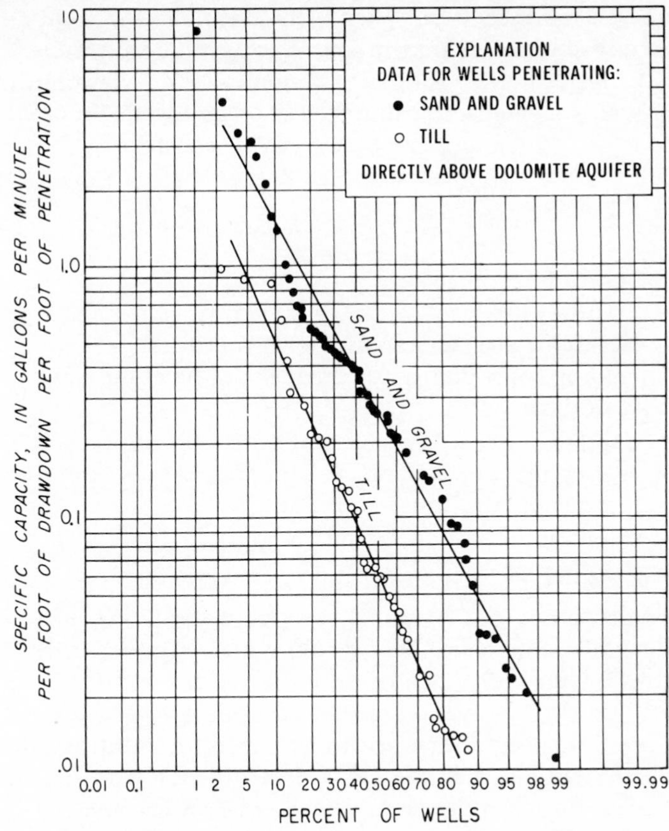

Figure 86[188] Relation between specific capacity and of glacial drift and till.

and that the productivity of the dolomite is primarily controlled by solution openings in the upper section of the aquifer.[493]

The above approach typifies one quantitative appraisal of specific capacity (safe yield). In practice, the appraisal is based on varying the permeability within different geohydrological units and the same units (or two units within a relatively thick aquifer).[187,196]

Well Spacing

The problem of spacing production wells in sedimentary formations is frequently encountered. The farther apart wells are spaced the less their mutual interference but the greater the cost of connecting pipeline and electrical equipment. The spacing of wells is often dictated by practical considerations such as property boundaries and existing distribution pipe networks. The following discussion is concerned only with the influence of aquifer characteristics and economics on the spacing of production wells.[650]

Theis[612] derived the following equation for determining the optimum well spacing in the simple case of two wells pumping at the same rate from a thick and areally extensive aquifer:

EQUATION 53

$$r_s = 2.4 \times 10^8 \frac{C_p Q^2}{kT}$$

where:

r_s = Optimum well spacing, in ft.
C_p = Cost to raise a gallon of water 1 foot, consisting largely of power charges, but also properly including some additional charges on the equipment, in dollars
k = Capitalized cost for maintenance, depreciation, original cost of pipeline, etc., in dollars per year per foot of intervening distance
Q = Pumping rate of each well (in gpm)
T = Coefficient of transmissibility (in gpd/ft)

For small values of T and Q, r_s (from Equation 53) is of no practical significance. Because the effects of partial penetration are appreciable within a distance of about twice the saturated thickness of the aquifer, 2m, from the production well, it is generally advisable to space wells at least a distance of 2m apart in aquifers 100 or less feet thick. Experience has shown that in the case of a multiple well system consisting of more than 2 wells the proper spacing between wells is at least 250 feet.

Production wells should be spaced parallel to and as far away as possible from barrier boundaries and as near to the center of a buried valley as possible. Wells should be spaced on a line parallel to a recharge boundary and as close to the source of recharge as possible.

Theis also derived the following equation to determine the permissible distance between production and disposal wells in an areally extensive isotropic aquifer:

EQUATION 54

$$r_d = \frac{2Q_d}{TI}$$

where:

r_d = Permissible distance between production and disposal wells to prevent recirculation of water (in/ft)
Q_d = Pumping and disposal rate, (in/gpd)
T = Coefficient of transmissibility, (in gpd/ft)
I = Natural hydraulic gradient of water table or peizometric surface, (in ft/ft)

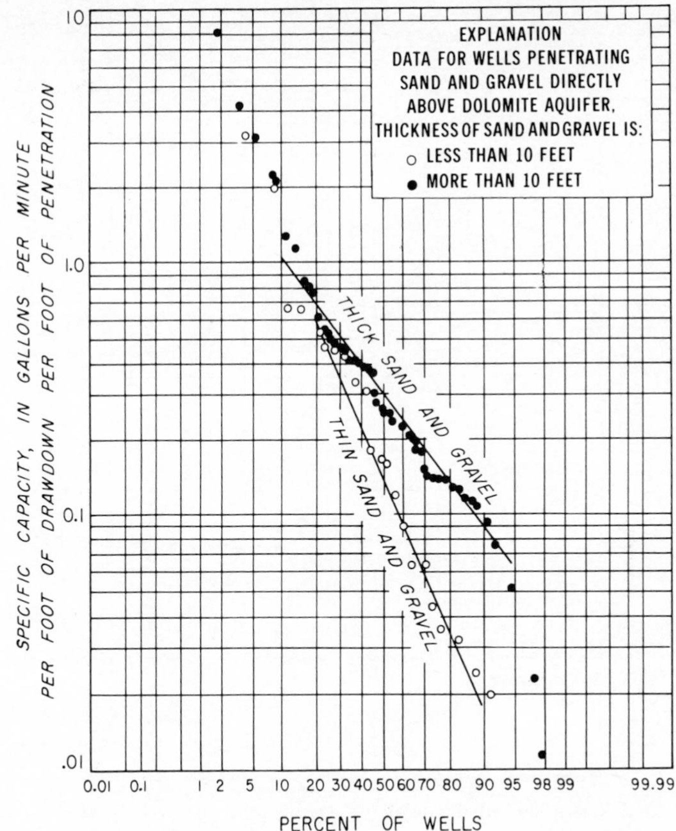

Figure 87[188] Relation between specific capacity and thickness of sand and gravel directly above a dolomite aquifer.

WELLS IN IGNEOUS AND METAMORPHIC FORMATIONS

Although an overwhelming majority of water wells are drilled in sedimentary rocks, consideration must also be given to drilling and development of water wells in igneous and metamorphic rocks since many geographical locations are predominantly underlain by "crystalline" rocks.[317,614] Igneous and metamorphic rocks are at or near the surface in approximately 20 percent of the world's land surface. The search for new supplies of water in these areas, particularly in Australia, India, and the Union of South Africa, has been responsible for extensive exploration drilling. This experience combined with isolated work in the United States, Sweden, Japan, and other countries has yielded knowledge of only the very basic drilling and development principles involved.[198]

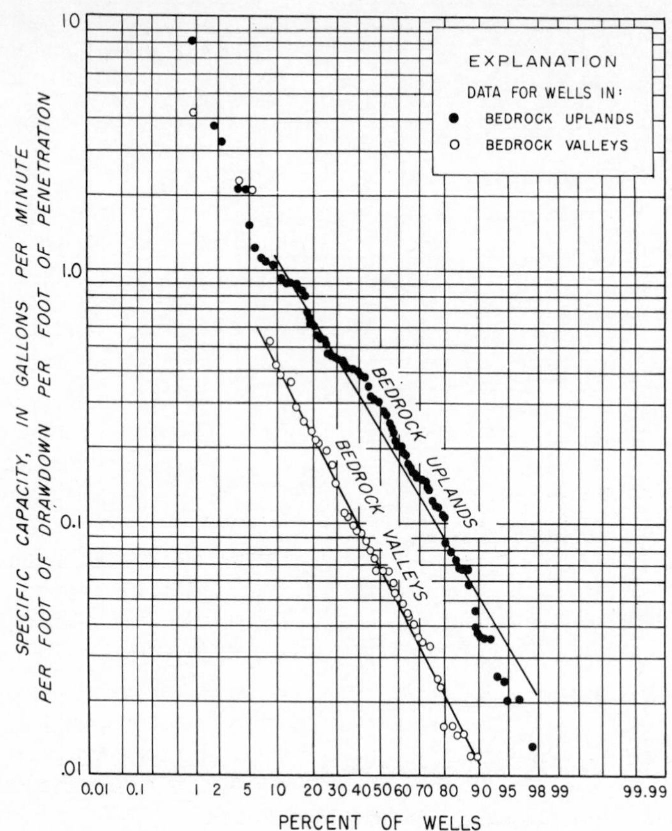

Figure 88[188] Relation between specific capacity and bedrock topography.

Drilling in hard crystalline rock is presently accomplished most effectively by the "down-hole hammer" air rotary method, although drilling efficiencies are low because of an inadequate knowledge of the drilling parameters involved. Many of the emerging drilling methods and related techniques offer particular promise for higher penetration rates.

Well design for such rocks varies much in the same areas as for sedimentary rocks. Factors such as variation in water quality within the same well, water-bearing fracture development, selective-cementing requirements, etc., combine to offer potential problem areas, which often require more specialized approaches than are necessary in sedimentary drilling. However, little detailed work has been completed to date on the problematical areas of igneous and metamorphic drilling or on completion and development of wells in such rocks. The mining industry will be leading in the development and application of solutions to the difficulties in drilling. Other aspects must be examined in detail by the ground water industry in the future.

Experience shows that yield per foot of well penetration in igneous and metamorphic (as well as creviced limestone/dolomite) aquifers generally decreases as depth increases. Drilling beyond an optimum depth may add to total yield, but the cost per unit of production is significantly higher.[197]

General observations of topography and positions of surface streams provide some guidance for drilling in regions where detailed studies of outcropping rock are not possible. Even under the best conditions, however, results of depth and yield of wells are difficult to predict. Under the worst conditions, extensive exploratory drilling must be employed, and results can only be generally predicted from statistical analysis of data gathered from previous drilling in comparable rocks situated in similar topographic locations. Table 23 summarizes data obtained from the Greensboro, North Carolina area, on the typical effect of topography on well yield. Table 24 presents data for industrial and municipal wells in the same area. Although specific rock types do not yield characteristic supplies over great distances, certain conclusions can be made on the potential of a specific rock type within an area when well information is available. The tables illustrate the potential of crystalline rock aquifers.

Water-bearing properties of these rocks depend upon the extent of weathering and the presence of joints and faults. Porosity, resulting from weathering, generally develops at depths less than 100 feet.

TABLE 23

Topographic Location	Number of wells	Average depth (feet)	Yield (gallons a minute)			Percent of wells yielding less than 1 gal a min
			Range	Average	Per foot of well	
Hill	282	147	0–125	7 1/2	.052	28
Flat	152	154	0–200	17	.113	3
Slope	228	127	0–120	14	.108	6
Draw	66	180	1/2–212	27	.148	3
Valley	74	212	0–150	28	.132	1
All wells	802	151	0–212	14 1/2	.097	12 1/2

Source: North Carolina Dept. Cons. & Develop., Div. Mineral Resources. Bull. 55, 1948.

TABLE 24

Type of Rock	Number of wells	Average depth (feet)	Yield (gallons a minute)			Percent of wells yielding less than 10 gal a min
			Range	Average	Per foot of well	
Granite	26	361	2–212	33 1/3	.093	34
Gneiss	44	286	1–125	34 1/2	.121	9
Greenstone	56	282	3–200	55 1/2	.197	3 1/2
Slate	8	315	5–32	22	.071	12 1/2
Triassic	19	291	8–150	35	.120	21
Diorite	3	181	10–25	17	.092	0
All wells	156	297	1–212	41	.138	13

Source: North Carolina Dept. Cons. & Develop., Div. Mineral Resources. Bull. 55, 1948.

Porosity, represented by joints or fractures, also is depth dependent since at depth, joints generally become smaller and less numerous. Openings along fault surfaces also tend to become tighter in deeper formations.

These geologic observations together with test data from wells indicate that rock permeability often decreases with depth. This, in turn, implies that unit cost of water from deep wells is likely to be higher than the unit cost of water from more shallow wells. Economic factors, therefore, largely determine optimum well depth in crystalline-rock aquifers.

Unweathered and unfractured crystalline rock normally has less than one percent porosity, with permeability so low as to be almost negligible. Porosities produced by weathering commonly range from 30 to 50 percent. Weathering causes differential expansion of various mineral grains due to partial hydration. Expansion and differential movement produce, in turn, inter-granular pore space. Circulating ground water may then increase local porosity by dissolving unstable minerals.

If the rock is originally coarse-grained, with a moderate abundance of stable minerals, such as quartz, a relatively high permeability may result from weathering. Wells completed in the zone of weathering commonly produce ten to 15 gpm.

Thickness of the zone of weathering depends on the geologic history of the area. When erosion is rapid, fresh rock may be exposed at the surface. In contrast, depths of almost 300 feet of weathered rock are common in areas with little apparent surface erosion, e.g., the north central region of Australia. Depths of ten to 100 feet, however, are more common. The transition from weathered to unweathered rock is relatively abrupt, and this transition may occur within a vertical distance of only ten to 20 feet.

Permeability is significant in unweathered rock only along fractures that result from volume changes or from externally applied forces that produce faulting. Regardless of how the openings may be formed, fractures and joints tend to be smaller and to decrease in frequency with an increase in depth.

However, mineral exploration drilling operations have demonstrated the existence of permeable zones at depths of several thousand feet. These occurrences appear to be associated with faulting. Even along vertical fault zones, however, slight amounts of rock creep tend to close the fractures, so that permeability is generally decreased in deeper zones. Low angle faults, in particular, tend to close as depth increases due to the weight of overlying sediments and rocks.

Records of water wells in granite and schist in the eastern United States were compiled from published descriptions.[197] In addition, data for wells in the Sierra Nevada area, California, mostly completed in granodiorite (a granite-like material) but some in metamorphic rocks, were examined.

Data used in a study by Davis and Turk[197] include well depth, diameter, rock-type penetrated, and final production. Records of some wells in the eastern United States and all wells in California show static water levels, casing lengths, and perforation intervals.

Figures 89 and 90 show some of the results of the investigation. In these diagrams the well yield per foot of well penetration in the aquifer is shown as a function of well depth. These figures clearly show that permeability generally decreases with depth. The scatter of the plotted points in the figures is large, which is common because natural variations of structure and rock type cause large differences in the permeability of individual fracture zones as well as in the overall frequency of these zones.

The analysis provides a basis for estimating optimum and economic depths for drilling water wells in crystalline rocks; and optimum depth depends largely upon cost factors.

Exceptions to this approach are, of course, situations in which geologic knowledge or surface geophysical data may predict the existence of productive zones at particular depths. Surface geophysical methods are especially applicable to areas underlain by crystalline rocks with highly fractured quartz, dikes, etc., e.g., Eastern United States, Wisconsin, etc.

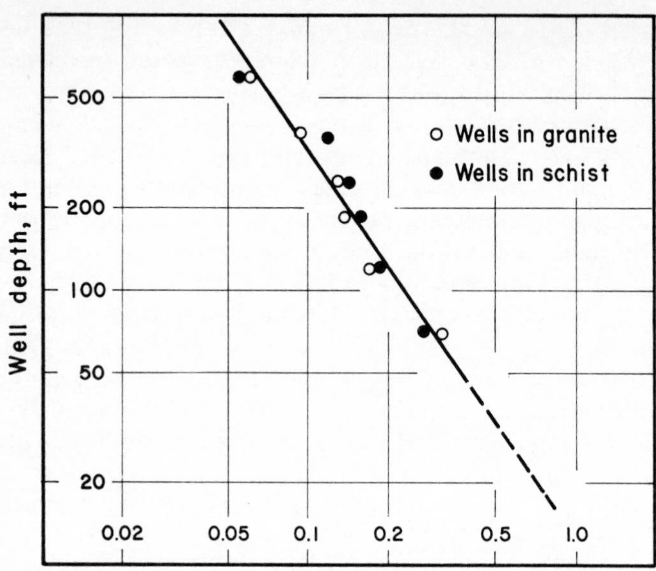

Figure 89[197] Average depth-yield relations for 814 wells in granite and 1522 wells in schist in eastern United States.

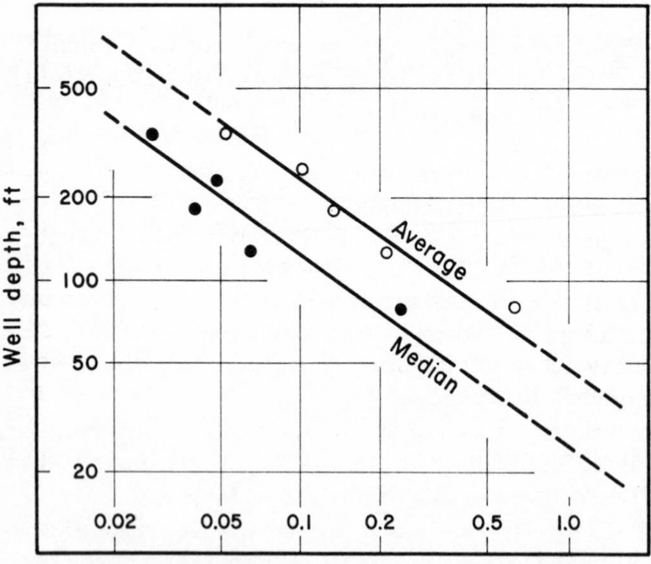

Figure 90[197] Depth-yield relations for 239 wells in granodiorite or closely related crystalline rocks in the Sierra Nevada area of California.

Aesthetic considerations, biological and chemical purity of water, local zoning regulations, and a number of other factors which cannot be treated in such a simple economic or geologic analysis are commonly important in making decisions concerning water well drilling. Concerning a problem with oil-well drilling, Grayson[278] states that although many steps will always remain qualitative, some type of formal approach is necessary to minimize the possibility of erroneous decisions. Davis and Turk[197] attempt to give some quantitative expression to only one decision-making step in one class of water-bearing material. The authors stress that the entire decision-making process should be continuously analyzed. The following conclusions can be made concerning igneous and metamorphic rock water-bearing characteristics:

(1) The water-bearing characteristics of most "crystalline" rocks are primarily controlled by weathering and structure. Rock type alone is commonly of secondary importance.

(2) In the absence of geological and geophysical guidance, when drilling in crystalline rocks, highly variable amounts of water are encountered. In unweathered rocks, from five to 15 percent of the wells are failures, median yields are less than eight gpm, and roughly ten percent of the wells will have yields of 50 gpm or more. (See Table 24.)

(3) Water production per foot of well decreases rapidly with an increase in well depth. This decrease is roughly ten-fold between depths of 100 and 1,000 feet.

(4) The optimum depth of water wells in crystalline rocks is determined largely by economic factors, unless the geologic structure is known in detail.

(5) Although extensive economic studies have not been made in many of the hard rock areas, rough estimates suggest that the depth of single domestic wells should be less than 150 to 250 feet, and wells of larger production should be less than 600 feet. In many places, the optimum depth of domestic wells will be less than 100 feet. Additional research is obviously needed on both drilling parameters and production factors involved in developing ground water from igneous and metamorphic rocks.

ALTERNATIVE WELL DESIGN CONCEPTS

The concept of horizontal collection, and hence production of ground water, was apparently understood and utilized by the early Egyptians, Greeks, and Romans. It was also applied as the "Maui" well of the prehistoric Hawaiians.

Early public water supplies in the United States also relied upon shallow "infiltration galleries" for their supply. Notable examples are the city of Des Moines, Iowa, which developed an extensive gallery adjacent to the Des Moines River and the city of Columbus, Ohio, which developed a smaller gallery near the confluence of the Olentangy and the Scioto Rivers at an early date. Many other small communities, such as Pella, Iowa, followed a similar course of action. Unfortunately, these early galleries were limited by several factors. For economic reasons they had to be shallow, usually 25 feet deep or less. Due to shallow depths, these galleries had very little storage and so were directly dependent upon constant flow in the adjacent stream for continual recharge. In addition, as a consequence of their limited depth they were more vulnerable to pollution. As pollution and contamination became an increasing factor, the desirability of water treatment became increasingly apparent. Thus, in many cases galleries were abandoned for either deep wells or a surface supply.

A modern concept of the infiltration gallery was introduced by a geologist, Leo Ranney, in 1933. Ranney had developed a system of horizontal boring of hard rock which met limited success in the production of petroleum from shallow sandstones in southeastern Ohio. He adapted his concept to the production of water from unconsolidated sand and gravel aquifers. The method involves projecting a horizontal screen, or "lateral," and simultaneously extracting fine materials from the gravel aquifer, thereby developing the gravel-pack. By combining this development with certain improvements in the method of installing deep, large diameter caissons by the "open-end" method, "galleries" could be economically installed at depths and capacities previously unattainable.[506]

The first "radial collector well" was installed at London, England, in 1933. The second, following soon thereafter, was installed for the Timken Roller Bearing Company in Canton, Ohio. Both of these were operational for over 35 years. Since that time, hundreds of collector wells have been installed throughout Europe, Asia, and North America. Individual innovations and modifications of the basic design have occurred over the past few years.

Collector Well Design

The typical radial collector well is illustrated in Figure 91. The central caisson is a minimum of 13 feet inside diameter with an 18-inch wall thickness. The typical caisson is between 80 and 130 feet in depth, although much deeper units can be and have been installed.

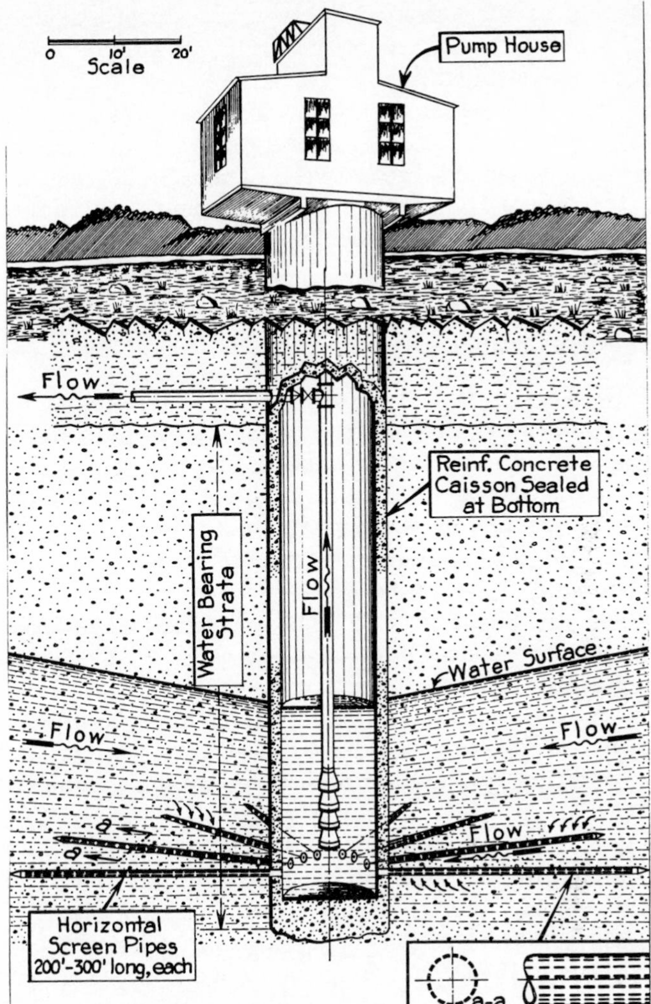

Figure 91 Typical Radial Collector Well (Ranney Water Systems, Inc.)

From near the bottom of the caisson, horizontal screen pipes, or laterals, are projected and developed. The diameter of these horizontal screens normally varies between eight and 24 inches, depending upon design velocities. Screen entrance velocities are often on the order of one foot per minute, after allowance for partial screen blockage by the gravel present in the aquifer.

The horizontal "laterals" can be projected in either a linear or radial pattern or in a combination of both. If the pattern is linear, the collector well produces ground water with a linear trough or depression that is geometrically analogous to the drawdown produced by a true gallery-type installation.

With a radial projection pattern, the drawdown resulting from production resembles the cone of depression of a standard vertical well with a diameter equal to between 60 and 80 percent of the total span of the laterals but with a slight "nosing" or "mounding" of water between laterals. This mounding is particularly noticeable at the outer extremities of the laterals.

Collector Well Potential

Thus, a collector well in an unconfined aquifer, with no adverse boundary conditions (with laterals averaging 240 linear feet in length in a radial distribution), can be expected to produce water at a rate equivalent to a vertical well with a diameter of [(2 x 240) + 16] x 0.80, or 397 feet. Obviously many factors enter into this comparative consideration. Lateral lengths may range to 300 feet or more. (See Figure 92.)

Collectors can be utilized to their maximum advantage when installed adjacent to a surface recharge source.[448] With laterals installed in a semi-radial pattern toward or beneath the surface recharge source, optimum conditions exist for inducing infiltration.

Capacities of collector wells have ranged from 700 gpm to 21,000 gpm, the variation resulting entirely from aquifer characteristics and well design.[247]

In general, the capital cost of a collector well system is similar to the capital cost of a comparably designed vertical well system. Over the past 20 years, amortized costs have averaged slightly less than one cent per thousand gallons of water produced, and the total costs of water to the wellhead including maintenance and lifting costs have averaged just under three cents per thousand gallons. (T. Bennett, pers. com.)

Principal advantages of the collector well systems are as follows:

(1) Favorable capital cost.
(2) Increased pump efficiency resulting in lower lifting costs.
(3) Minimal maintenance due to extremely low screen velocities.
(4) Ease of operation due to centralization of all pumping equipment and controls.
(5) Some degree of quality control due to individual operation of each lateral.
(6) Ability to completely dewater the entire aquifer due to the horizontal installation of the screens at the bottom of the aquifer.
(7) Safety and indestructibility of the supply.

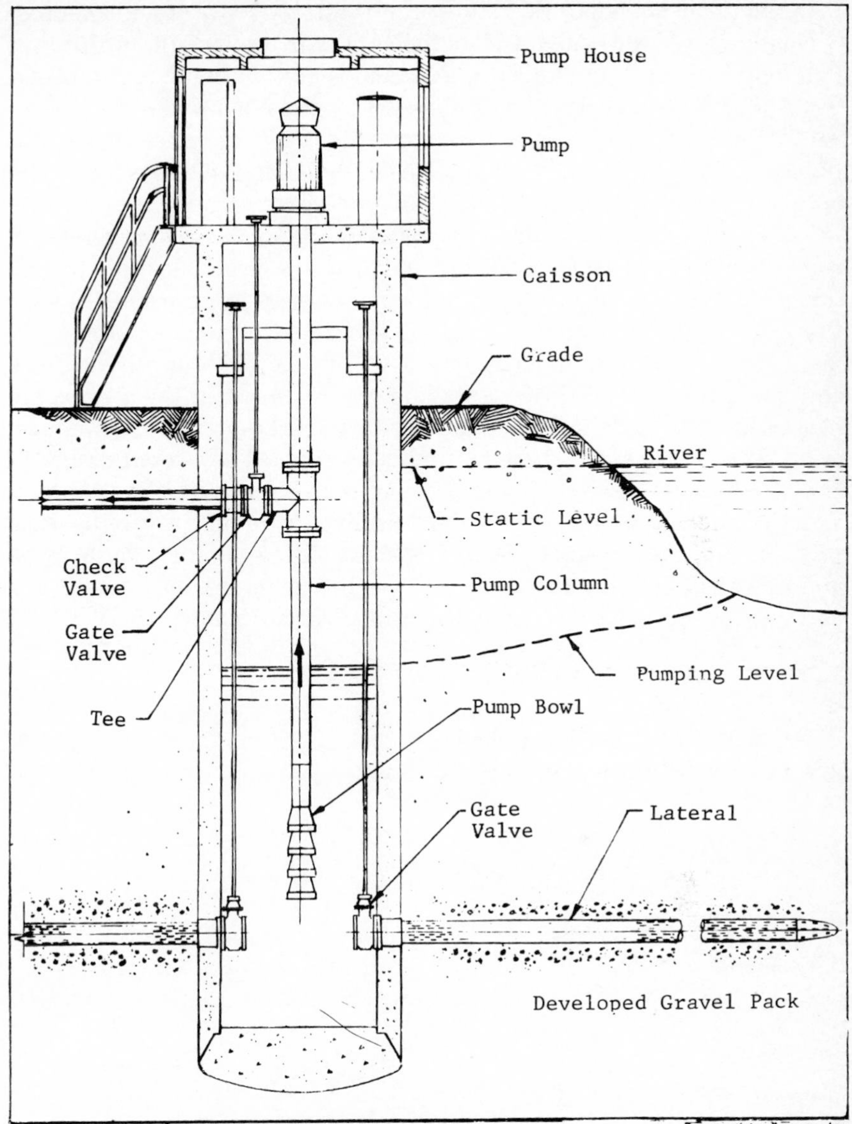

Figure 92 Section of the Radial Collector Well.

Principal disadvantages of collector well installations are as follows:

(1) Inadequate evaluation of recharge prior to construction resulting in declining water levels under conditions of long term heavy pumping.

(2) Increased potential for plugging of intake laterals due to corrosion/incrustation by products.

(3) Difficulty in well development.

Bennett[113] has examined the pump-test procedures and methods of analysis of the yield for radial-collector wells and infiltration galleries in general. In addition, Huisman is completing a text entitled *Groundwater Recovery* to be published by the MacMillian Press Ltd., which will comprehensively explore the flow of ground water to galleries, collector well systems, single wells, and well systems. The text is being produced specifically for the design engineer.

Well design may be most accurately defined as a progressive, quantitative evaluation using engineering principles of several scientific fields. This procedure must necessarily be coordinated with the most suitable methods of testing and construction. As previously mentioned, well design must be based on the geologic, hydrologic and chemical nature of the prevailing ground water in the area of the proposed well. Well yield and quality of the ground water, of course, are direct functions of these natural conditions. The overall design of a well embraces several phases. The preliminary survey of a proposed well location combined with the drilling of a test or pilot hole provide the basic design data of the prevailing geologic hydrologic and chemical characteristics of the specific ground water conditions and insures the most favorable operational performance.

12 Well Construction Operations

During well construction, certain operations and materials must be considered in detail. Gravel packing operations, for example, are commonly employed for high capacity industrial and municipal wells. The selection of proper well casing is also of vital concern to the structural integrity and longevity of the well. In addition, well cementing operations are without question ultimately responsible for short and long term sanitary protection of the well. Other operations, such as well sterilization, establishing well plumbness and alignment, fishing, etc., will also be considered on the basis of their relative importance to technologically sound well construction operations.

GRAVEL PACKING OPERATIONS

The demands of the rice industry gave rise to gravel packing water wells 70 years ago. In Arkansas and Kansas, irrigation wells were being gravel packed for the rice industry as early as the turn of the century. Popularity of the practice has continued to increase with the advent of reverse-rotary drilling. The gravel packing of drilled wells using an outer casing and wells drilled by either reverse circulation or the standard rotary method have met with equal success both in the petroleum and ground water industries.[7,46,181,221,416,452,558,576]

Two types of gravel packing are in general use — the uniform-grain size pack and the graded-grain size pack. The former has been widely accepted in recent years, especially when manufactured screens are used because the size of the openings can be controlled.[48]

In the case of a graded pack, the formation material may invade a graded pack at the gravel-formation interface, partly filling the pores and resulting in reduced permeability. With a well-sorted (uniform) gravel pack, the fines of the formation can travel between the grains and be pulled into the well during development, thereby increasing the formation permeability while retaining the highly permeable nature of the pack.

To prevent segregation of the graded pack during placement, special equipment is needed. The tremie, or ordinary four-inch pipe, is normally filled with pack material and allowed to settle four or five feet at each application. On the other hand, the uniform-grain-size pack can be shoveled into the well with acceptable results, although segregation can occur causing bridging, etc.

The pack material often must be processed on the site when the desired size is not readily available from local sources. Lack of availability is the big disadvantage of the uniform-grain-size pack. The most important physical property of uniform-grain-size material is the particle size as represented by the mean grain diameter which is the 50 percent grain size. To prevent the movement of formation material, it is necessary to provide a pack in which the mean grain size of the pack material bears a specific relationship to the mean grain size of the formation material.

Not all water-bearing formations require gravel packing; however, any formation can be successfully gravel packed, assuming special emphasis is placed on drilling fluid removal and development prior to gravel packing operations.[452] Generally, formations with an effective size (a size such that only ten percent of the formation is finer) of 0.01 inch and a uniformity coefficient of two or more can be safely developed without a gravel pack, providing there are few vertical changes in sizing in the formation. As the formation becomes coarser, the desirability of the gravel pack decreases; however, exceptions to the above are common.

Field Selection

Probably the four most common reasons for gravel packing are:

(1) to increase the specific capacity of the well

(2) to minimize sand flow through the screen in fine formations

(3) to aid in the construction of the well

(4) to minimize the rate of incrustation by using a larger screen slot opening where the formation is relatively thin but very permeable, and the chemical characteristics of the ground water suggests a potential for significant incrustation.

Figure 93 shows a typical sieve analysis curve of a water-bearing formation which is classified as medium sand. The other curve is that of the gravel pack generally used, classified as very coarse sand to fine gravel. The mean grain size of the formation material is 0.38 millimeter, and that of the gravel pack material is 1.8 millimeters, e.g., the size ratio of gravel pack to formation is 4.8. This ratio of mean sizes is called the "*Gravel-pack ratio*" (GpR).

The Gravel-Pack Ratio is usually obtained by:

EQUATION 55

$$\text{GpR} = \frac{50\%\ \text{gravel pack}}{50\%\ \text{formation}}$$

If a gravel pack is used, the V_c must reflect the permeability of the pack material. However, since the permeability is usually unknown, a value is assumed generally between three and five times the aquifer permeability. This factor is then generally used in obtaining a V_c value (critical velocity) for a gravel-packed well by the following equation:

EQUATION 56

$$V_c = \frac{V_{\text{formation}} + V_{\text{gravel pack}}}{2}$$

When the gravel-pack ratio is from four to five, wells generally have a high efficiency; whereas, wells having ratios of seven to ten were considerably less efficient.[576] When ratios are above ten, the wells produced considerable amount of sand, and a well with a ratio of 20 produced excessive sand and is considered a complete failure. The results of an Illinois Water Survey study show that ratios of four to five are satisfactory.[576]

The formation need not have a low uniformity coefficient to be effective when treated with uniform-gravel pack. In one case examined by the Illinois Water Survey, the formation uniformity coefficient was 8.3, and the well was gravel packed with material having a uniformity coefficient of 1.5. The gravel-pack ratio was 5.1, and this well produced no sand.

Probably the most common cause of sand pumping wells is the use of a gravel pack that is too coarse for at least part of the formation. A relatively thin interval of fine sand, sandwiched between coarser sand and gravel, may continue to sift through the pack indefinitely. The gravel (or sand) pack should be selected as a ratio of the finest

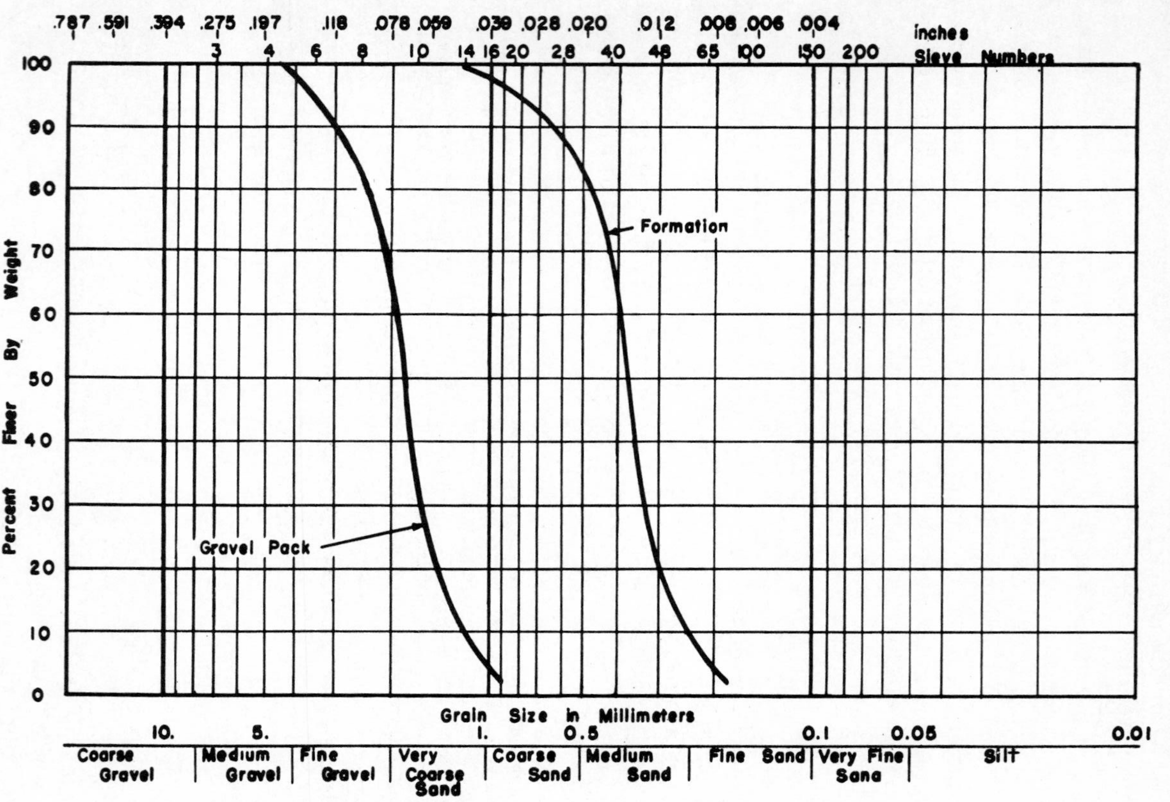

Figure 93[576] Example sieve analysis curve of aquifer material and of gravel pack material generally used for formations in Illinois.

part of the formation to be screened.[181] This problem is caused by two factors: (1) poor sampling and (2) lack of care in selecting the gravel pack size. (See Figure 94.)

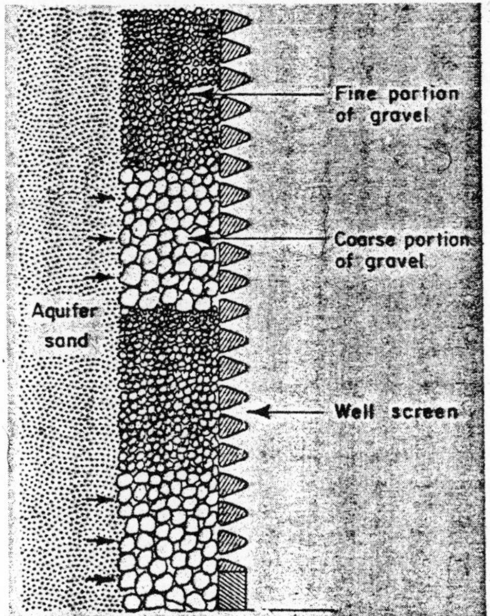

Figure 94[48] Potential segregation of graded-pack material. If placed by simply dumping into well, material can separate into layers. Formation sand may leak through coarser layers when well is put in service.

As previously mentioned, the pack can be of small grain size and still be effective. One well, in which the formation graded from medium to coarse sand, was constructed with a gravel-pack ratio of five (pack-grain diameter of 1/8 inch) and produced 2,100 gallons per minute with a ten-foot drawdown (a specific capacity of 210 gallons per foot).

Field Operations

Placement of gravel by a reverse-circulation system is generally accepted as being more effective than the "tremie" method.[66] The method is usable for wells of any depth; with certain modifications for relatively deep gravel pack applications. (See Figure 95.)

If the velocity of the descending stream in the annular space is about the same as the velocity at which a particle of "gravel" falls in a fluid, no separation of sizes can occur. In a light drilling mud, the velocity of the fall of the particle will be less than one-half foot per second.

The volume of fluid that must be circulated to attain a given stream velocity can be calculated from the area of the annular space in the wells. In practice, a fluid velocity less than the calculated fall velocity can be used to effectively prevent segregation. For example, with a light drilling fluid, circulating rates could be approximately 40 to 50 gpm for carrying 1/8-inch particles into place around the screen and 20 to 30 gpm for transporting 1/16-inch particles. In large diameter wells, practical considerations may require circulation rates lower than those ideally required to prevent total separation.

In gravel-pack installations incorporating the rotary drilling method, reverse-circulation of the drilling mud may be started at a relatively slow rate. The pumping rate will often depend upon the flow of mud through the well screen openings, and water should be added to thin the mud so that circulation can be increased. As soon as the desired circulation is attained, gravel may be put into the stream.

Thinning of the mud with water does not increase the risk of formation caving, since the descending fluid stream in the annular space tends to exert more borehole support even though the fluid weight per gallon has been reduced. Once the introduction of the granular material is started, the weight of the material builds up the effective weight per gallon of the fluid; therefore, caving is seldom a problem.

When circulating by the method illustrated in Figure 96, the two limiting factors of the pumping rate will be the head loss of the fluid flow within the well itself and the suction capability of the pump.

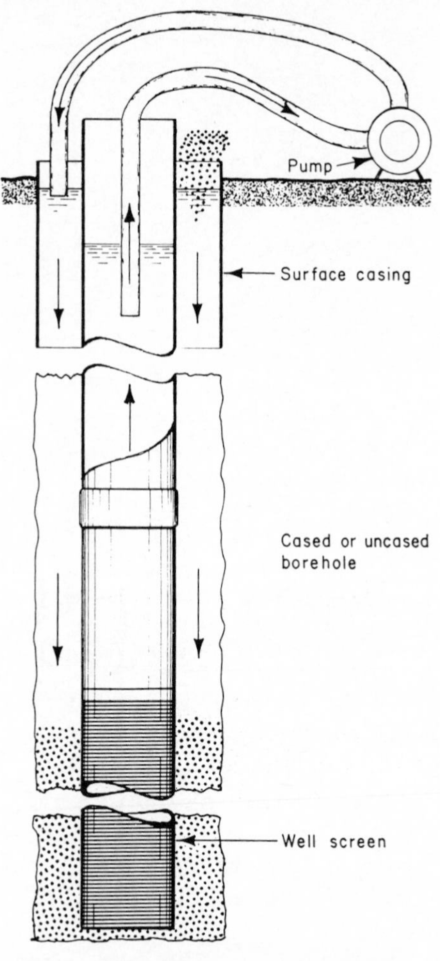

Figure 95[66] Simple water or mud circulation for gravel-pack placement can be arranged using available pumps for wells of moderate depth.

In relatively deep wells, total friction loss in the system quickly surpasses the suction characteristics of the pump. In such cases, the fluid must be pumped into the annular space under pressure introducing the granular material into the stream after proper circulation has been established. Figure 97 illustrates the methods employed in such deep gravel packing applications.

As sand and water are pumped into the annulus and around the screen in deep applications, the return flow of fluid passes through the well-screen openings, then up to the surface through the pipe used for suspending the screen assembly or the well. The lower end of the return flow pipe (the "stinger") almost reaches the bottom of the screen. During the emplacement operation, the pressure gauge is carefully monitored, with the flow of water-granular material slurry being controlled at a constant rate. As the material is emplaced to the top of the main screen, a slight pressure increase occurs, signaling that the level of the gravel pack has reached the top of the main screen.

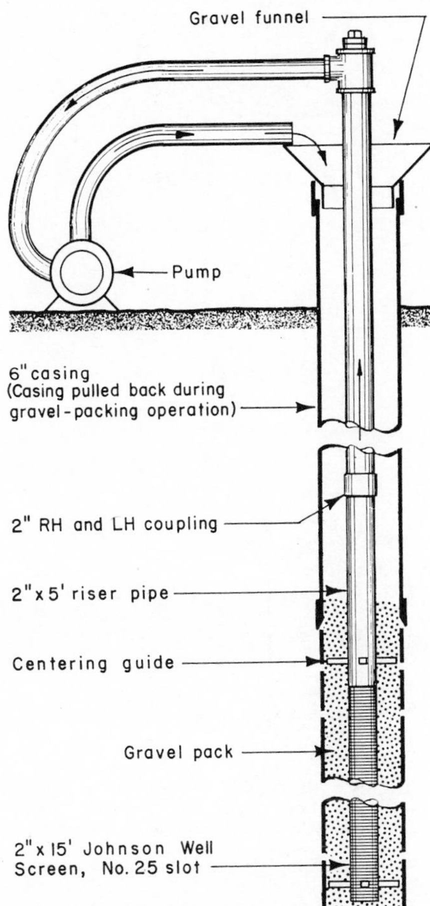

Figure 96[66] Details of 6-inch demonstration well with small centrifugal pump used to provide water circulation during placement of gravel pack.

As slurry emplacement continues, an abrupt pressure increase occurs when the material reaches and encloses the "tattle-tell" screen. (See Figure 97.)

The pumping method using the reverse-circulation techniques requires proper organization, an experienced crew, and reliable equipment.

Another method, originally developed by the oil industry is illustrated in Figure 98. This has been referred to as the "cross-over" method of gravel-packing.[48] Water, carrying the gravel-pack material, is circulated down the drill-pipe, into the "cross-over" tool, and out the two sides into the annulus around the extension pipe above the screen and upward from the bottom of the hole. Water circulates through the screen to the open end of

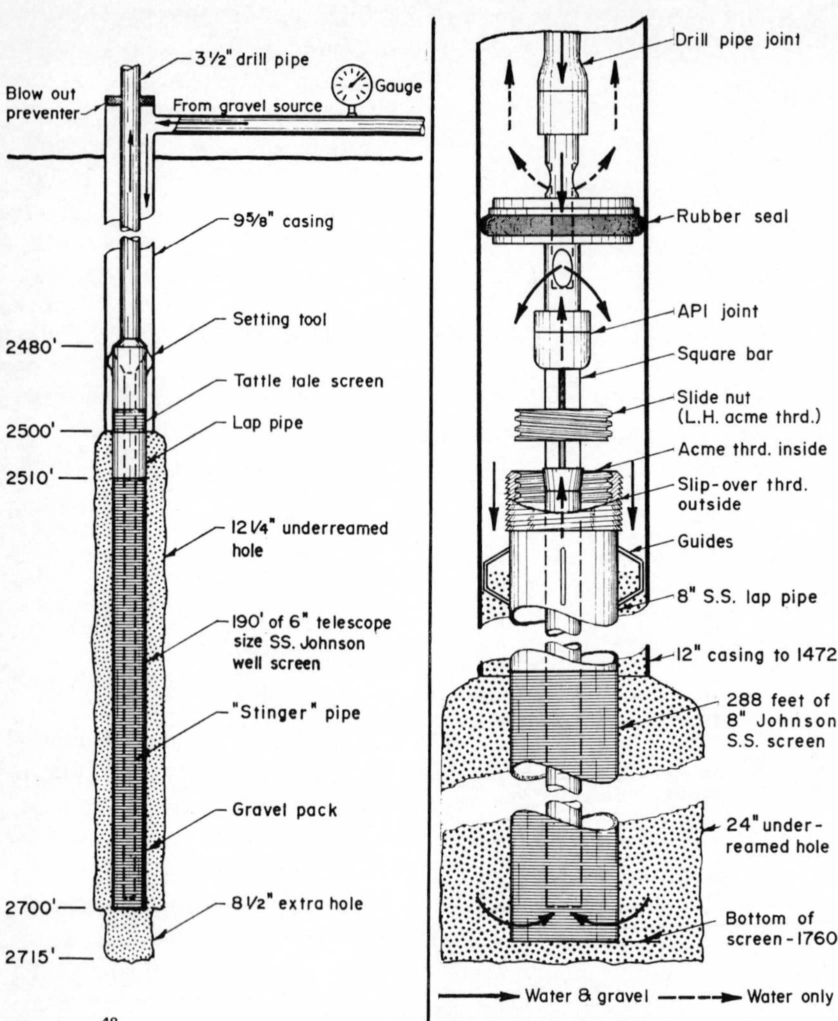

Figure 97[48] Gravel pack can be pumped into annular space with fluid circulating under pressure in deep wells.

Figure 98[48] Sketch showing the principal components used in a deep well when gravel-packing by the "cross-over" method of hydraulic placement.

the "stinger" pipe located near the bottom of the screen, then moves upward to the two openings in the "cross-over" tool above a rubber seal.

Although this method appears simple, it requires elaborate and relatively expensive equipment in addition to considerable skill and operating techniques in order to assure continuous flow without plugging or causing a blow-out. High pressures are required as the gravel

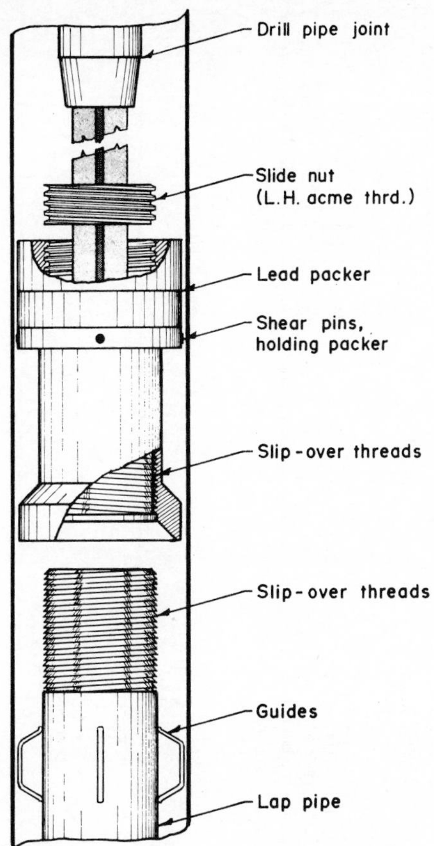

Figure 99[66] Special packer which was used to seal between the screen lap-pipe and the casing in the deep gravel-packed well.

is placed. When placement is completed, a special slip-packer is needed to seal the areas between the screen lap-pipe and the casing. (See Figure 99.)

WELL CASING

Casing is used in many types of well design, ranging in diameter from 1 to 1 1/2 inches in driven wells to 36 inches in some large irrigation, municipal, and industrial wells. Depths at which the casing is set range from ten feet or less to over 1,000 feet. Standard-weight pipe, with casing diameters up to six inches, is usually adequate for the depths at which these small sizes are set. For installations of greater depth, particularly in unconsolidated formations where collapse loads become significant, the wall thickness to diameter ratio becomes important, and heavier walled pipe is mandatory. Of interest for measuring casing lengths are papers by Frimpter[229] and Ross and Adcock.[539] A relatively new method of installing casing has been developed which allows casing emplacement during rotary drilling (see Figure 100.) This method can eliminate loss of rig time when drilling in unconsolidated formations.

There are five products produced for the water well and plumbing industries, as distinguished from the oil field products.[89] They are:

1. Standard Pipe.
2. Line Pipe.
3. Reamed and Drifted Pipe.
4. Drive Pipe.
5. Water Well Casing.

Standard pipe is pipe which comes in a range of sizes, from 1/8 inch to six-inch nominal inside diameter. The sizes are designated by the approximate inside diameters, although the pipe is not exactly these diameters. Each size is available in one wall thickness only, which is classed as "standard weight." The threads on all sizes of pipe are tapered 3/4 inch per foot in diameter, but the threads in the couplings of sizes two inch and smaller are untapered. The couplings are short and not recessed.

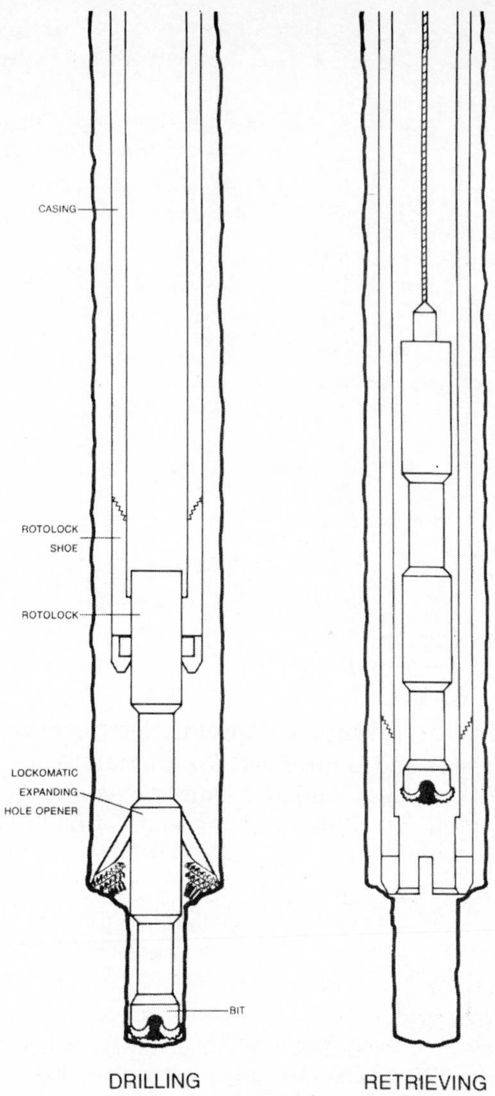

Figure 100 Landing Casing While Drilling (Baker Mining and Construction, Inc.)

Line pipe comes in sizes from 1/8 inch through 36-inch, larger sizes are also available but not to API specifications. Threaded and coupled line pipe are available up to and including a 12-inch size. Larger sizes are available with plain or beveled ends only on special order. As with standard pipe, the threaded sizes are designated as nominal inside diameters. In sizes to six inches (inclusive), the pipe has the same I.D. and O.D. as standard pipe, so the only difference is in the threads and the couplings. All sizes of line pipe have tapered threads on the pipe and the couplings; the couplings are long and recessed. For sizes above six inches, a choice of two or three wall thicknesses are available; therefore, these sizes must be indicated by weight or wall thicknesses in addition to diameter.

Reamed and drifted pipe is offered in sizes, both threaded and coupled, from one-inch to 12-inches inclusive. The pipe is the same as line pipe in most ways but is reamed and drifted, and the couplings are a little shorter (except the two-inch size), and tapped slightly deeper.

Drive pipe is similar to line pipe but is threaded to produce a joint where the ends of the pipe abut when twisted power tight.

Water well casing is a thin-walled, fine-threaded casing which differs in dimensions and threads from all other casing items. Water well casing is fitted with recessed couplings which are shorter and smaller in outside diameter than an equivalent diameter of line pipe. The thread is sharp and tapered 3/8 inch per foot in diameter.

When using any of these products for casing a well, particular attention should be paid to the threading specifications. Some of the thread dimensions are different for each type of pipe, so an improper joint and a possible casing failure can result, for instance, from attempting to join line pipe and standard pipe.

The specifications for casings are designated by letters and numbers. There are several such specifications, but the ones most likely to be used for water well casing are ASTM A-120 and A-53, AISI Standard for R&D Pipe and API 5-L.[81,82,83,92,93,94] ASTM A-120 covers pipe made by the welded or seamless processes, for normal use in water wells.

ASTM A-53 covers seamless and welded steel pipe manufactured by the furnace butt-weld, electric resistance weld, or seamless processes and is intended for normal use in water wells, as well as some other uses. Added requirements are included beyond those covered by A-120. Pipe of this specification is suitable for bending and general fabrication requirements which are not covered by ASTM A-120 testing procedures.

API 5-L is a somewhat more precise specification than ASTM A-53 or A-120 and covers seamless and welded-steel pipe for line pipe. Hydrostatic, tensile, flattening, and bending tests are included under this specification, as well as tight controls of the steel chemistry.

The threading of pipe and casing is an important and precise operation. When the matching threads of a pipe and coupling are cleanly cut and accurately aligned, the resulting joint is dependable. When the threads are improperly made, the resulting joint may be weak and unsatisfactory.

The threading process is monitored very closely at the mills. A poorly threaded joint will rarely come from a reputable modern mill. Unfortunately, threads cut outside of the mill are often of lower quality. Many things cause inferior threads, e.g., improperly dressed, dull, or incorrectly positioned chasers in the die, etc. Threads should be carefully inspected, and any imperfections corrected if good joints are required. The subject of thread-cuttings is complex and is not covered here in any detail.

About 50 years ago, the American Standards Association (ASA) established standard wall-thickness specifications for line pipe called "schedule numbers." Table 25 contains a selected group of sizes which shows the relationships of these schedule numbers to the class designations, e.g., "standard weight," "extra strong," and "double extra strong." There is not a direct correlation between the schedule numbers and the classes of pipe weights; however, the standard class (up to and including six-inch nominal) is the same weight (wall thickness) as Schedule 40. Standard class weight (Schedule 40) in these sizes is generally used because of strength, thickness to resist corrosion, and general availability. These sizes include all of the standard pipe sizes. In the larger sizes of line pipe, R&D pipe, and drive pipe, the nine-inch and ten-inch nominal sizes have the same wall thickness in Standard Class as Schedule 40; but in the 12-inch size, the standard class weight lies between that of Schedule 30 and Schedule 40.

Certain types of tubular products are considered to be oil industry goods as distinguished from water well items. Due to differences which can cause confusion and trouble, the separation of the two is desirable since the requirements in these two fields are different.

TABLE 25[89]

Nominal Size	Actual ID (in.)	OD (in.)	Wall (in.)	Class	Schedule No.
2	2.067	2.375	0.154	Std.	40
2	1.939	2.375	.218	X	80
2	1.687	2.375	.344	None	160
2	1.503	2.375	.436	XX	None
4	4.026	4.500	.237	Std.	40
4	3.826	4.500	.337	X	80
4	3.438	4.500	.531	None	160
4	3.152	4.500	.674	XX	None
8	8.125	8.625	.250	None	20
8	7.981	8.625	.322	Std.	40
8	7.625	8.625	.500	X	80
8	6.875	8.625	.875	XX	None
8	6.813	8.625	.906	None	160
12	12.000	12.750	.375	Std.	None
12	11.938	12.750	.406	None	40
12	11.750	12.750	.500	X	None
12	10.750	12.750	1.000	XX	120

Comparison of OD, ID and wall thickness with class and schedule numbers for elected sizes of line pipe.

TABLE 26[89]

Size OD (in.)	Nom. Wt. T & C (lbs.)	Wall Thickness (in.)	Inside Diameter (in.)
4 1/2	9.50	0.205	4.090
4 1/2	11.60	.250	4.000
4 1/2	13.50	.290	3.920
5 1/2	13.00	.228	5.044
5 1/2	14.00	.244	5.012
5 1/2	15.35	.275	4.950
5 1/2	17.00	.304	4.892
5 1/2	20.00	.361	4.778
5 1/2	23.00	.415	4.670
6 5/8	17.00	.245	6.135
6 5/8	20.00	.288	6.049
6 5/8	24.00	.352	5.921
6 5/8	28.00	.417	5.791
6 5/8	32.00	.475	5.675
7 5/8	20.00	.250	7.125
7 5/8	24.00	.300	7.025
7 5/8	26.40	.328	6.969
7 5/8	29.70	.375	6.875
7 5/8	33.70	.430	6.765
7 5/8	39.00	.500	6.625

Four of the AP1 casing sizes available. Note variation in wall thickness and resultant variation in inside diameter.

API Casing

API casing is always designated by the nominal weight of the outside diameter and the wall thickness. Sizes range from 4 1/2 inches to 20-inches inclusive. Table 26 shows four sizes and the weights available in these sizes.

On certain specifications of API casing, two lengths of couplings — long and short — are available. There are several steel grades available because of different requirements. The threads used are called the API standard-round thread. This thread differs materially from the threads used on the water well products previously described.

As will be seen from the sizes in Table 26, the size and weight of the casing must be designated to assure that items, such as tools, or well screens and pumps can be fitted correctly into these casings since there is a range of inside diameters. When API standard-round thread is needed, care must be exercised to avoid confusion of this thread with other pipe threads. Mixing the threads could result in a misfit which could lead to a casing failure.

Another oil field product which occasionally wanders into water well use is "non-upset" API tubing. This is made in sizes ranging from 1.900-inch O.D. to 4 1/2-inch O.D. Sizes below 4-inch O.D. are threaded, ten threads per inch, and the 4-inch and 4 1/2-inch sizes have eight threads per inch. All these threads have 3/4-inch taper and are of the API round-thread form. Failure to distinguish these items from regular water well pipe has frequently caused misfits and trouble. The use of oil field products should be avoided in water well design, unless there are very good reasons, aside from expediency.

Casing Life

In a water well, the casing is generally received from the supplier in a black condition as a result of an oil coating which protects the casing from atmospheric corrosion during shipment and for a reasonable length of time in storage. Because of objectionable tastes and odors, more durable coatings are seldom practicable in water wells.

Two important approaches to insure the extension of casing life are, first, to install a casing with a relatively thick wall and, second, to select the casing on the basis of its resistance to corrosion. Corrosion and incrustation, important features in casing longevity, are covered in detail later in this text.

Local Casing Requirements

The minimum recommendations or requirements for casing-wall thickness are of interest in some states. A few of the states specify standard weight, Schedule 30 or 40, or equivalent wall thickness with a minimum thickness of 0.250-inch for 6-inch diameter or larger, although a number of states have no requirements or recommendations. With the exception of some areas in the southwest United States, Schedule 30 or 40 casing is used. This casing, to 12 inches in diameter, has adequate strength for depths of 500 to 600 feet.

The AWWA[44] recommendations for heavy-walled pipe (diameters to 20 inches) are adequate for depths to 350 feet or more and the thinner-walled pipe for depths to 200 feet.

With the possible exception of Colorado, few of the state requirements or recommendations, based on experience and local acceptance, consider increased pressures with depth even though well life in some areas averages less than ten years. Many states follow AWWA Standard A100-66 (see Table 27). Casing, as a rule, is one of the major costs of a water well; however, an increase in wall thickness is a relatively minor cost factor, and one of the best means for extending well life.

TABLE 27[7]

STATE CASING REQUIREMENTS OR RECOMMENDATIONS

CALIFORNIA –	.1046 inch wall thickness up to 10 inches diameter
	.1345 inch wall thickness 12 to 14 inch diameter
	.1644 inch wall thickness 16 to 20 inch diameter
	.250 inch wall thickness 20 inches and larger diameters
COLORADO –	Minimum casing diameter nominal 4 inches. Minimum wall thickness .188 inches but wall thickness must be adequate to resist stresses to which it may be subjected.
NEVADA –	.1345 inch wall thickness up to 8 inches diameter
	.1644 inch wall thickness 10 to 12 inches diameter
	.250 inch wall thickness 14 to 20 inches diameter

Casing Strength

There are a number of equations available for estimating the collapse resistance of casing, but the one most applicable in water wells is that developed by Cleindeinst.[174] His equation for elastic collapse is:

EQUATION 57

$$P_c = \frac{2E}{1-U^2} \quad \frac{1}{\left(\frac{d}{t}\right)\left(\frac{d}{t}-1\right)^2}$$

where:
- P_c = Critical collapse pressure (psi)
- E = Modulus of elasticity
- U = Poisson's ratio
- d = O.D. of the pipe (in.)
- t = Wall thickness of the pipe (in.)

Tests on pipes indicate that minimum collapse strength may be as little as 75 percent of the estimated critical strength. (See Table 28.) Casing specifications permit a wall thickness 12 percent less than the nominal, and the casing may be as much as one percent out-of-round. Both factors act to reduce collapse resistance. Axial tension and bending reduce and axial compression increases collapse resistance. In addition, threading or poor welding at joints may reduce the tensile strength at these points to 40 percent. Because of these factors and the assumed average value of soil pressure, Equation 57 is used with a design factor of three for wells in unconsolidated formations, and of 1.8 to two for wells in consolidated rock. The estimated required wall thickness is seldom the same as standard, nominal wall thickness, so the nearest thick-walled casing is used.

TABLE 28 [44]

Theoretical Collapsing Pressure for 2-Ply Casing

	Casing Gage							
	12		10		8		6	
Casing Diameter *ft*	Excav. Depth *ft*	Pressure *Psi*	Excav. Depth *ft*	Pressure *Psi*	Excav. Depth *ft*	Pressure *Psi*	Excav. Depth *ft*	Pressure *Psi*
8	694	300	1540	670	––	––	––	––
10	353	153	780	339	––	––	––	––
12	204	88	450	196	780	339	––	––
14	127	55	280	122	500	216	––	––
16	85	37	188	81	336	146	552	240
18	60	26	130	60	239	104	388	168
20	43	18	96	41	170	73	280	122
22	––	––	72	31	127	55	212	92
24	––	––	56	24	96	41	163	71
26	––	––	43	18	78	34	127	55
30	––	––	28	12	50	21	83	36

Field Operations for Cable-Tool Drilling

Moss[471] states that the cable tool or percussion method of drilling is used in the construction of most high capacity wells (in excess of 500 gallons per minute) in the southwest United States. The formation characteristics in deep, unconsolidated alluvial basins require that the casing be installed by driving or jacking coincident with drilling. Following installation, the casing is either selectively perforated or screened in the desired aquifers.

Consideration must be given to the following factors in selecting casing to be installed in wells drilled by the cable-tool method:

(1) The casing must have a smooth exterior to minimize seizing by the formation.

(2) The casing must have a smooth interior to permit passage of drilling tools.

(3) The inside diameter must be large enough to accommodate the pump.

(4) The casing should be manufactured from steel having physical properties that resist the severe stresses from drilling and handling.

(5) The casing must have sufficient wall thickness to resist stresses from placement and subsequent production.

(6) Within economic limits, the chemistry of the steel should promote long life of the casing.

(7) The joint design must provide reliable coupling with high joint efficiency.

One solution to these requirements has been achieved with two-ply, slip-joint well casing which has been developed over the past 70 years. Double-walled casing is manufactured from high strength copper-bearing steel, in four-foot lengths. The inside and outside joints are offset two feet. The inner casing fits to the outer casing with a tolerance not exceeding 10/1000ths of an inch. Strength and durability are thus enhanced from inherent structural qualities rather than from the efficiency of the welded joint.

For field assembly, the inside joints are flush and the outside joints are separated, approximately 1/8 inch so that a circumferential weld can join the outside joints to the inside joints. In order to withstand severe driving forces, the bottom 20 feet of the casing is reinforced and is welded to a forged, heat-treated drive shoe.

The inside diameter of well casing is generally two inches larger than nominal pump-bowl diameter. There are instances, however, when greater clearance between pump bowls and column-pipe well casing may be desirable. Table 29 shows minimum inside diameter of the casing as a function of nominal bowl diameter and yield range. Minimum wall thickness is selected primarily from the following considerations:

(1) Required durability.
(2) Maximum drawdown during production.
(3) Physical requirements for installation.

Generally, casing durability can be anticipated from accumulated data in other wells of the same area. If highly corrosive ground water conditions exist, casing durability can be extended by increasing the wall thickness.

Moss[471] states that during production under ideal conditions, a cone of depression surrounds the casing with a negligible pressure differential imposed on the casing. Under normal field conditions, unless the casing is perforated above the pumping level, lowering the water level inside the casing produces a pressure differential on the casing equal to the difference between the standing water level and the pumping level. Wall thickness must be adequate to sustain this pressure. The critical collapsing strength of double-walled casing can be computed by the following equation:

EQUATION 58

$$P = \frac{(62.5 \times 10^6)(0.65)}{\left(\frac{Dm}{t}\right)\left(\frac{Dm}{t-1}\right)^2}$$

Where: P = collapsing pressure (psi)

$D_m = \frac{D_1 + D_2}{2}$

D_1 = Casing inside diameter (in.)

D_2 = Casing outside diameter (in.)

$t = \sqrt{(T_1^2 + T_2^2)}$

T_1 = Thickness of inside joint (in.)

T_2 = Thickness of outside casing joint (in.)

Note: Two thirds of the computed values should be used to provide an adequate safety factor.

Physical requirements for installation depend primarily on the nature of the formation and the total depth. Jacking forces in excess of 250 tons have been required to install some types of casing. Tables are given by Moss[471] for double-walled casing for collapsing pressures and weights, axial crushing strength and recommended minimum

TABLE 29[471]

Nominal Bowl Diameter (inches)	Operating Pump Speed (RPM)	Yield (gpm)	Minimum Casing I.D. (in)
8	3500	200 – 1200	10
	1800	100 – 600	
	1200	160 – 400	
10	1800	200 – 1500	12
	1200	370 – 670	
12	1800	400 – 2300	14
	1200	250 – 1500	
14	1800	1000 – 4500	16
	1200	700 – 3000	
16	1800	2000 – 5200	18
	1200	1300 – 3400	
18	1800	3200 – 4100	20
	1200	2200 – 4000	
	900	2800 – 3000	
20	1200	3100 – 4400	24
	900	2300 – 3600	
22	1200	7500	24
	900	5600	

wall thickness as a function of well depth and diameter, See Tables 30, 31, and 32. Although the use of double-wall or "stove pipe" is not in general use today in the United States because of readily apparent deficiencies, the foregoing discussion briefly summarizes its design features for purposes of comparison with the commonly used single wall casing.

With standard cable-tool methods of well construction, the requirement for medium-weight casings, particularly with large diameters, precludes the use of relatively expensive ferrous and non-ferrous metals in fabrication. Experience and test data indicate

TABLE 30[471]

Collapsing Pressures & Weights of Double Wall Casing.

	12 Ga.		10 Ga.		8 Ga.		6 Ga.	
Dia.	Coll.	Wgt.	Coll.	Wgt.	Coll.	Wgt.	Coll.	Wgt.
8	694	19.6	1540	25.3				
10	353	24.4	780	31.3				
12	204	28.0	450	37.3	780	45.9		
14	127	33.7	280	43.3	500	55.0		
16	85	38.3	188	49.9	336	62	565	75
18	60	43.2	130	55.5	239	70	388	86
20	43	47.8	96	61.5	170	78	280	96
22			72	67.3	127	85.6	212	105
24			56	73.2	96	93	163	114
26			43	80.5	78	98.5	127	121

NOTE: The collapsing values shown above are in feet of water. Weights are in pounds per lineal foot.

TABLE 31[471]

Approximate Axial Crushing Strength of Double Wall Casing in Tons.

Dia.	12 Ga.	10 Ga.	8 Ga.	6 Ga.
8	65	80	100	120
10	80	105	125	150
12	95	125	150	180
14	110	145	175	210
16	125	165	200	235
18		185	225	265
20		205	250	295
24		245	300	355

that well durability can be extended by the addition of 0.20 percent copper to steel. Recent tests with corrosometer probes indicate that, under minimal corrosive conditions, "splash" zone is higher than that in the submerged zone. The corrosive environment in the "splash" zone is equivalent to a moist atmosphere in which copper-bearing steel has twice the corrosion resistance of carbon structural steel.

Field Operations for Rotary Drilling

Most high capacity, gravel-packed water wells are constructed by the rotary method of drilling. A typical well installation consists of: conductor casing, grouted in place; blank-well casing; screen casing; and gravel packing, if applicable.

Consideration must be given to the following factors when selecting casing and screen to be installed when drilling by rotary methods:[471]

(1) The casing and screen must have a smooth interior to permit installation of the pump and the operation of development equipment.
(2) Casing inside diameter must be large enough to accommodate the pump. Screen inside diameter must be large enough to permit water flow without excessive head loss.
(3) Casing and screen must have physical properties adequate to permit installation and to resist stresses from development and production.
(4) Field connection joints must be equal to casing strength.
(5) Casing and screen material must withstand a potentially corrosive ground water environment.
(6) The shape of the screen openings should be designed to reduce plugging to a minimum, to retard the entrance of fine sands and silts, and to provide minimum resistance to water flow.

Casing and screen sections are subjected to tensile stresses during installation. Additional tensile stress is caused by the downward movement of the gravel pack material during consolidation and well development. Although these stresses cannot be calculated, Moss[471]

TABLE 32[471]

Suggested Minimum Thickness for Double Wall Casing

Depth of Casing, In Feet	Diameter, in inches								
	10	12	14	16	18	20	22	24	30
0 – 100	12	12	12	12	10	10	10	10	8
100 – 200	12	12	12	10	10	10	10	8	8
200 – 300	12	12	10	10	10	10	8	8	8
300 – 400	12	12	10	10	10	8	8	8	8
400 – 600	10	10	10	10	8	8	8	8	8
600 – 800	10	10	10	8	8	8	8	8	8
Over 800	10	8	8	8	6	6	6	6	6

Note: Values are United States standard gage.

indicates that a factor of safety of 2:1 to 3:1 over theoretical casing and screen weight is adequate.

The casing yield strength, applying only to a perforated casing, can be computed by the following equation:

EQUATION 59

$$Y_s = S_y\,(\pi\, D_m t)$$

Where:
Y_s = Yield strength (lb.)
S_y = Yield strength of metal (psi.)
$D_m = \dfrac{D_1 + D_2}{2}$ (mean diameter, in.)
D_1 = Casing inside diameter (in.)
D_2 = Casing outside diameter (in.)
t = Casing wall thickness (in.)

Screen-yield strength is a function of the casing cross-sectional area between the perforations. The above equation is for wire-wrapped screens. Equations for other types of screens can be obtained from various screen manufacturers. The columns critical strength must be considered an important factor in setting deep screens with casing or riser pipe.

Minimum wall thickness is determined primarily from the following conditions:

(1) Anticipated drawdown during production.
(2) Total well depth and cumulative effect of stresses imposed by movement of the gravel pack and formation materials during development.
(3) Required durability.

The critical collapsing strength of single-walled, steel casing may be estimated by the following equation:

EQUATION 60

$$P = 50.2 \times 10^6 \left(\frac{t}{D}\right)^3$$

Where: P = Collapsing pressure (psi.)
t = Casing wall thickness (in.)
D = Casing outside diameter (in.)

Moss[471] reports that the combined stresses imposed on the casing and screen during installation, development, and operation are difficult to estimate. Many wells have been constructed using casing with a collapsing strength of 40 pounds per square inch. As depth increases, greater allowance should be made for these stresses. Tables 33 and 34 give collapsing pressure and weight and recommended wall thickness as a function of depth and diameter for steel well casing.

TABLE 33[471]

Collapsing Pressures & Weights of Steel Well Casing.

	3/16		1/4		5/16		3/8	
Dia.	Coll.	Wgt.	Coll.	Wgt.	Coll.	Wgt.	Coll.	Wgt.
8	1070	17	2800	22	5500	28		
10	600	21	1430	28	3000	35		
12	378	24	900	33	1770	41		
14	242	28	570	38	1140	48	2000	56
16	162	32	390	43	760	54	1350	64
18	116	36	275	49	540	61	940	72
20	85	40	202	54	390	68	690	80
22	64	44	153	59	300	74	520	88
24	49	48	118	65	230	81	400	96
26	39	52	93	70	181	88	317	104
28	31	56	74	75	147	94	245	112
30	25	60	60	81	119	101	205	120

NOTE: The collapsing pressures shown above are in feet of water.
Weights are in pounds per lineal foot.

With thin-walled, large-diameter casing, welded construction for field assembly of both casing and screen sections is indicated for the following reasons:

(1) The intial cost of casing designed for assembly by welding is less.
(2) The uniformity of wall thickness at the joints.
(3) The increased efficiency of properly welded joints approaches 100 percent; whereas, standard threaded and coupled joints have an efficiency of approximately 60 percent. Either bell-ends or collars should be provided to facilitate welding and installation. Criteria for determining the inside casing diameter is the same as that previously described in the section on double-walled casings.

TABLE 34[471]

Suggested Minimum Thicknesses For Steel Water Well Casing.

Single Casing Depth of Casing In Feet	6	8	10	12	14	16	18	20	22	24	30
00 – 100	12	12	12	10	10	8	8	8	8	8	3/16
100 – 200	12	12	10	8	8	8	3/16	3/16	3/16	3/16	1/4
200 – 300	10	10	8	8	8	3/16	3/16	3/16	1/4	1/4	1/4
300 – 400	10	8	8	3/16	3/16	3/16	1/4	1/4	1/4	1/4	5/16
400 – 600	10	8	3/16	3/16	3/16	1/4	1/4	1/4	5/16	5/16	5/16
600 – 800	3/16	3/16	3/16	3/16	1/4	1/4	1/4	5/16	5/16	3/8	3/8
Over 800	3/16	3/16	3/16	1/4	1/4	1/4	5/16	5/16	3/8	3/8	7/16

note: upper in guage, lower in inches

The most commonly used metals for well casing and screen are listed in Table 35 with their approximate cost ratio using carbon-structural steel as the base unit.

The *cupro-nickel alloys* are most suited for sea-water production wells (with high sodium chloride combined with a high content of dissolved oxygen). Silicon bronze is used in wells with a ground water environment of high total hardness, high sodium chloride content, and low pH. Copper-bearing steel has about twice the corrosion resistance of plain carbon steel. High strength, low-alloy steel has at least three to four times the corrosion resistance of carbon-structural steel.

Carbon steel (High strength, low-alloy steel) and stainless steels in general have demonstrated a high resistance to soil corrosion. The relative merits and performance of these metals are covered in detail in the National Bureau of Standards Circular 579 (April 1957) which is based on extensive field-burial programs that determined the influence of various types of soils on metal corrosion.

TABLE 35[471]

Name of Metal	Material cost ratio	Casing cost ratio	Screen cost ratio	Approx. installed cost ratio
Carbon structural steel (ASTM A-7)	1.0	1.0	1.0	1.0
Copper bearing structural	1.0	1.0	1.0	1.0
Kai-well (high strength copper bearing, used in double well casing)	1.2	1.3	1.2	1.1
High strength, low alloy structural (ASTM A-242)	1.4	1.3	1.2	1.1
Cupro-nickel (10% nickel)	15.6	15.5	15.3	6.9 – 9.3
Silicon Bronze	15.0	14.9	14.7	6.6 – 8.9
Stainless Steel, type 304	6.8	6.6	6.4	1.5 – 2.0
Stainless Steel, type 316	9.7	9.5	9.3	2.2 – 3.0
Transite	–	1.3	2.5	1.4 – 1.9

Transite pipe has been used for casing water wells, although this pipe is unsuitable for the following reasons:

(1) Low joint strength.

(2) Insufficient joint alignment to permit the passage of close-fitting swabs.

(3) Inability to sustain abuse, due to brittleness.

(4) High cost of perforating.

(5) High cost of installation.

(6) Transite pipe cannot be perforated in a manner which will satisfy the requirements for high-capacity wells.

Where considerable durability and high reliability are desired, the *stainless steels*, having relatively high tensile and yield strengths and twice the modulus of elasticity of the copper alloys, offer the best solution. The cost of these metals has declined in recent years, and overall costs can generally be maintained within reasonable limits by predetermining the inside diameter and wall thickness required. Since the deterioration of these metals is minor over a period of time, the wall thickness and weight per unit area can be strictly limited to the required collapsing strength imposed by drawdown.

Non-Metallic Casing

The widespread use of *plastic casing* has increased in recent years. The quality of such casing has been improved because of more efficient extrusion and molding techniques and developments in thermoplastic materials.[524] The polyvinyl-chloride plastic well casing (PVC) in use today is NSF approved and usually is the schedule 40 or 80 type. Maximum installation depths are normally less than 200 feet, due to potential vertical-stress failures. Specifications for wall thickness and diameter based on ASTMD #1785 are given in Table 36. Another type of plastic well casing is constructed from virgin, white, high-impact, rubber-modified polystyrene material. This casing is not recommended for depths in excess of 300 feet. Pertinent specifications are given in Table 37. Plastic well casings are usually not larger than six inches in diameter for structural reasons, and generally all joints are other than screw type.

Wells in highly corrosive environments are sometimes cased and screened with ceramic tile, concrete, asbestos cement, or even wooden pipe. Many of the installations have been relatively successful, although the limited open area of the perforated sections produced low well efficiencies. However, most of the substitutes are

TABLE 36

POLYVINYL CHLORIDE (PVC) PIPE FOR WELL CASING

U. S. Dept. of Agriculture, Soil Conservation Service Standard

Nominal Size	Outside Diameter	Inside Diameter	Minimum Wall Thickness	Outside Diameter	Inside Diameter	Minimum Wall Thickness
	PVC 1220 SCHEDULE 40			PVC 1220 SCHEDULE 80		
inches	inches	inches	inches	inches	inches	inches
1½	1.900	1.610	0.145	1.900	1.500	.200
2	2.375	2.067	0.154	2.375	1.939	.218
2½	2.875	2.469	0.203	2.875	2.323	.276
3	3.500	3.068	0.216	3.500	2.900	.300
4	4.500	4.026	0.237	4.500	3.826	.337
6	6.625	6.065	0.280	6.625	5.761	.432

TABLE 37

HIGH-IMPACT, RUBBER-MODIFIED POLYSTYRENE PIPE FOR WELL CASING

U. S. Dept. of Agriculture, Soil Conservation Service Standard

Nominal Diameter	Minimum Wall Thickness
Inches	Inches
4	0.220
5	0.250
6	0.300

heavy, structurally weak, requiring special care and equipment for installation, and are not readily available in many areas. Concrete and asbestos-cement pipe are subject to rapid deterioration in high-sulfate soils and water. If incrustation is a problem, the acidizing of the concrete, asbestos-cement, wooden casing, and screen usually results not in the rehabilitation of the well but in the destruction of the well.

The development over the last 30 years of plastic pipe which is dielectric, light weight, inexpensive, and corrosion resistant has provided an excellent substitute for use in small diameter shallow wells, but such materials have a tendency to creep and yield under sustained stress. These materials, therefore, have not proven satisfactory in large diameter deep wells in unconsolidated sediments.

The recently developed, fiberglass-reinforced, *epoxy pipe* has more desirable characteristics than pure plastics. Epoxy plastic has been used extensively in the petroleum industry in corrosive environments for small diameter installations in deep wells and as pressurized distribution pipe. Eight- and ten-inch diameter 0.20-inch walled pipe has been used extensively to depths of about 300 feet for water well casing in Pakistan extensively and in the United States to some extent. A few large-diameter installations with 0.50-inch wall thickness or heavier have been installed in wells for special purposes in the United States.

Epoxy plastic pipe has a high resistance to corrosion and reduces incrustation problems. However, the 0.20-inch and thinner walled pipe is reported to collapse under the stress of normal development, and the slotted pipe, used as screens, has a relatively low percentage of open area. There has been a reluctance to experiment with this type of pipe in large diameter deep wells in unconsolidated materials regardless of the advantages. The collapse resistance could be improved by increasing the wall thickness — diameter ratio, at which time, however, stainless steel and other proven alloys become economically competitive. Epoxy plastic pipe is now made in nearly all sizes. As knowledge about properties of this pipe expands, technology improves, and competition becomes more effective, there will be an increased use of such pipe as casing in the water well industry on the basis that potential corrosion can be all but eliminated.[678]

DRILL PIPE

The petroleum industry has done excellent work on the allowable hook load and the torque combination of drill pipe as well as studies on collapse strength.

Numerous drill pipe companies can supply additional information on pipe strength, etc.[590]

FISHING OPERATIONS

Fishing operations can often be prevented by proper equipment care and attention. Loss, breakage, and the parting or collapsing of a string of casing is generally the result of carelessness and may be avoided with proper care of the equipment. A thorough periodic inspection of equipment will greatly reduce the frequency of fishing operations. Drilling cables should be carefully inspected for signs of weakness or unusual wear; drilling tools should be inspected for fatigue cracks particularly at welds, and tools should not be lowered into the well unless the equipment is in perfect condition. Fishing is the cause of annoying delays and financial loss in drilling operations.

Probably the majority of fishing jobs are the result of mechanical failures — either from overworking the tool or equipment or from the improper use of accessories in the drilling operation. The abuse of tools or the use of worn tools contributes to the need for a great number of fishing operations. Any work being done in the drilling, cleaning, or servicing of a gas, oil, or water well involves potential problems: (1) failure of the drilling tools or the casing or (2) erroneous procedure — either of which may necessitate fishing. Thus, the potential cost of fishing is frequently the cause of high drilling costs.[63,378,516]

Field Operations

Dipping or creviced formations, subsurface caverns, crooked or slanting holes, settling sands, boulders, etc. are conditions which contribute to drilling problems. Careful and deliberate assessment of these problems will greatly reduce the necessity of fishing operations.

When fishing is required, knowledge and past experience should be considered before recovering the "fish." In fishing, the operator must know what to do as well as what not to do, for many minor fishing jobs have become complicated because the problem was not properly handled early. Many simple jobs are made difficult when improper fishing tools and faulty judgment are employed. The first principle of fishing is never to run an unsuitable tool in the hole. If the proper tool is not available, fishing should not be attempted until the tool can be obtained or built. The vast majority of fishing jobs are simple and easily accomplished unless complicated by improper procedures.

If the operator is prepared for potential fishing problems, the exact dimensions of all downhole drilling and fishing tools should be recorded. Some of the important tool measurements are: outside diameter and length of rope socket; diameter and length of the neck of rope socket; diameter and length of drill stem, etc. The measurements should include size of joints and outside diameter of pin and box collars as well as body size and length of bits. A careful record of the depth of the hole and the overall length of the drilling string should be noted.

If fishing is difficult or becomes complicated, the cost of the tools, time, and labor may greatly exceed the cost of the hole. It may be more efficient to start a new hole. In any event, cost should be the main factor in deciding what action to take.

Fishing operations require skill, patience, and ingenuity. In many cases, the completion of the well depends upon success in recovering lost equipment. A great variety of special tools and procedures have been devised to assist fishing, some of which are described by Decker[199] for cable-tool equipment, but many of the techniques are widely used with other drilling methods. Only common fishing tools are discussed. Many of these tools are rarely used; in some cases, a tool will be made for a particular purpose and never be used again. Only the largest contractors can afford to own more than a limited assortment of fishing tools. Many operators make a practice of renting tools when needed from local rental agents.

No amount of measurement at the surface can exactly determine the position of the lost tool in the hole or, in some cases, whether the top is free from obstruction. *Impression block* is often used to obtain an impression of the top of the tool before attempting any fishing operations. This is particularly necessary in rotary-drilled, uncased holes. Impression blocks have many forms and designs, one of which is shown in Figure 101. A short block of wood having a diameter one inch less than that of the drilled hole fits tightly into the drill-pipe box collar. Warm paraffin wax, yellow soap, or other plastic material fills the sub and then cools and solidifies. The block is carefully lowered into the hole until the "fish" is encountered. The impression block is then raised to the surface where the impression made in the wax or soap can be examined. By careful interpretation of the impression, the position of the "fish" and the best means of retrieving it can be determined.

The recovery of drill pipe that has twisted off in the hole is a frequent fishing operation. The break may either be due to shearing of the pipe or failure of a threaded joint. An impression block can be used to determine the exact depth and position of the top of the pipe, whether there has been any caving, or whether the pipe has become embedded in the wall of the hole. If the top of the pipe is unobstructed, either the *tapered fishing tap* or *die overshot* can be effective if used before the cuttings in the hole settle and "freeze" the drill pipe. The *circulating-slip overshot*, which permits the circulation of drilling fluid, is the best tool to use if the pipe has been frozen by the settling of cuttings around it. These tools are all illustrated in Figure 102.

The *tapered fishing tap*, made of heat-treated steel, is tapered one inch per foot from a diameter smaller than the inside of the coupling to a diameter equal to the outside of the drill stem. The tapered portion is threaded and fluted to permit the escape of chips cut by the "tap." The tap is lowered slowly on the drill stem until it engages the lost pipe; the circulation is maintained at a low rate through the hole in the tap during this period. After engaging the lost pipe, the tap turns slowly by the rotary mechanism or by hand, and circulation is stopped until the tap is threaded into the pipe. Circulation should be re-established through the drill string before pulling the lost pipe.

The *die overshot* is a long tapered die of heat-treated steel designed to fit over the top of the lost drill pipe and cut thread when rotated. This tool is fluted to permit the escape of metal cuttings. Circulation cannot be completed to the bottom of the hole through the lost pipe since the flutes also allow the fluid to escape. The upper end of the tool has a box thread designed to fit the drill pipe.

The *circulating-slip overshot* is a tubular tool, approximately three feet long, with an inside diameter slightly larger than the outside of the drill pipe. The beveled lower portion of the tool helps to centralize and guide the top of the lost drill pipe into the slip that is fitted in the tapered sleeve. The slot cut through one side of the slip enables expansion of the slip as the tool contacts the drill pipe. When the tool is raised, the slip is pulled down into the tapered sleeve, thus, tightening the slip against the pipe. Circulation of fluid can then be established freeing the pipe for recovery.

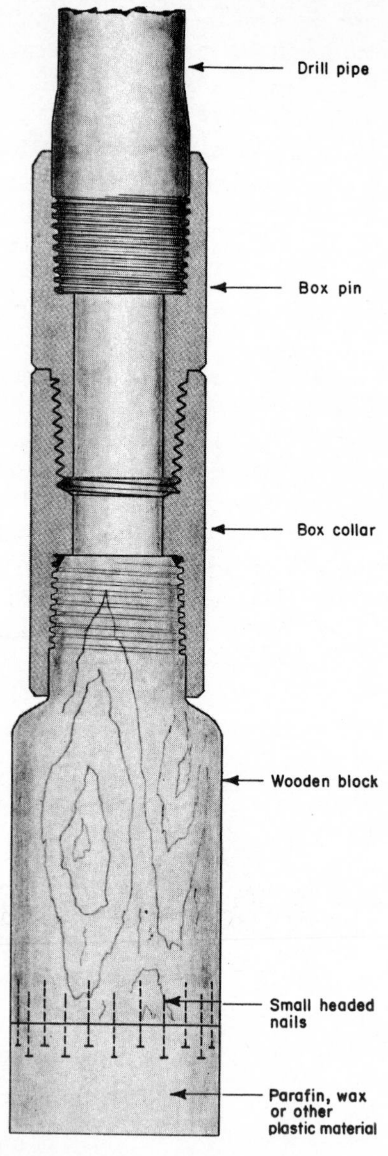

Figure 101[246] Impression Block

A *wall hook* is used to straighten the lost drill pipe in the hole in preparation for removal by the tap or overshot tools. The wall hook is a simple tool that can be made from steel casing, shaped with a cutting torch. A reducing sub connects the top end of the tool to the drill stem. The wall hook is lowered until it engages the pipe, then slowly rotated until the pipe is fully within the hook. The hook is slowly raised to set the pipe in an upright position and then it is disengaged from the pipe.

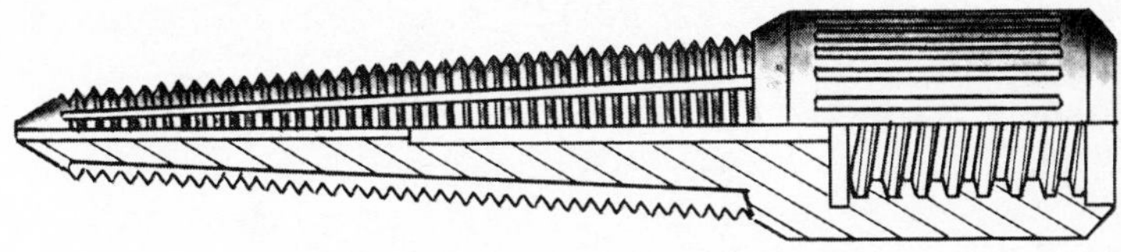

Tapered Tap

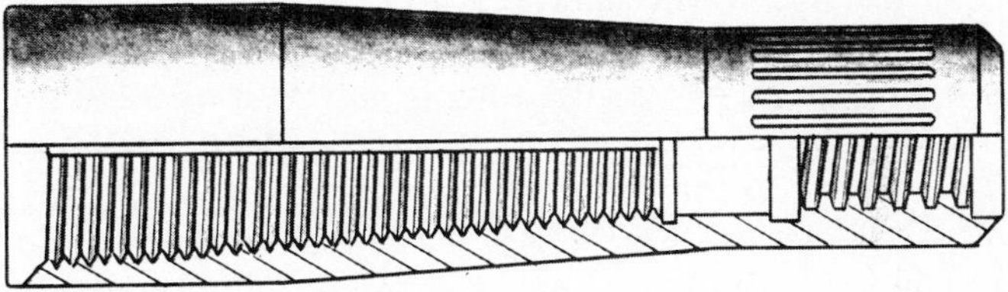

Die Overshot

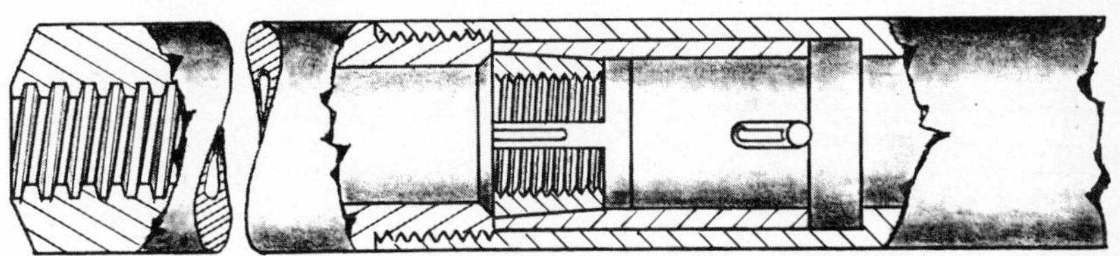

Circulating Slip Overshot

Figure 102[43] Tapered Tap and Overshots.

A tapered fishing tap can be attached to the upper portion of a wall hook. Using this combined tool, the hook can realign the lost drill pipe and then, while being lowered, guide the tap into the drill pipe to complete both operations in one run. This method is particularly desirable when the drill pipe leans against the wall of a much larger hole.

The loss of a downhole casing is perhaps the most difficult fishing problem. The loss results from freezing, collapsing, telescoping, parting, or splitting of the casing. Freezing results from the caving of the wall against the pipe, accumulation of cavings or sand around the casing collars, contact with the walls in a crooked hole, or an improperly reamed interval in the well. Collapse of the casing is due to external pressure — generally hydrostatic. Caving of the walls or loose boulders pressing against the pipe may deform it. Telescoping of a column of pipe can result from either dropping or driving the pipe carelessly when the lower end is frozen. A column of casing may part because of: extreme tensional strain, engendered by its own weight; an attempt to lift the casing when it is frozen; defective threads; or improperly coupled joints. The casing can split as a result of defective welding in the manufacturing process, drilling formation material which has "heaved up" into the casing, or improperly using a swedge, casing spear, or other fishing tools. Most of these difficulties can be avoided by proper selection and inspection of casing and by carefully coupling the joints before the casing is lowered into the well. Good judgment is also necessary in determining the depths at which a string of pipe can withstand strain. The condition of the walls of the well, the straightness of the hole, and the proper reaming of all sub-guage intervals have an important bearing on the success of a casing installation.

If the casing develops frictional contact with the formation (i.e., if it shows indications of being collar bound with mud or loose material from the walls) the casing can often be freed by being alternately raised and lowered a vertical distance of 20 to 40 feet, depending on the length of joints. If this procedure fails to relieve the friction, the well may be bailed down within the casing so that hydrostatic pressure can clear the space around the pipe.

If friction results from an attempt to lower a column of casing into a too-small hole, the best remedy is to pull the casing until the casing shoe is above the tight interval and to thoroughly under-ream the hole. Under such conditions, particularly if the casing has been driven, lifting the column against the friction with power available is often impossible without placing undue strain on the rig. In such cases, a combined pull and jar action of the rig is often successful. Another method is shown in Figure 103.

Many dangers are involved when pulling frozen or lodged casing. When the mechanical advantage of the reduction gear and hoisting block is considered, it is apparent that a sufficient force to pull the pipe apart or to collapse the rig mast can be generated. Men have been killed or injured by derricks that have collapsed and casing lines, hoisting blocks, or elevators that have snapped from the sudden release of tension when the casing pulls apart near the surface.

If friction is due to cavings in the bottom of the hole, or if friction is due to a restricted hole diameter, the casing can be mechanically driven at the surface to assure emplacement before freezing occurs. Alternate driving and pulling is also effective to free the casing from wall friction. When casing is being pulled from a caving hole, freezing can occur several feet from the bottom, but by driving a short distance, the casing often can be freed.

If all other means of pulling frozen casing have failed, a casing connector can be used with considerable success. A *casing connector* is a tool with a tool-joint box; a mandrel top (similar to a rope-socket neck, but solid); and a shoulder between the neck and box threaded to fit a casing coupling. The casing connector may also be used to pull broken casing that remains in the bottom of the hole. The fishing string may be the same size as the lost casing, and a short stem can be connected between the spear and the casing connector. Thus, if the top joint of the casing is split, the spear may secure a firm hold.

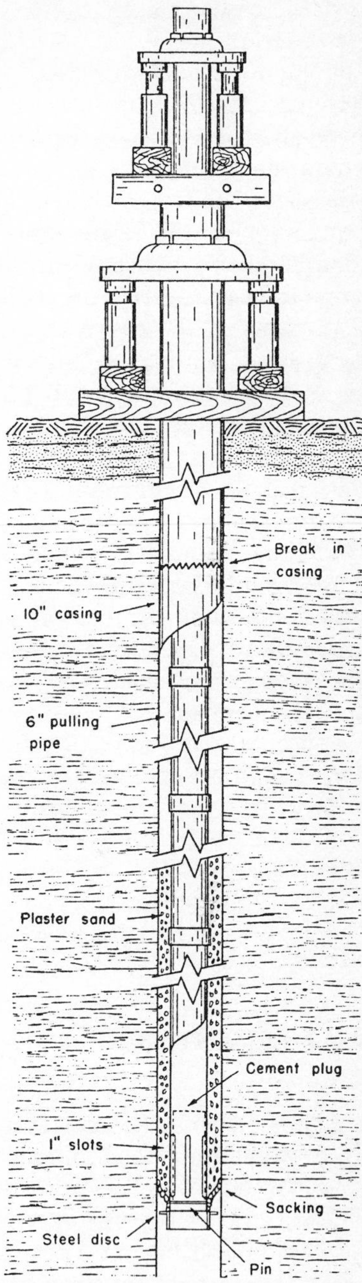

Figure 103[169] Rigging for pulling back 10" casing which parted 60' below ground.

If the casing can neither be pulled nor driven, it is possible to save that section which is not frozen. The freeze may be located by running a drive-down spear and testing for vibration. The spear is driven to the bottom of the casing and raised and jarred at regular intervals. No vibration will occur until the spear contacts the area where the casing is frozen. A *casing ripper* (or cutter) is then lowered just above the freeze and the casing is parted and removed. If tension is maintained, the casing should move when the parting operation is completed. To continue drilling, casing must be used that will telescope the casing left in the hole.

Collapse of the casing is caused by external pressure from hydrostatic head bearing on the pipe; by caving walls of the hole; by a loose boulder in the wall pressing against the pipe; or by sudden "heaving" of unconsolidated sand into the well from oil, gas, or water horizons. If casing is set to exclude ground water of undesirable quality, fluid accumulates outside the pipe to many feet above the casing shoe — at times even to the surface. The hydrostatic head which can collapse the casing will vary with the density of the fluid. Fresh ground water develops a pressure of 0.433 pound per foot of depth and saline ground water (containing 34,000 parts dissolved salt per million) develops a pressure of 0.444 pound per foot of depth. For average conditions in making approximate computations, the fluid is assumed to have a density of 1.15 and develops a static pressure of 0.5 pound per square inch per foot of depth. Conservation computations assume that fluid outside the pipe extends to the surface and that a collapsing pressure is developed on the casing equivalent to a fluid head of the full length of the casing. This collapsing pressure can be a force of great magnitude reaching as much as 3,000 pounds per square inch in a well 6,000 feet deep.

References on other aspects of fishing are included in the Annotated Bibliography.[31,63,378,516]

WELL CEMENTING

Petroleum industry cementing techniques have become highly specialized and apply to water well construction, although this transfer of cementing technology has not progressed as well as in other areas. As water wells are drilled deeper, this transfer, however, will become a necessity. An excellent report on oil well cementing practices was published in 1959 by the American Petroleum Institute[27] which is very well referenced and comprehensive in its treatment of oil industry techniques. Many case histories of various cementing techniques are given in the literature.[257,450]

The following is a summary of present cementing practices and salient features related to water well construction. The term *cementing* includes the entire operation of mixing and placing the grout. Grouting well casing involves filling the space around the pipe, usually between the pipe and the drilled hole with a suitable slurry of cement or clay and sand. Cement slurry, a mixture of only Portland cement and water, is commonly used, but puddled clay can also be used, if used at a depth where drying and shrinkage of the mud will not occur and where water movement will not wash away the clay particles. If the well construction includes both an inner and outer casing, grout is placed between and outside the two casings.[46, 61, 138, 144, 405, 406] Many defects in well construction can be traced to inadequate cementing practices. Figures 104 and 4 through 7 (in Chapter 2).

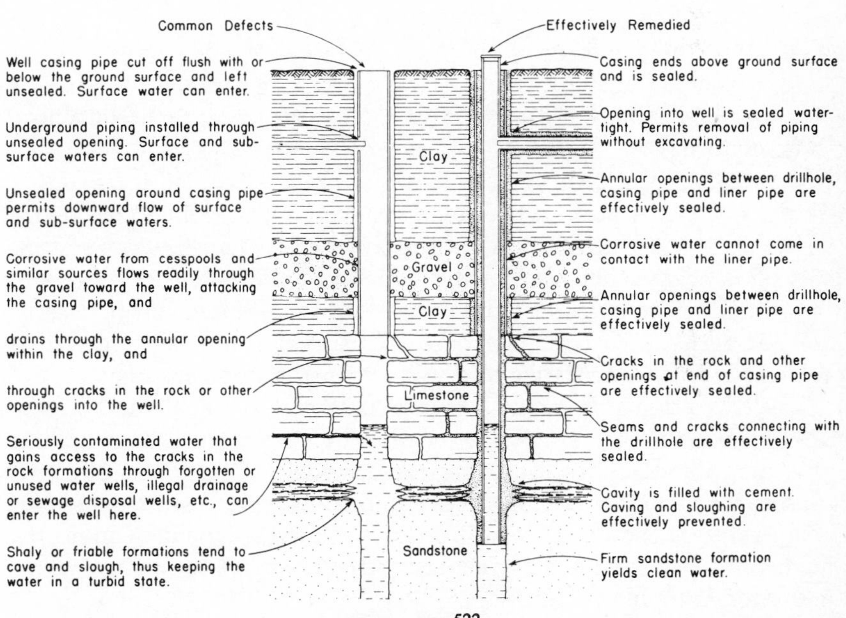

Figure 104[522]

The size of the grout space required when drilling the well usually depends upon the method of cementing. The area of the annular space around the casing frequently influences the success of the work because a complete sheath of cement around and along the casing is often necessary. Planning the diameter of the hole is important since tight areas and dead spots can occur when improperly centered casing contacts the wall of the hole causing the slurry to channel.

Isolation by grouting is desirable to protect the producing water-bearing formation from contamination by less desirable fluids or from the surface. The proper use of cementing materials will provide the most effective means of isolation, will form a seal to permit various techniques of well stimulation, and will help to protect casing from corrosion.[43] Types of materials used to cement casings are many, including the dry blending of several additives to meet specific well conditions.

Cementing Materials

Portland cement used for well cementing is used in general concrete construction and the water well industry since availability is not dependent upon other service industries.

Bentonite is a colloidal clay which requires a large volume of water. When added to cement in quantities varying from one to eight percent, a slurry lighter than the neat cement will result with an increased yield depending upon the percentage of bentonite. The addition of bentonite will generally reduce the compressive strength of the material.

Pozzolans are siliceous materials which, in a powdered form and in the presence of water, will react chemically with lime at ordinary temperatures to form compounds that have cement-like properties. These pozzolans when used with Portland cement react with the free lime liberated from cement and thus become solidified. Properties of such materials have high resistance to sulphate and salt water, low permeability of set material, and low expansion upon setting.

Perlite is a volcanic material which is mined, screened, and expanded by heat to form a cellular product of extremely low weight. A small amount of bentonite in Portland cement (two to six percent) helps to prevent segregation of the perlite particle from the cement slurry, reduces the density of the slurry, increases yield of material, and acts as a bridging material for porous or fractured formations encountered while cementing.

A special-graded *diatomaceous earth* is added to Portland cement to permit formulation of high-density material, increasing the weight of the slurry from 10 to 40 percent.

Gilsonite is a selected and graded asphaltite, having a specific gravity of 1.07, which will neither accelerate nor retard the setting time of the cement. Gilsonite reduces corrosion by waters or brines and resists the attack of acids or alkalies. Slurries containing

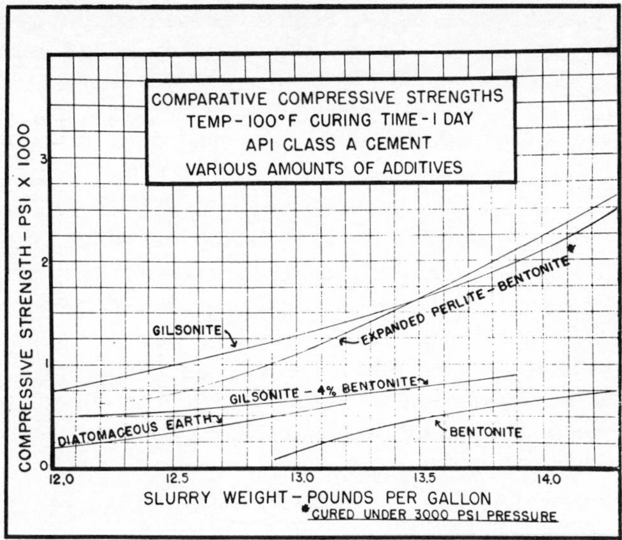

Figure 105[144] Compressive strengths of various materials used for well cementing work.

gilsonite are suitable for zones requiring low hydrostatic pressure, bridging aids, maximum compressive strengths, and minimum slurry density. There are various accepted minimum compressive strengths for cementing materials; however, the most popular minimum is 500 psi. (See Figure 105.)

Cementing Operations

The economic factors must be considered by the well operator in planning the well completion program. However, special materials are often required to fulfill certain specifications which preclude cost considerations.[406]

The normal ratio of water to cement for a suitable grout is five to six gallons per 94-pound sack of cement.[46] Laboratory tests indicate that 5.4 gallons of water will hydrate one U.S. sack of cement. Mixtures with more than six gallons of water per sack have been used as grouting foundation materials, but this is not acceptable for water well cementing. Shrinkage increases with water content. Water is removed from grouting mixtures by filtration through fine sand or other permeable formation materials. Cement can settle out of the slurry rather than remain in suspension if the water-cement ratio is greater than ten gallons per sack of cement.

One decided advantage of the proper water-cement ratio is that effective bridging of cement particles in the pores of permeable formations can be accomplished. Bridging prevents excessive penetration of the grout into the formation, although some penetration is desirable for adequate sealing.

Bentonite clay, as well as other additives, is commonly added in amounts from three to five pounds per sack of cement, in which case about 6.5 gallons of water per sack is required. Bentonite assists in the suspension of the cement particles which reduces shrinkage. The bentonite and water should be mixed before the cement is added to the clay-water suspension. Features of common additives and effects on the physical properties of cement are summarized in Table 38.

Additives are mixed with neat cement to increase yield per sack, to reduce cementing costs, and to alter slurry properties for special well conditions. Additives are utilized, generally, as the following:[414]

(1) Extenders to provide a greater yield, or slurry volume, for each sack of cement.
(2) Weighting materials to increase slurry density and overcome high formation pressures.
(3) Accelerators, such as calcium chloride, to reduce waiting time in shallow well completions, particularly in cold climates.
(4) Retarders, commonly used in wells 8,000 feet and deeper, to increase thickening time.
(5) Low-water-loss additives important to the "squeeze" cementing technique where control of water loss is critical.
(6) Lost-circulation materials to reduce loss of cementing fluids to a very porous or permeable formation, fractured zones, or weak formations hydraulically fractured by the pressure of mud and cement in the annulus.
(7) Special additives to change cement-flow behavior.

Water for grout should be free of oil or other organic material, and the dissolved mineral content should be less than 2,000 ppm — a high sulphate content is particularly undesirable. Exceptional conditions may call for the addition of sand or other bulky material to permit the grout to bridge larger openings without excessive loss of fluid; a bentonite additive serves some of these requirements.

In cases where an open borehole has been drilled below the depth to which the casing is to be grouted, the lower part of the hole must be backfilled or a bridge set in the hole to retain the slurry at the desired depth. Backfilling the hole to the proper level with a fine-grained sand is a common procedure. If the sand is fine, cement will not penetrate the sandy backfill material more than two or three inches. Material ordinarily sold as plaster sand or mortar sand is generally used with grain sizes of 0.012 inch to 0.025 inch. Investigations indicate that no significant cement penetration occurs in sand

TABLE 38[27]

Some Additives Commonly Used with Cement.

EFFECTS OF SOME ADDITIVES ON THE PHYSICAL PROPERTIES OF CEMENT		BENTONITE	PERLITE	DIATOMACEOUS EARTH	POZZOLAN	SAND	BARITE	ARSENOFERRITE	CALCIUM CHLORIDE	SODIUM CHLORIDE*	LIGNOSULFONATES	CMHEC†	DIESEL OIL	LOW-WATER-LOSS MATERIALS	LOST-CIRCULATION MATERIALS	ACTIVATED CHARCOAL
DENSITY	DECREASE	⊗	⊗	⊗	x											
	INCREASE					⊗	⊗	⊗	x	x	x					
WATER REQUIRED	LESS										⊗					
	MORE	⊗	x	⊗	x	x	x	x							x	x
VISCOSITY	DECREASE								x		⊗					
	INCREASE	x	x	x	x	x	x	x							x	x
THICKENING TIME	ACCELERATED	x					x	x	⊗	⊗						
	RETARDED			x						x	⊗	⊗	x	⊗		
SETTING TIME	ACCELERATED						x	x	⊗	⊗						
	RETARDED	x	x	x	x						⊗	⊗		x		
EARLY STRENGTH	DECREASED	x	x	x	x		x	x			⊗	⊗		x	x	x
	INCREASED								⊗	⊗						
FINAL STRENGTH	DECREASED	x	x	⊗	x		x					⊗		x	x	x
	INCREASED										x					
DURABILITY	DECREASED	x	x	x									x		x	
	INCREASED				⊗											x
WATER LOSS	DECREASED	⊗									x	⊗	x	⊗	x	
	INCREASED		x	x												

x DENOTES MINOR EFFECT.

⊗ DENOTES MAJOR EFFECT AND/OR PRINCIPAL PURPOSE FOR WHICH USED.

* SMALL PERCENTAGES OF SODIUM CHLORIDE ACCELERATE THICKENING. LARGE PERCENTAGES MAY RETARD API CLASS A CEMENT.

† CARBOXYMETHYL HYDROXYETHYL CELLULOSE.

with uniform grain sizes finer than 0.025 inch or in non-uniform sand with permeability less than 3,000 gallons per day per square foot.[144]

To assure that grouting provides a satisfactory seal, the slurry must be placed by a continuous operation before initial setting of the cement. Regardless of the method employed, the grout should be introduced at the base of the grouting interval to minimize contamination or dilution of the slurry and bridging of the mixture with upper-formation material.

Suitable pumps, air pressure, or water pressure is used to force grout into the space to be filled, although placement by gravity is practical and satisfactory in some cases. Gravity placement in a shallow borehole, for example, is accomplished by introducing the slurry to the bottom of the hole and by then lowering the casing into the slurry. The casing is centered in the hole by centering guides, and the bottom of the casing contains a tight, drillable plug. As the casing is lowered, the grout is forced upward and around the casing, filling the annular space.

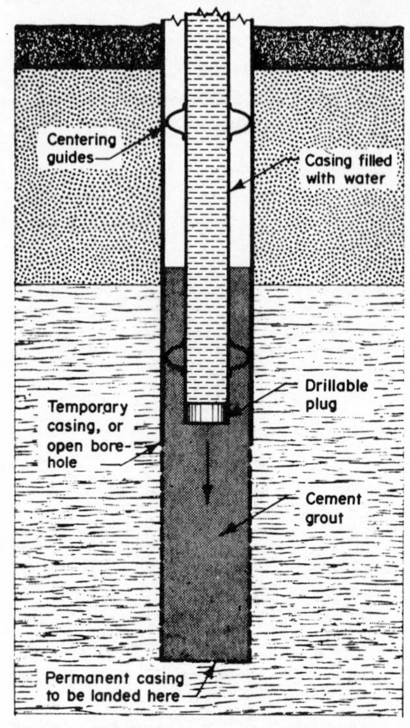

Figure 106[46] Cement grout being forced upward in annular space as plugged casing is lowered into slurry placed in bottom of borehole.

If the casing does not sink to the bottom, it is filled with water. More weight may be added in some instances. When a temporary casing is used to prevent caving of an oversized hole, this casing must be pulled before the grout has solidified so that the grout will make intimate contact with the wall of the hole. Figure 106 shows particulars of the procedure.

A pipe of two-inch to four-inch diameter should be used to conduct the proper quantity of slurry to the bottom of the oversized hole. If the hole is drilled by the rotary method and is filled with drilling fluid, the mud will rise up the hole since the cement mixture is the heavier slurry. The volume of grout placed must be adequate to fill the annular space around the permanent casing as it is sunk to the proper depth. Unless a caliper log has been run of the hole to estimate the volume of cement required, common practice is to estimate the required volume from the nominal-hole diameter and O.D. of the pipe. (See Appendix.) This volume is increased 20 to 25 percent to assure adequate quantities. Should there be an overrun, the extra grout is wasted at the ground surface as the pipe is lowered into the hole.

When the cement is set and has sufficiently hardened, the bottom plug is drilled out, and drilling is continued below the grouted section. A 72-hour setting period is normal for most Portland cement slurries. High early strength cement may be used to reduce the waiting period to about 48 hours.[46]

Where the annular space is of sufficient size, grout is placed through a string of small pipe from outside the casing. The casing with centering guides attached is lowered into the hole. The lower end of the casing is closed with a drillable plug, or the casing may be driven into the clay bottom of the hole, if present, so that the grout cannot enter. To prevent the casing from floating in the slurry, the casing is filled with water or is held down by the weight of the drilling rig.

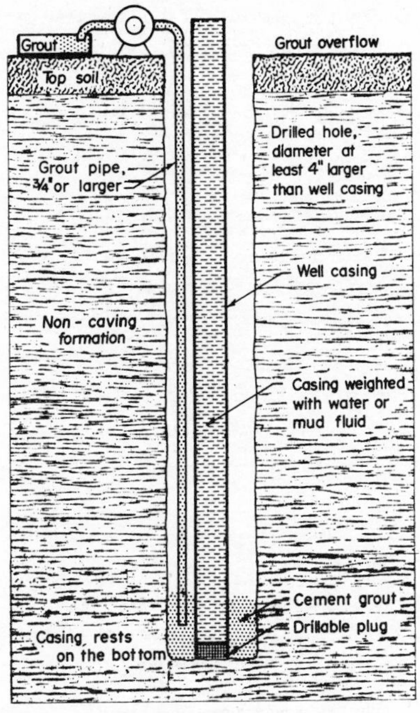

Figure 107[46] Cementing casing by pumping grout through a pipe lowered in the annular space outside the casing.

The 3/4-inch or 1-inch grout pipe, generally used, must be large enough to allow the required volume of grout to be placed in the time available. The oversized hole should be four to six inches larger than the casing in order to provide sufficient space to accommodate the grout pipe.

Grout is placed by gravity flow only when it is certain that the operation can be completed quickly. Pumping facilitates the rapid introduction of the required volume of grout and is preferred. A pump pressure equal to the hydrostatic pressure of the grout, plus the fluid friction in the grout pipe and annular space, is usually required and should be anticipated (See Figure 107).

The grout pipe is similar to a tremie pipe normally used by the construction industry for placing masses of concrete under water. Initially the pipe is extended to the bottom of the annular space and is submerged in the slurry during the entire grouting operation. The pipe must be removed soon after grouting emplacement has been completed. If the flow of grout is interrupted, the pipe must be raised above the grout level until all air and water have been displaced. The grout pipe must, however, be pulled prior to hardening, to allow the grout to fill the hole made by this pipe. Otherwise the hole left by the grout pipe will allow contamination from the surface or other sources.

When the use of a small pipe outside the casing is impractical, grouting is accomplished by means of a grout conductor pipe installed within the casing referred to as the *tubing method* of cementing in the oil industry. A suitable packer connection — a cementing or float shoe — is used at the bottom of the casing to regulate the flow of fluid from the grout pipe and to prevent reverse flow of the grout into the casing during and after the grouting procedure. A common type of packer connection, with a ball-type check valve, prevents reverse flow of the grout. The internal materials are drilled-out of the casing upon completion of the cementing after an adequate setting period.

Cementing can be accomplished without using the cementing plug or float shoe on the casing. The *inside casing method* requires a fluid-tight stuffing box which accommodates the grout pipe and a heavy cap to close off the top of the casing. The grout pipe is positioned three or four feet above the lower end of the casing which, in turn, is suspended from the bottom of the hole. The air inside the casing is released by a bleeder valve as the casing is filled with water or drilling fluid. The details of this method can be found in the literature.[46,144,406]

The *casing method* of grouting, by which the slurry is forced down the casing and into the annular space, is widely practiced in the oil well industry.[27,46,61] Either one or two spacer plugs are used in this method, which was developed and patented a number of years ago by the Halliburton Oil Well Cementing Company of Duncan, Oklahoma. One plug separates the cement slurry from the fluid in the casing, the other separates the slurry from water pumped in above the plug.

After pumping water or mud through the casing to circulate fluid in the annular space and clear any obstructions from the hole, the first plug is inserted and the casing capped. A measured volume of grout is pumped in, the casing is opened, and a second plug is inserted. A measured volume of water is then pumped into the casing until the second plug is pushed to the bottom of the casing expelling the cement slurry from the casing and into the annular space. The water in the casing is held under pressure to prevent backflow until the slurry has set.

A modification of this procedure is to use only the lower plug; then, after pumping a predetermined quantity of slurry, a volume of water sufficient to force most of the grout out of the casing is added. The usual practice is to leave 10 to 15 feet of grout in the pipe (See Figure 108).

Another modification is to use only the upper plug, the reasoning being that the part of the slurry which may be diluted by the fluid in the well will be expelled to waste at the surface. This assures sound, uncontaminated grout at the lower critical end of the casing.

Spacer plugs are necessarily made of materials which can be drilled-out. Wood and rubber-fiber plugs can be difficult to drill since these materials tend to elude the bit teeth. A plaster-type plug or equivalent is suitable.

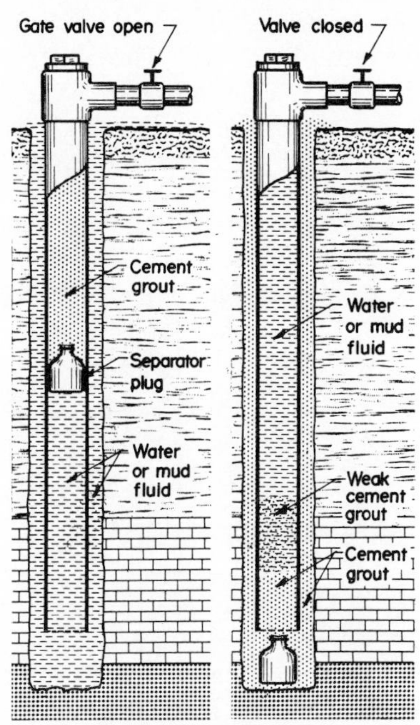

Figure 108[46] Cementing by casing method using single plug between grout column and drilling fluid. Plug and grout are forced out of bottom of pipe; grout then moves upward in annular space.

One problem which may arise during the cementing operation is that of mud contamination. An excellent procedure for minimizing mud contamination is the "two-plug" system. This operation consists of a fixed container for the movable valve plugs as an integral part of the casing (See Figure 109). Drilling or circulating fluids may be pumped through this container until the slurry is introduced. The bottom plug valve is closed, and the plug is pumped ahead of the slurry, fulfilling two functions: (1) providing a barrier between the slurry and drilling fluid and (2) cleaning the casing wall of drilling mud. When this plug reaches the float collar, the differential pressure opens the valve of the plug and allows the cement slurry to flow through the plug and floating equipment, around the casing, and up the annular space.

Upon completion of the grouting operation, the top plug is released from the container. This plug should seat effectively on the bottom plug resulting in a pressure build-up which prohibits grout backflow. The two-plug system requires knowledge of formation permeability characteristics because, if the annular space is not filled to the top, the slurry is partially wasted, proper cementing is not achieved, and more slurry cannot be added.

There are instances when circulation of slurry is a necessity, but information is lacking as to washout factor. In order to minimize contamination, the continuous method is more effective. The *continuous method* involves running tubing or drill pipe to the proximity of the float shoe or collar with a sealing element between the tubing and the casing. Grouting materials can be pumped until returns are noted on the surface.

The *plunger-type receptacle method* is an improved technique of the continuous method involving a special type of floating equipment designed with the receptacle to seat an adapter attached to the tubing or drill pipe (see Figure 109). This system minimizes contamination and provides the advantage of continuous mixing. The continuous method is a very practical approach for cementing of casing sizes 16 inches and above, since a two-plug system in many instances would require special plugs, plug containers or swedges (see Figure 109).

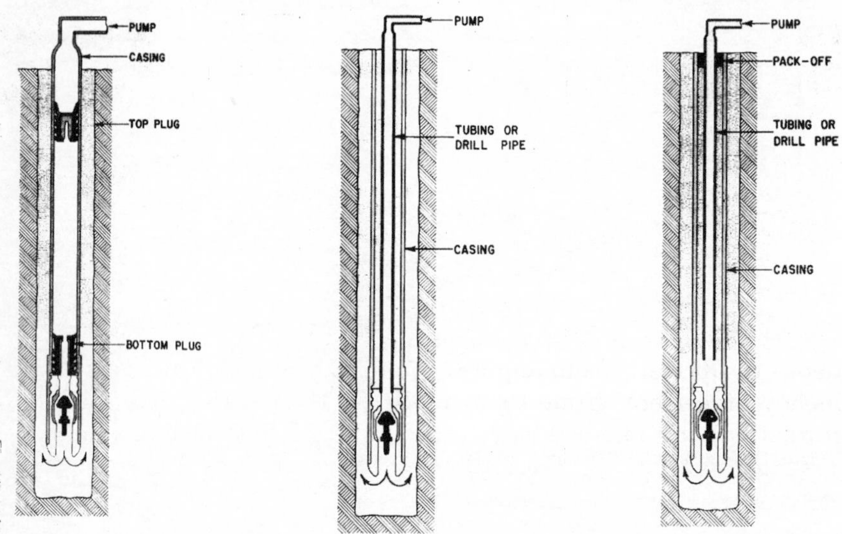

Figure 109[144] Techniques of cementing wells: Left, two plug system; center, plunger type receptacle; and right, continuous method.

A new grouting tool, developed for the petroleum industry, incorporates economic as well as operational features for water well completions. In the operation of this tool, a special plug is dropped from the surface, and by pressure of the pump, the sealing element is set, and circulating ports are opened. Grout material is mixed and pumped into the casing through circulating ports to above the sealing element. The top cementing plug seats on a special sleeve, and through application of pressure that shears the pins, plugs and sleeve are pumped out of the tool eliminating the drilling-out operation.

A problem encountered in cementing large diameter surface casing is that of flotation. To counteract the flotation factor, the density of the displacing fluids must occasionally be weighted-up or increased.

This operation is expensive; however, with knowledge of the properties of various grouting materials, the cost can be minimized. The required volume of grout cannot always be accurately determined. Irregularities in the size of the borehole and losses in fractured or highly permeable rock occur in many holes.

Ordinary concrete mixers are used for grout that requires 15 to 50 sacks of cement. For larger volumes, a jet-type mixer, similar to the type that prepares drilling fluid, is generally used. As volume requirements of grout increase, the facilities for mixing and placing a uniform slurry must be more dependable. Halliburton[298] has published a book of tables on various aspects of the cementing operation and should be consulted for detailed information. The literature on cementing practices is limited; however, the information available should be closely examined.

Local Grouting Requirements

An increasing number of states are establishing requirements on sealing and grouting of wells. Grouting is generally required as a sanitary protection, but the requirements from state to state are as widely divergent as casing requirements. Nevada, for example, requires a two-inch thickness of cement grout around the casing to a depth of 50 feet. Idaho requires a minimum annular thickness of 1.5 inches of cement grout to a depth of 18 feet (in a consolidated formation) and puddled clay to the same depth (sealed in an unconsolidated formation). California has no regulations for agricultural wells and a two-inch minimum thickness of grout to a depth of 10 to 50 feet for others or a driven outer casing only. Well casings in Colorado must have a minimum depth of ten feet and be grouted in for the entire length. No thickness is specified, and grout may be any impervious material. In Oregon, casing grouted with cement of five feet in a consolidated formation is required and in unconsolidated formations, a minimum thickness of 1.5 inches and 18 feet of depth of cement or bentonite slurry.

One state accepts grouting of the annular space between the pump housing casing and surface casing, while another requires filling the annular space with heavy, bentonite mud while the casing is driven, forming a seal. The former regulation is inadequate — the latter, only probably effective.

Grouting Seals

Abandoned and improperly cemented wells provide vertical openings or channels through which polluted water may gain entry to a valuable aquifer. A near-surface, upper-annular-space seal of an aggregate of fine sand and cement is commonly used.[46] Another seal of possible merit is the use of finely ground, rubber fiber that is sandwiched between two layers of grout (or gravel pack).[101] This would insure the protection of lower intervals from surface water contamination. The rubber fiber is not prone to shrinkage, as is cement, and makes an excellent seal between the well casing and the borehole wall. Foaming epoxy may replace cement in the future.

A driven casing does not form an effective seal even though the unconsolidated material is assumed to collapse against the casing! A seal might occur in a uniform sand, but the general effectiveness is highly questionable. Driven casing can have a drive shoe that is slightly larger in diameter than the couplings. As the casing is driven, the wall of the hole is compacted. Subsequently, the borehole will collapse forming an aggregate of compacted particles with open channels that offer relatively unimpeded paths for the flow of contaminated water from the surface.

When a temporary surface casing is used in well construction, the annular space between the surface casing and the working casing should be grouted as the surface casing is withdrawn. When a permanent surface casing is used, the annular space between the casing and the formation should also be grouted. For sealing, driven casing is usually pulled back a foot or so, the annular space is flushed with clear water, and the grout is pumped through the casing so that the water and grout flush and fill the annular space from the lower end of the driving shoe to the surface. The hole is occasionally filled with a thick bentonite slurry as the casing is driven, although this practice may not be universally acceptable.

When casing is set in a consolidated formation, a puddled clay or bentonite seal is usually effective. However, if the casing is seated in unstable material and not supported permanently at the surface, a clay seal may act as a lubricant and permit the casing to settle if a telescoped casing is used.

A cement or a concrete grout seal, although relatively stable, may have other drawbacks. Cement slurry or concrete shrinks upon setting. When the slurry is placed above the water table, much of the water may be drawn from the mixture by capillary action before setting occurs causing even more shrinkage. A narrow but permeable passage for water may be left between the grout and the wall of the hole. The more water that is in the mix, the greater the possible shrinkage. As previously mentioned, 5 1/2 to 6 gallons per cubic foot of mix (one U.S. Cement sack: 94 pounds) is about the maximum which should be used for neat cement grouting. Concrete grout should consist of equal parts by volume of cement and dry, clean sand. If greater adaptability is desired, a water-reducing, set-retarding agent such as hydroxylated carboxylic acid should be added instead of more water. The addition of three to five percent bentonite by weight also increases adaptability and reduces shrinkage and strength. The addition of about 1/2 teaspoon of aluminum powder per sack of cement causes the grout to swell on setting and produces an acceptable seal.

An adequate grout seal not only protects the well from possible contamination but also serves other purposes. The minimum thickness of the grout is generally 1 1/2 inches around the casing and couplings. This protects the casing in the event that corrosion is a problem. Steel casing can be completely corroded; however, the stability of the hole has been maintained solely by the cement-based grout seal in rare instances..

In many areas, formation materials above the water table consist primarily of relatively clean, fine sand with little or no cohesion or bridging capability. When a well is developed, sufficient formation material can cave and compact to form a void near the top of the screen. Frequently, such a void eventually causes progressive caving to the surface with consequent damage to surface installations of piping and pitless adaptor, etc.[300] A cement-based grout around a surface casing forms an adequate bond with the casing and with the surrounding soil. Soil often encourages a stable bridging or arching of the formation with the grout and stops upward caving.[5]

The methods of grouting-in casing given in AWWA[44] are the generally accepted specifications presently available for water wells, but field conditions and design should be considered in selecting the kind and quality of grout to use. In a paper by Moehrl,[450] well grouting and well sealing with Portland cement, neat-cement slurry, and special slurries in addition to causes of cementing failures are examined in some detail.

Surface-Apron Cementing

When considering cementing operations, surface-apron construction may present serious problems in casing failure. For example, a 24-inch concrete apron firmly grouted to an eight-inch casing subjected to 30-psi frost-heave pressures, would have the potential to transmit approximately 72,000 pounds of force to the casing. In some practices where the surface-cement apron thickens in cross-section towards the casing, considerable tension is transmitted to the casing. Radial-expansion joints in apron construction are used to minimize the potential compressive force to the casing. Hockstra, *et al.*[316] report on frost-heaving pressures which bear upon the above problem. In many cases, a surface platform of cement, serving as the pump house floor, is not required. The "pump house" should be located away from the well, itself, e.g., 10 to 15 feet. This allows convenient workover of the well without endangering pressure tanks and electric plumbing configurations located in the "pump house" — the enclosure should perhaps be considered a "service house."

PITLESS ADAPTERS

Mention should be made of a sanitary, underground, discharge assembly called the pitless adapter since some state well-construction codes call for its use.[100] In some areas, it is common practice to terminate the casing below the frost level in a pit. This protects the water-carrying pipe from freezing and facilitates the connection of underground-discharge piping. However, the U.S. Public Health Service does not recommend that pits be used because proper drainage cannot be established. Furthermore, the hazard of poisonous gases (*always* present in pits) heavier than air, tasteless, odorless and invisible is paramount and must be considered in pit operations. Wherever possible, equipment that can be operated from the surface should be employed, and if personnel enter the pit, supplemental oxygen should be provided. Many cases of well contamination have also been traced to flooded well pits.

Pitless-adapter units, which are installed as a permanent extension of the casing, allow termination of the casing above ground and connection of the discharge piping below ground (Figure 110). When a well cannot be terminated above the flood level of the area, a pitless-adapter unit with a water-tight cap should be used.[46,50,68]

WELL PLUMBNESS AND ALIGNMENT

A hole drilled into the earth to any substantial depth may not be either perfectly straight or perfectly plumb; although, some deviation is always present. To determine how much a well casing may be out of plumb, practical methods of measuring any deviations must be available. Many modern water well specifications require that plumbness be checked with a special plumb bob and that straightness be tested with a 40-foot cylindrical dummy, slightly smaller in diameter than the inside of the well casing.

Measurement of the deviation from plumbness and the deviation from straightness is possible by means of the plumb-bob test alone.[28]

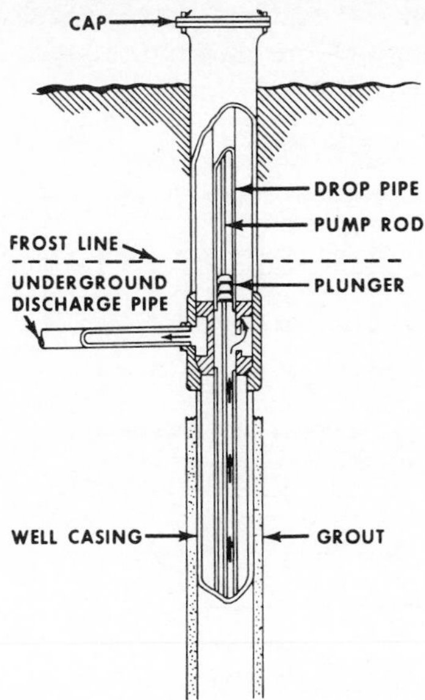

Figure 110[100] Pitless adapter units allow termination of the well casing above ground and connection of the discharge piping under ground.

Figure 111 illustrates a field plot of a well employing the plumb-bob test method. Methods have also been developed for testing deep-well alignment.[333] The conditions that cause wells to be out of plumb are:

(1) The character of the subsurface material penetrated during drilling.
(2) The trueness of the pipe used as a well casing.
(3) The pull-down force on the drill pipe during rotary drilling.

While drilling, gravity tends to make the drilling bit cut a vertical drill hole. Varying hardness of the materials being penetrated, however, deflect that bit from a truly vertical course. The edge of a boulder in a glacial drift, for example, will force either a cable-tool bit or a rotary bit to one side. In cable-tool drilling, the boulder may deflect the well casing as it is driven and cause the hole to slant as the well is deepened. Variations in the hardness of consolidated rock formations will also start a deflection which results in a continued drift of the hole from the desired alignment.

WELL STERILIZATION

The necessary final step in proper well completion is the sterilization procedure, which attempts to destroy all disease-producing organisms which were introduced into the well during the well construction operations.[29] Entry of these organisms into the well can occur through contaminated drilling fluids, via equipment, materials, or through surface drainage into the well. All newly constructed wells or existing wells subjected to any repair or workover should be disinfected before being placed into service.

A chlorine solution is the simplest and most effective agent for disinfecting or sterilizing a well, pump, storage tank, or piping system. Highly chlorinated water for this purpose may be prepared by dissolving calcium hypochlorite, sodium hypochlorite, or gaseous chlorine in water.

Periodic disinfection of a well during the drilling operation is recommended. The disinfectant can be added to the water in the well each day the drilling progresses to disinfect the casing and drilling tools. Theoretically, any water introduced into the well as a drilling fluid or for well stimulation should be of drinking water quality. Gravel-pack material should also be disinfected before being placed in the well.

The well should be cleaned as thoroughly as possible of foreign substances such as soil, grease, oil, joint dope, etc., before disinfection. These substances may harbor bacteria, especially oil-based materials since bacteria can be effectively protected from mild bacteriacides. Disinfection is most conveniently achieved by the addition of a strong solution of chlorine to the well. The solution should be introduced into the well in such a manner that all well surfaces above the static level will be completely flushed with the solution. The well contents should then be agitated to uniformly distribute the chlorine throughout the well.

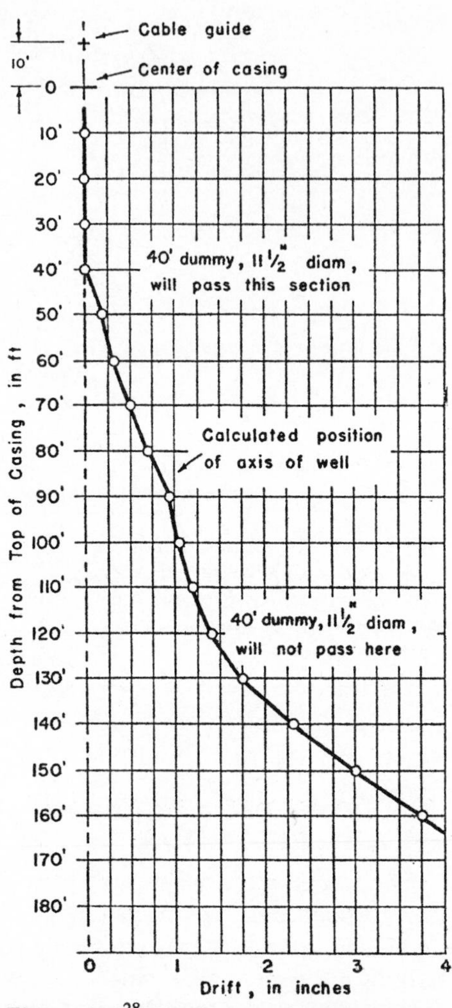

Figure 111[28] Typical graph of hole drift, based on field data. Trace of center line of a well that is out of plumb and crooked.

The most convenient method of preparing the strong chlorine solution is by dissolving calcium hypochlorite or sodium hypochlorite in a relatively small quantity of water. Gaseous chlorine may also be used but it requires greater care in handling and storage. Numerous tables exist in the literature which give the amount of the chlorine compound required to produce a desired chlorine strength, or concentration in a known volume of water.[29]

Calcium hypochlorite is a dry, white, granular material containing about 70 percent "available" chlorine by weight. This material has also been marketed in tablet form under the trade names Pit-Tabs, HTH Tablets, Chlor-Tabs, etc. The name high-test calcium hypochlorite distinguishes this chemical from chlorinated lime or bleaching powder.[29] When dissolved in water, one pound of calcium hypochlorite produces a solution that has the oxidizing potential of 0.70 pound of chlorine gas dissolved in the same quantity of water (or, 1.43 pounds of calcium hypochlorite is equivalent to one pound of chlorine gas in a water solution).

The strength of chlorine solutions is usually expressed in parts per million (ppm). Solution strengths of 50 to 200 ppm chlorine are commonly used for sterilizing wells and well construction materials. Table 39 gives the quantity of calcium hypochlorite required to make 100 gallons of disinfecting solution of various concentrations.

TABLE 39[29]

Desired Chlorine Strength	Dry Calcium Hypochlorite, lb.
50 ppm	0.06
100 ppm	0.12
150 ppm	0.18
200 ppm	0.25
300 ppm	0.4
400 ppm	0.5

Dry calcium hypochlorite is a fairly stable material. This chemical does lose some of the available chlorine slowly, but when properly packaged and stored at a cool temperature, 90 percent of the chlorine content will be retained 12 months after manufacture. The chemical is quite corrosive and loses chlorine rapidly if moisture is present.

Sodium hypochlorite, as laundry bleach, is widely obtainable in solution form. Sodium hypochlorite generally contains from 12 percent to 15 percent by weight of available chlorine although some common bleaches may contain only 3 percent to 5 percent available chlorine. These solutions are not very stable and, if more than about 60 days old, should not be considered to contain the full amount of available chlorine originally in solution.

The bactericidal efficiency of free chlorine is dependent on pH, temperature, chlorine concentration, microbial type and number, and time of contact. Because of the reaction of chlorine with organic matter, which results in free chlorine either being used up in oxidizing the organic matter or converted to less efficient combined chlorine compounds, the efficiency of the bactericide is also dependent on the chemical composition of the water being disinfected. The amount of chlorine which must be added to produce a chlorine residual of such a strength that, after a definite contact time, the desired degree of disinfection that may be attained is not an absolute value and cannot be expected to be the same under different environmental conditions. This uncertainty likely accounts for the wide range in chlorine concentrations and contact times required the various regulatory agencies. Generally, an initial concentration of 50 parts per million (ppm) with a residual of 1.0 ppm remaining at the end of two hours of contact will most likely inactivate any known disease-producing organisms. If no residual remains after two hours, the chlorination procedure should be repeated.

In relatively deep wells, with high water levels, special steps may be required to insure chlorination throughout the entire well depth. One practical method is to place dry calcium hypochlorite in a container made from a short length of perforated tubing capped on both ends and suspended by a cable. By dipping the container in the full column of water in the well, the chemical will be dispersed evenly. The same device may be lowered into a flowing artesian well and agitated near the bottom of the well. The natural upward flow will then carry the chlorinated water to the surface.

It is extremely difficult to test for specific disease-producing organisms, of which there may be several types present in water. Because of this, *coliform* bacteria are used as indicators of the possible presence of disease-producing organisms of human or animal origin in water. Disinfection is, therefore, considered to be complete only when sampling and testing of the water show the presence of no *coliform* bacteria. After the well and piping system have been thoroughly flushed to remove all traces of chlorine, the samples should be collected in a container supplied by a laboratory and in accordance with sanitary procedures. The samples should then be sent to the laboratory without delay. If negative coliform tests cannot be achieved, it may be necessary to equip the water system with a constant-feed chlorinator and storage tank to permit adequate contact time.

The "Multiple-Tube Fermentation," or MPN (most probable number), method was the only approved procedure for detecting coliform bacteria in water — officially the principal criterion of sanitary quality for public drinking water. The use of membrane filters for isolating and identifying coliform and other bacterial organisms is now well established. In many state and federal laboratories it is the preferred method owing to its simplicity, sensitivity, and the fact that the MPN method has a safer time factor (about four times).[444]

In many cases, the membrane filter method has made it possible to substitute field testing for laboratory analysis. Investigators have used it to isolate pathogenic bacteria for identification purposes when outbreaks of enteric illness were traced to water supply.

In addition, an "early warning" system can be developed by well contractors, consultants, etc., to establish the presence of most of the common bacteria, both before official testing of the water supply and at specified periods after the well has been in service for a prolonged period.[445]

Wells may also be subject to attack by iron, sulfate-reducing and other bacteria, normally harmless to health, with metabolic processes that result in incrustation of screens or aggressive corrosion. This topic is treated later in the text. Whether such bacteria are present in some aquifers or are introduced during construction is a problem which requires solution. Many of them are resistant to chlorine concentrations and contact times normally used for disinfection. For instance, the "disulpho-vibrio" type of sulfate-reducing bacteria may require 400 ppm chlorine concentration and six hours of contact time.

When concentrations of about 300 ppm or greater of available chlorine are required, gaseous chlorine or sodium hypochlorite should be used. When calcium hypochlorite is used, particularly in waters containing calcium bicarbonate, calcium carbonate and calcium hydroxide are precipitated. These precipitates may result in incrustation of the screen or blockage of the aquifer.

Since the complete elimination of these types of bacteria may not be possible by a single "shock" chlorination treatment, periodic, or even continual chlorination may be required to inhibit the growth of these bacteria. The use of other bactericides, more commonly used in the petroleum industry, that do not result in residuals that may be toxic to man or produce other undesirable effects should be investigated. For example, iodine is known to be more toxic to more organisms than chlorine and it does not significantly react with organic matter.[125] If iodine proves to be effective against iron and sulfate-reducing bacteria, even though it is more expensive than chlorine, it may result in overall savings if lower concentrations for shorter periods of contact time would be required. Also, the treatment may not have to be applied as frequently. Further investigation, however, of the effectiveness of iodine and other bactericides is necessary.

Well design, whether intended for sedimentary, igneous, or metamorphic applications, determines the future production characteristics and overall longevity of the well. Obviously then the best available technology should be applied to a particular well design. This considers, of course, the economic practicability. However, all too often wells are designed on the basis of expediency or a lack of appreciation of the many possible variations of design which offer incremental savings, either short term in the form of avoiding "overdesign" or long term in the form of avoiding future production problems, e.g., clogging of screens or sand production due to improper selection, well structure failure, due to improper selection of casing type, loss of head due to inter-aquifer pressure bleeding, water quality deterioration due to improper cementing or casing operations, etc. Well design and the desired production depends solely on the local geological hydrogeological conditions. A practical compromise between design requirements and the specific economic framework, however, can be achieved with the assistance of a thorough knowledge of well design alternatives.[674]

13
Well Efficiency and Maintenance

Water well maintenance is one of the most important features of water well technology. Primary completion and development techniques of original wells must be designed so that redevelopment and "workovers" can be accomplished. Many of the practices have been discussed in the previous chapters. A strict division among completion, development, redevelopment, and rehabilitation will not be made since it will be obvious where the various practices are applicable. There is considerable overlap, however, and several topics treated in this chapter may be applicable to *primary* completion and development practices.

Water well maintenance procedures begin with the techniques implemented during the drilling of the original well and continue with the type of materials used and with the extent of care, or the lack of it, that the well is given over the months and years of operation.[142]

It should be noted that under very special conditions water well installations can be a hazard. A most dramatic example was reported in Pennsylvania where a leak in a subsurface fuel tank, and a thunderstorm may have combined to produce an explosion which propelled a 200-pound pump up and over a 38-foot-high power line, landing about 180 feet away.[254] Natural production of potentially explosive gases could, under special conditions, also create problems which should be considered in areas where the geology is suspect.[224,436]

Under most conditions, a temperature survey can pinpoint the zone or zones from which natural gas is leaking. Figure 112 shows one gas leak occurring, which could be cemented off during well completion.

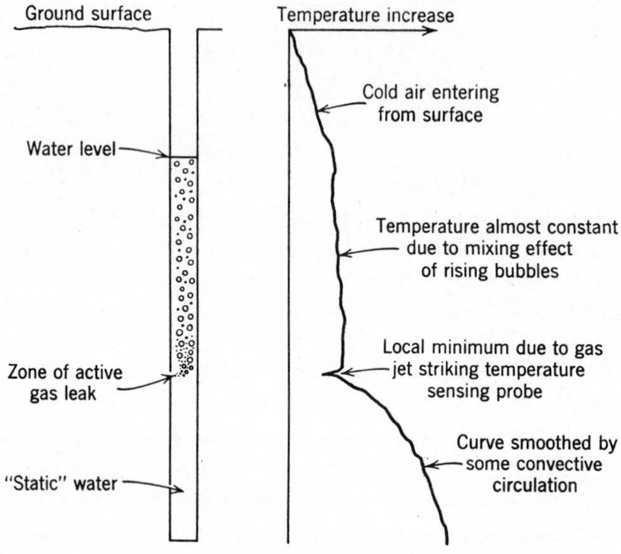

Figure 112[172] Schematic diagram showing temperature log used to pinpoint location of expanding gas entering a well.

As in previous chapters, special emphasis is placed on subjects which either the petroleum industry has developed, or which the water well industry has yet to fully incorporate,[114] e.g., topics such as: well efficiency, direct observation methods, corrosion-incrustation types, prevention, inhibition, well stimulation and others.

WELL EFFICIENCY

Well efficiency is formally defined as the actual specific capacity adjusted for well loss divided by the theoretical specific capacity.[556] The computed average efficiency of a typical newly constructed well, for example, is about 80 percent. An increase in yield of 20% can be accomplished by removing fine materials which have migrated into the formation during the drilling process. Many drilling fluids tend to inhibit the removal of fine materials, causing permanent formation damage in many cases.

Field Requirements

For many years, drillers of oil and gas wells were concerned with water wells only as a source of water. In recent years, the rapid growth of secondary recovery methods using waterflood and steam has required the construction of water wells for other uses. In many oil fields, there are surplus or unused wells, such as stratigraphic test holes or dry oil wells. "Dead" wells decorate many oil fields. Unfortunately, the first plan petroleum management is likely to follow for source water is to put these wells into service by perforating the casing opposite a water sand. Perforating a cased and cemented water sand obviously results in a grossly inefficient water well. Expense and anxiety result from breakdowns and delays caused by wells that pump sand, thereby, plugging filters, and distribution lines. The use of existing wells for source water may eventually cost more than properly designed and constructed new water wells.[365] The high cost of power because of hydraulic inefficiency compared to the new water wells is usually overlooked.[600]

The efficiency of oil wells is not generally considered because the existing bottom-hole pressure is high and/or allowable production is so low that the extra cost of producing or pumping an inefficient oil well is negligible. On the other hand, an inefficient water well becomes very expensive in terms of power costs and additional wells that may be required to meet the demand for source water. For example, a well producing 605,000 gallons of water per day with a drawdown of 720 feet (or specific capacity of 0.58 gpm − 840 gallons per day per foot) has only a 50 percent efficiency if the hydraulic constants of the water-bearing formation are such that the drawdown at that rate should only be 360 feet (or a specific capacity of 1.16 gpm − 1,680 gallons per foot). If the well is pumped continuously with an electric-powered pump, having a wire-to-water efficiency of 75 percent, the increased cost per year of the additional 360 feet of lift is $10,000 for an electric rate of three cents per kilowatt-hour and $20,000 for six cents per kilowatt-hour. This is a somewhat exaggerated example, but nonetheless, the situation is more common in the water well industry than in the oil industry.

Optimization of Well Efficiency

In practice, the efficiency of a well is the ratio of yield per unit of drawdown actually obtained to that theoretically possible. Stated in another way, the well efficiency is the theoretical drawdown at a given time at a constant rate of production divided by the actual drawdown for the same rate of production. Drawdown is the difference between the pumping fluid level and the static fluid level. Yield per unit of drawdown is called *specific capacity (productivity index*, in the petroleum industry's terminology). The PI (productivity index) declines with time; therefore, the duration of pumping or production should be recorded. Properly designed and constructed wells should have efficiencies as high as 80 to 90 percent. In fact, an increasing percentage of water well specifications stipulate that efficiencies less than 80 percent are a cause for rejection or reduced payment to the contractor.

Schreurs [556] presents an excellent example of the salient features of well efficiency. The hydraulic efficiency of a water well can be calculated if the hydraulic constants of the producing aquifer are known. The transmissibility (T) is the product of the average field permeability multiplied by the aquifer thickness. The *coefficient of storage* (S) of an aquifer is the volume of water released from or taken into storage per unit surface area of the aquifer per unit change in the component of head normal to that surface. S is a dimensionless index and ranges from 3.0×10^{-1} to 1.0×10^{-5}. These hydraulic constants are usually determined by pumping a water well at a constant rate and measuring drawdown and/or recovery in the pumped well and one or more observation wells. S and T are calculated by using the well-known Theis non-equilibrium formula or the Jacob modified non-equilibrium formula (as mentioned in a previous chapter).[392,393]

Terms that are familiar to the petroleum industry include the *skin effect* which is the reduction in permeability from damage of the bore face caused by drilling mud invasion of a producing stratum adjacent to a well, or by improper completion methods.[634] *Damage ratio* (DR) has been defined as the dimensionless ratio of transmissibility (T) to the instantaneous productivity index (or specific capacity).[204,600]

EQUATION 61

$$DR = \frac{T}{Q/s_a} \quad \text{or} \quad \frac{Ts_a}{Q}$$

Where: s_a = Actual drawdown (ft.)
Q = Rate of production (gpm)
T = Transmissibility (gpd)

The relationship of this ratio to well efficiency is apparent.

When using data from a drill-stem test, for example, the damage ratio and transmissibility are calculated by the following equations:

EQUATION 62

$$DR = 0.183 \frac{P_o - P_f}{\Delta p}$$

EQUATION 63

$$T_o = 162 \frac{Q}{\Delta p}$$

Where: P_o = Undisturbed formation pressure
P_f = Final flowing pressure
Δp = Change in pressure per log cycle of graph of pressure build-up from drill-stem test
T_o = Transmissibility (millidarcies)

"DR" relates the approximate quantity of fluid that might have been produced if mud damage had not occurred to the quantity of fluid actually produced. Although the "DR" is useful because the data is easily obtained from drill-stem tests, it only approximates the reciprocal of well efficiency. The factors of well diameter, coefficient of storage, and time of production are not taken into consideration. The damage ratio will approach well efficiency if the value for $\log_{10} \frac{0.3\,Tt}{r^2 s}$ is about 5.5, (where t = time of production in days, r = radius of the well in feet). This occurs if the time or production is less than two hours; if transmissibility is less than about 1,000 gpd per foot (48.8 darcy per foot); if diameter of the well is less than eight inches; and if the value for storage is in the artesian range (confined 10^{-3} to 10^{-5}. These conditions are often met in drill-stem tests, so that the reciprocal of the "DR" will approximate the well efficiency.

Several factors initiate well losses in inefficient water wells.[121] These may include: (1) losses caused by reduction in permeability from mud invasion or by compaction and smearing of the formation face from cable-tool or auger drilling and (2) losses from the turbulent flow of water adjacent to the well through the screen or well face and inside the casing to the pump intake. High well losses are unnecessary since they can be minimized by good well design and construction. Well loss can be calculated over a wide range of pumping rates.[454,650]

The permeabilities of water sands are commonly more than 500 times higher than oil sands; therefore, the potential rate of water production is many times that of most oil wells. It is obvious that the design of water wells should be different from the design of most oil wells. For example, experienced water well designers recognize that a critical well design feature is the inlet velocity through the screening device. Ahrens[6] suggests an average, ideal entrance velocity is from 0.05 to 0.125 foot per second. A screen manufacturer recommends a maximum inlet velocity of 0.1 foot per second.[35,46,67,75] Sceva[550] prefers 0.15 foot per second; Smith[580] 0.066 foot per second. Walton[650] suggests a varying inlet velocity from about 0.02 to 0.1 foot per second depending upon the average permeability. Experience demonstrates that maintaining low inlet velocity results in high efficiency, minimizing sand pumping and reducing the rate of corrosion and incrustation in wells.

Equally as important as entrance velocity on the effect of well efficiency is the distribution and shape of slot openings in the screening device, which has been briefly discussed in a previous chapter.

New materials and new design concepts are used in an attempt to construct longer lasting and, hopefully, more efficient wells. Likewise, screen design, gravel-pack selection, development techniques, and an overall understanding of ground water hydraulics are some of the factors that have been given serious consideration and study. The life of a well will be limited unless it is originally constructed to permit both a high level of initial efficiency and ready redevelopment and is pumped at the rate dictated by the well design. For example, a stainless-steel screen will be useless if initial and accumulated debris does not allow water to flow freely into the well.

Redevelopment of water wells can be a successful and economical practice because it is an important factor in reducing production costs. Redevelopment considerations must be included in the original design if the well is to function at top efficiency throughout its life. An understanding of redevelopment is essential in the water well industry, although these methods are only recently finding widespread use. Large sums of money can be wasted on unnecessary operating expenses, and greater capital invested because of inefficient wells.[556]

Optimization of Pump Efficiency

W. H. Walker[147] has developed a series of graphs relating pumping costs to changes in well and pump efficiency. The graphs are intended to be of use to the well owner, well contractor, technical consultant, planner, pump manufacturer, etc. When the investigation is completed and released, it will be an important contribution in determining when to rehabilitate or to replace inefficient pumps or wells.

A brief explanation of the graphs is presented here. For a detailed review of the applications of the approach, the technical report should be consulted when it becomes available from the Illinois State Water Survey. As support data, Figure 113 illustrates the relative head — discharge characteristics of steep and flat head pumps. Figures 114 and 115 give the individual flat head and steep head pump characteristics.

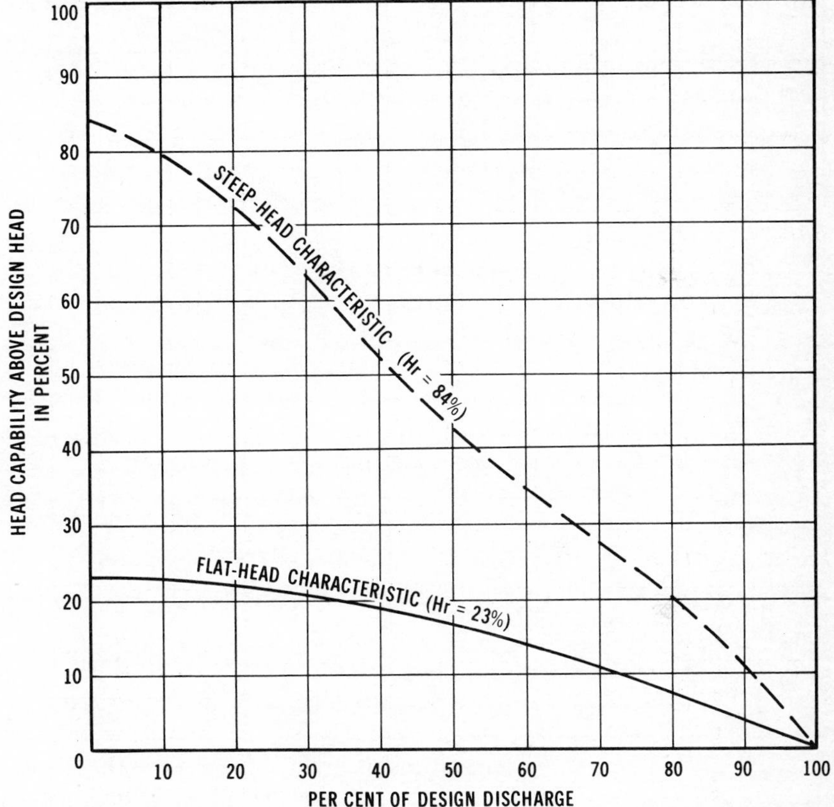

Figure 113[647] General comparison of flat-head and steep-head pump characteristics.

Figures 116 and 117 permit an approximation of the overall field efficiency of a particular pumping plant under a range in percent of full-load operation. Figure 118 approximates the increase in operating cost per unit pumped under a range of percent of designed Q load. Figures 119, 120, and 121 approximate the cost of pumping any quantity of water against any head with a pump of any given overall efficiency. The graphs are designed so that the user has only

$$\text{HEAD RATIO } (H_R) = \frac{\text{SHUTOFF HEAD} - \text{DESIGN HEAD}}{\text{DESIGN HEAD}} \times 100$$

$$H_R = \frac{80.5 - 65.5}{65.5} \times 100 = 23\%$$

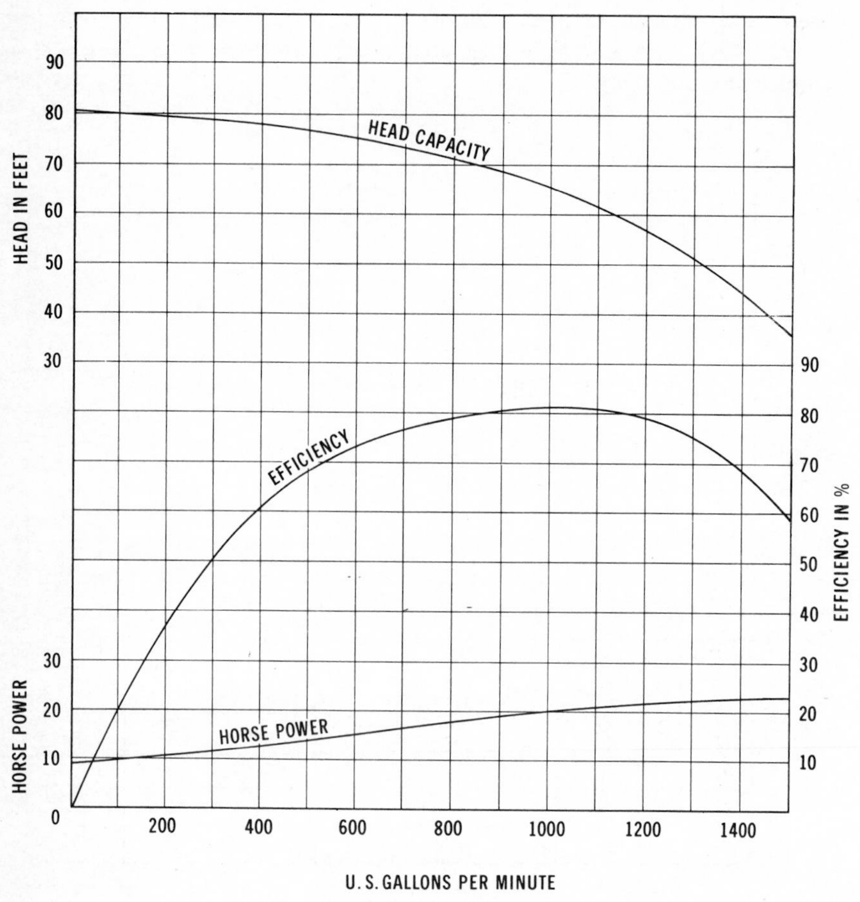

Figure 114[647] General characterics of Flat-head pumps.

to project straight lines and to multiply all results obtained by the users actual power-rate charge, if different from one cent per KWH, to obtain his actual cost of pumping water under a combination of conditions mentioned above.

The increase in pumping cost per year, at 100 percent efficiency can be determined from Figure 121, once the overall efficiency has

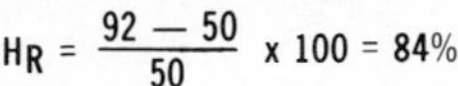

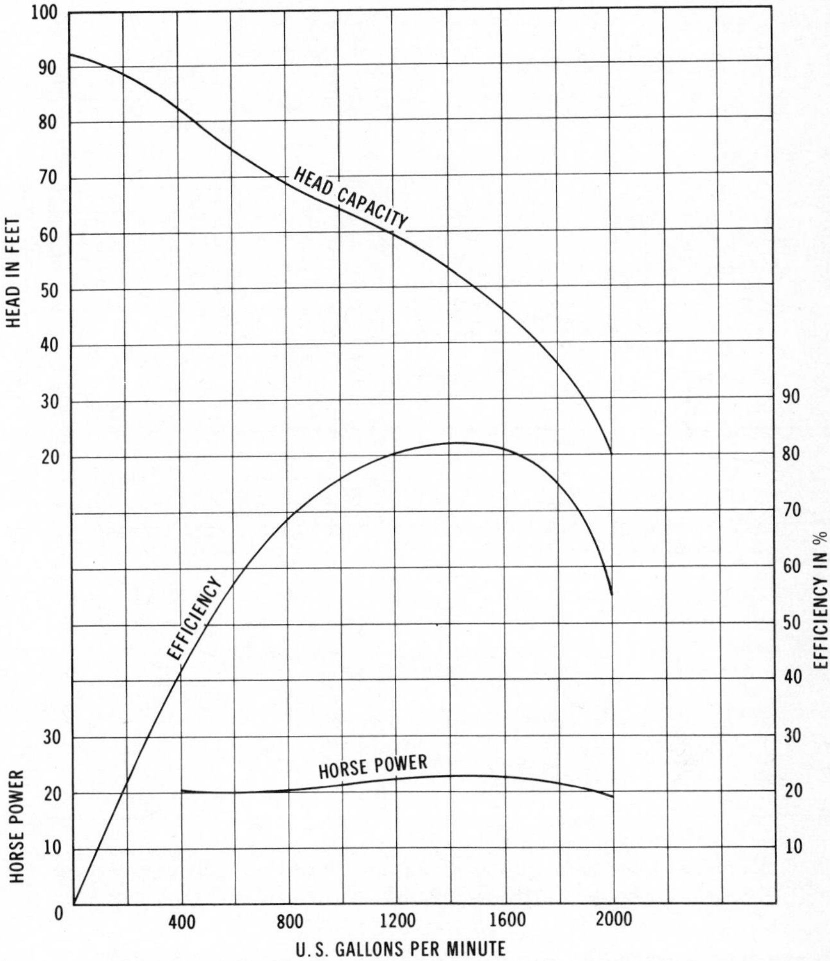

Figure 115[647] General characteristics of Steep-head pumps.

been determined from Figure 116 and 117, and the well efficiency has been established from pumping tests. In addition, by projecting the pumping cost value vertically in Figure 122 to the proper service life-line (20 years for a well or 5 years for a pump) at the interest rate of the capital investment, an equivalent improvement cost can be obtained — or the expense that can be economically justified

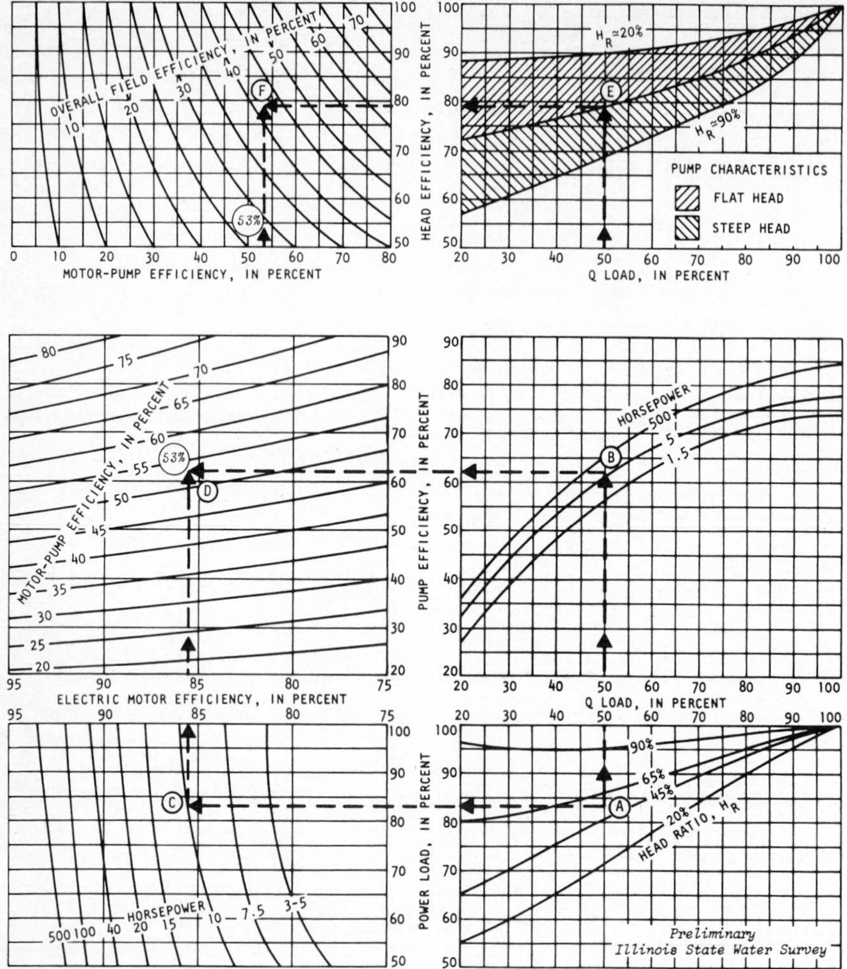

Figure 116[647] Overall field efficiency of electrically driven deep well turbine pumps operating at 1750-1800 rpm. An approximation of pumping system's efficiency under specific range of operation. Enter graph set at A.

during the service life of the installation to rehabilitate the specific system. If the C_{EI} (Equivalent Improvement Cost) is greater than the estimated cost of a new well or pump, well abandonment/pump replacement rather than repair should be considered.

Overall efficiency depends to an important extent on the efficiency of the specific pump and power plant. Excessive operational costs can result from one of the following causes or a combination of one with another:

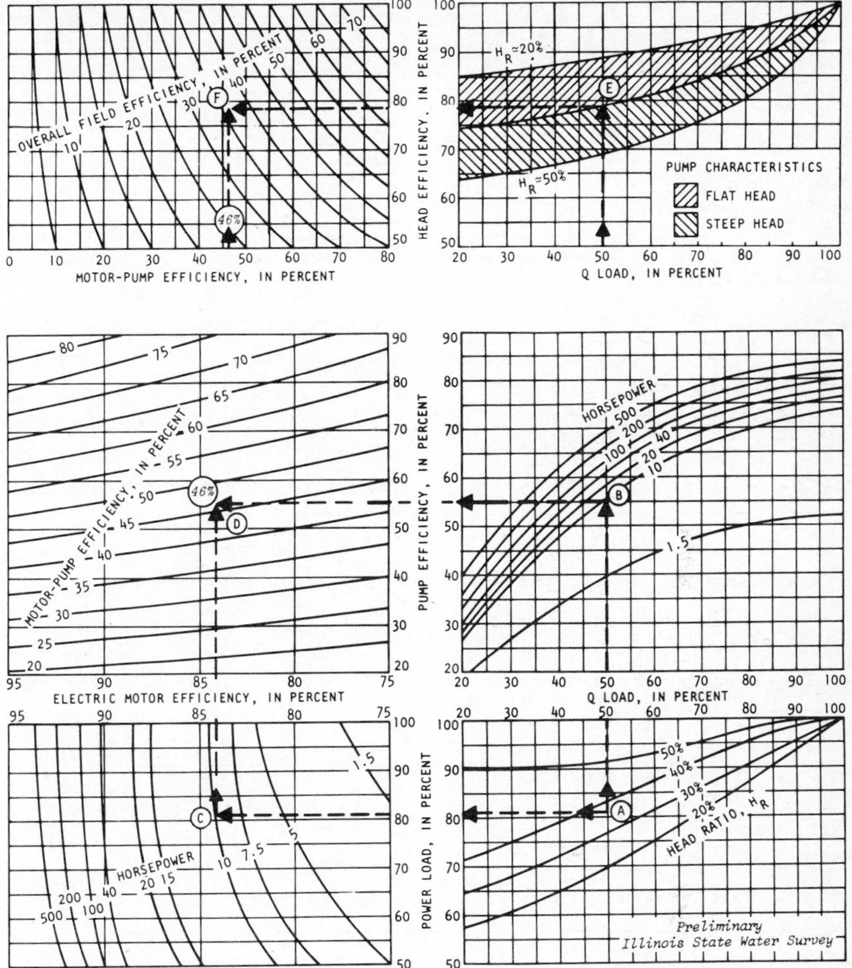

Figure 117[647] Overall field efficiency of electrically driven deep well turbine pumps operating at 3500-3600 rpm. Another approximation of pumping plants overall field efficiency under a range of full-load operation.

(1) Improper well design and/or construction.
(2) Improper selection of the pump and/or power unit.
(3) Inadequate monitorization of system performance.

Well characteristics which bear directly upon pump operations are:

(1) Drawdown.
(2) Total life for a specific pumping rate.
(3) Straightness of the well.
(4) Inability of well to produce sand-free water (Improper well development).

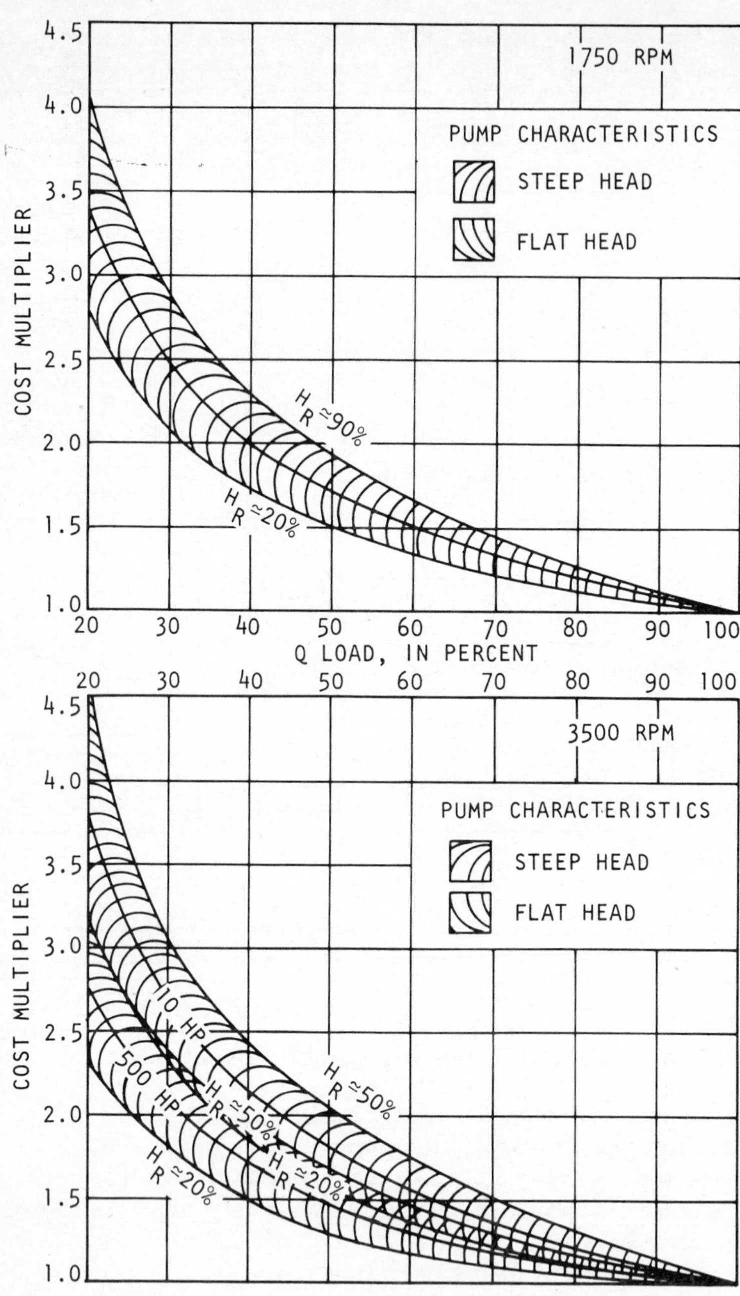

Figure 118[647] Increased cost of lifting water with pumps operating at less than full capacity. Increases in operating cost per unit pumped under a range of percent of designed Q load for flat-head pumps.

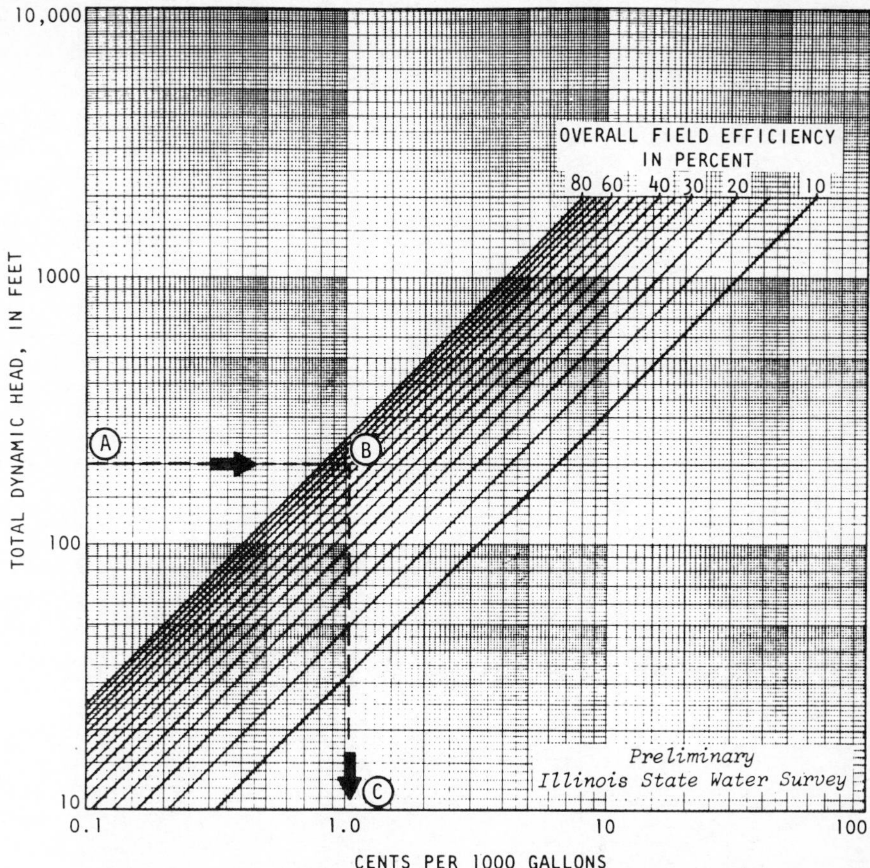

Figure 119[647] **Pumping cost in cents per 1000 gallons at power rate of 1 cent per KWH.**

To select the proper pump for a given well, certain data is required from a pumping test. The data required includes:

(1) Maximum quantity well can produce.
(2) Depth from which pump must lift water.
(3) Pump discharge pressure.
(4) Extent of sand content of producing well.

In a study on pumping plants, Hansen[304] of Kansas State University determined the ranges in overall efficiency of the power unit and pump combined. Figure 123 shows the combined efficiency of the electric power units and pumps for 21 irrigation well installations in Kansas. Electric motors convert electrical energy at about 88 percent efficiency, so overall efficiency figures are higher for electric

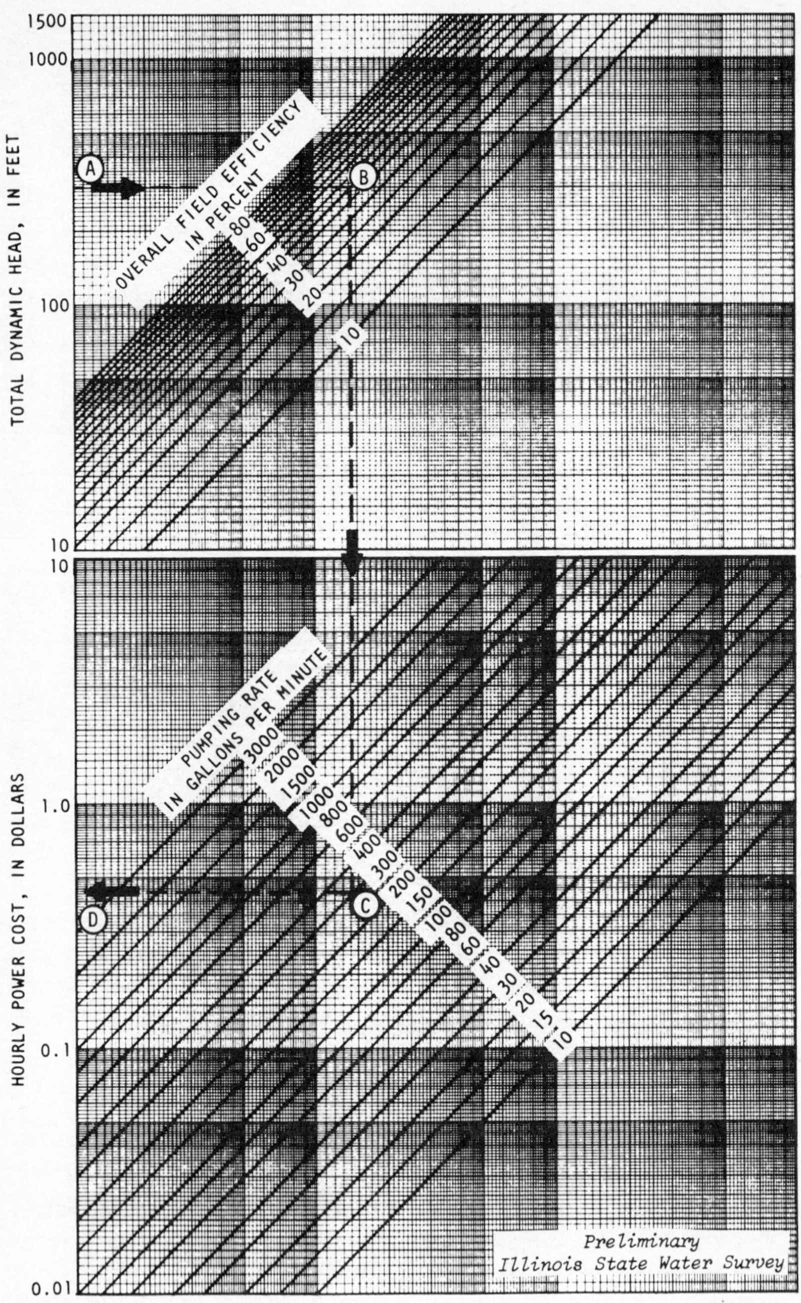

Figure 120[647] Pumping cost in dollars at power rate of 1 cent per KWH. A continuation of the approach for obtaining approximate cost of pumping costs, resulting in hourly power costs.

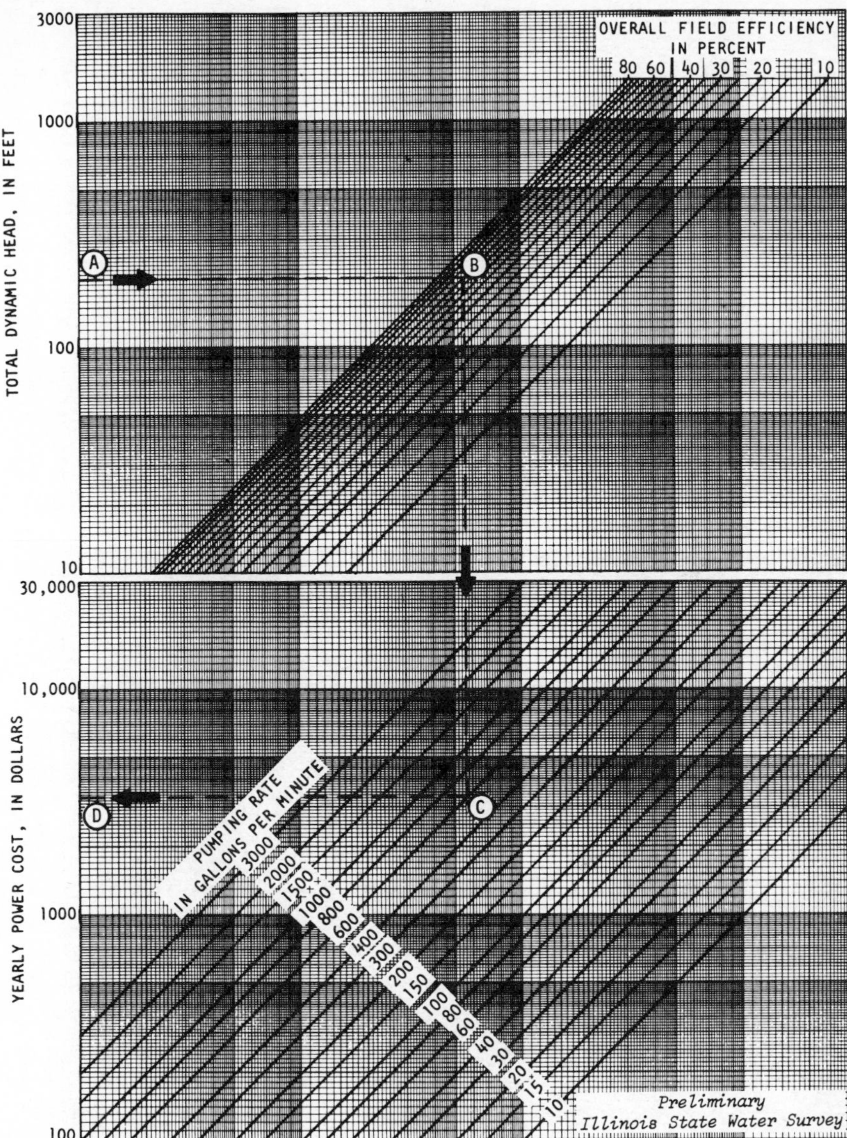

Figure 121[647] Yearly pumping cost in dollars at power rate of 1¢. A continuation of the approach for obtaining cost of pumping costs, resulting in yearly power costs at 100% pumping efficiency.

plants than for pumps driven by internal combustion engines. In referring to Figure 123, therefore, an overall efficiency of 40 percent means that only 40 percent of the electric energy input to the motor is converted to *water horsepower* in the operation of the pumping plant.

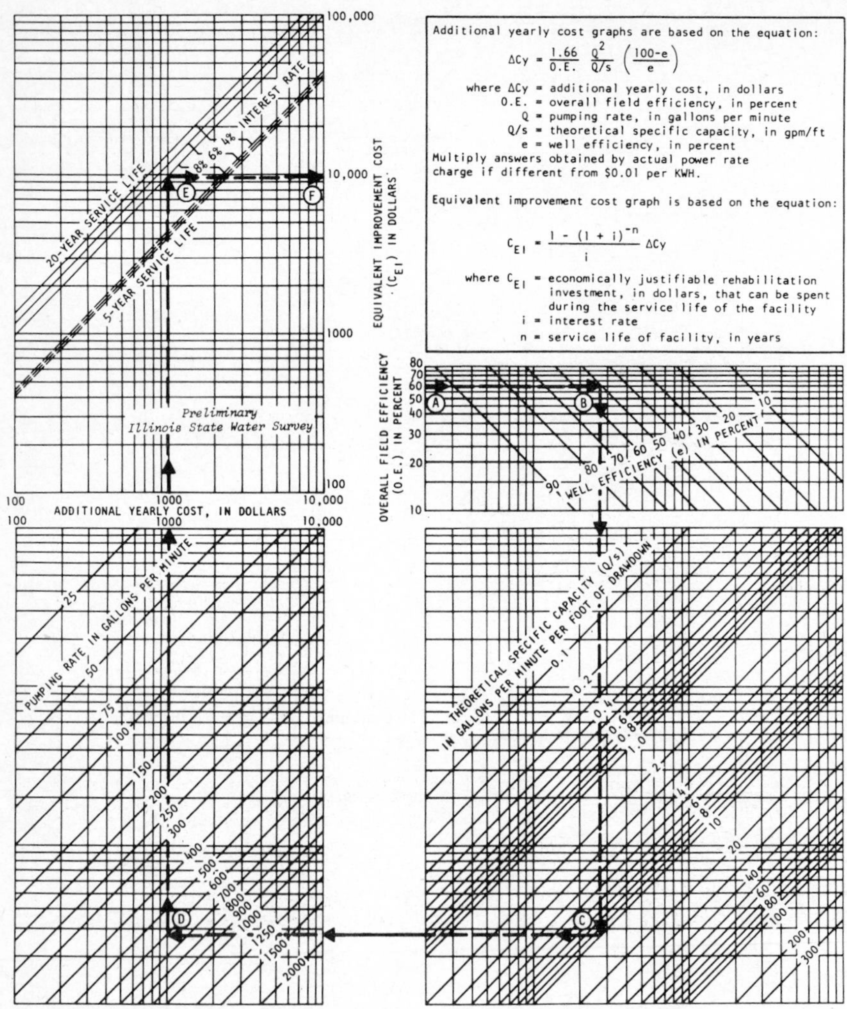

Figure 122[647] Economically justifiable rehabilitation investment to increase well efficiency. A graphical representation of an approach for obtaining equivalent improvement cost.

The power required to pump water at a given rate and against a given head is defined as water horsepower (W_{hp}).

EQUATION 64

$$W_{hp} = \frac{Q(\text{gpm}) \times \text{Head (ft.)}}{3960}$$

Total head is the lift, from pumping level to ground surface, plus the discharge head against which the pump is working. Total horsepower required is Water Horsepower plus allowances for power loss from inefficiencies in the pump, mechanical drive and power unit.

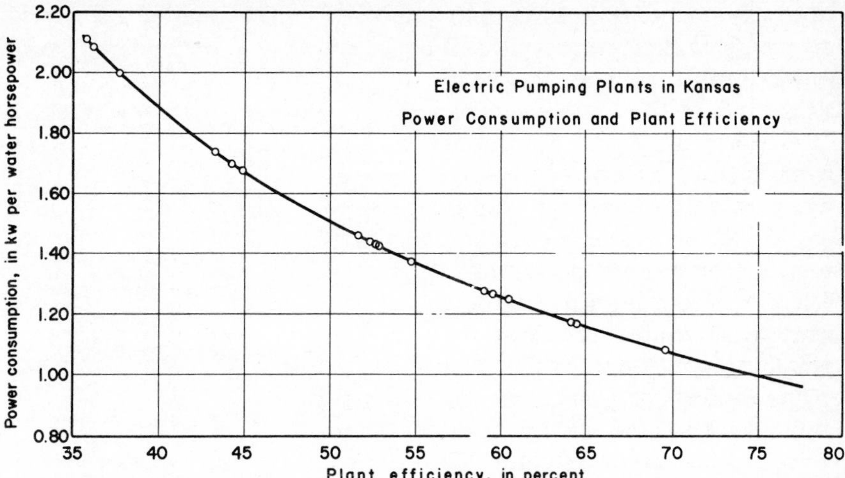

Figure 123[72] Field test results for 21 irrigation wells in Kansas with pumps driven by electric motors.

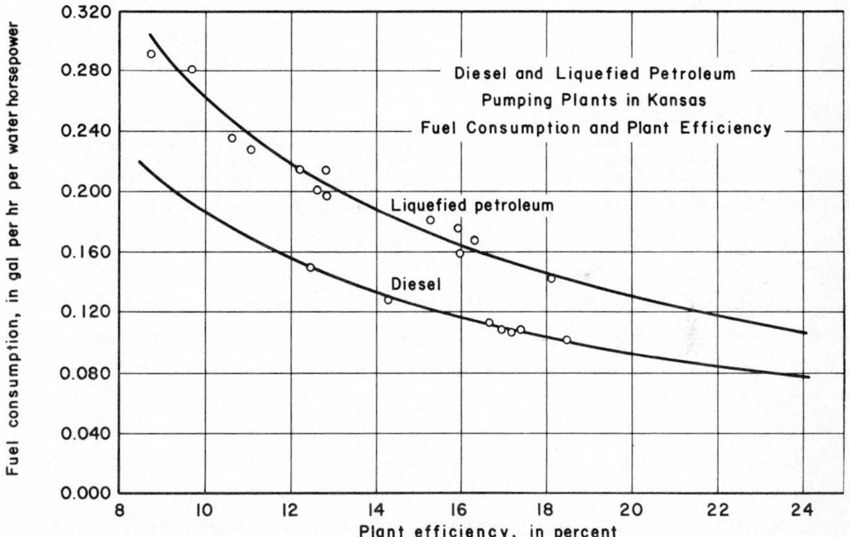

Figure 124[72] Field test results for 19 irrigation wells in Kansas with pumps driven by internal combustion engines.

Internal combustion engines are not particularly efficient modes to convert energy and fuel into mechanical horsepower; however, operating cost per horsepower may be less than that for electric motors. Figure 124 illustrates the combined efficiency for diesel and liquified petroleum pumping plants from the study made by Hansen of 19 units.

Average total cost for pumping plants with various types of power units vary with initial investment and the energy source. Total costs are fixed costs plus operating and maintenance costs. Figures 123, 124 and 125 represent conditions, fuel prices and electric energy prices prevailing in Kansas at the time the pumping plants were conducted. Since the cost factors included in the total cost structures expressed in the graphs vary independently, comparisons illustrated in Figure 125 should be interpreted on a relative basis, not in terms of present dollar value.

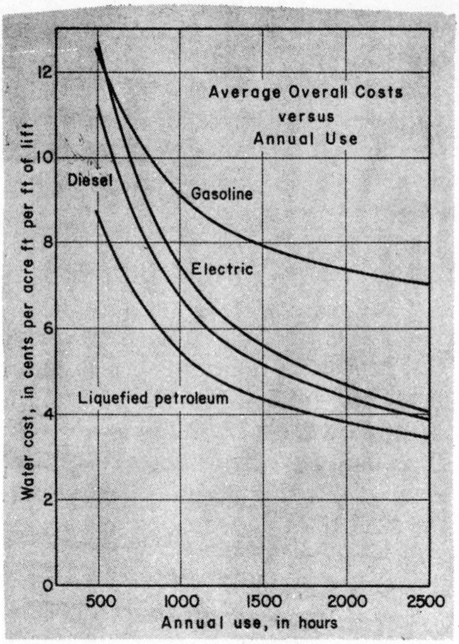

Figure 125[72] Comparison of annual costs for four energy sources of pumping plants.

Efficiency and Technical Criteria

Additional research on pumping plant efficiency (as briefly discussed previously in terms relating to the overall efficiency of the well system) is needed.

In considering pumping plant efficiency, the selection of any pump to a particular application must be in terms of efficiency. To accomplish this, pump performance factors must be considered. Fowler[226] with the support of the Water Systems Council, has proposed that pumps be rated on the basis of pumping capacity or output rather than in such abstract terms as motor horsepower.

Under Fowler's proposal, the nominal pump rating would be its discharge in gallons per minute when operating at the best efficiency point indicated by the head-capacity-efficiency chart for the pump. Figure 126 shows typical characteristic curves for a pump that would be rated as a ten gpm pump and assumes that the pump is driven by a submersible motor.

The efficiency curve always indicates the best value at a particular combination of head and pumping rate. The pumping rate represents the normal rating of the pump.

In addition to pump capacity in gallons per minute, however, output rating must also include a factor that includes head factors or pressure against which the pump must deliver its output. This factor must represent the water horsepower involved in the pumping

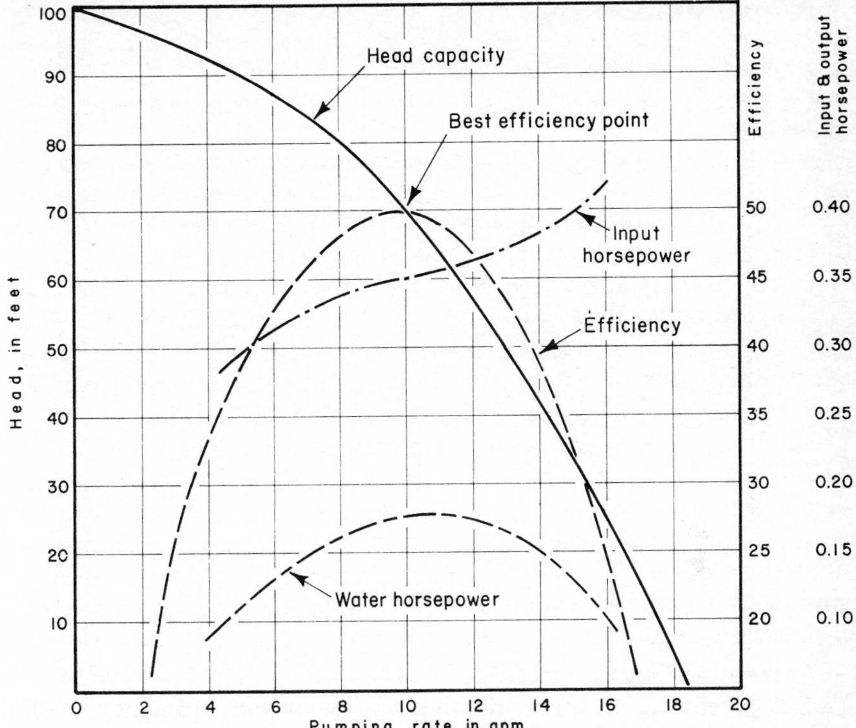

Figure 126[226] Proposed plan for rating water system pumps by pumping capacity and output rather than by pump motor horsepower. Typical performance curves for a pump that delivers 10 gpm against 70 feet of total head at its best efficiency point.

operation, since water horsepower varies directly with head or pressure at any given pumping rate.

Fowler further suggests using an "output factor" that would approximate 100 times the water horsepower at the best efficiency point. In practice, the output factors would be bracketed in groups or ranges that correspond to respective values of water horsepower. Of course, the output factor could be expressed in other terms.

The approach, briefly outlined above, if it is accepted by the industry, would more easily facilitate efficiency studies of water systems than the present use of the rather cumbersome pump motor horsepower.

PHYSICAL CONDITIONS OF WELLS

Prior to the development and use of downhole photographic closed-circuit television surveys and other downhole methods of observation,[621] water well problems were difficult to define directly.

Photographic Evaluation

With the expansion of photographic technology, photographs can now be produced to clearly show specific problems that might exist in a well.[336] Study and evaluation of these photographs assist in the selection of appropriate corrective methods. Well repair expenses can be significantly reduced if downhole photographs are employed. In some cases, photographs have indicated that wells should be abandoned since repair would be economically prohibitive or, if repaired, effective for only a short productive period.[335]

The photographic approach has been used during the past several years to inspect new wells or damaged portions of old wells, to locate various materials or equipment dropped into wells and to identify geologic formations. Such photographs show accurately and in detail the nature of conditions below the ground surface.

Photographs may be made from black and white or color film. The color photographs have been used in uncased holes to identify formation materials. The best color photographs of formation materials have been obtained in dry holes; however, color photographs of boreholes containing clear water can also be used for formation identification.

Stereo-photographic surveys made of new wells have provided a visual inspection of the entire well.[472] These surveys suggest to the regulatory agency and well owner that the well has been completed in accordance with guaranteed plans and specifications. However, the surveys cannot show detailed features of the grouting, gravel pack, screen placement versus formation changes, or many other aspects related to proper well construction and specifications. Such photographic records of a new well can be used for comparison with photographs made at a later date; thus, furnishing clues from which conclusions may be drawn for well rehabilitation. These comparisons also provide a basis for determining the endurance of the casing and screen materials. Subsequent surveys can also show the apparent effectiveness of the chemical rehabilitation of the well screen.

Photographs have been made in limestone and sandstone wells containing clear water and have assisted identification of porous and cavity zones. In limestone formations, photographs have been used to determine well conditions before and after acidizing. The stereo-photographic surveys have also been employed for many miscellaneous applications, such as: ventilation holes, mine shafts, brine wells, and others. A 360° camera for example for photographing an abandoned mineshaft via drilled holes near which pillars are constructed to prevent roof collapse. The photographic method is useful in observing the effectiveness of pillar construction.

Closed-Circuit Television Evaluations

In the past, it has been necessary to core the water-bearing formations to study their physical properties, such as the percentage and size of pore space but this method has not always been successful. In addition, some formations are too fragile to core, and examination of the outcrops of the water-bearing rocks many miles from the well site may not reveal the desired features of the formations. However, the use of a television camera in a well provides an instantaneous look at the *in situ* consolidated water-bearing formations.[143,153,214,258] A closed-circuit television system has obvious advantages over a photographic survey, with the exception of cost in some cases. Features that may be determined by this approach are: the nature of and depth to geologic contacts; the gross lithologic characteristics; the configuration of the hole wall; the location and condition of well casing and screens; and the precise location of debris in the well.[391,601]

Presently used television equipment permits either a vertical or horizontal view of downhole conditions.[143] An initial survey using a vertical-viewing probe is common, during which sections are noted where a horizontal view is desirable. When the initial survey is completed, the equipment is hoisted to the surface; the horizontal viewing mirror is attached to the cable and inserted in the well — a time-consuming process. Development of a surface-controlled mirror or prism which could be incorporated in the vertical-viewing camera and which could be mechanically activated to give a horizontal view would be a valuable contribution to ground water closed-circuit television.

The above surveys have not found widespread use to date in the water well industry even though significant cost-saving factors are involved.

A problem in photographic and television well surveys is the absence of an acceptable method to rapidly precipitate suspensions in water, particularly in older wells. The clarity of the pictures can be adversely affected or blacked-out entirely by suspensions in the water, e.g., colloidal material, bacteria and algae clumps, and debris from corrosion or incrustation. Various precipitants and coagulants such as alum, ferrous sulphite, etc., have been used occasionally with only limited success.

WELL CORROSION AND INCRUSTATION

Corrosion control has been of prime importance to well maintenance in the oil industry for many years because of the high cost of "workover" operations and of well replacement. However, practical, detailed corrosion control investigations have not been undertaken in the water well industry except for generalized commercializations. This neglect has resulted not only from a lack of detailed knowledge of the principles of corrosion control but from a lack of adequate economic impetus for employing any control methods.

Most ground water contains corroding or incrusting elements.[172] The difference between the effect of these elements on various waters is entirely a matter of degree and nature. The two are seldom found together because corrosion, naturally, tends to prevent the accumulation of incrusting materials; whereas, the accumulation of incrusting materials tends to protect or insulate the casing or screen against the action of corrosion. Corrosion only occasionally results in the formation of an incrustation — due largely to the erosion products that may be deposited "downstream." No one metal or alloy will resist all types of corrosion, but some have a greater tolerance to corrosion than others.

Research in metallurgy is of secondary importance in solving the problem of incrustation. Incrustation is caused in most cases by entirely different factors than those which cause corrosion. Metals can be used which will resist corrosion, but incrustation may take place on any kind of metal or material depending on the chemical character of the ground water and would require some "workover" method to remove the incrustation, although the physical condition of the metal in use must be able to withstand the "workover" techniques.

Corrosion

The subject of corrosion is complex. It is now generally accepted that metals are corroded for many reasons, under various conditions, and at different rates.[322] Any of the following agents may contribute to corrosion: oils, acids, oxidizers, ground water of various geochemical histories, brines, organic compounds, high temperatures, sulphur compounds, and alkalies. In both material and environment, a single causitive factor is not readily discernible in the variety of corrosion phenomena observed.

The oil industry has found considerable success, however, in various corrosion control methods.[496,636] Of these methods, most of the work deals with *cathodic protection* — an artificial electrical system constructed to redirect stray electrical currents that cause a specific type of corrosion in casing, transmission piping, and drill pipe.

The type of corrosion which commonly occurs may be any one of the following: direct chemical corrosion, dezincification, graphitization, galvanic or two-metal corrosion, concentration or solution cell corrosion, pitting, fatigue, or corrosion cracking. A visual inspection and/or a mechanical examination will usually indicate which form of corrosion is occurring.

Corrosion can severely limit the useful life of a well in three ways:

(1) Screen slot opening enlargement, followed by sand-pumping failure.

(2) Strength reduction, followed by collapse of well screen or casing.

(3) Redeposition of corrosion products, followed by screen blockage.

The rate at which corrosion takes place also depends on several environmental factors: (1) the ground water acidity (the pH), (2) the presence or absence of oxidizing characteristics of the ground water, (3) the rate of movement of ground water over areas being corroded (entrance velocity), and (4) the ground water temperature.

The extent of corrosion also depends on certain properties of the metals being corroded which may be inherent in the metals or may be acquired depending upon the form of corrosion taking place.[90,553] The important properties of metals that affect the extent of corrosion are: (1) the chemical affinity, (2) the oxidation passivity, (3) the properties of corrosion by-products, (4) the composition of the metal, (5) the changes due to strains from heating or cooling, (6) the original state of the metal surface, and (7) the galvanic relationship of two or more different metals.

Corrosion Potential

Moss[471] suggests that the following factors can be considered indicators of corrosive ground water:

(1) Acid pH — as the pH decreases (less than 7.0), corrosion activity increases.

(2) Dissolved oxygen — oxygen is usually present if serious corrosion takes place in ground water. Dissolved oxygen will greatly accelerate corrosion in acidic, neutral, or slightly alkaline waters.

(3) Hydrogen sulphide — as concentration in ppm increases, possibility of corrosion increases.

(4) Total dissolved solids (TDS) — in general, an increase in dissolved-solids content of water increases corrosivity. Serious corrosion may occur with dissolved solids in excess of 1,000 parts per million.

(5) Carbon dioxide — in excess of 50 parts per million, even in the absence of dissolved oxygen, accelerates the corrosion process.

(6) Chloride — in excess of 300 parts per million is capable under acidic conditions of penetrating and breaking down protective films and is a corrosion accelerator with most metals.

(7) Temperature — the chemical reactions of corrosion increase in rate with increase in temperature. Increased temperature decreases water viscosity, thus increasing the diffusion rate of oxygen. Electrical conductivity is increased at higher temperatures, thus, producing an effect similar to an increase in dissolved solids.

Highly mineralized ground water promotes scaling and deposition of calcium carbonate ($CaCO_3$) deposits which have a restraining influence on corrosion by shielding the metal surfaces. Langelier[376] devised a method of determining the incrustation or scale-forming parameters involved in $CaCO_3$ deposition which produces an indicator referred to as the Langelier Index (or Saturation Index). In order to derive this index, it is first necessary to know the following about the water in question:

(1) The alkalinity, which is a function of the carbonate ion concentration (ppm).
(2) The concentration (ppm) of ions present.
(3) The pH.
(4) The temperature.
(5) The total dissolved solids (TDS).

In simplified form, the Langelier Index (Saturation Index) for water within a pH range of 6.5 to 9.5 is the pH at which water reaches equilibrium with calcium carbonate:[543]

$$pH_s = (pK_2' - pK_s') + pCa + pA1k \qquad \text{EQUATION 65}$$

where: pK_2' and pK_s' = apparent constants computed from the true thermodynamic constants, pK_2 and pK_s, the values of which have been determined.

K_2' = second dissociation constant for carbonic acid.

K_s' = activity product of calcium carbonate. The term, $(K_2' - K_s')$ varies with the ionic strength, total dissolved solids (TDS) content, and temperature.

pCa = the negative logarithm of the calcium ion concentration mols per liter.

pAlk = the negative logarithm of the total alkalinity of the sampled water to methyl orange in terms of titratable equivalents per liter.

A major modification of the above has resulted in the following:

EQUATION 66

$$pH_s = \log\frac{K_s}{K_2} - \log(Ca^{++}) - \log(alk.) + 9.30 + \frac{2.5\sqrt{u}}{1 + 5.3\sqrt{u} + 5.5}$$

where: (Ca^{++}) and (alk.) are expressed in parts per million as Ca and $CaC0_3$, respectively.

An extended form of the Saturation Index is as follows:

EQUATION 67

$$SI = \text{Actual pH} - pH_s$$

where: The algebraic difference between the actual pH of a water sample and its calculated pH (see Equation 60) is a measure of the degree of $CaC0_3$ saturation.

A *plus* value for the Saturation Index (SI) indicates a tendency for the sampled water to deposit calcium carbonate; whereas, a *minus* value indicates a tendency to dissolve calcium carbonate. Ryznar,[543] however, reports that the Saturation Index is only qualitative. Langelier [376] emphasizes that his index is only an indicator of the directional tendency of the water to either deposit or dissolve calcium carbonate. Ryznar modifies Langelier's Index so that the approach is a more quantitative and, therefore, a more usable technique in defining a specific water's incrustation potential.

Ryznar's minor modification is called the Stability Index. [543] It was developed to eliminate the possibility of misinterpreting a positive Saturation Index as being non-corrosive or scale forming:

EQUATION 68

$$\text{Stability Index} = 2pH - pH_S$$

where: pH_s = (Equation 60)

A Stability Index value below 7.0 suggests scale formation or incrustation; when above 7.0 corrosion ("red water") of water systems becomes increasingly severe (see Figure 127). Considerable research on this approach is needed, since field results suggest that little is known of these processes. Further modifications are, of course, urgently needed. The Ryznar Stability Index is still widely used as a general indicator for the corrosive-incrustative nature of ground water.

Field Occurence of Corrosion

Only the most common forms of corrosion will be described here. Unusual types, of more importance to the oil industry, are treated by Ostroff.[488]

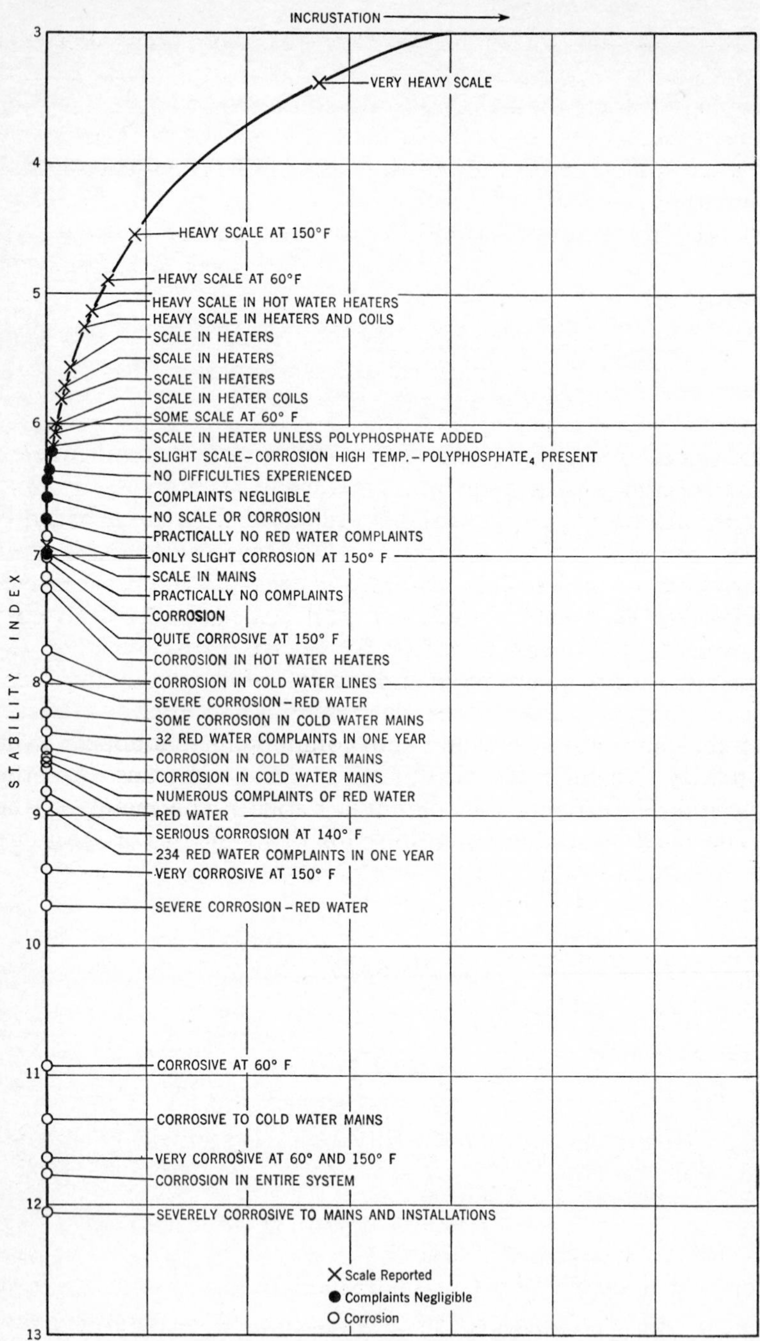

Figure 127[543] Case Histories of Using the Stability Index

The first form of corrosion, *direct chemical* , uniformly destroys the metal's surface generally leaving the body of the metal in its original shape. For example, this type of corrosion can attack the well screen, leaving the slots enlarged. The screen is weakened only to the extent that the thickness of the metal has been reduced by the corrosive action.

Case histories are given by Mogg[455] outlining examples of chemical corrosion by ground water which caused early failure of well screens. In many cases, the cause of failure can be related to particular chemical and physical factors or to a combination of both. Carbon dioxide (CO_2), oxygen (O_2), hydrogen sulphide (H_2S), pH, temperature (and in some special cases, unusual chemical compounds such as hydrochloric acid (HCl), chlorine (Cl_2) and sulfuric acid (H_2SO_4) are constituents in solution) which induce or inhibit the corrosion process.

The second type of corrosion, *dezincification* or *graphitization* is sometimes called *selective corrosion.* This type of corrosion consists essentially of one metal of an alloy being removed, leaving the structure in its original shape but in a spongy and weakened condition. Brass is such a metal, and this type of corrosion selectively removes the zinc from the copper-base alloy. Selective corrosion is deceptive in appearance because failures often suddenly occur in brass screens and other well and pump components that appear to be structurally sound. Dezincification results because of the electro-chemical difference in potential between the metals in the alloy. The most favorable conditions for this type of corrosion are those which combine (1) highly conducting solutions, such as a brine or slightly acidic ground water, (2) the presence of oxygen in solution, and (3) a two-metal alloy. Dezincification is often considered a special form of galvanic corrosion and is different than direct chemical corrosion since anodes and cathodes are separate. Graphitization of low carbon steel and cast iron is frequently associated with the presence of sulphate-reducing and other bacteria in a well.

The third type of corrosion, *galvanic*, results in the formation of a two-metal electrolytic cell. Galvanic corrosion occurs whenever two different metals are electrically connected in a conductive solution. An electric current is generated by the metals and results in the corrosion process — the section of the cell being corroded is the anode, and the protected section is the cathode.

As electrical current increases, the rate of corrosion increases. The rate is also influenced significantly by the scouring action of the electrolyte as it passes over the surface being corroded, by the relative size of the anode and cathode, and by the corrosion-resisting properties of the two metals involved. When both metals concerned are highly resistant to the corrosive condition, the rate of attack is low.

Most of the electrolytic corrosion encountered in well screens is of the galvanic type. It is recognizable by heavy accumulations of the products of the electrolysis and by the severe erosion to the most corrosion-prone metal (see Table 40). The order of the various metals tendencies to corrode galvanically is listed in Table 40.

As mentioned, when two dissimilar metals (some distance apart in the preceding table) are joined and immersed in an electrolyte, e.g., ground water, the metal higher on the list will be corroded. Metals

TABLE 40[46]

Electromotive Series of Metals and Alloys

Corroded End *(Anode)*

↑ DECREASING CORROSION RESISTANCE

Metals and Alloys
Magnesium Magnesium Alloys
Zinc
Aluminum 25
Cadmium
Aluminum 17ST
Steel or Iron Cast Iron
Chromium-Iron (Active)
Ni-Resist
18-8 Chromium-Nickel-Iron (Active) 18-8-3 Chromium-Nickel-Molybdenum-Iron (Active)
Led-Tin Solders Lead Tin
Nickel (Active) Inconel (Active)
Brasses Copper Bronzes Copper-Nickel Alloys Monel
Silver Solder
Nickel (Passive) Inconel (Passive)
Chromium Iron (Passive) 18-8 Chromium-Nickel-Iron (Passive) 18-8-3 Chromium-Nickel-Molybdenum-Iron (Passive)
Silver
Gold Platinum

Protected End *(Cathode)*

(After the International Nickel Co., Inc.)

located in close proximity on the list have little tendency to produce strong galvanic corrosion. The table applies for electrolytes such as common dilute water solutions — saline water, weak acids, and weak alkalies, all common in the ground water environment. Bimetallic corrosion takes place only when current flows between the metals through the liquid in which they are immersed.

Metals change relative positions in the previous table because of external influences. For example, the stainless steel alloys can change relative positions depending on the oxidizing characteristics of the electrolyte. This is the reason that the stainless steels are listed in the galvanic series as both "passive" (stable) and "active." Materials that possess this characteristic should be carefully evaluated for a particular environment to determine if an undesirable active state is possible.

Metals and alloys which have been found to be most resistant to corrosive agents that attack well screens under many conditions of soil, ground water, usage, and ambient pressure (graded in order of corrosion resistance) are listed below:

(1) Monel Metal (Approx. 70 percent Nickel, 30 percent Copper); (2) Stainless Steel, (74 percent Low Carbon Steel, 18 percent Chromium, 8 percent Nickel) (provided that soil ground water conditions and screen openings are such that the protective film is maintained); (3) Everdur Metal (96 percent Copper, 3 percent Silicon, 1 percent Manganese); (4) Silicon Red Brass (83 percent Copper, 1 percent Silicon, 16 percent Zinc); (5) Anaconda Red Brass (85 percent Copper, 15 percent Zinc); (6) Common Yellow Brass (Approx. 67 percent Copper, 33 percent Zinc); (7) Armco Iron; (8) Low Carbon Steel.

To establish the most common causes of corrosion, the following are the usual contributing factors:

(1) Low pH value, coupled with low alkalinity, low TDS, and high content of free carbon dioxide (high Ryznar Stability Index).
(2) A water with a high content of dissolved oxygen which by combination with liberated hydrogen, prevents the formation of a protective film.
(3) The presence of hydrogen sulphide, sulfur dioxide, or similar gases.
(4) The presence of organic acids.
(5) The presence of iron sulphate.

Corrosion Control

The following general guidelines were listed by Mogg[455] and apply to major factors involved in corrosion control:

(1) If low-carbon steel will corrode due to chemical corrosion in a particular chemical type of ground water, it will corrode more quickly if connected to a metal listed below it in the electromotive series (see Table 40). The corrosion rate is reduced considerably at a distance of about three diameters away from the connection. Stainless steel end fittings should be used on both ends of stainless steel screens, for corrosion is concentrated on a relatively small area of mild steel which is commonly welded to the stainless steel screens.
(2) Work-hardened material corrodes faster than the same metal in an annealed condition.
(3) Stressed parts are more likely corroded than unstressed parts.
(4) Higher temperatures increase corrosion rates which generally double for each 40° F increase.
(5) High fluid velocities increase corrosion rates.
(6) Oxygen in solution increases corrosion rates.
(7) Electro-chemical corrosion causes loss of material along portions of the casing or screen with corrosion products being redeposited (see Figure 128). This can occur in ground water with relatively high pH and high total dissolved solids (TDS) (usually $> 1{,}000$ ppm).

The most common type of corrosion is an electro-chemical phenomenon involving oxidation resulting in a loss of electrons and ionization of the metal. The chemical reaction and electron release leads to other reactions such as rust formation (see Figure 129). One form of rust is ferric hydroxide and is formed in the overall reaction:

$$4Fe + 3O_2 + 6H_2O \longrightarrow 4Fe(OH)_3$$

For iron to rust, the above oxidation reactions must occur and both oxygen and moisture must be present. Iron will not rust in an atmosphere containing only oxygen (assuming, of course, carbon dioxide, hydrogen sulphide, other salts, etc., are not present in significant quantities).[80]

Two points require illustration: (1) Metals have an oxidation potential, which is a measure of the energy required to remove electrons. (2) This electrode potential for production of ions and electrons depends on: (a) the nature of the metal or alloy and (b) the nature of the electrolyte or corrosive environment. The metallurgical factors which are involved are: (1) The nature of the metal or alloy, i.e., composition. (2) Alloy behavior — single phase, two phase, distribution of phases. (3) Imperfections — grain boundaries, grain size and shape, impurities or inclusions. (4) Mechanical factors — localized stresses, surface finish.

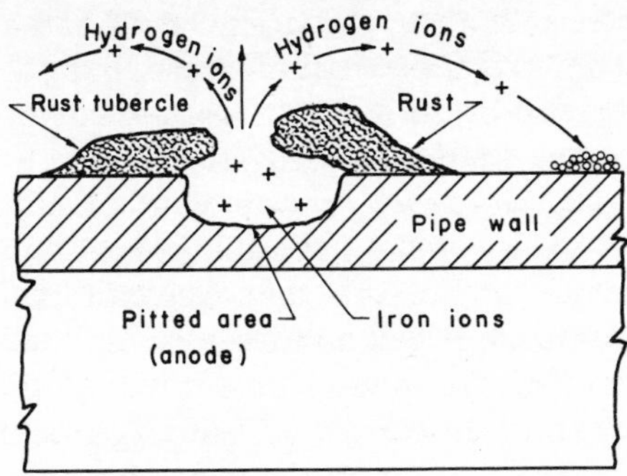

Figure 128[455] Anode and cathode can develop in nearby areas on a single metal surface, resulting in corrosion by local action.

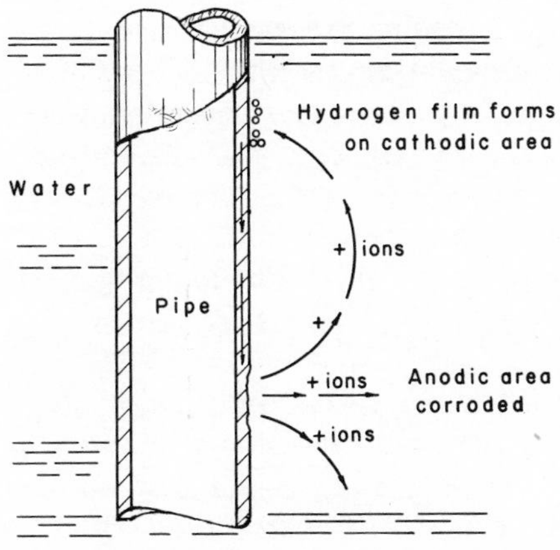

Figure 129[455] Corrosion of iron forms deposits of iron hydroxide and oxide rust at anodic areas.

Expensive corrosion-resistant alloy screens are essential in some areas of aggressive corrosion. Because of their construction and use, screens with large surface areas exposed to constantly renewed corrosion activity are particularly subject to attack. Casing with solid walls presents, on the other hand, much less surface area and is often protected to a large extent by redeposited corrosion products. Low-carbon steel casing is relatively inexpensive and by doubling the wall thickness, the life of the casing can be quadrupled.[471] However, if connected directly to a more resistant metal screen by welding or a coupling, failure of the casing in the vicinity of the connection is hastened. Development of a relatively inexpensive, dielectric coupling of adequate strength would contribute considerably to the problem of well design and to the extension of well life.

Complexity is the outstanding feature of corrosion on subsurface equipment in oil wells. Wells in an entire field thought to be reasonably free from corrosion can become a critical problem at any time and specific corrosion-control methods used in older fields may not be applicable in newly developed fields.

A well differs significantly from almost any other structure which may be subject to corrosion. Wells differ from pipe lines not only in the variety of exposure but in greater inaccessibility. This problem of access in water wells is important — it means that data to indicate the progress of corrosion cannot be readily obtained, and it means that protection methods, such as cathodic protection used in pipe lines, are not applicable. Cathodic protection, properly designed, theoretically resists external attack on the outerside of the casing, but the nature of the electrical-current distribution renders it useless for internal corrosion. An important area of needed research is on the problem of increased galvanic corrosion caused by induced currents from submersible pumps. Shallow wells, where induced ground currents from the pump or overhead power lines enhance galvanic corrosion, can be protected by the use of sacrificial anodes.

Internal corrosion on a pipe line can frequently be controlled by the continuous injection of one of the following modifying agents: a corrosion neutralizer or inhibitor or a preferential wetting agent. Application of this technique in a producing oil well involves at least two additional difficulties: first, the well is accessible at one end only, and there may or may not be a path through which a non-toxic chemical can be injected; second, the agent responsible for the corrosion is more variable both with respect to time and to depth. Nevertheless, the use of inhibitors and other treatments has proven to be a valuable tool in combating oil well corrosion.

The method of casing installation in wells causes a problem in the application of coating materials. The installation itself can scratch and break the coating on the inside of the casing and expose the metal to corrosion. The coating process has been modified by mill or factory application, using coatings that are thinner and, generally, more expensive than those used for the external protection of pipe lines.

In another important and relevant respect, the vertical steel structure used in a well differs from horizontal steel structures used in pipe lines. Since the vertical structure is more inaccessible per unit of exposed area, any failure is more costly per occurrence. It is not a simple matter of repairing or replacing the casing section which has failed; it is a major operation to recover the part for repair. Often, the final result may be the complete loss of the well. The significance of the inaccessibility of the vertical structure and the potentially high cost of a corrosion failure is that more expensive means of combating corrosion are fully justified. Therefore, special (and expensive) alloys for casing and other vulnerable metallic well components are readily adopted, once their superiority in corrosion control has been demonstrated.

A standard method of corrosion control has not yet been fully developed for pipe lines. Nevertheless, practical techniques have been devised which inhibit several types of pipe line corrosion. The probable cause of pipeline corrosion in different environments can be predicted with reasonable accuracy, and for this reason, protective measures can be taken for control. However, corrosion cannot be predicted with any accuracy in subsurface equipment of producing wells. Although the degree of distribution of corrosion is never known in advance, after sufficient corrosion experience has been accumulated on the first wells in the field, this experience may be applicable to older wells.

In well fields where corrosion is a critical problem, standard effective countermeasures are difficult to define. Rarely are the techniques from other fields transferable, and certainly no reliable set of simple measures guarantees retardation. Just as the attack is varied, so are the methods available for control.

Briefly, the control measures that are available are:

(1) The injection of chemicals to condition the corrosive fluid and to treat the surface of the attacked metal — various alkaline substances for pH control, inhibitors, and preferential wetting agents.
(2) The use of corrosion-resistant alloys, either for components or as platings and linings.
(3) Nonmetallic coating and/or lining materials such as plastic-lined tubing which has been widely adopted for corrosive environments.
(4) Cathodic protection, although not suitable for protecting internal surfaces (with a few exceptions), could be more widely used for the mitigation of external casing attack.

In recent years, cathodic protection has been used with increasing frequency throughout the oil and gas producing industry to protect the external surfaces of well casings against corrosion. Costly "workovers" of oil and gas wells, due to leaks in the casing caused by external corrosion, have prompted much research in cathodic protection systems.

If corrosive ground water conditions are anticipated or have been encountered in existing wells, corrosion measuring equipment can determine the relative passivity of available metals resulting in a substantial savings in costs.

Moss [471] reports that corrosion was a major problem in the El Kharga oasis region of Egypt's western desert where carbon structural steel failed in less than 12 months due to ground water with high concentrations of CO_2 and H_2S, and pH values ranging from 6.6 yo 7.0. Various materials, including aluminum and plastics, were used but also failed due to inadequate physical properties and high bottom-hole temperatures. However, some progress has been made in West Pakistan by using fiberglass-reinforced plastic casing where mild steel casing failed witin two years due to corrosion, and the initial results have been satisfactory.[575] Corrosion tests were made with corrosometer probes. It was noted that the performance of aluminum, although superior to the structural steels, was satisfactory only for specific applications and for a limited amount of time. Moss also observes that the corrosion rate of copper-bearing steel, although high, was less than one half that of mild steel materials. Stainless steel-type 304 was relatively passive to the ground water and, therefore, was selected for application in future wells of that region. Barnes and Clarke[109] have also done extensive work on this approach.

Detecting corrosion on external casing and determining its extent are two of the most difficult problems in the entire field of corrosion control. The inaccessibility of the external casing surface hampers examination, not only in assessing the problem but in determining the feasibility of cathodic protection. Advanced techniques to control corrosion of external casing have been developed in the oil industry, most of which involve sophisticated electrical equipment. Observation via photographs of closed-circuit television can aid in evaluatting the micro-characteristics of early corrosion.

Rising drilling costs have necessitated the oil and water well industries concern with the protection of existing facilities against deterioration by corrosion.[37] The importance of coatings in the control of corrosion is evidenced by the large expenditure of money for them by the petroleum industry. Preliminary results of a survey conducted by the National Association of Corrosion Engineers show that almost one billion dollars is spent annually for industrial paints and coatings within the oil well industry alone.[80]

> Protective coatings may be divided into two categories, organic and inorganic. Inorganic coatings are composed of metals or other noncarbon compounds. One example is the zinc used in galvanizing which controls corrosion by providing cathodic protection to the base metal until the zinc is consumed. Other types include plating with metals such as cadmium or nickel alloys which form a corrosion-resistant surface for the base metal.
>
> Organic coatings are those which are compounds of carbon and include wax, asphalt, coal tar, and plastics. This type of coating prevents corrosion by isolating the base metal from its surrounding environment by forming an impermeable and electrically insulating barrier.
>
> Organic and inorganic materials can be combined to form an effective coating as for example, a zinc-rich primer with an organic top coat. [The potentially-toxic nature of the coatings must, however, be evaluated.]
>
> In order to permanently prevent corrosion of the metal coated, a perfect organic coating would have the following properties:

(1) Electrically insulating.
(2) Complete and continuous.
(3) Unaffected by deterioration with time.

A coating possessing all of the above properties has not yet been developed.

Bacterial Corrosion Incrustation

The presence of sulphate-reducing bacteria has been found in numerous instances of corrosion damage as recorded by Postgate.[517] Von Wolzogen Kuhr[641] proposed the first widely accepted theory of the corrosion activity of sulphate-reducing bacteria. This thesis has been investigated by Starkey and Wight[593] and further discussed by Starkey[592] and Sharpley.[560] Although the importance of sulphate-reducing bacteria in the external corrosion of pipe lines is well known, the bacteria's activity within water distribution pipes is often overlooked. Black, red, or "rusty" water or offensive odors at isolated points in a water-supply system can often be attributed to sulphate-reducing bacteria. Trautenberg[623] treats the general features of aerobic and anaerobic bacteria in oil industry water flood operations. Ostroff,[488] in an important contribution, covers the salient features of oil field technology, including the chemical and bacterial aspects of water flood operations; and Case[161] covers many of the problems encountered in these operations.

Role of Bacteria

The aerobic genus, *Pseudomonas*, includes some 30 species found in fresh and sea water. They are non-spore-forming, small slender rods. *Pseudomonas sp.* frequently produce a water-soluble, bluish-green pigment. These micro-organisms produce enzymes and can cause, under certain conditions, the decomposition of cellulose and many types of hydrocarbons. Chloride and sulphate salts reportedly stimulate the growth of *Pseudomonas sp.*[54] This genus is one of the most difficult to control and is known to become resistant to the normally toxic, quaternary ammonium compounds and can cause severe corrosion.

On the other hand, much of bacterial corrosion of well structures is due to the metabolic activities of *Desulphovibrio desulfuricans.* These anaerobic, sulphate-reducing bacteria (which do not assimilate free oxygen) utilize sulphate (SO_4) as the preliminary energy source in the presence of enzymes, hydrogenase, and other naturally occurring compounds. The oxygen within the sulphate of naturally occurring organic matter and other mineral sources is assimilated by these organisms using hydrogen as an energy source.

King and Miller[361] evaluate several reasons why sulphate-reducing bacteria initiate the corrosion of iron and steel in oxygen-free conditions:[361]

(1) Stimulation of the cathodic part of the corrosion cell by the removal and utilization of the polarizing hydrogen by the bacteria. (2) Stimulation of the cathodic reaction by solid ferrous sulphides formed by the reaction of ferrous ions with sulphide ions produced by bacteria. (3) Stimulation of the anodic reaction, metal dissolution, by bacterially produced sulphide. (4) Local, acid-cell formation. (5) Formation of iron phosphide by reaction of the metal with bacterially reduced phosphates.

The last two reasons are not significant, except possibly in isolated cases, whereas, (3) is probably important only at the start of the corrosion process due to the eventual formation of protective sulphide films in the presence of the free sulphide ion. The generally accepted theory of corrosion by cathodic depolarization was explored by King and Miller[361] and was found to be lacking. A new theory was proposed — the cathodic reaction (hydrogen evolution) occurs on the ferrous sulphide produced by reaction of ferrous ion with the bacterially produced sulphide ion. The ferrous sulphide activity diminishes with time, possibly because of the bonding of atomic hydrogen within the ferrous sulphide crystals.

Certain aerobic bacteria can also utilize hydrogen. Slime material has been found to contain the enzyme hydrogenase and, therefore, can be involved in bacterial corrosion, e.g., *Aerobacter aerogens*, *Escherchia coli*, and *Proteus vulgaris* — all aerobic bacteria!

Iron bacteria include such filamentous genera as *Clonothrix*, *Crenothrix*, *Leptothrix*, and the "true" bacteria, *Gallionella*. The latter is a rod or kidney-shaped bacteria which secretes long, slender, twisted bands or stalks of ferric hydroxide. Figure 130 illustrates the relative size of most iron bacteria.

All of the iron bacteria oxidize ferrous iron to ferric iron utilizing the derived energy for the chemo-synthetic assimilation of carbon. In satisfying their energy requirements, these bacteria precipitate large quantities of ferric hydroxide.

Iron bacteria grow best at low temperatures and are commonly found in wells, springs, and rivers containing soluble iron salts. They, unfortunately, have a wide range of oxygen tolerances and will grow in water with 0.3 to 9.0 ppm dissolved oxygen.[54] The common types of slime-forming, aerobic (free oxygen assimilators bacteria) are generally responsible for bacterial incrustation problems in wells.

Iron bacteria may also cause pronounced corrosion. When a small amount of ferrous salt is formed (at some point where oxygen concentration is low) by anodic attack, iron bacteria convert the ferrous salt to hydrated iron oxide. This product encloses the metallic surface and corrosion becomes more vigorous.

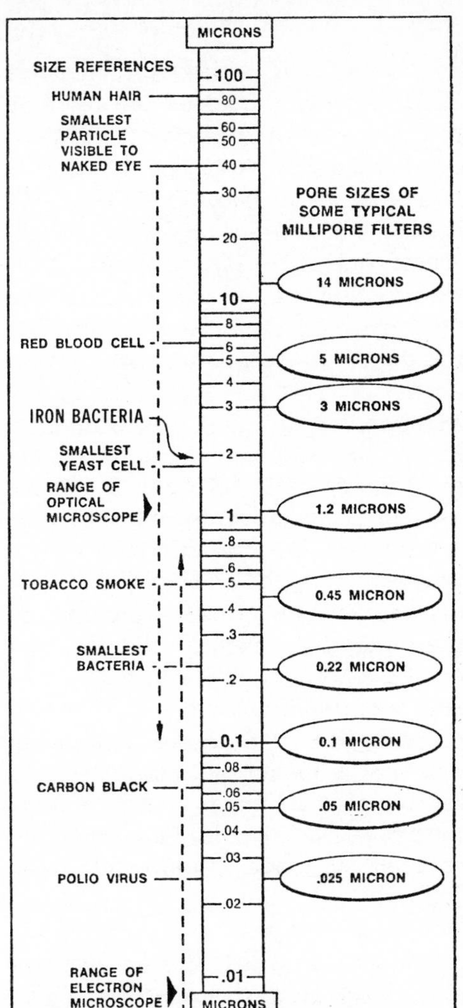

Figure 130[444] Scale showing relative size of microorganisms and other particulates.

The presence of aerobic bacteria usually makes an environment much less favorable for sulphate-reducing bacteria.[54] Sulphate-reducing bacteria require an anaerobic environment for optimum metabolic activity.

However, aerobic bacteria, growing as slime masses on metallic surfaces or in sediment pore spaces, create localized anaerobic microenvironments which promote maximum growth of species of *Desulforibrio* and others.

The anaerobic corrosion of ferrous metals is usually attributed to the activities of sulphate-reducing bacteria. Hydrogenase strains of these bacteria depolarize the metal surface by removing cathodic hydrogen. Hydrogenase systems are not confined to the sulphate- reducing bacteria but are possessed by many bacteria and microalgae, as mentioned. A strain of *Escherichia coli*, an organism which possesses a hydrogenase system, was found to be able to utilize nitrate as a hydrogen acceptor under anaerobic conditions in experiments performed by Mara and Williams.[423] The results indicate that the organism preferentially utilizes nitrate for the oxidation of organics, and also that when the cells are in a resting state, nitrate reduction occurs with attending corrosion if ferrous metals are present. In view of the diverse natures of nitrate-reducing bacteria and of their high numbers in soils, the corrosion caused by these organisms deserves additional study to resolve current problems, presently attributed to sulphate-reducing bacteria. Other types of bacteria have recently been isolated that may play important, yet presently undefined, roles in the complex corrosion-incrustation process.[419]

Bacteria Control

Updegraff[629] states that chlorination is of limited effectiveness against anaerobic bacteria that cause corrosion because the sulphide produced by the bacteria inactivates the chlorine. Lewis[387] suggests that the above statement is true only when there is an accumulation of sulphide material such as in a living tubercle. Because chlorine will kill the individual cells, the maintenance of residual chlorine throughout a distribution system is, however, one of the best presently available measures to prevent the formation of tubercles. Once the tubercles are formed, they must be physically removed and only then can the bacteria's activity be reduced to the formation of new tubercles.

The introduction of bacterial contamination into a well from drilling mud or fracture fluid has been misunderstood for many years in spite of the knowledge that bacteria are introduced during water-flooding operations. But the problem has recently been recognized, and remedial action can be taken if the proper chemicals are used early in an effective maintenance program. Wells drilled for the production of oil, gas, or ground water are all subject to contanimation because of contact with drilling fluid, cement filtrate, and hydraulic fracturing fluid.

Most trouble in contamination cases can be traced to one or two sources — sulphate-reducing bacteria or iron-oxidizing bacteria. Both of these are capable of reducing the permeability of a producing formation either through deposition of their by-products or from their presence in relatively large numbers. Having recognized the problem, the next step is to find a chemical that will reduce the contamination.

A cationic quaternary alkylbenzyl trimethyl ammonium chloride has been found to be an effective bactericide which also behaves as a corrosion inhibitor because of its film-forming properties.[38] This compound is a nonemulsifier and its surface active properties prevent clay swelling. However, this particular bactericide reportedly reacts with many clays and loses its effectiveness.

In the water-fracturing treatment used by the petroleum industry, water carrying minute quantities of bacteria is forced into the formation, thereby, contaminating the surrounding reservoir area. Sulphate-reducing bacteria and iron-oxidizing bacteria can adapt to the formation environment in a small amount of time, as previously mentioned. Formations containing lignite may aid in this adaptation since lignite or other carbonaceous materials can often be utilized as energy sources, e.g., pyrite oxidation produces hydrogen sulphide corrosion, etc.

Incrustation

Incrustation is defined as being "any clogging, cementation or stoppage of a well screen and water-bearing formation which is the result of the collection of material in and about the openings of the screen and the voids of the water-bearing formation."[46] Where proper conditions are present, stoppage, or incrustation will occur regardless of the material used in the screen or the type of well construction.

Field Occurrence of Incrustation

Incrustation may take the form of a hard, brittle, cement-like deposit and under different conditions may be soft and pasty like sludge or stiff jelly.[510] The causes of incrustation in the probable order of occurrence are as follows:

(1) Precipitation of materials carried to the screen in solution, such as carbonates of calcium and magnesium.

(2) Deposition of materials carried to the screen in suspension such as clays, silts, etc.

(3) Presence of iron bacteria in the ground water.

(4) Presence of slime-forming organisms other than iron bacteria (such organisms feed on ammonia and organic matter).

Incrustation is usually of the first type, i.e., incrustation which results from precipitation of minerals carried in solution in the ground water. In most of these cases, little or no organic material is present, and the chief incrusting agent is calcium carbonate which serves to cement the sand grains together. This agent is commonly found with aluminum silicate, iron sulphate, and a number of other minerals, contained in the openings between the sand and gravel grains. Although only a part of the total incrustation, calcium carbonate is, however, the fundamental binder.

There are a few geographic localities where the incrusting material is deposited entirely by direct mechanical means. Where the water-bearing formations contain considerable amounts of lignite which may partially decompose to form a slimy coating around the screen, both plastic- and metal-intake hardware are known to incrust.

The combination of carbon dioxide with water forms weak carbonic acid. Water can dissolve only a very small amount of calcium carbonate, but when appreciable amounts of carbon dioxide gas are present, water can dissolve calcium carbonate and other carbonate salts under special conditions of temperature and pressure.

Incrustation Control

Barnes and Clarke[109] examined the chemical properties and corrosive effects of ground water with reference to equilibrium conditions for 29 possible coexisting oxides, carbonates, sulphides, and elements. Of the 29 compounds considered, only calcite ($CaCO_3$) and ferric hydroxide ($Fe(OH)_3$) showed any correlation with the corrosiveness of the waters to mild steel (iron metal). Not one of the 39 waters tested was in equilibrium with the metal, but those waters in equilibrium or which are supersaturated with both calcite and ferric hydroxide were the least corrosive. Supersaturation with other solid chemical phases apparently was unrelated to corrosion. This report states that:

> . . . a number of solids may form surface deposits in wells and lead to decreased yields by fouling well intakes (screens and gravel packs) or increasing friction losses in casings. Calcite, $CaCO_3$; ferric hydroxide, $Fe(OH)_3$; magnetite, Fe_3O_4; siderite, $FeCO_3$; hausmannite, Mn_3O_4 (tetragonal); manganese spinel, Mn_3O_4 (isometric); two iron sulphides — mackinawite, FeS (tetragonal); greigate, Fe_3S_4 (rhombohedral) — copper hydroxide, $Cu(OH)_2$; and manganese hydroxide, $Mn(OH)_2$ were all at least tentatively identified in the deposits sampled.
>
> Of geochemical interest is the demonstration that simple stable equilibrium models fail in nearly every case to predict compositions of water yielded by the wells studied. Only one stable phase (calcite) was found to exhibit behavior approximately predictable from stable equilibrium considerations. No other stable phase was found to behave as would be predicted from equilibrium considerations. All the solids found to precipitate (except calcite) are metastable in that they are not the least soluble phases possible in the systems studied.[109]

The Barnes and Clarke study showed that equilibrium considerations are useful in establishing a reference for the actual properties and probably corrosive effects of naturally occurring ground water solutions. The detailed physical-chemical calculations are very useful in understanding the nature and for predicting behaviors of such solutions in water wells and other environments.

The practical results of the study are that the corrosive and incrustive nature of ground water upon mild steel may be predicted on the basis of rapid and inexpensive field measurements coupled with laboratory analyses made on appropriately treated water samples. The same data produce useful information on potential incrustation problems that may occur in water wells. Knowledge of the corroding and incrusting potentials of the ground water aids the selection of appropriate well construction materials. (See previous sections covering Langelier's Saturation Index and Ryznar's Stability Index.)

Water flow is created in a well because of an artifically induced difference in head. As pumping begins, the water in the well is pulled down, thus, creating a pressure differential between the water in the formation outside and the water level inside the well.

Because of the pressure change which is necessary to cause flow to the well, the dissolved carbon dioxide is released from the water. When this gas is released, carbonates, principally calcium carbonate ($CaCO_3$), precipitate from the water and deposit around the well-intake area. Because conditions for the release of the gas are most favorable in the well-intake area, the greatest amount of incrusting material is deposited as a cement-like material on the screen and in the sand or gravel around the screen. This incrustation can build up a cement-like wall, which is relatively solid for several inches around the screen. Partial incrustation may extend back for several feet into the water-bearing formation.

From the description just given, a reduction in the head or pressure drop may clearly reduce the extent of incrustation. To accomplish this, five factors are generally considered to play a part in the reduction of the incrustation process by the water well industry:[455]

(1) The well must be completed in such a manner which will permit the water to enter the well with the least resistance possible, and which will permit the well to be "developed" thoroughly. "Developing" a well is the application of one or more approved methods for removing forcefully the silt, fine sand, etc. from the formation around the screen, thus making the remaining sand or gravel more uniform in size and perfectly graded to prevent sand pumping, etc. Pumping is not an effective method of development and is insufficient unless combined with other means. When a well is properly developed, it will yield more water with less drawdown than is possible by any other procedure, with greatest security against failures such as sand pumping, except under very abnormal conditions.

(2) The pumping rate may be reduced and the pumping period increased. There are a number of cases on record where the trouble has been greatly reduced by pumping a smaller quantity of water per minute and pumping for a longer period of time. In one case a large part of the trouble was eliminated by using a smaller capacity pump which was operated more continuously than the larger pump which had been operated at rather infrequent intervals. It should be borne in mind that the most economical operation of both pumps and wells is obtained when pumping is as continuous as is practicable.

(3) The pumping load is spread among several wells instead of pumping one or a few large wells at excessive rates. A group of slightly smaller wells properly spaced can develop the full yield of the aquifer more efficiently. Velocities in the formation near each well are materially reduced, and the

drawdown in each well is decreased. These factors not only lessen the possibility of trouble from incrustation but, at the same time, reduce the cost of pumping.

(4) Since excessive available oxygen is involved in the incrustation process, downhole packers or vacuum seals are used to isolate the aquifer (from atmospheric oxygen) if the pumping level is considerably above the top of the well intake area (which would be applicable for water table wells in unconsolidated sediments).

(5) A periodic maintenance or cleaning procedure for each well is scheduled wherever local experience shows considerable difficulty from incrustation.

Corrosion/Incrustation Research

The accepted incrustation-corrosion theory generally states that the type and amount of dissolved minerals and gases in ground water determine the water's tendency either to corrode metals or to deposit minerals as incrustation. This theory has been challenged recently. A new approach, based on the electrokinetic theory, has been studied in an attempt to understand the underlying causes of incrustation and corrosion in water wells.[421]

From that investigation, the following conclusions were drawn:

(1) A saturated, water-bearing aquifer generates a streaming electrical potential when the water is flowing.
(2) The streaming potential is a catalyst in the formation of incrustation on water well screens (for additional information on streaming potential, see SP Log, Chapter 9).
(3) Incrustation forms on a water well screen only when it is negatively charged.
(4) No deposition forms on the water well screen when it is positively charged.

Mandal and Edwards[421] suggest that possibly direct current voltage can be generated in the aquifer to prevent the formation of incrustation on water well screens in a field installation. The required potential could be supplied by a DC source, operating coincidentally with the pump. Additional investigations should clarify the effectiveness of this approach. However, J. S. Fryberger (pers. com.) suggests that this solution could possibly result in greater damage by increased galvanic corrosion than would be alleviated by retarding incrustation.

W. H. Walker (pers. com.) suggests that many types of incrustation may be due to overpumping which in turn is related to poor well design, construction/development or the installation of a pump having excessive capacity for the hydraulic efficiency of the completed well. If Equation 52, i.e.,

$$L = \frac{Q}{A_e V_c \, 7.48}$$

is used and a pump is capable of producing only the Q computed in the above equation, the potential for significant incrustation should be reduced considerably. An inefficient well, Walker stresses, when pumped at an excessive rate, may cause incrustation at, or near, the holewall-screen interface, which in turn, causes a significant increase in the pressure drop at the interface.

Careful consideration should be given to all methods of preventing or retarding incrustation; but in a large number of cases, incrustation will occur sooner or later, so that well treatment becomes necessary. In general, three possible approaches for accomplishing well treatment are: (1) pull the incrusted screen, treat it to remove the incrusting material, and reset, (2) pull and replace with new screen, and (3) treat the screen and the water-bearing formation with inhibited acid or other solutions without pulling the screen. Agents such as sulfamic acid, hydrochloric acid, high concentrations of chlorine gas, and hydroxyacetic acid, etc. are in common use. Dispersing agents may also be used. The oil industry has made significant advances in the removal and inhibition of various types of scale formation.[573]

WELL STIMULATION

In view of the corrosion-incrustation processes described, an excellent analysis was made in 1959 of existing and emerging preventative techniques in crude-oil production in order to assess the extent of applicability to ground water production.[366,367,368,369] One of the more oustanding techniques capable of reducing the effects of corrosion and incrustation was well stimulation, known in the water well industry as well rehabilitation. More than 90 percent of the approximately 600,000 existing U.S. oil wells have been stimulated — some many times in order to increase their production. The frequency of stimulation of water wells in the United States could not be ascertained, although thought to be low but increasing.[218] Of the many publications reviewed by Koenig on oil well stimulation, very few references to the literature on water well hydraulics or water well stimulation were found. Conversely, in the scores of articles on water well hydraulics and stimulation reviewed, only a few authors, notably Mylander,[478] referred to oil industry techinques.[535] Koenig[369] observed that professionals in the two fields apparently do not read or contribute to any but their own technical journals, and this lack of inter-reliance was substantiated during this investigation.

The observed prevalence of this isolationist condition leads to a number of logistic propositions. Some of the techniques developed in one area may be applicable in others. Moreover, techniques which have been considered failures in one geographic region may be perfectly applicable in another. Research and development work is expensive, especially in the ground water field, since noncommercial research facilities are rare, and the chemical and mechanical industries will not risk funds without first predicting the potential return. This ordinarily involves an economic analysis of the total national market which indicates only a marginal return from the water well industry.

Stimulation was defined by Koenig's study as treatment of a well by mechanical, chemical, or other means for the purpose of reducing or removing an underground resistance to flow.[368]

> The term stimulation was chosen because it is the standard term of the crude-oil industry, and there is much more extensive literature on oil well stimulation than on water well treatment. Furthermore, the water supply field does not have a comprehensive term that denotes all that is meant by stimulation. Comparable terms in the water supply field are development, redevelopment, rehabilitation, or reconditioning. Development usually means some treatment applied upon completion of the well. Redevelopment means treatment applied at some subsequent time, ordinarily after a well has declined in production although rarely when a well has become unsatisfactory in production without having declined. Rehabilitation and reconditioning not only imply a treatment after completion of the well, but they may mean complete reconstruction of a well, for example, by removing and replacing the screen or casing, deepening the hole, or enlarging the well. Such operations, which are also part of what the oil man calls "workover," are not included in the definition of stimulation which is confined to treatment of an existing well as is, not its reconstruction. Among ground water technologists, the terms development and redevelopment also include treatment of a well for the purpose of clearing the produced water of turbidity, sand, sediment, or other impurities introduced during drilling. The term stimulation, [as used by Koenig] does not include these operations, except insofar as they are necessary for removing an underground resistance to flow.

The methods of stimulation fall into five major categories. They are, in order from the more familiar to the less familiar: surging, jetting, shooting or blasting, acidizing, and hydraulic fracturing.

Surging

Those methods which involve movement of water through the formation are included in surging. Numerous methods of surging, with and without the aid of chemicals, exist. There are many methods in use to flush out sand, silt, clay and fine materials which tend to reduce the permeability of the water-bearing formation. Typical surging, however, involves the movement of water back and forth through the formation, accomplished by using a surge block, air pressure, or by starting and stopping the pump. Included in the surging category are those treatments in which chemicals are placed in the well to penetrate the borehole wall — phosphates, chlorine, wetting agents, and acids. Surging should begin slowly at first, developing the aquifer at greater and greater distances into the formation.[46]

Jetting

Jetting, which is a method of high-velocity washing of completed intervals, effectively increases the well's yield and decreases the drawdown. The equipment consists of a relatively simple jetting tool together with a high-pressure pump and necessary hose and piping. The washing action of the high-velocity jets, working through the screen openings, agitates and rearranges the sand and gravel particles that surround the screen. The wallcake deposit on the borehole in the conventional rotary method of drilling is effectively broken up and dispersed so that the drilling mud can be removed. The procedure consists of operating a horizontal water jet inside the well in such a way that the high-velocity stream of water will be forced through the screen openings. (See Figures 131 and 132.) Fine sand, silt, and clay are washed out of the formation,

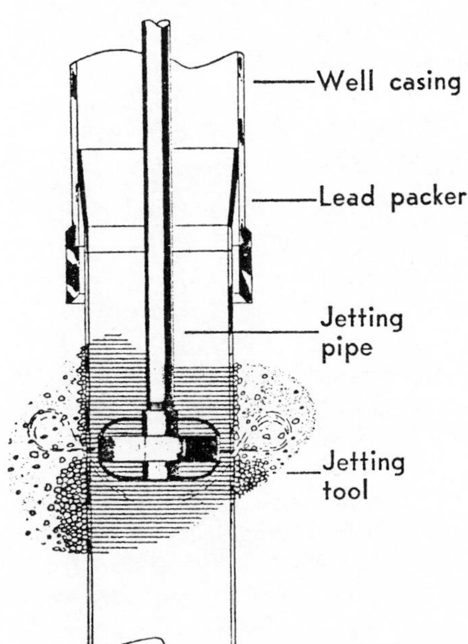

Figure 131[32] Simple high-velocity developing tool. Diagram showing the operating principle of the high-velocity, horizontal jet for vigorous agitation to remove silt and fine sand to develop well.

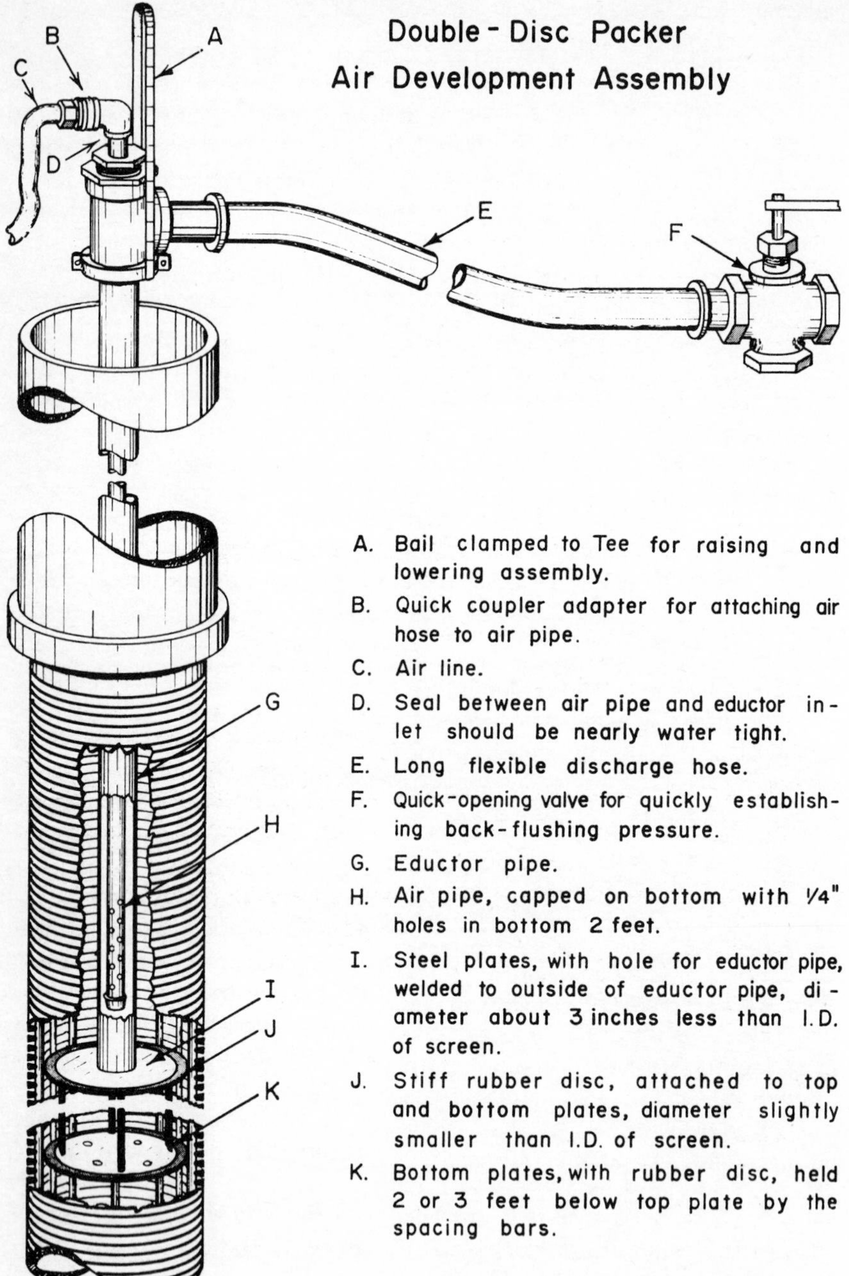

Figure 132[32] Jetting tool for screens requiring more intense development. The tool was designed in 1947 by Robinson, Roberts and Associates of Tacoma, Washington.

and the turbulence created by the jet will recirculate these fine materials into the well through the screen openings.[32,67]

For this method to be most effective, it is desirable to pump or air lift the turbid water from the well at a rate higher than water is being jetted into the formation. This causes a net movement of water through the screen and into the well, speeding the "cleaning" process. Furthermore, by simultaneous pumping, it is possible to continuously monitor the affect of the jetting by inspecting the pumped water.

There are many relatively unknown, but highly effective, well development methods, e.g., the double-packer air-lift developing tool. This tool, for example, incorporates two packers separated two to four feet apart by spacers (See Figure 133). The spacers isolate the water-bearing zone being developed. Effective surging is accomplished by alternately pumping with air and shutting the air off and allowing the water column in the educator pipe to surge back out between the packers.

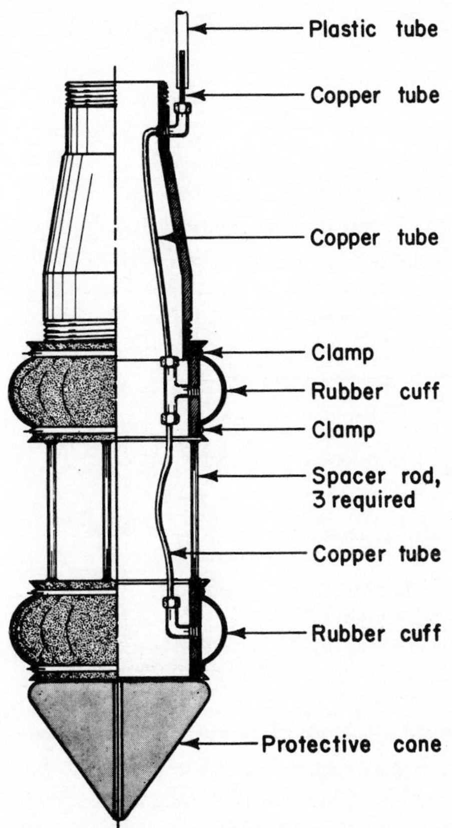

Figure 133[32] Essential features of packer modified for improved efficiency. Developed by L. L. Hindson, Hydrological Branch of the Ministry of Water Development, Salisbury, Rhodesia. An improvement of tool illustrated in Figure 132.

This method has several advantages: (1) only one power source is needed: the air compressor, (2) the fines are being continuously removed, (3) the fines do not cause wear on the pumping unit, (4) each segment of the formation is developed thoroughly, and by monitoring the discharge water, it is possible to establish which intervals of the aquifer require additional developing effort.

Some disadvantages of the above method are: (1) when the static level is very deep and the percent submergence is low, pumping with air is not practical, (2) when the static level is very shallow, i.e., less than ten feet, the degree of surging on the backwashcycle is not as effective as when the static (or pumping) level is deeper.

Blasting (Shooting)

Blasting (shooting) involves detonating massive charges of explosive in the well. Among the explosives used are various forms of dynamite, nitroglycerin, and TNT as well as a powerful, high-velocity explosive fabricated in the form of a string to produce a linear explosion as distinct from a point explosion. Mills,[446] Johnson, U.O.P.,[19,46] Spencer, *et al.*,[587] Dysart, *el al.*,[208] and Eakin, *et al.*[209] discuss shooting or blasting from various points of view for particular cases.

Vibratory explosion is a special form of shooting in which the explosive is divided into relatively small shaped charges and arranged to fire in rapid sequence, thus, producing a vibrating effect on the casing and the formation. At the same time, since the explosion releases a gas, a bubble is created which expands and contracts with an amplitude which gradually decreases. The effect of the vibratory explosive is to prolong the high-amplitude phases of the bubble cycle; the expansion and contraction produce a strong, surging action in the formation precisely during the period when the vibrations are freeing particles and deposits. Koenig[368] studied approximately 300 case histories in Southern California on the use of vibratory explosions and found the method to be particularly effective and applicable to other geographic areas.

Blasting Efficiency

Walton and Csallany[652] studied the effects of blasting deep sandstone wells. In most cases well yields were increased by blasting, because the hole is enlarged and fine materials deposited on the wall face during drilling are removed. According to this report, enlarging the effective diameter of the well by blasting results in an average increase in specific capacity of a well of approximately ten percent. The yield of a newly completed well is, on the average, increased 20 percent by removing fine materials from the well face.

Explosives have been used successfully to develop newly constructed deep sandstone wells or to rehabilitate old wells in many parts of the United States. Many wells are commonly shot with nitroglycerine (liquid or solidified) opposite several areas in the well bore. Shots of approximately 100 to 600 pounds of 80 to 100 percent nitroglycerine are usually exploded opposite the most permeable zones of a formation. Shots are commonly spaced vertically 20 feet apart. The explosions loosen quantities of rock varying from a few cubic feet to several hundred cubic yards that have to be bailed out of the well. Recently, a lighter method of shooting consisting of a string of high explosives (primacord) has been used. Smith[577] reports that the results of the primacord "shooting" exceeded expectations in several cases and that there was no evidence of any large quantities of loosened rock in the wells.

Careful study of the effects of shooting suggests that in many cases the yields of deep sandstone wells are increased because (1) the hole is enlarged and (2) fine materials and incrusting deposits on the face of the well bore and extending a short distance (perhaps less than in inch) into the formations are removed.[652] Caliper surveys of deep sandstone wells furnish valuable information concerning hole enlargement as the result of shooting. Examples of caliper logs showing the effects of shooting are presented in Figure 134. Hole enlargement depends largely upon the size and number of shots and upon the physical characteristics of the sandstone.

During the construction of most deep sandstone wells, very fine drill cuttings invariably infiltrate a short distance into the formation

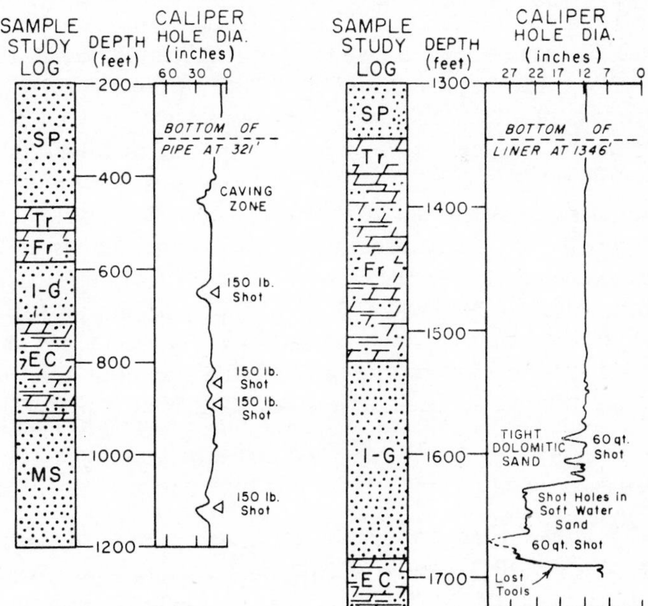

Figure 134[652] Caliper logs of selected wells showing effects of shooting.

as a result of the vertical oscillation and the frequent withdrawals of the bit. The permeability of the aquifer in the immediate vicinity of the well bore is thus reduced. In addition, the face of the well bore is often partially clogged with very fine drill cuttings and mud derived in part from layers of shale. For these reasons, a newly completed well is seldom 100 per cent efficient and the yields of wells are almost always less than what would be predicted based on the hydraulic properties of the aquifer. The yield of a newly completed well, as previously mentioned, can be increased about 20 percent by removing fine materials which have migrated into the formation during construction.

Explosives have also been used to rehabilitate old wells. Under heavy pumping conditions, the specific capacities of wells sometimes decrease as a result of well deterioration. The well face and well wall become partially clogged, commonly with calcium carbonate. When wells are operated at high rates of pumping, the pressure of the water in the aquifer is greatly reduced, carbon dioxide is liberated, and the water is unable to hold in solution its load of mineral salts. Consequently some of these mineral salts are precipitated in the openings of the well bore. This clogging is particularly noticeable in multi-aquifer wells, where waters have moved through the well from one formation to another. The yields of many clogged wells have been restored to the original values through shooting. The specific capacities of rehabilitated wells are sometimes three times the specific capacities before shooting. The driller's problem in rehabilitation is mainly a matter of increasing the permeability of the well wall to its original value and not a matter of hole enlargement.

Unfortunately, well-production tests are not generally made before and after shooting so that the effects of shooting cannot be directly evaluated. Increases in specific capacities of newly constructed wells due to shooting were inferred from specific-capacity data from wells which have not been shot and the performance of wells, after shooting.[652] Well-production test data are of critical importance if results are to have any relative meaning. Cost factors in blasting are treated in the section covering well construction cost.

Field Operations

Mills[446] states that on detonation, one pound of 100 percent blasting gelatin (BG) yields, at standard atmospheric pressure, 6.3 cubic yards of expanding gases. The detonation velocity averages 27,000 feet per second (18,500 mph) compared to about 9,000 feet per second for construction grades of 40 percent dynamite. The gas velocity or impending force at the face of the well wall is about 7,000 mph. Furthermore, 100 percent BG develops at the instant of explosion, a working pressure of approximately 3,600,000 pounds per square inch, which rapidly decays to a steady pressure of 600,000 psi, if totally confined (nearly impossible).

Another report suggests that 60 to 80 percent BG is in general use for water well development.[209] Charges of 100 to 500 pounds are reported as standard, the size varying with the rock strengths and depths at which the charge is to be detonated.

By contrast, the energy for mechanical-hydraulic fracturing is supplied by high-pressure pumps with upper limits around 17,000 psi. The energy transmission ranges from 6,000 to approximately 10,000 psi (the upper failure limit of common steel tubing although packer failure may occur sooner).

Mechanical-hydraulic fracturing is a very successful procedure, when the compressive strength of the rock is not excessive. Compressed-air fracturing is limited to about 1,000 psi. Many types of granite, for example, have compressive strengths far in excess of 30,000 psi, and only high explosives can fracture these rocks.

A review of common rock strengths, among other indices, shows many rocks with strengths in excess of 8,000 psi. The previous section on "Rock Drillability" (Chapter 3) has discussed many of the features relating to rock strengths and potential behavior under high pressure, encountered not only under drilling conditions but also under similar conditions relative to blasting.

High-detonation-velocity 100 percent BG is required for under-reaming or enlarging a well bore face in very consolidated rock. The use of a relatively slow-velocity explosive can dislodge large pieces of rock from the wall which, after the explosion and during the ejection process, can plug the hole excavating a crater at the surface or expelling the surface casing. High-velocity 100 percent BG can finely pulverize the rock which, when mixed with formation water, becomes a slurry that reduces the possibility of bridging or plugging the hole during the venting process.

Blasting Research

A particularly important blasting technique has been described by Miller and Howell at the Fourth Annual Oil Shale Symposium held in Denver during April, 1967. The method involves the forcing of liquid explosives under pressure into fine cracks. Research on this approach could prove to be of considerable value to effectiveness of well stimulation.

Another approach to increasing the effectiveness of blasting is reported by J. S. Fryberger (pers. com.) A 12-inch production well in a sandstone formation with a declining yield is blasted according to commonly accepted procedures. Four or more slim shotholes are drilled in two or three directions from the production well, spaced 10 to 20 feet apart. Each of the shotholes is then blasted. The results simulate a very large, effective well diameter and increase the short-term yield by a factor ranging from two to five. Irrigation wells using this method range from 100 to 400 feet in depth.

A combined electrical blasting method has been reported by J. F. Mann, Jr., a consultant in LaHabra, California (pers. com.). This method was developed by the City of Los Angeles. The method involves the instantaneous release of large amounts of stored electrical energy into the well bore through a discharge across a spark gap (See Figure 135). This reportedly produces a very effective double acting shock wave which dislodges incrustation from well screens and perforations. This electrohydraulic device is not only reportedly more effective in restoring well capacity than conventional methods but is faster and more economical than methods in general use today.

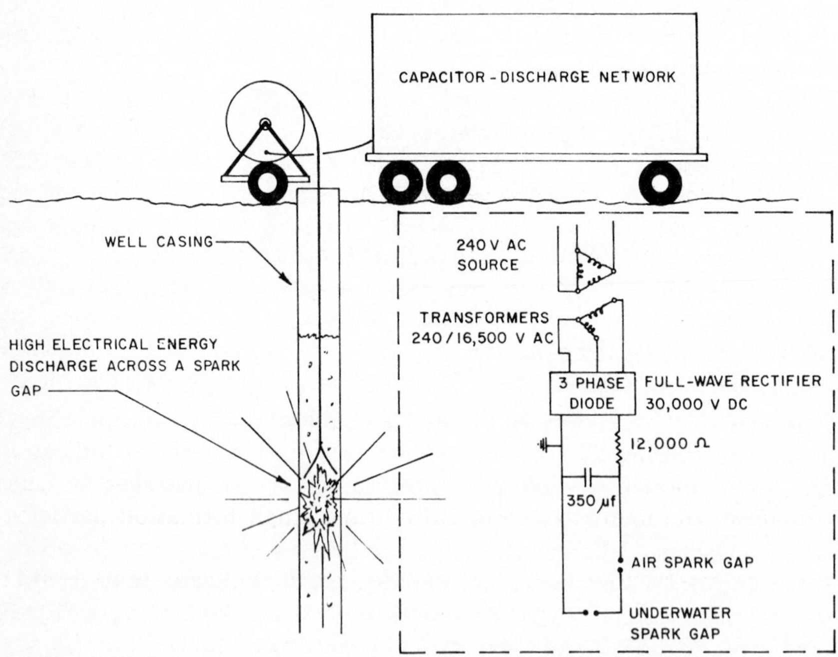

Figure 135[662] Well rehabiliation tool incorporating a high energy electric arc.

The approach was originally developed by C. S. White, and D. E. Williams[662] has reported on the application of the method.

In summary, usage of 100 percent BG is EXTREMELY DANGEROUS in the hands of an amateur and even in the hands of an experienced explosives engineer. A blasting checklist procedure should be followed for each project including alternatives for emergency conditions, e.g., non-fire situation, upper-hole pack failure, etc. Blasting caps SHOULD NEVER be carried in the same vehicle with gelatin. For details on placement and firing procedures, information should be sought from explosive manufacturers.[58] (See Appendix for summary information.)

Acidizing

Acidizing, an accepted method of water well stimulation for many years, usually brings significant production increases. However, the results occasionally do not meet expectations. Bielek[120] reports that the process consists of placing a nontoxic-acid solution in contact with the water-bearing formation. The acid-soluble parts of the formation are dissolved, permitting higher flow rate of water into the well bore.

Two major problems have been encountered as more and more acidizing treatments are performed. In some wells, water production after acidizing did not meet expectations; in other wells, production increased but declined rapidly after several months.

Bielek's study of treated wells resulted in certain general conclusions. Production from all wells in the area was found to be limited by:[120]

(1) Falling water table.
(2) Fouling of screens and/or formations by:
 (a) Carbonate deposits.
 (b) Iron deposits.
 (c) Silts and fine sands.
 (d) Living or dead micro-organisms

Research in acidizing has been directed primarily toward eliminating the fouling of screens and formations. Experience indicates that even though acidizing creates larger water passages in the formation, the increased-water flow brings small formation particles against the screen. The accumulation of these deposits reduces the water flow. Furthermore, laboratory results indicate that hydrochloric acids and others cause certain silicates to swell, expanding the individual particles to as much as five times their original size. Such a reaction could, of course, completely plug the formation and offset the increased permeability caused by the acid.

As the formation is dissolved, minerals containing iron are dissolved and form iron chloride. The iron chloride remains in solution until the acid spends to a pH of 3.5 or higher. At this time, the iron precipitates as iron hydroxide, a jelly-like material having remarkable plugging properties. Iron sequestering agents assist in the control of the formation of secondary iron compounds.

Field Operations

Acid treatment has been used successfully to rehabilitate old shallow carbonate wells and to develop newly constructed wells.[188,650] Many wells have been treated with inhibited 15 percent hydrochloric acid in quantities ranging from 100 to 4000 gallons. The pump and discharge column are usually removed from the well during the treatment period. Acid is introduced through a temporary line extending to a position near the bottom of the well. The solution is allowed to stand under pressure in the well for periods ranging from one-half hour to four days and averaging about one day. The pump is then reinstalled, and the spent acid is removed from the well during pumping periods ranging from one to eight hours.

Well-production tests were made on a few wells before and after acid treatment. The results of acid treatment of two wells are shown graphically in Figure 136.

When wells are operated at high rates of pumping, the pressure of the water in the shallow carbonate aquifer is greatly reduced in the immediate vicinity of the well bore and in fractures extending some distance from the well bore. Because of the decline in pressure, carbon dioxide is liberated and the water is unable to hold in solution its load of mineral salts. Consequently calcium carbonate is precipitated in the openings of the well face and well wall, and the permeability of the well face and well wall is greatly reduced. This clogging is especially noticeable in those wells with pumping levels below the top of the carbonate. The yields of clogged wells can often be restored to their original values by acid treatment.

During the construction of many shallow wells some very fine drill cuttings invariably infiltrate a short distance into the water-yielding openings of the aquifer and reduce the permeability of the well wall. A newly completed well is, as mentioned, less than 100 percent efficient because of the partial clogging of openings. With acid treatment the yield of a new completed well can often be increased by removing the fine materials which have migrated into the openings of the carbonate aquifer.

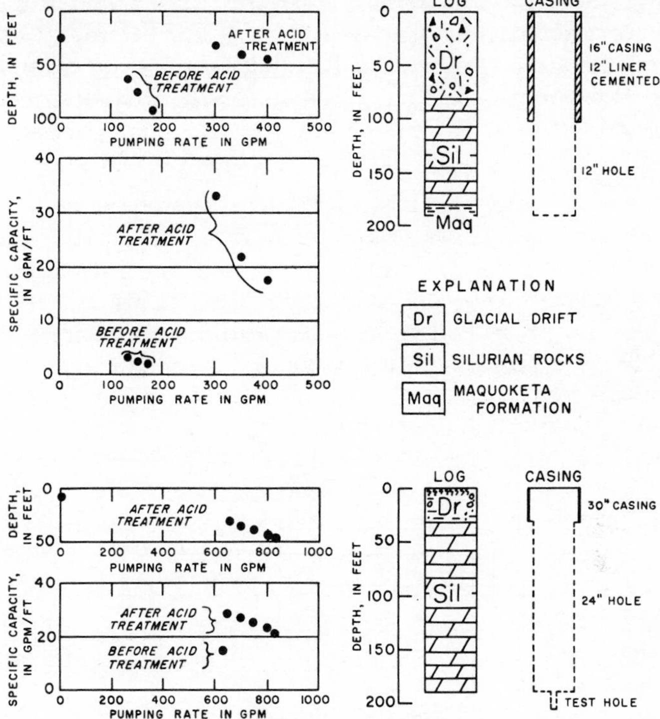

Figure 136[188] Effects of acid treatment on yields of selected wells.

Acid introduced into a well tends to flow into, and widen, fractures leading into the well bore. Also the acid reacts with drill cuttings in openings and the carbonate of the well wall. The effect of the reaction with the massive carbonate of the well wall is to increase the radius of the well bore. Large increases in the well bore result in comparatively small increases in specific capacity. Several thousand gallons of acid cannot dissolve in a day enough massive carbonate, for example, to substantially increase the radius of the well bore. Thus, large increases in the yield of a shallow carbonate well cannot be attributed to well bore enlargement. However, the acid will penetrate considerable distances along the fractures and will widen openings and increase their permeability. In addition, the acid will dissolve drill cuttings in openings and increase the permeability of the well wall.

The effect of treatment will vary according to the permeability of the well wall before treatment. A tight carbonate formation with narrow openings will respond differently than one with openings of appreciable width. Furthermore, a formation which has been partially clogged during drilling will respond differently than one which has not been clogged. Acid should be removed from the well before it is entirely spent; if acid remains in the well after it has been spent, clogging due to iron falling out of solution may occur.

Specific capacities per foot of penetration of newly completed wells before and after acid treatment have been tabulated in order of magnitude, and frequencies computed, during investigations made by Csallany and Walton in Illinois. Values of specific capacity per foot of penetration before and after acid treatment have been plotted against percent of wells on logarithmic probability paper as shown in Figure 137. There was some improvement in the yields of

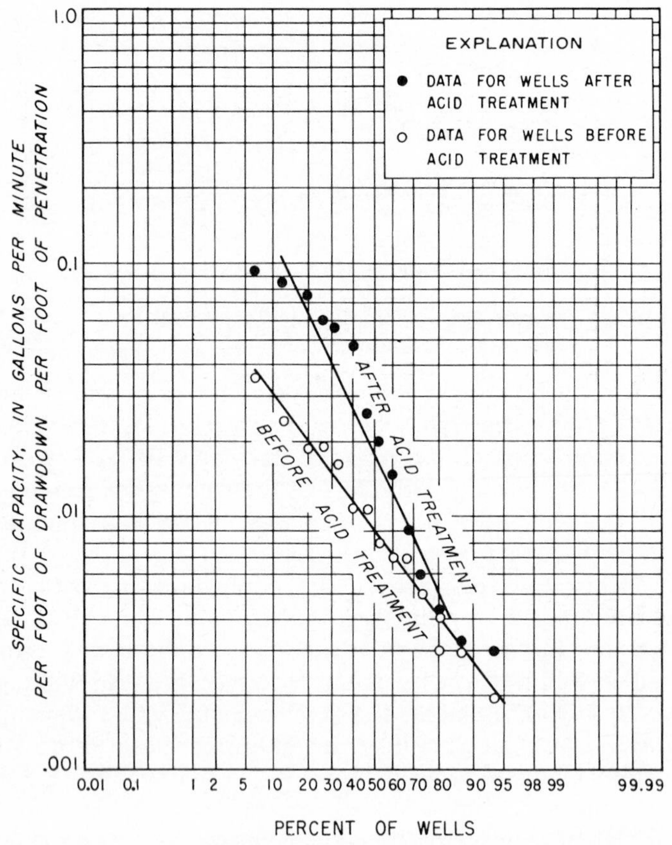

Figure 137[188] Effects of acid treatment on specific capacities of newly constructed wells.

80 percent of the wells indicating that in most cases acid treatment will increase the initial yields of newly completed wells. Improvement increases as the specific capacity increases. In 50 percent of the wells tested, improvement averaged over 100 percent.

Frequency graphs for rehabilitated wells are shown in Figure 138. There was some improvement in all wells indicating that in most cases the specific capacity of a well whose yield has deteriorated because of partial clogging of water-yielding openings by incrustation can be greatly increased with acid treatment. In 50 percent of the wells tested, improvement averaged over 150 percent.

However, acidizing to increase yield has proved to be an uneconomical method of well development or rehabilitation in some

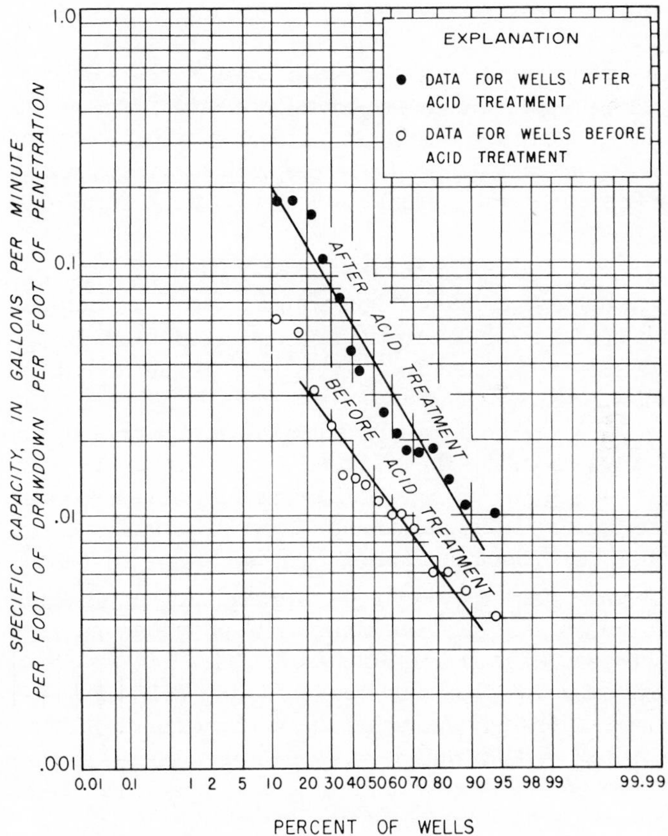

Figure 138[188] Effects of acid treatment on specific capacities of old wells.

cases where the aquifer permeability was low and normal operating head great, in wells affected by methane gas, or where the practical sustained yield of the aquifer had been exceeded. Illinois studies show that controlled pumping tests before, during and after development are necessary to accurately evaluate the effectiveness of rehabilitation methods employed in the area.[644] Geologic definition of aquifers and records of aquifers and records of pumpage and water-level information are invaluable aids.

For many years, chemical additives, such as surface-tension decreasing agents (surfactants) to increase penetrating ability, have been included in the acid solution.[285] Other chemical agents are added to the acid solution in an effort to stimulate and sustain production, such as silicate-controlling agents to minimize formation swelling and stabilizing agents to prevent the precipitation of iron.[574]

Not only improved acid solutions but engineering methods have been established in order to improve well production. Engineering experience indicates that a formation should be redeveloped after being treated with acid. One process consists of backwashing and surging. This agitation of the acid solution redistributes the loose formation solids and traps the large particles at the screen. These large particles, in turn, prevent the small particles from clogging the screen.

The generally accepted treatment essentially consists of two acid stages together with as many "redeveloping stages" as are required. The first stage of this method consists of an application of enough acid to cover the screen or "pay zone." The acid is allowed to remain in place for a sufficient length of time to become spent. During the soaking period, the acid usually is surged several times by pumping. The spent acid solution then is pumped to waste until clear water is discharged which assures that no foreign material will be carried into the formation by subsequent acid stages. The well usually is then surged and backwashed until production no longer increases.

The second stage requires a quantity of acid equivalent to five times the screen volume. Four times as much water as acid is used to displace the acid into the formation. The acid is kept in the formation from two to four hours. During this time, the acid is surged every thirty minutes using the well pump. At the completion of this stage, the well is pumped to waste until the pH of the water is normal. Surging and backwashing are continued until water production fails to increase. After these redeveloping stages, the acidizing treatment is complete.

Certain advantages have always been characteristic of acidizing. These include:

(1) Immediate production increase and flatter decline curve
(2) Rapid and inexpensive method of well workover.
(3) Method adaptable to a wide variety of well conditions.
(4) Effective aid in combating fallen water tables, e.g., the efficiency or specific capacity may need to be increased because of a decline in the water level in order to maintain the desired yield.
(5) Method for sustaining peak production.
(6) Minimizes lifting costs.

As a result, effectiveness of the new acidizing techniques incorporating backwashing and surging has been established. These techniques have proven effective for use in preventing rapid fall of pumping levels and steep decline curves resulting from blocking of screens and flow channels by formation particles. However, the economic factors of acidizing and well rehabilitation should be considered. Acidizing costs can be considered within the approach expressed in Figures 136, 137 and 138.

For maximum success in an acidizing treatment, Bielek[120] suggests that all pertinent data should be studied and evaluated. The well history, water analysis, geophysical logs, and past experience are valuable guides in adapting proven techniques to the individual well. Even if the treatment fails to produce the desired production increase, sufficient information may be obtained to formulate a successful workover program or to prove the well economically "dead." Detailed case histories of well acidizing in particular geographical areas are available in the literature for review.[218,574] A reasonable knowledge of this literature will promote more effective acidizing programs. However, acidizing often treats the effect and not the cause of declining productivity. Poor well development during the original construction does not remove the fine materials around which incrustation of various intermatrix minerals can often occur.

Matrix acidizing is an oil-field-proven technique found effective in correcting producing formation damage incurred while drilling or cementing sandstones.[572] This method is the result of two significant modifications in conventional treatments to overcome "skin effects" around the well bore. The key to this acidizing method is a granular synthetic polymer that swells upon contact with acid and serves as an effective, fluid-loss, control additive. The other principal difference in this "mud" acidizing technique is in the application of low injection pressures.

Mud acid contains a combination of hydrochloric and hydrofluoric acids, specifically intended for low acid solubility formations. This acid relieves formation damage in the following ways:

(1) Dissolves and disintegrates residual mud cake.
(2) Dissolves insoluble clay or cement particles which may have filtered into the formation rock.
(3) Dissolves silicate particles in the matrix which may have been swollen by contact with mud or cement filtrates.

In addition to restoring damaged permeability, "mud acid" has often been used to improve the permeability of sandstone formations. Many rocks showing only five to ten percent solubility in hydrochloric acid reveal from 15 to 50 percent solubility in mud acid because the hydrofluoric-acid component attacks and dissolves the silicate-cementing materials in the sandstone. Special handling is required for mud acid application.

Acetic acid has been widely used as an aid in overcoming many of the problems encountered in well completion, stimulation, and reconditioning. Even though this acid has been used in the past for well stimulation, factors such as economics, handling, and the lack of technical data have limited the use over the past few years.

Acetic acid does not present many of the operational difficulties often associated with other acids.[285] The corrosive action of acetic-acid mixtures can be greatly minimized even at temperatures in excess of 240°F. Contact time of acid and pipe can now be extended for days with proper inhibition. The mixtures currently being used in the oil industry have not caused electrolytic corrosion, hydrogen embrittlement, or stress cracking of metals. Unlike hydrochloric acid, acetic acid can be effectively inhibited for most types of steel.

The type of corrosion caused by acetic acid differs from that caused by hydrochloric acid, the latter tends to "pit corrode" tubular goods with extended exposure. This action accelerates with increasing temperatures. By comparison, acetic acid at equivalent test conditions of time, temperature, and pressure will produce only a minor removal of steel from pipe. Furthermore, when the acid was allowed to spend completely, serious damage did not occur to laboratory and field test samples, and pronounced pit-type corrosion did not form. A special inhibitor is always added to the acid to permit storage within the casing or tubing for many hours so that full-strength acid will be present when needed.

Acetic acid inherently reacts more slowly than hydrochloric acid. When acting on carbonates, acetic-acid-reaction rates are influenced greatly by pressure. While temperature does exert an influence on rate of reaction, the rate is not accelerated as much as with hydrochloric acid.

Like hydrochloric acid, acetic acid will chemically react on clay particles to form hydrogen-activated clay. With an acidic environment, these clay particles will remain in the least-swollen condition.

Because of the distinct difference of reaction rates between hydrochloric acid and acetic acid on limestone under identical conditions, Harris[307] utilized this difference to obtain a significant retardation by blending the acids. In field applications, the acids were blended and excellent results were obtained. The acetic acid was pumped as a first-stage treatment immediately followed by the conventional hydrochloric-acid treatment, a procedure which has been performed remarkably well in many instances. In this method, the theory is that the leading edge of the acid will be considerably retarded, thus lessening the possibility of pumping spent acid. An understanding of this retardation is considered by the oil industry to be one of the primary factors governing the success or failure in acidizing a limestone formation.[308]

A three to five percent solution of acetic acid has been used as a "kill" fluid for petroleum workover operations. This fluid because of its acidic characteristics is highly suitable. If fine-grained sediments are exposed to fresh water, or infrequently to brines, clay particles will absorb water and swell several times their original size, thus blocking the formation permeability. Acids can limit this undesirable condition. An acetic-acid solution will limit clay-swelling more efficiently than other types of fluids. Properly inhibited acetic acid does not damage metals during this type of acidizing.

Inhibited acetic acid has been used frequently as a carbonate-scale remover in pumping wells containing chromium-plated and heat-treated pump parts. Even though this acid has only about 60 percent the dissolving ability on carbonates as hydrochloric acid, inhibited acetic acid has performed effectively without harm to the pumps.

Acetic acid can be made into a high-viscosity, gelled and/or emulsified acid, thus becoming a suitable carrier for producing temporary bridging agents during the treatment of a well, where a stronger hydrochloric acid may have a harmful effect on the formation.

Pressure Acidizing

Koenig summarizes the effectiveness of the pressure-acidizing techniques which are in common use in the petroleum industry.[368]

> Pressure acidizing, the first of the advanced techniques borrowed from the crude-oil industry, may be applicable to water wells. Pressure acidizing is treatment with large volumes of acid solution at high rates of injection into the formation and at correspondingly high pressures producing deep, radial penetration. The acid typically used is a 15 percent solution of muriatic acid to which an inhibitor has been added to prevent the acid from attacking the metal parts of the well structure while allowing it to dissolve the iron, carbonate, and other acid-soluble minerals. Gelling agents are commonly added to modify the flow properties, and retarders may be used to slow down the action of the acid so that at least some of it reaches the extreme depth of penetration of the fluid without losing its dissolving power. Pressure acidizing can be applied only when the formation to be treated is sealed-off from overlying formations. If the formation is not sealed, the acid being pumped under pressure into the well will simply rise through the overlying formations and will not penetrate radially far from the well. The well itself must also be sealed, usually by cementing the annulus between the casing and hole.

Grohskopf[383] collected and studied data on 36 cases of pressure acidizing of limestone wells in 15 Missouri counties. Grohskopf and his associates were pioneers in the field of pressure acidizing of water wells. Walker,[644] Ross, *et al.*,[539] Bielek,[120] Harris,[307] Harris, *et al.*,[308] and Stow and Renner[598] discuss the various features of the normal and pressure methods of acidizing.

Hydraulic Fracturing

Hydraulic fracturing is the most advanced of the stimulation techniques used in the petroleum industry. Hydraulic fracturing, used on thousands of oil wells,[320] consists of injecting liquid into the well under a high pressure so that the formation is actually parted or fractured. The approximate pressure required is one psi or less for every foot of overburden. Special gelling agents and other additives are used to reduce the flow in order to generate the required pressure. When the pumping pressure is released, the formation usually returns to the original position and the fractures close. When sand, or other propping material such as special beads, are forced into the cracks along with the liquid, the propping agent keeps the cracks from closing. The method is called *sand fracing*. The fractures may be only a few millimeters wide but may extend for hundreds of feet and be either horizontal or vertical. The permeability of a propped fracture may be 1,000-2,000 gallons per day per square foot.

Hydraulic fracturing has been used to accomplish four basic functions in the oil industry.[320]

(1) To overcome well/bore damage.
(2) To create deep-penetrating, reservoir fractures which improve the productivity of a well.

(3) To aid in secondary recovery operations.
(4) To assist in the injection of disposal of brine and industrial waste material.

Modern drilling operations generally use fluids that will assist in achieving the most efficient drilling system. However, many of these fluids and contained solids invade the matrix and damage formation permeability. This situation frequently causes a marked reduction in the ability of the formation to produce oil, gas, or ground water.

Fracturing that was practiced early in the history of this stimulation procedure consisted of making very short, radial fractures. In past instances, penetration into the surrounding rock was limited to ten to 20 feet. The oil production increases of ten-to fifty-fold over pretreatment rates result from fractures that penetrate a damaged zone immediately adjacent to the well. The success of multi-fracturing (numerous short, radial fractures at various elevations in the well created by small batches of fracturing fluid) further illustrates the advantage of penetrating a damaged well bore.

Numerous studies indicate that by utilizing deep-penetrating, high-flow-capacity fracture systems, a low-resistance flow path to the well can be established in large portions of the reservoir. The deeply invading fracture promotes production of large volumes of oil and possibly ground water. In addition, the fracture system provides large drainage areas into which formations of very low permeability can slowly transmit a liquid, thus utilizing available reservoir energy to the maximum; consequently, many areas that previously had been considered non-oil-productive were, in fact, subject to commercial exploitation.

The deep-penetrating reservoir fracture with the ability to rejuvenate and extend the economic life of an oil well and to open new areas to commercial production is responsible in large measure for increased oil reserves. The deep-fracture technique may soon find widespread applicability in ground water stimulation since more efficient aquifer development is possible.

In the area of secondary oil recovery, fracturing has played two important roles: (1) increasing the capacity of the water-injection well to accept fluid at a predetermined pressure and (2) establishing high-capacity flow channels into the producing well, thus increasing the efficiency of the gas or waterflooding project. However, the use of hydraulic fracturing within the economic framework of the ground water industry may be limited.

The adaptation of the hydraulic fracturing process for the disposal of industrial waste has gained wide acceptance by industry. The Atomic Energy Commission has made extensive studies of this technique to dispose of radioactive material.

The concept of fracturing, or formation breakdown, recognized many years ago played a very important role in well stimulation, acidizing, water injection, and cementing in the oil industry. While little understood, a means was established for fracturing a formation by displacing a fluid into a matrix at rates and volumes greater than the rock could accept.

Hydraulic fracturing has found application today as a well stimulation procedure for overcoming damaged-matrix permeability surrounding a well bore, and for creating deep-penetrating fractures that provide high-capacity channels from within the producing formation to the well. Water injection and waste disposal are also important applications incorporating hydraulic-fracturing techniques. This procedure has become the most widely used well-stimulation method developed to date by the oil industry.

Well Stimulation Effectiveness

The technical-performance factors defined by Koenig[368] in water well stimulation, drawn from approximately 900 stimulation cases, are very encouraging. Cumulative, frequency-distribution curves were produced for three criteria of technical performance:

(1) The ratio of specific capacities immediately after and before treatment.

(2) The ratio of specific capacities immediately after treatment and at the time of constructing the original well.

(3) The ratio of the increase in specific capacity as a result of treatment to the specific capacity of the original well.

Koenig summarizes as follows:[368]

> For all types of treatments in all types of formations, the median ratio indicated a 97 percent improvement over specific capacity immediately before treatment and a 20 percent improvement over the original production of the well. Failures to achieve any improvement over the treated well were 11 percent; whereas, failures to achieve improvement over the original well were 43 percent. The median increment of specific capacity, achieved by treatment, was 43 percent of the specific capacity of the original well.
>
> When considered by type of formation, the data show that success has been definitely greater in consolidated formations than in unconsolidated formations. Based on improvement in the well before treatment, consolidated formations show a median of 141 percent improvement; whereas, unconsolidated formations show only 45 percent improvement.
>
> When analyzed by treatment type, the data show a quite consistent pattern by all three criteria; the methods, arranged in ascending order of effectiveness, are surging, shooting, vibratory explosion, pressure acidizing, and fracturing.

> Water well stimulation has definitely been technically successful; some of the more advanced techniques that have already been adopted as conventional practice in crude-oil production technology have shown higher performance records than the conventional techniques in the water industry. Judged by the wide geographic coverage of successful stimulation treatments, stimulation could probably be much more widely applied to improve ground water production.

From analyses of the technical and economic data of that survey, water well stimulation has shown highly favorable results, not only in technical performance but also by the economic criterion chosen. In the majority of instances, stimulation has significantly increased the specific capacity (at a unit cost less than that produced by the original well). The median cost of improvement of even the least favorable stimulation method by this economic criterion occurs at a unit cost less than half the unit cost of the original specific capacity. This approach is not only important in regions of existing ground water exploitation but equally applicable (and possibly more important) in the large areas with adequate quantities of ground water in aquifers of such low permeability that ground water production from wells has not been economical. Extensive research and application of water well stimulation techniques can potentially result in a greater and more efficient utilization of ground water resources.

However, a number of limitations on the benefits of water well stimulation exist. The well production is not only limited by the specific capacity and available drawdown but other factors, such as interference with other wells on adjacent properties, intrusion of undesirable waters, and legal restrictions. "Once such limits are reached," Koenig states, "water well stimulation can be of no further influence in reducing investment cost per unit of production although it can still function to reduce operating costs."

If a water supply consists of only one well, then the choice between stimulation and construction of an additional well should be influenced not only by considerations of investment cost but by the general desirability of having alternative water supply facilities for flexibility and emergency use. However, the investment cost for the original well in Koenig's survey does not include the pump or additional equipment that is necessary before the well can become a water supply. To that extent, Koenig's comparisons, specifically those of unit-cost ratios, are conservative. If an additional well is contemplated, the actual economic advantage of stimulating an existing well may be advantageous.

Furthermore, serious consideration should be given to the possibility of well stimulation:

(1) Before a new well is constructed.
(2) Before a proposed well field is abandoned because of economics.
(3) Before a newly constructed well is abandoned because of low production.
(4) Before an unsatisfactory old well is abandoned or is subjected to drastic reconstruction.

Water well stimulation should also be carefully evaluated for well fields, where the desired total production may be efficiently increased from fewer wells, thus achieving lower capital investment and maintenance costs.

Mining Industry and Well Stimulation

The mining industry receives considerable assistance from water well technology, particularly from research involving well stimulation. With the increasing economic impetus provided for the mining companies, significant advances in well stimulation technology should be forthcoming from that industry. Relatively severe economic pressure, low-grade ore, great mining depth, and remote locations in mining certain minerals (e.g., uranium, potash, sulfur, phosphate, etc.) often necessitate *in situ* leaching and solution mining in which water well technology is vitally important. Other mining operations, such as dewatering for open-pit mining and disposal of waste fluids, are also discovering the importance of water well technology.[79,479]

WELL ABANDONMENT

The proper plugging of abandoned wells is a fundamental practice in the preservation of high ground water quality.[633] One of the more important aspects of well construction is proper abandonment or permanent sealing because serious ground water pollution can occur from either contaminated surface water or inter-aquifer transfer of ground water from a saline or otherwise contaminated aquifer into a potentially productive aquifer. Thousands of abandoned wells, either water wells, shallow seismic holes, or shallow mineral exploration holes, exist throughout the United States; too many of which are contaminating potential ground water sources as a result of improper abandonment techniques. Short wooden logs, or "hole-caps," covered by a few inches of soil hide abandoned mineral exploration holes, especially in the western United States. These holes are intentionally left open at depth, in the event that additional geophysical logging is necessary at a later date. Natural caving eventually closes the hole preventing future logging but creates highly permeable avenues for vertical ground water flow and possible contamination. In addition, many of these holes are in areas where artesian water may "bleed" without proper containment.

The basic concept in the proper sealing of an abandoned well is the restoration, as far as possible, of the previously existing geologic conditions. Sealing or abandonment is usually achieved by grouting with puddled clay, cement, or concrete. When grouting below the water table, the cementing material should be placed from the bottom up by methods that would avoid segregation or excessive dilution of the material, as previously described. (Chapter 12).

Gibson and Singer[246] suggest that, in some cases, well casings in water-bearing zones should be pulled to assure an effective seal. In addition, if the upper 15 to 20 feet of the surface casing was not carefully cemented during the original construction, this section of the casing should be removed before final grouting for abandonment. Downhole geophysical probes will be helpful to determine the effectiveness of previous cementing operations, as mentioned in the section on borehole geophysics (Chapter 9).

Shallow underground mining operations (200 to 400 feet) have discovered many exploration holes while mining, demonstrating the impermeable nature and excellent sealing character of bentonitic mud.

The presence of a high rate of artesian flow of water from a borehole causes problems during the sealing operation. To facilitate sealing of such a borehole, a cement plug is set near the top of the well. A fluid-pressurized, borehole plug has been developed for use in the sealing of wells to stop artesian flow.[111]

To seal an abandoned well properly, the type of ground water occurrence must be considered. If under water table conditions, the objective is to prevent the percolation of the surface water through the well structure or along the outside of the casing to the water table. If under artesian conditions, the sealing operations must confine the water to the aquifer in which it occurs. This prevents loss of artesian pressure that will result from uncontrolled flow from the aquifer. Removal of liner pipe from wells maybe necessary to assure placement of an effective cement seal. However, the local geologic conditions must be assessed in detail via geophysical logs, i.e. caliper, SP, resistivity, etc., or a detailed driller's log to determine the most effective method of well plugging, e.g. extent of cementing, use of puddled clay, etc.

14
Well Cost Analysis

Although any cost analysis has the problem of discriminating proper and realistic data input, general guidelines can be established. Studies completed by the Illinois State Water Survey are an excellent example of one approach for the analysis of well construction costs in sedimentary rocks, both domestic and industrial-municipal. The guidelines to be put forth are not all inclusive, since well costs depend to a large extent on: (1) geographical location, (2) labor supply, and (3) geological environment and other factors. A particular well cost will fluctuate since general cost factors can combine to either simplify or complicate well construction.

The use of the following graphs need not necessarily be limited to Illinois, although they are based on wells drilled in sedimentary rocks of that state. In those parts of the United States where geologic conditions are similar and where labor costs are comparable, the results are applicable to other parts of the country. However, well costs in predominantly igneous or metamorphic rocks, may be considerably greater since the cost per foot of drilling could be significantly higher.

COST ANALYSIS (SEDIMENTARY ROCKS)

No system for estimating costs ever provides 100 percent accuracy. However, historical data and analysis of the construction cost of many wells provide an excellent means of estimating future costs in any given geographical area.[77]

Domestic Wells

Cost information for 345 wells of various types, constructed in sedimentary rocks throughout Illinois during 1967, 1968, and 1969, was collected for a study conducted by Gibb.[244] All data were adjusted to a common 1969 economic level by using a domestic and farm well index, developed as a part of the study. This index, shown in Figure 139, indicates the increase in the costs of farm and

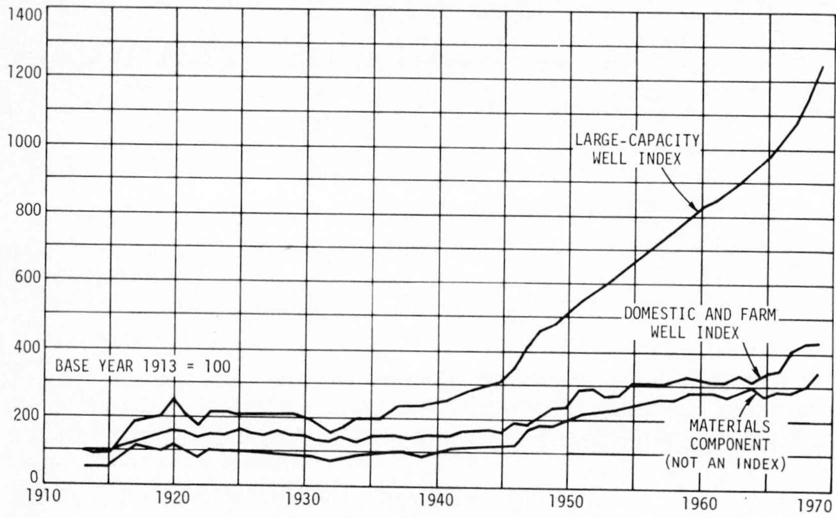

Figure 139[244] Well construction cost indices.

domestic wells in Illinois from 1913 through 1969. Figure 139 also illustrates the increase in the costs of large-capacity wells and the cost of materials, such as steel and concrete normally used in construction (Engineering News Record, 1970).

Figure 139 indicates that the recent costs of domestic-type wells are approximately four times the 1913 base costs. On the other hand, the costs of large-capacity wells are more than 12 times the 1913 prices. Figure 139 also suggests that the costs of domestic wells are primarily determined by the costs of the construction materials although the profit margin has been low in the past.

After the reported cost information was adjusted to the 1969 economic level, well-cost data were divided into three categories by Gibb,[244] according to the aquifer tapped and the type of well construction.

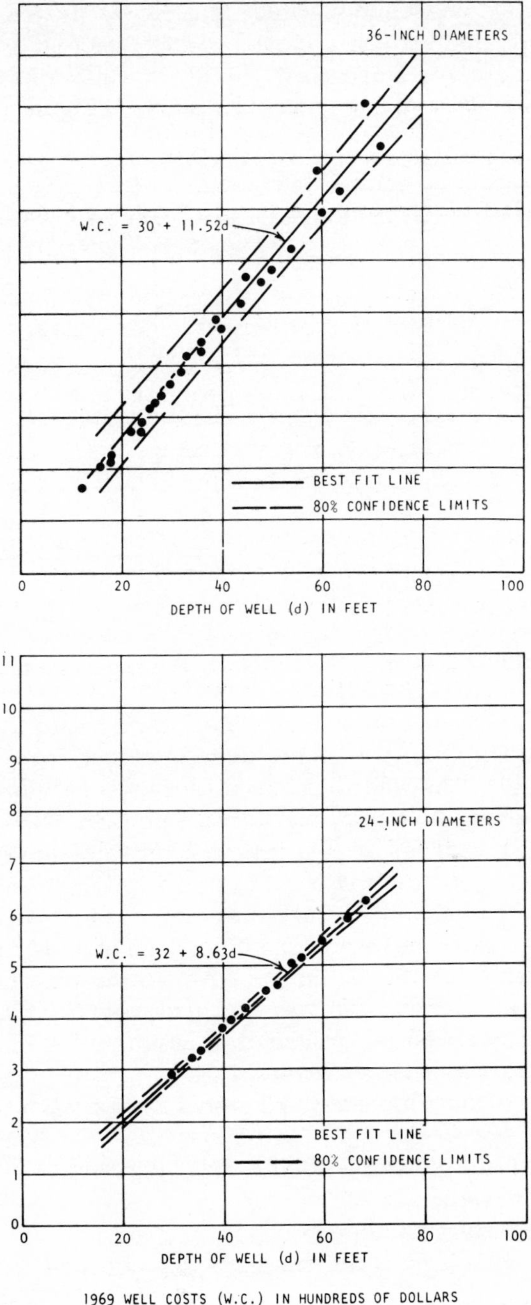

Figure 140[244] **Cost of 24- and 36-inch augered wells.**

(1) Large-diameter, concrete-cased, augered wells completed in unconsolidated materials above bedrock (see Figure 140).
(2) Commercially screened, drilled wells completed in water-bearing sand and gravel deposits (see Figure 141).

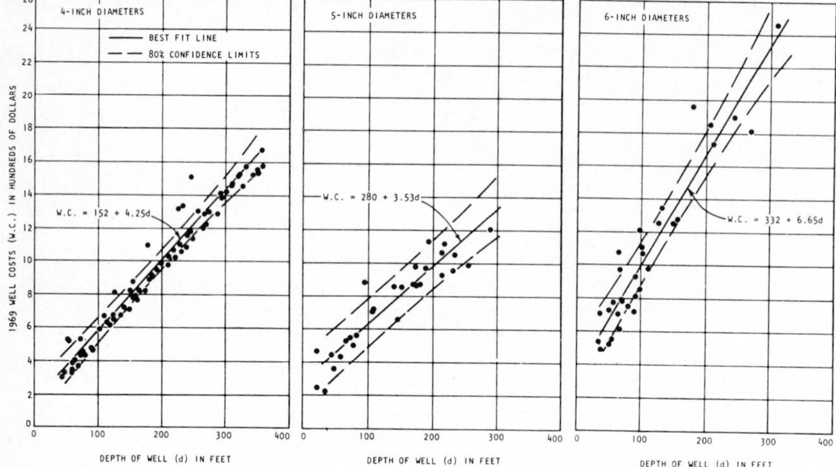

Figure 141[244] Cost of 4-, 5-, and 6-inch sand and gravel wells.

(3) Drilled wells completed in water-bearing sandstone, limestone, and dolomite of consolidated bedrock formations (see Figure 142). The wells in each category were grouped according to the inside casing diameters or nominal-well diameters.

Gibb examined each set of cost information to eliminate data for wells with unusual construction features or obvious inconsistencies in pricing. Data for wells where a subcontractor was involved generally indicated higher values and were, therefore, eliminated from further study. Information on wells equipped with slotted pipe rather than commercially made screens also was excluded from the final analysis because this type of well is considered undesirable for modern well construction.

The final sets of data include the material and/or labor cost of the following:

(1) Setting up and removing the drilling equipment.
(2) Drilling the well.
(3) Installing casings and liners.
(4) Grouting and sealing the annular spaces between casings and boreholes.
(5) Installing well screens and fittings.
(6) Developing the well.

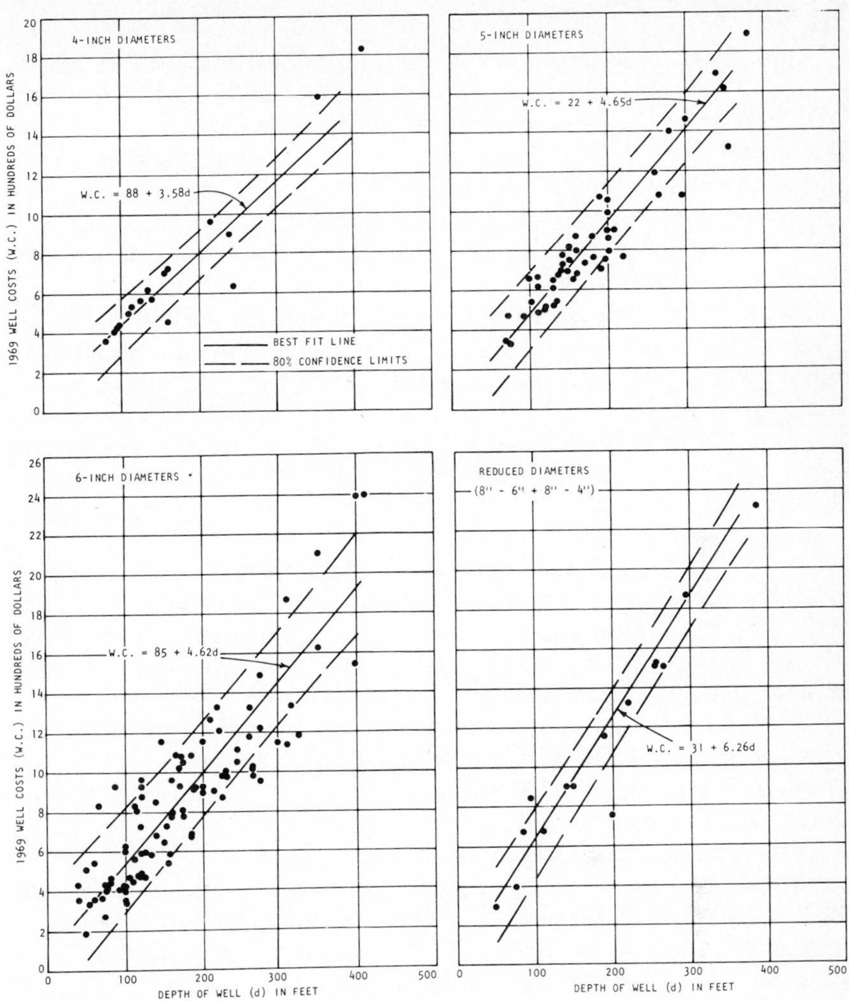

Figure 142[244] Cost of 4-, 5-, and 6-inch, and reduced diameter bedrock wells.

These well costs should, therefore, represent the actual cost to the consumer.

During the cost analysis for the three types of wells in sedimentary rock considered in the study, no local variation in costs was noted by Gibb.[244] Even though different well contractors reported varying drilling prices (cost per foot), the fixed costs applied by each well contractor tended to minimize the difference in the final consumer well cost.

Pump Systems

Completed cost information for approximately 200 domestic pumping systems installed during 1967, 1968, and 1969 was collected from 19 well contractors cooperating in the study conducted by Gibb.[244] The data include the costs of the installed pump, the pitless-adapter unit, the pressure tank, and all associated piping and wiring. All pumps are of the submersible type (indicating a recent trend away from the use of well houses and pits). Pump capacities range from one to over 20 gallons per minute.

Pumping-system cost data also are adjusted to a 1969 economic level using the farm- and domestic-well index. Preliminary analysis of the adjusted cost data reveals unexpectedly large variations in the costs of these systems. No significant relationships between the pump costs or pumping-system costs and the pump capacity and depth of setting could be developed by Gibb.[244] Instead, costs appear to vary in different areas of the state which may or may not apply to other parts of the United States.[1,2]

To minimize the large variation in pumping-system costs and to permit a more detailed presentation, data for the systems equipped with 10-gpm pumps were selected for further analysis because the average pump capacity of all systems was about 10 gpm, and more than 50 percent of all systems were equipped with 10-gpm pumps.

The final set of pumping-system cost data represents the consumer cost (labor and materials) of the following:

(1) A 10-gpm submersible pump.
(2) A pitless-adapter unit.
(3) An adequately sized pressure tank.
(4) All associated piping and wiring to deliver water to the house.

The range and distribution of costs of the selected systems are illustrated in Figure 143 and 144.

If sufficient geohydrologic data are available to determine the probable type and size of well required to furnish an adequate water supply at a given location, the data presented in the foregoing discussion can be used to estimate the initial costs of that well and a typical pumping system. Once the initial costs have been determined, these values can be reduced to a more meaningful equivalent monthly cost.

The equivalent monthly cost of a given well and pumping system is obtained by amortizing their initial costs over the expected life span of each unit. Previous studies by the Illinois State Water Survey have found the median service lives of domestic wells and pumps to be approximately 20 and 10 years, respectively. Figure 144 shows the equivalent monthly costs obtained using these life spans at an

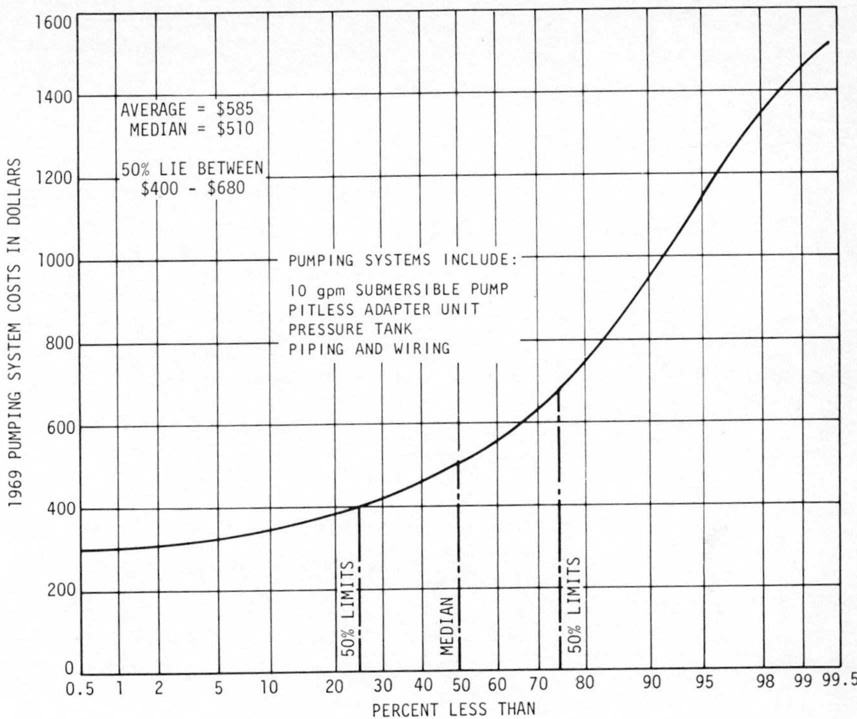

Figure 143[244] **Cost of pumping systems.**

interest rate of 8 percent, plus a combined annual maintenance expense of $10, for several combinations of initial well and pumping system costs.

The equivalent costs shown represent the monthly costs required to retire a loan on a newly constructed well and pumping system. The loan period is the life of the well (20 years), and it is assumed that two pumping systems will be required over this period (10 years for each system). Therefore, payment on the well and pumping systems is required for the entire 20-year period.

These equivalent monthly values represent the cost of raw water from a selected well and pumping system regardless of the quantity of water pumped. The electrical power costs of pumping water, however, are not included.

Water Treatment Systems

Desirable water for general domestic use should contain no objectionable or dangerous concentrations of minerals or gases and

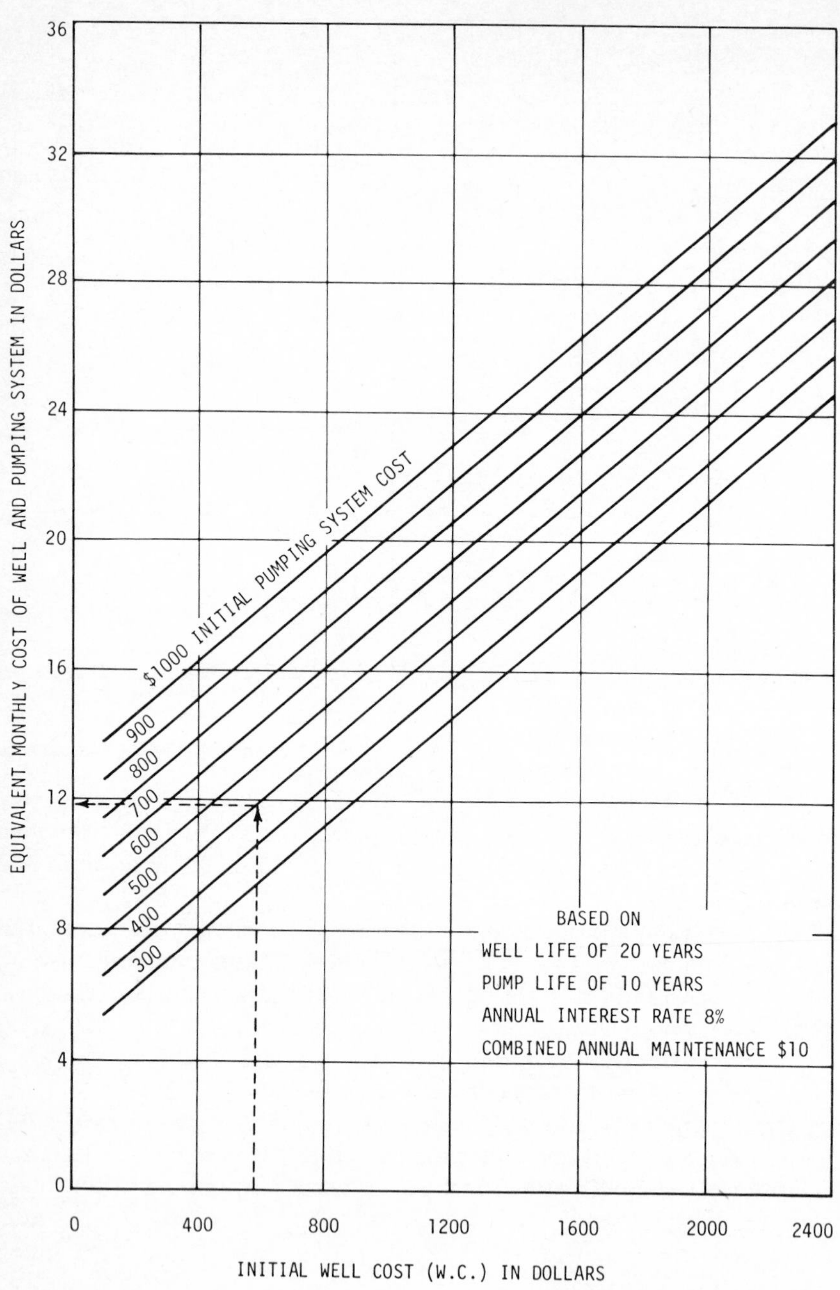

Figure 144[244] **Monthly cost of wells and pumping systems.**

should have a safe sanitary quality. However, objectionable concentrations of hardness-forming minerals and iron are common in many aquifers. In the past, tolerance rather than treatment of the high concentrations of minerals was the common practice. Today, an increasing number of private water-supply systems are being equipped with home water-treatment units to soften the water, remove the iron, and, where necessary, chlorinate. The general quality of water from many private water-supply systems could be improved with the installation of properly selected water-treatment equipment. Figures 145 and 146 show typical costs of softening and iron removal in the

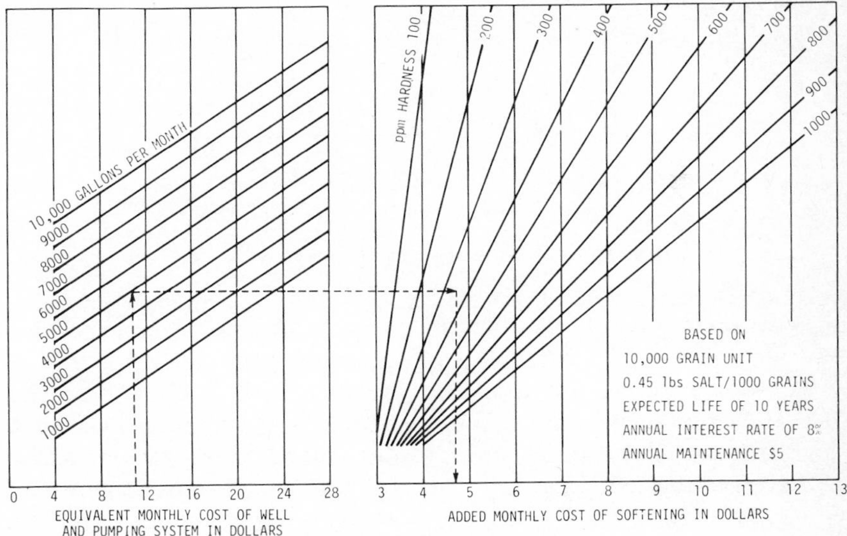

Figure 145[244] Monthly cost of softening.

Midwest. Figure 147 shows summary costs of a typical, domestic, ground water supply in Illinois. Similar analyses in other sections of the United States would be a vital contribution to the ground water industry.

Municipal and Industrial Wells

An excellent report by Gibb and Sanderson[245] of the Illinois State Water Survey, presents a summary of data on the initial cost of 143 municipal and industrial wells constructed throughout Illinois during 1964, 1965, and 1966. These data were obtained primarily from questionnaires, voluntarily returned by 78 consulting engineers, 48

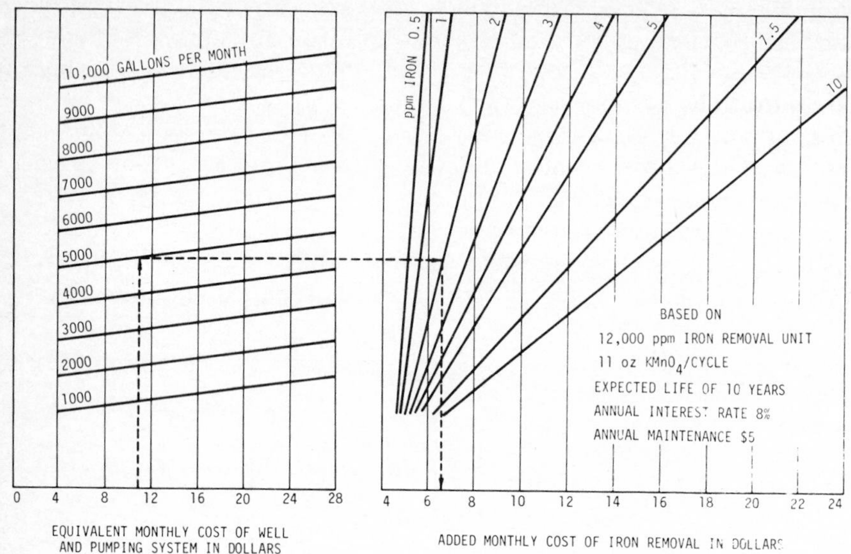

Figure 146[244] **Monthly cost of iron removal.**

water well contractors, 38 industries, and 30 municipalities. Excellent correlations were obtained from the data collected.

All cost data were adjusted to a common 1966 level by increasing the 1964 and 1965 cost figures ten and five percent, respectively. Furthermore, Gibb (pers. comm.) reports that "costs have increased approximately 60 percent since 1966." All cost figures of municipal-industrial wells should be so adjusted for an approximation of 1972 cost levels.

Well-cost data were divided into three categories covering four basic well types, according to the aquifers tapped: (1) sand and gravel (see Figures 148 and 149), (2) shallow bedrock (see Figure 150) and (3) deep sandstone (see Figure 151). Fully cased and artificially gravel-packed wells completed in the glacial materials above bedrock were considered separately in the "sand and gravel" category.

After separating all well data into respective groupings, each set of cost information was examined to establish a relative standard for acceptable data. Cost figures for wells having unusual construction features or obvious inconsistencies in pricing were immediately omitted from further analysis. Information for wells where the exact cost components of the reported sole contractor bids could not be separated was also discarded from the final analysis. Data from

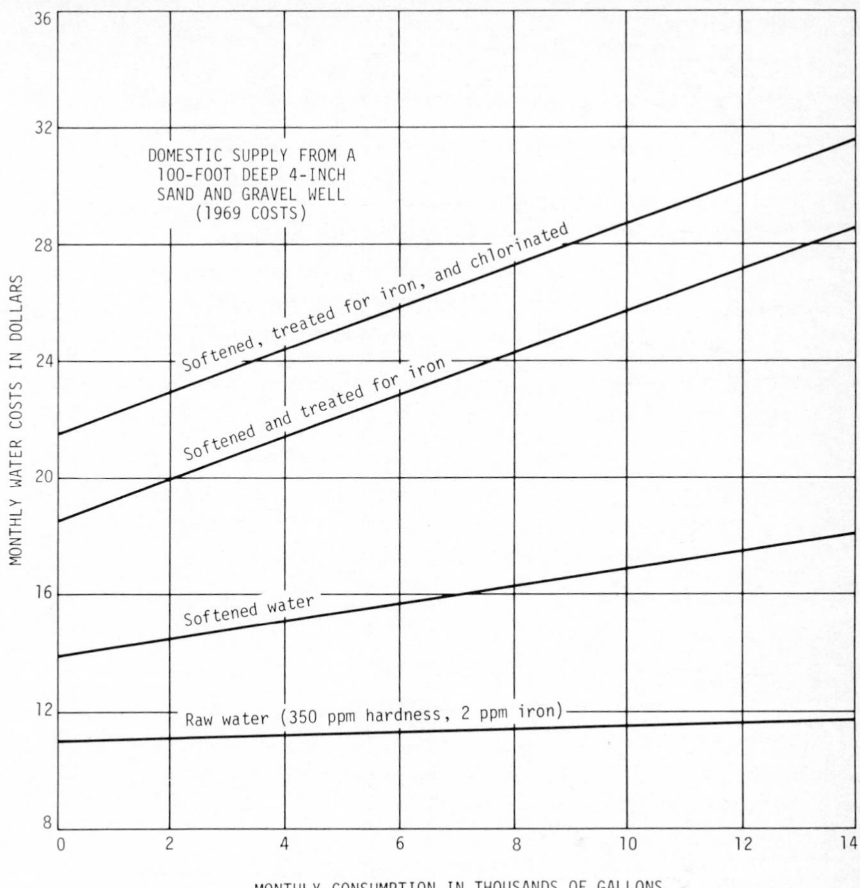

Figure 147[244] Summary costs of a typical domestic water supply in Illinois.

joint-bid prices for two or more wells were deleted, because in all cases of this type the resulting individual-well costs were below normal.

Additional information omitted from the analysis of sand and gravel wells included information pertaining to wells equipped with slotted pipe rather than commercially made screens (as in the previously discussed domestic-well analysis) and information for gravel-packed wells with a gravel-pack annulus greater than ten inches in thickness. Data for bedrock wells, or wells completed in consolidated sediments with unusual casing configurations, were also excluded from the final analysis.

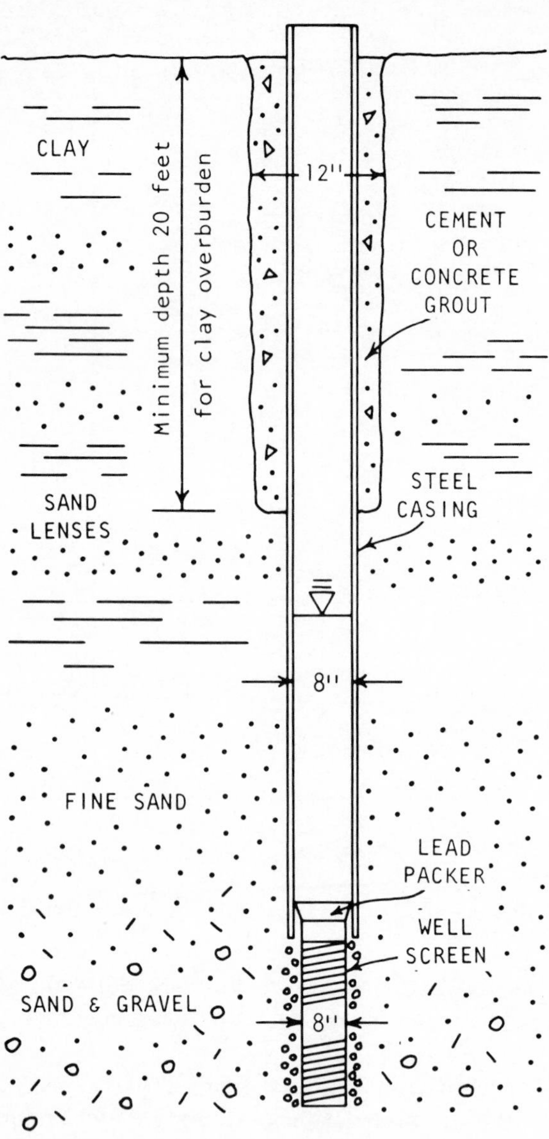

Figure 148[245] Well completed in unconsolidated sediments, e.g., sand and gravel. (See Figure 152 for cost analysis.)

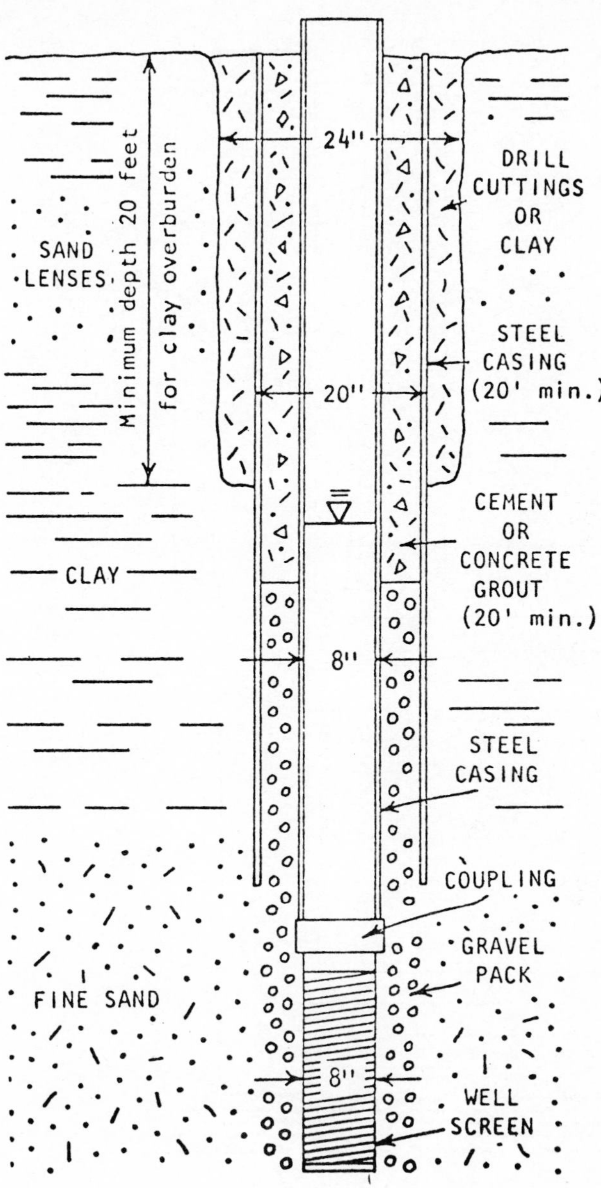

Figure 149[245] Gravel-packed well completed in sand and gravel — unconsolidated to partly consolidated sediments. (See Figure 153 for cost analysis.)

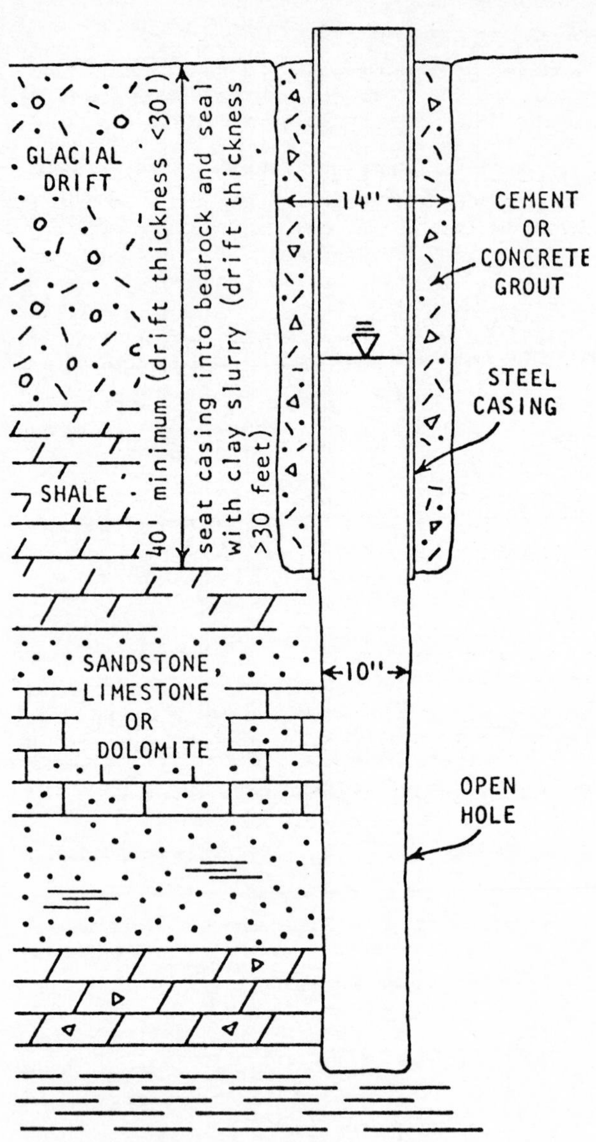

Figure 150[245] Shallow well completed in consolidated sediments of sandstone, limestone or dolomite (bedrock well). (See Figure 154 for cost analysis).

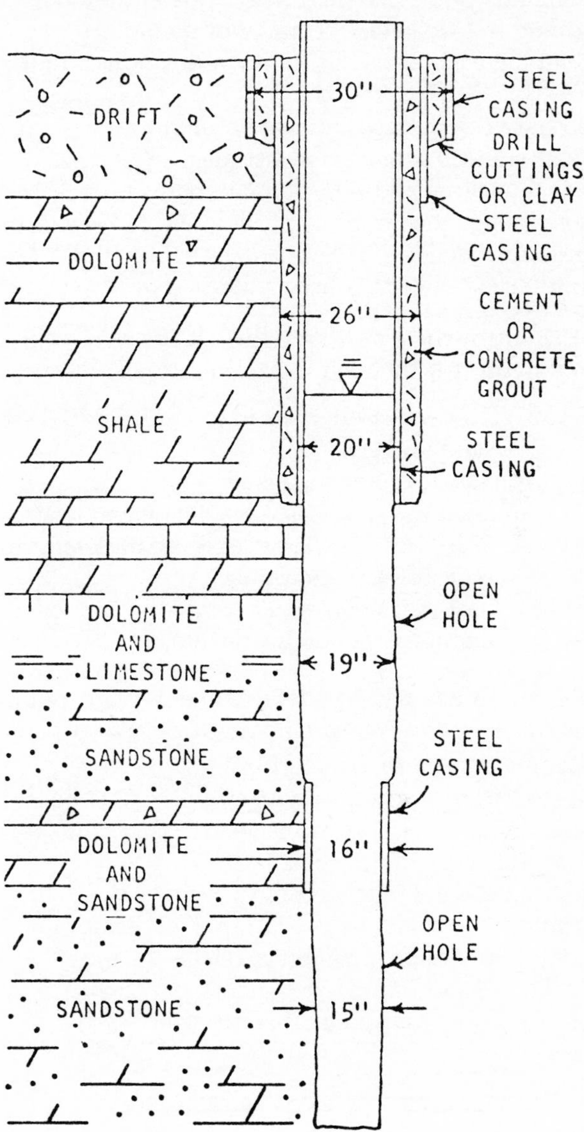

Figure 151[245] Deep well completed in consolidated sediments, e.g., sandstone, etc. (See Figure 155 for cost analysis.)

The material and labor costs of the following were represented:

(1) Setting up and removing the drilling equipment.
(2) Drilling the well (test drilling not included).
(3) Installing casings and liners including construction casings.
(4) Grouting and sealing the annular spaces between the casings and between casings and the borehole.
(5) Installing well screens and fittings.
(6) Gravel-packing and placing materials.
(7) Developing the well excluding blasting (shooting) of deep sandstone wells.
(8) Conducting one eight-hour pumping test.

Figures 152 through 155 show the analysis of well costs plotted against well depth for the four, basic well types:

Figure 152: Cost of cased wells in unconsolidated sediments, e.g., sand and gravel.
Figure 153: Cost of gravel-packed well completed in sand and gravel.
Figure 154: Cost of shallow well completed in sandstone, limestone, or dolomite.
Figure 155: Cost of deep well completed in consolidated sediments, e.g., sandstone, etc.

Well diameters are usually determined by the pump needed and the minimum clearance requirements of casing to facilitate removal of the pumping system for inspection and repairs. Some variations in the pump sizes for a given rated capacity were noted when comparing pumping installations dealing with small and large pumping lifts. These variations coupled with the rapid changes in pump design prohibit a relationship between minimum well diameters and the proposed pumping capacities. Table 41 represents typical well diameters used in Illinois.

TABLE 41[245]

Well Diameters Used in Illinois

Pumping rate (gpm)	Diameter of Well (inches)
125	6
300	8
600	10
1200	12
2000	14
3000	16

Pump Systems

Industrial-municipal pump cost data were divided into two categories according to the type of pumping installation: (1) vertical turbine pumps and (2) submersible turbine pumps. Each piece of

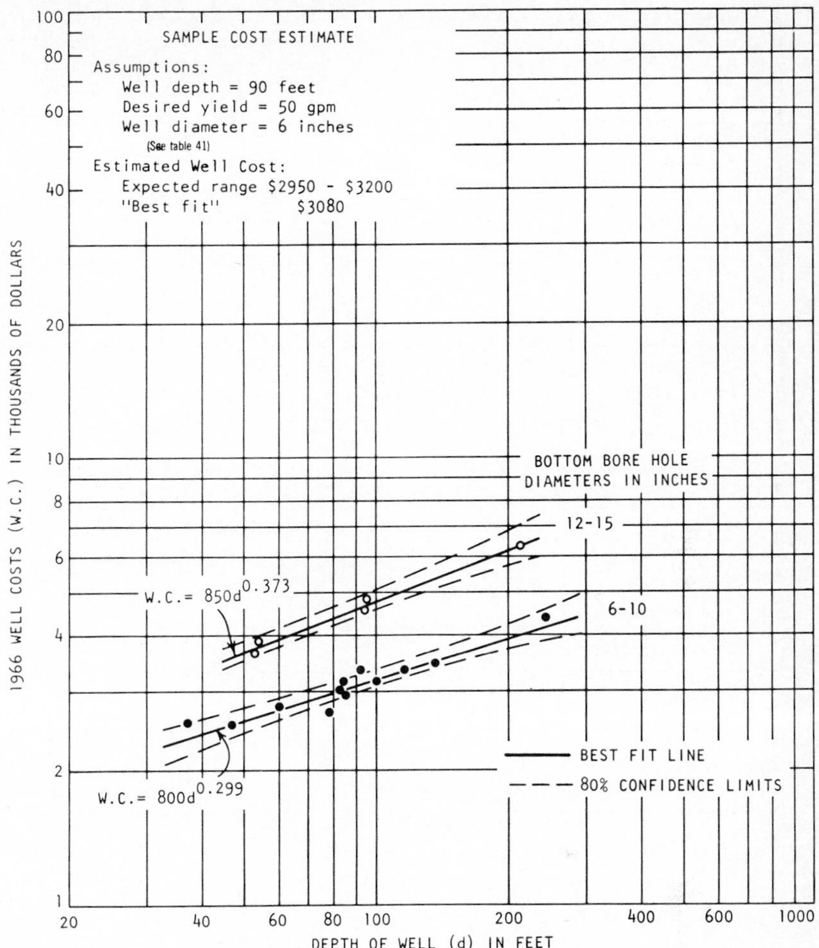

Figure 152[245] Cost of cased wells finished in sand and gravel.

pump-cost data was examined to achieve a common base for comparative purposes. Detailed investigations into individual cases revealed that unusually low pump costs could normally be attributed to the installation of used pumps or materials. Excessively high pump costs normally resulted from the inclusion of sophisticated control systems and, in some cases, the addition of the pump house in the reported installed pump costs. All of Gibb and Sanderson's data that deviated greatly from the norm were, therefore, investigated more thoroughly and judged on merit alone.

The sets of data represent the direct expenses involved in furnishing and installing a basic pumping system of a given size and

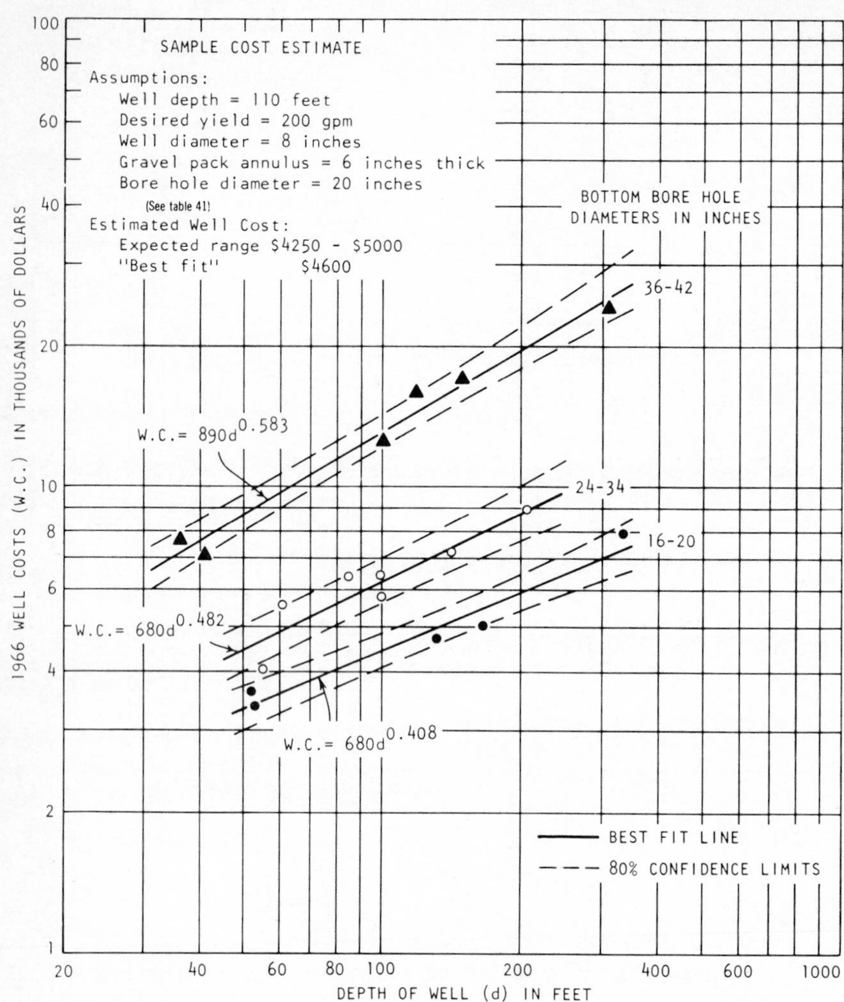

Figure 153[245] Cost of gravel-packed wells finished in sand and gravel.

type. The costs of well houses and control systems are not included in the installed-pump costs.

Adequate, detailed cost information was reported for 108 pumping systems installed in the wells surveyed. Of this total, about 45 percent were vertical-turbine pumps, and the remainder were of the submersible-turbine type. Because of the marked difference in the cost/pump capacity relationships for various heads for the two types of pumps, the data have been treated separately. See Figure 156 (Cost Analysis of Vertical Turbine Pumps) and Figure 157 (Cost Analysis of Submersible Turbine Pumps).

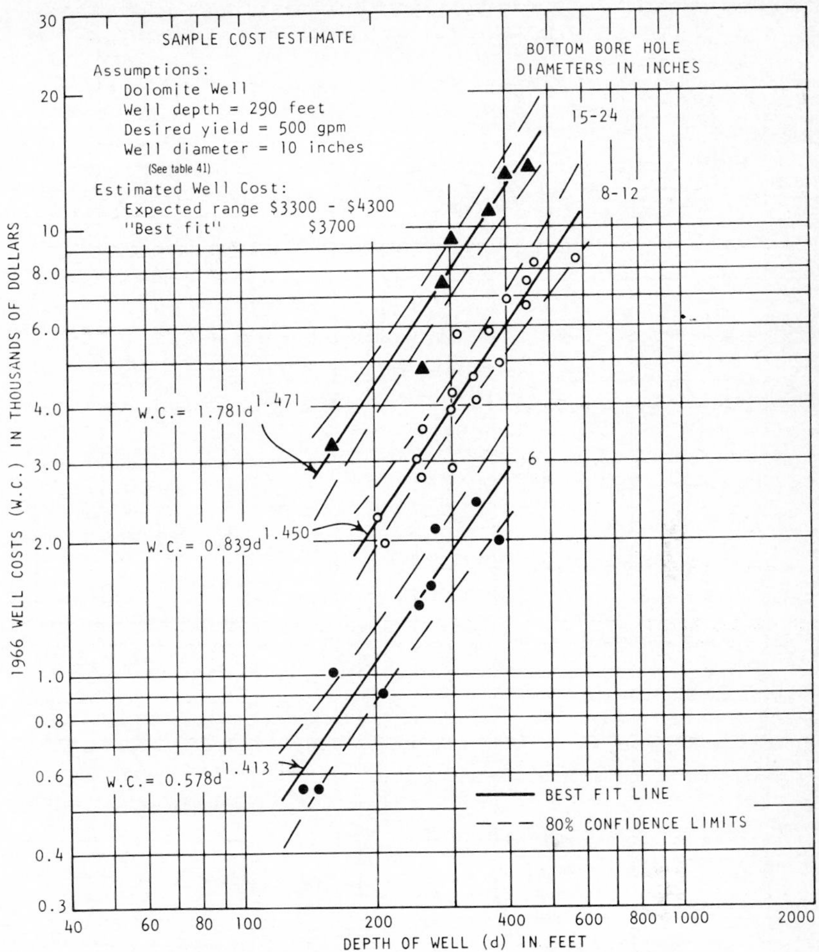

Figure 154[245] Cost of shallow sandstone, limestone, or dolomite bedrock wells.

Blasting (Shooting)

In examining the cost factors of deep sandstone-type wells, data obtained by Gibb and Sanderson[245] indicated a wide range in the blasting and cleaning expenses of some deep sandstone wells; therefore, the cost was excluded in the final analysis. Completed cost information for 15 wells separated shooting and cleaning costs from the total well costs. Figure 158 illustrates the relative costs of the wells which were blasted in comparison with those developed by other methods. The relationship of cost to yield for rehabilitated wells has not been established.

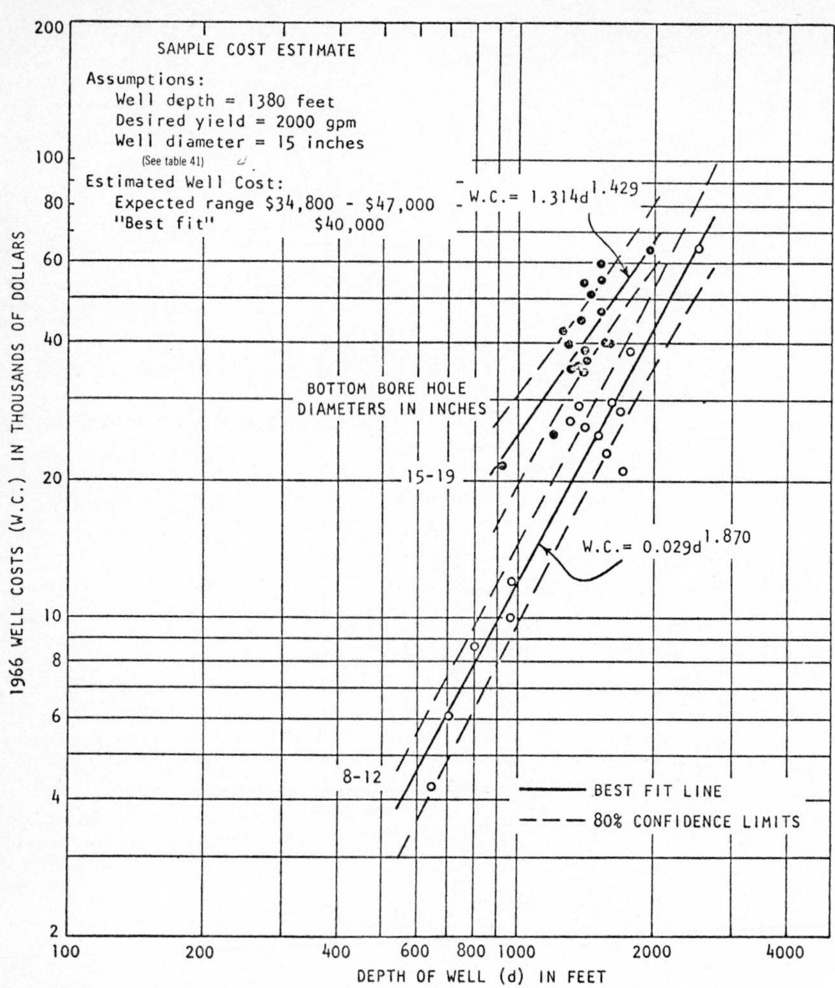

Figure 155[245] Cost of deep sandstone wells.

COST ANALYSIS (IGNEOUS AND METAMORPHIC ROCKS)

Although the previous approach to well cost analysis is only occasionally applicable to drilling in igneous and metamorphic rocks, Davis and Turk[197] have developed an analytical approach, basing well yield on cost of water in igneous and metamorphic rocks. Other features dealing more with well design have been explored in a previous chapter: (Chapter 11, "Well Design and Yield,") on igneous and metamorphic rocks, and some sedimentary rocks which have similar characteristics such as fractured dolomite, limestone, etc.

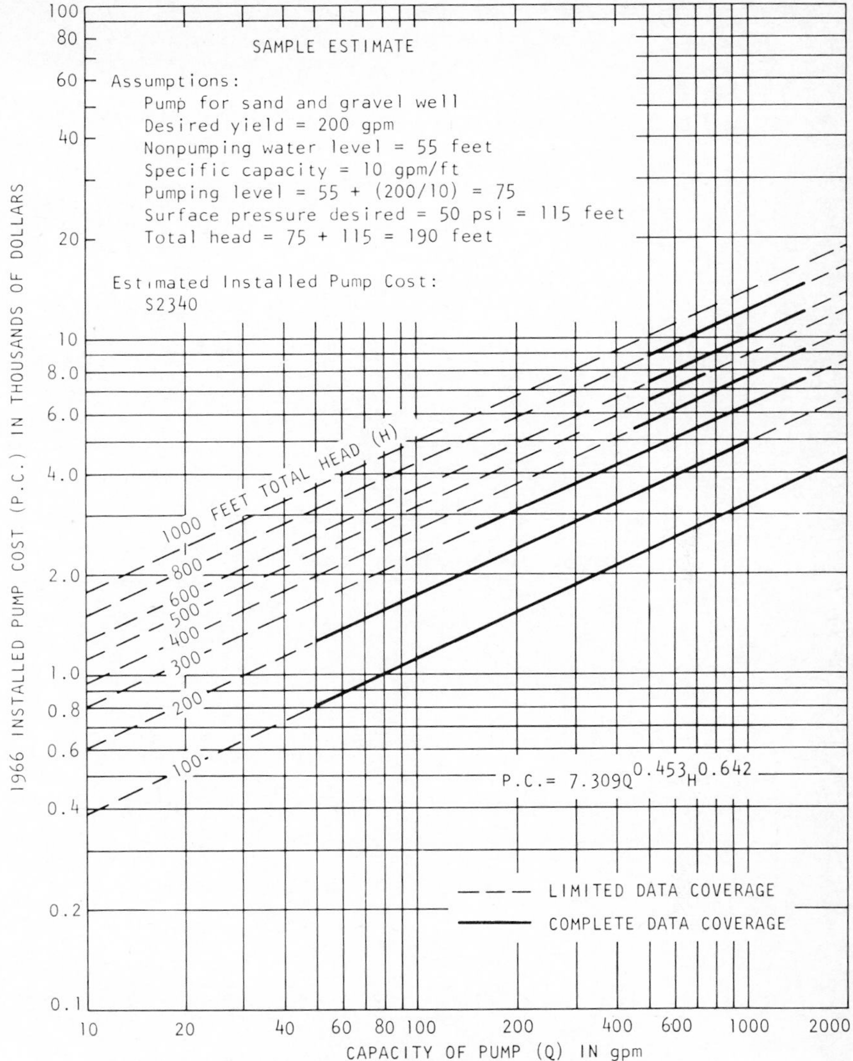

Figure 156[245] **Cost of vertical turbine pumps.**

Exceptions to this particular approach include situations where geologic knowledge or geophysical data may predict the existence of productive zones at particular depths.

Since the yield per foot of well diminishes with depth, explained in the previously mentioned section, the unit cost of water increases with well depth. This is shown in Figure 159. Cost of water per month at various pumping rates as shown for comparison is based on

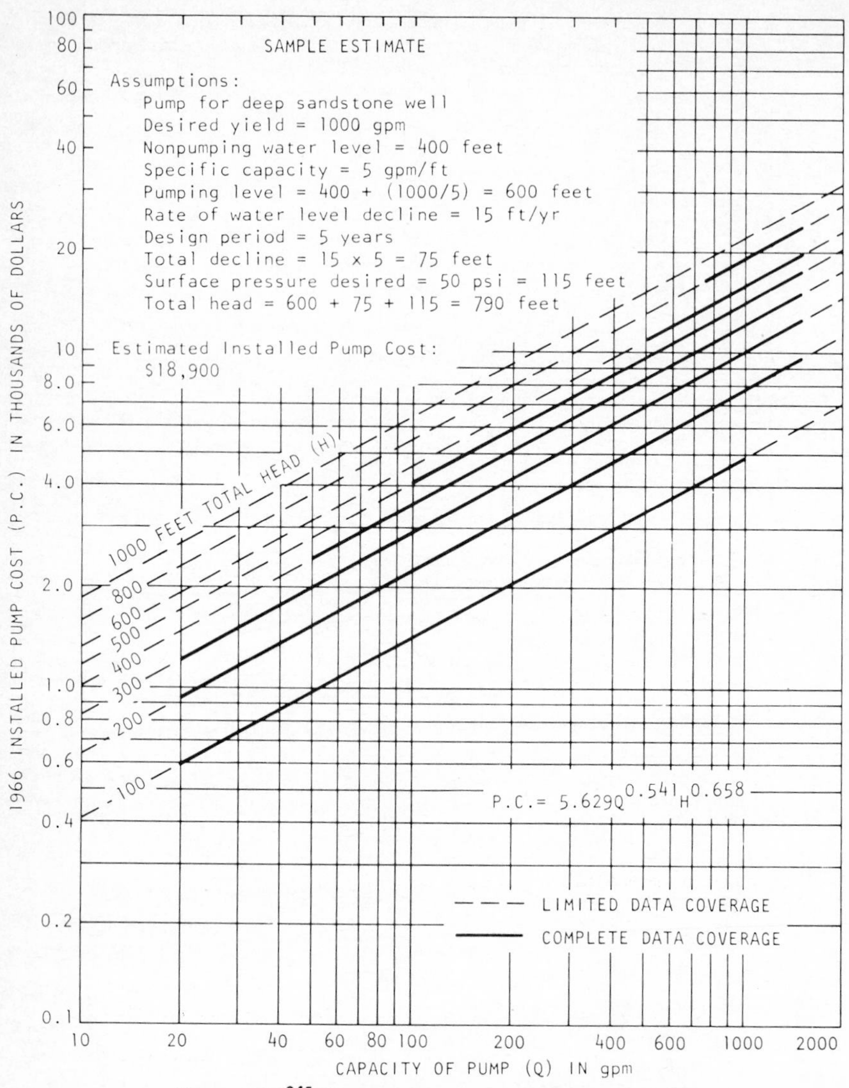

Figure 157[245] **Cost of submersible turbine pumps.**

assumptions of well cost at $10.00 per foot of depth, interest at six percent, useful well life of 30 years, and power cost at two cents per KWH.

The curves in Figures 159 are based on the depth-yield relationships expressed in Figures 89 and 90 with a static water level at 30 feet. Other curves can be developed for other aquifers which exhibit other economic and geologic factors, but the general slope of all such

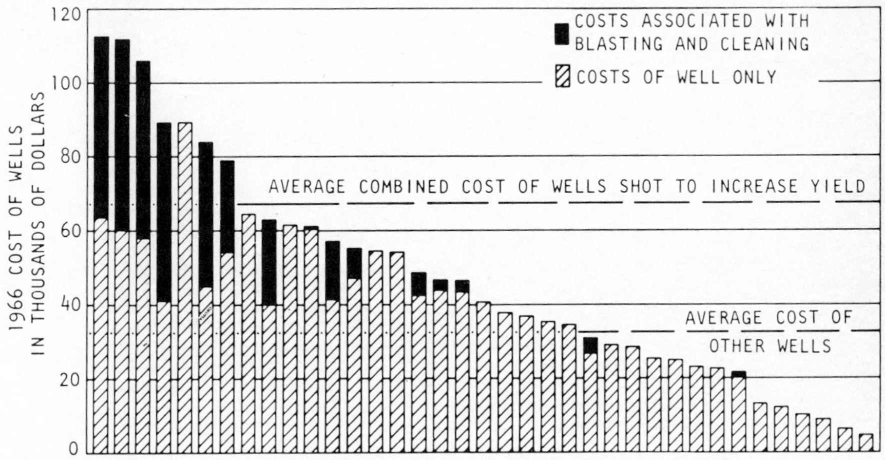

Figure 158[245] Cost of blasting deep sandstone wells.

Figure 159a[77]

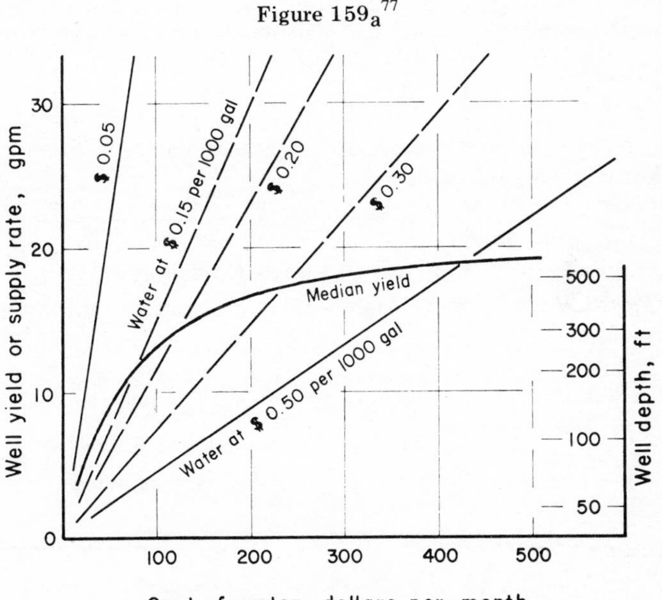

A — Curve showing cost of well water based on average yield of wells in granite and schist, eastern United States, compared to water bought at fixed prices; use assumed to be continuous at well yields or supply rates indicated.

Figure 159b[77]

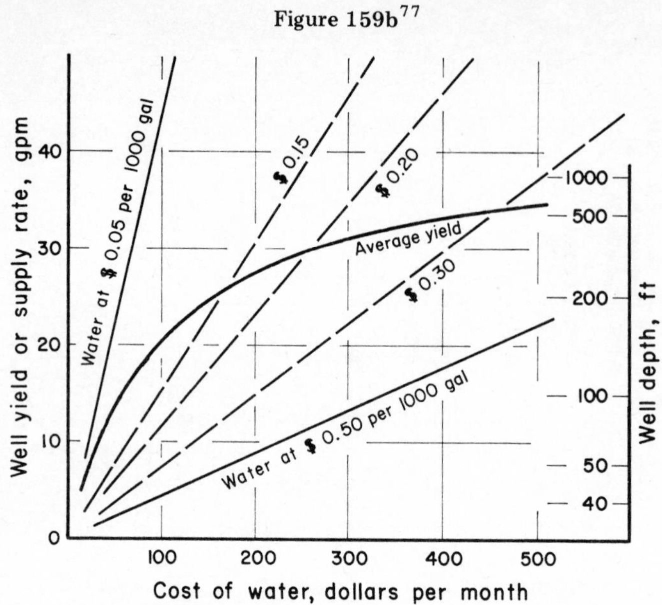

B — Curve showing cost of well water based on median yield of wells in granite and schist, eastern United States, compared to water bought at fixed prices; use assumed to be continuous at well yields or supply rates indicated.

curves should be similar to those in Figures 89 and 90. See Appendix for other cost-related guides.

The cost analysis briefly examined here provides a basis for determining the more specific economic framework for wells in "crystalline" rocks, which include many highly fractured rocks, e.g., igneous metamorphic, etc., where the general optimum depths were considered.

15 Summation and Outlook

Water well construction technology has made considerable progress in the past 15 years, and the petroleum and mining industries have played a vital part in this advancement. In accelerating the development of our ground water resources, however, the practical application of well construction technology has remained relatively low primarily because of the lack of economic impetus inherent in the oil and mining industries.

The mining and oil industries have been highly efficient for many years in locating and developing oil and other high-value minerals; hence, the technology of these industries has reached elevated levels of technological sophistication. The economic impetus of drilling for oil and developing mining properties is obviously greater than drilling for water; however, sophistication of operations does not necessarily imply that these techniques are too expensive to employ in the ground water industry. In fact, many oil and mining practices have now been simplified, assimilated, modified, and put into field use by the ground water industry. The results have been technically successful, and the costs, unquestionably justified. As the development of ground water continues, the number and depth of water wells will certainly increase, especially in the irrigation regions of the western United States. New techniques and concepts have, for many years, been adapted by the water well construction industry after practical applications were realized by numerous service companies, equipment manufacturers, and technical personnel in the petroleum, mining, and ground water industries. Furthermore, the ground water industry, in recent years, has developed many concepts and techniques which have found subsequent use in the petroleum and mining fields.

DRILLING TECHNOLOGY

Because of technical sophistication, the petroleum industry is translating the drilling process into a set of controlled variables and predictable results. The mining industry is developing this process, although the degree of complexity of this industry's drilling parameters is more diverse and perhaps more difficult to define than those of the petroleum industry because mineral exploration is usually conducted under less difficult hole conditions, e.g., less depth, less pressure, etc., but under relatively more severe economic pressure than oil exploration. However, the water well industry must drill under the most severe economic pressure; at present, millions of dollars in research are not available to control the partially unique drilling variables inherent in relatively shallow drilling. By increasing technical communication, drilling variables which the petroleum industry have already defined can be transferred to the water well drilling industry. All variables lie within a system which has been defined for the water well industry but awaits further industrial development.

ROTARY DRILLING

More technical knowledge is available on the rotary system than on any other system of drilling. This results from a wide acceptance of the method by the oil industry whose extensive research toward more efficient and, therefore, less costly systems has promoted the natural evolution within the water well industry to the rotary system. More recently, not only have the tools become available to the shallow drilling industry, but advanced rotary techniques have begun to find application in water well construction, although this transition is slow because cable-tool drilling is still in widespread use in the water well industry. The accent today in water well technology is on efficiency, and the rotary drilling system generally offers considerable advantages.

"Optimization of operations," a method of maximizing efficiency, has been an important concern of the oil industry for a number of years because better drilling practices mean higher profits. One of the most important aspects of an information transfer from the oil industry is a concern for detail, and if this concern is emphasized in the shallow drilling industry, the effect will be a substantially increased level of efficiency in water well drilling.

More than 40 variables are generally assumed to influence drilling, and many of the relationships of these variables must be established in the laboratory where drilling conditions can be controlled. Mere simulation of field conditions in the laboratory is inadequate since field conditions must be more closely duplicated to provide meaningful results. A parameter or criterion which describes both field and laboratory behavior is needed, and the results of any study must then be applied to practices in the mining, petroleum, and ground water industries.

DRILLIBILITY

In the broadest sense, the rate of rock drilling is directly proportional to rock drillability. If all the factors which affect drilling rate were possibly isolated and expressed mathematically, some function of rock drilling strength would certainly be included. Fluctuations in drilling rate are due to a difference in the drilling resistance of each rock type, and this relationship translates as the inverse of rock drillability.

Only after a correlation has been established between some measurable rock properties and drillability can ideal conditions be defined. Because the water well drilling industry operates under relatively severe economic pressure, the need for such a definition is urgent. Actual performance can then be expressed in terms of the ideal, and the relative merits of various approaches toward improvement in drilling efficiency can be evaluated. Improvement in shallow drilling efficiency is a prerequisite for healthy industrial growth.

DRILLING ECONOMICS

The staggering cost of deep rotary drilling, especially offshore, has concentrated the drilling research effort of the petroleum industry to reducing the time spent on deep wells. Such efforts have been concerned with increasing penetration rates and minimizing a variety of hole problems. All of the developments that have been made do not offer the same economic rewards to shallow rotary drilling operations, but many of the principles and practices can be beneficially adapted to shallow drilling conditions. Certainly the shallow hole drilling industry must progress within its economic structure, but knowledge of the developing technology is a necessary prerequisite to growth in any industry.

Sizeable cost reductions have been made by the application of improved rotary-drilling technology. Air rotary drilling promises additional cost-reductions. Also, through education and motivation, the best current technology can be rigidly applied in all drilling operations to continue the cost-reduction trend. For example, by placing drilling engineers on a rig around the clock for the drilling of several wells in one area, drilling costs have been reduced 5 to 15 percent or more in the oil industry. A cost reduction could also be realized in the water well industry if the size of a project permits the addition of qualified technical supervision.

Furthermore, as the industry gains additional fundamental knowledge about the rotary-drilling process, such knowledge should produce even better technology and equipment than that currently available, some of which will be directly applicable to shallow drilling technology. In addition, some items developed for new drilling systems are applicable and will further enhance the potential of rotary drilling.

In spite of efforts to discover new and improved drilling systems, rotary drilling maintains an economic leadership. Undoubtedly, rotary-drilling costs will continue to be reduced by rigid application of the best available technology and by development of new rotary technology. In view of the extensive past development programs, however, significant long-range improvement appears to be a research, not a development problem. Research must postulate and prove theories and principles governing various subsurface rock-failure processes pertinent to both rotary and new systems as well as physical and engineering data. Major improvements in rotary drilling can then be expected, and the systematic evolution of an improved drilling method can be initiated — with a strong probability for success.

The oil industry is developing drilling methods other than rotary drilling as a potential means of reducing drilling costs in the future. Major cost-reduction efforts, however, have been centered on engineering development work aimed at incremental improvements in rotary drilling. A considerable effort has been expended to establish basic physical principles and engineering data pertaining to the earth-boring process which can serve as a foundation for the development of equipment that can reduce drilling cost.

OTHER DRILLING METHODS

The downhole hammer, or air percussion drilling method can be modified to combat excessive ground water influx by incorporating foam. Furthermore, drilling for ground water in igneous, metamorphic, or other "hard" rocks is drawing particular interest and will no doubt increase as this method of drilling improves.

Reverse circulation offers the most inexpensive method of drilling large-diameter holes in soft, unconsolidated formations. Where geologic conditions are favorable, the cost per foot of borehole increases little with increase in diameter.

NOVEL DRILLING METHODS

The most extensive drilling development over the past ten years is the turbodrill. The basic concept of the turbodrill is to transmit relatively large amounts of power to the bit through the drilling fluid, thereby overcoming the limitations of transmitting power by means of a long rotating drill string. With currently available turbodrills, the increased power is presented to the bit in the form of much higher rotary speeds than can be used safely with conventional rotary practice.

In the United States, turbodrill test and development work has been reduced, and the effort in this area is being directed toward the development of a suitable bit for use with the turbodrill. Recent work indicates that the turbodrill bit problem is related more to cutting structure configuration and basic rock-failure principles than to bearing failures.

Other novel drilling methods of interest are: forced-flame drills, plasma drills, laser drills, modified turbine drills, spark drills, and numerous others including rotary. Considerable development work is now being done in the United States and other countries on at least several of the novel drills having the highest potential drilling rates: primarily the erosion drills (with and without abrasives), modified turbine drills, and the explosive drills.

In recent years, publications relating to oil-field drilling emphasize progress in the study and development of novel drills, while continued and steady improvements in rotary-system performance have been less dramatically treated. But rotary performance is moving ahead steadily. As the power outputs of lasers and other drilling concepts increase, drilling rates and potential applicability will increase. For this reason, these devices should be continually re-evaluated as new equipment and better techniques are developed.

DRILLING ECONOMICS AND FUTURE DEVELOPMENTS

Many economic factors must be considered when evaluating the potential of new drilling methods. Oil-field drilling is currently accomplished with very large expensive rigs, requiring large crews. New drilling methods that would employ smaller rigs, smaller crews, or flexible drill pipe to increase trip speed could be economical even though penetration rate would be lower. In oil-field drilling, the cost of the power transmitted to the rock by the bit is relatively insignificant compared to the other costs, so novel drilling systems that require considerably more power could be economical if the drilling rate increased or the operating costs decreased.

Consideration should be given to combining two or more of these novel drilling methods. For example, if the laser system, used to cut slots in rocks, is combined with the spark or erosion jet systems, the unsupported rock can be crushed and removed. Novel systems could also be combined with conventional systems, e.g., lasers, plasmas, or electron beams could be used to heat and degrade rock, and roller or drag bits could be used to remove the weakened rock. Obviously, some of these new drilling methods can theoretically drill large-diameter holes faster than conventional methods in some rocks.

A considerable amount of research on new drilling techniques is in progress throughout the world, and some of these drilling systems may soon find initial field application. These systems should be tested for special applications where drilling rate is more important than unit drilling cost. Once tested and developed, efficiencies, power outputs, and drilling rates of these systems will be increased until more widespread use in industry is merited, which would mean reduced costs and possible application to the shallow drilling operations of the water well industry.

Many years of research may be required before a novel drilling concept can become sufficiently simplified and, therefore, economically feasible for the water well industry to adapt. However, the development of a novel or an entirely new drilling concept may be more applicable to particularly shallow drilling than to the petroleum industry, especially in the early stages of research. Therefore, the petroleum industry's drilling research must be closely monitored for a potential breakthrough.

COMPLETION AND DEVELOPMENT TECHNOLOGY

The petroleum industry has supplied much of the present knowledge of formation evaluation techniques which have found applicability in the water well industry. However, aspects, such as detailed well design features, aquifer productivity evaluations, pump-test methods, etc., have nearly all evolved from the field-developed efforts over the past 25 years of the water well industry. Furthermore, a close similarity exists between the technology of crude oil production and the technology of ground water production, both of which deal with the flow of fluids in porous underground formations. The physical laws governing the flows are identical (although the mathematical expressions for the flow of crude oil are more complicated), and the wells used are nearly the same for both technologies and are constructed by similar methods. With the increase of water-flooding oil recovery operations, both for primary and secondary completions, the petroleum industry can draw from the available technology of the water well industry. Both industries, in fact, can prosper by working together closely and by overtly supporting the interdisciplinary transfer of technology.

FORMATION EVALUATION METHODS

The most important technological feature under considerable research and development in the petroleum and mining industries is in the area of formation evaluation, e.g., borehole geophysics, drill-stem testing, etc. Other features, such as cementing techniques, casing designs, formation stimulation techniques, etc., have made excellent advances in the past 15 years. The petroleum industry is responsible for the development of almost all of the logging equipment and interpretive techniques in use today. Borehole geophysics might never have reached the present state of development without the economic impetus provided by the world market for oil. Ground water may never reach this economic level, but the industry has benefitted from the existing techniques financed by the oil companies. A thorough knowledge of developments in petroleum well logging is essential to avoid duplicating basic research that can be applied to other fields of mineral exploration including ground water. These developments function to bring equipment and service costs in line with the value of the product.

The total consumer investment in ground water development is very high, however, the unit value of water is rarely equal to the unit value of oil. The cost of logging equipment and commercial logging services must be compatible with well costs and the value of the product. Shallow, small-diameter domestic wells in the East have a low average cost; however, in the western United States municipal, irrigation, and disposal wells are comparable in cost to some oil wells. The general trend is toward deeper, more expensive wells which have required and have economically justified more advanced knowledge of the ground water environment and better well construction and completion practices. As a result of recent developments in electronics, small and relatively inexpensive logging units which can be operated by one man are now available.

A basic problem in the widespread use of geophysical well logs in ground water is that few geologists, hydrologists, or engineers are working on water well drilling projects, and many of these professionals do not have the knowledge and experience necessary for detailed interpretation of geophysical logs. Therefore, interpretations are usually provided along with logs by commercial well logging companies. If logging is to have practical application to the solution of hydrologic problems, not only must the usefulness and concept of logging be accepted by the well contractor, but more detailed knowledge of the interpretation of the logs must be understood by the professional.

The many water wells being drilled throughout the world provide an invaluable fund of geological knowledge if the proper data are obtained. By interpreting geophysical logs with a single objective, such as discovering the presence of water or oil, valuable mineral deposits in the area may be overlooked. The well contractor should be aware of the fundamental features of mineral exploration since the same geophysical logs needed for superior water well construction can be used to evaluate the mineral potential of prospective aquifers.

Geophysical logs are often used in conjunction with drilling rate information to aid in identifying formation characteristics. During drilling, a record of bit penetration rate can be kept, either quantitative or qualitative. After the common suite of logs are run, the penetration rate can be matched with the logs. The combination of data helps to interpret more rock properties than can be obtained from the individual logs or the drilling rate alone.

In much of the present water well logging, SP, single point resistivity, and often gamma logs are coordinated. They form a package that gives the most information for the lowest cost to the well contractor. A few state geological surveys and some federal projects log water wells for their own information. Copies of these logs generally are available to the driller or well owner. Oil well logs can occasionally supply general lithologic information.

An accurate differential temperature survey technique is successfully used to locate precisely a downhole-water loss in a waterflood-injection well, which was previously undetectable by any other means. The differential temperature log accurately measures fractional-temperature anomalies associated with fluid movements downhole. Besides being used to detect fluid communication downhole in water injection wells, the technique is being used to find tubing-casing leaks, gas communication, productive zones, lost circulation zones, oil and water production profiles, and gas/oil/water contacts in addition to some other water well applications.

Of the many logging tools available today, the acoustic log offers considerable promise in determining the density of cement behind casing, in analyzing fluid saturation, in locating cement tops, and in determining the presence and extent of formation fractures. The recent development of a superior acoustic log instrument has made possible the collection of porosity and lithology data from cased wells. At the same time, a valuable evaluation of the cement condition can be made from the data recorded. Recent advancements in electronic technology, as well as improvements in acoustic transducers, made the design of this acoustic equipment feasible.

Several logging and interpretive techniques used experimentally in petroleum logging have considerable potential for ground water investigations. These sophisticated tools and others that will be developed in the future will be widely used in hydrology when the equipment is perfected for routine use and the size and cost are reduced. Currently the techniques with the greatest potential appear to be: (1) neutron life-time logging, (2) several types of spectral logging, (3) nuclear magnetic logging, and (4) computer analysis of logs. The petroleum and mining industries will, in the future, continue to refine the various features of borehole geophysics. An increasingly aware ground water industry will take advantage of these developments and perhaps make special modifications which the oil industry can also use.

One area in which the petroleum industry offers considerable technological assistance is the general area of permeability-damage repair. However, most if not all of the work done by the oil industry on formation investigations for water sources to be used for secondary recovery purposes incorporates little of the technology available in the ground water industry. With the increase of water-injection practices, the petroleum industry should rely on the completion practices of the water well industry, not on oil-biased completion practices which are grossly inefficient when applied to water well completion.

Well design for igneous, metamorphic, and sedimentary formations is much the same. Factors such as variations in water quality within the same well, water-bearing fracture development, selective-cementing requirements, etc., combine to offer potential problem areas in "hard rock" drilling which often require more specialized approaches than are necessary in sedimentary drilling. However, little detailed work has been completed to date on the problematical areas of igneous and metamorphic drilling or on completion and development of wells in such rocks. The mining industry will be leading in the development and application of solutions to the difficulties in drilling. Other aspects must be examined in detail by the ground water industry in the future.

WELL CASING

Improvements in the past few years in metal alloys have made some metallic casing increasingly corrosion resistant; however, basic inherent problems remain. The advantages of plastic well casing, being non-metallic and, therefore, corrosion-proof should be considered. Furthermore, the collapse resistance of plastic can be improved by increasing the wall thickness to diameter ratio. As knowledge about properties of plastic casing expands, the technology improves, and the competition becomes more effective, this material will find increased application in the water well industry.

WELL CEMENTING

Ground water contamination can many times be a direct result of improper or inadequate grouting of wells both in the original well construction and in well abandonment. Grouting procedures in the petroleum industry have been reasonably standardized, and although a direct transfer is not practical, a transfer of the basic approach to cementing would be advantageous. Petroleum industry cementing techniques have become highly specialized and apply to water well construction although this transfer of cementing technology has not progressed as well as in other areas. As water wells are drilled deeper, this transfer, however, will become a necessity. An increasing number of states are establishing requirements on sealing and grouting of wells. Grouting is generally required as a sanitary protection, but the requirements from state to state are as widely divergent as casing requirements. One state accepts grouting of the annular space between the pump housing casing and surface casing, while another requires filling the annular space with heavy, bentonite mud as the casing is driven, forming a seal. The former regulation is inadequate, the latter, only probably effective. Obviously, the cementing technology should be delineated and eventually standardized in order to protect the ground water resources from contamination.

WELL EFFICIENCY

Since the ground water industry is under severe economic pressure, well efficiency is vitally important. As construction costs and population rise, the efficient use of each well becomes a necessity both in the conservation of the water resource and in the utilization of available space.

Several factors initiate well losses in inefficient water wells. These may include: losses caused by reduction in permeability and losses from the turbulent flow of water adjacent to the well through the screen or well face and inside the casing to the pump intake. High well losses are unnecessary since they can be minimized by good well design and construction.

The permeabilities of ground water aquifers are commonly more than 500 times higher than oil sands; therefore, the potential rate of water production is many times that of most oil wells. It is obvious that the design of water wells should be different from the design of most oil wells. For example, experienced water well designers recognize that a critical well-design feature is the inlet velocity through the screening device.

Breakdowns and delays are often caused by wells that pump sand, eventually plugging filters and distribution lines. The use of existing wells for source water in the petroleum industry may cost more than properly designed and constructed new water wells. The high cost of power, because of hydraulic inefficiency, compared to cost of constructing new water wells is usually overlooked. The efficiency of oil wells is not generally considered because the existing bottom-hole pressure is high and/or allowable production is so low that the extra cost of producing from an inefficient oil well is negligible. On the other hand, an inefficient water well becomes very expensive in terms of power costs and additional wells that may be required to meet the demand for source water.

The water well industry as well as the petroleum industry loses many thousands of dollars each year to corrosion which affects the well casings, the well screens, and the quality of the ground water produced by the well. True well efficiency depends to an important extent on the events which occur after the well has been drilled, developed, and completed. The nature of the casing, the well screen, the pump, the aquifer characteristics, and the original ground water quality all play a vital role in determining the longevity of the well. The petroleum and water well industries have begun to deal with these problems.

CORROSION/INCRUSTATION CONTROL

Corrosion control has been of prime importance to well maintenance in the oil industry for many years because of the high cost of "workover" operations and of well replacement. However, corrosion control has been neglected in the water well industry. This neglect has resulted not only from a lack of knowledge of the principles of corrosion control but from a lack of adequate economic impetus for employing any control methods.

The oil industry has found considerable success in various corrosion control methods. Of these methods, most of the work deals with cathodic protection — an artificial electrical system constructed to redirect stray electrical currents that cause a specific type of corrosion in casing, transmission piping, and drill pipe. However, corrosion cannot be predicted with any accuracy in subsurface equipment of existing wells. Although the degree or distribution of corrosion is never known in advance, after sufficient corrosion experience has been accumulated on the first wells in the area, this experience may be applicable to older wells.

In well fields where corrosion is a critical problem, standard effective countermeasures are difficult to define. Rarely are the techniques from other areas transferable, and certainly no reliable set of simple measures guarantees retardation. Just as corrosion is caused by many factors so are the methods available for control.

Briefly, the control measures that are available are: the injection of chemicals to condition the corrosive fluid and to treat the surface of the attacked metal; the use of corrosion-resistant alloys either for components or as platings and linings; and nonmetallic coatings or plastic casings have been widely adopted for corrosive environments.

Detecting corrosion on external casing and determining the extent are two of the most difficult problems in the entire field of corrosion control. The inaccessibility of the external casing surface hampers examination, not only in assessing the problem but in determining the feasibility of cathodic protection. Advanced techniques to control corrosion of external casing have been developed in the oil industry, most of which involve sophisticated electrical equipment. Rising drilling costs have necessitated the oil and water well industries' concern with the protection of existing facilities against deterioration by corrosion. The importance of coatings in the control of corrosion is evidenced by the large financial expenditure of the petroleum industry.

Incrustation is caused in most cases by entirely different factors than those which cause corrosion. Metals can be used which will resist corrosion, but incrustation may take place on any kind of metal or material depending on the chemical character of the ground water and would require some "workover" method to remove the incrustation. However, the physical condition of the metal in use must be able to withstand the "workover" techniques.

Careful consideration should be given to all methods of preventing or retarding incrustation; but in a large number of cases, incrustation will occur sooner or later so that well treatment becomes necessary. In general, two possible approaches for accomplishing well treatment are: pull, treat, or replace the screen or treat the screen and the water-bearing formation without pulling the screen. Dispersing agents or acids may also be used in the treatment. The oil industry has made significant advances in the removal and inhibition of various types of scale formation although generally the chemicals used are toxic. Observation via photographs or close-circuit television can aid in evaluating the micro-characteristics of early corrosion and incrustation.

WELL EVALUATION METHODS

In the past, physical properties such as the percentage and size of pore space could be studied only by coring a water-bearing formation, but this method has not always been successful. In addition, some formations are too fragile to core, and examination of the outcrops of the water-bearing rocks many miles from the well site may not reveal the desired features of the formations. However, the use of a television camera or photographs in a well provides a look at the *in situ* consolidated water-bearing formation.

Photographic techniques can be used as a tool in water well construction completion/development and maintenance methods to facilitate many aspects of well efficiency and to reduce overall cost factors. Photographs have been made in limestone and sandstone wells and have assisted identification of porous and cavity zones. In limestone formations, photographs have been used to determine well conditions before and after acidizing. Stereo-photographic surveys have also been employed for many miscellaneous applications; such as ventilation holes, mine shafts, brine wells, and others.

A closed-circuit television system has obvious advantages over a photographic survey with the exception of cost in some cases. Features that may be determined by this approach are: the nature of and depth to geologic contacts; the gross lithologic characteristics; the configuration of the hole wall; the location and condition of well casing and screens; and the precise location of debris in the well.

Presently used television equipment permits either a vertical or horizontal view of downhole conditions. An initial survey using a vertical-viewing probe is common during which sections are noted where a horizontal view is desirable. When the initial survey is completed, the equipment is hoisted to the surface, the horizontal viewing mirror is attached to the cable and inserted in the well — a time-consuming process. Development of a surface-controlled mirror or prism which could be incorporated in the vertical viewing camera and which could be mechanically activated to give a horizontal view would be a valuable contribution to ground water closed-circuit television.

Photographic and television surveys are not widely used to date in the water well industry even though significant cost-saving factors are involved. One problem is the absence of an acceptable method to rapidly precipitate suspensions in water, particularly in older wells. The clarity of the pictures can be adversely affected or blacked-out entirely by suspensions in the water, colloidal material, bacteria and algae clumps, and debris from corrosion or incrustation. Various precipitants and coagulants have been used occasionally with only limited success.

WELL STIMULATION

In an existing well with declining production, water well stimulation techniques can be employed to increase aquifer productivity and overall well efficiency. Modern drilling operations generally use fluids that will assist in achieving the most efficient drilling system. However, many of these fluids and contained solids invade the matrix and damage formation permeability. This situation frequently causes a marked reduction in the ability of the formation to produce oil, gas, or ground water.

The frequency of stimulation of water wells in the United States could not be ascertained, but it is thought to be low and increasing. Of the many publications reviewed on oil well stimulation, very few references to the literature on water well hydraulics or water well stimulation were found. Conversely, in the scores of articles on water well hydraulics and stimulation reviewed, only a few authors referred to oil industry techniques.

Numerous studies by the petroleum industry indicate that by utilizing deep-penetrating, high-flow-capacity fracture systems, a low-resistance flow path to the well can be established in large portions of the reservoir. The deeply invading fracture promotes production of large volumes of oil and possibly ground water. In addition, the fracture system provides large drainage areas into which formations of relatively low permeability can slowly transmit a liquid, thus utilizing available reservoir energy to the maximum; consequently, many areas that previously had been considered non-oil-productive were subject to commercial exploitation. Water well stimulation should also be carefully evaluated for water well fields where the desired total production may be efficiently increased from fewer wells, thus achieving lower capital investment and maintenance costs.

The mining industry receives considerable assistance from water well technology, particularly from research involving well stimulation. With the increasing economic impetus in the mining industry, significant advances in well stimulation technology should be forthcoming. Relatively severe economic pressure, low-grade ore, great mining depth and remote locations in mining certain minerals (e.g., uranium, potash, sulfur, phosphate, etc.) often necessitate *in situ* leaching and solution mining in which water well technology is vitally important. Other mining operations such as dewatering for open-pit mining and disposal of waste fluids are also discovering the importance of water well technology.

WELL ABANDONMENT

One of the more important aspects of well construction is proper abandonment or permanent sealing because serious ground water pollution can occur from either contaminated surface water or inter-aquifer transfer of ground water from a saline or otherwise contaminated aquifer into a potentially productive aquifer. Thousands of abandoned wells, either water wells, shallow seismic holes, or shallow mineral exploration holes, exist throughout the United States, too many of which are contaminating potential ground water sources as a result of improper abandonment techniques.

GROUND WATER CONTAMINATION

All wells, regardless of their use, pose a potential threat to the ground water environment. The part that oil, water, and gas wells play in local ground water contamination has been well documented, but obviously neither oil/gas wells nor water wells will be prohibited in the foreseeable future since contamination is usually a result of inadequate well construction techniques and practices and is totally unnecessary. Petroleum and potable water are not of equal importance to society's needs for there are energy source substitutes for petroleum and related oil/gas products, e.g., uranium, geothermal, and solar energy, etc., but there is no known alternative for potable water. The development of petroleum and ground water is competitive only to the extent that the latter must be protected more jealously than the former. Both can be developed simultaneously, if technology can serve to protect one from the other.

The ground water and oil industries must make a reasonable profit but also must act as the day-to-day guardian of the ground water resource. By incorporating the most up-to-date techniques of well construction, completion, and maintenance, these goals can be achieved. There is no tenable excuse for poor well construction practices or for inadequate profit in an open economy. Improved technology, designed to increase profit, also increases the quality of well construction and serves as the final answer to water well-induced contamination of the ground water resource.[676]

Monitoring equipment is needed in the petroleum and water well industries to continuously assure that all materials, used in the construction of wells, remain intact throughout the years of operation. Furthermore, abandonment techniques bear heavily on potentially widespread ground water pollution. These techniques should be under continuous development, the most effective of which should be used consistently in the field as a matter for the common good not as a matter of economics. Based on an examination of well construction techniques and the potential aquifer productivity, the extent of permeability damage attributable to drilling fluids, drilling muds, well-treatment chemicals, and drill cuttings needs evaluation.[677]

Ground water contamination is always present and will become intensified with increased use of ground water and concentration of facilities, especially in areas with high-population growth rates. Once recognized, however, the problem can be solved by the development of new and improved well construction techniques, materials, and equipment and by appropriate statutory and administrative measures. Appropriate monitoring tools are also needed to assure the sanitary protection of the water well and the valuable ground water resource.

CONCLUSIONS: RESEARCH AND DEVELOPMENT

Ground water will, in the near future, achieve broader status as an economic mineral, deserving all the rights of scientific consideration not only by professionals within the field but also by those in the petroleum and mining industries. Early signs of this acceptance are becoming obvious, and the respective terminologies are being merged although technical isolation is still apparent. Professionals in the two fields apparently rarely read or contribute to any but their own technical journals, and this lack of interreliance was substantiated during this investigation. The observed prevalence of this isolationist condition leads to a number of logistical problems. Some of the techniques developed in one area may be applicable in others. Moreover, techniques which have been considered failures in one geographic region may be perfectly applicable in another. Research and development work are expensive, especially in the ground water field since noncommercial research facilities are rare, and the chemical and mechanical industries will not risk funds without first predicting the potential return. An economic analysis of the total national market indicates only a marginal return from the water well industry.

A serious and consistent effort must be made by the ground water industry to develop and then to maintain the avenues of communication between the three industries: petroleum, mining, and ground water. The communication must be both in general and selected areas of interest, e.g., bit design, drilling fluid technology, borehole geophysics, casing, and cementing practices, etc.

Since geothermal resources are directly related to ground water development and to oil drilling practices and since the petroleum industry is heavily involved in deep-well disposal of their wastes and in ground water development for their water flooding recovery operations, a well-planned effort by the ground water industry to develop and maintain communications with the petroleum and mining industries would result in a twofold return. First, pertinent technology can be transferred more rapidly to the water well industry, and second, an information transfer could be made to the petroleum industry of the technical information presently available on water wells for the water-driven oil recovery operations. The assistance would produce a wide dissemination of water well construction technology including the inherent growing awareness of ground water pollution prevention and control. Each industry would have the benefit of the other's technology, thereby promoting the objectives of rapid technological advancement in water well construction and ground water pollution control, in addition to promoting professional cooperation which will be required in the difficult years of natural resource conservation ahead.[679]

The water well construction industry is operating under added disadvantages. The petroleum industry organized technical information and research centers in the 1930's that made significant contributions toward solution of the fundamental technological problems of that era, thus eventually benefiting the water well industry. However, since the water well industry today confronts many unique problems, the industry must solve its technological problems — the wealthy oil industry cannot provide these answers, only the methods of approach. Without viable centers of research, the rate of the vitally needed progress in all aspects of water well technology will not be sufficient to meet the dual demands of pollution control and economic profitability in the years of ground water development ahead.

Communication is the key to an increased rate of progress, and better communication within the industry itself must be developed. The oil industry has been able to develop a working relationship between the geologist and the well contractor since a geologist or engineer has been historically involved in oil drilling operations. This is not the general case in the water well industry due to, of course, the economic framework. This lack, however, has fostered division, and few possibilities exist that the economics will change to allow this potential relationship to develop in the near future. This contractor-geologist relationship provided educational possibilities for the training of the well contractor. Drilling experiences have yet to be fully understood or explained in the classroom without field observations. Furthermore, without a financially sound equipment industry capable of developing better tools and a better knowledge of how to use them more profitably, the contractor cannot be expected to radically improve his own economic position.

During the course of this investigation, the broad technological features of well construction were reviewed in conjunction with the literature of the petroleum, mining, and ground water industries. In order to accelerate the rate of advance in water well construction technology, priorities must be established for the most pressing technological problems which relate to the development of the ground water resource.

Well construction is vitally important to potential ground water pollution control. The investigation has found that certain areas of well construction technology have a potential major impact on ground water pollution control and aquifer productivity and deserve detailed study. Of these, the following subject areas are of high priority:

(1) *Water Well Cementing Practices and Technology*, i.e., detection of effectiveness, sanitary protection, longevity features, etc.
(2) *Water Well Corrosion and Incrustation Control Technology*, i.e., early detection, advanced techniques in control, materials resistance, etc.
(3) *Water Well Rehabilitation and Stimulation Technology*, i.e., downhole inspection, aquifer stimulation techniques, economic justification, etc.
(4) *Water Well Construction Specifications*, i.e., geographically pertinent requirements, rational standards for well construction, industrial acceptance, etc.[674]

(5) *Water Well Drilling Cost Analysis Technology*, i.e., optimization of operations, economic environment of cost-profit structure, industrial cooperation, etc.
(6) *Water Well Industry Assimilation of New Technology*, i.e., optimum transfer, psychological framework of transfer potential, economic framework of transfer potential, etc.

In addition, the following are pressing research and development needs in the field of well construction technology:

(1) New and improved methods of drilling to reduce introduction of potentially contaminating materials into aquifers.
(2) New and improved techniques for aquifer sampling to facilitate proper well design.
(3) Reliable design criteria to reduce sand pumping, well collapse, vertical flow, and similar factors that contribute to contamination.
(4) Design features specifically directed toward facilitating eventual abandonment and destruction.
(5) Methods of treating drilling equipment and fluids to reduce bacterial contamination and eventual possible corrosion/incrustation and to determine the effectiveness of the treatment during drilling.
(6) New and improved materials for casing, liners, and screens especially for use in highly corrosive ground water environments.
(7) New and improved methods and materials for grouting, especially (a) inexpensive, expendable packers, (b) simple grout pumping equipment, (c) nonstructural grouts and (d) in-place grout detection and evaluation equipment.
(8) Base legislation and administration on established principles of ground water occurrence and well hydraulics.
(9) Shift the regulatory emphasis from protection of the water user to protection of the aquifer.
(10) Apply the same sanitary standards to all wells, regardless of the water use; however, make the standards functionally rigid but structurally flexible enough to enable use under a wide range of hydrogeologic conditions.
(11) Require prior approval by the regulatory agency of the drilling method and design and construction features of each well before issuing a drilling permit, and submission of a log upon completion.
(12) Require notation of the exact location of each well on a deed, plat, or other permanent land record.

(13) Provide for the right of the regulatory agency to inspect and sample each well by implied consent as a condition of issuance of a permit.
(14) Require that notice be given to the regulatory agency upon disuse of a well for a stated period of time. If periodic follow-up indicates permanent disuse, destruction could be prescribed.
(15) Establish the right of the regulatory agency to correct or abandon an offending well.
(16) Establish a continuing fund from permit fees or other sources to be used for abandonment procedures.
(17) Require licensing of well contractors based on an appropriate examination.
(18) Establish appropriate curricula at a vocational school and/or college level for those engaged in water well contracting and drilling. For example, Indiana Vocational Technical College, South Bend, Indiana, in cooperation with the National Water Well Association, has established one of the first curricula of this type, and Salem Technical College, Oregon, has been involved in this type of educational program.

Detailed research, treating each of the above areas should have direct applicability to practical field problems. Since the magnitude of the above problems is great and of widespread importance to the nation's short- and long-term ground water development, both industry and the federal government should co-sponsor research programs on high-priority problems. The water well industry is operating under much more severe economic pressure than the petroleum industry, but is making progress in assimilating technological advances. The water well industry will, however, require financial assistance from the government to accelerate this progress to a rate consistent with the increased usage and long-term, pollution-free development of the ground water resource. Furthermore, an adequately funded technical/industrial information and research center is needed to monitor and develop research programs and to provide consistent industrial/governmental guidance and continuity.

16
Annotated Bibliography

This annotated bibliography is keyed, where possible, to the computerized accession numbering index incorporated in *Selected Water Resources Abstracts,* a semimonthly publication of the Water Resources Scientific Information Center (WRSIC), Office of Water Resources Research, U. S. Dept. of the Interior.

1. Ackermann, W. C., 1969, "Cost of Pumping Water," *Ground Water,* Vol. 7, No. 1, January-February, pp. 38-39.

Use of this material will assist in the determination of cost of pumping water, given the quantity of flow required, the total pumping head, the wire-to-water efficiency, and the unit cost of power. A table of conversions is presented to aid in reducing theoretical equations to simplified equations, and a figure is provided for graphical solutions of the equations. WRSIC #W71-09731

2. ———, 1969, "Cost of Wells and Pumps," *Ground Water,* Vol. 7, No. 1, January-February, pp. 35-37.

The use of this material will give an estimate of the well and pump costs for projects requiring a given capacity, but it does not substitute for detailed engineering studies. Well cost data were analyzed for three categories according to the aquifer tapped: sand and gravel, shallow bedrock, or deep sandstone. WRSIC #W71-09730

3. Adams, J. H., 1957, "Air and Gas Drilling Cuts Costs 22 Percent," *Petrol. Eng.*, Vol. 29, No. 5, May, pp. B30-B34.

Rotary drilling with air or gas as a circulating medium has been highly successful in drilling four wells in the extremely hard-rock area of eastern Oklahoma and northwest Arkansas. An over-all savings of 22 percent was realized by utilizing air or gas rather than conventional rotary mud. One application of the exhaust-gas drilling technique was not as successful from an economic standpoint, and this paper discusses both the advantages and disadvantages of this drilling technique.

4. Ahmad, N., 1969, *Tubewells Construction and Maintenance,* Scientific Research Stores, 4 Abkari Road, Lahore, West Pakistan, 236 pp.

The private interest in the lifting of ground water by tubewells started in West Pakistan in 1950. Since 1969, more than 80,000 tubewells installed by farmers mainly from their own resources are in operation. This book covers the science of tubewell construction; the technique of designing; installing; and operating a tubewell. It is a textbook for engineering students, a guide for the practicing engineers engaged in installing tubewells and exploiting ground water, and a reference book for the engineers of West Pakistan in general.

5. Ahrens, T. P., 1957, "Well Design Criteria," *Water Well Journ.*, Vol. 11, Nos. 9 and 11, September and November, 8 pp.

Procedures and well design criteria dealing with irrigation wells; the criteria are applicable to domestic, industrial, and municipal water wells but not applicable to wells for ground water observations, ground water recharge, drainage, excavation de-watering, drilling water supply, or for other temporary or special purposes. Preliminary investigations, drilling requirements, sample analyses, justification for packs, selection and design of packs, placing of pack materials, well screen slot size, length and diameter of screens, screens and packs in composite aquifers, percent penetration and drawdown, well development, selection and installation of casing, cementing-in and sealing, well completion.

WRSIC #W71-08931

6. ———, 1958, "Water Well Engineering, More on Well Design Criteria," *Water Well Journ.*, Vol. 12, Nos. 9, 11 and 12, September, November, and December, 8 pp.

Benefits of measurements, water level, and log notations made while drilling a water well; of formation samples to estimate yield of well; of perforated casing or slotted pipe instead of screens when entrance velocities permit; of limiting minimum diameter of screen and casing below pump chamber to one inch less than the column pipe Bureau of Reclamation pack criteria compared to others; estimation of specific capacity for design purposes; influence of well dimensions on yield; effective well diameter; fractional penetration and fractional open hole; anisotropy.

WRSIC #W71-08666

7. ———, 1970, "Basic Consideration of Well Design," *Water Well Journ.*, Vol. 24, Nos. 4, 5, 6, and 8 (Parts I-IV), 17 pp., 32 refs.

The four-part article deals with water well design, pilot holes, screen size determinations, drilling specifications, principles of well hydraulics, analysis of well casings and grouting, ground packing and well development and testing.

8. Alger, R. P., 1966, "Interpretation of Electric Logs in Fresh Water Wells in Unconsolidated Formations," *Soc. Professional Well Log Analysts Symposium,* Houston, 25 pp., 12 refs.

Evaluation of wells drilled for fresh water presents special problems in log analyses. In such analyses the problem is to determine the quality and quantity of water that may be obtained from various strata. The SP is used for determination of fresh water quality, but firm empirical data for the locale are required. Measurements of resistivity provide the means for determining relative productivities of fresh water sands. Unconsolidated sands generally exhibit uniformly high porosities; however, a surface conductance effect in fresh water sands causes the formation resistivity factor to vary with both R_W and grain size. Because permeability is related to grain size, resistivity values indicate relative productivity. WRSIC #W71-06947

9. American Association Petroleum Geologists and the U.S. Geological Survey, 1971, "Underground Waste Management and Environmental Implications Symposium," December 5-9, Houston, Texas; Proceedings, *Amer. Assn. Petrol. Geol. Bull.*, Vol. 55, Nos. 11 and 12 (Parts 1 and 2).

Symposium covered a variety of topics embracing the area of deep-well waste disposal.

10. Ammann, C. B., 1960, "Case Histories of Analyses of Characteristics of Reservoir Rock from Drill-Stem Tests," *Journ. Petrol. Tech.*, Vol. 12, No. 5, p. 27.

The presented case histories of analyses of characteristics of reservoir rock from drill-stem tests illustrate that drill-stem test analyses may be used to evaluate a formation within the vicinity of a wellbore. The examples presented show that an analysis will determine to a practical degree such properties as true pressures and potentiometric surfaces, transmissibilities, actual and *in situ* producing capacities and permeabilities, productivity index, damage ratio, production rates, distance to any apparent barrier interferences near the wellbore, and the approximate drainage radius during a test.

11. Anderson, K. E., 1967, *Water Well Handbook,* Scholin Brothers Printing Corp., 2nd Edit., St. Louis, 281 pp.

Drillers, engineers, and geologists working with water wells have occasion to refer frequently to charts, tables, and other data which are scattered throughout many books and catalogs. At present, there has been no convenient reference source available which brings much of this information together. This booklet has been prepared for that purpose.

Anderson, K. E. . . . . (con't)

An attempt has been made to bring together only the information most frequently referenced by persons working with ground water supplies. It is well indexed to assure useability. WRSIC #W71-06919

12. Angel, R. R., 1957, "Volume Requirements for Air and Gas Drilling," *Trans. SPE of AIME,* Vol. 210, pp. 325-330, 5 refs.

Air requirements for air drilling are developed which have applicability to foam drilling. Resultant tables are based on gage of hole and are valid only when dry formation is drilled. In wet formations, the variable and complex flow conditions that develop within the well require more air.

13. ———, 1968, *Volume Requirements for Air-Gas Drilling,* Gulf Publishing Co., Houston, Texas.

An expansion of Angel's earlier work – see above.

14. Annis, M. R. and P. H. Monaghan, 1962, "Differential Pressure Sticking – Laboratory Studies of Friction Between Steel and Mud Filter Cake," *Journ. Petrol. Tech.,* May, pp. 547-553, 12 refs.

The control of mud properties affords two practical means of mitigating pipe sticking caused by differential pressure: (1) reducing mud weight and, therefore, differential pressure and (2) reducing the friction between the pipe and mud cake. This paper describes investigation of the second of these – the friction between the pipe and the mud cake.

15. Anonymous, undated, "Turbine Bit," Christensen Diamond Products Co., *Technical Brochure,* 12 pp.

Describing details of the Christensen turbine bit state-of-development, operational requirements, and applications.

16. ———, 1952, "Rotary Drilling, Setting Screens and Developing," *Johnson Driller's Journ.,* Vol. 26, No. 4, July-August, pp. 1-5.

Rotary drilling requires special methods of screen emplacement and well development. The effect of mud invasion and removal is explored as well as the "bail-down" and "wash-down" methods of screen installation.

17. ———, 1953, "Physical Properties of Some Typical Foundation Rocks," U.S. Bur. Reclamation, *Concrete Laboratory Report SP-39.*

Describes selected physical properties of common foundation rocks; gives compressive strengths under uniaxial loading and other properties.

18. ———, 1954, *Baroid Drilling Mud Data Book,* Baroid Div., National Lead Corp., 471 pp.

A complete text on drilling fluid theory and application despite the commercial intent. Topics: history, function, and types of drilling mud; clays weighing material; chemical treatments; oil-mud use; basic mud-mixing techniques; cost appraisals; field testing; trouble shooting; lost circulation problems and their solutions. Illustrated, referenced, and updated since 1954 with glossary. WRSIC #W71-05294

19. ———, 1955, "Shooting Rock Well Involves Risks," *Johnson Driller's Journ.*, Vol. 27, No. 6, November-December, pp. 3-6.

Reasons and methods for shooting rock wells; common types of rock aquifers and their water bearing properties; cases and results.

20. ———, 1955, "The Well Screen Envelope: Judging Proper Gravel-Pack Thickness," *Johnson Driller's Journ.*, Vol. 27, No. 2, March-April, pp. 1-4.

Discusses general aspects of well development of gravel-pack operation and screen installation. Gravel-pack thickness is explored for determination of optimum thickness under specific conditions. Function of gravel pack, permeability ratio, and case histories are explored.

21. ———, 1956, *Effects of Drilling Fluid on Penetration of Rock Bits*, Battelle Memorial Institute, Sponsored by the Amer. Assoc. of Oil Well Drilling Contractors.

Describes an early investigation on penetration rates of rock bits in various drilling fluids. By lowering bit nozzles closer to bottom of hole, results indicate that approximately two-thirds less fluid flow was required to move cuttings from hole into fluid stream than was required on a standard jet bit.

22. ———, 1956, "Russian Turbodrill . . . How Good are They?" *World Oil*, December, pp. 151-158.

This article includes orientation tests of 40 turbodrills imported to the United States from the Soviet Union, comparison of turbodrill and conventional rotary drilling rates as experienced in the Soviet Union. WRSIC #W71-05306

23. ———, 1957, "Development Work is Essential," *Johnson Driller's Journ.*, Vol. 29, No. 6, November-December, pp. 1-2.

Water development and usage, well design needed, methods of development, improvement of permeability, and porosity of water bearing sand for some distance around the well screen from development operations are included.

24. ———, 1957, "Methods of Setting and Pulling Johnson Well Screens," *Johnson, U.O.P. Bull. 933*, Revised, 16 pp.

Where conditions permit its use, the simplest and best method of setting screens is called the "standard" or "pull-back" method. The only requirement for this method of installation is that the casing must be of such kind that it can be sunk down to the point where the bottom of the screen is to be set and then pulled back the length of the screen. When a screen cannot be set by the "standard" method, the "bail-down" method may be employed. The principles upon which all

Anonymous, 1957 . . . . (con't)

bailing-down processes are based is that by maintaining a constant weight on the screen and removing the sand and gravel which lies directly below the bottom of the screen, the screen will settle down by force of gravity to take the place of the material which is withdrawn. The two essentials are, therefore, provision of proper weight upon the screen and a means whereby the material beneath the screen may either be removed or displaced so that the screen can settle through the water bearing formation. WRSIC #W71-08468

25. ———, 1958, "Development Work is Essential," *Johnson Driller's Journ.*, Vol. 30, No. 1, January-February, pp. 1-3.

Describes the general aspects of well development and its effect on optimum production and screen selection. Various geological conditions require various development procedures.

26. ———, 1958, "Jet Cleaning of Wells Described," *Johnson Driller's Journ.*, Vol. 30, No. 6, November-December, pp. 1-3.

Description of a technique of horizontal jetting used in conjunction with various types of chemical treatment.

27. ———, 1959, *Oil-Well Cementing Practices in the United States,* A.P.I., New York, 297 pp., 2,356 refs.

This book has been written about the process of placing oil-well cement and covers how this and related operations are carried out. The compositions used in cementing are basically portland cements or cements prepared by supplementing portland with additives or by variations in grinding or other manufacturing processes; however, portland-cementing methods cannot invariably be used with other compositions. Some parts of the process of "cementing a well" are closely similar wherever they might be carried on, while other parts may vary widely depending upon geographical location. Nomenclature of terminology particularly vary from one area to the next. This book attempts to fill this gap. WRSIC #W71-09710

28. ———, 1960, "Checking Alignment and Plumbness of Wells," *Johnson Driller's Journ.*, Vol. 32, No. 1, January-February, pp. 8-11.

Discusses arrangement for checking the straightness and plumbness of a well.

29. ———, 1960, "Disinfecting Wells and Water Piping," *Johnson Driller's Journ.*, Vol. 32, No. 5, September-October, pp. 1-3.

A step in well completion; thorough disinfection tables give chemical quantities required.

30. ———, 1961, "American Standard for Vertical Turbine Pumps," *A.W.W.A. E101-61,* 2 Park Ave., New York, 49 pp.

Standard published by A.W.W.A. and approved by Amer. Standards Assoc. on January 31, 1961. Guide for users of the line shaft vertical turbine pump and for selecting new equipment. Six sections: scope and purpose; definitions; nomenclature; general specifications; engineering data; factor inspection; and tests. Formulas, tables, and charts.

31. ———, 1961, "Here's Ingenious Way to Fish for Pump," *Oil and Gas Journ.*, Vol. 59, No. 42, October, p. 94.

During removal of submerged water well pump motor unit from well, casing parted, and pump dropped back to its operating level, 280 feet below grade; fishing tool was designed to slip over pump's top and then grip it as tool moved upward; tool, attached to length of four-inch drill pipe, was lowered into well opening; one at a time, 22-feet lengths of drill pipe were added until contacting pump; pump was removed in two days.

32. ———, 1961, "Jet Development Does the Work," *Johnson Driller's Journ.*, Vol. 33, No. 6, November-December, pp. 1-5.

Jetting, which is a method of high velocity washing of completed intervals, effectively increases the well's yield and decreases the drawdown. The equipment consists of a relatively simple jetting tool, together with a high-pressure pump and necessary hose and piping. The procedure consists in operating a horizontal water jet inside the well in such a way that the high velocity stream of water will shoot out through the screen openings. Fine sand, silt, and clay are washed out of the formation, and the turbulence created by the jet will bring these fine materials back into the well through the screen openings.

33. ———, 1962, "Charts Show Air-Lift Performance," *Johnson Driller's Journ.*, Vol. 34, No. 3, May-June, pp. 3-4.

Illustrated charts for general performance of an air-lift under different operating conditions and air compressor requirements. The performance of an air lift cannot be precisely predicted because minor factors can have more effect upon results than is generally the case with other types of pumps.

34. ———, 1962, "Sand Studies Can Improve Well Design," *Johnson Driller's Journ.*, Vol. 34, No. 3, May-June, pp. 8-10.

Test sieves for sand analyses; geological processes in formation of unconsolidated sedimentary deposits; size frequency curves.

35. ———, 1962, "The Yield of Water Wells," *Johnson, U.O.P. Bull. 1238*, Revised, 8 pp.

The potential capacity of a water-bearing sand or sand-gravel formation to yield water is governed primarily by the permeability, extent, and thickness of the formation. The presentation of the fundamental principles in this bulletin serves as a reliable guide for choosing the various elements of the well structure to get the desired end results – maximum yield and efficient operation. WRSIC #W71-08923

Anonymous, 1963 . . . . (con't)

36. ________, 1963, "Basic Principles of Water Well Design," *Johnson Driller's Journ.*, Vol. 35, No. 4, (Part I) pp. 1-3, 12-13; No. 5, (Part II) pp. 1-3; No. 6 (Part III) pp. 4-5, 8; 1964, Vol. 36, No. 1, (Part IV) pp. 6-7, 10.

Describes the general aspects of water well design for municipal, industrial, and irrigation wells since highest efficiency is required, e.g., high yield and high specific capacity. Well casing selection, total well depth, and screen length is explored.

37. ________, 1963, "Review of Survey Made on Oil and Gas Well Corrosion Costs," *Mats. Protection,* Vol. 2, No. 12, December, pp. 85-89.

Interim Report of NACE Technical Unit Committee T-1H reviews data on economics of protecting oil and gas wells from corrosion in ten regions of United States; information obtained from questionnaire indicates uses of inhibitors, cathodic protection and coatings, and cost of corrosion protection per month per well; tabular data give ranges of reported costs for inhibiting oil and gas wells in various regions; cost comparisons are made between wells with and without corrosion protection programs.

38. ________, 1964, "Don't Pollute A Clean Reservoir," *Oil and Gas Journ.*, Reprint, November 16, 2 pp.

Producing formations in wells drilled for the production of oil, gas, or water are all subject to bacterial contamination because of contact with drilling fluid, cement filtrate, and hydraulic fracturing fluid which can be traced to one or two sources – sulfate-reducing bacteria or iron bacteria. Both of these are capable of reducing permeability of a producing formation through deposition of their by-products or their presence in extremely large numbers. After testing many chemicals, it was found that a cationic quaternary, alkylbenzyl trimethyl ammonium chloride was most effective in solving the problem. It is a bactericide. It behaves as a corrosion inhibitor because of its film-forming properties. Its detergent properties aid in preventing clay swelling, and it is a nonemulsifier. WRSIC #W71-06943

39. ________, 1964, "Engineering a Hole in the Ground," Hughes Tool Company, Reprint, *Bull. EH-656,* Houston.

Characteristic problems related to drilling relatively small diameter holes. Covers the general characteristics of shot-hole drilling problems in some depth – figures and few references. The basic objective in drilling small diameter holes is fundamentally economic – to produce the maximum hole at minimum cost for equipment and labor, to provide a reasonable footage price to the customer, and reasonable footage profit to the contractor. WRSIC #W71-07203

40. ________, 1964, "Industry Asked to Report Bit Wear in Eighths," *Drilling,* Vol. 25, No. 4, February, pp. 35-39.

Through accurate and uniform evaluation of dull bits, the petroleum industry is seeking to develop bit selection guides and drilling practices which will make for a faster, more profitable hole – by rock-bit manufacturers, drilling contractors and producing company engineers at AAODC meetings.

41. ———, 1965, *Lessons in Rotary Drilling, Unit I, Lesson 3: The Drill String,* Petroleum Extension Service, University of Texas, Austin, 58 pp.

The drill string consists of the drill pipe, the tool joints, and the drill collars together with certain accessories which are commonly used while drilling, although not always required. These accessories include stabilizers, reamers, and subs (short for "substitutes"). Although no string of drill pipe is complete without tool joints, the two items are treated separately in order to more clearly describe each and in order to simplify discussion of maintenance of each. In the oil world today, an overwhelming majority of all drill pipe is made to standards approved by the American Petroleum Institute. The important thing is that when the purchase is made on the basis of API standards, the buyer is sure of what he is buying. WRSIC #W71-09925

42. ———, 1965, *Water Well Symposium for Professional Engineers,* Sponsored by Mississippi Soc. Prof. Eng. and Mississippi Board of Water Commissioners, 102 pp.

Proceedings of symposium. Papers included cover (1) professional responsibilities, (2) Mississippi water resources, (3) water information availability, (4) considerations of water quality, (5) test holes in Louisiana, (6) well logging, (7) pumping tests, (8) well design in the Gulf Coastal Plain, (9) sanitary requirements for well construction, and (10) water well specification development.

43. ———, 1965, "Well Drilling Operations," *Dept. Army Tech. Manual TM 5-297,* U.S. Gov't. Printing Office, September, 249 pp.

Provides instruction and guidance for military personnel engaged in or responsible for establishing water sources by the use of wells; providing geological data through the medium of exploratory drilling, core drilling and sampling; and providing demolition drilling support. Military applications of construction techniques, including dug, bored, driven, jetted, and drilled wells. Use of equipment for developing ground water sources and exploratory drilling, core drilling, sampling, and demolition drilling support. Methods of well construction, subsurface soil exploration, and drilling and tunneling for demolition emplacement.
WRSIC #W71-05293

44. ———, 1966, *A.W.W.A. Standard for Deep Wells,* A.W.W.A., A100-66, 2 Park Ave., New York and N.W.W.A., 88 East Broad St., Columbus, Ohio, 57 pp.

This standard and accompanying appendix are intended as a guide in the preparation of contract documents governing well construction. Because of the broadness of well drilling requirements and the fact that this standard covers construction rather than materials, it cannot be used verbatim and must be supplemented by the user to fit the particular needs and conditions.

Anonymous, 1966 . . . . (con't)

45. ———, 1966, *Cable Tool Drilling Manual,* Sanderson Cyclone Drill Co., Orrville, Ohio, 50 pp.

A commercial booklet on cable tool drilling, e.g., for set-up through drilling to bit sharpening and trouble shooting.

46. ———, 1966, *Ground Water and Wells,* Johnson, U.O.P., 440 pp.

Technical aspects of ground water occurrence, ground water movement, well hydraulics, well design, and ground water geology together with the practical aspects of well drilling, well screen selection, well maintenance, and well operation. An excellent source of information for the driller. WRSIC #W71-06920

47. ———, 1966, *Lessons in Rotary Drilling, Unit I, Lesson 2: The Bit,* Petroleum Extension Service, University of Texas, Austin, 50 pp.

The bit and the way it performs is what drilling is all about. Main consideration in this series is limited to those bits which are turned by the rotary table. There are many variables in the performance of bits with respect to any given formation. These all resolve into questions of economics, that is, into the selection of the one that seems most likely to make the greatest contribution to the progress of drilling.
WRSIC #W71-09924

48. ———, 1966, "Procedures Used in Gravel Packing Wells," *Johnson Driller's Journ.,* Vol. 38, No. 2, pp. 1-4, 6 figs.

Gravel packing is defined as a process of placing a selected size grading of gravel (or sand) to fill the annular space between a centrally positioned well screen and the borehole drilled into an aquifer. Conditions are described which must be met if gravel packing is to be economically practical and successful, e.g., gravel size, size-grading, gravel emplacement, etc.

49. ———, 1966, *Rotary Drilling Bits,* Hughes Tool Co., Houston, 37 pp.

The introduction of the rolling-conical cutter bit greatly increased the economic feasibility of drilling hard formations and led to the general acceptance of the rotary method of drilling oil and water wells. To obtain maximum footage with lower drilling costs, the type of bit should be designed for the specific type of formation being drilled. Various types of bits, several different broad classifications of formations, and the types of rock bits best suited to drill them are examined. Other topics such as importance of bit records, importance of using proper drilling muds with bits, and selection and operation of rolling cutter bits are discussed. WRSIC #W71-05298

50. ———, 1966, *Water System and Treatment Handbook,* Water Systems Council, 4th Edition, 112 pp., 51 refs.

This handbook describes the various features of the private water supply system, e.g., water sources, well construction, system fundamentals, selection and sizing, jet pumps, submersible pumps, positive displaced pumps, pressure tanks and controls, electric supply, servicing, treatment of domestic water, etc.

51. ______, 1966, *What's New With Mud?* Baroid Div., National Lead Corp., Special Report, 35 pp.

Updates previous work on drilling mud developments and techniques, including improvements in drilling rates using new mud techniques, hole stability techniques, solids control, downhole mud temperature stability, and corrosion. Illustrated with new experimental studies.

WRSIC #W71-05301

52. ______, 1967, "Accuracy and Speed of Sampling Improved – Drive-Core Drilling Gets Results," *Johnson Driller's Journ.*, Vol. 39, No. 5, September-October, pp. 1-3, 13.

A type of sampling procedure particularly suited to cable-tool equipment is that known as drive-core sampling. The procedure is not new, certainly, but as yet it is not in common use. Simply, the procedure involves taking samples of unconsolidated material by driving a hollow tube into the material and withdrawing it while retaining the material inside the tube. The procedure especially lends itself to cable-tool drilling because the rigging necessary is almost always a part of the normal inventory of equipment. The tool string necessary for drive-core sampling involves a rope socket, a short heavy stem, a set of long stroke fishing jars, a pin substitute with a straight pin turned on the lower end, a core barrel adapter box with a thread to fit the lower pin on the sub, and a sampling tube welded to the adapter. A typical string of tools is illustrated. The stroke of the jars must be longer than the stroke of the machine. This is necessary to avoid jarring upward with each stroke of the machine.

53. ______, 1967, "Aquifer Evaluation with Radioisotope Well Logs," *Amer. Water Resources Assn. Proc.*, Ser. No. 4, pp. 319-328.

Descriptions of equipment and specific applications of logging techniques in this report refer to a U.S. Geological Survey, Water Resources Division, research project on borehole geophysics as applied to geohydrology. Radioisotope logging is emphasized because it is the only presently available technique for obtaining information from cased holes, and to date it has been little used in ground water hydrology. The radiation logs utilized in these studies include natural gamma, gamma-gamma, neutron-epithermal neutron, and radioactive tracer. Information on well construction is necessary for the interpretation of logs.

54. ______, 1967, "Bacteria Cause Many Problems," *Johnson Driller's Journ.*, Vol. 39, No. 4, July-August, pp. 1-3, 13.

Methods are discussed for detecting the presence of bacteria in water samples and for controlling bacterial contamination of water wells during drilling. A common procedure for detecting water pollution is to test for the presence of coliform bacteria, not generally considered a harmful organism, however, indicating the presence of other types of organisms which can be harmful. When drilling water supply wells, it is essential to use the drilling fluid and only chlorinated water (if not available, the drilling water should be treated with either calcium or sodium hypochlorite). All gravel packs should be chlorinated before

Anonymous, 1967 . . . . (con't)
installation in a well. Inexpensive test kits are available for finding the presence of chloride or nitrate in a water sample – usually an indication of pollution.

55. ———, 1967, "State of the Art Report: Fishing Tools and Services," *Petrol. Equip. Serv.*, Vol. 30, No. 5, September-October, pp. 30, 32, 34.

This report of fishing tools considers new tools, new materials of construction, and the increasing demand for competent fishing tool operators. One of the most important changes in fishing tools is the changeover to tungsten carbide-tipped mills, overshots, washover shoes, and similar tools. Because of the recent trend in the oil industry toward smaller diameter well bores and completions, slim-hole tools have been developed for nearly every purpose. Newly developed fishing tools are tabulated according to company names. Tool handling problems have been alleviated with the development of such equipment as hydraulic vises and portable tool-servicing units which can be taken to the field, to offshore platforms, or pipe yards.

56. ———, 1967, "Taking Representative Formation Samples," *Johnson, U.O.P. Bull. 638-S,* December.

Much of the success of a completed water well depends upon the degree of care exercised in obtaining formation samples for analysis – a responsibility which lies directly with the driller. The right well screen for a given water-bearing formation can be recommended and provided by the screen manufacturer only when the samples are representative of the various strata in the formation. Correct labeling of samples, full and accurate information describing water levels, depths, methods of construction to be used, and the amount of water wanted depend on care and good judgment plus observance of a few simple rules.
WRSIC #W71-06945

57. ———, 1967, "Testing Water Wells for Drawdown and Yield," *Johnson, U.O.P. Bull. 1243,* (Revised April, 1967), 8 pp.

It is the purpose of this bulletin to describe the most accurate and inexpensive methods of determining the actual yielding capacity of a finished well. If a well is to be equipped with a permanent pump, it should be accurately tested for drawdown and yield before the pump is purchased to operate at highest efficiency. If a well is important enough to be tested for capacity on completion, it is essential that it be tested accurately by the use of approved measuring devices and standard methods.
WRSIC #W71-06949

58. ———, 1968, *Blasters Handbook,* E.I. DuPont DeNemours, Explosives Dept., Wilmington, Delaware, 479 pp.

Describes procedures, fundamentals, principles, layout, ignition, safety precautions, types, effectiveness, and application of various blasting techniques.

59. ———, 1968, *Lessons in Rotary Drilling, Unit II, Lesson I: Making Hole,* Petroleum Extension Service, University of Texas, Austin, 55 pp.

The business of making hole is often a long series of compromises. To make the right decision, the driller and pusher have to be aware of the factors involved in drilling. The most significant of these are: (1) the geological section, (2) the capability of the rig, (3) the drilling fluid program, (4) the bit, (5) three critical factors: weight on the bit, rotary operation, the hydraulic factor. WRSIC #W71-09927

60. ———, 1968, *Lessons in Rotary Drilling, Unit II, Lesson 2: Drilling Mud,* Petroleum Extension Service, University of Texas, Austin, 98 pp.

The drilling fluid is now recognized as one of the major factors involved in the success of the drilling operation. Speed, efficiency, safety, and cost of drilling depend upon the performance of the drilling fluid used. The term "drilling fluid" properly includes gases as well as liquids and suspensions of liquids and solids in liquids. This lesson is limited to a discussion of suspensions of solids in a liquid, "muds," and of solids and a liquid in a second liquid, "emulsion muds." The cost of maintaining effective drilling mud is now necessarily a major item of the overall expense of drilling in many areas. The purpose of this lesson is to discuss the need for, the methods employed in, and the most economical maintenance of effective mud control. WRSIC #W71-09928

61. ———, 1968, *Lessons in Rotary Drilling, Unit II, Lesson 4: Casing and Cementing,* Petroleum Extension Service, University of Texas, Austin, 81 pp.

Drilling for oil or gas involves two main objectives: (1) to bore a hole to the mineral accumulation and (2) to install a pipe from the reservoir to the surface. The pipe is called "casing" and has six important functions: (1) to prevent caving of the hole, (2) to prevent contamination of fresh water in upper sands by fluids from lower zones, (3) to exclude water from the producing formation, (4) to confine production to the well bore, (5) to provide a means of controlling well pressure, (6) to permit installation of artificial lift equipment for producing the well. The cost of casing is often the greatest single item of expense of a well. Selection of casing sizes, weights, grades, and types of threaded connections for a given situation constitutes an engineering and economic problem of considerable importance. WRSIC #W71-09929

62. ——— 1968, *Lessons in Rotary Drilling, Unit II, Lesson 5: Testing and Completing,* Petroleum Extension Service, University of Texas, Austin, 64 pp.

From the very earliest days, it was recognized that testing the porous zones to obtain samples of the fluid contents in a well was an essential part of the drilling job. The cable tool method of drilling made possible a continuous test prior to completing a well. The introduction of coring tools, drill stem testing, and electric logging changed the picture completely for rotary drilling by making drill stem testing and electric log analyses a practical field tool. The development of these tools has progressed to the point that formation evaluation can be accomplished without setting casing. WRSIC #W71-09930

Anonymous, 1968 . . . . (con't)

63. ———, 1968, *Lessons in Rotary Drilling, Unit III, Lesson 2: Fishing,* Petroleum Extension Service, University of Texas, Austin, 31 pp.

The term, "fishing," stems from the early cable-tool practice of dangling a homemade hook or spear in the hole in an attempt to snag a broken drilling line and, thereby, retrieve the tool attached. On modern drilling rigs, the fish may be anything from a part (or all) of the drill string, to smaller pieces of equipment such as bit cones, pieces of tools, or any material accidentally dropped into the well bore. Basic techniques and tools for fishing in open hole are discussed in this paper.
WRSIC #W71-09931

64. ———, 1968, *Manual of Water Well Construction Practices,* Oregon Well Contractors Assn., Portland, 85 pp.

The first purpose of this manual is to introduce young drillers to the natural resource with which they will be dealing, to the earth's crust in which water is found, to the machinery and tools drillers employ, to the practicable and permissible techniques of drilling, and to the terminology of the trade. The second purpose is to suggest sources where answers may be found to the questions likely to be encountered in Oregon's State Examination for a Drilling Machine Operator's license and a Drilling Contractor's license. WRSIC #W71-06957

65. ———, 1968, *Operation and Maintenance Manual for Mission Hammerdrill,* Mission Mfg. Co., Houston, 38 pp.

The Mission Hammerdrill is a pneumatically operated, bottom-hole drill that efficiently combines the percussive action of cable-tool drilling with the rotary action of rotary drilling. The Hammerdrill is to be used on any standard rotary rig with the necessary air-compressor capacity. Used for fast and economical drilling of medium-hard formations in quarry, construction, water well, oil well, and geophysical work. Fast penetration results because Hammerdrill is a bottom-hole tool, the piston blows are transmitted directly to the bit without losing percussive energy through the drill string. Thus, the air is fully utilized for the rapid impacting action and for blasting and lifting cuttings from the hole. Continuous hole cleaning bares new formation to the bit so that practically no energy is wasted in chewing up cuttings. Straight hole is assured by short rapid blows that minimize the effect of dipping and broken formation. WRSIC #W71-05289

66. ———, 1968, "Place Gravel Pack Properly for Best Results," *Johnson Driller's Journ.,* Vol. 40, No. 6, p. 1-4, 13, 6 figs.

Two keys to successful gravel packing operations are: (1) proper selection of the pack material, and (2) proper placement of the material around the well screen. Avoidance of bridging of the gravel in the annular space is an important feature as well as is proper development after installation.

67. ———, 1968, "Principles and Practical Methods of Developing Water Wells," *Johnson, U.O.P. Bull. 1033,* Revised, 24 pp.

Water well development and the importance of well screen selection have occupied the attention of many water well drillers for the past several decades. Well screen openings of the right size permit development to the proper extent. The development of wells by "overpumping" and backwashing" and with surge plungers, compressed air, and high velocity jetting is also discussed in detail.

WRSIC #W71-06944

68. ———, 1968, *Private Water Systems,* Midwest Plan Service, Iowa State University, Ames, Iowa, 60 pp., 76 figs., 46 tabs.

This handbook will be of value to many people with an interest in private water systems. It will be an aid to the owner of such a system faced with the problems of operation and maintenance. It will be valuable to well contractors, pump installers, plumbers, and farmers. Even those individuals on community or municipal systems, but who have problems with quality of municipal systems, will find this handbook useful.

69. ———, 1968, "Procedures for Installing Well Screens," *Johnson Driller's Journ.,* Vol. 40, No. 2, pp. 1-5, 5 figs.

Procedures for installing well screens vary with the design of the well and with the method in drilling the well. Problems encountered in the drilling operation may also dictate a particular method of installation differing from that which may have been originally planned. Various standard methods of screen installation are explored.

70. ———, 1968, "Survey Reveals Averages in Drilling Operations," *Ground Water Age,* Vol. 2, No. 5, January, pp. 8-9.

Based on the results of a survey recently conducted by "Ground Water Age," the average depth of water wells drilled in the United States was established. The survey also revealed that the average respondent to the questionnaire drilled 75 water wells a year.

71. ———, 1969, "A.P.I. Specifications for Oil-Well Drilling Fluid Materials," *A.P.I. Standard 13A,* 5th Edit., February, 12 pp.

This specification, under the jurisdiction of the API Committee on Standardization of Drilling Fluid Materials, provides Standards for materials used in oil-well drilling fluids and covers physical properties and test procedures for materials used in oil-well drilling fluids.

WRSIC #W71-09715

72. ———, 1969, "Finding Ways to Save Pumping Costs," *Johnson Driller's Journ.,* Vol. 41, No. 1, p. 1-4, 3 figs., 2 tabs., 3 refs.

Equipment manufacturers, power suppliers, agricultural universities, agricultural extension services – state and federal – have conducted numerous field tests and studies on irrigation pumping plants. These investigations serve in a practical way to point out where money can be saved and overall efficiency increased.

Anonymous, 1969 . . . . (con't)

73. ———, 1969, "How Many Individual Wells Are Contaminated?" *Ground Water Resources Institute Quarterly,* Vol. 2, No. 1, pp. 11-12.

A Ground Water Resources Institute survey was made covering 1,500 county sanitarians throughout the country. 345 survey questionnaires were returned, giving a response of 23 percent. G.W.R.I. survey shows that the extent of well contamination has been grossly overstated, although a problem, and that mandatory well sample testing is needed to isolate the problem wells and protect public health, and that sound well-pump legislation is needed to control well construction methods and misuse of private disposal systems. WRSIC #W71-13967

74. ———, 1969, *Jet Bit Hydraulics,* Petroleum Extension Service, University of Texas, Austin, 141 pp.

The major requirement of a hydraulic system for a drilling rig is to provide adequate hydraulic power at the rock face, where the mechanical forces are at work fracturing and removing the rock. The second requirement of the system is to provide a means of returning cuttings to the surface for identification in such a way as to maintain clean, stable hole conditions for drilling and for the passage of bit and drill stem during round-trips. It is in the interest of those wishing to reduce footage drilling cost to maintain the hydraulic system on their rigs at a high overall hydraulic efficiency to offset the cost of supplying the energy.

WRSIC #W71-09923

75. ———, 1969, "Johnson Screens," *Johnson, U.O.P. Cat. 169,* St. Paul.

This catalog briefly summarizes the use of screens for water well installations, discusses well screen accessories, and application engineering. A well screen serves as the water intake portion of the well and must allow sand-free water to flow into the well in ample quantity and with minimum loss of head to supply the pump at full capacity. The screen affects the following factors: (1) capacity, (2) drawdown and pumping lift, (3) production of sand-free water, and (4) maintenance cost and life of well. WRSIC #W71-07207

76. ———, 1969, *Principles of Drilling Fluid Control,* Edited by A.P.I. and University of Texas in cooperation with the Amer. Assoc. of Oil Well Drilling Contractors, Dallas, 215 pp.

The fluid used in rotary drilling, once regarded only as a means of bringing rock cuttings to the surface, is now recognized as one of the major factors involved in the success or failure of the drilling operation. This manual discusses the well conditions which occasion the need for control of drilling fluid properties and the most economical means of gaining and maintaining this control. The term, "drilling fluid," includes air, gas, water, and mud. The term "mud" refers to a suspension of solids in water or oil, or of solids and droplets of one of these liquids dispersed in the other. This training manual discusses each of these drilling fluids including the modification of muds for use as packer fluids but deals primarily with those fluids used most often in the field,

namely, suspensions of solids in a liquid, "muds," and suspension of solids and droplets of a liquid in a second liquid, "emulsion muds."
WRSIC #W71-09921

77. ———, 1969, "Research is Guide for Estimating Well Costs," *Johnson Driller's Journ.*, Vol. 41, No. 5, September-October, pp. 6-8.

Illustrated graphs and other data of use in estimating costs of different types of wells. Data from Illinois were divided into three categories: sand and gravel wells; shallow bedrock wells; deep sandstone wells; including costs for: setting up and liners; grouting and sealing the annular space between different sizes of casing and between the casing and the borehole; well screens and fittings; gravel pack and installation; developing the well and conducting an eight-hour pumping test.

78. ———, 1969, "Well Point Systems," *Johnson, U.O.P. Bull. 467D,* Revised, 9 pp.

Well-point systems, or suction-lift wells, can supply large quantities of water economically where conditions are favorable. Well-point systems find extensive practical use in four fields: (1) to obtain water for irrigation; (2) to obtain water for municipal or industrial purposes; (3) to temporarily dewater construction sites in wet ground, and (4) to permanently lower the water table over an area for special reasons. The first two of these applications put the water to beneficial use. The third and fourth are drainage measures where the ground water is carried away and wasted after being pumped from the wells. The operating principles, the installation details or well-point systems to supply water for irrigation or other uses, and the fundamentals of dewatering are discussed.
WRSIC #W71-06948

79. ———, 1969, "Well Technology Serves the Mining Industry," *Johnson Driller's Journ.*, Vol. 41, No. 2, March-April, pp. 1-4.

Ground water technology serves the mining industry as (1) dewatering for open-pit mining, (2) solution mining, (3) wells for extracting mineralized water as a raw material, and (4) disposal wells for difficult-to-handle waste fluids. Lowering the water table over an area for an open pit operation involves creating a composite cone of depression by pumping from a series of properly spaced wells. The article also covers solution mining of underground uranium ore in formations; wells for bringing salt and sulfur to the surface from deep deposits; mineralized ground water pumped from which magnesium metal is extracted; salt water disposal underground through unused oil wells.

80. ———, 1970, *A Digest of the Proceedings of the University of Oklahoma Corrosion Control Course,* Norman, in cooperation with the National Assoc. of Corrosion Engineers, 35 chapters.

Describes basic metallurgical considerations in corrosion control, application of cathodic protection, corrosion detection and control monitoring, protective coatings, corrosion inhibitors, oil field chemistry, scale treatments, chemical selection, etc.

Anonymous, 1970 . . . . (con't)

81. ———, 1970, "A.P.I. Bulletin on Performance Properties of Casing, Tubing and Drill Pipe," *A.P.I. Standard 5C2,* 17th Edit., April, 65 pp.

The purpose of the bulletin is to provide minimum performance properties on which the design of casing, tubing, and drill pipe strings may be based. The performance properties cover the grades, sizes, and weights of casing, tubing, and drill pipe as given in API Std. 5A, 5AC, and 5AX as well as values for various performance properties of casing, tubing, and drill pipe conforming to API Std. 5A, 5AC, and 5AX (not including factors of safety). WRSIC #W71-09716

82. ———, 1970, "A.P.I. Specifications for Grade C-75 and C-95 Casing and Tubing," *A.P.I. Standard 5AC,* 7th Edit., April, 52 pp.

The purpose of this specification is to provide standards for casing and tubing with restricted yield-strength ranges for use in drilling and producing operations. This specification covers Grade C-75 casing and tubing and C-95 casing in the sizes and wall thicknesses shown in the standard lists and the dimensional tables. It also includes requirements for couplings and thread protectors. Dimensional requirements on threads and thread gages, stipulations on gaging practice, gage specifications, and certification as well as instruments and methods for inspection of threads are given in API Std. 5B and are applicable to products covered by this specification. WRSIC #W71-09712

83. ———, 1970, "A.P.I. Specifications for High Strength Casing and Tubing," *A.P.I. Standard 5AX,* 8th Edit., April 45 pp.

This specification provides standards for high-strength casing and tubing suitable for use in drilling and producing operations. High-strength seamless steel casing and tubing; casing and non-upset tubing in the sizes and wall thicknesses shown in the standard list and the dimensional tables; requirements for couplings and thread protectors are included. Dimensional requirements on threads and thread gages; stipulations on gaging practice; gage specifications and certification; as well as instruments and methods for inspection of threads are given in API Std. 5B and are applicable to products covered by this specification.
WRSIC #W71-09713

84. ———, 1970, *Bit Tips,* Hughes Tool Company, Houston, 33 pp.

In designing rock bits, space is the controlling factor because the sizes of the parts are determined by the size of the hole to be drilled. Proper selections of the right proportions for the various parts, such as tooth size, cone shell thickness, and bearing sizes are required. It is known that rock bits perform best when rotating about their own axis. Bearing life, because of reduced thrust loading, is extended when the bit rotates about its own center – especially under maximum loading. The advent of jet bit was a major breakthrough in obtaining greater penetration rates. Properly directed jet streams combined with correct nozzle sizes clean the cones of the bit and provide efficient removal of cuttings from bottom. Other topics: bit stabilization, well logs and bit selection, and optimum bit weight and rotary speed. WRSIC #W71-05302

85. _______, 1970, *Diamond Drilling Handbook,* Christensen Diamond Products Co., Salt Lake City, Utah.

Authoritative handbook on diamond-drilling by the major manufacturer of diamond drilling equipment. Covers most aspects of diamond drilling, e.g., bit selection, operational requirements, case histories of various applications, etc. Various drilling rig models are described.

86. _______, 1970, "Ground Water Pollution," A special issue, *Water Well Journ.,* Vol. 24, No. 7, pp. 31-61.

Is a United States contribution to the International Hydrological Decade. Describes ground water pollution problems, use of aquifers, desalination, subsurface waste storage, classification of pollutants, sources, purifications, role of federal legislation and management.

87. _______, 1970, "Guidelines for Optimum Diamond Bit Performance," *World Oil,* Vol. 171, No. 4 (Part I) September, pp. 63-66; No. 5 (Part II) October, pp. 94-96; No. 6 (Part III) November, pp. 113-116; No. 7 (Part IV) December, pp. 75-77.

Diamond bits can provide economic and operating advantages when capabilities and limitations of these tools are understood. Describes the determination of economics of diamond drilling and selection of proper bit for specific drilling situation.

88. _______, 1970, *Jet Bit Mechanics,* Petroleum Extension Service, University of Texas, Austin, 141 pp.

Probably the most important single realization in rotary drilling is that rock must not only be fractured on bottom but also must be removed from the rock face instantly and efficiently to provide for further fracturing and drilling progress. For this purpose, two types of energy sources are brought from the surface to the rock face and applied as efficiently as possible: (a) mechanical energy applied through rotary and drill string and converted into its most useful form through the mechanism embodied in the bit cones and teeth and (b) hydraulic energy applied through pump and drill string conduit and converted into its most useful form through correctly positioned jet nozzles aimed at the rock face just ahead of the on-coming tooth. Both types of energy play a part in both rock fracture and rock removal. Although no progress can be achieved without rock fracture, these studies are largely concerned with the more complex mechanism of rock cutting removal. The hydraulic analysis here is concerned principally with the manner in which the hydraulic energy assists and complements mechanical energy at the rock face. WRSIC #W71-09922

89. _______, 1970, "Pipe and Casing in Water Well Drilling," *Johnson Driller's Journ.,* Vol. 42, No. 1, January-February, pp. 1-6, 6 refs.

Five products for the water well industry (e.g., standard pipe, line pipe, reamed and drifted (R & D) pipes, drive pipe, and water well casing) are distinguished from oil-field products. Because of confusion between oil-field products and water well items, the similarities and differences

Anonymous, 1970 . . . . (con't)
in these products which are used as well casings are pointed out. It is best practice to avoid using oil-field products in water wells. Two means of adding life to water well casing are to use a relatively thick wall and to galvanize by adding zinc coating over the entire surface. To prevent contamination from ground water, surround the upper portion of the casing with a neat cement grout of adequate thickness. An important reason for using a relatively thick casing wall is that the rate of corrosion tends to become slower as the penetration of corrosion below the surface increases.

90. ______, 1970, "Second Corrosion Study of Pipe Exposed to Domestic Waters," *Mats. Protection,* Vol. 9, No. 6, June, pp. 34-37, 3 refs.

Committee examination and analyses of the specimens studied resulted in the following conclusions: corrosion of galvanized steel and wrought iron were equal in resistance to cold water; copper and chloride were predominant factors causing increased corrosion of galvanized metals. Relative difference in degree of corrosion was a problem. Cases are reported where completely softened water (Zeolite treated) increased corrosion of all metals as compared to unsoftened water. Other results included relate to other factors causing corrosion, e.g., oxygen, copper and chloride, relative velocities, and silicate treatment of high-chloride-sulfate water resulted in inhibition of corrosion of galvanized metals.

91. ______, 1970, "Seal-Sealing Packer Saves Time on Job," *Johnson Driller's Journ.* Vol. 42, No. 6, November-December., pp. 1-3.

Describes a self-sealing packer consisting of a flexible, Neoprene sealing ring, securely mounted on a heavy metal filling attached to the top of the well screen on extension pipe above the screen. No swedging or lead packer expanding operation is nécessary. The new design is not intended for screens which require circulation outside of the screen, for casing in poor condition, or in an open borehole.

92. ______, 1971, "A.P.I. Specifications for Casing, Tubing, and Drill Pipe," *A.P.I. Standard 5A,* 31st Edit., April, 70 pp.

The purpose of this specification is to provide standards for casings, liners, tubing, work tubing, and drill pipe suitable for use in drilling and producing operations. This specification covers casing, tubing, work tubing, drill pipe, and casing liners in the sizes and wall thicknesses applicable to the various grades of pipe as shown in the standard lists and in the dimensional tables. It also includes casing and tubing couplings and thread protectors. Dimensional requirements on threads and thread gages, stipulations on gaging practice, gage specifications, and certification as well as instruments and methods for inspection of threads are given in API Std. 5B and are applicable to products covered by this specification. WRSIC #W71-09711

93. _______, 1971, "A.P.I. Specifications for High-Test Line Pipe," *A.P.I. Standard 5LX,* 18th Edit., April, 60 pp.

The purpose of this specification is to provide standards for more rigorously tested line pipe having greater tensile and bursting strengths than for pipe manufactured under API Std. 5L. This specification covers high-test line pipe in grades X42, X46, X52, X60, X65, and grades intermediate thereto. The chemical composition and certain physical properties of intermediate grades are subject to agreement between the purchaser and manufacturer. The agreed upon requirements must be consistent with the corresponding requirements for grades X42, X46, X52, X56, X60, and X65. Grade X60 or higher pipe shall not be substituted for pipe ordered for grades X52 and lower without purchaser approval. American Petroleum Institute (API) specifications are published as an aid to procurement of standardized equipment and materials. WRSIC #W71-09719

94. _______, 1971, "A.P.I. Specifications for Line Pipe," *A.P.I. Standard 5L,* 26th Edit., April, 59 pp.

The purpose of this specification is to provide standards for pipe suitable for use in conveying gas, water, and oil in both the oil and natural gas industries. This specification covers seamless and welded steel line pipe. It includes standard-weight and extra strong threaded pipe; and standard weight plain-end, regular weight plain-end, special plain-end, extra-strong plain-end, and double-extra-strong plain-end pipe. The sizes and wall thicknesses applicable to the various processes of manufacture under this specification are indicated. Dimensional requirements on threads and thread gages, stipulations on gaging practice, gage specifications and certification as well as instruments and methods for inspection of threads are given in API Std. 5B and are applicable to products covered by this specification. Grade X60 or higher pipe (API Std. 5LX) shall not be substituted for pipe ordered to this specification without purchaser approval. American Petroleum Institute (API) specifications are published as an aid to procurement of standardized equipment and materials. WRSIC #W71-09718

95. _______, 1971, "Drilling Fluids File," *Petrol. Equip. Serv.,* March-April, pp. 19-45.

The mud-system classifications and product functions employed in the Fluids File are defined. These definitions reflect general industry practice and terminology and to a great extent incorporate descriptions adopted by the American Petroleum Institute and the American Association of Oilwell Drilling Contractors, although no endorsement nor approval is claimed or intended.

96. _______, 1971, "Geophysics and Ground Water, An Introduction – Part I; Applied Use of Geophysics – Part II," *Water Well Journ.,* Vol. 25, No. 7; Vol. 25, No. 8, 35 pp.

This is a two-part series dealing with surface geophysical techniques. The first section deals with an introduction to the basic principles and features of all geophsics. The second part considers the applied aspects of ground water geology.

Anonymous, 1971 . . . . (con't)

97. ———, 1971, "Geothermal Resources Gather a Head of Steam," *Eng., New-Rec.*, Vol. 186, No. 18, May, pp. 30-31, 35-36.

The answer to the problem of more power without pollution may lie in geothermal energy. Geothermal is the only potential new energy resource that is both economically and technically feasible now. A geothermal resource is essentially the heat generated from the interior of the earth and recoverable in some medium, such as hot ground water or existing as steam. Any minerals obtained in the process are considered geothermal resources. Estimates of the energy potential in geothermal resources range from a conservative 60-million-kw capacity for 50 years in the U.S. to an unlimited world potential. There are only four known dry steam fields in the world: The Geysers in N. California, the Larderello field in Italy, and two Japanese fields. Because The Geysers is a dry steam field, it solely produces power. After ten years of power production, The Geysers drives four Pacific Gas and Electric Co. turbines with a total capacity of 82,000 kw. By the end of this year, Pacific Gas and Electric will add two 55,000-kw units and expects to have almost 600,000 kw of power available by 1975. The success of the present wells and analysis of the reservoir suggests a total steam reserve in excess of one to three million-kw capacity per year. The steam produced at The Geysers is almost ideal with little brine and only 0.5% of noncondensable gaseous materials.

98. ———, 1971, "Standard Procedure for Testing Drilling Fluids," *A.P.I. Report 13B*, 3rd Edit., A.P.I., Dallas, February, pp. 3-22.

The purpose of this recommended practice is to provide standard procedures for the testing of drilling fluids, mud weight, viscosity, gel strength, filtration, sand, pH, resistivity, chemical analysis, bridging materials, etc.

99. ———, 1971, "Virus and Water Quality: Occurrence and Control," *Proceedings 13th Water Quality Conference*, University of Illinois, College of Engineering and Illinois Environmental Protection Agency, 224 pp.

Contains 15 papers on the biologic parameters involved in the transmission of viruses via water, e.g., detection, control, human resistance, etc.

100. ———, 1971, "Water Supply Source for the Farmstead and Rural Home," *Farmer's Bull. 2237*, U.S. Dept. of Agriculture, p. 13.

Produced by the Agricultural Engineering Research Division as a general guide for the development of water-supply sources for the farmstead and rural home. Explores briefly water requirements, wells, well disinfection, etc.

101. ———, 1972, "Rubber Fiber Well Seal," *Water Well Journ.*, Vol. 26, No. 6, p. 48.

A field report on a possible solution to the problem of polluted surface water seepage down the side of the well casing. A rubber fiber made from ground tires reportedly is, when compressed, water proof and "rot"-proof making a seal between the casing and borehole wall.

102. Appl, F. C. and D. S. Rowley, 1968, "Analysis of the Cutting Action of a Single Diamond," *Trans. SPE of AIME,* Vol. 243, pp. II-269.280, 16 refs.

Assuming that rock behavior during cutting with a single diamond may be approximated by that of a rigid Coulomb, plastic material, a theory of single diamond cutting action has been developed. Using this theory, the stresses on the diamond cutting surface and the components of the cutting force have been determined. Theoretical results agree reasonably well with available experimental data.

103. Aron, G., *et al.*, 1967, "Cyclic Pumping for Drainage Purposes," *Ground Water,* Vol. 5, No. 1, January-February, pp. 35-38, 3 refs.

Criteria and formulas are presented for scheduling cyclic pumping operations to maintain a given minimum drawdown as required for drainage purposes. Solutions of the equation for the relative length of the shutoff periods are presented in a family of curves, which greatly simplifies the scheduling procedure. WRSIC #W71-09907

104. Atkins, E. R., Jr., 1961, "Techniques of Electric Log Interpretation," *Journ. Petrol. Tech.,* Vol. 13, No. 2, February, pp. 118-123, 46 refs.

Techniques of interpretation, including determination of formation water resistivity values from log data, water saturation from electric logs and formation porosity from resistivity data; 12 assumptions commonly made in log interpretation are emphasized.

105. Ault, D. C. and R. H. Bethart, 1970, "Tools for Well Development," *Water Well Journ.,* Vol. 24, No. 3, March, pp. 45-47.

Describes tools and techniques involved in increasing well yield in unconsolidated aquifers, e.g., solid surge block, compressed air, high velocity jet washing tool, overpumping acid treatment.

106. Bade, J. F., 1963, "Cement Bond Logging Techniques – How They Compare and Some Variables Affecting Interpretation," *Journ. Petrol Tech.,* Vol. 15, No. 1, January, pp. 17-22.

250 cement bond logs have been run in Cedar Creek anticline, Mont. continuing investigation has produced several significant tool design changes; relative merits and shortcomings of each design are noted; variables affecting amplitude of measured signals include gating arrangement, tool sensitivity, centering, cement density, thickness of cement sheath, casing size, logging fluid and formation velocities; field examples illustrate these factors.

107. Bader, J. S., 1966, "Device for Removing Debris from Wells," *U.S. Geol. Surv. Water Supply Paper 1822,* pp. 43-46.

Grappling device similar to pair of ice tongs has been designed and effectively used to remove debris from wells; rehabilitating existing wells by using this device eliminates cost of drilling new ones.

108. Barbish, A. B. and G. H. F. Gardner, 1969, "The Effect of Heat on Some Mechanical Properties of Igneous Rocks," *Petrol. Eng.,* December, pp. 395-402, 5 refs.

Elastic moduli of most igneous rocks are greatly reduced by heating and depend at room temperature on the highest temperature to which they have been previously heated. Gabbro samples from French Creek, Penn., were measured for velocity, attenuation, bending strength, and point penetration as functions of moisture content, pore pressure, confining pressure, and highest previous temperatures. Conclusion: heating introduces a system of macrocracks between grains distinct from a system of microcracks that exists within grains. Thus, bending strength is reduced by the introduction of macrocracks, but resistance to indentation, which depends on microcracks, is not. It is suggested that from the measurement of velocity and attenuation alone, the macro and microcracks can be detected and distinguished.

109. Barnes, I. and F. E. Clarke, 1969, "Chemical Properties of Ground Water and Their Corrosion and Encrustation Effects on Wells," *U.S. Geol. Surv. Prof. Paper 498-D,* 58 pp., 47 refs.

Certain practical problems can be identified before extensive well construction and unnecessary well failure. Well waters in Egypt, Nigeria, and W. Pakistan were studied for chemical properties and corrosive or encrusting behavior. From the chemical composition of the waters, reaction rates with reference to equilibrium were tested for 29 possible coexisting oxides, carbonates, sulfides, and elements. Only calcite, $CaCO_3$, and ferric hydroxite, $Fe(OH)_3$, showed correlation with the corrosiveness of the waters to mild steel (iron metal). All 39 of the waters tested were out of equilibrium with iron metal, but those waters in equilibrium or supersaturated with both calcite and ferric hydroxide were the least corrosive. Supersaturation with other solid phases apparently was unrelated to corrosion conditions leading to either corrosion (related to lack of supersaturation with protective phases) or encrustation (supersaturation with phases that were found to precipitate) or both and can be identified.

110. Basham, R. B. and C. W. Macune, 1952, "The Delta-Log, A Differential Temperature Surveying Method," *Petrol. Trans., AIME,* Vol. 195, p. TP 317.

Very small anomalies in oil well temperatures are detected and measured by recording the difference in temperature existing between two thermally sensitive elements which are spaced several feet apart on a small diameter carrier and lowered into the borehole. The system is highly sensitive to small changes in thermal gradients caused by gas or fluid movements and to the boundaries between beds of different thermal conductivities. Typical logs show successful applications of the process for locating gas and water leaks in casing, gas entry, and gas-oil contact.

111. Bassani, P., 1971, "Seismic Borehole Plug," *U.S. Patent 3,* 613, 784, October.

Boreholes often strike water and become artesian wells. The presence of a large flow of water in a borehole causes problems during the sealing operation. To facilitate sealing of such boreholes, a cement plug is set near the top of the borehole. A fluid-pressurized borehole plug can be adapted for use in sealing off seismic boreholes to prevent the flow of subsurface water from such holes.

112. Bates, R. E., Jr., 1965, "Economics of Percussion Versus Air Rotary Drilling," *World Oil,* February, pp. 51-59, 11 figs., 3 refs.

Percussion drilling tools and technique are gaining acceptance and moving out of specialty application. The extremely rapid hammer blows delivered with relatively light weight on bit increase bit life, decrease drill collar wear, and help control deviation while maintaining penetration rates. The technique is limited, however, to systems using circulating fluids which are compressible, such as air, natural gas, and foam (mist). Percussion rotary drilling is the rotary technique in which the main source of energy for fracturing rock is obtained from a percussion machine connected directly to the bit. The circulating system, drill string, and hoisting system are unchanged from that used in the conventional air rotary method. Field performance results and cost-analysis nomographs determine when to consider using percussion drill. WRSIC #W71-05687

113. Bennett, T. W., 1970, "On the Design and Construction of Infiltration Galleries," *Ground Water,* Vol. 8, No. 3, May-June.

Reliable prediction of the yield of infiltration galleries can be made only on the basis of careful field testing combined with evaluation of surface flow-duration data. Pump test procedures and methods of analysis are outlined. Design and maintenance suggestions were made from successful operations.

114. Bennison, E. W., 1953, "Fundamentals of Water Well Operation and Maintenance," *A.W.W.A. Journ.,* Vol. 45, No. 3, March, pp. 252-258.

Common causes of well failures, incrustation, and corrosion; construction and hydraulic characteristics of deep rock wells; measures to maintain and protect ground water supply. WRSIC #W71-13968

115. Bentall, R., 1963, "Methods of Collecting and Interpreting Ground-Water Data," *U.S. Geol. Survey Water Supply Paper 1544-H,* pp. 1-97.

This report consists of six papers, each describing a different phase of ground water investigations. The subjects discussed are: test drilling; the installation of shallow observation wells; the phenomenon of reverse water-level fluctuations; underground temperatures as an index to ground water velocity; the development of water supplies by induced streambed percolation; and the development of a water budget for an artesian aquifer.

116. ———, 1963, "Shortcuts and Special Problems in Aquifer Tests," *U.S. Geol. Survey Water Supply Paper 1545-C,* pp. 1-117.

Bentall, R. . . . . (con't)

Special methods of solving fundamental formulas used in analyzing aquifer test data and solutions for particular ground water problems are described in 16 short papers by various authors. Some of the papers present shortcut methods for the solution of the general nonequilibrium formula; some extend the equilibrium straight-line methods for purposes of obtaining more information with less work; some analyze specific boundary problems; and one discusses hydrologic and economic factors in spacing multiple wells.

117. Bentson, H. G., 1964, "New Tools and Procedures for Better Drilling Operations," Technical Manual Reprint, *Oil and Gas Journ.*, Tulsa, 56 pp.

This booklet describes in considerable detail the various aspects of rock-bit design, selection, and evaluation. Variables affecting rotary-drilling penetration rate are examined. Some of the more recognizable are the effects of drilling-fluid properties, hydrostatic pressure, hydraulic factors, rotary speed, and weight on bit. Also involved are personnel-rig efficiency, straight-hole requirements, and the extreme variance of rock properties and other subsurface conditions in different areas. The various features of the turbodrill are discussed with emphasis on the present lack of adequate bit design features and supporting mud pump capacities. WRSIC #W71-05316

118. Berube, S. C. and R. N. Young, 1966, "Bottom-Hole Percussion Tools – Where and How to Use Them," *Rock Products,* October, pp. 79-81.

When other drilling methods are uneconomical due to slow penetration rates, high bit costs, or high rig maintenance costs, bottom-hole percussion tools and bits are used. Requirements to insure successful application of the bottom-hole percussion drilling methods are: (1) adequate volume of compressed air to efficiently clean the hole; (2) proper size drill pipe; (3) sufficient pull down on the drill stem to keep the bit engaged against the rock face to be drilled; (4) sufficient torque to turn the drill stem; (5) an adjustable speed rotary mechanism; (6) means of tool lubrication; (7) adequate hoisting capacity to lift the tool and drill pipe from the deepest hole the driller expects to drill; (8) proper breakout equipment to service the tool; and (9) means of sharpening bits. WRSIC #W71-05309

119. Beyer, M. G., 1966, "Ground Water Production from the Bedrock of Sweden," *Ground Water Problems,* Pergamon Press, Chapter 10, pp. 161-179.

The ever-increasing fresh water quantities needed for the supply of growing communities' recreation areas and industries in Sweden has greatly increased the amount of water drawn from the mainly Precambrian bedrock. Drilling methods by compressed air, now being introduced, are many times more efficient than conventional cable-tool drilling. Increased output and closer spacing of wells, however, demand closer over-all control of bedrock ground water resources.

WRSIC #W71-06923

120. Beilek, I. R., 1956, "Successful Acidizing," *Water Well Journ.*, Vol. 10, No. 8, August, pp. 9, 24-26, 29, 4 refs.

Acidizing is the process of placing a nontoxic acid solution in contact with the water-bearing formation. Properly administered, this process usually results in significant increases in well production. Acid treatment and effects are discussed.

121. Bierschenk, W. H., 1964, "Determining Well Efficiency by Multiple Step-Drawdown Tests," *International Assoc. Sci. Hydrol.*, Vol. 64, pp. 493-507.

Graphical solution of multiple step-drawdown test data permits an approximate determination of the two components of drawdown in a pumped well; that due to formation loss and that due to well loss.

122. Billington, S. A. and K. A. Blenkarn, 1962, "Constant Rotary Speeds and Variable Weight for Reducing Drilling Costs," *A.P.I. Drill. and Prod. Pract.*, pp. 52-63, 1 ref.

Presented in this paper are theoretical charts for use in calculating the optimum constant-speed and variable-weight program to be followed during a bit run to obtain minimum drilling costs. Field tests have been conducted with this calculating system, which may be used on rigs with fixed-speed rotary drives; and other tests with a similar variable-weight and variable-speed system. Results of these tests show that both systems lead to reductions in calculated direct drilling cost in the range 12 to 51 percent, as compared to conventional drilling in the greater Anadarke Basin area.

123. Bingham, M. G., 1965, "A New Approach to Interpreting Rock-Drillibility," Technical Manual Reprint, *Oil and Gas Journ.*, Tulsa, 93 pp.

Rotary drilling is an anachronism. It stands as the last remaining art in an industry where virtually all other operations have been reduced to a science. A single link remains to be forged before drilling also can elevate itself to scientific status. That vital link is a thorough understanding of what happens on the bottom of the hole as the bit rolls over it. This plan of action measures only a few square inches, but it is the location of the only "useful" work performed in drilling. This manual relates rotary drilling to rock strength, or rock drillability. The aim of establishing such a relation is to allow reproduction of the behavior of a drill bit from one well to the next. To accomplish this objective mathematical equations must be derived which describe the interaction of bit and rock. The author firmly believes that equations can be written to describe and predict drilling in the field. More than 26 variables have been identified which influence drilling. Without understanding their effects, maximum drilling efficiency and minimum costs can never be reached. WRSIC #W71-05313

124. Bird, J. M., 1954, "Interpretation of Temperature Logs," *A.P.I. Drill. and Prod. Pract.*, pp. 187-195.

Bird, J. M. . . . . (con't)

The movement of fluids in a borehole is a problem of interest to petroleum engineers. There are several tools available to detect and record the direction and extent of this movement. The use of temperature surveys for this purpose is the subject of this paper. The general shape of temperature logs in water-injection wells is discussed in the light of the principles of heat transfer involved.

125. Black, A. P., *et al.* 1965, "Use of Iodine for Disinfection," *A.W.W.A. Journ.,* Vol. 57, No. 11, November, pp. 1401-1421, 18 figs., 9 tabs., 24 refs.

The physical and chemical properties of iodine make it particularly suitable for use as a water disinfectant. This paper explores the use of iodine in dilute-aqueous solutions, the effect of pH, the formation of tri-iodide ion, the formation of the iodate ion, the feed of elemental iodine to water, a medical assessment and physiologic response to iodine intake. In addition, cost features are also discussed.

126. Blankennagel, R. K., 1968, "Geophysical Logging and Hydraulic Testing," *Ground Water,* Vol. 6, No. 4, July-August.

The U.S. Geological Survey, on behalf of the U.S. Atomic Energy Commission, performs a broad range of geological studies related to underground nuclear explosions at the Nevada Test Site. One phase of the studies involved extensive geophysical logging and hydraulic testing in deep exploratory holes at Pahute Mesa, Nevada Test Site. Various geophysical logs and hydraulic testing techniques used in the study are described in this paper.

127. Bleakley, W. B., *et al.,* 1965, "The Sidewall Epithermal Neutron Porosity Log," *Soc. Petrol Eng., A.I.M.E.* Preprint No. 1180, 20 pp.

Describes the development and application of a special logging tool for determination of porosity and potential reservoir characteristics.

128. Bobo, R. A., 1966, "The Significance of Current Practices in Drilling Hydraulics," *Spring meeting of the Pacific Coast District,* Division of Prod., A.P.I., Los Angeles, May.

Today's practice in drilling hydraulics is to operate under a prescribed set of surface conditions. The purpose is to get the maximum scavenging at the bit for a program that is affixed for the pump. These methods are presented herein in a manner intended to clarify the similarities, differences, and limitations.

129. ———, 1971, "Surface Mud Systems," *Oil and Gas Journ.,* Vol. 69, January, February, and March, pp. 72-74, 78-87, 47-51, 77-84.

The functions performed by any drilling fluid system are essentially the same. The way these tasks are accomplished depends on several factors, including economy, safety, drilling conditions, adaptability of rig and equipment. Too often, needed changes and modifications to accommodate new equipment are not made properly. This paper explores some of the necessary procedures.

130. ______, *et al.*, 1958, *Keys to Successful Competitive Drilling,* Gulf Publishing Company, Houston, 133 pp.

Describes, in part, the solids in unweighted drilling fluid that are derived from two sources: (1) those which are added to the active mud storage to give some specific property to the mud and (2) those which are obtained in the process of drilling. The first classification refers generally to purchased products, and the second pertains to drilled solids. Water is the continuous phase in water-based muds. When water is added in sufficient quantities to reduce the concentration of solids, the viscosity of the mud is reduced.

131. Bond, L. O., *et al.*, 1969, "Well Log Applications in Coal Mining and Rock Mechanics," *Soc. Mining Eng.* Preprint No. 69-F-13, 16 pp., 12 refs., 8 figs.

The paper presents the initial results of a logging program applied to coal prospects evaluation. Particular applications discussed are: (1) identification of coal seams, (2) measurement of coal seam thickness, (3) evaluation of lithology and moisture index of formations surrounding coal seams, (4) determination of a "Strength Index" for formations penetrated by the drill hole, and (5) analysis of coal quality.

132. Bonham, C. F., 1955, "Engineering Analysis of Cable Tool Drilling," *Petrol. Eng.,* December, pp. B-93 to B-100.

Covers the general aspects of cable-tool operations, effectiveness of the method, and limitations.

133. Boulton, W. S., 1954, "The Drawdown of the Water Table under Non-Steady Conditions near a Pumped Well in an Unconfined Formation," *Proceedings of the British Institute of Civil Eng.,* Pt. 3, Vol. 3.

A classic in the field. Description of drawdown of a pumped well in an unconfined formation under non-steady conditions. Mathematic expansion of hydraulics involved.

134. Bourgoyne, T., 1969, "A Critical Examination of Rotary Drilling Hydraulics," *Fourth Conference on Drilling and Rock Mechanics,* SPE No. 2386, Austin, Texas, January, pp. 41-54, 20 refs.

A new computer technique has been developed for providing information on rotary-drilling hydraulics. The computer program determines the proper bit nozzle size and pump operating conditions for all combinations of depth and mud density for a given hole size, making repetitive calculations for each well unnecessary. The computer program presented was used to compare various published optimization procedures over a typical range of field conditions. The comparison includes differences among the procedures in pump cost as well as the differences in the hydraulic horsepower, jet impact, and nozzle velocity of the bit. WRSIC #W71-08916

135. Bowman, I., 1911, "Well Drilling Methods," *U.S. Geol. Surv. Water Supply Paper 257,* p. 139, 25 figs.

The first comprehensive American contribution to well drilling technology. Explores all important aspects of the state-of-the-art of drilling as of 1911, e.g., aquifers, oil and gas bearing structures, early history, drilling methods, well construction, water well contamination, well costs, etc.

136. Brantly, J. E., 1961, "Hydraulic Rotary-Drilling System," *History of Petroleum Engineering,* A.P.I., Boyd Printing Co., Dallas, pp. 273-452.

Chapter 6 traces the development of rotary-drilling equipment and rotary-drilling practices from prehistoric times to 1960. The historical record is divided into seven time intervals. The second interval begins with the first reference to a drilling machine using a rotating tool, hollow drill rods, and circulating fluid to remove cuttings – Robert Beart's patent of 1844. The rotary method spread rapidly with the developing oil industry following the discovery of oil at Spindletop, Texas, in 1901. The development of the principal features of the rotary rig is described and is illustrated by reproductions of drawings and photographs.

137. ———, 1961, "Percussion-Drilling System," *History of Petroleum Engineering,* A.P.I., Boyd Printing Co., Dallas, pp. 135-269.

Historical treatment of percussion-drilling system. Follows development, operation, and application of system.

138. ———, 1961, *Rotary Drilling Handbook,* 6th Edit., Palmer Publications, New York, 825 pp.

This book is one of the few comprehensive classics on rotary drilling produced by the oil industry. This sixth edition has been up-dated to include most of the common techniques in use today. Its value to the ground water industry is not found in the many petroleum and natural gas oriented techniques discussed, but rather in the philosophy of development over the past years. Some of the techniques are, however, clearly applicable to shallow drilling with only minor modifications or extrapolations, especially the chapters on bits and cementing as well as the glossary which contains a multitude of semi-pertinent tables, formulas, and graphs. WRSIC #W71-09920

139. Bredehoeft, J. D., 1964, "Variations of Permeability in the Tensleep Sandstone in the Bighorn Basin, Wyoming, as Interpreted for Core Analysis and Geophysical Logs," *U.S. Geol. Surv. Prof. Paper 501-D,* pp. 166-170.

An examination of the Tensleep Sandstone in the Bighorn Basin, and the permeability via drill-stem testing, core analysis, and geophysical logs.

140. ———, 1965, Drill-Stem Test: Petroleum Industry's Deep Well Pumping Test," *Ground Water*, Vol. 3, No. 3, July, pp. 31-36, 8 refs.

Drill-stem tests provide the petroleum industry information on three critical properties of subsurface formations – pressure head, permeability, and water chemistry – that the ground water hydrologist also seeks in making pumping tests of water wells. As it is increasingly necessary to study the hydraulic and geochemical properties of deep-lying rocks in order to understand the behavior of ground water, data from these tests made by methods currently in use are highly useful in water studies.

141. Briggs, G. F., *et al.*, 1967, "Water Well Hydraulics – Part 2," *Industrial Water Eng.*, Vol. 4, No. 6, pp. 24-28.

A review of the principles involved in tapping ground water resources.

142. ———, 1967, "Maintaining Water Well Yield – Part 3," *Industrial Water Eng.*, Vol. 4, No. 7, pp. 28-32.

Reviews practical considerations in evaluating and correcting problems which affect well production.

143. Briggs, R. O., 1964, "The Downhole T.V. Camera," *Soc. Prof. Well Log Analysts*, 5th Annual Logging Symp., Midland, Texas, May, p. N-1.

This paper describes the design and construction of a deep well inspection tool, based upon the use of closed circuit television techniques. The tool was developed and patented by Shell Development Corporation of Houston, Texas. A license to manufacture and lease or sell the tool has been granted to Oceanographic Engineering Corporation of La Jolla, California. The surface equipment includes a van with a winch and sufficient cable to operate the camera to the desired depth, a control console with a television screen and controls for the light source, the camera, and the rotating mirror. After describing the tool in general terms, this paper concludes with a film presentation showing the tool, the van, and associated equipment and a demonstration of the use of the tool in an experimental well logging application.

144. Brown, B. D., 1959, "Cementing Water Wells," *Pub. Works*, Vol. 90, No. 9, September, pp. 99-100.

Tools and materials to aid in obtaining maximum control of isolating producing zone; isolation protects water-bearing formations from contamination from strata or from surface; techniques: two-plug system, plunger-type receptacle, and continuous method; materials, such as: Portland cement, bentonite, pozzolans, perlite, diatomaceous earth, and gilsonite.

145. Brown, K. G., 1956, "Mud-Scow Drilling," *Water Well Journ.*, Vol. 10, No. 6.

A special kind of cable-tool drilling, known as mud-scow drilling, is practiced in the alluvial valley fills of the Southwest, primarily California and Arizona. Experience has refined the method and

Brown, K. G. . . . . (con't)

developed a special type of drilling machine, generally known as the California Sky Beam. The problem which led to the development of this method of drilling stems from the fact that most ground water obtained in California and Arizona is found in the alluvial valley fills in the inland areas and in the sand, silt, and delta deposits of the seacoast area.

146. Brown, R. H., 1964, "Hydrologic Factors Pertinent to Ground-Water Contamination," *Ground Water,* Vol. 2, No. 1, pp. 5-12, January.

Covers a few principles of fluid movement in porous media for ground water flow systems of simple geometry.

147. Bruce, W. E., 1968, "Bureau of Mines Conducts Diamond Drilling Experiments," *Mines Mag.,* Vol. 58, No. 4, pp. 17-21.

Interior Department's Bureau of Mines' indices for predicting rock drillability, based on the applied drilling forces, the physical properties of the rock and associated parameters. Studies involve 26 rock formations drilled in the field with a commercial diamond drill as well as bits similar to those used in the laboratory tests. The 26 rocks tested ranged from soft limestone to hard, dense taconite. Diamond drilling rates can be predicted from the machine characteristics rotational speed, thrust, and torque in conjunction with the compressive strength of the rock being drilled.

148. Brun, A., 1969, "Turbodrill Solves Tough, Costly Operating Problems," *World Oil,* August, Vol. 169, No. 2, pp. 35-38.

Compagnie Francaise des Petroles has confirmed, after extensive use of the turbodrill, the drilling efficiency is upgraded when this tool is used for special purposes, and if it is applied by properly trained personnel, the tool is ideal for increasing energy on bottom; to break and remove formation and to drill faster at more economical rates. In one case, use of a turbodrill from a drill ship operating in the Mediterranean Sea increased penetration rates threefold. On land and at depths of 12,000, 13,500, and 18,000 ft. respectively, running a turbodrill after conventional tools increased the average feet drilled per hour from 2.6 to 5.7; from 3.1 to 10.6; and from 2.6 to 4.25. These improved drilling rates result because a turbodrill rotates faster on bottom and torque buildup is eliminated in the drill string; drills faster in deviated hole with less weight; drills surface hole through hard formations when heavy weight cannot be transmitted through the drill string; and can operate when circulation is not possible. WRSIC #W71-02983

149. Bugbee, J. M., 1953, "Lost Circulation – A Major Problem in Exploration and Development," *A.P.I. Drill. and Prod. Pract.,* pp. 14-27, 18 refs.

The major drilling problem of lost circulation can be avoided or minimized by the study of possible losses and the programming of treating procedures in advance of drilling. Not only must there be openings in the formation of sufficient size to accept and store the lost mud, but excessive pressures must act to force it away. Careful study of the nature of loss zones, their formation-fluid pressures and the hydrostatic and mechanical pressures that can be imposed is explored.

150. Burt, E. M., 1970, "Well Development: A Part of Well Completion," *Principles and Application of Ground Water Hydraulics Conference,* Michigan State Univ., East Lansing, December, 15 pp., 34 refs.

Available tools and various techniques of well development are discussed. Illustrates elements of the basic types of wells found in Michigan wherever water bearing unconsolidated formations rest on water-bearing consolidated formations. Well development accomplishes the design goal of maximum, sandfree capacity during construction; and the selection of a screen, the possible necessity of artificial gravel packing, and the selection of the method of drilling during the design phase.

151. Cahill, J. M., 1967, "Hydraulic Sand-Model Study of the Cyclic Flow of Salt Water in a Costal Aquifer," *U.S. Geol. Surv. Prof. Paper 575-B,* pp. 240-244.

Model studies demonstrate visually that "cyclic flow" takes place in the salt water when fresh water flows seaward over intruding ocean water. When tidal fluctuations are simulated in the model by alternately raising and lowering the level of the free body of salt water, additional mixing of the two fluids occurs, and additional salt water is transported into the fresh-water region. The fluid movement was observed by following colored tracers that were injected into both fluids. These studies indicate that the degree of salt-water intrusion is controlled primarily by the flow of fresh water rather than by tidal effects.

152. Calabresa, T. A., 1963, "Statutes and Regulations Governing Private Water Well Construction and Pump Installations," *Ground Water,* Vol. 1, No. 2, April, pp. 25-32, 2 refs.

This paper discusses the need for laws and regulations governing construction of wells for private water supplies, including: (1) why laws and regulations are necessary, (2) need for registration and licensing of drillers and pump installers, (3) how laws and regulations affect uniformity of construction and the desirability of such uniformity, (4) how they encourage the manufacture of new materials and group organization, (5) what a law and code should contain, (6) need of judgment in administering laws and the regulations adopted pursuant thereto, and (7) the difficulties in enforcing laws and regulations including budgetary reasons.

WRSIC #W71-06953

153. Callahan, J. T., 1963, "Television – A New Tool for the Ground Water Geologist," *Ground Water,* Vol. 1, No. 4, October, pp. 4-6.

"Subaqueous" studies made by means of a television camera in water wells at Brunswick and Savannah in the coastal area of Georgia. In these evaluations, a Polaroid and a 35-mm camera were used to photograph the results shown on a monitor. Observed were areas of smooth and rough walls, ledges of hard material (some of which had rock fragments on them), scratches made by a caliper logging device, drilling mud deposits, small openings and cracks and color changes. Porous zones and the differences between hard and soft limestone could be detected.

154. Campbell, J. M. and B. J. Mitchell, 1963, "Effect of Tooth Geometry on Tooth-Wear Rate of Rotary Rock Bits," *Mid-Continent District Spring Meeting,* A.P.I., Div. of Prod. Paper 851-887-I.

Laboratory studies on wear-rate of simulated rotary bit teeth show that the shape of the bit teeth characterize the wear rate. Equations for wear rate based on theoretical concepts can be applied to field observations on bit wear and to the selection of optimum bit weight and rotary speed through the development of equations for each type of bit.

155. Campbell, M. D., 1972, "Ground Water Aids Mineral Exploration," (In prep.)

Reviews applicability of hydrogeological aspects of ground water to exploration for uranium, phosphate, silver, copper, and other "strata-bound" minerals. Explores field use of quantitative methods, problems encountered in applications and interpretations. Paper applies knowledge available in ground water geology to mineral and mining exploration industry.

156. Cannon, G. E., 1957, "Development of a High-Speed, Low-Torque Drilling Device," *Fall A.I.M.E. Meeting,* Dallas, October.

Describes development of a turbine tool capable of high speed and low torque drilling. Various characteristics of power and effectiveness are explored.

157. Carlson, C. W., 1943, "Early History of Water-Well Drilling in the United States," *Econ. Geol.,* Vol. 38, No. 3, March, pp. 119-136, 30 refs.

Standard cable-tool drilling rig was invented and developed in drilling salt wells following successful completion of first drilled well in 1808 near Charleston, W. Va., auger boring for artesian water, invented in 1884, became chief method of sinking artesian wells in Atlantic and Gulf Coastal Plain by end of century.

158. Carr, J. R., 1969, "Boring Equipment is One More Tool for Water Well Contractor," *Johnson Driller's Journ.*, Vol. 41, No. 5, pp. 1-3.

In drilling wells with well-boring equipment, both bucket-type and flight-type augers are used. Flight augers are used less frequently than the bucket-type, because flight augers do not excavate as rapidly in most materials as do the bucket-type augers. However, for penetrating relatively hard strata and under special conditions, such as where large roots are encountered, flight augers must be used. Under good conditions, the bucket fills rapidly and is then brought to the surface to be dumped. Most machines include complete hydraulic control systems including hydraulic cylinders for raising the derrick, for positioning outrigger arms, and for operating leveling jacks.

159. Carstens, J. P. and C. O. Brown, 1971, "Rock Cutting by Laser," *46th Annual SPE of A.I.M.E. Fall Meeting,* Preprint SPE-3529, New Orleans, October, 11 pp.

A high-power, electric discharge, closed cycle, CO_2 convection laser developed at the United Aircraft Research Laboratories was used for the rock cutting tests. State-of-the-art multikilowatt high-beam-quality electric discharge CO_2 closed-cycle gas lasers are capable of cutting deep slots or drilling deep holes in hard rock. Kerf depth to width ratios of up to twelve in basalt have been achieved and hole depths of over six cm have been achieved in taconite. Although all rock types examined are cut by the laser beam, some rock types are more easily cut than others. Penetration in granites, in particular, is hampered by the formation of a viscous glassy melt region. Laser beam energy requirements to form kerfs in rock can for appropriate cutting conditions closely approach the energy required to just melt the affected volume of rock. Laser power level, peak-power density, focusing system "f" number, traverse speed, and the presence of a blowing jet are all important parameters in the rock cutting process.

160. Carter, V. B., 1966, "Supplementing Sample Logs," *Ground Water,* Vol. 4, No. 3, July, pp. 49-51.

In the past 20 years there was been a substantial increase in the application of electric well logging to water wells. The well drilling industry is becoming more and more aware of the benefits that can be derived from use of the electric log, and there will almost certainly be an increase along these lines in the years ahead.

161. Case, L. C., 1970, *Water Problems in Oil Production,* Petroleum Publishing Co., Tulsa, 133 pp., 42 refs.

First definitive volume ever published on oil-field water problems. It embraces more than 40 yrs. of intensive experience with a number of petroleum companies and in a variety of situations. Following an uncomplicated discussion of water chemistry, the book covers solid deposits and scale formations and corrosion causes, prevention, and treatment. Subjects such as plugging or fouling deposits, filtration,

Case, L. C. . . . . (con't)

identification of common water problems, and injection-water specifications are examined in detail. Attention is given to avoiding pollution, combatting unjust claims, and to establishing criteria for evaluating qualified services for water problems.

WRSIC #W71-05295

162. Chambers, K., 1967, "What's Up In Air Drilling and Workover Operations?" *Petrol. Equip. Serv.*, May-June, pp. 14-18.

Air drilling concepts are due for a major overhaul, especially if present knowledge supports only one type of air drilling. There are at least five basically different forms of air drilling. Each type has its own areas of application, its own related techniques, and its own limitations. This paper explores these types.

163. Cheatham, J. B., Jr., 1963, "Rock-Bit Tooth Friction Analysis," *Trans. SPE of AIME,* II-327-332, 4 refs.

The influence of friction on the force required for an idealized bit tooth to penetrate a "plastic" rock is analyzed. The rock is assumed to obey the Coulomb-Mohr yield criterion, and the tooth is represented by a two-dimensional sharp wedge. At the tooth-rock interface, the stress field not only satisfies the yield condition but also is such that the ratio of the shearing stress to the normal stress is equal to the coefficient of friction.

164. ———, 1967, "Strain Hardening of a Porous Limestone," *Trans. SPE of AIME,* II pp. 229-234, 8 refs.

Applications of the mathematical theory of plastic behavior promises to lead to the solution of many rock mechanics problems. Tests were made to develop an analytical model which fits the initial and subsequent limit surfaces. Torsion and extension tests examined to define limit surfaces, etc.

165. Chenevert, M. E., 1970, "Shale Alteration by Water Adsorption," *Journ. Petrol. Tech.,* September, pp. 1141-1148, 14 refs.

An adsorptional isotherm technique makes it possible to define quantitatively the hydrational tendencies of any formation, thereby eliminating the cumbersome rock classification schemes presently used in industry.

166. ———, 1970, "Shale Control with Balanced-Activity Oil-Continuous Muds," *Journ. Petrol. Tech.,* October, pp. 1309-1316, 7 refs.

Field tests have shown that properly designed oil-continuous drilling fluids require little maintenance and significantly simplify drilling operations by maintaining extremely stable well bores. The laboratory data explain why well bores are unstable when drilled with water base and untreated oil-continuous muds.

167. Cherry, R. N., 1965, "Portable Sampler for Collecting Water Samples from Specific Zones in Uncased or Screened Wells," *U.S. Geol. Surv. Prof. Paper 525-C,* pp. 214-216.

Water sampler for use in uncased or screened wells which tap more than one water-bearing zone consists of two inflatable packers and of submersible pump which removes water from section of well isolated by packers; auxiliary instruments can be used to measure temperature, specific conductance, or other characteristics of water in isolated interval during pumping; sampler is easily handled and is readily moved.

168. Chombart, L. G., 1960, "Well Logs in Carbonate Reservoirs," *Geophysics,* Vol. 25, No. 4, pp. 779-853.

Modern well logs can play an important, often decisive role in the evaluation of carbonate reservoirs and in well completions therein. To do so, however, the logs must be selected and interpreted with due regard for the specific rock "types" and pore structures encountered by each well. To cope with such a reservoir, an evaluation and logging program adhering to certain principles is most likely to yield valid results and insure better completions and greater ultimate recovery at minimum cost. First, in every well, the cuttings or cores should be described precisely as to rock types and depths. Second, any techniques used should permit the largest possible number of determinations through the reservoir, so that any existing relationships between pore size distribution, porosity, and water saturation may be established on a sound statistical basis. Among logging devices, "focusing" tools meet this requirement best. Third, starting very early in the development of the reservoir, the latter should be cored and logged in key wells, the cores subjected to capillary pressure and other petrophysical tests, and all potentially diagnostic logs run and analyzed in the light of all other data. Fourth, in non-key wells, the logging program should include only those logs proved most reliable in the key wells for the pore structures encountered and the data desired (usually porosity, water saturation, net ft. of pay).

169. Church, M., 1960, "Mud Pressure Aids Cable Tool Drilling," *Johnson Driller's Journ.,* Vol. 32, No. 3, pp. 4-5, 2 figs.

A field report on a method which allows easier driving of a pipe while cable-tool drilling. Hydraulic pressure is maintained on an envelope of drilling mud surrounding the well casing as it is being driven. The mud lubricates the casing and seals the annular space.

170. Clancy, G. E., *1967,* "Water from Wells," *Heating, Piping and Air Conditioning Journ.,* Vol. 19, No. 11, November, pp. 75-79.

Factors in using well water in air conditioning, refrigeration, and heat pump installations; types of supply and disposal wells; data on drilling wells; charts for power required for pumping water from various depths and compressor power for various quantities of water for given condenser.

WRSIC #W71-13713

171. Clark, E. H., Jr., 1956, "A Graphic View of Pressure Surges and Lost Circulation," *A.P.I. Drill. and Prod. Pract.*, pp. 424-438, 7 refs.

Much has already been published to bring to light the fact that pressure surges can be a major factor in lost circulation. It is the intent of this paper to clarify when and how these pressure surges take place and to point out the important factors which must be controlled in order to reduce them.

172. Clarke, F. E., 1966, "The Chemistry of Water Well Development," *Industrial Water Eng.*, Vol. 3, No. 9, pp. 19-25.

Chemistry must be considered a necessary partner with hydraulics in well field planning. Chemical processes determine compatibility of construction materials with ground waters, whether troublesome deposition will occur on critical well parts, and what treatments can be applied to combat objectionable processes.

173. Clausing, D. P., 1959, "Comparison of Griffith's Theory with Mohr's Failure Criteria," *Third Symposium on Rock Mechanics*, Colorado School of Mines, Golden.

A mathematical comparison of Griffith's and Mohr's criteria, with modifications and clarifications.

174. Cleindeinst, W. V., 1939, "A Rational Expression for Critical Collapse Pressure of Pipe Under External Pressure," *A.P.I. Drill. and Prod. Pract.*, pp. 383-391.

An examination and derivation of collapse pressures for pipe under external pressures. A classic paper which is used as the foundation for subsequent evaluations of pipe collapse.

175. Cobb, J. B., 1962, "In Hard-Rock Country, Percussion Tool Increases Drilling Rate, Cuts Costs," *World Oil*, Reprint, October, pp. 116-120.

Southern Natural Gas Company has successfully used the percussion air drilling method to increase penetration rates four to ten times, and reduce over-all drilling costs as much as $16,000 on wells drilled in the hard rock country of San Juan and Grand Counties, Utah. Other advantages of the percussion air drilling technique follow: (1) wells may be spudded easily in extremely hard formations, (2) straight hole conditions are maintained more easily because of reduced bit load requirements, (3) fewer drill collars are required, (4) less bit wear in some formations extends bit life, reduces required trips to change bits, and (5) larger cuttings are obtained for geologic analysis. Pneumatic percussion drilling now is becoming fully established as an aid to rotary drilling where air or gas can be used as the circulating medium.

WRSIC #W71-05308

176. Collier, S. L., *et al.*, 1965, *Know Your Mud Pump*, Mission Mfg. Co., Houston, Reprint, 64 pp.

Drilling costs can be reduced through a better understanding of the mud pump and accessory equipment. The mud pump is the main element of the surface mud system, but it is only a link in the three element chain consisting of the suction system, the mud pump itself, and the discharge system. To know what is required of the suction system, both the normal and abnormal situations that occur during and as a result of the suction stroke of the piston must be understood. In the discharge stroke, the piston's pattern of motion is molded into a completely different pattern by the complexity of the discharge system. Each element of the system should be operated for the maximum effectiveness by controlling the problems of each through the best use of the piping and the mud reserve or when necessary, the proper use of the auxiliary equipment. WRSIC #W71-05300

177. Collings, B. J. and R. R. Griffin, 1960, "Clay-Free Salt Water Muds Save Rig Time and Bits," *Oil and Gas Journ.*, April, pp. 115-117.

Field trials of a newly introduced mud are bringing outstanding results in west Texas drilling. This promising fluid contains extremely few solids – some two percent compared to the six-twelve percent of conventional muds of the same weight. Through use of guar gum and starch, water loss is held below ten cc. This "clay-free" fluid has shown many advantages over other muds with higher solids content, e.g., increases penetration rate, lengthens bit life, and reduces costs.

178. Collins, A. G., 1966, "Here's How Producers Can Turn Brine Disposal into Profit," *Oil and Gas Journ.*, Vol. 64, pp. 27, 112.

Discusses a general plan for and against the salvage of minerals from oil-field brines. Such an operation may not result in profit for all producers, but precipitating such minerals in a brine-gathering system might be cheaper than providing brine disposal facilities in the field.

179. ________, 1971, "Oil and Gas Wells – Potential Polluters of the Environment?" *Journ. Water Pollution Control Fed.*, Vol. 43, No. 12, December, pp. 2383-2393, 34 refs.

Discusses potential of oil-gas well ground water pollution via drilling fluids, chemical treatment of oil-gas wells, oil industry corrosion inhibitors from products of oil and gas and from oil industry waste disposal. WRSIC #W72-04069

180. Corey, G. L., 1949, "Hydraulic Properties of Well Screens," unpubl. M.S. thesis, Colorado A. and M., Fort Collins.

The problem for which an answer is sought in this thesis may be stated as follows: What are the hydraulic properties of well screens when surrounded by gravel envelopes containing various sizes of gravel? Problem analysis: What is the loss of head through each well screen operating in clear water with no gravel or sand surrounding the screen? What effect does placing gravel envelopes around the screen have on the loss of head through the well screen? What effect does size of particles in the gravel envelope have on the loss of head through the well screen? What effect does size of particles in the gravel envelope have on the loss of head through the gravel envelope? What effect does variation in

Corey, G. L. . . . . (con't)

discharge have on the above losses of head? What effect does length of screen have on the loss of head through the screen?

181. Coulter, A. W. and D. G. Gurley, 1971, "How to Select the Correct Sand Control System for your Well," *Soc. Petrol. Eng., A.I.M.E.,* Reprint Paper No. SPE-3177, 6 pp., 10 refs.

There has long been a need for a guide to the selection of the sand control system needed for a particular well. There are four types of systems in use today: (1) screens, (2) gravel packing, (3) sand consolidation, and (4) a resin-coated particulate such as sand. Although universal agreement may not be possible at this time, there are certain guidelines that are followed. The factors involved include formation permeability, clay and silt content of formation, interval length, number of zones, homogeneity, well history, type completion, depth, expense of rig time, and others. This paper will consider these factors and their influence on the selection of sand control systems. WRSIC #W71-05320

182. Craft, B. C., *et al.,* 1962, *Well Design – Drilling and Production,* Prentice Hall, Inc., Englewood Cliffs, New Jersey, pp. 571.

A text describing various features of oil well design both for drilling and production stages. Text's development parallels other texts covering same general topics. Well referenced and illustrated.

183. Cregeen, D. J. and H. Moir, 1961, "Evaluation of Limestone Formation Characteristics from Well Logs," *Journ. Petrol. Tech.,* Vol. 12, No. 11, November, pp. 1087-1092, 4 figs., 3 tabs., 4 refs.

The evaluation of limestone formations by well logs has brought many interesting problems to light, some of which are as yet unresolved. While a major purpose of well logging is to insure that zones containing producible hydrocarbons are not bypassed, this paper is more concerned with the further information that can be derived from the logs to assist in more clearly defining reservoir characteristics, fluid content, and performance. In the hard limestones of Southeast Arabia, sufficient evidence has not yet been gathered to allow any large-scale tailoring of the logging program; a number of logs are still run which ostensibly can be used to compute the same parameter. The validity of some of these measurements is discussed, and some tentative interpretations are made concerning their significance.

184. Crouch, R. L., 1964, "Investigation of Alleged Ground-Water Contamination Tri-Rue and Ride Oil Fields, Scurry County, Texas," *Texas Water Comm.,* LD-0464-MR.

Discusses case history of ground water contamination as a result of oil production of Scurry County, Texas.

185. Csallany, S., 1966, "Graphical Method for Determining Coefficient of Transmissability," *A.W.W.A. Journ.,* Vol. 58, pp. 628-634.

Method is an attempt to determine accurately the coefficient of transmissability from specific capacity data. Specific limitations are noted on the use of the graphs.

186. _______, 1966, "Yields of Wells in Pennsylvanian and Mississippian Rocks in Illinois," Illinois State Water Survey, *Report of Investigation 55*, 43 pp.

During the period 1920 to 1963 about 250 well-production tests were made on more than 200 wells penetrating Pennsylvanian and Mississippian rocks. Statistical analysis of specific-capacity data provided a basis for comparing the productivity of individual formations. Several wells show marked improvement in yield as the result of shooting. Yields are increased because: (1) the hole is enlarged and (2) fine materials and incrusting deposits on the well face and in the well wall are removed.

187. _______, 1967, "The Hydraulic Properties and Yields of Dolomite and Limestone Aquifers," *International Assoc. Sci. Hydrol.*, No. 73, pp. 120-138.

Geology, hydraulic properties, aquifer tests, and yields on dolomite and limestone aquifers in Illinois.

188. Csallany, S. and W. C. Walton, 1963, "Yields of Shallow Dolomite Wells in Northern Illinois," Illinois State Water Survey, *Report of Investigation 46*, pp. 43-46, 31 figs., 6 tabs, 16 refs.

About 1000 well-production tests were made, 1921-1961, on more than 800 shallow dolomite wells. Statistical analysis of specific-capacity data provided a basis for determining: (1) the role of individual shallow dolomite aquifers or formations, uncased in wells, as contributors of water; (2) whether or not significant relationships exist between the yields of wells and geohydrologic controls; and (3) the effects of acid treatment on the productivities of wells.

189. Cunningham, R. A., 1960, "Laboratory Studies of the Effect of Rotary Speed on Rock-Bit Performance and Drilling Cost," *A.P.I. Drill. and Prod. Pract.*, pp. 7-14, 9 refs.

This paper presents laboratory data which show relationships between rock-bit bearing life and rotary speed, drilling rate and rotary speed, rock-bit tooth wear and rotary speed. Discussion is given to show that these relationships combine in such a manner as to result in decreased rock-bit footage and increased costs at high rotary speeds.

190. Cunningham, R. A. and J. G. Eenink, 1959, "Laboratory Study of Effect of Overburden, Formation, and Mud Column Pressures on Drilling Rate of Permeable Formations," *Trans. A.I.M.E.*, Vol. 217, pp. 9-17, 10 refs.

One phase has been completed of a laboratory investigation of formations with relatively high permeability under conditions of overburden, formation, and mud column pressures. Drilling rate decreased when mud column pressure was greater than formation pressure. The decrease was primarily due to a layer of cuttings and mud particles held to the hole bottom by the difference in pressure. Adequate jet velocities helped clean away the filter cake and chips and resulted in increased drilling rate – the higher the jet velocity, the faster the drilling rate. Drilling rate increased slightly when formation pressure was greater than mud column pressure. The increase resulted from cleaning the hole

Cunningham, R. A. and J. G. Eenink . . .(con't)
bottom as formation fluid flowed into the borehole. Overburden pressure had practically no effect on drilling rate.

191. Cunningham, R. A. and W. C. Goins, Jr., 1957, "Laboratory Drilling of Gulf Coast Shales," *A.P.I. Drill. and Prod. Pract.,* pp. 75-85, 10 refs.

Based on laboratory results, Gulf Coast shales exhibit drilling characteristics different from harder, more brittle formations which under certain conditions tend to become gummy and cause bit balling. Oil added to lime-base mud tends to reduce balling and increases drilling rate. Reduced fluid loss resulted in a reduced drilling rate.

192. Darby, R. L. and W. H. Veazie, 1968, "Writing a State-of-the Art Report," *Mats. and Research and Standards,* MTRSA, Vol. 8, No. 5, pp. 28-32, 5 refs.

A state-of-the-art report is a comprehensive analysis of available knowledge (published and unpublished) on the status of a particular subject area or mission frequently written for the use of a specific reader audience. This paper suggests some guidelines for their preparation. These guidelines are based on the experience gained in assisting many technical specialists to write state-of-the-art reports. While most of these reports were prepared as a function of information analysis centers, others were written on subjects for which such specialized assistance was not available. In these cases informational resources and assistance normally available through the library were used. In addition, many authors across the U.S. have contributed to this paper and to the guidelines through their response to survey questions and their personal experiences in writing such reports. WRSIC #R203002X68A

193. Darley, H. C. H., 1969, "A Laboratory Investigation of Borehole Stability," *Journ. Petrol. Tech.,* July, pp. 883-892, 6 refs.

Establishes basic mechanisms underlying various types of borehole instability and determines the influence of drilling fluids thereon. Study of swelling and dispersion of shales in water and in mud filtrates; two simple tests were developed for evaluating these properties. A model study of borehole stability was made in a machine in which shale specimens were subjected to triaxial stresses and fluids were circulated through an axial hole. Conclusions: (1) water absorbed by surface hydration causes unstable hole conditions in montmorillonitic shales. This phenomenon occurs even if an inhibited mud is used. The mode of failure is fracturing into relatively hard fragments. If a fresh water mud is used, additional water is adsorbed by osmosis and the mode of failure is plastic deformation. (2) Caving in brittle shales is caused by water penetration into rebonded fractures and parting the fractures by hydration of its surfaces. (3) Spalling is to be expected in areas where the horizontal earth stresses are high. WRSIC #W70-01107

194. ———, 1969, "Model Borehole Helps Solve Shale Problems," *Petrol. Eng.,* September, pp. 76-96, 5 refs.

A study has been made by Shell E & P Research Center of hole stability in shales by means of a small-scale model. Object was to establish principles underlying various types of shale problems, and to show effect of such variable as filter loss, electro-chemical conditions, etc.

195. da Silva, F. A. M., 1965, "Safe Way to Clear Plugged Bits," *World Oil,* Vol. 161, No. 2, August, pp. 114, 118.

Plugged bits, common problems in the drilling industry, usually are plugged inside by cuttings, mud cake, or sand. This can be caused by differences in mud densities (higher in annulus than in the drill pipe) when a connection is made. Mud flows back into the drill string and some cuttings, mud cake or sand enter through bit nozzles. After the connection is made, the pump is started, causing an abrupt increase of pressure which compacts the cuttings in the bit. Many times, it is impossible to resume circulation. A safe solution to the problem lies in progressively increasing the pressure differential at the bit.

WRSIC #W71-08505

196. Davis, S. N., 1966, "Initiation of Ground-Water Flow in Jointed Limestone," *Natl. Speleol. Soc. Bull.,* Vol. 28, pp. 111-118.

Review of the limited amount of data available suggests that jointed limestone under normal conditions is quite impervious to ground water flow. Joints adjacent to existing bodies of surface or subsurface water are affected by differential movement along the joint surfaces. The main effect will be to pump water in and out of the joint and start incipient solution. Once solution enlarges the joint, water will be diverted from surrounding rocks and follow the headward advance of the solution opening.

197. Davis, S. N. and L. J. Turk, 1969, "Best Well Depth in Crystalline Rocks," *Ground Water,* Vol. 2, No. 2, April, pp. 6-11, 10 refs., 10 figs.

Water quality, local regulations, aesthetic considerations, and other factors also are commonly important in deciding upon well depth in addition to simple economic or geological analyses. The authors give some quantitative expression to just one decision-making step in one type of water-bearing material, e.g., water-bearing characteristics of crystalline rocks, drilling indicates highly variable amounts of water, production per foot of well depth decreases rapidly, optimum well depth controlled by cost and geological factors. Estimates are given for probable depths of domestic wells and of larger production wells.

WRSIC #W71-10420

198. Davis, S. N. and R. J. M. DeWiest, 1970, *Hydrogeology,* John Wiley and Sons, Inc., New York, 463 pp. (third printing).

A comprehensive text covering the following topics: Hydrologic Cycle; Physical and Chemical Properties of Water; Water Quality, Radionuclides in Ground Water; Elementary Theory of Ground Water Flow; Applications of Ground Water Flow; Exploration of Ground Water; Ground Water in Igneous and Metamorphic Rocks; Ground Water in Sedimentary Rocks; Ground Water in Nonindurated Sediments; and Ground Water in Regions of Climatic Extremes.

199. Decker, M. G. 1968, *Cable Tool Fishing,* Water Well Journ. Publ. Co., Box 29168, Columbus, Ohio, 70 pp.

Fishing is the cause of annoying delays and financial loss in drilling operations. Therefore, it is essential to take every precaution with drilling tools and to exercise care in drilling in order to avoid as many fishing jobs as possible. However, in anticipation of the inevitable fishing job, it is a good plan to record the exact dimensions of everything used about or in the well so that information will be at hand for designing or selecting a suitable fishing tool. Any work being done in the drilling, cleaning out, or servicing of a gas, oil, or water well involves the risk of (1) failure of the drilling tools or the casing or (2) erroneous procedure – either of which may result in a fishing job. Thus, the hazards of fishing are frequently the cause of high drilling costs.

WRSIC #W71-06921

200. Denney, J. P. and J. L. Shannon, 1968, "New Way to Inhibit Troublesome Shales," *World Oil,* July, pp. 111-117.

Drilling problems encountered in Bridger Lake Utah wells prompted operator to analyze effect of varied drilling fluids on collected core samples. Mud containing di-ammonium phosphate and a special poly-anionic cellulose polymer was indicated as effective for inhibiting shales.

201. Deutsch, M., 1963, "Ground-Water Contamination and Legal Controls in Michigan," *U.S. Geol. Surv. Water Supply Paper 1691,* 79 pp.

Discusses ways that chemical and bacteriological wastes have contaminated aquifers in Michigan and reviews many actual cases of contamination. Also describes cases of contamination by natural saline waters because of overpumping, unrestricted flow from artesian wells, vertical movement through open holes, dewatering operations, and deepening of stream channels. Schematic diagrams illustrate hydrologic principles involved. The activities and legal controls of state agencies in protecting and conserving Michigan's ground water resources are discussed.

202. Deutsch, M., 1965, "Natural Controls Involved in Shallow Aquifer Contamination," *Ground Water,* Vol. 3, No. 3, pp. 37-40.

Shallow aquifers, commonly the most important sources of ground water, are also the most susceptible to pollution and contamination. The paper explores the mode of entry of contaminants to shallow aquifers, the major factors involved in aquifer contamination cases and the typical hydraulic-environmental relationships. Well illustrated.

203. Dewan, J. J., *et al.,* 1961, "Chlorine Logging in Cased Holes," *Journ. Petrol. Tech.,* June, pp. 531-537.

A new chlorine logging tool, developed primarily for the detection of oil saturation behind casing, is described. The principle involved is the selective detection of gamma radiation from neutron capture in chlorine. The tool is relatively insensitive to porosity, responding mainly to the concentration of chlorine in the interstitial liquid rather than to the total chlorine content in the formation.

204. Dolan, J. P., *et al.*, 1957, "Special Application of Drill-Stem Test Pressure Data," *Petrol. Trans., A.I.M.E.*, Vol. 210, pp. 318-324.

Case histories of applying drill-stem test data, with respect to reservoir evaluations, e.g., permeability, transmissibility, etc.

205. Doll, H. G., 1955, "Filtrate Invasion in Highly Permeable Sands," *Petrol. Eng.*, Vol. 27, No. 1, January, pp. B53-B66.

Salt-water-bearing sands of high permeability drilled with fresh-water mud show on resistivity logs that the invasion is deep in the upper part of the sand section and shallow in the lower part. This observation is explained by the upward migration of the filtrate due to the difference in density between mud filtrate and connate water.

206. Dreeszen, V. H., 1959, "Rotary Drilling Muds," *Water Well Journ.*, Vol. 13, No. 9, September, pp. 18-19, 26, 29-30, 10 refs.

The methods for protecting water-bearing formations when the rotary method of drilling is used are discussed. The methods are: remove all the cuttings from the hole, do not recirculate cuttings (keep the drilling mud clean), and use a low filter-loss mud to build a thin filter cake opposite permeable formations and to prevent sloughing of silt and clay formations. The drilling mud must be removed from the drilled hole and from the hole walls in order to develop a producing well satisfactorily. WRSIC #W71-09910

207. Duran, R. J., 1961, "How to Evaluate Mud Desanding," *Petrol. Eng.*, Vol. 33, No. 9, August, pp. 39-44.

An evaluation procedure to determine exactly how much money desanding saves was developed in a year-long program and proved out on a 12,000-ft. well in Mississippi. By applying this procedure, which also takes operational aspects into consideration, a total of $8,112 was saved through use of a desander on the Mississippi well. The savings consisted of $6,420 in mud materials and $1,692 in water costs. Desanding drilling fluid from top to bottom made it possible to carry less mud weight (from ½ to 1 lb. per gal. less) than offset wells. Hydraulics were improved, penetration rates were increased, and the hole was maintained in good shape.

208. Dysart, G. R., *et al.*, 1969, "Blast-Fracturing," *A.P.I. Drill. and Prod. Pract.:* Completion and Re-Completion Treatment, pp. 68-76, 19 refs.

In December 1967, El Paso Natural Gas Company, in conjunction with the U.S. Government, detonated a thermonuclear device underground (Project Gasbuggy) to create extensive fracturing in the Pictured Cliffs formation. This form of explosive fracturing is a continuation of one of the first well-stimulation techniques – nitro shooting. Nitro shooting was a dangerous process in all aspects – storage, handling, and application. In recent years new high-energy explosives have been developed which are safe to handle, temperature-insensitive, and can be produced in various plasticities, so they can be made to conform to well-bore or fracture surfaces. Experience may show conventional high-explosive fracturing to be more convenient and less costly than hydraulic

Dysart, G. R. . . . . (con't)

fracturing for certain applications and, in some cases, the only way a reservoir can be stimulated. WRSIC #W71-01418

209. Eakin, J. L. and J. S. Miller, 1966, "Explosives Research to Improve Flow through Low Permeability Rock," *Soc. Petrol. Eng., A.I.M.E.,* Preprint Paper No. SPE-1715, 10 pp., 16 refs.

Preliminary to underground testing, surface and near-surface tests with liquid explosives showed that explosives in sheetlike layers simulating underground fractures would detonate and propagate through the layers. The successful surface tests used layers of explosive confined between glass plates and explosive-saturated sand confined in small diameter metal tubes. In addition, it was shown that explosions propagated through the pores of Berea sandstone saturated with a liquid explosive. Effects of the explosion were observed by coring, caliper logs, downhole camera surveys, elevation measurements, and air-flow tests. WRSIC #W71-05322

210. Eckel, J. R., 1954, "Effect of Mud Properties on Drilling Rate," *A.P.I. Drill. and Prod. Pract.,* pp. 119-125, 6 refs.

Field observations indicate that the drilling rates obtainable with muds may vary from 30 to 70 percent of those obtainable with water under same conditions. The causes of this reduction based on laboratory and field experiences are discussed. Fluid viscosity appears to be the significant factor affecting drilling rate through cleaning action.

211. ———, 1958, "Effect of Pressure on Rock Drillability," *Trans. SPE of AIME,* Vol. 213, pp. 1-6, 2 refs.

A laboratory drilling rig has been devised and placed in operation which permits the application of hydrostatic, terrastatic, and formation pore pressures to a rock sample for drilling under controlled conditions. Calibrated tests of this equipment indicate qualitative agreement with field results on the effect of weight on bit, rotary speed, and rate of circulation on drilling rate. Tests under pressure, using water and air as the drilling fluids, indicate that when drilling saturated limestone samples, a pressure differential between hydrostatic and formation pressure is the only pressure which affects drilling rate. A differential in either direction causes a reduction in rate, but the same differential from the wellbore into the formation causes the greater reduction. In general, the magnitude of the changes due to the pressure effect were small compared to changes in drilling rate obtainable by changing either bit weight or rotary speed.

212. ———, 1967, "Microbit Studies of the Effect of Fluid Properties and Hydraulics on Drilling Rate," *Trans. SPE of AIME,* Vol. 240, pp. 541-546.

It has long been known that mud properties affect drilling rate and that drilling with water is as much as six to seven times faster than with mud. However, it is not known why drilling rates with mud are lower. Recent studies show what fluid properties govern microbit drilling rate and also provide a quantitative correlation of these fluid properties and hydraulics with microbit drilling rate.

213. Eckel, J. R. and D. S. Rowley, 1957, "How Rotary Speed Affects Penetration," *Oil and Gas Journ.*, Vol. 55, No. 47, November, pp. 86, 87.

An A.A.O.D.C. committee composed of personnel from drilling and equipment companies have made a joint study of the factors that enter into the effect of weight and speed of rotation on penetration. The test site was in Dora Roberts field near Odessa, Texas. An exploration drilling company rig was used to drill The Texas Gulf Producing Co.'s well, Roy Parks B-23. This is a report of the rotary-speed study.

214. Eddy, J. E., *et al.*, 1967, "Description and Use of an Underwater Television System on the Atlantic Continental Shelf," *U.S. Geol. Surv. Prof. Paper 575-C,* pp. 72-76.

Relatively inexpensive underwater closed-circuit television cameras and a monitor, which were adapted first for use in boreholes, were used to obtain a 500-line-resolution picture of the Continental Shelf off the Georgia coast.

215. Edwards, J. H., 1964, "Engineering Design of Drilling Operations," *A.P.I. Drill. and Prod. Pract.*, pp. 39-55, 17 refs.

The drilling character of the formations to be penetrated is the only non-variable quantity in a drilling operation. If parameters of these formations are known, the operation can be designed for lowest cost by varying rig equipment and drilling practices. The wells can be "drilled" on paper several times, and the best available drilling equipment and practices can be preselected.

216. Edwards, J. M. and S. G. Stroud, 1966, "New Electronic Casing Caliper Log Introduced for Corrosion Detection," *Journ. Petrol. Tech.*, Vol. 18, No. 8, August, pp. 933-938.

The electronic casing caliper tool measures the average inside diameter of pipe over a length of one to two inches. Holes as small as 3/4-inch diameter in casing are detectable and vertical splits or cracks are readily evident. The tool may be run with the casing wall-thickness tool to provide simultaneous records of inside diameter and wall thickness.

217. Emerson, D. W. and S. S. Webster, 1970, "Interpretation of Geophysical Logs in Bores in Unconsolidated Sediments," *Australian Water Resources Council,* Research Project 68/7 – Phase I, 212 pp., 378 refs.

This report covers the first phase of a two-phase project to investigate the interpretation of geophysical logs of water wells in unconsolidated sediments. Phase I of the project was designed to (1) appraise the well logging practices presently in use for ground water investigation in Australia, (2) evaluate the geophysical well logging method in its application to unconsolidated sediments, (3) interpret selected well logs to estimate the possibility of this research project producing a practical

Emerson, D. W. and S. S. Webster . . . (con't)
quantitative analysis procedure. The investigation above showed that the well logging practices are inadequate for a full evaluation of aquifers. Present data presentation formats and radiation intensity units are shown to have little uniformity throughout Australia, while calibration practices are inadequate for the provision of accurate logs. Procedures to eliminate these deficiencies are outlined.
WRSIC #W71-07685

218. Erickson, C.R., 1961, "Cleaning Methods for Deep Wells and Pumps," *A.W.W.A. Journ.*, Vol. 53, No. 2, February, pp. 155-162.

Methods of well stimulation by surging, vibratory explosion, acidizing treatment, pressure acidizing, polyphosphate treatment, shooting treatment and chlorination; pump cleaning by dismantling, by treatment of inhibited muriatic acid, or with polyphosphate compounds when cost of additional water supply by well stimulation is less than cost of this additional water supply by constructing new wells, justify well stimulation.
WRSIC #W71-07685

219. Estes, J. C., 1971, "Selecting the Proper Rotary Rock Bit," *Journ. Petrol. Tech.*, November, pp. 1359-1367, 14 refs.

This paper was written for an engineer who plans casing and drilling programs and the tool pusher who must select his rock bits at the rig. Its primary purpose is to offer guidelines for the use of the latest rock-bit technology so as to avoid obsolete bits and drilling techniques, hence getting better returns on the dollar invested.

220. Fairhurst, C. and W. D. Lacabanne, 1966, "Some Principles and Developments in Hard Rock Drilling Techniques," *6th Annual Drilling and Blasting Symposium,* University of Minnesota, Minneapolis, October, pp. 15-25, 10 refs.

This paper describes the general mechanism of rock breakdown in the two most used drilling methods, namely percussion and drag-type rotary. Most other systems can to a large extent be classified within these types. Consideration of the mechanics reveals the inherent limitations of each and indicates the reasons why the comparatively recent method of rotary-percussive drilling offers hope for improved drilling performance, especially in the harder rocks.
WRSIC #W71-09912

221. Fawcett, A. E., 1963, "Hydraulic Gravel Packing for Deep Water Wells," *Ground Water,* Vol. 1, No. 1, pp. 16,24.

Discusses a method of well packing.

222. Feenstra, R. and J. J. M. Van Leeuwen, 1964, "Full-Scale Experiments on Jets in Impermeable Rock Drilling," *Trans. SPE of AIME,* Vol. 231, pp. 329-336, 12 refs.

The effect of jets on bit penetration has been investigated by means of a 50-ton drilling machine and 8½-in. commercial jet bits drilling under representative bottom-hole conditions. The conclusions apply to the drilling of impermeable rock only. Penetration rate is in all cases primarily hampered by a differential pressure effect, known as dynamic hold-down. Jet action can reduce this slightly. Major gain in penetration rate up to 25 or 50 percent is already obtained at medium-high jet power.

223. Ferris, J. G., 1948, "Ground Water Hydraulics as a Geophysical Aid," *Techn. Report No. 1,* State of Michigan, Department of Conservation, Geol. Surv. Div., East Lansing.

Explores general features of hydraulics in relation to geophysical application.

224. Fisher, L. W. and W. H. Sawyer, Jr., 1951, "Methane Gas in Water Wells," *Science,* January, pp. 7-8, 1 ref.

An account of discovery of methane gas in a 510-foot cable-drilled water well near Winthrop, Maine. Suggests as possible explanation, migration of the gas from near-by swamps. WRSIC #W71-13717

225. Fournier, R. O. and A. H. Truesdell, 1971, "A Device for Measuring Down-Hole Pressures and for Sampling Fluid in Geothermal Wells," *U.S. Geol. Surv. Prof. Paper 750-C,* pp. C146-C150.

Long flexible stainless-steel tubes can be used to measure downhole pressures in geothermal wells by balancing gas pressure in the tube against the hydrostatic pressure external to the tube. If water will flow naturally from a well (wells with positive wellhead pressure), water samples may be collected from any desired depth by allowing water to flow up through the steel tube (provided boiling does not occur during ascent). WRSIC #W71-13457

226. Fowler, K., 1970, "Rating Pump Performance by Capacity," *Johnson Driller's Journ.,* Vol. 42, No. 2, p. 4-5, 1 fig.

Rating water system pumps by pumping capacity on output rather than by motor horsepower has been proposed by the Water System's Council. This article explores the justifications for the proposed change.

227. Fox, F. K., 1960, "New Pipe Configuration Reduces Wall Sticking," *World Oil,* December, pp. 83-87.

When pipe in the open hole is freely movable at any depth, the hydrostatic pressure of the drilling fluid is distributed uniformly over the entire surface area of the pipe. If this mud pressure is restricted and not uniformly distributed around the pipe by the wall of the borehole and its mud cake, an unbalanced hydrostatic pressure condition exists. This results in the pipe being held against the wall of the hole by the differential or unequalized hydraulic pressure.

228. Franzoy, C. E. and C. D. Busch, 1966, "A Portable Airline to Measure Water Level," *Agr. Eng.*, Vol. 47, pp. 86-87.

An instrument using an airline and manometer to detect the water surface in a well is claimed to have an accuracy of 0.1 ft. and to be unaffected by oil on the water surface or cascading water.

229. Frimpter, M. H., 1969, "Casing Detector and Self-Potential Logger," *Ground Water*, Vol. 7, No. 5, September-October, pp. 24-27, 5 refs.

Rapid method of determining casing length and permeable zones in wells tapping bedrock. A device consisting of a galvanometer, a reel of insulated wire, and a copper electrode locates the casing depth, changes of lithology, and permeable zones. The small-diameter electrode permits measurement through well-seal access ports avoiding the procedure of removing the seal access ports and drop pipes to measure casing depth with a magnet. The measured electromotive force changes rapidly when the electrode passes the end of casing. Thus, the depth of casing is easily determined from the length of wire paid out.
WRSIC #W71-09724

230. Galle, E. M. and H. B. Woods, 1960, "Variable Weight and Rotary Speed for Lowest Drilling Cost," *20th Annual Meeting AAODC*, New Orleans, September.

This paper includes the best available empirical equations for the effects of weight, rpm, and tooth dullness on drilling rate, tooth wear, and bearing wear. These equations apply to many formations and drilling conditions. From these equations, useable methods to determine the way to vary weight and rotary speed for lowest drilling cost have been developed.

231. ———, 1963, "Best Constant Weight and Rotary Speed for Rotary Rock Bits," *A.P.I. Drill. and Prod. Pract.*, pp. 48-73, 3 refs.

Empirical equations for effects of weight, rpm, and tooth dullness on drilling rate, tooth wear, and bearing wear apply to many formations and drilling conditions; sample field problems are given to show how to get bit weight and rotary speed which result in lowest cost.
WRSIC #W71-05297

232. Galle, E. M., *et al.*, 1967, "New Equipment, Optimum Drilling Techniques Make for Better Drilling Practices," *Oil and Gas Journ.*, Reprint, 69 pp.

For each particular drilling situation and rig, there is some "best" combination of bit weight and rotary speed. If this weight and speed are held constant throughout the bit run, they produce the lowest possible cost. This booklet lays out in cook-book fashion a method of figuring the best constant bit weight and rotary speed. Other topics covered include: Packer terminology and recommended standardization, tool joints, and drill pipe, ways of avoiding drill collar damage and new ideas on electric drive rigs. WRSIC #W71-05318

233. Gallus, J. P., *et al.*, 1958, "Use of Chemicals to Maintain Clear Water for Drilling," *Trans. SPE of AIME*, Vol. 218 pp. 70-75, 5 refs.

Fresh water or brine drilling fluids may be kept free of suspended drilled solids by the addition of a water soluble acrylamide-carboxylic acid co-polymer at the flowline. Addition of from 0.01 to 0.2 lb/bbl of the polymer solution to drilling fluids containing less than five percent clay solids by weight causes solids to flocculate and settle to bottom of pits.

234. Garber, M. S., 1968, "Methods of Measuring Water Levels in Deep Wells," *U.S. Geol Surv. Misc. Rept.*, Chapt. A1, Book 8, 23 pp.

Describes methods developed by the U.S. Geological Survey's Nevada Test Site and New Mexico staffs for obtaining accurate measurement of water levels at depths in excess of 1,000 feet. Equipment and field and calibration procedures are described. Devices developed for automatically recording water level fluctuations are also described. The application of pressure-sensing transducers is discussed.

235. Garcia, V. M., 1958, "Physical Properties of Mine Rock and Their Effect on Percussive Drilling," *M.S. Thesis T-882,* Colorado School of Mines.

This investigation was performed to study the effect of the physical properties of rocks and their resistance to penetration when drilling with pneumatic machines with the objective of finding an easy method of classifying rock as to drillability. Five kinds of natural rock and one kind of synthetic rock were carefully selected for the tests. A "Drop Tester" was used for the drillability test. Crater volume has been established by previous investigators as the most useful criterion of drillability. In this investigation, volumes of more than fifteen craters for each rock were averaged. Twelve physical and elastic properties of the rocks were determined. Drillability varies linearly with the reciprocal of the unit cohesive strength of the rocks; consequently, this property can be used as a practical method of classification. The Schmidt Impact Hammer test seems to give an indication of drillability. The other physical properties studied show little direct correlation with drillability.

236. Gardner, J. D., 1960, "Engineering Fundamentals Involved in Bottom-Hole Tool Drilling," *Water Well Journ.*, Vol. 14, No. 3, March, 3 pp.

Fundamentals of "bottom-hole drilling"; the major equipment requirement change when rotary drilling rigs using mud were introduced to the water well industry. Previously, only the accessory equipment was needed to add air as a drilling fluid and increase drilling rates. With the introduction of "bottom-hole-tools," the air compressor manufacturers have kept up with their needs. WRSIC #W71-08905

237. Garner, L. L., 1969, "New Concepts in Rotary Drilling Bits," *A.P.I. Drill. and Prod. Pract.*, Vol. 31, November, pp. 148-155.

When drilling specific rock types, the basic design concept principles of three-cone rotary drill bits in recent years has remained relatively constant with no significant changes. New applications, developments, and improvements utilizing the same design principles are changing

Garner, L. L. . . . . (con't)

some of the old concepts, creating many new and interesting trends. Most design developments improved in steps. Replacement of roller bearings with friction bearings in rotary rock bits has seen limited success until recently, due to improved seals. Tungsten carbide bits in the past have been used, generally to drill only hard formations. Design changes following the same general concept of milled tooth designs are now using tungsten carbide bits to drill even softer rock. Bottom hole jets will be utilized more as recent tests indicate their usefulness in conjunction with downhole mud motors. Mud motor bits or rotary bits, modified in recent years to resist the skidding and abrasion wear created when used with high speed downhole motors, are being improved and will be more widely used. WRSIC #W71-07205

238. Garner, N. E., 1967, "Cutting Action of a Single Diamond Under Simulated Borehole Conditions," *Trans. SPE of AIME,* Vol. 240, pp. 937-942, 4 refs.

Laboratory experiments were conducted to investigate the cutting of a single diamond on limestone and shale under simultaneous downhole conditions. Weight on diamond, geometry of stress, and different time pressure across the face were varied.

239. Garner, N. E. and C. G. Gatlin, 1963, "Experimental Study of Crater Formation in Plastically Deforming Rocks," *Trans. SPE of AIME,* Vol. 228, pp. 1025-1030, 24 refs.

Results of impulsive wedge penetration tests on two synthetic, plastically deforming rocks are presented. Basic data obtained were force-time, displacement-time, and force-displacement curves for the impacts, plus the crater geometry. Wedge geometry and blow frequency were varied over a considerable range. The synthetic rocks consisted of wax-sand mixtures; two waxes of different degrees of ductility were used to provide variable "rock" characteristics. Conventional triaxial tests showed that these synthetic rocks exhibited force-deformation curves and Mohr envelopes quite similar to real rocks except that strengths were much lower.

240. Garrison, E. P., 1966, "Field Results with The "Dyna-Drill" in Directional Drilling," *Amer. Assn. Oilwell Drilling Contractors Rotary Drilling Conf.,* Dalas, February, pp. 31-36.

The Dyna-Drill provides an economical system of controled deviation that yields full-gage, continuously deviated directional hole in all types of formations and at any depth. Normal, as well as unusual, directional problems are being solved daily by the Dyna-Drill, with directional runs being made under all types of conditions. The Dyna-Drill's penetration rate not only equals that of rotary drilling in medium to hard formations, but even surpasses it in softer formations and unusually hard formations. Controlled deviation is obtained by means of a bent sub, run directly above the motor. This enables the tool to carry its own bending force along with it as it drills. Instead of a series of abrupt doglegs, the Dyna-Drill produces a full-gage corrective deviation along the continuous smooth arc of a circle.

241. Gary, J. H., Jr., 1969, "Engineered Diamond-Bit Drilling in South Louisiana," *A.P.I. Drill. and Prod. Pract.*, pp. 156-159, 4 refs.

This paper discusses: factors affecting the performance of diamond bits; engineering the proper bit selection; summary of performance of diamond bits for 1967 and 1968 in Humble's LaFayette District; and surrounding economic incentives.

242. Gatlin, C. G., 1957, "Effect of Rotary Speed and Bit Weight on Drilling Rate," *Oil and Gas Journ.*, May, Vol. 55, No. 20, pp. 193-198, 12 refs.

Variables affecting rotary-drilling penetration rate are exceedingly numerous. Some of the more recognizable are the effects of drilling-fluid properties, hydrostatic pressure, hydraulic factors, rotary speed, and weight on bit. This is to say nothing of personnel and rig efficiency, straight-hole requirements, and the extreme variance of rock properties and other subsurface conditions in different areas. In hard-rock areas, penetration rate is limited by the weight on bit or rotary speed which may be applied or both. Normally these limits are imposed by either crooked-hole or equipment considerations. This discussion will deal with the individual and combined effect of these two factors as indicated from various data in the literature.

243. ———, 1960, *Petroleum Engineering, Drilling and Well Completions,* Prentice-Hall, Inc., Englewood Cliffs, New Jersey, 341 pp.

A complete text on the fundamentals of petroleum engineering. Covers nature of petroleum, concepts of petroleum geology, basic rock properties, exploration methods, drilling, drilling hydraulics and penetration rates, coring, well logging, formation damage, drill-stem testing, cementing, and oil well completions. WRSIC #W71-05296

244. Gibb, J. P., 1971, "Cost of Domestic Wells and Water Treatment in Illinois," *Illinois State Water Survey Circ. 104,* 23 pp., 4 refs, 11 figs.

This study provides cost information for private home ground water supply systems in Illinois. Relatively accurate cost predictions for different types and depths of wells, ranging in cost from about $150 to $2,400, can be made from the graphs presented. A typical domestic well in Illinois may be expected to cost about $575. Cost data for pumping systems equipped with ten-gpm submersible pumps (approximately 50 percent of all collected data) show that the average cost of these systems is about $585 with 50 percent ranging between $400 and $680. The cost of treating water for domestic use also is summarized. Two graphs illustrate the monthly costs of softening and removing iron at varying monthly consumption rates and concentrations of hardness-forming minerals and iron. The monthly cost of continuous chlorination is calculated. WRSIC #W72-02753

245. Gibb, J. P. and E. W. Sanderson, 1969, "Cost of Municipal and Industrial Wells in Illinois, 1964-1966," *Illinois State Water Survey Circ. 98.*

This study of the cost of water wells and pumps is based on information obtained for 143 municipal and industrial water-supply wells drilled in Illinois during 1964, 1965, and 1966. Regression analyses, using the

Gibbs, J. P. and E. W. Sanderson . . . . (con't)

method of least squares, show that the cost of wells is directly related to depth, and the cost of pumps is directly related to capacity. A series of cost-depth relationships which plot as straight lines on log-log graph paper are developed for wells tapping sand and gravel aquifers; shallow sandstone or dolomite aquifers; and deep sandstone aquifers. Similar graphs relating pump cost to capacity are also presented for various operating heads.

246. Gibson, U. P. and R. D. Singer, 1971, *Water Well Manual,* Agency for International Development (Small Wells Manual-1969), Premier Press, 156 pp.

Paperback book covering most aspects of small water well construction; includes fundamentals of geology, ground water exploration, well design, well construction, well completion, pumping equipment, and sanitary protection.

247. Gidley, H. K. and J. H. Millar, 1960, "Performance Records of Radial Collector Wells in Ohio River Valley," *A.W.W.A. Journ.,* Vol. 52, No. 9, September, pp. 1206-1210, 1 ref.

Twenty-four radial collector wells for extraction of water from unconsolidated deposits in Ohio River Valley in W. Va. have been successfully used; accurate predictions on well yield at specific locations were made in advance by hydrologic studies; table shows mineral characteristics of water from wells, decline of yield and its causes; restoration of well capacity was not attempted. WRSIC #W71-13720

248. Giefer, G. J., 1963, "Water Wells – An Annotated Bibliography," *Archives Series Report No. 13,* Water Resources Center Archives, University of California, Berkeley, April, Indexed, 141 pp.

An annotated bibliography on water wells covering the period 1940-1961. Includes papers covering well design criteria, well construction, and well operation and maintenance. References to material on well hydraulics limited to those pertaining to theory of flow in the vicinity of wells. Literature on pumps and pump tests not included.

249. Glover, R. E., 1967, "The Effect of Pumping Over an Area," *Proc. Natl. Symp. Groundwater Hydrol.,* Am. Water Assoc., San Francisco, pp. 149-156.

A generalized chart was developed giving the drawdown at the corner of a rectangular area in which pumping can be idealized as being uniformly distributed over the area. The method of using the chart is illustrated by solving examples.

250. Gnirk, P. F., 1966, "An Experimental Study of Indexed Single Bit-Tooth Penetration into Dry Rock at Confining Pressures of 0 to 7,500 PSI," *Proc. First Congress Int. Rock Mechanics,* Lisbon, pp. 121-129.

The optimum distance between successive bit-tooth penetrations required for maximum rock damage and chip formation decreases substantially with increasing confining pressure above the brittle-to-ductile transition pressure of a particular rock, but remains approximately constant for a variation in confining pressure below the transition pressure. At a given confining pressure, the bit-tooth force required for chip formation is constant for indexing distances greater than optimum but decreases linearly, in general, with decreasing indexing distance for distances less than optimum. WRSIC #W71-05303

251. Gnirk, P. F. and J. B. Cheatham, Jr., 1963, "Indentation Experiments on Dry Rocks Under Pressure," *Trans. SPE of AIME,* Vol. 228, pp. 1031-1039, 5 refs.

An experimental investigation has been made to study the effects of tooth angle and confining pressure on the force required for indentation of dry rock samples under confining pressure. In these experiments static single-tooth indentations were made on two sandstones, a limestone, a schist, and a slate under confining pressures of 5,000 to 15,000 psi. In all cases the force-displacement curves are approximately linear for rocks tested under confining pressure, as contrasted with the discontinuous curves obtained for similar rocks at atmospheric conditions. Photomicrographs of thin sections through the craters indicate very little evidence of chip formation during indentation.

252. Gnirk, P. F. And J. A. Musselman, 1967, "An Experimental Study of Indexed Dull Bit-Tooth Penetration into Dry Rock under Confining Pressure," *Journ. Petrol. Tech.,* pp. 1225-1233, 7 refs.

Consideration is given to indexed penetrations by single drill bit-tooth under statically applied loads into rock subjected to confining pressures of atmospheric to 5,000 psi and atmospheric pore pressure; experimental results obtained with 45° wedge tooth over above range of pressures are presented for two limestones and sandstone, variety of indexing distances, and two degrees of tooth dullness.

253. Goins, D. L. and T. B. O'Brien, 1970, "Recent Improvements in Bit Performance – West Texas and New Mexico, *A.P.I. Drill. and Prod. Pract.,* Div. Production, A.P.I., Odessa, Texas, March.

There have been significant reductions in drilling cost due to improved bit performance. Improvements are the result of evolutionary changes in a variety of drilling parameters. Casing point selection and drilling fluids have been refined, and more recently, changes have been made in bit design. Case histories are given, with discussions of bit design changes, running techniques, and cost analyses.

254. Gold, D. P., *et al.,* 1970, "Water Well Explosions," *Earth and Mineral Sciences,* Penn. State University, University Park, Penn., Vol. 40, No. 3, December, pp 1, 18-21.

Reports on explosion in a water well after a storm which caused extensive damage to surrounding structures. Causes associated with leak in underground gasoline storage tank and favorable climatic conditions for rod lightning (or spark of pump motor.)

255. Gondouin, M. and C. Scala, 1958, "Streaming Potential and the SP Log," *Journ. Petrol. Tech.*, Vol. 213, No. 8, August.

Reports that streaming potential (E_k) of deep shales do occur, but SP deflection from shale line is, however, unaffected by E_k.

256. Gondouin, M., *et al.*, 1956, "An Experimental Study on the Influence of the Chemical Composition of Electrolytes on the SP Curve," *Joint Meeting of Rocky Mtn. Petrol. Sections,* Schlumberger Well Surveying Corp., Casper, Wyoming, May, T.P. 4455.

An investigation has been made of the influence of HCO_3^-, $SO_4^=$ Ca^{++}, and Mg^{++} on the amplitude of the SP deflection which in particular included determinations of the activity coefficients of Ca++ and Mg++. The theory, the experimental techniques, and qualitative method for applying the results are described.

257. Goolsby, J. L., 1969, "Proven Squeeze-Cementing Technique in Dolomite Reservoir," *Journ. Petrol. Tech.,* Vol. 21, No. 10, October, pp. 1341-1346.

Case history of secondary recovery operations in West Texas. It became apparent through injectivity profiles that many primary cement jobs on injection wells were in need of repair – and because of reservoir characteristics; such as thick "pay" with low porosity and permeability, vertical fracturing, and desire to prevent premature dehydration of slurry at perforations, it was necessary to design unique cementing system and technique. Details of technique are analyzed.

258. Gorder, Z. A., 1963, "Television Inspection of Gravel Pack Well," *A.W.W.A. Journ.*, Vol. 55, No. 1, January, pp. 31-34.

The city of LaCrosse salvaged a unit well station worth an estimated $27,000 at a cost of less than $10,000. A closed-circuit television system was used to make a visual probe of a problem well. It was possible not only to observe ruptured or encrusted areas, but also to estimate degree of efficiency of mechanical brushing or cleaning operations, and whether such procedures had been worthwhile.
WRSIC #W71-05695

259. Gordon, R. W., 1958, *Water Well Drilling with Cable Tools,* Bucyrus-Erie Co., South Milwaukee, Wisconsin, 230 pp.

Written for inexperienced people interested in owning or operating a cable-tool rig – selection and use of tools and machinery of the water well drilling industry, emphasizing cable tools and percussion drilling machines, history of water well drilling; types of wells; machines; cable tool; drilling tool string, fishing tool string, miscellaneous tools and pipe handling equipment; cable equipment; pipe and casing; cable tool drilling machine and its selection; drilling with cable tools; prospecting; sample and core taking; test work; witching; drilling problems and how to overcome them; installation of screens; well development and recovery; mud scow drilling; bit dressing; pumping equipment; well failure; maintenance and care; includes tables, illustrations, and a glossary. WRSIC #W71-05292

260. ———, 1959, "Development of Wells with Cable Tools," *Water Well Journ.*, Vol. 13, No. 2, Reprint, February, 2 pp.

Sand and gravel wells; increasing the intake; screened wells; solid surge method; compressed air development. WRSIC #W71-03845

261. Gordon, R. W., *et al.*, 1957, "Water Well Drilling Methods." *Water Well Journ.*, Reprint, 4 pp.

The material in this book is not set down as a rule but rather as one generally accepted method of handling percussion drills and cable tools. Operators have developed drilling and fishing tools to meet their needs. As a result of much ingenuity and experimentation, many effective tools are on the market today. Only tools which are in widespread use are discussed. WRSIC #W71-05292

262. Gorelik, A. M., *et al.*, 1962. "Experiment in the Application of Micrologging for the Study of Water Wells," *Razved Okrana Nedr,* No. 6, June, pp. 54-56 (in Russian).

The micrologging method has been used with success. It is shown that combined with other geophysical methods, micrologging can be used in both hydrological and geological engineering projects. The relationship of logging results to geologic sections in water wells is shown by illustrations.

263. Gosselin, J. C., 1971, "IFP's Computerized Optimization System for Turbo Drilling," *Trans. SPE of AIME,* 3509, October.

During the drilling process, the downhole drilling rate may be quite different from the surface feeding speed due to drill pipe elasticity. An accurate measurement of the downhole drilling rate has been developed and successfully tested by the Institut Francais du Petrole.

264. Gossett, D. C. and H. Lauman, 1959, "Development of Reverse-Rotary Wells," *Water Well Journ.*, Reprint, March, 2 pp.

Describes cleaning and development requirements of the reverse-rotary well, principally municipal and industrial well water supplies. WRSIC #W71-03844

265. Graham, J. B., 1966, "New Tools for New and Old Wells," *A.W.W.A. Journ.*, Vol. 58, No. 10, October, pp. 1278-1284.

Many of the tools currently available are not new in concept but most have undergone recent improvements or modifications that greatly increase their usefulness. For example, an up-to-date, detailed, and reliable map showing the geology, topography, and at least some hydrologic features is the most important single tool that the ground water investigator can have. Progress in field mapping in recent years has been outstanding. By means of aerial photography coupled with improved capability in getting into difficult areas to make the necessary ground checks, high caliber maps can be provided in a relatively short time. Other tools are briefly examined. WRSIC #W71-10422

266. Graham, J. W. and N. L. Muench, 1959, "Analytical Determination of Optimum Bit Weight and Rotary Speed Combinations," *Trans. SPE of AIME,* 1349-G, October, 8 pp.

This paper presents an analytical method for the selection of optimum combinations of rotary speed and bit weight to minimize total drilling cost. A mathematical analysis of the cost to drill any depth interval is the basis of the method. In the analysis, it is assumed that the total drilling cost can be expressed as the sum of rig cost for making round trips, rig cost for drilling, and bit cost. It is further assumed that bit life is limited by bearing failure, that bit weight is not limited by considerations such as hole deviations, and that drilling hydraulics are adequate.

267. Grandone, P. and L. Schmidt, "Survey of Subsurface Brine-Disposal Systems in Western Kansas Oil Fields," *U.S. Bur. Mines Rept. of Inv. 3719.*

Case history of brine-dispoal wells in western Kansas oil fields.

268. Gray, G. R.,1943, "Discussion of Filtration of Salt-Water Muds at Elevated Temperatures," *Trans. AIME,* Vol. 151, pp. 251, 252.

The use of gums and starches in drilling muds requires attention to the prevention of fermentation of these organic products. Fermentation can be prevented by: (1) maintenance of high pH (pH 11-12); (2) saturated salt solution; or (3) adequate concentration of bactericide, such as formaldehyde.

269. ———, 1970, "Chemicals in Oil Well Drilling Fluids," 159th National Meeting, American Chemical Society, February, abridged as "Where the Industry's Mud Money Goes," *Oil and Gas Journ.,* April, pp. 157, 159.

Quantities of chemicals used in drilling muds cannot be correlated directly with footage or number of wells drilled because both depth and location of the wells are major factors affecting mud consumption. Annual expenditure for mud products is about $160 million. Barite, at nearly a million tons, accounts for nearly one-third of the total cost. Bentonite is second in tonnage at about one-half million. Chrome-lignosulfonates, used for control of both flow and filtration properties, almost equal in amount all of the other chemicals added for these purposes. Emphasis on faster drilling rate has promoted the use of acrylic and biopolymers and decreased the consumption of thinners. Nearly 100,000 tons of common inorganic chemicals are added to muds yearly.

270. ———, 1970, "Mud Know-How Helps Exploration Drilling," *Mining in Canada,* May and June, 6 pp.

Modern drilling fluids technology is making substantial contributions to mineral exploration through reduction in time spent in coring and improvement in core recovery. The enormous capital investment needed to develop low-grade ores requires full information on the prospect, hence, the demand for complete core recovery. The mud should be used as a tool, not as a last resort when the situation otherwise appears hopeless. Recognition of the limitations and capabilities of the personnel, the equipment, and the drilling fluid is essential to the effective planning of a coring program. WRSIC #W71-05686

271. _______, 1971, "Plan the Mud Program to Reduce Exploration Costs," *Western Miner,* Vol. 44, No. 4, April, pp. 36-44, 8 refs.

A well planned exploratory-drilling program must consider such factors as: the primary objective, water supply, lithology, drill site, drilling equipment, climate, and manpower. The desired end-product is maximum information at minimum cost. Examples are cited of the close relationship between the drilling fluid and other factors affecting the success of exploration drilling programs.

272. Gray, G. R. and S. Grioni, 1969, "Varied Applications of Invert Emulsion Muds," *Journ. Petrol. Tech.,* Vol. 56, No. 3, March, pp. 261-266, 11 refs.

Invert emulsion muds, compounded to have specific properties, can solve such divergent problems as are encountered in producing from water-sensitive pay zones; in drilling formations that dissolve, disperse, and disintegrate in water and in alleviating pipe corrosion.

273. Gray, G. R. and E. E. Huebotter, 1957, "Drilling Fluids," *Encyclopedia of Chemical Technology,* 1st Supplement Vol., The Inter-Science Encyclopedia, Inc., New York, pp. 246-258, 88 refs.; see also, *loc. cit.,* 1965, 2nd Edit. Vol. 7, pp 287-307, 80 refs.

The technology of drilling fluids has kept pace with the advances in drilling equipment and techniques. In 1845, a French engineer named Fauvelle for drilling water wells used water pumped through a hollow boring rod to bring the drilled particles to the surface. P. Sweeney in 1866 received a United States patent on a "stone drill" that displayed many features of today's rotary drilling rig including the swivel head, rotary drive, and roller bit. The role of mud as a sealing agent was recognized about that time. The primary function of drilling fluid is to remove bit cuttings from the hole. The drilling fluid is pumped down through the drill pipe and out the ports in the bit. The drilling fluid carries the cuttings up the annular space between the drill pipe and the wall of the hole or casing and out at the surface. Although gas or air is used to carry the cuttings out of some wells, by far the greater number of wells are drilled with mud.

274. Gray, G. R. and F. S. Young, 1971, "Drilling Fluids for Deep Wells in the United States," *8th World Petroleum Congress,* Preprint, Panel Discussion 7: Problems, Techniques, and Possible Solutions in Drilling and Producing Very Deep Wells, Elsevier Publishing Co., London, 17 pp., 75 refs.

The major problems associated with the drilling fluid in current drilling to below 20,000 feet (6 km.) are: poor penetration rate, high pressure and temperature, lack of hole stability, soluble and dispersive substances in the formations drilled, and lost circulation. Total cost is least

Gray, G. R. and F. S. Young . . . . (con't)
when time spent on the well is shortest; attention is focused on ways to maximize penetration rate and minimize non-drilling time.

275. Gray, G. R., *et al.,* 1942, "Control of Filtration Characteristics of Salt-Water Muds," *Trans AIME,* Vol. 146, pp. 117-125, 11 refs.

The wall-building properties of salt-water drilling muds can be improved markedly by the addition of: (1) natural gums, such as tragacanth, karaya, and ghatti; (2) seaweeds, such as Irish moss; or (3) gelatinized starch. Results of laboratory tests have been confirmed by field trials. For drilling operations in which the mud becomes contaminated by salt, either through penetrating salt-bearing rocks or by addition of sea water to the mud, the added cost resulting from the use of one of the substances listed above is more than offset by freedom from stuck drill pipe and ease in running casing.

276. Gray, K. E., 1967, "Some Current Rock-Mechanics Research Related to Oil-Well Drilling," *A.P.I. Drill. and Prod. Pract.,* pp. 82-99, 46 refs.

Some highlights of recent and current basic rock-mechanics research pertinent to the drilling problem are discussed. Results clearly show the complex nature of rocks and rock behavior, the local and short-term nature of the phenomena at and near a drilling bit, and the potential usefulness of such basic research results to the drilling problem. Specifically, items of discussion include: (1) pressure gradient below a drilling bit and penetration rate; (b) formation damage during drilling; (3) "effective stress" and rock behavior below a single bit tooth; and (d) directional strength of rock and implications with regard to hole deviation. It is concluded that rocks and rock behavior are very complex but that basic research into the nature of rock and its interaction with a bit under dynamic downhole conditions is essential to an understanding of the drilling process and, therefore, to any systematic approach to improvements in drilling.

277. Gray, K. E., *et al.,* 1962, "Two-Dimensional Study of Rock Breakage in Drag-Bit Drilling at Atmospheric Pressure," *Trans. SPE of AIME,* Vol. 225, pp. 93-98, 22 refs.

This paper presents some preliminary results of two-dimensional cutting tests of dry limestone samples at atmospheric pressure. Cutting tips having rake angles of +30°, +50°, 0°, −15° and −30° were used to make cuts on Leuders limestone samples at six depths of cut ranging from .005 to .060 in. at cutting speeds of 15, 50, 109, and 150 ft./min. The vertical and horizontal force components on the cutting tips were recorded with an oscilloscope equipped with a Polaroid camera. Motion pictures of the cutting process as camera speeds of 5,000 to 8,000 frames/sec. were taken at strategic points in the variable ranges.

278. Grayson, D. J., 1960, *Decisions Under Uncertainty, Drilling Decisions by Oil and Gas Operators,* Harvard Business School, Boston, 33 pp.

Explores the basic philosophy of decision making in drilling operations. This article is for oil and gas drilling operators but approach is applicable to shallow drilling industry.

279. Green, B. Q., 1959, "New Tests Show Which Lost-Circulation Materials to Use and How to Use Them," *Oil and Gas Journ.*, March, pp. 110-115, 170-173.

This article describes the results of lab tests made on materials used to prevent and cure loss of circulation of drilling mud: size of particles necessary to seal is proportional to the size of the openings to be sealed; the seal must take place inside the opening, not on its face; 15 lb. lost-circulation material per barrel of mud is optimum; if 15 lb./bbl. does not seal, larger particle size must be used on a different distribution of particle sizes; waiting period is vital because it permits solids to accumulate in the bridge through filtration.

280. ________, 1963, "Eight Steps to Stop Lost Circulation," *Petrol. Eng.*, Vol. 35, No. 3, March, pp. 76-87.

Eight ways are discussed to stop lost circulation. All lost-circulation problems in rotary drilling can be prevented if the effective downhole pressure can be reduced to equal the formation pressure. This success was due to considerations of (1) better casing programs, (2) improved mud-weight control, (3) better control of flow properties, and (4) improved rig techniques. WRSIC #W71-02999

281. Green, C. F., 1965, "How to Cut Costs in Exploratory Drilling," *World Oil,* Vol. 161, No. 1, July, pp. 103-105.

Working knowledge of drilling contracts can help geologists and other exploration personnel to assist in reducing cost of exploratory holes; excessive requests for unnecessary special equipment and drilling contracts with severe deviation limitations are two areas of potential savings. The geologist can assist operational management in selecting and specifying terms of drilling contracts by predicting drillability of various formations, especially in areas where penetration rates are difficult to predict. The engineer should also be familiar with the different types of drilling contracts in order to plan drilling programs properly. WRSIC #W71-02995

282. Griffith, W. H., 1952, "Discussion of Sprengling and Stephenson Paper," *World Oil,* September, p. 72.

Discusses merits of Sprengling and Stephenson's paper with regard to drilling-time logs, penetration rates, and equipment repairs; drilling system as ideal elastic system is explored on basis of tool motion. Criticism of approach.

283. Grohskopf, J. G., undated, "Summary of Results of Acidizing Water Wells," (unpubl.), Missour Geol. Surv., Rolla.

A review of the effectiveness of acidizing water wells conducted by the Missouri Geol. Survey – based on the files of an unpublished report.

284. Gronemeyer, W., 1956, "Increased Drilling Rates with Turbodrills," *World Oil,* Vol. 143, No. 7, December, pp 144-146.

A paper describing turbodrilling and its early development by the U.S.S.R. explores early models of turbodrills, e.g., T 12 M10, etc.

285. Grubb, W. E. and F. G. Martin, 1963, "A Guide to Chemical Well Treatments," *Petrol. Eng.*, Vol. 21, No. 5, 6, 7, 8, 9, 10 and 11, Parts 1 to 7, June to November.

A comprehensive coverage of well treatment by the use of various chemicals. Explores, in part, acid usage and effectiveness, surfactants and their uses as dispersants. Chemicals which assist in corrosion control are also described.

286. Gstalder, S. and R. Raynal, Jr., 1966, "Measurement of Some Mechanical Properties of Rocks and Their Relationship to Rock Drillability," *Trans. SPE of AIME,* Vol. 237, pp. I-991-996, 3 refs.

Consideration was given to simple tests which could be performed on rocks to give a measure of rock drillability. Various methods of breaking rocks were considered, and the standard hardness test was selected for this study. The test involves application of an increasing load on the rock face through a flat-faced cylindrical punch until rupture occurs. Correlations of drilling results with rock hardness measurements were good. It was concluded that hardness, as measured in lab or deduced from sonic logs, could be used for predicting rock drilling performance.

287. Guardino, S. T., 1958, "Developments in the Design and Drilling of Water Wells," *A.W.W.A. Journ.,* Vol. 50, No. 5, June, pp. 769-776, 3 refs.

Gravel filter; test bore; surface seal; casing installation; well development; reverse-circulation method; re-development and repair.

WRSIC #W71-13715

288. ———, 1964, "Progress in the Water Well Construction Industry," *Ground Water,* Vol. 2, No. 1, January, pp. 31-34.

Contractor's skills and accumulated knowledge should be used as the basis for design and water well specifications. Engineering a water well should have the primary objective of obtaining an adequate supply of potable water, with equal importance placed upon the protection of the source of supply. WRSIC #W71-06954

289. Guerrero, E. T., 1968, *Practical Reservoir Engineering,* Petroleum Publishing Co., Tulsa, 266 pp.

The practical methods of the petroleum industry for reservoir engineering are presented in 102 parts. Practical applications are emphasized and enough theoretical concepts and references are included to give interested readers the fundamentals of the methods used by the petroleum industry. This book demonstrates most of the practices and concepts used by the industry which overlap similar practices used in the ground water industry. WRSIC #W71-05290

290. Guyod, H., 1946, "Temperature Well Logging," *Oil Weekly,* Seven Parts (Oct. 21, 28; Nov. 4, 11; Dec. 2, 9, 16).

Reviews early development of temperature logging, application, interpretation, and problems encountered.

291. ———, 1952, *Electrical Well Logging Fundamentals,* Well Instrument Developing Co. (W.I.D.C.O.), Houston, 164 pp., 301 refs.

Describes fundamentals of resistivity and SP logging as well as applications and interpretations of the logging method.

292. ———, 1963, "Use of Geophysical Logs in Soil Engineering," Amer. Soc. of Testing and Materials, *Special Tech. Publ. No. 351,* pp. 74-85, 8 refs.

The measurement of some of the physical properties of the soils and rocks penetrated by boreholes provides information of value to the soil engineer. Electrical resistivity, radiation, and acoustical properties are particularly useful. The measurements continuously recorded in terms of depths give a detailed log on which geological and physical changes are indicated. By correlating the logs and the core data, the geology of the area and some of the rock properties can be mapped.

WRSIC #W71-05310

293. ———, 1966, "Interpretation of Electric and Gamma Ray Logs in Water Wells," *The Log Analyst,* January-March, 16 pp., 48 refs.

A typical inexpensive logging program for water wells consists of a single-electrode resistivity, the SP, and sometimes gamma ray measurements. Some of the principles underlying the analysis of these logs are reviewed. The usual logging situations are classified into three groups and simple interpretation procedures are outlined for each.

WRSIC #W71-05688

294. Hackett, J. E., 1965, "Ground-Water Contamination in an Urban Environment," *Ground Water,* Vol. 3, No. 3, pp. 27-30.

Ground water contamination may occur from surface or near-surface sources and from subsurface sources. Of all potential sources, the disposal of human and industrial wastes at or near the surface is the most widespread hazard for ground water in northeastern Illinois. Pollution hazards result from individual sewage disposal systems, refuse disposal sites, ponds or settling basins for disposal of sewage and other liquid waste products, and sewage effluent and industrial wastes that have been discharged into streams. The role of the hydrogeologist in assessing ground water contamination hazards and in providing information on the mechanism of contamination is stressed.

295. Haden, E. L. and G. R. Welch, 1961, "Techniques for Preventing Differential-Pressure Sticking of Drill Pipe," *A.P.I. Drill. and Prod. Pract.,* pp. 36-41, 3 refs.

In recent years differential-pressure sticking of drill pipe has been a common occurrence in the Gulf Coast area. Failure to recognize the problem and lack of preventive techniques have made this type sticking costly. A laboratory investigation revealed that in addition to the factors suggested by earlier investigators, mud type and composition influenced the severity of sticking. Tests with surface-active agents and modified collar designs demonstrate these methods to be useful in reducing sticking, but a combination of techniques is necessary for maximum protection.

296. Hagmaier, J. L., 1971, "The Relation of Uranium Occurrences to Ground Water Flow Systems," *Wyoming Geol. Assoc. Earth Science Bull.*, Vol. 4, No. 2, June, p. 19-24, 31 refs., 4 figs.

The relation between uranium and ground water in the sedimentary basins of the Rocky Mountain region is a function of the geologic history of the basins and the past ground water flow systems and their hydrochemistry. Along the regional flow path from recharge to discharge areas, the following chemical facies will theoretically develop for reasonably heterogenous silicate sediments: HCO_3 SO_4 Cl. Because uranium can be easily carried in solution as a carbonate complex, it would likely be leached and transported in the bicarbonate ground water facies and precipitated in the transition zone between the bicarbonate and sulfate facies where a number of other chemical changes occur. The bicarbonate ground water facies is generally associated with regional ground water recharge areas; it follows that the likely location for uranium exploration is in these recharge areas. The ground water and uranium geology in the Powder River Basin, Wyoming, supports this conclusion. WRSIC #W72-00893

297. Hall, H. N., *et al.*, 1950, "Ability of Drilling Mud to Lift Bit Cuttings," *Trans. SPE of AIME,* Vol. 189, pp. 35-46, 10 refs.

Removal of bit cuttings is an important function of drilling muds. In an effort to obtain better understanding of the factors influencing the removal of cuttings, an extensive series of laboratory tests were made in which slip velocities of various sizes and shapes of particles were measured in muds of different physical properties. Empirical equations were then derived from these experimental data. These equations show that slip velocity is dependent on cutting size and shape, mud flow constants, and flow state of the mud. Applicability of these equations for field use is demonstrated by comparing computed slip velocities with slip velocities obtained from field tests.

298. Halliburton Services, 1969, *Halliburton Cementing Tables,* Halliburton Services, Duncan, Oklahoma, 75 pp.

Contains tables on (1) volume and height between tubing and hole and casing and hole, (2) pipe displacement, (3) pipe dimensions and strengths, (4) pipe capacity, (5) volume and height between tubings, between tubing and casing, and between casings, (6) other technical data on cement additives.

299. Halloran, T., 1966, "What to Look for in Selecting Air-Drilling Foaming Agents," *Oil and Gas Journ.*, Vol. 67, No. 23, August, 3 pp.

To get the most out of air drilling, operators looked for better foaming agents. Most surfactants available for oil-field use in the late 1950's were by-products of detergents made for other industrial or even household uses. These products were not versatile enough for the wide range of oil-field conditions. Few soaps and ordinary detergents can give good results in every type of fluid to which they are exposed.
WRSIC #W71-08924

300. Ham, H. H., 1971, "Water Wells and Ground Water Contamination," *Bull. Assn. Eng. Geol.*, Vol. 8, No. 1, Spring, pp. 79-90.

Ground water in urban and other areas of concentrated usage is subject to contamination from a number of sources, among which is the faulty water well. Entrance of contaminants can take place while drilling, during the operational life, or following abandonment. Because of the complex environment, such contamination is often insidious and difficult to rectify. Principal contributing factors are: (1) public and water well industry misconceptions regarding ground water, (2) lack of appropriate statutory regulation and effective enforcement, and (3) the out-of-sight, out-of-mind nature of wells. Common sources of, and avenues for, contamination are examined in light of typical well construction features. Corrective statutory and administrative measures are suggested and research and development needs listed.

301. Hancock, J. C., 1963, "Public Health Aspects of Individual Water Wells," *Ground Water,* Vol. 1, No. 3, July, pp. 27-29, 4 refs.

Local government in Michigan, and perhaps in other states, must try to solve health problems which are within their legally designated area of responsibility, such as the construction and operation of private individual water well systems. Monroe County, Michigan, and Toledo, Ohio, recognized in 1961 the need for regulation of such water well systems. The subsequent development of a Code, and its enforcement have made us realize that we are just beginning to fully understand a subject which is much broader in scope from a public health standpoint than just construction details and bacterial tests. WRSIC #W71-08470

302. Handin, J.,*et al.,* 1963, "Experimental Deformation of Sedimentary Rocks under Confining Pressure: Pore Pressure Tests," *Amer. Assn. Petrol. Geol. Bull.,* Vol. 47, No. 5, pp. 717-755. 55 refs.

Berea sandstone, Marianna limestone, Hasmark dolomite, Repetto siltstone, and Muddy shale have been subjected to triaxial compression tests in which the external confining pressures and internal pore pressures (to 2 kilobars) are applied and measured independently. The interstitial water pressure is maintained constant throughout the test, and porosity changes are determined. The ultimate strength and ductility of porous rocks are found to depend on effective confining pressure – the difference between external and internal pressures when the pore fluid is chemically inert, the permeability is sufficient to insure pervasion and uniform pressure distribution, and the configuration of pore space is such that the interstitial hydrostatic (neutral) pressure is transmitted fully throughout the solid framework.

303. Handin, J. and R. V. Hager, Jr., 1958, "Experimental Deformation of Sedimentary Rocks Under Confining Pressure, Tests at High Temperatures," *Amer. Assn. Petrol. Geol. Bull.,* Vol. 42, No. 12, pp. 2892-2934, 17 refs.

Short-time triaxial compression tests of dry anhydrite, dolomite rock, limestone, sandstone, shale siltstone, slate, and halite single crystals under pressure-temperature conditions simulating depths down to

Handin, J. and R. V. Hager, Jr. . . . . (con't)

30,000 feet reveal the following: (1) Invariably an increase of pressure at constant temperature increases the yield stress, but an increase of temperature at constant pressure reduces it. (2) An increase of pressure at constant temperature always enhances the ultimate strength. However, heating at constant pressure may raise the ultimate strength by increasing the ductility of work-hardening rocks. More commonly, heating lowers the ultimate strength by eliminating work-hardening, so that even though a rock is more ductile, it is weaker because its yield stress is reduced. (3) In any event,for all materials tested except the halite, the ultimate strength at any simulated depth exceeds the crushing strength at atmospheric conditions. Below about 15,000 feet the strength of halite is less than that at the surface. (4) The effect of heating up to 300°C. on the strength and ductility of anhydrite, dolomite, sandstone, and slate is small. However, the strength of limestone, shale, and siltstone at room temperature exceeds that at 300°C. by about 50 percent, and of halite by nearly seven times.

304. Hansen, H. J., 1967, "The Electric Log: Geophysics' Contribution to Ground Water Prospecting and Evaluation," *Maryland Geol. Surv. Inform. Circ.*, No. 4, 11 pp.

Within the last five years, the use of small portable electric logging units within the coastal plain areas of Maryland has accelerated until it now has been classified as a routine operation for evaluating all new water wells. In water well work most logging machines are of the single-electrode type. The single-electrode logger produces a two-curve log consisting of one resistivity and one SP curve. Logs produced by this type of unit should normally be used only for qualitative interpretations. To obtain electric logs useful for quantitative evaluation, a unit having a multiple-electrode probe that can produce three resistivity curves is needed.

305. Hantush, Mahdi, S., 1964, "Hydraulics of Wells," *Advances in Hydroscience,* Academic Press, Vol. 1, 442 pp., New York.

A 150-page collection of information on hydraulics of wells. A full mathematical theory and discussion of the fundamental equations and basic principles is given. Vector calculus is used in the formulation of the fundamental equations. Formulae are applicable to a wide variety of problems. Full treatment is given to: flows to artesian wells; wells partially penetrating artesian aquifers; flow to water table wells; well fields and spacing of wells; drainage wells; flow to collector wells and pumping tests. The methods used are analytical.

306. Harris, H. D., 1967, "Rotary Air-Drilling of Phosphate Deposits in Southeastern Idaho," *Anatomy of the Western Phosphate Field,* 15th Annual Field Conference, Intermountain Assn. of Geol., Salt Lake City, pp. 167-171.

A rapid, low cost method of obtaining samples for the evaluation of phosphate deposits in southeastern Idaho is provided by the use of a truck-mounted rotary drill rig utilizing compressed air as the circulating medium. The seismograph-type drilling rig commonly equipped for mud or air drilling is compact, fast, portable and is suited for operation in the rugged terrain in which the phosphate beds occur in southeastern Idaho. Air-derived cuttings represent a complete and relatively uncontaminated sample of the drilled interval within certain limitations.

307. Harris, O. E., 1961, "Applications of Acetic Acid to Well Completion, Stimulation and Reconditioning," *Journ. Petrol. Tech.*, Vol. 13, No. 7, pp. 637-639.

Acetic acid has been used successfully many times in the past few months in treating various mixtures and in a number of different applications. It has been used as (1) a perforating fluid, (2) a retarded acid without viscosity, (3) a treatment for removal of carbonate scale in the presence of aluminum metal at elevated temperatures, (4) a "kill" fluid for wells, (5) a weak aqueous solution for carrying surfactants to remove emulsions and water blocks in the presence of water-sensitive clays, (6) a first-stage treating fluid ahead of hydrochloric acid for a greater drainage pattern, and (7) a transitory true gel or emulsion for placement of temporary bridging agents. WRSIC #W71-09917

308. Harris, O. E., *et al.*, 1966, "Why Strong Acid Pays off in Carbonate Reservoirs," *World Oil*, Vol. 163, No. 7, pp. 76-80, 6 refs.

Hydrochloric acid in concentrations greater than the 15 percent normally used will significantly improve fracture acidizing results from carbonate reservoirs. Acid properties at concentrations above 15 percent change with increasing concentration. By understanding these changes and how and why they occur, high-strength acid can be advantageously used to provide: (1) greater formation conductivity resulting in sustained production and injection increases. This is particularly true in reservoirs of high water saturation which make weaker solutions of acid ineffective because of rapid dilution. (2) greater penetration because of more rapid initial decrease in area-to-volume ratio. (3) greater volume of CO_2 gas per barrel of reacted acid. This has proved extremely effective in aiding well clean-up. (4) greater penetration due to retardation produced by concentration increases of the acid itself and of the reaction products in solution in the partially spent acid. WRSIC #W71-09916

309. Hartman, H. L. and I. Chao, 1963, "The Simulation of Percussion Drilling in the Laboratory by Indexed-Blow Studies," *Journ. Petrol. Tech.*, September, pp. 214-226, 9 refs.

The drop tester has proved an invaluable tool for the investigation of percussion drilling in the laboratory in "slow motion." It has allowed the process of rock penetration by impact to be studied a single blow at a time. Indexed blows form craters which are influenced by others adjacent to them. Two unique effects are involved: (1) the provision of additional free faces in proximity to the point of impact and (2) the creation of subsurface damage by the previous blows. Both have a pronounced influence on the volume of rock removed per blow which

Hartman, H. L. and I. Chao . . . . (con't)

governs the rate of penetration in actual drilling. A single blow approach also has value in cable-tool drilling. The results of this study are: (1) the optimum index distance on a damaged surface is greater than on a fresh, undamaged rock surface; (2) the volume of rock broken at optimum index distance with wedge is about the same for both surfaces; (3) the optimum index distance and maximum crater volume are proportional to the energy level; (4) at a given energy level, the volume of rock broken by dies generally exceeds that by wedges; and (5) maximum crater volume varies, approximately, inversely with width of die but is nearly independent of included angle of wedge.

WRSIC #W71-02997

310. Heim, A. H. and H. W. True, 1956, "Low Resistivity on Electric Log Could Mean 'Chert,'" *World Oil,* Vol. 142, No. 4, pp. 126, 129-130.

Low resistivities frequently shown on electrical logs in "Mississippi chert" sections have been attributed to a combined result of: (1) high porosities, (2) high formation water salinities, (3) possible interbedded shales. Cores from the Mississippian chert zones of some Oklahoma and Kansas wells were analyzed to find the causes for low resistivities. All cores were a heterogeneous mixture of microcrystalline quartz material and siliceous shale. The quartzitic material (chert) varied from a dense, flint-like form to a chalky and porous form; porosities up to 40 percent were not uncommon and formation factors were as low as five. Cores contained no conductive minerals. WRSIC #W71-09906

311. Helmick, W. E. and A. J. Longley, 1957, "Pressure Differential Sticking of Drill Pipe and How It Can Be Avoided or Relieved," *A.P.I. Drill. and Prod. Pract.,* pp. 25-60, 2 refs.

Investigation of the mechanism of the sticking of drill collars, now called pressure-differential sticking, was initiated after the acute observation was made that spotting oil would only free pipe that had stuck while remaining motionless opposite a permeable bed. Laboratory investigations established that the force acting to hold the pipe against the side of the hole was proportional to the differential pressure acting across the drill pipe and the area of the pipe isolated from the hydrostatic pressure by a thickening mud cake and that the primary mechanism of release by oil is wetting action.

312. Hicks, T. G. and T. W. Edwards, 1971, *Pump Application Engineering,* McGraw-Hill Book Company, New York.

This book was written for the design engineer who must select and apply pumps in any of numerous fields – power, petroleum, chemical process, nuclear, marine, food, etc. – and who must assess the operating conditions for the pump needed, determine the best of received bids and evaluate alternative arrangements, so that if there is a problem, a solution can be found.

313. Highbarger, W. J., 1962, "Obtaining Improved Bit Performance in Abruptly Changing Formations," *22nd Annual Meeting of the A.A.O.D.C.,* Denver, October.

A procedure for drilling in abruptly changing lithologies is discussed. This procedure includes the use of electrical and radioactivity well logs and five foot interval drilling times to arrive at operating conditions for the bit and to aid in bit selection. The results of a number of field tests have shown a significant improvement in bit performance.

314. Hill, H. J., *et al.*, 1955, "Effect of Clay and Water Salinity on Electrochemical Behavior of Reservoir Rocks," *Petroleum Branch Fall Meeting.*

In quantitative interpretation of electrical logs the presence of clay minerals introduces an additional variable. Six typical sandstone formations having a wide variety of petrophysical properties were selected for the study. An empirical equation has been developed which quantitatively relates formation resistivity factor to saturating solution resistivity, porosity, and "effective clay content." This relation is indicated to be uniformly applicable to clean or shaly reservoir rocks.

315. Hill, H. R., 1966, "New Technique in Rotary Shot Drilling with Reversed Circulation," *Water and Water Eng.*, Vol. 70, No. 843, May, pp. 195-199.

Drilling of water wells by chilled shot rotary methods including the use of the old forward circulation and the new reversed circulation is described. The use of reversed circulation has been noted as a major advance in rotary drilling with significant improvements in well bore and completed well conditions. Some of the disadvantage of this method are stuck tools, vertical drifts, the extent of permanent partial sealing taking place in formation fissures, and fluid sampling. The more important outstanding advantages of this new technique are listed. Two diagrams are provided that illustrate the construction and operation. Some of the calculation criteria are also considered.

316. Hockstra, P., *et al.*, 1965, "Frost Heaving Pressures," *U. S. Army Mat. Command,* Cold Regions Research and Eng. Lab., Research Report 176, Hanover, New Hampshire.

Explores frost heaving pressures and effect on radial-expansion joints with respect to potential compressive forces. An engineering study.

317 Hodges, A. L. Jr., 1968, "Drilling for Water in New England," *Journ. New England Water Works Assoc.* Vol. 82, pp. 287-315.

Considerations in ground water development; general features of various types of wells; drilling equipment and construction of drilled wells, particularly rock wells most common in New England; well development; and the use of water well pumps.

318. Hooper, W. F. and J. W. Earley, 1961, "Compositional Logging of Air-Drilled Wells," *Amer. Assn. Petrol. Geol. Bull.*, Vol. 45, No. 11, November, pp. 1870-1883.

Mineralogical compositions of representative samples of air-drilled cuttings taken at ten-ft. intervals from two wells in the Punxsutawney

Hooper, W. F. and J. W. Earley . . . . (con't)

Driftwood gas field of central Pennsylvania have been determined by routine methods of X-ray analysis; distinct mineralogical zones were found to correspond closely with formations and formational contacts; although wells are 18 mi. apart, correlations can be made.

319. Hopkin, E. A., 1967, "Factors Affecting Cuttings Removal During Rotary Drilling," *Journ. Petrol. Tech.*, Vol. 19, No. 6, June, pp. 807-814, 10 refs.

Laboratory tests and field experience in Canada indicated the magnitude of some of the factors affecting ability of drilling mud to clean the hole. A correlation was observed between funnel viscosity and particle slip velocity. A relationship was observed between the Bingham yield value of the mud and the particle slip velocity. Increasing the mud density, creating laminar annular mud flow, or rotating the drill pipe may improve the carrying capacity of mud. WRSIC #W71-02991

320. Howard, G. C. and C. R. Fast, 1970, "Hydraulic Fracturing," Monograph, *Soc. Petrol. Eng., A.I.M.E.*, Vol. 2, 210 pp., 185 refs.

Hydraulic fracturing is a method for increasing well productivity by fracturing the producing formation and thus increasing the well drainage area. This monograph is a thesis on hydraulic fracturing covering the state-of-the-art from the theory and technique of hydraulic fracturing to the application of nuclear energy as a means of cracking the reservoir rock and forming rubble. WRSIC #W71-06950

321. Howard, G. C. and P. P. Scott, Jr., 1951, "An Analysis and the Control of Lost Circulation," *Petrol. Trans., A.I.M.E.*, Vol. 192, pp. 171-182, 41 refs.

During the drilling of wells, fractures which are created or widened by drilling fluid pressure are suspected of being a frequent cause of lost circulation. A study of the variables which are believed by the authors to be related to fracturing led to the premise that the presence of an effective lost circulation material in a drilling mud stream would serve to plug small fractures at the moment they are encountered or created and, thereby, eliminate a possible cause of lost circulation by preventing fluid pressure from widening and extending the fractures. A large number of lost circulation materials were classified and tested in simulated fractures and in shallow wells. Granular type materials with a gradation of particle sizes were found to be the most effective for plugging. Within the limits of pumpability, the concentration in mud of each lost circulation material tested was found to be a controlling factor in determining the maximum size fracture which could be sealed. A new type centrifugal sorter was given preliminary tests and found to be capable of concentrating cuttings for rejection and reclaiming lost circulation materials.

322. Hoxie, E. C., 1964, "Corrosion of Well Points and Piping in Domestic Water near A.E.C. Plant," *Ground Water*, Vol. 2, No. 3, July, pp. 45-48, 1 ref.

Report of investigation to determine cause of excessive corrosion of construction materials of private wells supplying domestic water to several communities bordering Atomic Energy Commission's Savannah River Plant; it was found that corrosion resulted from natural corrosiveness of soft water with high carbon dioxide and oxygen content, no effect from atomic energy plant was indicated. WRSIC #W71-09726

323. Hubbert, M. K., 1969, *The Theory of Ground-Water Motion and Related Papers,* Hafner Publishing Co., 311 pp.

The early analytical treatments of ground water flow have mostly been founded upon the erroneous concept, borrowed from the theory of the flow of the ideal frictionless fluids of classical hydrodynamics, that ground water motion is derivable from a velocity potential. This conception is in conformity with the principle of conservation of matter but not with that of the conservation of energy. In the present paper it is shown that a more exceptionless analytical theory results if a potential whose value at a given point is defined to be equal to the work required to transform a unit mass of fluid from an arbitrary standard state to the state at the point in question is employed. The remainder of the work is devoted to deducing the consequences of Darcy's law with particular regard for the practical problems of ground water hydrology.

324. Hudeck, R. R., 1969, "How to Select and Use Drill Bits Profitably," *World Oil,* Vol. 169, No. 4, September, pp. 73-77.

Although rock bit shapes are similar, the angle and length and design of the teeth must change for soft through hard formations. This article describes how to recognize, understand, and apply these varied bit designs to best advantage. By selecting and using the proper bit and by programming rig performance to match bit capabilities, the contractor can also: (1) drill straighter and safer hole, (2) reduce fishing jobs, (3) improve design of bottom hole assemblies, (4) reduce trips, and (5) apply optimum weight and rpm if formation characteristics are known. WRSIC #W71-02985

325. Hurr, R. T., 1966, "A New Approach for Estimating Transmissibility from Specific Capacity," *Water Res. Research,* Vol. 2, pp. 657-664.

Shows a method for estimating transmissibility from the single-point observation by a manipulation of the Theis nonequilibrium formula.

326. Ingham, E. C., 1955, "The World's Deepest Cable Tool Well," *Drilling,* Vol. 16, No. 10, August.

News Release of record making total depth of cable-tool rig, e.g., 11,145 feet in New York in 1953.

327. Ionnesyan, R. A. and Y. R. Ioanesyan, 1971, "Development of Deep Well Turbodrilling Technique," *Neft. Khoz.,* No. 5, May, pp. 23-26, (in Russian).

Work on the development of new turbodrill designs for turbodrilling rationalization is discussed. The old turbodrill's flow rate was artificially

Ionnesyan, R. A. and Y. R. Ioanesyan . . . . (con't)

increased, and this made them less competitive than rotary drilling. The use of large-size bits makes turbodrilling more efficient. The turbodrilling practice in the U.S.S.R. showed that the rubber bearings of old turbodrills do not allow them to run at low rpm. At present, the turbodrills with ball bearings and packing glands operable at pressure drops up to 60 atm are under production in the U.S.S.R. New design of a turbodrill with rotating casing for diamond drilling also is discussed.

328. Jackson, G. O., 1970, "Economic Application of Insert Bits," *30th Annual Meeting of Amer. Assn. Oil Well Drilling Contractors,* September, 12 pp.

Sealed bearing insert bits have proven to be economical in a wide variety of applications previously drilled by sealed and nonsealed milled tooth bits. A practical approach to evaluating the possible economy of a sealed bearing insert bit is to examine past performance of milled tooth bits, the logs, and the mud programs which produced that performance. A cost analysis of the bit record and a study of the bit record in conjunction with the log and recorded dull condition have proven to be a good approach. Long intervals of similar drillability and cost per foot usually lend themselves to the economical application of a sealed bearing insert bit. Drilling problems, such as deviation, undergage sections and lost circulation, are added cost incentives.

WRSIC #W71-09904

329. Jacob, C. E., 1944, "Notes on Determining Permeability by Pumping Tests under Water-Table Conditions," *U.S. Geol. Surv. Special Report.*

Discusses case histories of pumping tests of wells under water-table conditions with a view toward determining permeability. Report is one of earlier classic papers which brought forth new understanding of basic pump test principles.

330. Jacob, C. E., 1967, "Research Trends in Groundwater Movement and Well Hydraulics," *Proc, Natl. Symp. Groundwater, Hydrol.,* Am. Water Res. Assoc., San Francisco., pp. 135-137.

Explains the studies and research in subsurface geotechnology and geoscience in which the New Mexico Institute of Mining and Technology is interested. Areas of needed research are discussed.

331. Jaeger, J. C., 1960, "Shear Failure of Anisotropic Rock," *Geological Mag.,* Vol. 97, No. 1, pp. 65-72, 6 refs.

The two-dimensional theory of two simple generalizations of the Coulomb-Navier criterion for shear failure is developed. A material with a single plane of weakness which has a different shear strength and coefficient of internal friction from the remainder of the material may fail, according to circumstances, either in the plane of weakness or in planes cutting across it. A layered material whose shear strength varies continuously from a maximum in one direction to a minimum in the perpendicular direction has only one possible plane of failure.

332. Jaeger, J. C. and N. G. W. Cook, 1969, *Fundamentals of Rock Mechanics,* The Chaucer Press, Bungay, Suffolk, England, 513 pp.

The results of investigations of the mechanical properties of rocks are described in a great number of papers scattered over many scientific and engineering journals. The authors list over 600 references in presenting the mathematical and experimental foundations of the mechanical behavior of rock. Following an analysis of stress and strain, the theories of elasticity, perfectly plastic and other rheological materials, Coulomb aggregate and porous elastic materials are treated mathematically. Laboratory studies on the behavior of rock specimens are linked to observed rock masses in the field. Applications are made to geology, mining, excavations, and slope stability.

333. Jann, R. H., 1966, "Method for Deep Well Alignment Tests," *A.W.W.A. Journ.,* Vol. 58, No. 4, April, pp. 440-445.

This article describes a convenient method for plotting and interpreting the results of deep well alignment tests. It is common in making alignment tests for persons to obtain accurate field data and to reduce the readings with mathematical precision and then to interpret the results with vague or inadequate methods. It describes methods used in making an alignment test of a deep well. WRSIC #W71-13969

334. Jensen, O. F., Jr., 1961, "General Industrial Water Well Design Factors," *Industrial Water and Waste Conference,* Texas Water and Sewage Works Assn., Rice Univ., Houston, March, 30 pp.

Industrial water well design may best be defined as a progressive quantitative evaluation using engineering principles of several scientific fields, coordinated with the methods of testing and construction. The design must be based on the natural geologic, hydrologic, and chemical conditions of the ground water prevailing in the area of a proposed well. The pumping rate, or yield, and quality of water produced are direct functions of these natural conditions. The design of a water well system embraces several phases. The preliminary survey of a proposed well location together with the drilling of a test hole or pilot hole survey provide basic data of the prevailing geologic, hydrologic, and chemical quality of the ground water conditions in a proposed area. These data together with application of water well testing and construction methods provide the basis for determining a well design that will result in the most favorable operational performance. The preliminary survey, test hole, or pilot hole survey of a proposed well location, construction methods, and well hydraulics are discussed. WRSIC #W71-05691

335. Jenson, O. F., Jr. and W. Ray, 1964 "Photographic Evaluation of Water Wells," *A.W.W.A. 53rd Annual Conference,* Southwest Section, Houston, 20 pp.

Photographs of wells in stereoscopic, three dimensional pairs are used for evaluation of well conditions. The pictures provide accurate detail pictures from which determination can be made for remedial procedures or for inspection of a well from top to bottom. These stereoscopic photograph surveys are used for evaluation of new well completions, damaged

Jenson, O. F., Jr. and W. Ray . . . . (con't)
well, junk in wells, and formation material identification.
WRSIC #W71-05299

336. Jensen, O. F., Jr., and W. Ray, 1965, "Photographic Examination of Wells," *A.W.W.A. Journ.*, Vol. 57, pp. 441-447.

Stereographic photography and associated camera equipment enables accurate examination of wells as small as 5 in. ID to a depth of 10,000 ft.

337. Johnson, A. S., 1967, "The Development of Techniques and Equipment for Deeper U.S. Mineral Exploration," *12th Symposium on Exploration Drilling,* Univ. of Minnesota, Sch. of Min. and Metal. Eng., Minneapolis, October, pp. 229-242.

Reviews history of related industry (petroleum), and its effect on the minerals industry, techniques of coring, equipment, coring economics, etc.

338. Johnson, C. R. and R. A. Greenkorn, 1963, "Description of Gross Reservoir Heterogeneity by Correlation of Lithologic and Fluid Properties from Core Samples," *Bull. International Assoc. Sci. Hydrol.,* Vol. VIII, No. 3, pp. 52-63, September.

The paper presents a method for the geologic assessment of the distribution and character of heterogeneity for reservoir studies.

339. Johnson, H. P., 1964, "Meter for Measuring Flow Discharge from Pipes," *Agr. Eng.,* Vol. 45, pp. 378-379.

Device is constructed whose principal application is measurement of discharge from wells. Designed to be commonly used, it is simple in principle, small and light in weight, convenient to use, reasonably accurate and inexpensive.

340. Jones, E. E., "Well Construction and Water Quality," *1971 Winter Meeting of Amer. Soc. Agricultural Eng.,* Chicago, December, Paper No. 71-703, 15 pp., 3 refs., 2 tabs., 15 figs.

Wells lacking adequate sanitary protection serve as unauthorized uncontrolled ground water recharge points. Common easily detected sanitary defects are described. A detailed investigation of a contaminated well apparently having adequate sanitary protection was made. Design recommendations are not included.

341. Jones, P. H. and T. B. Burford, 1951, "Electric Logging Applied to Ground Water Exploration," *Geophysics,* Vol. 16, No. 1.

A method is described for the determination of the quality of ground water in granular aquifers penetrated by rotary-drilled holes electrically logged. Conventional techniques of electric-log interpretation to determine true-bed resistivity from apparent resistivity values are briefly described; and a method for converting water-resistivity values into hypothetical chemical analyses is explained. The objective of the method is to narrow the limits of error in quality-of-water estimates based upon electric logs. The paper deals also with methods of determining formation porosity *in situ,* which is an important factor in salt-water-encroachment problems.

342. Jorgensen, D. G., 1969, "Field Use of Orifice Meters," *Ground Water,* Vol. 7, No. 4, July-August, pp. 8-11, 4 refs.

A well designed and calibrated orifice meter is an accurate and inexpensive measuring device for flow. End-line orifices can be calibrated at work sites by solving an equation that interrelates easily measured dimensions of the orifice and outflow. WRSIC #W69-08702

343. Kading, H. W. and J. S. Hutchins, 1969, "Temperature Surveys: The Art of Interpreting," Acid Treating, *A. P. I. Drill. and Prod. Pract.,* pp. 1-20, 5 refs.

Temperature logs are in common use in both drilling and production efforts. Because of their basically empirical experience-based development and semi-quantitative nature, the industry was slow to accept the use and results of this instrument. There are specific applications where temperature logs provide comparable or superior information to other investigative logging tools. In each case, however, qualitative results depend strongly on the proper commitment of equipment capability to well condition (before and during logging) and to down-hole logging methods. Six accepted temperature survey uses which are covered in detail are: cement tops, gas entry or channels, water-reduction logs, fluid-injection profile (shut-in temperature profiles, water-injection wells), fracture-evaluation logs, acid-evaluation logs.
WRSIC #W71-10427

344. Kastrop, J. E., 1971, "Automatic Drilling Data Acquisition, Processing and Telemetry," *Petrol. Eng.,* Vol. 43, No. 9, August, pp. 39-62.

The host of available, complex drilling systems present a twofold problem: available systems offer so much capacity that is difficult for the average drilling organization to utilize it efficiently and productively, and because of the high degree of sophistication, cost is a significant factor. This paper is a "mini-symposium" of automatic Drilling Data Aquisition, Processing, and Telemetry (ADAPT) supplied by companies to solve these problems.

345. Keech, D. K., 1970, "Ground Water Pollution," *Principles and Applications of Ground Water Hydraulics Conference,* Kellogg Center, Michigan State Univ., East Lansing, Michigan, December, 20 pp., 8 refs.

The purpose of this paper is to show, mainly by example, how the valuable ground water resource has been contaminated primarily through negligent practices. Pollution of ground water results in degradation of the water quality in one of three categories: microbiological, chemical, or physical. WRSIC #W71-05693

346. Keller, G. V., 1956, "Development of Well Logging in the U.S.S.R.," *7th Annual Soc. of Prof. Well Log Analysts Symposium,* Tulsa, May.

Keller, G. V. . . . . (con't)

In Soviet practice, a widely used set of logs for interpretation consists of a sequence of six or more lateral-spacing logs. With these, resistivity variations are studied in detail and conclusions drawn about fluid saturations. A combination neutron-neutron and neutron-gamma log is widely used to determine porosity. Pulsed neutron logs are used extensively to determine water saturations in cased wells. Soviet well logging groups are currently faced with the familiar problem of upgrading their borehole tools and cables to survive under the higher temperatures and pressures encountered in newly opened fields. Well logging methods are also used widely in non-petroleum boreholes, particularly in evaluating coal beds and metallic mineral deposits.

347. Kelley, D. R., 1969, "A Summary of Major Geophysical Logging Methods," *Penn. Geol. Surv. Bull. M 61,* 88 pp., 98 refs.

The major subsurface well-logging methods available domestically are reviewed. The principles upon which these methods are based, the effects to be considered from the evaluation of these methods, and the uses of these methods are summarized in text, figures, and tables. The type of subsurface information is stressed rather than the geophysical principles upon which the methods are based.

348. Kempe, W. F., 1967, "Core Orientation," *12th Annual Drilling and Blasting Symposium;* Univ. of Minnesota, School of Min. and Metal. Eng., Minneapolis, pp. 161-182.

There is nothing more valuable to authoritative decision making than reliable data in sufficient quantity. This is equally true in the realm of formation evaluation. For purposes of soil and substrata investigation several means are available. Those selected depend on the kind of information sought and on the type of formation to be studied. Each one of these methods accomplishes certain functions. Most of them have serious limitations and all, except the seismic, need a borehole to receive the logging instruments. Coring is the only direct method of formation evaluation. Coring is explored in some detail, with emphasis on *in situ* orientation of rocks.

349. Kendall, H. A., 1965, "Application of SP Curves to Corrosion Detection," *Journ. Petrol. Tech.,* Vol. 17, No. 9, September, pp. 1029-1032.

A correlation was obtained between active internal electrochemical corrosion occurring in a producing oil well and a standard SP log. The observations were confirmed by extensive laboratory tests in which selected sections of tubing were placed in a corrosive environment. A potential profile was obtained for each tubing joint after which the joint was sectioned longitudinally and visually examined for active corrosion cells. The anodic areas, as determined by a ferroxyl test, were correlated with the potential profile.

350. Kendall, H. A. and W. C. Goins, 1960, "Design and Operation Jet-Bit Programs for Maximum Hydraulic Horsepower, Impact Force or Jet Velocity," *Trans. SPE of AIME,* Vol. 219, pp. 238-250, 17 refs.

This paper shows the maximum obtainable bit horsepower, impact force, and jet velocity at all depths, taking into account the limitations of the pump, piping, hole, and minimum circulating rate for adequate cuttings removal. Ranges of operation are developed; and flow rates, surface pressure, and bit pressures are specified for each range to provide a maximum of any one of the desired effects.

351. Kennedy, J. L., 1968, "Preshearing, Pelletizing Improves Asbestos Fibers for Drilling Mud," *Oil and Gas Journ.*, October, pp. 93-95.

Asbestos materials used as drilling mud additives provide one method of increasing the carrying capacity and suspending ability of clear fluids, without sacrificing the high penetration rates possible when using clear water. But asbestos products have presented two main problems in the past, e.g., high bulk and slow yield. An improved asbestos is described which reduces the major previous problems.

352. ______, 1970, "Drilling Porosity Log Proves Accurate," *Oil and Gas Journ.*, August, pp. 53-55.

Continued refinement in the use of drilling data has been a big factor in keeping oil drilling costs within reasonable limits as materials and labor costs continue to climb. Drilling data can be gathered on a given bit run and analyzed in time to determine the optimum drilling conditions to be used on the succeeding bit run. The use of computers is the basis for this advancement. It has been concluded that drilling response is related to the shear and compressive strength of rocks, and these properties are used to indicate porosity in several logging techniques. Thus, rock properties used initially for predicting bit performance and drilling optimization are now used for formation evaluation. One of the resultant new logs is the drilling porosity log (DPL). It interprets drilling data that are related to formation properties providing an early indication of formation type and porosity.

WRSIC #W71-03850

353. ______, 1971, "Computer Drilling System Can Provide Optimization, Rig Control," *Oil and Gas Journ.*, Vol. 69, No. 19, May, pp. 61-64.

In the last few years, a trend has developed toward more rig instrumentation, and more sophistication in the analysis of drilling variables has resulted. Development of the hardware has been slow, application of the technologies that are available has been even slower.

354. Kerr, J. W., 1970, "Consider Lost Circulation in Well Planning; Rig Selection," *Oil and Gas Journ.*, Vol. 68, No. 38, September, pp. 92-94.

Lost circulation continues to be a serious drilling problem. Sufficient occurrences can double the cost of an oil well. The ability to calculate formation pore pressures and fracture gradients has done much to reduce the frequency of lost circulation problems. Using the minimum mud weight required to control the formation pore pressure will not only speed drilling but will also result in less mud loss. Lost circulation can be classified as: (1) seepage loss to highly permeable, uncemented,

Kerr, J. W. . . . . (con't)

shallow sands, gravel or shell beds, or vugular limestone, (2) sudden complete loss to cavernous formation, and (3) partial or complete loss to natural or induced fractures. WRSIC #W71-03834

355. Keys, W. S., 1967, "Aquifer Evaluation with Radioisotope Well Logs," *Proc. Natl. Symp. Groundwater Hydrol,* Am. Water Res. Assoc., San Francisco, pp. 319-328.

Equipment and specific applications of radioisotope logging techniques as used in a research project on borehole geophysics as applied to geohydrology. Radiation logs utilized include natural gamma, gamma-gamma, neutron-gamma, neutron-epithermal neutron, and radioactive tracer. Information provided by these logs include nature and character of ground water movement, aquifer properties, etc.

356. ———, 1967, "Borehole Geophysics As Applied to Ground Water," *Mining and Ground Water Geophysics/1967,* Geol. Survey Canada, Econ. Geol. Report 26, pp. 598-614, 20 refs.

The Water Resources Division of the U.S. Geological Survey has conducted research on the application of borehole geophysics to ground water investigations. The various logging techniques include: natural gamma, gamma-gamma, neutron-gamma, neutron-neutron, spontaneous potential, single point and short and long normal resistivity, caliper, sonic velocity, flowmeter, radioactive tracer, fluid resistivity, and gradient and differential temperature. An inexpensive system for recording logs on magnetic tape also has been developed. Quantitative interpretation of these logs will provide numerical values for the elements necessary to construct analog models of ground water systems, and data from logs are of considerable value in the design and interpretation of surface geophysical surveys. Geophysical logs can be interpreted in terms of the geometry, resistivity, bulk density, porosity, permeability, moisture content, and specific yield of water-bearing rocks and the source, movement, and chemical and physical characteristics of water. WRSIC #W70-09400

357. ———, 1968, "Well Logging in Ground Water Hydrology," *Ground Water,* Vol. 6, No. 1, January-February, pp. 10-18, 12 refs.

In 1966, more than 50 billion gallons of water was pumped daily from an estimated 10 to 15 million water wells in the United States, one-sixth of the national withdrawal of water. On the basis of past rates of increase, a much greater future use of ground water is suggested. Although most petroleum well logging techniques may be utilized in hydrology, modification in equipment and interpretation are necessary because of basic economic and environmental differences between petroleum and ground water evaluation. If logging is to be widely applied to ground water exploration and evaluation, the expense of equipment and services must be reduced. This can be accomplished, because most water wells are not as deep as oil wells, and the temperatures and pressures are lower. The U.S. Geological Survey is conducting research on the application of borehole geophysics to ground water hydrology. WRSIC #W71-02988

358. Keys, W. S. and L. M. MacCary, 1971, "Application of Borehole Geophysics to Water-Resources Investigations," *U.S. Geol. Surv. Tech. Water Resrouces Inv.*, Book 2, Chapt. E1, 133 pp., 120 refs.

This manual is intended to be a guide for hydrologists using borehole geophysics in ground water studies. The emphasis is on the application and interpretation of geophysical well logs, and not on the operation of a logger. Most of the logs described can be made by commercial logging service companies, and many can be made with small water well loggers. Geophysical well logs can be interpreted to determine the lithology, geometry, resistivity, formation factor, bulk density, porosity, permeability, moisture content, and specific yield of water-bearing rocks, and to define the source, movement, and chemical and physical characteristics of ground water. WRSIC #W72-04203

359. Killiow, H. W., 1966, "Fluid Migration Behind Casing Revealed by Gamma Ray Logs," *The Log Analyst,* Vol. 6, No. 5, pp. 46-49.

Briefly explores the use of the gamma log in deflecting fluid migration behind casing. Describes problems involved in interpretation and identification of gamma responses to fluid shield of gamma producing formation materials.

360. King, G. R., 1959, "Why Rock Bit Bearings Fail," *Oil and Gas Journ.*, Vol. 57, No. 47, November, pp. 166-182.

This report sums up years of testing work done on rock-bit bearings. Many of the most common materials in drilling mud drastically shortened bearing life. Explores the sealed bit versus common bit.

361. King, R. A. and J. D. A. Miller, 1971, "Corrosion by the Sulphate-Reducing Bacteria," *Nature,* Vol. 233, No. 5320, December, pp. 491-492, 13 refs.

There are several explanations for the aggravation of corrosion of iron and steel in oxygen-free conditions by sulfate-reducing bacteria: (1) stimulation of the cathodic part of the corrosion cell by the removal and utilization of the polarizing hydrogen by the bacteria; (2) stimulation of the cathodic reaction by solid ferrous sulfides formed by the reaction of ferrous ions with sulfide ions produced by bacteria; (3) stimulation of the anodic reaction, metal dissolution by bacterially produced sulfide; (4) local acid cell formation; and (5) formation of iron phosphide by reaction of the metal with bacterially reduced phosphates. A new theory is proposed such that the cathodic reaction (hydrogen evolution) occurs on the ferrous sulfide produced by reaction of ferrous ion with bacterically produced sulfide ion. It has been found that the ferrous sulfide activity is diminished with time, possibly owing to the bonding of atomic hydrogen within the ferrous sulfide crystals.

362. Klaer, F. H., Jr., 1963, "Bacteriological and Chemical Factors in Induced Infiltration," *Ground Water,* Vol. 1, No. 1, pp. 38-43, January.

Discusses general processes by which natural deposits of sand and gravel serve as large natural filter beds.

363. Klapka, K., 1967, "What is Going On and What May be Coming in Diamond Bit Design," *12th Symposium on Exploration Drilling,* Univ. of Minnesota, Sch. of Min. and Metal. Eng., Minneapolis, pp. 79-93.

The manufacturers of industrial diamond products as well as those engaged in the design and manufacture of drilling equipment realized that the new technological advances must also be applied to this industry. Whether man-made diamonds were developed to a point where they would become a factor in drilling was studied. New metals were evaluated. Many changes have been made in diamond bits, and the proper selection of a diamond bit is described in this paper. Years of performance studies have given manufacturers the information necessary to recommend the changes required in a bit so that it performs with maximum efficiency. Consideration is given to the correct selection of bit specification items, if the formation to be drilled is known, other aspects of diamond bit design are explored.

364. Klug, M. L., 1966, "The Ground Water Consultant Looks to the Driller," *Ground Water,* Vol. 4, No. 4, October, pp. 20-22.

The consultant expects the driller to provide good data and competent performance. Contacts between the consultants and the drillers have a tendency to upgrade the quality of the basic data collected. As a result, everybody associated with the water well industry – drillers, consultants, and clients – stands to gain by these contacts.

WRSIC #W71-09908

365. Klug, M. L. and W. F. Guyton, 1960, "Wells and Well Fields in Texas Gulf Coast Sediments," *Water Well Journ.,* Vol. 14, No. 1, January, pp. 14, 27-30.

Geologic and hydraulic characteristics of the aquifers in the vicinity where a ground water supply is to be developed must be carefully evaluated to obtain the most economical and efficient wells and well field. This article deals primarily with wells in the Gulf Coast sediments and much of it is applicable to other parts of the United States.

WRSIC #W71-07204

366. Koenig, L., 1960, "Economic Aspects of Water Well Stimulation," *A.W.W.A. Journ.,* Vol. 52, No. 5, May, pp. 631-637.

From the analysis of technical and economic data of the survey reported in this article, it is concluded that water well stimulation has shown highly favorable results, not only in its technical performance, but also by the economic criteria chosen. In the majority of instances, stimulation has added specific capacity at a unit cost less than the unit cost of the specific capacity produced by the original well. The median cost of improvement of even the least favorable method by this economic criterion occurs at a unit cost less than half the unit cost of the original specific capacity. Based on these analyses and results, the author recommends that ground water well stimulation be much more widely considered as a technique for developing ground water supplies.

WRSIC #W71-13719

367. _______, 1960, "Effects of Stimulation on Well Operating Costs and its Performance on Old and New Wells," *A.W.W.A. Journ.*, Vol. 52, No. 12, December, pp. 1499-1512, 4 refs.

Economy through stimulation is determined by "payout time" which refers to days of operation between time of stimulation and time when cumulative savings in pumping energy costs at same production rate equal cost of stimulation treatment, hydraulic fracturing, pressure acidizing, shootings, surging and vibratory explosion. Other methods of stimulation are evaluated for various types of water use; formula for payout-time calculation. WRSIC #W71-13710

368. _______, 1960, "Survey and Analysis of Well Stimulation Performance," *A.W.W.A. Journ.*, Vol. 52, No. 3, March, pp. 333-350, 11 refs.

Data from a survey of wells throughout the U.S., presented in tables and graphs, demonstrate the success of well stimulation (defined as the reduction or removal of underground resistance to flow by mechanical, chemical, or other means, but not including reconstruction); the production of crude oil has produced more advanced techniques of well stimulation than the conventional methods used in the water well industry. Water-producing wells and wells for the disposal of waste and flood waters were studied; five categories of stimulation (in ascending order of efficiency) are surging, shooting, vibratory explosion, pressure acidizing, and hydraulic facturing. WRSIC #W71-03849

369. _______, 1961, "Relation Between Aquifer Permeability and Improvement Achieved by Well Stimulation," *A.W.W.A. Journ.*, Vol. 53, No. 5, May, pp. 652-670, 22 refs.

Variation in permeability of aquifers has certain bearing on improvement achieved by well stimulation; relation is defined in numerical terms; it is concluded that improvement ratios achieved by every treatment type decrease with increases in permeability factor, methods to increase permeability in case of various types of well stimulation; some of the techniques used in the petroleum industry should be adapted to water wells. WRSIC #W71-03840

370. Koopman, F. C., *et al.*, 1962, "Use of Inflatable Packers in Multiple-Zone Testing of Water Wells," *U.S. Geol. Surv. Prof. Paper 450-B*, pp. B108-B109.

Packers have been used for many years in the testing and completing of oil wells, but have not been used extensively in testing water wells. Recent developments make the use of inflatable straddle packers feasible for testing sandstone and siltstone aquifers. One such test of aquifers of Permian to cretaceous age near Los Animas, Colorado is described in this paper. WRSIC #W71-13971

371. Krueger, R. F., *et al.*, 1967, "Effect of Pressure Drawdown on Clean-up of Clay or Silt Sandstone," *Journ. Petrol. Tech.*, Vol. 19, No. 3, pp. 397-403, March, 13 refs.

Laboratory experiments have shown that restriction of oil flow rates during clean-up of water sensitive sandstone invaded with fresh water or

Krueger, R. F. . . . . (con't)
filtrate from fresh water-base drilling fluid can materially reduce the amount of permeability damage. High rate clean-up of similar sandstone cores resulted in significantly higher damage. Attempts to repair permeability damage in cores through use of an oil backflush followed by low-rate reverse oil flow were only partially successful. Clean-up methods were usually ineffective for removing permeability damage in the Berea Sandstone. Clean-up of temporary permeability damage in clastic aquifer supplying domestic water supplies can be done via similar reasoning.

WRSIC #W71-02992

372. Kunnemann, E. A., 1967, "Dual String Drilling Method Yields Improved Samples," *12th Symposium of Exploration Drilling,* Univ. of Minnesota, Sch. of Min. and Metal. Eng., Minneapolis, pp. 195-201.

The dual string drilling concept and the merits of this system as an exploration tool has been recognized for some time. New developments and applications with the system in the last few years have created considerable interest in the industry. The special equipment needed to convert a standard drill rig to *Con-Cor* drilling will vary somewhat depending upon the particular job and the hydraulic system selected. Normally, the dual string pipe, a power swivel with dual swivel connections for rotating the string, and with large radius goose neck along with the rotating head and *Con-Cor* diamond core head or drill bits with some modifications required to the hydraulic system, is all that is required to adapt to this method. In the case of air or aerated water drilling, a suitable air compressor, mist pump for chemical injections, and a sample catcher chamber are also required. The drilling process is described in detail. The *Con-Cor* method has been employed to obtain continuous core samples as well as formation chip samples for analysis and has considerable application to the mineral and ground water exploration industries.

373. ________, 1970, "Geometry Considerations in Application of New Sealed-Bearing Carbide Insert Bits," *Drilling Contractor,* September-October, pp. 49-52.

This paper reviews oil well drilling bit geometry of steel tooth bits and the new generation carbide insert sealed bearing bits with limitations. Progress has been made in improved oil well drilling bits. Primarily, these improvements have centered around the sealed bearing carbide insert type bits.

374. Kutter, H. K., 1969, "The Electrohydraulic Effect: Potential Application in Rock Fragmentation," *U.S. Bur. Mines Rept. of Inv. 7317,* Bur. Mines, Twin Cities, Mining Research Center, St. Paul.

Cavitation was first considered as a possible new energy source for fragmenting rock, but small-scale tests indicated that the pulses from the collapse of tiny vapor bubbles lacked sufficient energy to produce effective fracturing. This fact led to an investigation of the underwater spark discharge which is a more effective generator of collapsible gas or vapor bubbles. The secondary collapse pulse, however, turned out to be negligible compared with the primary pressure pulse caused by the sudden expansion of the spark channel. The explosive expansion of the spark channel, or the electrohydraulic effect, thus became the prime interest of this study.

375. La Moreau, P. E. and W. J. Powell, 1963, "Stratigraphic and Structural Guides to the Development of Water Wells and Well Fields in a Limestone Terrain," *International Assoc. Sci. Hydrol. Comm. Subterranean Waters Pub. 52,* 13 pp.

Explores the problems involved in well-site selection and well development in limestone terrains. Local geologic controls on ground water availability is also treated.

376. Langelier, W. F., 1936, "The Analytical Control of Anti-Corrosion Water Treatment," *A.W.W.A. Journ.,* Vol. 28, p. 1500.

Examines the factors affecting $CaCO_3$ solubility equilibria. In pH range of 6.5 to 9.0 develops formula for Saturation Index (Langelier Index).

377. Langston, J. W., 1966, "A Method of Utilizing Existing Information to Optimize Drilling Procedures," *Journ. Petrol. Tech.,* June, pp. 677-686, 13 refs.

Describes a way to use existing information under day-to-day competitive drilling circumstances. Pitfalls when striving for optimization are noted, e.g., if one procedure increases drilling rate, this facet of the operations should not be over emphasized to the neglect of others.

378. Lankford, B. B., Jr., 1970, "Solving Four Common Drilling Problems: Keys to Milling, Washover Operations," *Oil and Gas Journ.,* Vol. 68, No. 38, September, pp. 97-98, 103-104.

There are many methods of retrieving "junk" from the hole, and each one has its own special application. In almost all cases, a fishing expert should be on location to supervise the operations. The details of the operation which are handled by the fishing expert are not discussed but some of the basic decisions which must be made before fishing is begun are considered. Washover operations are used to retrieve stuck drill pipe on drill collars, and in some instances, casing. Before washover operations are begun, several factors must be considered, i.e., (1) drilling fluid condition, (2) method and depth of sticking, (3) washpipe size and length, and (4) type of rotary shoe. WRSIC #W71-03833

379. Lauman, R. H., 1963, "Large Diameter Hole Drilling," *Ground Water,* Vol. 1, No. 2, pp. 9-10, 14, April.

Describes construction features of 24- to 42-inch diameter holes.

380. Ledgerwood, L. W. Jr., 1960, "Efforts to Develop Improved Oil Well Drilling Methods," *Trans. SPE of AIME,* Vol. 219, pp. 61-74, 73 refs.

Efforts to develop better oil well drilling methods have included automation of drilling rigs; continuous coring with reverse circulation;

Ledgerwood, L. W., Jr., ... (con't)

reelable drill pipe; retractable rock bits; magnetic waves; electric currents; chemical attack; erosion by high velocity gases; abrasive jets; grinding wheels; arc flame; high velocity pellets; explosives; shock waves; bottom hole machine with power output up to 400 hp.; electric and hydraulic bottom hole means of rotating bits; bit rotary speeds up to 2,000 rpm as well as electrical, mechanical, and hydraulic means of activating percussors; and impact at frequencies ranging from 6 to 300 cycles per sec. In spite of these efforts, rotary drilling has maintained its economic leadership. Application of existing and development of new technology will continue to reduce the cost. WRSIC #W71-05305

381 Le Grand, H. E., 1964, "Management Aspects of Groundwater Contamination," *Journ. Water Poll. Control Fed.,* Vol. 36, pp. 1133-1145.

Outlines sources of ground water contamination, classification of contamination problems, and hydrogeologic factors and their interrelation in waste management and ground water contamination problems.

382. ———, 1965, "Environmental Framework of Ground-Water Contamination," *Ground Water,* Vol. 3, No. 2, pp. 11-15, April.

Defines contamination and explores its ramifications.

383. ———, 1967, "Role of Ground Water Contamination in Water Management," *A.W.W.A. Journ.,* Vol. 59, pp. 557-565.

Differentiates between water supply from streams; water supply from wells; disposal in streams; and waste disposal in the ground. Separate study of these components and then their dependence must be examined in planning. Ground water contamination is shown to be a distinct planning alternative in long range integrated community plans.

384. Lehr, J. H., 1969, "A Study of Ground Water Contamination Due to Saline Water Disposal in the Morrow County (Ohio) Oil Fields," *State of Ohio Water Resources Center,* Ohio State Univ., Project No. A-044-Ohio OWRR, March, 81 pp.

The purpose of the project was to study the effects of pollution of the ground water in Morrow and Delaware Counties, Ohio, due to the introduction of saline oil field wastes through evaporation pits. The investigation was directed toward the determination of the source, severity, areal extent, and probable future movement of the pollution. Emphasis was placed upon methods of detection of such contamination. WRSIC #69-05749

385. ———, 1971, *Master Library of Water Well Equipment and Maintenance,* Micro-Graphix, Pennsylvania, 1, No. 104 pp.

Contains miniature reproduction of 347 different manufacturer's catalogs. Each catalog is preceded by a directory identification frame which lists the manufacturer's NWWA identification number, name, address, telephone number, and the name of the individual responsible for the sale of products in ground water systems and development. The

Master Library represents the most significant advancement in product information storage and retrieval ever accomplished in the ground water industry. Through the new science of micropublishing, the NWWA has been able to compile and reproduce actual catalog representation of most manufactures to the ground water industry. Each one has been cross-referenced in a comprehensive index. The contents of this library in original size would fill more than six filing cabinet drawers. A 7X handviewer is supplied with the Library. WRSIC #W71-05700

386. Lennox, D. H. and A. Vanden Berg, 1967, "Drawdowns due to Cyclic Pumping," *Amer. Soc. Civil Eng. Journ.*, Hydr. Div., Vol. 93, No. 6, pp. 35-51, November.

General expressions are derived and methods of computation discussed for the maximum drawdowns in a nonleaky artesian aquifer. Results can be applied to drawdown calculations for the pumped well or for any observation well and for either simple or complex cycling schedules.

387. Lewis, R. F., 1965, "Control of Sulfate-Reducing Bacteria," *A.W.W.A. Journ.*, Vol. 57, No. 8, August, pp. 1011-1015, 20 refs.

Although the importance of sulfate-reducing bacteria in the external corrosion of pipelines is well known, their activity inside water mains is often overlooked. Rusty water or offensive odors at isolated points in a distribution system are often attributable to sulfate-reducing basteria. Tubercles usually form where the oxygen concentration becomes very low at times – that is, at dead ends or in areas of low flow in mains. The maintenance of residual chlorine throughout a distribution system is one of the best preventive measures against the formation of tubercles. Once the tubercles are formed they must be removed by a reamer, and only then can steps be taken to prevent their recurrence. The use of specific chemical inhibitors for the sulfate-reducing bacteria would be very limited in drinking-water supplies, but can be of benefit in preventing corrosion of sewage mains or in controlling the "rotten eggs" odor in sewers or ponds containing waste waters. While much has been learned about sulfate-reducing bacteria in the last few years, control is still a major problem.

388. Linck, C. J., 1963, "Geophysics as an Aid to the Small Water Well Contractor," *Ground Water,* Vol. 1, No. 1, January, pp. 33-37, 5 refs.

Geophysical techniques as used in ground water exploration are subdivided into "borehole" and "surface" methods. The former include the commonly used electrical and gamma-ray logging and the less commonly used hole calipering and current meter logging. Included with this classification is the field of water-level measurements. The surface techniques discussed include electrical resistivity and refraction seismograph exploration. Real economic advantages, in many cases, are gained by use of one or more of the borehole methods. Geophysical methods properly used can do much

Linck, C. J. . . . . (con't)

to guide the water well contractor. It is extremely important, however, that their use be carefully directed because in the past, where geophysical methods have failed, it has often been due to the incorrect application of the technique, rather than a failure of the technique. WRSIC #W71-06946

389. ———, 1970, "Pumping Test Procedures for Aquifer Analysis," *Principles and Applications of Ground Water Hydraulics Conference,* Michigan State Univ., East Lansing, Michigan, December, 8 pp., 4 refs.

Obtaining reliable data which are amenable to mathematical analysis is one of the purposes of a pumping test. Such data must be analyzed with available techniques. The limiting factors and design of pumping and observation wells must be known to be sure the data can be analyzed with available techniques.

390. Lindsey, H. E., Jr., *et al.,* 1971, "A New Tool for Core Recovery of Soft, Unconsolidated Formation," *46th Annual SPE of AIME Fall Meeting,* Preprint SPE 3603, New Orleans, October, 8 pp.

This study describes a new core-barrel device for cutting and retrieving subterranean cores, particularly from soft or unconsolidated formations: it utilizes a fluid-pressure collapsible sleeve above the core bit to receive and hold the cored material and permit its retrieval in its original subterranean condition. Pressure responsive valves maintain a fluid pressure behind the collapsible sleeve to retain it in the collapsed position until the core is received in the sleeve and also permit fluid behind the sleeve to exhaust as the core enters the sleeve while the sleeve continues to offer lateral support for the core and also remains collapsed above the core.

391. Lloyd, D. P., 1970, "Down-the-hole TV, the Greatest Show on Earth," *Ground Water Age,* Vol. 8, No. 6, June, pp. 28-35.

For many years, contractors from various industries, particularly the well drilling industry, have questioned the nature and quality of work in areas not accessible to man. Down-the-hole TV gives this opportunity. And it is fast becoming a more useful and important tool – especially with the technological advancements of the electronics age. Its technical name is CCTV (Closed Circuit Television). but is more popularly known as Down-the-hole TV. Its basic function is to record for above-ground observers the intricacies and problems which go along with bore holes, well screens, pumps, caissons, and sewers. The system is described in some detail. Advantages are emphasized.

392. Lohman, S. W., 1972, "Definitions of Selected Ground-Water Terms – Revisions and Conceptual Refinements," *U.S. Geol. Surv. Water Supply Paper 1988.*

For many years there has been a need for redefinition or more precise definition of certain ground water terms used in publications by members of the U.S. Geological Survey. Another problem has been the expression of the coefficient of permeability (herein redefined as "hydraulic conductivity") and the coefficient of transmissibility (herein redefined as "transmissivity") in inconsistent units that included the U.S. gallon, the foot, and in some expressions, the mile. Such inconsistent units and the attendant confusing numerical conversion factors used in flow equations make it unnecessarily difficult for hydrologists, especially in foreign countries, to follow and use U.S.G.S. published results.

393. ———, 1972, "Ground-Water Hydraulics," *U.S. Geol. Surv. Prof. Paper 708,* 70 pp.

The principal method of analysis in ground water hydraulics is the application, generally by field tests of discharging wells, of equations derived for particular boundary conditions. Prior to 1935, such equations were known only for the relatively simple study flow condition, which incidentally generally does not occur in nature. The development by Theis (1935) of an equation for the nonsteady flow of ground water was a milestone in ground water hydraulics. Since 1935 the number of equations and methods has grown rapidly and steadily. These are described in a wide assortment of publications, some of which are not conveniently available to many engaged in ground water studies. The essence of many of these are presented and briefly discussed.

394. Lubinski, A., 1957, "Improved Rate-Weight Curve Over a Very Short Interval," *Oil and Gas Journ.,* Vol. 55, No. 47, November, pp. 91-93.

The purpose of this discussion is to propose a method which should permit determination of an entire rate-versus-weight curve over a very short interval, generally of a few inches. The method consists of either recording or reading and plotting weight versus time during drilling-off periods.

395. ———, 1961, "Maximum Permissable Dog-Legs in Rotary Boreholes," *Trans. SPE of AIME,* Vol. 222, pp. 175-194, 7 refs.

In drilling operations, attention generally is given to hole angles rather than to changes of angle, in spite of the fact that the latter are responsible for drilling and production troubles. The paper presents means for specifying maximum permissible changes of hole angle to insure a trouble-free hole using a minimum amount of surveys. It is expected that the paper will result in a decrease of drilling costs not only by avoiding troubles but also by removing the fear of such troubles.

396. Lummus, J. L., 1965, "Chemical Removal of Drilled Solids," *Drilling Contractor,* March-April, pp. 50-67, 6 refs.

Chemical methods are required for economical removal of solids smaller than ten microns. Of the various chemical methods investigated, flocculation uses force most effectively and as a result, flocculants are now widely used for complete removal of all solids or for maintaining the solids at a low level.

Lummus, J.L. . . . (con't)

397. ———, 1967, "A New Look at Lost Circulation," *Petrol. Eng.*, November, pp. 69-73, 9 refs.

Approximately twenty-two million dollars were spent for lost circulation materials in 1965. Additional sums were lost where the occurrence of lost circulation contributed to sealing a productive horizon, when blow-outs resulted, where drill pipe was stuck, or when formations caved. Lost circulation is one of the greatest expenses faced by the industry in drilling a well. A fair appraisal of the lost circulation picture is that there is a general understanding of the types of lost circulation and the conditions under which these occur, but with the exception of several squeeze techniques, there is an understandable lack of confidence in materials commercially available for sealing off these zones. Theories and observations voiced by the people who have had vast experience fighting lost circulation can be boiled down to these pertinent points: (1) a lost circulation material should contain high strength particles with a definite size distribution, (2) it should form an effective seal under both low and high differential pressure conditions, and (3) it should be equally effective in sealing unconsolidated formations and fractures or vugs in hard formation.

398. ———, 1968, "Squeeze Slurries for Lost Circulation Control," *Petrol. Eng.*, September, pp. 59-64.

It is estimated that the U.S. Petroleum Industry spends over $20,000,000 each year for lost criculation materials in drilling wells. This estimate does not reflect additional expenses incurred by loss of rig time, caving formations, stuck drill pipe, etc., all of which are possible side effects of the lost circulation problem. Consensus of engineers who have spent years fighting lost circulation is that materials used for its control should: (1) contain high strength particles with a definite size distribution; (2) form an effective seal under both high and low differential pressure conditions; (3) be equally effective in sealing unconsolidated formations and fractures or vugs in hard formations.

399. ———, 1970, "Drilling Optimization," *Journ. Petrol. Tech.*, November, pp. 1379-1389, 7 refs.

The development of rotary drilling can be divided into four distinct periods: Conception Period – 1900 to 1920; Development Period – 1920 to 1948; Scientific Period – 1948 to 1968; and Automation Period which began in 1968. The major accomplishments of the first three periods, and a prediction of what lies in the future for the Automation Period are given. In reviewing these development periods, the question naturally arises as to the reason for the approximate 30-year lapse between the end of the Conception Period and the start of the Scientific Period. There are a number of reasons that can be given, but undoubtedly the most significant is that major oil field equipment firms, mud service companies, and operators did not start appropriating the large amounts of money it takes to do high quality drilling research until about 1948.

400. Lummus, J. L., *et al.*, 1961, "New Low Solids Polymer Mud Cut Drilling Costs for Pan American," *Oil and Gas Journ.*, December, pp. 87-91, 4 refs.

Polymer is an inexpensive means of keeping mud solids at a low level, thus lengthening bit runs, speeding drilling rate, and cutting costs. While it extends the yield of bentonite, the polymer flocculates native drilled solids. Result is that solids content can easily be kept between three and six percent by volume.

401. Lynch, E. J., 1962, *Formation Evaluation,* Harper and Row, New York, 422 pp., 11 chapters, 4 app. and index.

The purpose of this text is to study the methods and the tools that are available for this very important task, to review the theory and principles of formation evaluation, to point out the uses and limitations and finally to show how they may be used to complement one another in providing a useful and coordinated system of evaluation at a reasonable cost.

402. MacCary, L. M., 1971, "Resistivity and Neutron Logging in Silurian Dolomite of Northwest Ohio," *U.S. Geol. Surv. Prof. Paper 750-D,* pp. D190-D197, 17 refs.

Data from resistivity and neutron logs of the Lockport Dolomite in NW. Ohio may be used in a semiquantitative empirical manner to estimate water quality and formation porosity if sufficient supporting data are available. Zones in the dolomite that appear to function like cemented granular aquifers can be analyzed for water quality and porosity on the basis of log response. No useful relationships between formation factor and permeability can be established for the Lockport Dolomite owing to the great range in its estimated porosity. Further study is planned on the applicability of these geophysical methods to carbonate rocks in other hydrologic environments. WRSIC #W72-04369

403, McCaslin, L. S., Jr., 1951, "Southwest U.S.A. – Birthplace of Rotary Drilling," *Oil and Gas Journ.,* Anniv. No., May, pp. 222-251.

Although rotary drilling may have come into the oil industry in Texas with the Baker brothers, water well drilling contractors from South Dakota, many items of equipment and much of the operating practice have developed in the Southwest. Drilling from a water location began at Caddo Lake, Louisiana, in 1911. The first submersible barge rig was introduced in Plaquemine Parish, Louisiana, in 1932 and offshore drilling developments have followed rapidly.

404. MacCracken, R. A. and H. D. Nickeroon, 1964, "Ground Water Contamination in Two New England Communities," *Public Works,* Vol. 95, No. 2, p. 162, February.

Describes the contamination of two wells due to lack of control over the use of contiguous lands.

405. McCray, A. W. and F. W. Cole, 1958, *Oil Well Drilling Technology,* Univ. of Oklahoma Press, Norman, Oklahoma, 492 pp., 76 refs.

A thorough technical discussion of the geological and geophysical principles governing oil accumulation and the drilling of wells from the standpoint of the driller petroleum engineer, geologist, and student is presented. From a survey of the geological formations likely to contain deposits of oil, it proceeds to methods of discovery, including some radioactive prospecting methods, then to the various types of drilling and drilling tools, to cementing operations, and to costs and ways of paying drilling charges. Practical problems are posed and methods of solving them demonstrated. The experienced oilman can find practical answers to many of his problems and questions, the student a sound text for basic principles of oil well drilling. The text provides complete information on drilling techniques, whether the well is 100 or 20,000 feet deep. WRSIC #W71-05698

406. McEllhiney, W. A., 1955, "Cementing Small Wells," *Water Well Journ.,* Vol. 9, No. 1 and 2, January-February, 5 pp.

This paper treats water well cementing techniques in practical detail. The author is well known in the field for his outstanding knowledge and experience in water well construction techniques. It has considerable emphasis on the sanitary aspects of well construction. WRSIC #W71-068667

407. ________,1960, "Application of Rotary Drilling to Water Wells," *A.W.W.A. Journ.,* Vol. 52, No. 3, March, pp. 351-355.

The use of rotary drills in the water industry both for drilling test holes and for constructing large-bore gravel-packed wells; the possibility of using rotary drills for constructing and enlarging wells in the deep sandstone of Wisconsin. WRSIC #W71-05307

408. McGhee, E. D., 1959, "Surface Mud System," *Oil and Gas Journ.,* Vol. 57, October, pp. 107-122.

Describes complete mud system at surface; pits; tanks; mud screens and de-sanders; mud agitators and mixers; centrifugal concentrations; degasser and other mud instruments; mud storage and handling.

409. ________, 1960, "What Weight and Speed Produce the Lowest Drilling Costs?" *Oil and Gas Journ.,* March, pp. 84-87.

There is a widely held belief that rotary drilling has gone about as far as it can go. Many have concluded we must find some drastically new method if drilling times are to be shortened appreciably. This is not true. Field tests now under way have greatly reduced drilling times without adding a single new piece of equipment to the rig. Key to this success is in those two commonplaces: bit weight and rotary speed.

410. McLamore,R. T., 1966, "The Mechanical Behavior of Anisotropic Sedimentary Rocks,"*Journ. Eng. Industry, Trans. ASME,* Vol. 89, Series B, No. 1, pp. 62-76.

Highly anisotropic strength characteristics are shown by rocks that have bedding, cleavage, or schistocity planes, such as shales, laminated sandstones, and finely interbedded sand and shale. In contrast to isotropic rocks, the compressive strength of anisotropic rocks is highly dependent upon the orientation angle between the applied load or principal stress causing failure and the plane of anisotropy. The minimum strength usually occurs at an angle of 30°.

411. ———, 1971, "The Role of Rock Strength Anisotropy in Natural Hole Deviation," *Journ. Petrol. Tech.,* November, pp. 1313-1321, 6 refs.

The influence exerted by the interaction of rock and bit is a major factor preventing a comprehensive solution to the problem of natural hole deviation. Described here is an analysis of the bit-tooth chipping mechanism in dipping laminated anisotropic rock. There are encouraging indications that the tendency of a bit to deviate in such rock can be diminished by proper bit-tooth design.

412. McLean, R. H., 1964, "Crossflow and Impact Under Jet Bits," *Trans. SPE of AIME,* Vol. 23, pp. 1299-1306, 31 refs.

Jet impingement produces two mechanisms to clean the bottom of a borehole during jet-bit drilling operations. One is an impact-pressure wave in the immediate area of jet impingement. The other is crossflow which spreads across the bottom away from the pressure wave. Equations derived from current jet technology describe the wave in this simple system for a wide range of conditions.

413. ———, 1965, "Velocities, Kinetic Energy, and Shear in Crossflow under Three-Cone Jet Bits," *Trans. SPE of AIME,* Vol. 234, pp. 1443-1448, 8 refs.

Velocity, kinetic energy, and shear in crossflow beneath three-cone jet bits may influence cleaning of the bottom of the borehole and the depth of the bit. Investigations show that each of these parameters is a function of the diameter of the borehole and the product of the volume rate of flow and velocity through the nozzles. These functions provide means for predicting the magnitude of each parameter and of sealing the cleaning forces.

414. McLeod, H. O., 1971, "Oil Well Cement, Pt. 3, Here are the Cement Additives," *Oil Gas Petrochem. Equip.,* Vol. 16, No. 6, April, pp. 4-5, 11 refs.

Additives are mixed with neat cement to increase yield per sack, to reduce costs or to alter slurry properties for special well conditions. These additives are utilized, generally, for the following purposes: (1) as extenders (an extender is any additive which will provide a

McLeod, H. O. . . . . (con't)

greater yield, or slurry volume, for each sack of cement); (2) weighting materials to increase slurry density and overcome high formation pressures; (3) accelerators, such as calcium chloride, to reduce time waiting in shallow well completions, particularly in cold climates; (4) retarders commonly used in walls 8,000 ft. and deeper to increase thickening time; (5) low-water-loss additives most important in squeeze cementing where control of water loss is more critical; (6) lost-circulation materials where cementing fluids may be lost to a very porous or permeable formation, fractures zones, or weak formations hydraulically fractured by the pressure of mud and cement in the annulus; and (7) additives to change cement-flow behavior.

415. McMillion, L. G. and J. W. Keeley, 1968, "Sampling Equipment for Ground Water Investigations," *Ground Water,* Vol. 6, No. 2, March-April, pp. 9-11.

Portable pumping equipment for sampling of wells described in this paper by written text, photographs, and a detailed drawing has been constructed by the Robert S. Kerr Water Research Center. The equipment can sample to depths of 300 ft. at pumping rates ranging between seven and 14 gpm with rate variation dependent upon sampling depth. The unit is convenient in size and easy to operate because only one line has to be handled during its operation. This is a wire-reinforced rubber hose that supports the submersible pump, contains the electrical cable, and conveys water. WRSIC #W68-00512

416. McReynolds, P. S., 1958, "Gravel Packing Controls Unconsolidated Sand in Venezuela Field," *Journ. Petrol. Tech.,* Vol. 10, No. 12, December, pp. 21-24, 5 refs.

Sand control has been the major problem encountered in producing oil from the unconsolidated Miocene sands in the Lagunillas area of Venezuela. One of the first completions developed to prevent excessive sand production was the combination oil string in which pre-perforated casing was employed which was replaced in 1951 by a special slotted liner completion. The slotted liner technique has been effective in controlling sand and in obtaining high volume producers. However, this technique will soon become obsolete because of the declining reservoir pressure in the Bachaquero field. A program to evaluate gravel packing as a replacement for the slotted liner technique and to provide a sand control method for use inside existing preperforated completions was initiated in April, 1956. Solutions to these problems have been found and procedures for both open-hole and inside gravel packs have been developed that can be applied with confidence to the sand control problems that exist in the Bolivar coastal fields. WRSIC #W71-08501

417. Maher, E. J., 1966, "Testing for the Development of Ground Water Supplies," *Journ. New England Water Works Assoc.,* Vol. 80, pp. 326-330.

Methods and procedures used in New England states in the well testing and exploration program for locating and evaluating ground water resources of any area, and in planning a permanent well.

418. Maher, J. C., 1964, "Logging Drill Cuttings," *Oklahoma Geol. Surv. Guide Book XIV,* 2nd. Edit., 48 pp. 60 refs.

This report has been prepared primarily to provide a reference, a manual, and a set of descriptive standards for subsurface investigations in the Midcontinent region. Specifically this report describes the preparation of composite interpretive logs and presents a system of description including terminology and definitions, abbreviations, and symbols. General discussions of drillers logs, drilling-time logs, electric logs, radioactivity logs, and sample logs are included to explain their supporting role in preparing composite interpretive logs. It is also of use to drillers to help them appreciate the primary problems of the geologist and to offer guidance in the preparation of a better driller's log. WRSIC #W71-06955

419. Malashenko, Yu, R., *et al.,* 1971, "The Isolation of Pure Cultures of Obligate Methane-Oxidizing Bacteria," *Mikrobiologiya,* Vol. 40, No. 4, July-August, 15 refs., (in Russian).

Either the strains of Nocardia and Mycobacterium oxidizing higher gaseous hydrocarbons ($C_2 - C_4$), or obligate methane-oxidizing strains of Methylococcus and Methylemonas were isolated from natural substrates, as a rule from soils. The method for the isolation of the methane-obligate bacteria into pure cultures is described. The methane-oxidizing bacteria are recommended to be cultivated and stored in the U-shaped tube containing the medium: the methane-obligate bacteria grow in one branch of the tube, the accompanying culture in the other branch. Some aspects of physiology of nutrition of the methane-obligate bacteria and the effect of some toxic components of the medium on these bacteria are discussed.

420. Mallory, H. E., *et al.,* 1960, "Low-Solids Muds Resist Contamination," *Petrol. Eng.,* July, p. B-25.

Through research and field experimentation, the well contractor has improved penetration rates by improved hydraulic programs and carefully planned mud programs. Field results indicate that overall costs can be reduced and more accurate geological information obtained by planning ahead.

421. Mandal, T. C. and D. M. Edwards, 1971, "The Effects of Electrokinetics upon Incrustation in Water Wells," *Amer. Soc. of Eng. Trans.,* May-June.

The results of this study were: (1) a saturated, water-bearing aquifer generates a streaming potential when the water is flowing; (2) the streaming potential is a catalyst in the formation of incrustation on water well screens; (3) incrustation forms on a water well screen

Mandal, T. C. and D. M. Edwards . . . . (con't)

only when it is negatively charged; and (4) no deposition forms on the screen when it is positively charged. WRSIC # W72-04870

422. Manuel, B., 1970, "100 Years of Ground Water History," *Ground Water Age,* Vol. 8, Nos. 11, November (Part 1), December (Part 2) and January (Part 3).

This series of articles covers the past 100 years of ground water history from dug wells to state and national associations including some of the first legislation and codes for the ground water industry. The discovery of the relationship of bacteria to disease in man during the 1880's and the connection of sewage with typhoid and cholera was the beginning of "well sanitation" and increased interest in salt water supplies.

423. Mara, D. D. and D. J. A. Williams, 1971, "Corrosion of Mild Steel by Nitrate Reducing Bacteria," *Chem. Ind.,* No. 21, May, pp. 566-567.

The anaerobic corrosion of ferrous metals is usually attributed to the activities of sulfate-reducing bacteria. Hydrogenase strains of these bacteria depolarize the metal surface by removing cathodic hydrogen. Hydrogenase systems are known not to be confined to the sulfate reducers, but are possessed by many bacteria and microalgae. Using a strain of *Escherichia coli,* an organism which possesses a hydrogenase system and is able to utilize nitrate as a hydrogen acceptor under anaerobic conditions, corrosion experiments were performed. The results indicate that the organism perferentially utilizes nitrate for the oxidation of organics, and also that when the cells are in a resting state, nitrate reduction occurs with attending corrosion if ferrous metals are present. In view of the diverse nature of nitrate reducing bacteria and of their high numbers in soils, the corrosion caused by these organisms should be investigated in greater detail.

424. Marsh, C. R. and R. R. Parizek, 1963, "Induction-Tuned Method to Determine Casing Lengths in Hydrogeologic Investigations," *Ground Water,* Vol. 6, No. 6, pp. 11-17., November-December.

Describes a device designed and tested under a variety of well casing conditions. Detects casing lengths to within 0.4 to 2.0 inches.

425. Massarenti, M., 1964, "Percussion-Reverse Circulation Water Well Drilling System Designed and Developed in Italy," *Ground Water,* Vol. 2, No. 2, April, pp. 25-27.

This paper describes the new percussion-reverse circulation system designed and developed in Italy for water well drilling. In a special reverse circulation rig, the percussion bit, consisting of a tube with external welded blades, slides up and down on the outside of the hollow drill pipes, while the hollow drill pipes remain still. The field of application of the reverse circulation method is thus enormously enlarged; either hard cemented formations or soft unconsolidated deposits as well as big boulders of any size are bored much faster and more efficiently than with any other method of drilling, provided that a sufficient source of water is available.

426. Mathis, H., 1965, "Mechanical Desanding of Drilling Fluids," *Drilling Contractors,* Vol. 21, No. 3, March-April, pp. 65-67.

The mechanical desanding of drilling fluids is the removal of sand, shale, and other solid particles 74 microns (200 mesh) in size and larger. The term "sand" or "API sand" as used in this paper refers to all such solids. Solids smaller than this are technically referred to as "silt."

427. Matlock, W. G., 1970, "Small Diameter Wells Drilled by Jet-Percussion Method," *Ground Water,* Vol. 8, No. 1, January-February, pp. 6-9.

Observation wells and access holes for neutron probe were drilled by a jet-percussion drill rig in coarse alluvial material. The method combines the jetting action which effectively removes loose materials with the percussion necessary to break up tighter formations and large particles. A unique feature of the equipment is the provision for simultaneous drilling and driving the casing to keep the hole open in loose formations. Washed samples of the material being drilled can be obtained from the recirculating water system. Drilling rates were from seven to ten ft. per hr., and costs including casing were less than $1.50 per ft. WRSIC # W70-07438

428. Maurer, W. C., 1959, "Impact Crater Formation in Sandstone and Granite," *M. S. Thesis T-887,* Colorado Sch. of Mines.

Impact craters formed in granite and sandstone by small, high velocity spherical steel projectiles have been studied. The sphere penetrates the rock by crushing and pushing aside the material in front of it, thus forming the burrow which in granite is completely obliterated as the cup is formed. As the projectile penetrates, fracturing is initiated in progressive stages. The fractures propagate along trajectories of maximum shear, each new fracture acting as a new free surface. The successive fractures are parallel and of equal length. In both sandstone and granite, it appears that the stress initiated at impact dislodges the rock from the cup. When a certain minimum stress level is reached on the surface of this fracture, the material is no longer removed from the crater.

429. ———, 1962, "The Perfect-Cleaning Theory of Rotary Drilling," *Trans., SPE OF AIME,* Vol. 225, November, pp. 1270-1274, 21 refs.

Drilling rate formula for roller-cone bits is derived from rock cratering mechanisms; this formula holds for "perfect cleaning," which is defined as condition where all of rock debris is removed between tooth impacts; under these conditions, drilling rate is directly proportional to rotary speed and to bit weight squared – and inversely proportional to bit diameter squared and to rock strength squared; good correlation of rate-weight-speed data obtained under perfect cleaning conditions to derived drill-rate formula is obtained. WRSIC #W71-05304

Maurer, W.C. . . . (con't)

430. ———, 1965, "Bit-Tooth Penetration Under Simulated Borehole Conditions," *Trans. AIME,* Vol. 234, pp. 1433-1442, 13 refs.

Crater tests in unconsolidated sand subjected to differential pressure showed that high friction was present in the sand at high pressures. Similar friction between the cuttings in craters produces the transition from brittle to pseudoplastic craters.

431. ———, 1966, " The State of Rock Mechanics Knowledge in Drilling," *Symposium on Rock Mechanics,* Univ. of Minnesota, Sch. of Min. and Metal. Eng., Minneapolis, pp. 355-395, 71 refs.

A better understanding of the drilling mechanisms would enable a more scientific approach to the design and use of drills and bits. The paper reviews several models which have been proposed to explain various aspects of the different drilling methods. These models are described as well as their limitations and areas of application. Drilling rate relationships are developed and related to the basic rock failure mechanisms. WRSIC #W71-13716

432. ———, 1968, *Novel Drilling Techniques,* Pergamon Press, New York, 114 pp., 111 refs.

Considerable research is going on throughout the world to develop new methods for drilling and excavating rock. Considerable research is being done to develop bottom hole drilling motors such as electric drills and turbodrills. Although the devices described in this study are called novel, many of the concepts involved are quite old as evidenced by the fact that the first flame drill was patented in 1853, and the first electric arc drill was patented in 1874. Novel devices remove rock by four basic mechanisms: mechanically induced stresses, thermally induced stresses, fusion and vaporization, and chemical reactions. Novel rock destruction methods will find initial application where rate of rock removal is more important than unit cost or where restrictions are imposed on the rock destruction device (such as weight limitation in space exploration or size limitation in deep oil wells. WRSIC #W71-05291

433. Maurer, W. C. and J. K. Heilhecker, 1969, "Hydraulic Jet Drilling," *Soc. Petrol. Eng., A.I.M.E.,* Preprint No. SPE 2434, pp. 213-224, 15 refs.

As a result of a detailed survey of over 25 drilling techniques, a laboratory study of hydraulic jet drilling was instigated. In initial tests, a cannon was used to fire 1.45 gallon water pulses at rocks at pressures up to 25,000 psi. These tests showed that a threshold nozzle pressure must be exceeded before hydraulic jets will drill rock and that water jets can effectively drill sedimentary rocks. Other tests showed that a full-scale hydraulic jet drill (3,000 hp.) should drill eight-inch diameter holes in average-strength sedimentary rocks at rates of 200 to 300 ft./hr. These high drilling rates show that hydraulic jet drills have high potential for drilling oil wells economically. WRSIC #W71-06922

434. Maxey, G. B., 1965, "Hydrogeologic Factors in Problems of Contamination in Arid Lands," *Ground Water,* Vol. 3, No. 4, pp. 29-32, October.

The ideal hydrologic system in arid lands includes a recharge area in mountains and a discharge area in lowlands modified by geologic and physiographic factors. Population and agricultural activity concentrates in valleys, usually in zones of ground water discharge. Most water-supply, contamination, and disposal problems arise from this combination of features. The suitability of hydrogeologic units for any function of operations involving water supply or waste disposal depends primarily on their position within the hydrologic system and secondarily on physical properties. At the Nevada Test Site, the ground water flow system is used to good advantage, whereas at Las Vegas, 70 miles away, the methods of disposal practiced are in direct conflict with the system.

435. Mead, J. L. and C. A. Reid, 1969, "Instrumentation and Analysis for an Optimized Drilling Program," *Drilling Contractor,* November-December, pp. 43-44, 6 refs.

During the past several years, West Texas operators have made impressive progress in deep drilling operations. The most significant benefit is reduction of overall time on location. It is now quite routine to drill a 20,000 foot well in 180 to 250 days. Drilling rates of this order are achieved in part through proper selection of casing points so that all intervals can be drilled with minimum mud weights while avoiding lost circulation. Continuous and current recording, analysis and interpretation of down-hole information done at the drilling location can insure optimum selection of casing points and drilling fluid density.

436. Meents, W. F., 1960, "Glacial-Drift Gas in Illinois," *Illinois State Surv. Circ. 292,* pp. 58.

Glacial-drift gas in Illinois occurs mainly in the northeast fourth of the state in some 60 areas in 27 counties. There are about 460 producing gas wells of which 250 are flowing pressure wells and the remainder are vacuum pumped. Some 172 pressure wells have been tested for open-flow gas volume and several dozen vacuum-pumped wells have been tested for formation vacuum. The gas is believed to be derived from buried soil zones and from organic matter in deep buried valleys. The glacial end moraines control the accumulation of drift gas by providing a cover of glacial till thick enough to prevent escape of the gas.

437. Melrose, J. C. and W. B. Lilienthal, 1951, "Plastic Flow Properties of Drilling Fluids–Measurement and Application," *Petrol Trans. A.I.M.E.,* Vol. 192, pp. 159-164, 10 refs.

The application of Bingham's law to the behavior of drilling fluids in a rotational viscometer permits the expression of viscometric data in terms of plastic viscosity and yield value, the flow properties of a plastic fluid. A commercially available rotational viscometer is described. Other data obtained by the device are shown to be useful in defining mud control problems relating to chemical treatment and to the hydro-behavior of muds.

438. Mesaros, J., 1957, "The Application of Flow Properties to Drilling Mud Problems," *A.P.I. Drill. and Prod. Pract.,* Bull. 243, pp. 83-93, 11 refs.

Field engineers have seldom used the flow properties of drilling muds to overcome drilling problems because: (1) unfamiliarity with flow-property concepts, (2) lack of use of flow-property equations, and (3) a viscometer to determine flow constants has not been available. Proper use of drilling-mud flow properties makes control of the mud more certain and increases penetration rate. It aids greatly in pump-liner selection, hole fill-up, gas cutting, and lost circulation.

439. Messenger, J. U., 1968, "How to Combat Lost Circulation," *Oil and Gas Journ.,* Vol. 66, Nos. 20, 21, 22, May, pp. 71-76, 90-97, 94-98, 21 refs.

The aim of this paper is to outline more effective procedures of lost circulation prevention and control; different types of loss zones have been classified on basis of severity; preventive and corrective measures and causes of failure to control are discussed; five specific remedial techniques discussed include pull up and wait technique, plug of bridging agents in mud, higher filter-loss slurry squeeze, cement, and diesel oil-bentonite-cement slurry squeeze; also discussed are four techniques using surface-mixed soft plugs, down-hole-mixed soft plugs, special tools, and drilling blind or with aerated mud and set pipe.
WRSIC #W71-05317

440. Meyer, G. and G. G. Wyrick, 1966, "Regional Trends in Water Well Drilling in the United States," *U. S. Geol. Surv. Circ. 533,* pp. 1-8, 8 refs.

Analysis of estimates of number of water wells drilled in United States during the period 1960 to 1964 reveals national and regional trends in water well construction: approximately 435,700 wells were drilled in 1964 – average of 1,700 wells starts each working day; for the nation as a whole, net change between 1960 and 1964 was an increase of 14 percent in drilling activity. WRSIC #W71-13972

441. Meyer, W. R., 1962, "Use of a Neutron Moisture Probe to Determine the Storage Coefficient of an Unconfined Aquifer," *U. S. Geol. Surv. Prof. Paper 450-E* pp. 174-176.

The coefficient of storage of an unconfined aquifer is approximately equal to the specific yield of the material that is drained when water is pumped from the aquifer. Under field conditions the storage coefficient can be determined by several methods. One of these, the aquifer-test method, consists of measuring the rate of water-level decline (or recovery) in one or more observation wells near a well from which water is being pumped (or has been pumped) at a steady rate. Another method makes use of a neutron moisture probe to determine the difference between the moisture content of saturated material and the moisture content of the same material after it has been drained.

442. Miesch, E. P. and J. C. Albright, 1967, "A Study of Invasion Diameter," *8th Annual Logging Symposium of the Soc. Prof. Well Log Analysts,* Denver, June, pp. 1-14.

This paper presents empirical correlations for the diameter of filtrate invasion and for the true formation resistivity as a function of mud properties, exposure times, and formation properties. Equations were obtained which predicted the diameter of filtrate invasion with an average absolute deviation of 41.8 percent, which in turn yielded an average absolute deviation of 13.1 percent for formation resistivity when good values are available for the deep induction log readings. Though these results are not as accurate as hoped for, they are significantly better than those obtained using our data in the equations of earlier investigators. The correlations are not oriented for hand calculations, but are readily adapted to a computer logging program.

443. Miller, L. M., 1957, "Design and Rating of Wells and Well Fields," *A.W.W.A. Journ.*, Vol. 49, No. 4, April, pp. 439-449, 6 refs.

Factors to be considered in the design of wells and well fields; effect of particle size on its porosity and permeability; coefficients of transmissibility and storage are defined. Examples of the calculation of these coefficients, using observations of the water level in test wells drilled in the region of the pumping well and of calculations of the soft yield and well losses. Practical application of these data in the design of a well field, and choice of well diameter and screen length are discussed. It is recommended that detailed records be kept of all test-drilled and operational observations in the well field. WRSIC #W71-03836

444. Millipore Corp., 1969, "Microbiological Analysis of Water," *Application Report AR-81.*

Detailed in this application report are a variety of specific techniques for isolating and identifying bacterial organisms in raw and treated water, swimming pools, treated sewage and industrial process, rinse and cooling water.

445. ———, 1971, "Coli-Count Water Tester," *Bulletin MB407.*

Field tests conducted on many different water sources in the U. S. show that, for any given water sample, the average results are the same for the Coli-Count and the standard MF technique.

446. Mills, A. E., 1970, "Stimulating Hard Rock Water Well Production with High Explosives," *Water Well Journ.*, Vol. 24, No. 2, February, pp. 39-42, 3 refs.

Shooting (blasting) is considered in the hard mountain granites and other types of clean, hard, elastic, well consolidated, brittle rock aquifers that fracture like glass. The first oil well shot with high explosives to stimulate greater production was around 1863 in the shallow, hard rock oil sands of Pennsylvania. When the wells were shallow, the pulverized rock debris was self-cleaned from the hole along with the water stemming (hole full of water for momentary confinement purposes) by the explosive gases. This article is concerned with only shooting shallow hard rock water wells or low cost construction of large cisterns. Deep wells resulted in long, expensive cleanouts of the fine shattered, pulverized rock debris. Thus, mechanical (pump) hydraulic fracturing was developed. WRSIC #W71-13718

447. Mills, K. N., 1952, "Mechanics of Cable Tool Drilling," *World Oil*, September, p. 123.

The cable tool process has been used to drill wells for centuries. Since its inception, the equipment used to produce the drilling motion has been constantly improved. However, the basic principles have not changed and will not change, because they are evolved around basic mechanical laws. As this process uses the principle of percussion (chipping) to produce the drilling action, it is well suited to drilling extremely hard formations, and for this reason, it enjoys wide popularity in "hard rock" country. The rates of penetration attained are dependent on the skill of the driller and the basic design of the drilling machine. The drilling practices are used on practical experience, and in general, they represent good practices for the equipment used to drill the well.

448. Milojevic, M., 1963, "Radial Collector Wells Adjacent to the River Bank," *Amer. Soc. Civil Eng. Journ.*, Hydr. Div., Vol. 89, No. 6, pp. 133-151, November.

Experimentally obtained formulas are given for a single well in the free water table aquifer and for a single well in a confined homogeneous and isotropic aquifer of limited thickness and unconfined side expansion.

449. Mitchell, J. A., 1957, "How are Turbodrills Performing?" *Oil and Gas Journ.* Vol. 55, No. 27, July, pp. 106-108, 111-112.

Turbodrills imported from the Soviet Union have logged over 13,000 ft. testing of various turbodrills to establish operating procedures and verify turbodrill characteristics in American conditions; hole sizes have ranged from 7 7/8 to 17 1/2 inches; penetration rates are 2 1/2 to 3 times those for comparable rotary drilling; under some conditions, lower drilling costs result with turbodrill.

450. Moehrl, K. E., 1964, "Well Grouting and Well Protection," *A.W.W.A. Journ.*, Vol. 56, No. 4, April, pp. 423-431.

Cementing and sealing of wells by Portland cement, neat-cement slurry, special cement slurries, and causes of failures are covered in this report.

451. Mogg, J. L., 1962, "Well Performance Shown by Simple Test," *Johnson Driller''s Journ.*, Vol. 34, No. 3, May-June, pp. 1-3.

Describes "water-input test" to check development work test performed by filling well casing with water, then noting the rate at which the water level drops; charts.

452. ———, 1963, "The Technical Aspects of Gravel Well Construction," *Journ. New England Water Works Assoc.*, Vol. 77, No. 2, pp. 155-164, June.

Discusses advantages and disadvantages of wells in sand and gravel formations, types of gravel wells, naturally developed wells, gravel packed wells, the steps in design, thickness of the gravel envelope, and vertical flow in the gravel pack.

453. ______, 1966, "Maximum Yields from Minimum Aquifers," *Ground Water,* Vol. 4, No. 2, April, pp. 11-12.

Maximum yields from minimum aquifers can be accomplished only by strict adherence to best known practice procedures while following the three necessary steps to well completion. These steps are well design, well construction, and well development.

454. ______, 1968, "Step Drawdown Test Needs Critical Review," *Ground Water,* Vol. 7, No. 1, January-February, p. 28.

This article discusses the results of well tests performed by the step-drawdown procedure. The expanded methods of Rorabaugh and Walton have been applied. WRSIC #W69-05813

455. ______, 1971, "What Experience Teaches us about Corrosion," *Johnson Driller's Journ.,* Vol. 43, No. 2, March-April, pp. 1-3.

Corrosion can severely limit the useful life of water wells in three ways: (1) screen slot opening enlargement followed by sand-pumping failure; (2) strength reduction followed by collapse of well screen or casing; and (3) redeposition of corrosion products that block screen slot openings and reduce yield to a point of failure. Processes have been observed by which these three types of failure can occur. One, chemical corrosion occurs when a particular constituent such as carbon dioxide, oxygen, hydrogen sulfide, hydrochloric acid, chlorine, and sulfuric acid is present in water in sufficient concentration to cause rapid material removal. The second, electrochemical corrosion, the corrosive attack on a metal, is accompanied by a flow of an electric current. Two conditions are necessary – a difference in electric potential on a metal surface and water containing enough dissolved salts to be a conducting fluid or electrolyte. Evidence of this type of corrosion problem often shows up through reduction in well yield which results from blocking of the screen slot openings by the redeposition of corrosion products.
WRSIC #W71-12474

456. Mondshine, T. C., 1966, "New Fast-Drilling Muds also Provide Hole Stability," *Oil and Gas Journ.,* Vol. 65, March, pp. 84-89, 2 refs.

Low-solids drilling muds designed to increase penetration and provide hole stability are being developed. Many shales, once thought to require high-solids muds, are being drilled faster and at less cost with these water-base, low-solids muds. Recent developments in mud technology are discussed.

457. ______, 1969, "New Technique Determines Oil-Mud Salinity Needs in Shale Drilling," *Oil and Gas Journ.,* July, pp. 70-75, 12 refs.

The osmotic pressure of an oil mud in contact with shale is a function of the differences in the salt concentration of the oil-mud water base and of the water in the subsurface shale. In shale drilling, the dissolved salt concentration of oil mud needs to be adjusted to provide a predetermined osmotic pressure to oppose the surface-hydration force of shale.

Mondshine, T.C. . . . (con't)

458. ———, 1970, "Drilling-Mud Lubricity: Guide to Reduced Torque and Drag," *Oil and Gas Journ.*, Vol. 68, No. 49, December, pp. 70-72, 77.

Laboratory studies were initiated to investigate drilling mud lubricity which primarily involves the rubbing of drill pipe and bit against the borehole during rotation and tripping. A mud lubricity tester simulates drill-pipe rotation and load and measures frictional force. The force needed to initiate movement (static) was measured, and the coefficient of friction was measured. Results are given in tabular data.

WRSIC #W71-03831

459. Mondshine, T. C. and J. D. Kercheville, 1966, "Successful Gumbo Shale Drilling," *Oil and Gas Journ.*, March, pp. 194-205.

Water wet shales (gumbo) have plagued drilling operations for many years. These shales called "plastic" are soft and mushy in ditch sample. Laboratory analysis of core samples of the shale indicates that silt and quartz are the predominant fractions and that these are bonded by montmorillonite clay, which contain an abnormally large quantity of water (30% weight). A well planned mud program assists in drilling clays of this type.

460. Moore, E. J. and J. M. Bird, 1962, "Cement Bond Logging, An Aid to Better Completion Practices," *The Log Analyst,* Vol. 3, No. 1 August, pp. 21-28.

Describes cement-bond logging and its application to well completion practices. The paper also explores problems of interpretation and identification of cement characteristics, bonding effectiveness, and theory of procedure.

461. Moore, P. L., 1958, "Five Factors that Affect Drilling Rate," *Oil and Gas Journ.*, Vol. 56, No. 40, pp. 141-170, 27 refs.

Increases in drilling rate as a result of higher bit weight and rotary speed depend on proper use of hydraulic horsepower. This means there must be enough hydraulic horsepower through the bit to remove formation cuttings as they are generated so that no regrinding occurs. At a given bit weight and rotary speed, drilling rate increases with increases in hydraulic horsepower until the point is reached where there is complete cleaning below the bit. After this, increases in hydraulic horsepower have no effect on drilling rate.

462. ———, 1961, "How to Apply Hydraulic and Bit Horsepower," *Oil and Gas Journ.*, January, pp. 77-80.

There are four basic points while using bit horsepower as the design criteria for jet-bit hydraulics programs. Laminar and turbulent flow, bit horsepower, annular velocity, compressors, nozzle sizes, case histories are discussed in the above paper.

463. ———, 1965, "Problems Associated with Design Programs for Air and Gas Drilling," *Drilling Contractor,* May-June, pp. 47, 48, 64, 5 refs.

This discussion of air and gas drilling is directed toward problems associated with program design. Problems of predicting pressure, solids accumulation, and critical solids concentration are discussed. To reduce repetition, the terms "air" and "gas" have been replaced by the term "gas."

464. ———, 1966, "Drilling for the Man on the Rig," Technical Manual Reprint, *Oil and Gas Journ.*, 72 pp.

This series of articles incorporates the latest in technology for use by the man in the field. The first step in planning a new well should be a review of all the available information on past wells in the area. This information should be evaluated to determine practices that should and should not be used in the proposed well. This means that second guessing is an accepted as well as a necessary practice. Information from the bit to use to the best cementing procedures can save considerable drilling time and trouble. WRSIC #W71-05314

465. Moore, P. L. and C. Gatlin, 1960, "Effect of Bit Weight on Drilling Rate," *Oil and Gas Journ.*, Vol. 58, No. 21, May, pp. 90-93, 8 refs.

After 100 years of oil well drilling, a complete theory does not exist for explaining the effect of bit weight on drilling rate. A review of recent empirical approaches to writing equations for bit weight as a function in oil drilling rate.

466. ———, 1960, "Effect of Rotary Speed on Drilling Rate," *Oil and Gas Journ.*, August, p. 170, 2 refs.

A review of recent empirical approaches on effect of rotary speed on drilling rate.

467. ———, 1960, "Six Variable Factors that Affect Penetration Rate," *Oil and Gas Journ.*, Vol. 58, No. 15, April, pp. 118-120, 9 refs.

The six factors are explained: bit weight, rotary speed, drilling-fluid properties, drilling-fluid hydraulics, rock properties, and bit selection.

468. Moore, W., 1968, "How to Dull a Bit for Fun and Profit," *Drilling*, March, pp. 64-65.

It has been demonstrated in the laboratory and verified by field tests that penetration rates generally increase directly in proportion to the vertical force applied to the bit so long as good cleaning and other factors remain constant. It is also shown that the rate of penetration increases in proportion to bit rpm raised to some power less than one and sometimes as low as 1/2.

469. Morris, R. B., 1968, "Sample Catcher Yields Clean Cuttings, Retains Fines," *World Oil*, Vol. 166, No. 6, May, pp. 97-98, 1 ref.

A simple machine has been developed that provides selected, screened, and washed subsurface rock cuttings. The machine's operation, advantages, and limitations are discussed. Primary design features are: (1) catch, sieve, wash, and collect aggregate or representative well cuttings in predetermined quantities; (2) provide a thorough but gentle washing action; (3) retain fines even when drilling carbonate, and (4) exclude most cavings.

470. Morris, R. I., 1969, "Rock Drillability Related to a Roller Cone Bit," *Fourth Conference on Drilling and Rock Mechanics,* A.I.M.E., SPE 2389, August, pp. 79-86.

The basic bit penetration mechanism of a roller-cone rotary bit has been used in the development of a drillability index for hard rock mining bits. A 1/8-in. radius, hemispherical drill bit element is pressed into a flat surface of a hand sample with a hydraulic pump and ram until a distinct crater is formed. The crater depth divided by the ram load constitutes a penetration or drillability index, P'/E. From this index, bit type, drilling weight, average penetration rate, and approximate bit life may be determined. Field results have been well within the degree of accuracy consistent with sampling error.

471. Moss, R., Jr., 1964, "Design of Casings and Screens for Water Production and Injection Wells." *A.P.I. Pacific Coast District Biennial Symposium,* Treatment and Control of Injection Wells, Anaheim, California, December, 25 pp., 8 refs., 8 tabs., 8 figs.

One of the most important contributing factors to development of the semi-arid southwestern United States has been the exploitation of the region's vast underground reservoirs. This paper discusses casing and screen practices applied to the two basic well designs most commonly employed in the area. Consideration is given to installation, operation, and economic criteria for both cable-tool and rotary-hydraulic methods of well construction.

472. Mullins, J. E., 1966, "Stereoscopic Deep Well Photography in Opaque Fluids," *7th Annual Symposium, Soc. Prof. Well Log Analysts,* Trans., Tulsa, Okla., May, pp. N1-N9.

The paper discusses the photography of deep wells in 3-D stereoscopic pairs as used for evaluation of well conditions or formational analysis. The photographs are obtained in wells and boreholes containing opaque fluids such as crude oil and drilling muds by an isolation displacement technique. The discussion covers the value of such visible examination of wells and boreholes for structural evaluation of materials, geologic studies of in-situ conditions, and recovery of lost equipment. Incorporated in the paper are photographs illustrating points of discussion and new techniques developed.

473. Murray, A. S. and R. A. Cunningham, 1955, "Effect of Mud Column Pressure on Drilling Rates," *Petrol. Trans., A.I.M.E.,* pp. 196-204, 10 refs.

If a dense fluid such as mud or water is used for circulation, the formation drilled is influenced by hydrostatic pressure which depends on hole depth and drilling fluid density. Laboratory tests indicate that drilling rates in many formations are decreased with increased pressure – in some cases as much as 90 percent. A comparison between laboratory and field tests indicates that drilling fluid head affects drilling rates in the field approximately the same as in the laboratory. Drilling rates of many formations are increased by reducing drilling fluid head. WRSIC #W71-05311

474. Murray, A. S. and J. E. Eckel, 1961, "Foam Agents and Foam Drilling," *Oil and Gas Journ.*, February, pp. 125-128, 2 refs.

Foam drilling is only one modification of air drilling, but its importance is growing. Successful foam drilling depends on an efficient foaming agent -- one that has the properties needed to get the job done. This article deals with these requirements of a foaming agent and other aspects of foam drilling. Foam drilling is used to penetrate competent water-bearing formations; aeration drilling is used to penetrate less competent formations that can be drilled with foam and to lift large flows of formation water; chemicals are used to shut off flows from water bearing formations. WRSIC #W71-05685

475. Muskat, M., 1949, *Physical Principles of Oil Production,* McGraw-Hill Book Co., Inc., New York, 922 pp.

Most of the content of this book appeared in the technical literature of the subject within the last fifteen years. Much of it is being currently extended, developed, and clarified. It has been the immediate purpose of this work to formulate and correlate what appears to be known now about the physical principles and facts underlying the mechanics of oil production. It has been a more serious purpose to stimulate and encourage further research and study of the subject to fill in the many gaps in our present knowledge, to clarify the many aspects that are still subject to speculation and conjecture, to generalize the simplified and idealized treatments of special problems, and to improve the correlation between laboratory theories and field observations.

476. Myers, G. M. and J. V. Funk, 1967, "Fluids Dynamics in a Diamond Drill Bit," *Trans. SPE of AIME,* Vol. 240, pp. 347-354, 7 refs.

The two-phase flow of a drilling fluid in a diamond drill bit is investigated by deriving steady-state continuity and momentum equations. Parameters in this analysis are rotational speed of bit, local rate of addition of drill solids, non-Newtonian nature of drilling fluid at drilling rate. This paper shows that under normal drilling conditions, certain parameters are neglected.

477. Myers, G. M. and K. E. Gray, 1968, "Rock Failure during Tooth Impact and Dynamic Filtration," *Trans. SPE of AIME,* Vol. 243, pp. 163-173, 9 refs.

Results of single-blow bit tooth impacts on saturated rocks at various stress states were reported. This paper extends these earlier works to include study of bit impact tests on salt water-saturated Berea and Bandera sandstone samples under conditions of elevated confining and pore pressure. During the tests, dynamic filtration and deposition of a mud cake were occurring due to the presence of drilling mud in the borehole and a borehole-to-formation pressure differential.
WRSIC #W69-09927

478. Mylander, H. A., 1956, "Oil Field Techniques Useful in Water Well Drilling," *Water Well Journ.,* Reprint, November and December 1955; January and March, 1956, 11 pp., 8 refs.

Mylander, H. A. . . . . (con't)

Some of the most useful oil field techniques are adaptable for water well drilling – electric logging; radioactive logging; magnetic detection; instruments for obtaining temperature, pressure, and flow data within well. WRSIC #W71-05696

479. Nagy, J., *et al.*, 1964, "Sealing a Coal-Mine Passageway through a Borehole," *Bur. Mines Rept. Inv. No. 6453,* 13 pp.

Five methods for remote sealing are discussed.

480. National Water Well Association, 1971, *Water Well Driller's Beginning Training Manual,* N. W. W. A., Columbus, Ohio, 84 pp.

The *Water Well Driller's Beginning Training Manual* is designed for use by water well contractors in the training of new men. It reduces to the barest essentials the knowledge a man must have in order to learn how to be a water well driller. The manual, which consists of 15 chapters, ranges from the simplest outline of ground water hydrology through the use of various types of drilling equipment for the construction of wells to the reasons for records and reports. The need for such a guidebook is well documented by a statistical analysis of the manpower shortage in the water well industry. Surveys show that the most widespread source of new workers are the contractors' own on-the-job training programs. This manual is designed for use with just such a program no matter how informal it may be. The information presented represents a foundation on which each contractor can build the knowledge of new employees to suit his own operation. But it is a foundation only. Other more comprehensive books and manuals are designed to cover a wider range of water technology. The ruling principle here has been to envision a man with no prior knowledge of water well work at all. Emphasis is on strict-introductory material which the employing contractor can supplement according to his own desires and local conditions. WRSIC #W71-13973

481. National Water Well Association and the Environmental Protection Agency, 1971, "Proceedings, National Ground Water Quality Symposium," Denver, August, *Ground Water,* Vol. 9-10, No. 6-1, November-December, January-February.

Symposium covered a variety of water quality problems including deep well waste.

482. Nikonov, G. P., 1971, "Research into the Cutting of Coal by Small Diameter, High Pressure Water Jets," *12th Annual Rock Mech. Symposium,* Rolla, Mo., November, Proc., pp. 667-680.

Results of preliminary research on the use of small diameter water jets at high pressure for cutting rocks have demonstrated the possibility of effective cutting of anthracite. Detailed research has been conducted to study the dynamics and structure of the small water jets and the mechanism by which they cut coal. The productivity of water jet mining machines depends on the hydrodynamic parameters of the jet.

483. Noble, D. G., 1963, "Well Points for Dewatering," *Ground Water,* Vol. 1, No. 3, July, pp. 21-26.

Techniques in use of well-point equipment for dewatering operation in Australia are described. Construction features of equipment used successfully in drying out trenches for pipe laying, excavation for underground structures, and dewatering base area for low level pumping stations are discussed. Typical layouts of well-point equipment and typical pumping equipment features are given. WRSIC #W71-09905

484. Obert, L. and W. I. Duvall, 1967, *Rock Mechanics and the Design of Structures in Rock,* Wiley & Sons, Inc., New York, 650 pp.

Explores the methods and procedures for measuring the mechanical properties of rock, and a consideration of mechanisms of failure; describes instruments and procedures for measuring stress, strain, deformation, and other related quantities together with results from both laboratory and field investigations; discusses procedures based on both theoretical and empirical results for designing, analyzing, and evaluating the stability of underground structures.

485. Obert, L., *et al.,* 1946, "Standardized Tests for Determining the Physical Properties of Mine Rock," *U. S. Bur. Mines Rept. of Inv. R. I. 3891.*

Fifteen physical tests are described which were adapted to and developed for the determination of the following physical properties of the rock. The tests were so designed that the test specimens could be obtained by a diamond-drill core, supplied either from field drilling or from rocks cored in the laboratory. With the exception of the abrasive hardness test, all tests gave results that were independent of the diameter and length of the diamond core specimen, provided that the length/diameter ratio was unity for the specimens used in the compression and impact-toughness tests. Although the moisture content in the rock changed some of the physical properties significantly, the air-dried state gave results which in terms of the reproducibility were as good as or better than those for other moisture states. It is anticipated that correlations can be made between the physical properties as determined by the tests described herein and various mining processes, such as drilling, blasting, and crushing. It is recognized, however, that successful application of laboratory results to actual mining problems depends on determination of the properties of the rock in place as distinguished from the properties of the drill-core specimens.

486. Oliver, P., Jr., and E. D. Michael, 1968, "Shallow Water Wells in Coastal Waters," *Mil. Engineer,* Vol. 60, pp. 409-411.

Various criteria for the design of shallow wells in coastal areas, and construction procedures, well-spacing, and discharge calculations for conditions in the Coastal Plain of South Vietnam.

487. Ormsby, G. L., 1965, "Desilting Muds with Hydroclones," *Drilling Contractor,* Vol. 21, No. 3, March-April, pp. 55-65.

Many drilling benefits have been reported from "desilting" drilling

Ormsby, G. L. . . . . (con't)

muds. The term and the process are not generally understood. The author explains terminology, lists benefits reported, outlines equipment requirements, and gives general installation and operation pointers.

488. Ostroff, A. G., 1965, *Introduction to Oil Field Water Technology,* Prentice-Hall, Inc., Englewood Cliffs, New Jersey, 412 pp.

This book provides up-to-date information on oil field water treatment techniques. It covers the chemical, physical, and biological problems encountered with these waters and their causes and correction.
WRSIC #W71-09721

489. Ostrovskii, A. P., 1962, *Deep-Hole Drilling with Explosives,* Consultants Bureau, New York, 133 pp., 67 refs.

This book discusses the new trend in worldwide application of explosives in technology and the national economy; it presents considerable experimental material on the effect of explosions on solid media, material that is not only interesting to specialists in mining but also to physicists. WRSIC #W71-09914

490. Otts, L. E., Jr., 1963, "Water Requirements of the Petroleum Refining Industry," *U. S. Geol. Surv. Water-Supply Paper 1330-G, pp. 287-340.*

Describes how water is used in the refining of crude oil and summarizes the findings of a survey of the sources, quantities, and chemical quality of makeup water used by 61 refineries in various parts of the United States. Estimates of future water requirements are included.

491. Outmans, H. D., 1958, "Mechanics of Differential-Pressure Sticking of Drill Collars," *Trans., A.I.M.E.,* Vol. 213, pp. 263-274, 11 refs.

Drilling progress is often delayed by sticking of the drill string. The development of preventive and remedial methods has been hampered by incomplete understanding of the sticking mechanism. A recent laboratory investigation has indicated that one type of sticking may be attributed to the difference in pressure between the borehole and formation. This paper shows by means of soil mechanics that the primary cause for differential pressure sticking is cessation of pipe movement; whereas, differential pressure and standing time determine the severity of the sticking.

492. Page, H. G., *et al.,* 1963, "Behavior of Detergents (ABS), Bacteria, and Dissolved Solids in Water-Saturated Soils," *U. S. Geol. Surv. Prof. Paper 450-E, Art 237, pp. 179-181.*

The comparative effectiveness of a coarse-grained, a fine-grained, and a colloid-coated soil in removing ABS, dissolved solids, and bacteria from sewage effluent was determined under saturated-flow conditions. The apparatus used in the study is described and sketched. Coarse sand and sandy loam removed about 90% of bacteria within a few feet of travel, but additional travel does not necessarily remove all the remaining bacteria. Dissolved solids and ABS were virtually unaffected by filtration of saturated flow. Because bacterial clogging occurs quickly in fine-grained soils, coarser sand is preferred as a pollution filter for the removal of bacteria. The use of colloidal alumina to remove bacteria or ABS seems to be economically unfeasible at present.

493. Parizek, R. R. and L. J. Drew, 1966, "Random Drilling for Water in Carbonate Rocks," Penn. State Univ., *Water Res. Research Pub. 3-66,* Univ. Park, Penn., March.

Assumes that geologic factors influencing order of magnitude differences in well yields are known in advance of drilling.

494. Parizek, R. P. and S. H. Siddiqui, 1970, "Determining the Sustained Yield of Wells in Carbonate and Fractured Aquifers," *Ground Water,* Vol. 8, No. 5, Sept.-Oct., p. 12-20, 7 figs., 2 tables, 11 refs.

Carbonate aquifers with highly developed anisotropic permeabilities and other fractured rocks under water-table or semi-water-table conditions present complex hydrologic settings in which to predict the sustained yield of individual wells or groups of wells. Yields of wells in these settings are particularly responsive to the position of the water level and its relationship to one or more producing zones. Often a well's total capacity may be accounted for by one or more openings encountered in drilling which are separated by varying thicknesses of essentially nonproductive rock. A well's yield is determined more by the position of the water table with respect to these openings than to the proportion of saturated rock penetrated by the well bore.

495. Park, A., *et al.*, 1960, "Chemical and Mechanical Means of Maintaining Low-Solids Drilling Fluids," *Oil and Gas Journ.*, Vol. 58, May, pp. 81-84, 6 refs.

Using a new chemical to replace part of the bentonite in a drilling mud can speed up drilling rate. All other factors equal, the fewer the dispersed solids in a mud, the higher the drilling rate. This chemical inhibits the hydration and dispersion of cuttings into the mud. The powdered free-flowing polymer can double yield of bentonite when used with soda ash.

496. Parker, M. E., 1960, "Corrosion and Its Control," *Oil and Gas Journ.*, Reprint, Tulsa, 52 pp.

The operating manual gives, in a series of easy-to-read, one page installments, a large variety of technical information and helpful pointers in the field of corrosion. WRSIC #W71-05315

497. Patchick, P. F., 1967, "Estimating Water Well Specific Capacity Utilizing Permeability of Disturbed Samples," *A.W.W.A. Journ.*, Vol. 59, No. 10, pp. 1292-1302, October, 19 refs.

A practical method, using little data, is illustrated by determining approximately the specific capacity of a well that is to be drilled. The method emphasizes use of any relevant hydrogeologic records from nearby wells.

Patchick, P.F. . . . (con't)

498. ________, 1967, "Predicting Well Yields —Two Case Histories," *Ground Water,* Vol. 5, No. 2, pp. 41-53, April, 26 refs.

Two case histories are presented which illustrate that by analyzing drill cuttings or bailed samples, by knowing total depth of a test hole and position of the static water level, and by studying the driller's log not only can a well's yield be predicted but also drawdown may be predicted for any well in advance of a pumping test.

WRSIC #W71-09733

499. Patten, E. P., Jr., 1963, "Application of Electrical and Radioactive Well Logging to Ground-Water Hydrology," *Pa. Geol. Surv. Fourth Series No. W19,* 60 pp., Harrisburg, Pa.

Discusses in detail several problems pertaining to the interpretation of electrical and radioactive well logs in ground water hydrology. Emphasis is placed upon situtations in which interpretation departs from the practices common in petroleum engineering. Certain interpretive methods of the oil industry are demonstrated to be unsatisfactory for use in ground water hydrology. An effort has been made to analyze the interpretive method in terms of underlying theory.

500. Patten, E. P., Jr., and G. D. Bennett, 1963 "Application of Electrical and Radioactive Well Logging to Ground Water Hydrology," *U. S. Geol. Surv. Water Supply Paper 1544-D,* pp. 1-60, 23 refs.

Discusses resistivity logging, spontaneous-potential logging, fluid-conductivity logging and gamma-ray logging in relation to selected ground water problems. A graphical method of combining radioactivity and resistivity data is proposed as an aid in qualitative comparison of sands penetrated by a well. Relates application of interpretive methods in ground water hydrology to methods used in the petroleum industry.

WRSIC #W71-05697

501. Paul, B. and D. L. Sikarskie, 1965, "A Preliminary Theory of Static Penetration by a Rigid Wedge into a Brittle Material," *Trans., AIME,* Vol. 232, pp. 372-383.

A theory is presented for the static penetration of a single rigid wedge into brittle material. The material considered is one which exhibits both crushing and chipping phases in the penetration process. If the wedge angle and three parameters are specified, the theory predicts forces and associated penetrations during both the crushing and chipping phases. For certain ranges of the parameters, agreement with the limited experimental data available is promising except for the initial phase of the penetration process where refinements on the proposed theory are required. The theory also predicts that for certain values of the wedge angle and other known parameters, the chipping process does not occur and penetration is due entirely to crushing.

502. Payne, L. L. and W. Chippendale, 1953, "Hard Rock Drilling," *Drilling Contractor,* June, pp. 58-62.

In drilling hard formations, it is necessary to know the characteristics of the formation being drilled which cause it to fall into the category of

"Hard Rock." Inasmuch as the physical properties of rock as examined at the surface may not represent those properties as presented to the rock bit cutters, it is important to study the nature of these changes in properties. An important advancement of recent years in the drilling of hard abrasive formations has been the introduction of a bit employing the use of sintered tungsten carbide inserts as cutting elements. This type of rock bit has shown remarkable results in West Texas, New Mexico, Oklahoma, and the Rocky Mountain areas when used in formations which react to a crushing action. The use of heavy weights and relatively slow rotary speeds is recommended when using this type of bit. Some formations, however, which are classified as "hard" due to their high abrasive properties may have relatively weak binders and will not respond readily to the crushing action of this type of bit.

503. Peacock, D. R., 1965, "Temperature Logging," Soc. Professional Well Log Analysts Trans., *6th Annual Logging Symposium,* Dallas, Vol. 1, May, pp. F1-F18.

The use of new, highly sensitive and stable temperature logging systems has produced data which correlate with earlier theoretical predictions and have stimulated the application of temperature logging into areas other than the conventional cement top location. Innovations lie chiefly in the concepts of combining gradient with differential logs to get different views of a given temperature anomaly. Also, the techniques of changing well conditions during a series of logs are effective in creating temperature anomalies which can reveal considerable information about downhole conditions. These techniques are particularly effective in producing injectivity profiles in secondary recovery programs and in evaluating fracturing operations.

504. Pennington, J. V., 1953, "Some Results of DRI Investigations, Rock Failure in Percussion," *A.P.I. Drill. and Prod. Pract.,* pp. 329-336.

Drilling Research, Inc., as one phase of an evaluation of unconventional methods of drilling deep holes investigated the fundamentals of rock fracture under percussion. Rock failure was studied by observing crater formation by static loading and by dropping a chisel edge on the rock. The relation of the energy of impact, the velocity of the bit, and the momentum of the percussive blow was examined. The energy of the blow determined the amount of rock drilled.

505. Peret, J. W., 1967, "Effect of Research of Present Drilling Costs," *5th Annual Rotary Drilling Conference,* A.A.O.D.C., Houston, pp. 39-52

The objective in this study is to present the various trends that have developed in drilling, along with measurement of the money involved. The basic factors used include the footage penetrated, holes drilled, and rigs used. An incomplete list of innovations includes improvements in weight-speed programs; hydraulics; muds low in viscosity, density and solids; better design and metallurgy in drill strings, bits, pumps, and rotary connections; better practices in inspection classification and grading of used drill strings and bits along with many other advancements.

WRSIC #W71-06958

506. Peter, Y., 1970, "Model Tests for a Horizontal Well," *Ground Water,* Vol. 8, No. 5, September-October.

This paper deals with the ground water flow into a drilled well consisting of a vertical shaft and radially spaced horizontal intake pipes. It is based on model tests devised and executed by the author in the Hydraulic Laboratory of the Haile Selassie University in Addis Abada, Ethiopia. The author tested the assumption of linearly distributed inflow into the pipes based on the steady-flow theory of Darcy-Thiem and found satisfactory agreement between theory and experiment. WRSIC #W71-00198

507. Peterson, D. F., *et al.,* 1952, "Hydraulics of Wells," *Ag. Exper. Station Bull. 351,* Utah State College, Logan.

New developments for the study of steady flow or equilibrium ground water flow condition. Includes flow of ground water into wells; seepage surface for unconfined systems; unconfined flow replenished by vertical percolation effect of replenishment; significance of the discharge number; effectiveness of wells; nonsteady flow; and zones of flow in well hydraulics.

508. Peterson, F. L. and C. Lao, 1970, "Electric Well Logging of Hawaiian Basaltic Aquifers," *Ground Water,* Vol. 8, No. 2, March-April, pp. 11-18, 10 refs.

An extensive program of electric well logging has been conducted in Hawaii during the past three years to determine the applicability of this tool to the volcanic environment which exists in Hawaii. Electric well logging techniques were found to be useful in Hawaiian basaltic aquifers; however, interpretation of both spontaneous potential logs and resistivity logs varies from the conventional interpretation of electric logs in sediments. WRSIC #W70-05480

509. Peterson, J. S., *et al.*, 1955, "Effect of Well Screens on Flow into Wells," *A.S.C.E. Trans.,* Vol. 120, pp. 562-584.

The hydraulics of wells involves flow (1) in the surrounding aquifer, (2) through the well screen, and (3) inside the well. This paper is concerned with the flow through the screen and inside the well. Head loss, screen selection, etc., are explained.

510. Piatek, A., 1967, "Preventing Filamentous Scale in Well Water," *Water and Wastes Eng.,* Vol. 4, No. 12, December, pp. 54-55, 9 refs.

Systems of chemical treatment for preventing formation of filamentous scale in well water containing iron bacteria and colloidal clay are reviewed. Laboratory experiments indicate that chlorine dioxide has ability to destroy iron bacteria. The treatment was divided into two parts: (1) well surging and cleaning and (2) line cleaning. The system of treatment has been successful for a year. It also appears that corrosion is being held at a lower level. WRSIC #71-03835

511. Pickett, G. R., 1960, "The Use of Acoustic Logs in the Evaluation of Sandstone Reservoirs," Geophysics, Vol. 25, No. 4, pp. 250-274.

It is shown that acoustic velocities in sandstones are primarily dependent on porosity, shaliness, and pressure differential between overburden and fluid pressures. Although there are undoubtedly other variables which have some effect on acoustic velocities in sandstones, usable porosity predictions can be made from acoustic borehole logs if measured velocities are corrected for effects of pressure differential and shaliness.

512. _______, 1966, "Prediction of Interzone Fluid Communication Behind Casing by Use of Cement Bond Log," *7th Annual Logging Symposium,* Soc. Professional Well Log Analysts Trans., Tulsa, May, pp. J1-27.

Quantitative criteria for predicting by the use of the cement bond log whether interzone fluid communication will take place through the casing-formation annulus are derived for cementing and logging conditions on the Cedar Creek Anticline. These criteria consist of a critical amplitude above which communications will take place and critical interzone distance below which communication will take place. These criteria were derived by correlation in 28 wells of conclusive communication tests with cement bond logs which meet certain quality control and calibration requirements. It is shown that other phenomena which significantly affect cement bond log response, but which are not associated with log quality control or with the presence and strength of the cement in the annulus, are "early arrivals" opposite high velocity formations, the apparent presence of a "micro-annulus" between casing and cement sheath, and a change in apparent bonding with time when gilsonite cement was used. The significance of the micro-annulus to interzone communication has not been established and is still under study.

513. Pigott, R. J., 1941, "Mud Flow in Drilling," *A.P.I. Drill. and Prod. Pract.,* pp. 91-103, 4 refs.

This paper is a revision of the study completed in 1931 on the flow of mud in pipe, well bores, and mud pits and on the lifting and releasing properties for cuttings. Tests made by Gregory on 4-in. pipe; by Ambrose and Loomis on capillary tubes; also on 1/2- and 1-in. pipe and a modified McMichael absolute viscosimeter were used to demonstrate that the flow of mud can be predicted in the viscous region by utilizing the variable apparent viscosity in place of the usual constant vicosity for a liquid.

514. Pirson, S. J., 1963, *Handbook of Well Log Analysis for Oil and Gas Formation Evaluation,* Prentice-Hall, Inc., Englewood Cliffs, New Jersey.

The aim of this handbook is to make available sufficient technical information together with an adequate number of well log interpretation charts in order to perform satisfactory evaluations of the productive capacities and of the reserves of prospective as well as of existing oil and gas fields. Practicing well log analysts will find this handbook

Pirson, S. J. . . . . (cont'd)

sufficiently complete; they no longer need to refer to the voluminous trade literature in order to find the required charts appropriate to the solution of their problems. Practicing geologists will find Chapters 6 to 11 of particular significance as by training they are prone to rely primarily on qualitative evaluation of electric logs in their subsurface correlation for stratigraphic and structural studies of sedimentary basins. The utilization of the qualitative principles involved will allow them the determination of lithology, of formation tops and bottoms, of porosity development, and of the petroliferous qualities of sediments with much more reliability than previously. Well completion and production engineers should especially welcome this book as they also rely a great deal on qualitative evaluation of well logs in designing casing and cementing programs, perforation, and formation testing by drill stem tests, formation tester and others, and in designing and choosing the appropriate well stimulation techniques: hydraulic fracturing, acidizing, water block removal, and the like.

515. ———, 1970, *Geologic Well Log Analysis,* Gulf Publishing Co., Houston, 370 pp.

Series of discussion topics proposes to use well logs to the fullest extent for geological studies (sedimentation, fluid migration, tectonic deformation, etc.) by deriving from logs all sorts of mapping parameters and indices of significance from the point of view of mineral accumulations -- primarily in sediments.

516. Porter, E. W., 1970, "Solving Four Common Drilling Problems: Fishing is More Art than Science," *Oil and Gas Journ.,* Vol. 68, No. 38, pp. 95-96.

There are precautions and practices which can result in eliminating fishing jobs or in reducing the time and money spent on these jobs. The causes of fishing jobs fall into three main categories: mechanical failure of equipment, downhole problems related to hole conditions, and human errors. Each cause can be minimized, but it is important to realize that economics and time must always be considered.

WRSIC #W71-03832

517. Postgate, J. R., 1963, "Versatile Medium for Enumeration of Sulfate-Reducing Bacteria," *Applied Microbiology Journ.,* Vol. 11, p. 265.

Explains the various environmental aspects of sulphate-reducing bacteria and occurrences of corrosion damage to water distribution piping. Cites numerous case histories of corrosion and presence of sulphate-reducing bacteria.

518. Pryor, W. A., 1956, "Quality of Ground Water Estimated from Electric Resistivity Logs," *Ill. State Geol. Surv.,* Circ. 215, 15 pp.

Comparison of chemical analyses of formation water to calculated values of NaC1 solution equivalent for 94 resistivity logs shows that chloride content and total solids content of water in Pennsylvanian sandstones of Illinois Basin can be determined within limited range by use of resistivity logs.

519. Purswell, G. M., 1967, "Drilling Assemblies Key to Crooked Hole Problems," *World Oil,* Reprint, Mid-Continent District Meeting of A.P.I., Oklahoma City, March, 5 pp.

New bottom-hole drilling assemblies and increased knowledge of forces acting on drill strings are overcoming crooked-hole problems. Various stabilizer and drill collar configurations are providing stiffness and controllability to bottom-hole assemblies and result in straighter holes, higher penetration rates, longer bit runs, and lower drilling costs.

520. Rabe, C. L., 1956, "A Relation Between Gamma Radiation and Permeability of Denver-Julesburg Basin," *Petrol. Trans., A.I.M.E.,* Tech. Paper 403, Vol. 210, pp. 358-460.

The Muddy and Dakota sands, more commonly known as the "D" and "J" sands, respectively, of the Denver-Julesburg Basin are correlative over a large area. It is reasonable to believe, therefore, that depositional environment throughout the Basin was comparable. Local variations in sand development do occur, however, and productivity is often governed by permeability as well as structural position. Experience indicates permeability is commonly a function of cementing material. While cases of siliceous or calcareous cementing have been observed, the bonding material is for the most part clay. Under the latter circumstance, clay should bear some relation to permeability. Similarly, gamma radiation should bear some relation to the amount of clay. This study was undertaken to determine if a cross relation exists between gamma radiation and permeability.

521. Rackley, R. I., *et al.,* 1968, "Concepts and Methods of Uranium Exploration," *Wyoming Geol. Assn. Guidebook.* 20th Annual Field Conference, Black Hills Area, pp. 115-124.

The major exploration guides for uranium concern the source and method of transportation of the uranium, the relationship of the deposits to structurally positive areas, the physical-chemical nature of the host rock, the character of the concentration process and its capabilities to produce the minerals in the quantities observed, the amount and distribution of carbonaceous matter, and the shape of the orebodies. The concept elucidated in this paper is a genetic hypothesis used to guide exploration for sandstone-type uranium deposits. The hypothesis may be applicable broadly in the search for vanadium, copper, silver, and other metaliferous sandstone-type deposits. Diagram (Figure 68) subsequently modified and published by B. Rubin, 1970, "Uranium Roll Front Zonation in Southern Powder River Basin, Wyoming," Wyoming Geol. Assoc. Earth Science Bull., Vol. 3, No. 4, December.

522. Rau, J. L., 1970, *Ground Water Hydrology for Water Well Drilling Contractors,* National Water Well Assn., 88 East Broad Street, Columbus, Ohio, 259 pp.

The objective of this book is to present the basic geologic factors which control the occurrence, availability, and quality of ground water. In an

Rau, J. L. . . . . (con't)
introductory treatment of this material, the author believes it is more important to consider the principles of geology as they relate to the discovery and exploitation of ground water than to develop the quantitative aspects of ground water hydrology. Ground water does occur in a geologic framework, and unless the nature of the framework is understood, an analysis of equations or computing procedures for developing various types of aquifers is meaningless. Further, this book is designed primarily for the water well drilling contractor who has had no formal training in geology. Therefore, it probably goes more deeply into geologic principles, identification of rocks and minerals, interpretation of geologic maps and crosssections, and geologic characteristics of aquifer systems than most introductory hydrology books.

WRSIC #W71-05699

523. Read, V., 1963, "Full Core Recovery in Unconsolidated Formations," *Drilling,* Vol. 24, No. 11, August, pp. 41-46.

Unique among new techniques of subsurface sampling and testing is one that achieved near full recovery and virtually undisturbed samples from near surface layers of unconsolidated boulder-strewn alluvium long recognized as one of the most difficult ground formations to core. This approach involves a combination of oil field rotary and mining diamond core equipment – heavy drill collars, rotary stabilizers, and thick vibration-reducing oil emulsion drilling mud, together with the precise hydraulic feed control, i.e., fast – over 300 rpm bit rotation, light drilling weight, hard-rock core barrel, and multi-step diamond bits, that are common in mining industry diamond drilling. In a 755 ft., 8 3/4 in. diameter exploration hole at the Nevada Test Site, 5 7/8 in. diameter cores were taken every 50 ft. producing cores up to 10 ft. long with an average core recovery of 92%. It is felt this core recovery was due to fast rotation, light weight, multi-step diamond bits, and precise controls.

524. Reinhart, F. W., 1968, "Recent Developments in the Thermoplastic Piping," *A.W.W.A. Journ.,* Vol. 60, No. 12, December, pp. 1404-1410.

The increase in quality of thermoplastic pipe and fittings has been due to: (1) improved extrusion and molding machinery and techniques used to form plastic pipe and fittings, (2) improvements in thermoplastic piping materials, (3) policies and procedures used to develop recommended hydrostatic design stresses, (4) the National Sanitation Foundation (NSF) inspection and certification programs, and (5) the increase in quality and quantity of the standards for plastic piping. The total effect of these developments has resulted in higher quality pipe and fittings, lower costs, increased confidence on the part of users and engineers and improved performance.

525. Reinke, J. W. and D. L. Kill, 1970, "Modern Design Techniques for Efficient High Capacity Irrigation Wells," *Winter Meeting, Amer. Soc. Agricultural Engineers,* Paper 70-732, December, 23 pp., 4 refs.

Common irrigation well design problem areas are reviewed. Modern design criteria that are presented insure sand-free water, overall high efficiency and long well life. The benefits of suggested procedures are economical, low cost irrigation wells. The long-term economics of modern irrigation well design is in the interest of adding much needed profits to agriculture. WRSIC #W71-09913

526. Riley, E. A., 1967, "New Temperature Log Pinpoints Water Loss in Injection Wells," *World Oil,* Vol. 164, No. 1, January, pp. 69-72, 1 ref.

A new, extremely accurate differential temperature survey technique has been used successfully to precisely locate a downhole water loss in a waterflood injection well which was previously undetectable by any other means. The differential temperature log accurately measures fractional temperature anomalies associated with fluid movements downhole applicable for finding tubing-casing leaks, gas communication, productive zones, lost circulation zones, production profiles, and gas-oil-water contacts. WRSIC #W71-02990

527. Robeck, G. G., 1969, "Microbial Problems in Ground Water," *Ground Water,* Vol. 7, No. 3, May-June, p. 33.

This article contains a summary of a few disease outbreaks caused by ground water contamination, and the difficulties in designing and monitoring for effective quality control. There is a discussion of problems associated with large basin recharge with treated sewage. Chlorination of well water withdrawn for domestic use is advocated as good insurance for microbial control. WRSIC #W69-06426

528. Roberts, B. J. and C. H. Mohr, 1971, "Down-Hole Motors for Improved Drilling Method," *SPE of AIME Rocky Mountain Reg. Meeting,* SPE 3343, Billings, Montana, June, 12 pp.

The concept of downhole motors as a drilling device is not new. Almost 100 year development has resulted in their recent increased utilization as an alternative to conventional rotary drilling practice. Already the success of downhole drilling motors has resulted in their routine use on performing directional drilling, sidetracking, and other special drilling operations. This achievement has resulted in the recognition that downhole motors can be used to improve normal drilling programs. Increased penetration rates as well as other drilling improvements for these programs can be achieved with proper application of the present development of downhole motors. This must be considered in respect to the specific drilling equipment and particular hole environment. Present successful utilization of this type of drilling method has resulted from the selective programming of the motor characteristics.

529. Robinson, L. H., Jr., 1959, "Effects of Pore and Confining Pressures on Failure Characteristics of Sedimentary Rocks," *Trans., AIME,* Vol. 21, pp. 26-32, 16 refs.

Triaxial compression tests have been performed to determine the strength characteristics of limestone, sandstone, and shale rocks subjected to controlled stress conditions. This control was exercised by varying the liquid pressures within and around a plastic-encased rock specimen. The pressure in the pores of the rock was varied throughout

Robinson, L. H. . . . . (con't)

the range from atmospheric to 15,000 psi; the external pressure was changed over the same range with various positive pressure differences between it and the internal pressure. The data shows that the rock strength increased and the mode of failure changed as the pressure surrounding the rock became greater than the pressure in the pores of the rock. These observations and the results of microbit drilling experiments indicate that the increased rock strength under pressure may be an important effect in reducing drilling rate but that other factors are probably of even greater importance.

530. Rogers, W. F., 1963, *Composition and Properties of Oil Well Drilling Fluids,* 3rd Edit., Gulf Publishing Co., Houston, Texas, 818 p.

The technology of oil-well drilling fluids (as of 1963) is treated in detail with numerous references to publications and brief abstracts of 556 United States patents on the subject. Following a brief history of the development of the technology, drilling equipment related to the drilling fluid is described. Chapters deal with test equipment, procedures, and specifications; calculations for mud treatments; manufacture of drilling fluid materials; flow and filtration properties; various compositions of water-base muds and their applications; aerated drilling fluids and loss of circulation.

531. Rollins, H. M., 1963, "Straight-Hole Drilling," *World Oil,* March and April, pp. 7-77, 113-119, 6 refs.

A three to five degree contract will in no way assure an operator of a hazard-free hole and rigid deviation clauses can result in drilling operations that will be slower, more costly, and less competitive. The prime objective should be to obtain a "useful" hole bottomed in the target sand containing a minimum of abrupt changes in angle, free of "keyseats," in the most efficient drilling time that will accommodate the casing program and production operations.

532. ________, 1970, "How to Drill a Better Hole," *World Oil,* January, pp. 49-52, February, pp. 33-36, March, pp. 66-69, 72, April, pp. 73-76, 15 refs.

Properly designed wall contact drilling assemblies drill straight, smooth, full gage, and trouble free holes. As a result, drilling costs are lower, fishing problems are minimal, and casing running is easier. This first installment of a four-part series explains how wall contact assemblies can be designed for optium deviation control and bit performance.

533. Romero, J. C., 1970, "The Movement of Bacteria and Viruses through Porous Media," *Ground Water,* Vol. 8, No. 2, March-April, p. 37.

This report is the result of a request by the State Board of Examiners of Water Well and Pump Installation Contractors. The Board is in the process of formulating a more reasonable set of guidelines which control the location of wells designed to produce water for human consumption and/or food processing with respect to potential or existing sources of ground water pollution. WRSIC #W70-05477

534. Rorabough, M. I., 1953, "Graphic and Theoretical Analysis of Step Drawdown Tests of Artesian Wells," *A.S.C.E. Proc.*, Hydraulic Div., Vol. 79, No. 362, September.

Drawdown in an artesian well resulting from the withdrawal of water is made up of head loss resulting from laminar flow in the formation, and head loss resulting from turbulent flow in the zone outside the well through the well screen and in the well casing. A graphical method for the empirical determination of laminar and turbulent head losses is given for computing the head-loss distribution outside the pumped well for various pumping rates. Analysis is made of the variation of specific capacity with discharge and of the importance of well radius in well design.

535. Rose, N. A., 1944, "Report on Research in the Field of Ground Water being Conducted by Oil Companies: Appendix C, Report of the Committee on Ground Water," *Amer. Geophy. Union Trans.,* Part II, January, pp. 420-421.

Explores the similarity of some of the problems of the petroleum engineer with those of the ground water geologist – cites case histories.

536. Rose, N. A., *et al.,* 1944, "Exploratory Water Well Drilling in the Houston District, Texas, "*U. S. Geol. Surv. Water Supply Paper 889-D,* pp. 291-292.

In the spring and summer of 1939, a program of exploration drilling was undertaken in the Houston district, Tex., in conjunction with a general investigation of the water resources of the district. The main purposes of the program were to determine the thickness and character of the water-bearing sands down to a minimum depth of 2,000 feet, the chemical character of the water at different depths, and the artesian pressures, and to provide additional observation wells for studying fluctuations in artesian pressure and possibilities of intrusion of salt water from the direction of the Gulf. On the whole, the results of the test drilling from Houston tend to show that although electrical logs give much information that is useful in the development of water wells, for the present at least, these logs should be used in conjunction with driller's logs and drill-stem samplings of both sand and water in all the more promising sand horizons.

537. Rosenburg, M. and R. J. Tailleur, 1959, "Increased Drill Bit Life through Use of Extreme Pressure Lubricant Drilling Fluids," *Trans., AIME,* Vol. 216, pp. 195-202.

Oil well drilling muds prepared with lubricating aids such as oil, graphite, or mica will not produce a sufficiently strong protective lubricant film for the bit bearing surfaces under high load conditions encountered in drilling operations. Consequently, the damage to the bearing surfaces resulting from the absence of such a film frequently results in a shortening of the drill bit life.

538. Ross, G. H. and G. Adcock, 1969, "Direct Conductance Methods of Measuring Casing Length," *Ground Water,* Vol. 7, No. 4, July-August, pp. 26-27.

The standard water-level indicator can be easily and inexpensively modified so that it also measures casing lengths. The semiquantitative method works well with most metal casings including non-magnetic ones and can be used to detect most casing reductions.
WRSIC #W71-09725

539. Ross, W. M., *et al.,* 1963, "Matrix Acidizing Corrects Formation Damage in Sandstones," *Petrol. Eng.,* Vol. 35, No. 12, November, pp. 64-69, 3 refs.

Matrix acidizing is a relatively new but field-proven technique to correct producing formation damage incurred while drilling or cementing sandstones. It is a result of two significant modifications in conventional mud acid treatments to overcome "skin effects" around well bores. Another principal change in mud acidizing is in the application of controlled injection pressures. WRSIC #W71-03000

540. Rowley, D. S., 1970, "Rotaries to Play Big Role in Future Rock-Drilling Methods," *Oil and Gas Journ.,* Vol. 68, No. 44, pp. 82-87, November, 10 refs.

Because rotary drilling rates continue to increase, rotary is likely to retain its place of leadership among drilling methods for the predictable future, possibly evolving from the conventional jet-rotary system. Improvements in rotary performance provide a moving target for the developers of novel drills. WRSIC #W71-03830

541. Rowley, D. S. and F. C. Appel, 1969, "Analysis of Surface Set Diamond Bit Performance," *Petrol. Eng.,* Vol. 9, No. 3, September, pp. 301-310, 5 refs.

With the assumption of perfect cleaning, a theory of the drilling performance of surface set diamond bits has been developed. The analysis is based on the previously developed theory of the cutting action of a single diamond in which it was assumed that rock behavior during cutting may be approximated by that of a rigid-plastic, Coulomb material.

542. Rowley, D. S., *et al.,* 1961, "Laboratory Drilling Performance of the Full Scale Rock Bit," *Trans. SPE of AIME,* pp. 71-81, 24 refs.

Laboratory drilling tests with 4 3/4 inch hard-formation rock bits were made under rock pressure and borehole fluid pressures simulating a 3,000-ft. borehole. The effects of bit weight and rotary speed on drilling rate and bit rotary power were determined in a hard, impermeable dolomite. With added bit weight, the drilling rate and rotary power both increased at an increasing rate; but with added rotary speed, the rate and power both increased at a decreasing rate. The rock volume removed per unit of energy increased as weight was raised or as rotary speed was reduced. Empiric quadratic equations for drilling rate and rotary power were obtained, and bit mechanical efficiency was calculated.

543. Ryznar, J. W., 1944, "A New Index for Determining Amount of $CaCo_3$ Scale formed by Water," *A.W.W.A. Journ.*, Vol. 36, No. 4, April, pp. 473-486.

The purpose of this investigation was to obtain a formula that will give a quantitative index of the amount of calcium carbonate scale that would be formed by a water at any temperature up to 200°F; and to predict, if possible, the corrosiveness of waters that are non-scale forming. The Langelier Index is modified for more quantitative use.

544. Saines, M., 1968, "Map Interpretation and Classification of Buried Valleys," *Ground Water,* Vol. 6, No. 4 pp. 32-37, July-August.

Discusses buried valleys cut in bedrock.

545. Sarma Jagannadha, V. V. and V. Bhaskara Rao, 1962, "Variation of Electrical Resistivity of River Sands, Calcite and Quartz Powders with Water Content," *Geophysics,* Vol. 17, No. 4, August, pp. 450-479.

Electric resistivity variations of samples of graded river sands, calcite and quartz powders are studied for different moisture contents of varying salinities. The variations exhibit a general hyperbolic trend. For the same grain size, the critical saturation index of a sample is correlative with its retentive capacity, and it is shown, from studies of the quartz samples that for grain sizes of the order of clay particles the critical moisture would reach 100 percent.

546. Sartain, B. J., 1960, "Drill Stem Tester Frees Stuck Pipe," *Petrol. Eng.*, October, pp. B87, B90.

A pipe stuck by differential hydraulic pressure was released with drillstem testing tool which might lead to a new hydraulic fishing tool. This tool accomplished in one run what $13,000 worth of conventional fishing equipment did not.

547. Sasman, R. T., 1957, "Reverse Rotary Well Construction," *Public Works,* Vol. 88, No. 4, April, pp. 117-118.

Construction drilling and testing of water well in Urbana, Ill., described; drilled using hydraulic rotary drilling methods; requiring large volume of water flowing from supply pit into well outside drill stem; water and drill cuttings return to surface by suction through drill stem.

WRSIC #W71-03847

548. Savins, J. G. and W. F. Roper, 1954, "A Direct-Indicating Viscometer for Drilling Fluids," *A.P.I. Drill. and Prod. Pract.*, pp. 7-22, 17 refs.

The understanding and efficient control of drilling-fluid behavior requires a fundamentally sound method of flow-properties analysis which

Savins, J. G. and W. F. Roper . . . . (cont'd)

is practical to use in the field. The concept of plastic flow may be used in describing the flow behavior of many drilling fluids requiring a suitable field viscometer.

549. Scanley, C. S., 1959, "Acrylic Polymers as Drilling Mud Additives," *World Oil,* July, pp. 122-128, 5 refs.

Acrylic polymers have been used as fluid loss control agents in drilling muds for about eight years. Article discusses (1) general background on polymers, history, and development, (2) chemical nature and preparations, (3) polymer properties, (4) the mode of interaction with clay, and (5) specific effort of polymer variables on drilling mud properties.

550. Sceva, J. E., 1966, "The Champoeg Park Demonstration Well, Design and Testing of Water Wells," *Ground Water Report No. 8,* State Engineer, Oregon.

This article briefly describes well construction and testing methods of wells in specific lithologies.

551. Schaff, S. L., 1950, "Geology and Water Well Construction," *A.W.W.A. Journ.,* Vol. 42, No. 4, May, pp. 475-478, 1 ref.

The author of contract proposals for a water well is confronted by geologic considerations from the start. Location, depth, casing, screen, size of pump, testing, prevention of pollution, elimination of sand, and turbidity – in each of these items, some element of geology is involved. As the geologic conditions below any given point on the earth's surface generally cannot be predicted with complete accuracy, approximations to the right answers must suffice, but without them, the writing of contract document founders on uncertainties. Knowledge of the geology of a proposed well site including facts on the thickness, character, and structure of rock formations and on the movement of water in them is fundamental to the preparation of water well contract proposals. The more accurate the geologic information, the better the proposals can be made to fit the conditions and the more economically the work can be accomplished. WRSIC #W71-13711

552. Scheidegger, A. E., 1957, *The Physics of Flow Through Porous Media,* MacMillan Co., New York, p. 309.

This book explores the wide spectrum of the porous media, fluids, hydrostatics in porous media, Darcy's law, and some solutions, physical aspects of permeability, general flow equations, elementary displacement theory, immiscible multiple phase flow, and miscible displacement. The guiding principles in the preparation of the text were as follows: (1) emphasis was laid on the general physical aspects of the phenomena rather than on particular cases applicable, to special engineering problems, (2) of the many solutions available for some of the basic differential equations, only one was chosen for presentation in each case. The theory of differential equations is a well established discipline of mathematics and has been considered of interest in the present context only if pertinent physical concepts were revealed, (3) the theoretical aspects have perhaps been stressed somewhat more than the experimental ones. However, descriptions of such procedures which enable one to determine theoretical "constants" have always been supplied in order to establish the proper logical sequence.

553. Schiefelbein, G., 1970, "Performance of Alloys Against Erosion–Corrosion Attack," *Mats. Protection,* Vol. 9, No. 6, June, pp. 11-13, 1 ref.

The high frequency of pump repair and replacement in the pulp and paper industry promoted a research program to develop an alloy that would resist the attack of abrasive slurries in corrosive media. The comparative performance of alloys commonly used in paper mills is described. WRSIC #W71-03838

554. Schmidt, L. and C. J. Wilhelm, 1938, "Disposal of Petroleum Wastes on Oil Producing Properties," *U. S. Bur. Mines Rept. of Inv. 3394.*

The attention of operators of oil-producing properties long has been directed toward methods of disposing of petroleum wastes as well as the disposal of oil-field brines. The problem of disposing of petroleum wastes is not easy to solve, and although many operators are to be commended for the efforts they have made, much work remains to be done before the problem can be considered solved. Petroleum wastes include emulsions of petroleum and brine mixtures of petroleum and sand that cannot be "broken down" or treated-economically by known methods of separating petroleum from deleterious matter; crude petroleum that escapes from pipe lines, fittings, and tanks because of holes in the lines, leaky connections or some accident; and oil-saturated materials that accumulate accidentally or otherwise in the normal operation of an oil-producing property.

555. Schmitz, G., 1955, "Slip Velocity of Bit Cuttings," *Canadian Oil and Gas Ind.,* July, pp. 47-55, 8 refs.

Inaccurate determination of net rise velocities of bit cutting in the uphole traveling mud stream is responsible for errors in sample-to-formation coordination. In addition, sample contamination takes place from intervals being drilled immediately before. This contamination is caused by deviations in slip velocities of the bit cuttings from mean values. The article covers actual sequence of time required for return of sample cuttings.

556. Schreurs, R. L., 1966, "Source Water Well Design and Efficiency," *Third Biennial Symposium on Microbiology,* Div. of Production, A.P.I., Anaheim, California, November, 18 p., 16 refs.

Several methods of determining the efficiency of wells as hydraulic structures are described with illustrated calculation procedures. The damage ratio of a well is approximately related to the efficiency. Many cheap-source water wells actually can be expensive if they are inefficient and are produced for long periods of time. It is important to have efficient water wells so as to reduce required number of wells and to minimize power cost. WRSIC #W71-06951

557. Schulz, D. B., 1971, "The Straight Hole Turbodrill," *Soc. Petrol Eng. A.I.M.E.*, Preprint No. SPE-3230, 8 p.

Diamond bits used with turbodrills must be built with large water courses and low pressure drop across the face. Cleaning and cooling is accomplished by changes in the flow characteristics of the fluid as the rotating speed of the bit is increased. Horsepower output of a turbodrill is affected by changes in the number of turbine stages in flow rate and in volume of fluid. A summary of drilling records shows some applications of the turbodrill. WRSIC #W71-05323

558. Schwartz, D. H., 1969, "Successful Sand Control Design for High Rate Oil and Water Wells," *Journ. Petrol Tech.*, Vol. 21, No. 9, September, pp. 1193-1198.

Presented here is a method for designing a gravel flow packed liner completion based on the following considerations: (1) formation analysis, (2) gravel-to-sand ratio, (3) formation sand uniformity, and (4) velocity through slots. The technique is developed from the technology and experience of the water well and petroleum industries, noting that continual improvement in well completions is necessary to achieve or maintain profitable secondary recovery projects. WRSIC #W69-09649

559. Scott, K. F., 1971, "A New Practical Approach to Rotary Drilling Hydraulics," *46th Annual Fall Meeting, SPE of AIME*, SPE 3530, New Orleans, October.

This paper presents a simple, practical, and accurate method of selecting jet bit nozzle sizes. Rather than calculating friction losses in the drilling system, a measure of total friction loss is taken from the standpipe and a graphical solution of pressure loss across the bit is subtracted from the standpipe reading resulting in the actual measurement of system pressure losses. Drilling hydraulics are based on maximum bit horsepower, hydraulic impact, or jet velocity and are all dependent on a predetermined friction loss in the system.

560. Sharpley, J. M., 1962, "Microbiological Corrosion in Water Floods," *Corrosion*, Vols. 17, 18, pp. 386, 247.

It is commonly observed that most bacterial corrosion in water-floods occurs as pit corrosion. A hypothesis has been advanced to explain a portion of the mechanism underlying bacterial pit corrosion, and a suggested method has been presented for determining the possible relationship between the general microbial flora and microorganisms capable of participating in pit corrosion. Topics discussed include role of sulfate-reducing bacteria, correlation between bacteriological examinations and corrosion damage, cultural techniques, laboratory, field evaluation techniques, and detection of sessile microorganisms.

561. Shell, F. J., 1969, "New Mud Improves Productivity," *Drilling*, November, pp. 47-50.

A new mud was developed for use in Utah where wells were being drilled in a National Forest. This mud was designed to prevent pollution of surface waters and at the same time to stabilize shale formations. The mud was based on the use of polyanionic cellulose (PAC) and diamonium phosphate (DAP). Both of these materials are good shale inhibitors, but neither affects electric logs as do other shale inhibitors.

562. Siddiqui, S. H. and R. R. Parizek, 1971, "Hydrogeologic Factors Influencing Well Yields in Folded and Faulted Carbonate Rocks in Central Pennsylvania," *Water Res. Research,* Vol. 17, No. 5, October, pp. 1295-1312, 31 refs.

Hydrogeologic factors influencing well yield in folded and faulted Cambro-Ordovician carbonate rocks and shales were investigated in central Pennsylvania. Fracture trace wells were more productive than nonfracture trace wells. Accidentally located fracture trace wells were as productive as intentionally located fracture trace wells because the accidentally located wells were clustered in more productive rocks. The success ratio of accidentally locating a fracture trace well is 4:6. Wells in sandy dolomite and coarse-grained dolomites were the best producers; wells in valley bottoms were more productive than those in valley walls and uplands; anticlinal wells were better producers than synclinal wells; and wells in beds dipping at less than 15° had higher yields than others.

563. Simon, A. L., 1960, "Production Analysis of Artesian Wells," *A.W.W.A. Journ.,* Vol. 52, No. 11, pp. 1438-1444, November, 6 refs.

Waste of water and energy can be eliminated by proper investigation discharge drawdown analysis; interconnection between aquifers; determination of characteristics for each aquifer by rheometering; role of thermal dilation and effect of temperature on head; problems of corrosion caused by hydrogen sulfide and carbon dioxide.

WRSIC #W71-03837

564. Simon, R., 1956, "Theory of Rock Drilling," *6th Drilling and Blasting Symposium,* Proc., Univ. of Minneapolis, Minneapolis.

A theory of rock drilling is presented, based on an experimental and analytical study of the elementary physical action involved, the impact of a chisel on the surface of the rock. A formula is developed for the rate of penetration of a bit into brittle rock which indicates that the significant parameters are the mechanical power input to the rock per unit area of the hole and a constant termed the drilling strength of the rock.

565. Simpson, J. P., 1967, "New Developments and Ideas in Drilling Mud Technology," *12th Symposium on Exploration Drilling,* Univ. of Minnesota, Sch. of Min. and Metal Eng., Minneapolis, pp. 203-216.

Reviews fresh water muds, salt water mud, oil muds for shale stability, solids control, beneficiated clay, L.S.G. mud, fresh water mud, visifiers, X-C polymer muds, asbestos fines, and seawater mud.

566. ———, 1967, "What's New in Mud Engineering?" *World Oil,* April, pp. 135-139.

Rapid changes in drilling mud technology make necessary a periodic review of these developments. Deeper drillings along with increased temperature and pressure has brought about new applications and problems related to complex drilling fluids. Reports on application of lignite/ligno sulfonate muds, low weight non-dispensive muds, and simple salt muds.

567. Simpson, J. P. and H. V. Sanchez, 1961, "Inhibited Drilling Fluids – Evaluation and Utilization," *Pacific Coast District Meeting,* A.P.I. Paper No. 801-37C, Los Angeles, May.

Examples of field usage of inhibited muds utilizing ferrochrome ligno sulfonate for both thinning and filtration control are cited to show that these muds exhibit excellent properties and that their use has resulted in reduced mud costs on improved hole conditions.

568. Singh, D. P., 1969, "A Review of Variables Involved in Drag Bit Drilling" *Australian Mining,* Vol. 61, No. 8, pp. 54-58, 32 refs.

In suitable strata, drag bit drilling is the fastest method of drilling. Design and operating conditions which affect the performance of drag bit drilling have been extensively investigated during the past few years. Drag bit drilling should have wide application if proper laboratory and field studies could separate those rocks suitable for this type of drilling from those which demand another type of drilling. The basic variables affecting the rate of penetration of a drag bit may be classified as follows: rock material, bit geometry, and material of the tip, thrust, rotational speed, torque, flushing conditions, and hardness reducers.

569. Singh, M. M. and A. M. Johnson, 1965, "Static and Dynamic Failure of Rock under Chisel Loads," *Second Conference on Drilling and Rock Mechanics,* SPE 1407, Univ. of Texas, January, 24 refs.

The mechanism of failure under a drill bit is still improperly understood in spite of several investigations on the subject. Generally the cratering process under static loading conditions is regarded as being similar to that achieved by dynamically applied impact. This paper attempts to indicate that although the sequence of events in the two cases appear identical, at least some dissimilarities exist. Thus, the characteristics of the crater vary to some extent, such as the width-to-depth ratio. The amount of energy consumed per unit volume of rock is also observed to be unequal. The energy aspect of the phenomenon may be explained by regarding rock as a viscoelastic substance. Estimations based on Maxwell and Kelvin-Voigt models have been made, but a Burgers model with a plastic component should provide better simulation.

570. Singh, M. M. and P. J. Huck, 1970, "Correlation of Rock Properties to Damage Effected by Water Jet," *12th Annual Rock Mech. Symposium,* Proc., Rolla, Mo., November, pp. 681-695.

Investigates the potential of high pressure water jets for rapid rock excavation. During this study, an attempt was made to correlate the extent of breakage effected to the mechanical properties as determined for each rock variety. Although tests were conducted to specific pressures of 35 p.s.i., no minimum in the specific energy consumption was noted.

571. ———, 1971, "Effects of Specimen Size on Rock Properties," *46th Annual Soc. Petrol. Eng., A.I.M.E., Fall Meeting,* Reprint No. SPE 3528, New Orleans, 12 p., 12 refs.

Triaxial tests were conducted on two rock types using specimens ranging from 2 to 32 inches in diam. to determine the influence of specimen size on strength and mechanical properties. While the influence of specimen size on mechanical properties was small, the effect on strength and character of failure was pronounced. Although sufficient data to verify statistical failure hypotheses was not collected, it is hoped that additonal work now in progress will help in the development of more realistic models.

572. Smith, C. F. and R. Hendrickson, 1965, "Hydrofluoric Acid Stimulation of Sandstone Reservoirs," *Journ. Petrol Tech.,* Vol. 17, No. 2, February, pp. 215-222.

Hydrofluroic acid (HF) has specific reactivity with silica which makes it more effective than HCl for use in sandstone; kinetics of reactions of HF have been studied to determine related effects of reservoir composition, temperature, acid concentration, and pressure on spending rate of HF; secondary effects from byproduct formation; predictions concerning improvement in productivity resulting from HF treatment of skin damage.

573. Smith, C. F., *et al.,* 1968, "Removal and Inhibition of Calcium Sulfate Scale in Water Flood Projects," *Journ Petrol Tech.,* Vol. 20, No. 11, November, pp. 1249-1256, 21 refs.

The problem of preventing calcium sulfate scale deposition has become increasingly important in the last few years due to the increasing use of waterflood as a means of secondary recovery. This paper describes the results of a laboratory testing program that evaluated potential scale inhibitors and scale-removal agents. The paper also describes a field testing program in which various removal methods and inhibitor placement techniques were evaluated in 19 wells, and it compares field and laboratory results.

574. ———, 1969, "Secondary Deposition of Iron Compounds Following Acidizing Treatments," *Journ. Petrol. Tech.,* Vol. 21, September, pp. 1121-1129, 17 refs.

There has been increased interest in using chemical additives in acid to prevent secondary precipitation of iron compounds following the acidizing treatment. Dissolved iron will remain in solution in the acid until the acid is spent. The precipitation of ferric hydroxide or other iron-containing compounds can seriously damage the flow channels recently opened by the acid reaction in the formation. Like many

Smith, C. F. . . . . (con't)

chemical additives for acid iron-control agents can be misused and overused with damaging results. Some agents precipitate if the expected downhole sources of iron are not present. Thus, the effective use of iron sequestering agents depends upon the chemical conditions existing downhole during acid reaction.

575. Smith, D. K., 1969, "Fiberglass Plastic Casing Overcomes Corrosion Problems in Water Wells in West Pakistan," *Soc. Min Eng., A.I.M.E., Trans.,* Vol. 224, No. 1, March, pp. 24-28.

Reclamation Program for Northern Zone of Indus Plains in West Pakistan involves construction of 30,000 irrigation wells to serve 20 million acres with annual pumpage of nearly 40 million acre-ft; within two years after construction mild steel casing in about ten percent of wells began to fail due to corrosion; fiberglass-reinforced plastic pipe has been tried and results of its initial use have been highly satisfactory.

576. Smith, H. F., 1954, "Gravel Packing Water Wells," *Water Well Journ.,* Vol. 8, No. 1, pp. 31-34, 2 figs., 5 refs.

Explores gravel packing methods employed in U.S. Midwest, e.g. gravel-pack ratio, formation grain sizing, sand production remedies, etc.

577. ———, 1959, "Well Rehabilitation with Primacord," *Water Well Journ.* December, 1 p.

Reports on the use of primacord in well rehabilitation. Case history.

578. Smith, L. R., 1969, "Development and Field Testings of Large Volume Remedial Treatments for Gross Water Channeling," *Journ. Petrol. Tech.,* Vol. 21, No. 8, August, pp. 1015-1025, 12 refs.

This paper describes laboratory development and field testing of plugging agents for large volume treatments of water-flood channeling problems which arise from fluids passing through interwell fractures of high permeability strata. Low-cost bentonite-nutshell slurries were developed and proved effective in field testing for improving sweep efficiency. Effective treatment using only a few thousand barrels of these gels are possible. Field testing demonstrated that the alkaline silica gels are effective in reducing flow through by-pass zones.
WRSIC #W71-02982

579. Smith, R. C., 1959, "The Well Driller and the Ground Water Hydrologist, Competitors or Allies," *Water Well Journ.,* Vol. 13, No. 7, July, 3 p.

In this day of growing water demands, of increasing conflicts between users, of individuals' demand for quantities that were almost unknown not too many years ago, ground-water is an extremely valuable resource. We must use, develop, and manage our vast ground water supplies to the fullest extent and in the most efficient manner. This is the aim of the water well contractor. It is also the aim of the ground-water hydrologist. This paper attempts to bridge the gap. WRSIC #W71-08408

580. ______, 1963, "Relation of Screen Design to the Design of Mechanically Efficient Wells," *A.W.W.A. Journ.*, Vol. 55, No. 5, May-June, pp. 609-614.

Mechanical efficiency of wells is discussed; investigation and development of a new Canton, Ohio, Sugar Creek well field is described; critical particle size, entrance velocity, open screen design factors; excessive entrance velocity due to improper well design is most critical design factor; by using mechanical efficiency and safe entrance velocities as basis for designing wells, kind of well needed can be easily specified.

581. Smith, W. C., 1965, "Asbestos in Drill Water Helps Cut Drilling Cost," *World Oil*, September, p. 116.

Asbestos added to drill water has been found effective in retaining good water penetration rates along with recovery of good samples. Asbestos can be added to fresh or salt water directly through the mud hopper. It has been found compatible with all mud additions used in West Texas drilling operations.

582. Soliman, M. M., 1965, "Boundary Considerations in the Design of Wells," *A.S.C.E. Journ.*, Proc., Irr. and Drainage Div., I.R.I.

The hydraulic movement of ground water toward wells can be divided into two parts: (1) the flow in the well and its boundary and (2) the flow through the aquifer. The foremost concern herein is with the flow in the well and its respective boundary. A theoretical treatment of entrance velocity distribution around the well and inside it is presented. A new development of well screen design has been achieved. In support of the theoretical development, a laboratory investigation has been made. Derived from the results attained theoretically and experimentally for the entrance velocity distribution around a well screen, a design was developed providing the most economical screen length in keeping with the amount of water to be withdrawn from the well.

583. Somerton, W. H., 1959, "A Laboratory Study of Rock Breakage by Rotary Drilling," *Journ. Petrol. Tech.*, Vol. 219, No. 5, May, pp. 92-97, 10 refs.

The effects of drilling variable on rotary drilling rates and efficiencies have been studied by a series of laboratory drilling tests. Two-cone 1.25 in. diameter bits were used to drill vertically upwards into rock samples at controlled weights and rates of rotation. Comparison was made with theoretical energy requirements and with energy requirements for size reduction by comminution methods. WRSIC #W71-08845

584. Somerton, W. H. and S. El Hadidi, 1970, "Well Logs Predict Drillability, Aid Computers," *Oil and Gas Journ.*, November, pp. 78-86, 15 refs.

Optimal scheduling of drilling operations and computerized drilling control require knowledge of the resistance of rock to drilling. Past practice has involved application of drilling data from previous wells drilled in same area to perform short-interval formation-drillability tests. Sonic-log-transit times correlate well with rock drillability.

585. Somerton, W. H., *et al.*, 1969, "Further Studies of the Relation of Physical Properties of Rock to Rock Drillability," SPE 2390, *4th Conference on Drilling and Rock Mechanics,* Austin, Texas, January, pp. 87-96, 9 refs.

In the present study, the correlation of physical properties of rocks with drilling performance has been extended and some new methods of measuring rock drillability have been proposed. The conclusion that sonic velocity is a good indicator of rock drillability has been substantiated, provided that a mineralogical factor is taken into account.

586. Speer, J. W., 1958, "A Method for Determining Optimum Drilling Techniques," *A.P.I. Drill. and Prod. Pract.*, pp. 130-147, 12 refs.

This report develops a simple method for determining the combination of weight-on-bit, rotary speed, and hydraulic horsepower which produces minimum drilling cost. Empirical relationships are developed to show the influence on penetration rate of weight-on-bit, rotary speed, and hydraulic horsepower. Optimum weight-on-bit is shown in relation to formation drillability, and optimum rotary speed is related to weight-on-bit. These five relationships are then combined into a chart for determining optimum drilling techniques for a minimum of field test data.

587. Spencer, A. M., *et al.*, 1970, "Powerful Borehole Slurry Passes Field Tests," *World Oil,* Vol. 171, No. 6, November, pp. 86-89, 8 refs.

The most powerful commercially available explosive is used in a versatile and safe slurry form to release high energy in the wellbore. Key features of the method are wellbore loading with bagged or high viscosity explosive and solid tamping with cement and gravel. Advantages of the system are: (1) good explosive-to-formation contact gives maximum energy transfer, (2) high stresses near wellbore cause desirable random fracturing, (3) concentrated explosive generates highest volume of gases for fracture extension, (4) casing is not damaged with protective tamping, (5) cost compares with hydraulic fracturing, and (6) the method has applicability to water well development.

WRSIC #W71-08476

588. Sprengling, K. and E. A. Stephenson, 1940, "Cable Tool Drilling," *A.P.I. Drill. and Prod. Pract.*, pp. 64-72.

Because drilling by present cable-tool methods is dependent to a large extent upon a driller's judgment of tool action transmitted to his sense of touch of his "feel" of the line, a mathematical analysis of the drilling system has been attempted in order to improve the efficiency of drilling. Examination of "drilling-time logs" by successive depth intervals or zones of 500 feet indicates that several zones are characterized by a decreased rate of penetration and an increased amount of time shut down for repairs of the drilling equipment. A comparison of the drilling action of a standard cable-tool system with an ideal elastic system shows that maximum tool motion is developed.

589. Stagg, K. G. and O. C. Zienkiewicz, 1968, *Rock Mechanics in Engineering Practice,* John Wiley & Sons, London, p. 442.

At the time of the first Congress of the International Society for Rock Mechanics held in Lisbon in 1966, it became clear that this subject had reached a stage where some *consolidation* was required. Over 300 papers were presented, showing interest and activity stemming from professions as varied as geology, mining, oil technology, and civil engineering. However, a demonstration was given that a common language was absent, that much duplication of effort was current, that a divergency of views often existed on basic matters, and that the practicing engineer was left without guidance on how to proceed with his work. An obvious need at such a stage in the development of a subject is for a text or treatise summarizing the progress already achieved, indicating promising investigation and current practice. This text attempts to serve these purposes.

590. Stall, J. C. and K. A. Blenkarn, 1962, "How Much Can you Pull on Stuck Drill Pipe", *World Oil,* Vol. 155, No. 6, November, pp. 91-96, 6 refs.

Calculations based on fundamental relationships are given on allowable hook load and torque combinations for commonly used drill pipe, including tool joints, subjected to combined tension, torsion and internal stress pressures. Strengths are given for both new and used pipe, and tables are based on minimum yield strengths. These calculations allow for normal circulating pressures, but do not allow for weakening effects of fatigue cracks, corrosion, or hydrogen embrittlement. Calculations and fundamental relationships have application to shallow exploration drilling for water or other economic minerals.

WRSIC #W71-03002

591. Stallman, R. W., 1967, "Examples of Recent Advances in Analytical Techniques for Evaluating Aquifers," *Proc. Natl. Symp. Groundwater Hydrol.,* Am. Water Res. Assoc., San Francisco, pp. 126-134.

A critical appreciation of recent advances in analytical techniques for evaluating aquifers. More complicated flow systems can now be analyzed with increased accuracy of descriptions of ground water resources and with reduced cost and risk in the development of water supplies.

592. Starkey, R. L., 1958, "The General Physiology of the Sulfate-Reducing Bacteria in Relation to Corrosion," *Producers Monthly,* Vol. 22, No. 8, p. 12.

Corrosion of metals is a natural phenomenon in that most metallic elements are unstable chemically and exist in nature as compounds. Therefore, effective use of metals depends not only on formation and their physical properties but also about the factors responsible for their disintegration. Corrosion is responsible for destruction of large quantities of metals. Wessel recently stated, "the yearly national cost of corrosion is estimated between $5 and $6 billion. The loss in buried pipe alone is placed at from $50 to $600 million. It has further been estimated that about 2.5 percent of all steel and iron in use corrodes

Starkey, R. L. . . . . (con't)

away every year. This amounts to 37.5 million tons, or approximately one third of our national output." Only a portion of this loss can be ascribed to micro-organisms, but they compose one of the important corrosive agents. In the early phases of studies of the relationships of microorganisms to corrosion of steel pipe lines in soil, Hadley stated that the importance of corrosion associated with sulfate reducing bacteria was second only to stray-current electrolysis.

593. Starkey, R. L. and K. M. Wight, 1945, "Anaerobic Corrosion of Iron in Soil," *Amer. Gas Assoc. Proc.*, Vol. 27, p. 307.

This article briefly describes anaerobic corrosion of iron materials in the soil environment with excess oxygen and explores the micro-environment of various soils.

594. Stevens, P. R., 1963, "Examination of Drill Cuttings and Application of Resulting Information to Solving of Field Problems on the Navajo Indian Reservation, New Mexico and Arizona," *U.S. Geol. Surv. Water Supply Paper 1544-H,* pp. H3-H13, 4 refs.

Only by drilling test holes and wells can detailed information on subsurface geologic and hydrologic conditions be obtained. The value of the information thus collected depends on the care exercised in sampling the rock cuttings, the thoroughness of the microscopic examination, and the accuracy and completeness of the lithologic and hydrologic descriptions. The more subsurface information compiled for an area, the better the understanding of the stratigraphy and geologic structure, and the greater the success in locating water supplies from ground water sources. WRSIC #W71-13970

595. Stone, V. D. 1964, "Low Silt Mud Increases Gulf's Drilling Efficiency, Cuts Costs," *Oil and Gas Journ.*, Vol. 62, No. 41, pp. 136-142.

A more effective means of removing abrasive solids from unweighted drilling fluid has helped make a further significant improvement in drilling efficiency. Mechanical solids-removal system has been developed which not only eliminates all sand from the drilling fluid but also reduces the content of finer silt particles. The ratio of native colloidal clays to remaining silt remains at a high level to produce slick, thin wall cakes.

596. Stoner, R. F., 1969, "Comments on Mogg's Criticism of Step-Drawdown Test," *Johnson Driller's Journ.*, Vol. 41, No. 1, pp. 10-12, 6 refs.

Mogg's analysis, printed in the July-August, 1968, issue of Johnson Driller's Journal is commented upon in this paper.

597. Stoner, R. F. and D. H. Lennox, 1969, "Step Drawdown Test Needs Critical Review: Discussion," *Johnson Driller's Journ.*, Vol. 41, No. 1, January-February, pp. 10-12.

Step drawdown tests are examined in detail, and it is concluded that they are limited in value. There was trouble both with the concept and with the practice. Small errors in discharge measurements affected the results drastically. The method is not recommended at all for determined well losses. The article agrees with most of J. L. Mogg's conclusions (PA 102,667), but not those concerning (1) superiority of individual, constant-rate step tests and (2) limitation of the Jacob step drawdown method to determine well efficiency as defined by UOP-Johnson. Step drawdown results should always be compared with constant rate results. If they agree, constant rate results when incorporated with the step drawdown results can improve the reliability of the step drawdown analysis.

598. Stow, A. H. and L. Renner, 1965, "Acidizing Boreholes," *Instn. Water Eng. Journ.*, Vol. 19, No. 8, November, pp. 557-572.

Use of hydrochloric acid on limestones; acids may be used in water well work to remove debris (from drilling) from test pump discharge; remove slurry from walls of well or borehole; develop and increase yield; and reduce friction loss to give more economic pumping; placing procedure for acid is discussed.

599. Stow, G. R. S., 1963, "Modern Water Well Drilling Techniques in Use in the United Kingdom," *Ground Water*, Vol. 1, No. 3, July, pp. 3-12.

The main water well drilling systems used in the UK with some details of the tools used and the principles of operation are described. Limitations as regards diameter and depth for various systems and their suitability for drilling different strata are given. The materials customarily employed for lining wells and for making sand screens and the usual method of measuring the deflection from the vertical are discussed. Methods of developing or improving the yields of wells and inspections by television cameras are outlined. There is no universal economic drilling system. Particularly on repetitive work, it is important to use the system appropriate to the site conditions. The most interesting developments of late have been (1) reverse-circulation rotary drilling for soft unstable sediments and for hard rocks; and (2) the use of air circulation instead of mud for drilling hard stable formations at relatively small diameters. Development of wells by acidizing and other means to improve the yield is worth more consideration. Many of the water well drilling techniques are applicable to and have been used for bored piles and caissons. WRSIC #W71-06952

600. Stramel, G. J., 1965, "Maintenance of Well Efficiency." *A.W.W.A. Journ.*, Vol. 57, No. 8, August, pp. 996-1010, 11 refs.

Problems associated with well design and development of successful programs for maintenance of high production efficiency are discussed; well construction techniques, improved well design, plugging, or well loss phenomena and better methods of well development are covered. WRSIC #W71-10432

601. Sturges, F. C., 1967, "Underground Surveys with Borehole Cameras," *12th Symposium on Exploration Drilling*, Univ. of Minnesota, Sch. of Min. and Metal. Eng., Minneapolis, pp. 39-52.

Sturges, F. C. . . . . (con't)

This paper discusses one of the newer developing phases of exploration – the use of borehole cameras – to supplement information obtained by drilling. The diamond core drill ranks as the outstanding tool in the exploratory field, as the cores it produces are subject to visual, physical, and chemical tests that provide specific and accurate data. But it is not always possible to get 100 percent core recovery and, more frequently than not, core loss occurs just where the information is needed most. Nevertheless, engineers have to make decisions, and when these are based on inference instead of specific information, they can be wrong. With modern core drilling techniques, it is possible to get good core recovery in any type of formation. Core loss is not caused by the type of rock, it is caused by the condition of the rock. And condition needs to be seen to be evaluated. Borehole television serves this purpose and is explored in some detail in this paper.

602. Sufall, C. K., 1960, "Water Shutoff Techniques in Air or Gas Drilling," *A.P.I. Drill. and Prod. Pract.*, pp. 74-77, 6 refs.

Water-shutoff techniques have reached a point where they are generally accepted for field usage if economic gain can be realized. The success, both economical and that of shutting off permeable zones, is a tribute to the planning of a job and the fine work done by the research and service organizations. Several types of shutoff material are now readily available through the service companies as are the tools and techniques needed to complete a water shutoff job.

603. Suliman, M. M., 1965, "Boundary Flow Considerations in the Design of Wells," *Amer. Soc. Civil Eng. Journ.*, Irrig. Div., Vol. 91, No. 1, pp. 159-177, March.

A theoretical treatment of entrance velocity distribution around and inside a well is presented. From theoretical and experimental results, a new well screen design was developed providing the most economical screen length consistent with the water withdrawn from the well.

604. Sullivan, E. T., 1960, "Drilling Shallow Salt-Water Source Wells," *A.P.I. Drill. and Prod. Pract.*, pp. 56-59.

Water flooding was initiated in the Wilmington Field during 1953 by the Long Beach Harbor Department through their contractor, the Long Beach Oil Development Company. During the launching of this pilot flood, the question of water supply was thoroughly examined. Of the three general sources available – fresh, ocean, and produced – the fresh water was unusable. The remaining source, ocean water, was available directly or from shallow sands which were ocean fed. The shallow sands were chosen because water taken directly from the bay was high in oxygen content and would also require filtering. The "water-well" approach using methods, equipment, and casing which were designed for shallow wells has brought an economical, trouble-free solution to the salt water requirements in the Wilmington Field water flood operations. WRSIC #W71-13714

605. Summers, W. K., 1971, *The Annotated Indexed Bibliography of Geothermal Phenomena,* New Mexico Institute of Mining and Technology, Socorro, New Mexico.

This book contains more than 14,000 references and provides more than 95 percent of the available references on thermal-water topics through December 31, 1969.

606. ———, 1972, "Horizontal Wells and Drains," *Water Well Journ.,* Vol. 26, No. 6, June, pp. 36-38.

Briefly summarizes early history of the horizontal well and basic operation of the equipment presently in use. Includes selected case histories in Southwestern United States based on U.S. Bureau of Mines Information Circular 8392: "Horizontal Boring Technology: A State-of-the-Art Study."

607. Summers, W. K. and L. A. Brandvold, 1967, "Physical and Chemical Variations in the Discharge of a Flowing Well," *Ground Water,* Vol. 5, No. 1, pp. 9-10, January.

Observation of temperature, pH, specific conductivity, and discharge, and concentrations of calcium, magnesium, potassium, lithium, and sodium in the discharge of a flowing well will reveal that geochemistry of aquifers should be based on sample wells which have been allowed to discharge for an extended period.

608. Sutcliffe, H., Jr., and B. F. Joyner, 1966, "Packer Testing in Water Wells near Sarasota, Florida," *Ground Water,* Vol. 4, No. 2, March, pp. 23-27.

During February and March, 1964, the U.S. Geological Survey ran caliper, conductance, and temperature logs on several wells in the Sarasota area. One of the tests produced excellent results indicating that the various producing zones were effectively isolated by the packers and that different quality of water, quantity of water, and a different head was available from each zone. The other test did not show sharp differences but did indicate the extent of contamination of the producing horizons in a flowing well which had been capped for approximately two years. These tests indicate that packer testing can measurably add to knowledge of the separation of permeable zones.

WRSIC #W71-09734

609. Sutko, A. A. and G. M. Myers, 1971, "The Effect of Nozzle Size, Number, and Extension on the Pressure Distribution Under a Tricone Bit," *Journ. Petrol. Tech.,* Vol. 23, No. 11, November, pp. 1299-1304, 8 refs.

A laboratory study was made of the pressure on the hole bottom as affected by nozzle velocity, size, extension, and number. Conclusions were that hole cleaning and drilling rate should improve as the number of nozzles is decreased from three to one and as the length of the nozzle is increased. Nozzle size and velocity and fluid density can be incorporated in an analysis of bottom-hole pressure.

610. Tailleur, R. J., 1963, "Lubricating Properties of Drilling Fluids," *World Petrol. Cong.*, 6th Proc., Section 2, Paper 36, PD 5, pp. 1-18, 11 refs.

The field application of this fluid has produced the results expected of any extreme pressure drilling fluid; torque reduction; wall sticking prevention; and assistance in direction drilling. The combination of E.P. properties with inhibitory properties has proved it to be a valuable tool as a preventative measure.

611. Taylor, S. S., *et al.*, 1940, "Study of Brine Disposal Systems in Illinois Oil Fields," *U.S. Bur. Mines Rept. of Inv. 3534.*

The discovery of many oil fields in Illinois since 1936 and their rapid development have brought the problem of brine disposal to the foreground in that state. The ultimate magnitude of the problem is not yet known, but already it is great enough to warrant technical investigation. The Bureau of Mines, therefore, has made a general survey of the methods of handling oil-field brines in Illinois and has studied in detail three subsurface disposal systems. The perplexities of the problem of brine disposal depend in general on the characteristics of the oil-producing formations in each particular locality. In Illinois some formations undoubtedly will produce large quantities of brine with the oil. Definite natural water drives apparently are present in the Devonian limestone in the Sandoval field and in the Benoist sandstone in the Dix field of Illinois. However, in the Benoist and Aux Vases sandstones in the Salem-Lake Centralia field no evidence of effective natural water drives has yet been noted. Many operators in Illinois plan to install adequate brine-disposal systems as soon as brine appears with the oil at the producing well. In that way the cost of such installations can be charged off during the period of "flush production," and abandonment of the property will not be forced prematurely by excessive costs of oil production and brine disposal. Furthermore, damage claims for contamination of domestic water supplies may be avoided if brines are disposed of satisfactorily.

612. Theis, C. V., 1935, "The Relation Between the Lowering of the Piezometric Surface and the Rate and Duration of Discharge of a Well Using Ground Water Storage," *Amer. Geophys. Union Trans.*, Pt. 2.

When a well is pumped or otherwise discharged, water levels in its neighborhood are lowered. Unless this lowering occurs instantaneously, it represents a loss of storage either by the dewatering of a portion of the previously saturated sediments if the aquifer is non-artesian or by release of stored water by the compaction of the aquifer due to the lowered pressure if the aquifer is artesian. The mathematical theory of ground water hydraulics has been based, apparently entirely, on a postulate that equilibrium has been attained and, therefore, that water levels are no longer falling. In a great number of hydrological problems involving a well or pumping district near or in which water levels are falling, the current theory is therefore not strictly applicable. This paper investigates in part the nature and consequences of a mathematical theory that considers the motion of ground water before equilibrium is reached and, as a consequence, involves time as a variable.

613. Thiery, J. R., *et al.*, 1969, "Slow Rotating Turbo-Drill Broadens Rock Bit Use," *World Oil*, Vol. 169, No. 5, October, pp. 83-86, 95.

Article discusses a modified turbodrill which operates effectively with rock bits between maximum rotary table speed (about 250 rpm) and standard turbodrill rpm. (near 600 rpm). To fill this gap is important because it allows use of rock bits at higher optimum speeds. Diamond bits also can be used within the 250 to 600 rpm range.

614. Thomas, H. E., 1967, "Igneous and Metamorphic Rocks," *Proc. Natl. Symp. Groundwater Hydrol.*, Am. Water Res. Assoc., San Francisco, pp. 103-106.

Discusses only the water in the igneous and metamorphic rocks that are no longer subjected to the head and pressure of origin, ignoring the "volcanic," "magmatic," and "metamorphic" waters.

615. Timur, A., 1968, "An Investigation of Permeability, Porosity and Residual Water Saturation Relationships," *9th Annual Logging Symposium*, Soc. Professional Well Log Analysts, June.

A reasonably accurate relationship for estimating permeability of sandstones from *in situ* measurements of porosity and residual fluid saturation would be instrumental in eliminating the expense of coring. To establish such a relationship, several possibilities were tested through laboratory measurements of permeability, porosity, and residual water saturation on 155 sandstone samples from three different oil fields in North America.

616. Tiraspolsky, W., 1957, "Theoretical Approach to Turbo Drilling," *World Oil*, Vol. 145, No. 6, 7, November, pp. 119, 122-124, 126, December, pp. 140, 144, 146-151.

Mechanical comparisons, illustrated with graphs of turbo and conventional drilling methods, indicate increased possibility of applying turbodrilling to industrial needs. Use and manufacture of turbodrills are discussed as well as their field application via favorable penetration rates compared with conventional rotary drilling methods.

WRSIC #W71-03003

617. Titkov, N. I., *et al.*, 1971, "Use of an Acoustic Method to Study the Longevity of Cement," *Burenie*, No. 3, pp. 24-26, (in Russian).

This study shows the advantage of using a pulsed acoustic method to study the longevity of cement mixtures. Resistance of cement to deformation, velocity of sound, and attenuation of sound in cement were determined and petrographic, thermal, and X-ray examinations of samples were made. This method was used to determine the longevity of six cement samples.

618. Tittman, J., *et al.*, 1966, "The Sidewall Epithermal Neutron Porosity Log," *Journ. of Petrol. Tech.*, Vol. 18, No. 10, October, pp. 1351-1362, 12 refs.

A directionally sensitive epithermal neutron detection system is

Tittman, J. . . . . (con't)

incorporated into a side-wall source-detector skid. The effects of variations in borehole size and shape, mud type, temperature, and salinity are greatly reduced. Residual borehole effects are corrected in the surface control panel, and a direct recording of neutron-derived porosity on a linear scale is provided. The SNP log can be made in uncased holes, whether filled with liquid or gas. Calibration and correction data are given along with examples of interpretation methods.

619. Tixier, M. P. and R. P. Alger, 1971, *Log Evaluation of Non-Metallic Mineral Deposits,* Schlumberger Well Services, Houston, 23 p., 18 figs., 2 tabs., 1 Appendix, 17 refs.

Well logs can be used to locate and evaluate deposits of various commercially important minerals. It is only necessary that the mineral of interest represent a significant fraction of the formation bulk volume, and that it exhibit characterizing properties measurable by logs. Because modern logging methods measure electrical, density, acoustic, radioactivity, and certain nuclear characteristics of formations, they may be used to identify many minerals, e.g., (1) for evaluation of sulphur deposits either density or sonic logs provide good resolution when compared with porosity computed from neutron or resistivity logs; (2) trona beds are identified by a sonic reading of approximately 65 microsec./ft. neutron porosity index of about 40 percent, low natural radioactivity, and pronounced hole enlargement; (3) gamma ray logs provide important information in the location, identification and evaluation of potash mineral deposits. Neutron, sonic, and density logs in various combinations, augment the gamma ray data in such studies; (4) coal beds are characterized by high resistivities and by high apparent porosities on sonic, neutron, and density logs; (5) density logs are particularly suited for evaluation of yield from oil shales. In all such explorations for non-metallic mineral deposits, well logging methods provide a fast, detailed and economical reconnaissance of the entire length of drilled hole. Results compare well with core assays.

620. Todd, M. D., 1967, "How to Complete Water Source Wells," *World Oil,* Vol. 164, No. 4, March, pp. 86-91, 11 refs.

This article discusses the experience-derived gravel packing techniques and other related completion practices. Carefully planned and executed completion procedures developed especially for water source wells have provided an adequate, reliable, and economical supply of water for three major Canadian fluid injection projects. Modified and simplified gravel packing techniques are primarily responsible for sand-free water production at rates to 4,000 bpd per well. WRSIC #W71-08830

621. Trainer, F. W. and J. E. Eddy, 1964, "A Periscope for the Study of Borehole Walls, and its use in Ground-Water Studies in Niagara County, New York," *U.S. Geol. Surv. Prof. Paper 501-D,* pp. 203-206.

A periscope made of aluminum tubing with a mirror and light at one end and a telescope at the other was used to study fractures in dolomite in shallow drill holes in western New York. Correlation of fractures, observed with the periscope, with inflections in temperature profiles in a well where downward flow occurs during periods of no pumping confirmed that the inflections coincide with water-bearing fractures in the rock. The periscope has also been used for examining well casings, well screens, and pump columns, and, in some places, for determining the texture and composition of the wallrock.

622. Traugott, M. V., 1971, "Use of SP Log in Water Flood Surveillance," *Soc. Petrol. Eng., A.I.M.E.*, Special Paper, SPE-3570.

An update on using the SP log in water flood operations for estimating permeability from streaming potential under several well-head pressures.

623. Trautenberg, G. A., 1965, "Don't Underestimate Role of Aerobic Bacteria," *Oil and Gas Journ.*, Vol. 63, No. 1, January, pp. 85-87.

In examining water sample data from a southern California produced water injection system, a close correlation between the aerobic bacterial counts and the sulfate-reducing bacteria counts was observed. The biocide treatment previously in use had little effect against the aerobic population and, consequently, its activity against the sulfate-reducers was negligible. In contrast, bacterial counts from the same sampling points being treated with the same level of a different biocide were reduced significantly. The aerobic bacteria isolated from this system were species of Flavobacterium, Alcaligenes, Proteus, and Bacillus.

WRSIC #W71-02996

624. Trescott, P. C., and G. F. Pinder, 1970, "Air Pump for Small-Diameter Piezometers," *Ground Water,* Vol. 8, No. 3, May-June.

The engine of a field vehicle can easily be used as an air pump by inserting a commercially manufactured check valve into a spark-plug socket. Small-diameter piezometers can be pumped with compressed air from this source. With sufficient air-line submergence, the air lift method can be used to pump one-inch diameter piezometers where the pumping lift is less than 70 feet, and 1.25-inch diameter piezometers where the pumping lift is less than 50 feet. In cases where this method will not work, some of the water contained in a piezometer can be pumped out by pressurizing the piezometer if the formation transmissivity is low and the pumping lift is not excessive. The maximum pumping lift is a function of the rate at which pressure in the piezometer is increased and the rate at which water flows into the formation with increasing head in the piezometer.

625. Tschirley, N. K. and K. D. Tanner, 1958, "Wetting Agent Reduces Pipe Sticking," *Oil and Gas Journ.*, November, pp. 165, 168, 2 refs.

The quality of an emulsion mud is normally described only in terms of volume percent of oil in the mud and by the presence or absence of free oil in the mud filtrate or free oil on the surface of the mud in the pits. The performance of an emulsifier is judged by its ability to keep oil emulsified in a given mud at the lowest cost. This report describes the use of a wetting agent which, when added to a mud that is already considered a good emulsion, alleviated conditions of tight hole.

626. Turcan, A. N., Jr., 1962, "Estimating Water Quality from Electrical Logs," *U.S. Geol. Surv. Prof. Paper 450-C,* pp. C135-C136.

The large number of electric logs on file at the Louisiana Geological Survey and the availability of a large number of chemical analyses of water make it possible to estimate, for selected aquifers, the relation of the concentration of certain chemical constituents to the resistivity reading. As a result, an empirical relation has been established for some of the major aquifers in Louisiana.

627. ———, 1963, "Estimating the Specific Capacity of a Well," *U.S. Geol. Surv. Prof. Paper 450-E,* Art 222, pp. 145-148.

Because the performance of a well is not readily predicted by the direct application of any one formula, the intent of the article is to present some of the equations derived and discussed by Theis, Muskat, Jacob and others, together with simple examples, which aid in predicting the specific capacity of a 100-percent-efficient well that taps an isotropic homogeneous artesian aquifer of large areal extent. The computations of the theoretical specific capacity of a well that is screened opposite the full thickness of the aquifer and of a well that is screened opposite only part of the aquifer are considered.

628. ———, 1966, "Calculation of Water Quality from Electrical Logs – Theory and Practice," La. Dept. of Publ. Works, *Water Res. Pamphlet No. 19,* Baton Rouge, La., May.

Three major aquifers – the Chicot, Evangeline, and Jasper – and two aquicludes – the Burkeville and one as yet unnamed – are recognized on an interstate basis in part of the Louisiana-Texas Gulf Coast region. Wells tapping the aquifers supply about 1 bgd of water for industrial, agricultural, and municipal use.

629. Updegraff, D. M., 1955, "Microbiological Corrosion of Iron and Steel," *Corrosion,* Vol. 11, p. 442.

A critical review of the literature on microbiological factors involved in the corrosion of iron and steel is presented. A brief account of the historical aspects of the subject is given along with a discussion of the mechanisms by which microorganisms affect corrosion, a description of some typical examples of microbiological corrosion and a discussion of methods used to prevent microbiological corrosion. It was found that while microorganisms do not corrode iron or steel, they often produce major physical and chemical changes in the environment. These changes may influence the electrochemical processes responsible for corrosion and, thus, markedly accelerate (or, under different conditions, decelerate) the corrosion rate.

630. Upp, J. E., 1966, "The Use of the Cement Bond Log in Well Rehabilitation," *7th Annual Logging Symposium,* Soc. Professional Well Log Analysts Trans., Tulsa, May, pp. XI-XII.

Since 1962, Michigan Consolidated Gas Company has been in the process of rehabilitating some 200 wells in the 25-year old Reed City

Oil Field. The project was undertaken for the purposes of gas storage and the secondary recovery of oil by gas injection and recycling. Since integrity of the wells is of prime importance in a project of this nature, a Cement Bond log was run as a matter of routine. This program was supplemented by running Gamma Ray-Neutron logs. The various aspects of a Cement Bond logging program involving 25-year old rotary and cable-tool completions, as well as reentering abandoned wells and evaluating liner jobs, are discussed.

631. Vaadia, Y. and V. H. Scott, 1958, "Hydraulic Properties of Perforated Well Casings," *Proc. A.S.C.E.* Irrigation and Drainage Div., Paper 1505.

Hydraulic performance of well casings of varied perforation characteristics; commercially perforated casings were tested in combination with various gravel envelopes; results presented and compared with criteria already available for well screens.

632. Van der Leeden, F., 1971, *Ground Water, A Selected Bibliography,* Water Information Center, Port Washington, New York, 116 p., 1,500 refs.

A bibliography of selected important papers in hydrogeology. Most papers concerning local hydrogeologic conditions have been omitted. All references are listed under topics. This book fulfills the need for a handy reference to publications in the ground water field. Because there has been a rapid increase in number of scientific articles and publications in the field and a need to refer to "classic" contributions, this quick reference guide was prepared.

633. Van Eck, O., Jr., 1971, "Optimal Well Plugging Plugging Procedures," *Public Information Circ. No. 1,* July, Iowa Geological Survey, p. 7.

The proper plugging of abandoned wells is a fundamental practice in the preservation of ground water quality. Any well, whether it be for domestic, municipal, or industrial use, is a potential site for pollution, either as an egress for liquids from the surface or as an avenue for the mingling of different quality water from the various aquifers penetrated by the well. Before the details of the differences in water quality in the many aquifers in Iowa were known, it was common practice to drill and case wells in a manner that permitted water from many, if not all, of the penetrated aquifers to enter the well bore and hence to move freely from aquifer to aquifer. As the water in some of these aquifers is naturally of a poorer quality than others, contamination resulted. Modern well construction is aimed at preventing this type of contamination. However, failure of the commonly used steel well casing after a number of years is normal and will result in contamination as in an uncased well. Therefore, it is important that all wells, regardless of depth or construction, be properly plugged when no longer in use.

634. Van Everdingen, A. F., 1953, "The Skin Effect and Its Influence On the Productive Capacity of a Well," *Petrol. Trans., A.I.M.E.,* Vol. 198, pp. 171-176.

The pressure drop in a well per unit rate of flow is controlled by the

Van Everdingen, A. F. . . . . (con't)

resistance of the formation, the viscosity of the fluid, and the additional resistance concentrated around the well bore resulting from the drilling and completion technique employed and perhaps from the production practices used. The pressure drop caused by this additional resistance is defined in this paper as the skin effect, denoted by the symbol S. This skin effect considerably detracts from a well's capacity to produce. Methods are given to determine quantitatively (a) the value of S, (b) the final buildup pressure, and (c) the product of average permeability times the thickness of the producing formation.

635. Van Lingen, N. H., 1962, "Bottom Scavenging – A Major Factor Governing Penetration Rates at Depth," *Trans. SPE of AIME,* Vol. 225, pp. 187-196.

A laboratory study has been made to determine what factors affect the penetration rate of roller bits, diamond bits, and drag bits in rock drilling with clay/water muds. The rather simple relations that exist when pressures in and around the borehole are equal become more complicated when under downhole conditions the penetration rate is hampered by the existence of a pressure differential between the mud at the hole bottom and the pore liquid at cutting depth. Expressions that have been derived for both the penetration rate and the magnitude of the pressure differential in permeable rock together fully account for operating, rock, mud, and bit variables. In impermeable rock, a similar pressure differential is caused by the bit action itself.

636. Van Nouhuys, H. C., 1962, "Master the Tools that Combat Corrosion," *Pipeline Eng.,* Vol. 34, No. 3, March, pp. 286, 288, 292, and 294.

The corrosion engineer has five basic tools with which to combat corrosion: galvanic anodes, rectifier-ground bed units, coatings, bonds, and insulators. How and where the corrosion engineer uses each method or combination of methods can spell the difference between an effective or an ineffective job of corrosion mitigation. This paper also has applicability to water well installations. WRSIC #W71-09915

637. Van Olphen, H., 1963, *Clay Colloid Chemistry,* John Wiley & Sons, New York, p. 301.

In almost every field of clay study, one has to deal at one time or another with dispersions of clay in water or in another fluid. Such dispersions which are characterized by the large interfacial area between the extremely small clay particles and the surrounding liquid are colloidal systems. Colloid chemistry, therefore, enters to some degree every technological problem involving clays and liquids, such as problems of soil consolidation, plant nutrition, molding of ceramic objects, and the circulation of drilling fluids in an oil well. The purpose of this book is to familiarize those engaged in some phase of clay technology in sedimentary geology or in soil science with the modern views of colloid science and its application to clay systems.

638. Vedder, J. G., *et al.*, 1969, "Geology, Petroleum Development, and Seismicity of the Santa Barbara Channel Region, Calif.," *U.S. Geol. Surv. Prof. Paper 679.*

The Santa Barbara blowout was the first significant oil-pollution experience resulting from drilling or working some 7,860 wells under federal jurisdiction on the Outer Continental Shelf since 1953. As a consequence of this event, federal operating and leasing regulations have been strengthened and additional safeguards have been added in all federally supervised oil and gas exploration and production operations. This report presents specific well construction information that could have effectively coped with the geology of the channel area and with the potential of oil seepage in general. WRSIC #W70-00836

639. Vidrine, D. J. and E. J. Benit, 1960, "Field Verification of the Effect of Differential Pressure on Drilling Rate," *Journ. Petrol. Techn.*, July, pp. 676-681.

A field study was conducted on eight South Louisiana wells to determine the effect of differential pressure on the instantaneous rate of penetration in shale. Drilling rate is affected significantly by changes in differential pressure and may be reduced as much as 70 percent as the differential fluid pressure is increased from 0 to 1,000 psi. When the formation pressure becomes greater than the mud column pressure, drilling rates continue to increase, sometimes at an increasing rate. The sensitivity of drilling rate to differential pressure depends upon the magnitude of the bit load. Overburden pressure and the hydrostatic head of the mud column over the intervals studied had no detectable effect on penetration rate.

640. Vonhof, J. A., 1966, "Water Quality Determination from Spontaneous-Potential Electric Log Curves," *Journ. Hydrology*, Vol. 4, pp. 341-347.

Spontaneous-potential electric logs from test holes drilled in the Pleistocene deposits of the province of Saskatchewan have been used successfully to determine the water quality of the aquifers encountered.

641. Von Wolzogen Kuhr, C. A. H., 1961, "Unity of Anaerobic and Aerobic Iron Corrosion Process in the Soil," *Corrosion*, Vol. 17, p. 293.

This paper on the unity of the anaerobic and aerobic iron corrosion process in soil was originally presented at the Fourth National Bureau of Standards Soil Corrosion Conference in 1937. It covers such topics as corrosion from an electrochemical point of view, the anaerobic iron corrosion process, iron corrosion by the sulfate reduction process, iron corrosion by carbonic acid, reduction process, iron corrosion by nitrate, reduction processes, iron corrosion due to chemical substances in the soil, and the aerobic iron corrosion process. A statement is made to the effect that the hydrogen acceptor which is invariably present with all forms of iron corrosion considered constitutes the unity in the anaerobic and aerobic iron corrosion process in the soil.

642. Wait, K. L. and J. T. Callahan, 1965, "Relations of Fresh and Salty Ground Water Along the Southeastern U.S. Atlantic Coast," *Ground Water,* Vol. 3, No. 4, pp. 3-17.

Reports on the relationship and on the different types of occurrence of salty water and shows the regional picture.

643. Walker, R. E., 1964, "Practical Oil Field Rheology," *A.P.I. Paper No. 926-9D,* March, abridged as "How to Predict Mud Performance," *Oil and Gas Journ.,* March, pp. 67-70.

A qualitative technique is described to predict hole cleaning, solids removal, and the time required to lift cuttings. The technique, which applies equally to all mud types, requires standard A.P.I. mud measurements and an understanding of a few fundamental concepts of rheology. The approach can be particularly helpful to geologists, drilling engineers, and those without extensive field experience in mud technology. Only one rheological property, the shear stress-shear rate relation, controls laminar flow, the type of flow which generally occurs in pits, and annuli when hole cleaning or sample collecting present problems.

644. Walker, W. H., 1967, "When not to Acidize," *Ground Water,* Vol. 5, No. 2, April, pp. 36-40, 4 refs.

Controlled step and constant-rate pumping tests used by the Illinois State Water Survey to determine well efficiencies and hydraulic characteristics of aquifers showed that acidizing water wells to increase yield proved to be an uneconomical method of well development or rehabilitation in some cases where aquifer permeability was low and normal operating head great, in wells affected by methane gas, or where practical sustained yield of aquifer had been exceeded.

645. ———, 1969, "Illinois Groundwater Pollution," *A.W.W.A. Journ.,* Vol. 61, No. 1, January, pp. 31-40, 10 figs., 16 refs.

The increasing importance of water quality on the use and reuse of available water supplies has led to many recent studies by the Illinois State Water Survey concerning the quality of water as it is found in nature and the changes in quality made by man. Selected case histories of adverse changes in ground water quality presented in this paper describe some of the more serious incidents of ground water pollution or contamination that have occurred in Illinois aquifers. Most are of fairly common types of pollutants that have readily entered shallow sand and gravel, creviced limestone, or dolomite aquifers from surface-derived sources.

646. ———, 1970, "Salt Piling – A Source of Water Supply Pollution," *Pollution Eng.,* July-August, pp. 30-33.

The paper explores the chemical pollutants from surface sources that cause serious groundwater quality deterioration in many shallow aquifers throughout the United States. Instances of such pollution are increasing, and wide-spread contamination of many major aquifers may occur unless stringent measures are employed.

647. ———, 1972, "Comparative Costs of Producing Water," Ill. State Water Surv., Urbana, Illinois, (In prep.)

Appreciable variations in the cost of pumping groundwater occur as the efficiency of a well or its pumping equipment changes with use. Graphs of pumping cost related to well and pump efficiencies are presented to assist in determining when to rehabilitate or to replace inefficient wells or pumps.

648. Walsh, J. B. and W. F. Brace, 1964, "Fracture Criteria for Brittle Anisotropic Rock," *Journ. Geophysics Research,* Vol. 69, No. 16, pp. 3449-3456.

The McClintock-Walsh modification of Griffith theory is extended to treat brittle fracture of anisotropic rock. It is suggested that anistropy is due primarily to preferred orientation of cracks in rocks and that a satisfactory mathematical model for fracture analysis is an elastically isotropic medium which contains a nonrandom array of Griffith cracks. The one array considered in detail consists of a long crack having a preferred orientation superimposed on a field of randomly oriented small cracks. Fracture in tension and in compression is considered. The fracture criterion found for compression resembles a similar criterion suggested by Jaeger on a somewhat different basis; it predicts behavior which is in good agreement with observed fracture of a slate reported by Donath. Determination of the effect of anisotropy upon the ratio of compressive strength to tensile strength shows that it can vary widely with sample orientation with values both above and below that predicted from isotropic analysis.

649. Walton, W. C., 1960, "Application and Limitation of Methods Used to Analyze Pumping Test Data," *Water Well Journ.,* Vol. 4-14, No. 2, (Part I), pp. 22, 49, 50, 53, 56; No. 3 (Part II), pp. 20, 46, 48, 50, 52, February and March, 13 refs.

The nonequilibrium formula is widely used with pumping test data for determining the hydraulic properties of an aquifer. The most popular method for solving that formula is the straight-line method; however, the straight-line method is not applicable in some cases, and it supplements rather than supersedes the more complicated type curve method. Geologists and engineers have become familiar with the interpretation of time-drawdown graphs. Distance-drawdown graphs (radial profiles of cones of depression) have, on the other hand, received little publicity, and they are not used as much as they ought to be to determine the hydraulic properties of aquifers and to confirm the existence of geohydrologic boundaries. The gravity drainage of water through stratified sediments is not immediate and the unsteady flow of water towards a well in an unconfined aquifer is characterized by slow drainage of interstices. Data collected toward the end of a pumping test, one or more days in duration, can in many cases be used to determine the hydraulic properties of an aquifer under water table conditions.

WRSIC #W71-09911

650. ———, 1962, "Selected Analytical Methods for Well and Aquifer Evaluation," *Ill. State Water Surv. Bull. 49.*

Walton, W. C. . . . . (con't)

Describes an analytical process involved in well-pump testing, productivity of various types of aquifers; explores various methods, problems of interpretation, and application.

651. ________, 1963, "Micro-Time Measurements of Ground Water Level Fluctuations," *Ground Water,* Vol. 1, No. 2, April, pp. 18-19.

Water-level measurements made during the early seconds and minutes of aquifer tests may often be used to determine the hydraulic properties of an aquifer and to locate hydrogeologic boundaries. This paper describes a method using a modified stylus carriage and pen and a stop watch of recording microtime water-level measurements in a well equipped with a recording gage. Ordinary and microtime records of water-level fluctuations are simultaneously obtained. Microtime water-level measurements often aid in appraising the effects of slow gravity drainage under water-table conditions and anisotropic and heterogeneous conditions.

WRSIC #71-09732

652. Walton, W. C. and S. Csallany, 1962, "Yields of Deep Sandstone Wells in Northern Illinois," Illinois State Water Survey, *Rept. of Inv. 43,* pp. 43-47, 17 figs., 11 refs.

During 1906-1960 well-production tests were made by the State Water Survey on more than 500 deep sandstone wells. Specific-capacity data were used to determine the role of the individual bedrock aquifers or units uncased in deep sandstone wells as contributors of water and to appraise the effects of shooting bedrock wells.

653. Watt, H. B. and T. B. Akin, 1967, "Tricore, a Continuous Sidewall Core Cutter," *8th Annual Logging Symposium,* Soc. Prof. Well Log Analysts, Denver, June, pp. U1-U9.

A new wireline coring device has recently become available. It saws a 60-inch long triangular sample of the formation with a pair of cutters which move along the borehole wall parallel to the axis of the tool. The formation sample retrieved is undisturbed by the cutting operation and retains the mud cake in place on the surface adjacent the borehole. Cores may be taken in boreholes of 7-7/8 inch diameter or larger. This device makes it possible to retrieve continuous cores in selected zones for quantitative determination of rock properties and for qualitative examination of strata characteristics such as dip and fracturing. Correlation and comparison of these triangular cores with logging measurements provide a new means for studying the effects of mud cake and invasion. Results of field tests are discussed and the characteristics of retrieved rock samples and mud cakes are described.

654. Wen, H. L., 1954, "Interaction between Well and Aquifer," *A.S.C.E. Proc.,* No. 578, November.

The piezometric head along a well is not constant as usually assumed but must decrease towards the discharging end of the well. An investigation is made to find the effect of this variation of piezometric head along the well on the flow in the aquifer. It is found that the formation loss to the discharging head of the well involves an additional term roughly equal to the velocity head on the discharging end of the well. The loss through the well screen is usually negligible. What is usually considered to be the screen loss is actually the additional loss mentioned above caused by the variation of piezometric head along the well. A solution is obtained for the axially symmetric flow from a confined uniform aquifer into a fully penetrating well.

655. White, C. G., 1969, "A Rock Drillability Index," *Colo. Sch. Mines Quart.*, Vol. 64, No. 2, April, 94 p., 30 refs.

The rock drillability index proposed incorporates the three major drilling systems – rotary, percussive, and rotary-percussive – and considers the complete range of rock types from the very soft, highly altered rocks to the extremely hard taconites. The drillability index is based on the penetration rate achieved by a 3/4-inch diameter bit drilling a hole four inches deep. A rock abrasive index is also proposed and is based on the change in area of the bit cutting-edge profile. Drillability and abrasive indices have been determined for 98 rock types.

656. White, R. J., 1956, "Lost Circulation Materials and Their Evaluation," *A.P.I. Prod. Div. Pacific Coast Dist. Spring Meeting,* 801-32D, Los Angeles, May, 10 p.

The function of lost circulation materials is briefly discussed. The types of materials available are listed, classified, and described. The usual effective range of sealable openings is given for each type material. The types of apparatus which have been used for evaluating lost circulation materials are listed. The effect of variables in the testing method on the results obtained is discussed and typical data on several types of lost circulation material is shown. Collection and interpretation of data from laboratory evaluation tests of lost circulation are discussed.

WRSIC #W71-05689

657. Whitesides, D. V., 1970, "Common Errors in Developing a Ground Water Aquifer," *Ground Water,* Vol. 8, No. 4, July-August, pp. 25-28.

This paper presents some of the more common errors made in the development of ground water supplies in the alluvial aquifers along the Ohio River in Kentucky. Ample available literature on proper methods of development of the alluvial aquifers generally seems to have been ignored by the water users in the area. The more common errors made in the typical developments are singled out for discussion.

WRSIC #W70-05506

658. Whitesides, D. V., 1971, "Yields and Specific Capacities of Bedrock Wells in Kentucky," Kentucky Geological Survey, *Information Circ. 21,* 18 p., 5 figs., 1 tab., 57 refs.

Pump testing of bedrock wells reveals that a large amount of water for potential development exists in bedrock aquifers in Kentucky. Since

Whitesides, D. V. . . . . (con't)

new and better pumping and drilling methods have been developed along with a better understanding of the occurrence of ground water in bedrock, it is often possible to tap more plentiful water supplies at greater depths. Most previously available records on maximum yields of bedrock wells in Kentucky are of limited use in determining maximum well yields or specific capacities. Most records reflect pump capacities rather than well yields because water-level drawdown measurements and pumping rates are rarely available. This report presents specific-capacity and well-yield data from controlled pumping tests on 106 selected bedrock wells in 41 counties in Kentucky and discusses the occurrence and movement of ground water in bedrock aquifers. Depths of wells tested range from 21 to 1,015 feet and yields range from less than one to several hundred gallons per minute.

659. Wiley, C. X., 1965, "The Use of Near-Bit Stabilizer-Subassemblies for the Control of Hole Deviation," *A.P.I. Drill. and Prod. Pract.*, pp. 66-75, 6 refs.

This paper reports the application of a new tool designed in accordance with "packed-hole" principles to practically and economically eliminate doglegs and effectively minimize the rate of change in hole deviation. Additional functions such as the elimination of undersized spiral holes and ledges are discussed. The beneficial effect on rock-bit life is noted.

660. Wilhelm, C. J. and L. Schmidt, 1935, "Preliminary Report on the Disposal of Oil Field Brines in the Ritz-Canton Oil Field, McPherson County, Kansas," *U.S. Bur. Mines Rept. of Inv. 3297.*

The initial step in making a study of the brine-disposal problem in the Ritz-Canton field was a survey of the quantity of brine produced, the methods in use at that time in disposing of the brine, and the efficiency of these methods of disposal in preventing mineralization of fresh-water-bearing formations and surface streams. Incidental to this study, it was necessary to consider the surface geology of the area. Surveys of the fresh-water wells and surface streams indicated that many of them were being contaminated. Test wells drilled within the limits of the field indicated the variable character of the surface formation. It was found that contamination of the fresh-water supply by brine had several sources. In conclusion, this report recommends various methods depending on conditions by which this mineralization may be either entirely eliminated or considerably reduced.

661. Williams, C. E., Jr., and G. H. Bruce, 1951, "Carrying Capacity of Drilling Muds," *Trans., AIME.*, Vol. 192, pp. 111-120, 14 refs.

The trend toward deeper drilling together with the attendant increase in power requirements for circulation of the drilling fluid has emphasized the need for a critical examination of the factors affecting the removal of bit cuttings from the hole by the drilling fluid. The ability of drilling fluids to lift cuttings is called their carrying capacity. A series of laboratory and field experiments has been conducted to determine the minimum annular velocity necessary to remove cuttings and to investigate the effects of properties of drilling fluids on their carrying capacities.

662. Williams, D. E., 1970, "Ground Water Development and Management in the Owens Valley," *Annual Conference Los Angeles Department of Water and Power,* Los Angeles, California, 12 p., 5 figs., 3 refs.

New techniques of development and management of ground water reservoirs are currently being practiced by the city of Los Angeles in the operation of their Owens Valley ground water reservoir. These techniques include utilization of earthquake faults as "ground water" dams, down-hole photographic devices, well drilling using directional explosives, well cleaning using a high-energy electric arc, and various mathematical and viscous-flow modeling techniques.

663. Woods, H. B. and A. Lubinski, 1955, "Use of Stabilizers in Controlling Hole Deviation," *A.P.I. Drill and Prod. Pract.,* pp. 165-182, 10 refs.

Results of a theoretical investigation on use of a stabilizer in drilling crooked formations are given in charts which show: (1) how much more weight may be carried and (2) where the stabilizer should be placed. The use of reamers and more than one stabilizer is qualitatively discussed.

664. Woods, H. B. and E. M. Galle, 1957, "How Weight Affects Penetration Rate," *Oil and Gas Journ.,* November, Vol. 55, No. 47, pp. 88-91.

An AAODC committee composed of personnel from drilling and equipment companies have made a joint study of the factors that enter into the effect of weight and speed of rotation on penetration. The test site was the Dora Roberts field near Odessa, Texas. An exploration drilling rig was used to drill the well. This is a report of the weight study.

665. Weight, C. C., 1963, "Testing and Evaluating Lost Circulation Materials," *Drilling Contractor,* September-October, pp. 64, 84-97, 12 refs.

A lost circulation control program is aimed at drilling the well at minimum cost, both in time and materials. There are two basic approaches to lost circulation control: prevention and cure. Some limiting factors to be considered are bridging agents, cementing agents, gelling agents, oil soluble agents, mud compatibility.

666. Wuerker, R. G., 1969, Annotated Tables of Strength and Elastic Properties of Rock," *Petrol. Trans., A.I.M.E.,* Reprint No. 6, pp. 23-45, 75 refs.

Testing for strength and elastic properties of rocks has made rapid progress since the end of World War II. The perfection of standard methods and the awakening interest of all branches of the mineral industry resulted in the collection of a great amount of new physical data on the most common rocks. The purpose of this paper is to

Wuerker, R. G. . . . . (con't)

tabulate the results from standardized tests and to discuss the importance of each property and its usefulness to the mineral engineer as reflected in the existing literature. Twenty-one rock properties are listed in tables which have been encountered in mining and petroleum exploration. WRSIC #W71-05312

667. Wuth, D. E. and R. L. O'Shields, 1955, "New Mud Desander Cuts Drilling Costs," *Drilling Contractor,* Vol. II, No. 6, October, pp. 76-81.

The contamination of drilling fluids by sand, rock particles, and other solids is a serious problem with the drilling industry. (1) The new desander continuously removes sand from drilling mud without requiring supervision or operating adjustments. The maximum observed sand content in the treated mud was 0.2 percent A.P.I. (2) The use of de-sanded mud has resulted in direct rig savings of between $2,700 and $3,300 per well when drilling in Southwest Louisiana. This paper presents other benefits of the new desander.

668. Wyllie, M. R. J., 1960, "Log Interpretations in Sandstone Reservoirs," *Geophysics,* Vol. 25, No. 4, pp. 748-778.

By comparison with carbonate rocks sandstones are texturally homogenous; in consequence the interpretation from well logs of the fluid content and physical properties of sandstones is relatively simple. From an exploration standpoint one significant fact must be gleaned from the log: will the formation of interest produce oil or gas? The precise porosity and thickness of the formation and its exact hydrocarbon saturation are, initially, secondary considerations. Some of the principal difficulties involved in determining interstitial water resistivity, porosity, and formation resistivity are examined. It is confirmed that the so-called "low zone" significantly decreases the resistivity of an oil bearing sand when this resistivity is found from an induction log. Theory and laboratory experiment confirm that diffusion and convective mixing of the filtrate and interstitial water do little to mitigate the problem. It is suggested that future correction charts for induction logs recognize this fact.

669. Young, F. S., Jr., 1969, "Computerized Drilling Control," *Trans. SPE of AIME,* Vol. 246, pp. 483-496, 19 refs.

Previous laboratory and field experimentation has demonstrated the effect of several variables on drilling rate. These results have been incorporated into optimization theories for the purpose of reducing drilling cost. Humble has developed an on-site computer system to control bit weight and rotary speed and, thereby, implement the concepts of minimum-cost drilling. This system can (1) perform short interval drilling rate tests for formation evaluation, (2) use the results of these tests as input information to solve minimum-cost drilling formulas, and (3) control bit weight and rotary speed in accordance with these computed solutions. The purpose of this paper is to describe the computer control system and present results of initial field testing.

670. Young, R. S., Jr., and K. E. Gray, 1967, "Dynamic Filtration During Microbit Drilling," *Trans. SPE of AIME,* Vol. 240, pp. 1209-1224, 24 refs.

This paper reports results of an investigation to evaluate dynamic filtration beneath the bit during microbit drilling. Some obvious experimental limitations of previous works have been eliminated so as to quantitatively evaluate this filtration process and its effect upon drilling rate. These limitations include (1) complicated filtrate flow path both beneath the bit and through the borehole wall, (2) limited filtrate flow rate measurements beneath the bit for varying rock-mud properties, and (3) limited pore pressure history during the drilling operation.

671. Zangar, C. N., 1953, "Theory and Problems of Water Percolation," *Eng. Monograph No. 8,* U.S.D.I. Bur. Reclamation, Denver.

The flow of ground water through dams and their foundations and the accompanying pressures and gradients that exist are examined. Several methods for determining the permeability of soil by field tests are explored as well as flow problems via electric analog experiments.

672. Zdenek, F. F., 1969, "Revert Cuts Costs and Makes Better Wells," *Johnson Driller's Journ.,* Vol. 41, no. 6, November-December, pp. 1-3.

Revert is a food grade, organic material which makes a bright blue drilling fluid when mixed with water at a ratio of about six pounds per 100 gallons. After a certain period of time, the fluid reverts to the viscosity of water through enzymatic action. Being a low solid, self-destroying fluid, revert does not contaminate water-bearing sands with clay particles and can be removed completely from the well during development. WRSIC #W71-03846

673. Zoeller, W. A. and D. M. Muir, 1969, "Acoustilog-Nuclear Logging in Cased Wells," *A.P.I. Drill. and Prod. Pract.,* Cements and Cementing, pp. 61-67, 3 refs.

The development of a superior acoustilog instrument has made it possible to obtain porosity and lithology data from cased wells. At the same time, a valuable evaluation of the cement condition can be made from the data recorded. The versatility of the Acoustilog and the nuclear logging equipment makes it possible to simultaneously record the acoustic, gamma ray, neutron, caliper, and collar logs. These data are obtained routinely in both open and cased wells. The value and advantages of the acoustic-neutron through casing log are: obtaining valuable log information in wells that were cased prematurely because of bad hole conditions, obtaining valuable porosity information in wells that are to be recompleted for secondary-type production that were first completed before the advent of the acoustic log.
WRSIC #W71-10428

674. National Water Well Association, 1976, *Manual of Water Well Construction Practices,* U.S. Environmental Protection Agency, Office of Water Supply, EPA-570/9-75-001, 155 pp., 6 figs., 12 tabs., Appendices, 1 Plate

The standards have been prepared by NWWA for engineers and governmental personnel whose expertise is not in the field of well construction. They are presented in such a manner that appropriate descriptive paragraphs can be adapted for use in the preparation of water well specifications.

675. Campbell, M.D., and Lehr, J.H., 1975, "Engineering Economics of Rural Water Systems: A New American Approach," *Journal of American Water Works Association,* May, pp. 225-231, 3 figs., 20 refs.

The rural areas of the U.S. have always been plagued with the difficulty of receiving services from public or private utilities. This article presents a view of engineering principles with regard to the problem as applied by the National Demonstration Water Project and the Commission on Rural Water. Local conditions and system design are explored and defined in both quantitative and qualitative forms.

676. Lehr, J.H., et al., 1976, *A Manual of Laws, Regulations and Institutions for Control of Ground Water Pollution,* U.S. Environmental Protection Agency, Washington, D.C., EPA-440/9-76-006, 34 figs., 4 tabs., 160 refs.

This manual is for use on the state and local levels by people concerned with ground water quality. It covers ground water pollution problems broadly and supplies a range of suggestions for ground water pollution control primarily in the form of regulatory provisions.

677. Campbell, M.D., and Gray, G.R., 1975, "Mobility of Well-Drilling Additives in the Ground-Water System," U.S. Environmental Protection Agency, *Conference on Environmental Aspects of Chemical Use in Well-Drilling Operations,* May, EPA-560/1-75-004, pp. 261-288, 9 figs., 47 refs.

The purpose of this paper is to examine the factors involved in the movement of drilling fluid components in the ground water reservoir. Components of the drilling fluids are considered generally and no attempt is made to deal with specific substances. Examples of the important mechanisms involved in the movement of liquids, solids, and gases from the borehole into the formation are discussed.

678. American Society of Testing and Materials, 1976, "Standard for Thermoplastic Water Well Casing Pipe and Couplings Made in Standard Dimension Ratio (SDR)," 76 pp., 4 figs., 5 tabs., 6 refs.

This Standard has been developed to fill a need of federal and state agencies in administering laws and programs dealing with protection of ground water from pollution by preventing the use of inadequately designed and manufactured thermoplastic well casing. The Standard has been developed by a cooperative research program staffed by thermoplastic well casing manufacturers, users of such casing, and state and federal regulatory personnel. The Standard has become a part of the "Manual of Recommended Water Well Construction Practices" prepared by the National Water Well Association for the United States Environmental Protection Agency.

679. Campbell, M.D. (ed.), 1977, *Geology of Alternate Energy Resources in the South-Central United States,* The Houston Geological Society, Houston, Texas, 500 pp., 140 figs., 30 tabs., 1,200 refs.

A review of the state-of-the-art is presented on the geological aspects of uranium, lignite, and geothermal-geopressured energy of the South-Central United States. Four chapters on each resource are devoted to: (1) frontier exploration, (2) trend analysis, (3) utilization/development, and (4) environmental considerations.

17
APPENDIX: Miscellaneous Tables, Charts, Formulae and Technical Guides

CONTENTS

Table		Page
(1T)	Length (General)	578
(2T)	Area (")	578
(3T)	Volume (")	578
(4T)	Flow (")	578
(5T)	Weight (")	579
(6T)	Power (")	579
(7T)	Volumes and Weight Equivalents	579
(8T)	Friction Loss in Smooth Pipe	579
(9T)	Areas of Plane Figures	580

(10T) Volumes of Solids 580

(11T) Miscellaneous Water Data 581

(12T) Conversion Table (Units in Water Analyses) 581

(13T) Irrigation Table 582

(14T) Comparison of Units Used in Petroleum Industry with Units Used by Ground Water Industry . . . 582

(15T) Conversion Table 583

(16T) Water Well Casing— Dimensions, Weights and Test Pressures 583

(17T) Weight of Water at Various Temperatures 584

(18T) Friction Loss in Pipe and Fittings 584

(19T) Maximum Quantity of Water Through Pipe 584

(20T) Pounds of Material Needed to Make 100 Barrels of Mud 585

(21T) Conversion Table for Hardness. 585

(22T) Mud Weight Conversion Tables 586

(23T) Structural Data for Drill Pipe, Tubing and Casing 586

(24T) Capacities of Suction Pits (cu. ft/inch of depth) 587

(25T) Lifting Power 587

(26T) Ascending Velocity of Mud 588

(27T) Drive Pipe - Dimensions, Weights and Test Pressures 588

(28T) Volumetric Requirements - Air Drilling for Blast, Water and Seismographic Drilling 589

(29T) Volumetric Requirements - Air Drilling for Gas and Oil Well Production Drilling 589

(30T) Standard Fuel Requirements for Pumping Plants . . . 589

(31T) API Standard Round-Thread Tubing - Dimensions, Weights are Test Pressures - Non-upset . . . 590

(32T) Square Feet of Surface/Lineal Foot of Pipe 590

(33T) Pressure, Lb/Sq. Inches to Feet (Head) of Water (ft. = 2.31 x 15/sq. inches) 591

(34T) Pressure, Feet (head) of Water to Lb/Sq. Inch (lb/sq. inches = 0.4331 x ft) 591

(35T) General Methods of Specifying Casing 592

(36T) Capacity of Pipe, Tubing and Drill Pipe 592

(37T) Hole Capacity (gals/foot; Linear feet/gals., etc.) 593

(38T) Pipe Displacement 594

(39T) Suction-Lift Table 594

(40T) Advantages and Disadvantages of Various Types of Pumps 595

(41T) Volume and Height Between Casing and Hole-Wall . . . 596

(42T) Pressure in lb/sq. inches to feet of head (General) . . . 596

(43T) Air-Lift Pumps 597

(44T) Pipe, Cylinder or Hole Capacity 597

(45T) Air-Lift Pumps (Con't) 598

(46T) Pumping-Plant Performance Standards 598

(47T) Deep-Well Turbine Pumps and Ordering Pumps 599

(48T) Windmill Pumping Capacity 599

(49T) Determining Cost of Pumping Water 600

(50T) Pump Efficiency 601

(51T) Permissible Velocities in Open Channels,
Erosion and Deposition of Silt 601

(52T) Kilowatt/Hours for 100 Feet of Field Head with
Various Overall Efficiencies 602

(53T) Velocities of Water Which Will Cause Sand to Rise . . . 602

(54T) Estimating Flow from Vertical Pipe or Casing 603

(55T) Standard Pipe - Mill Practices 603

(56T) Estimating Discharge from a Horizontal Pipe or
Casing (General) 604

(57T) Explosives (Properties of Straight and Ammonia
Dynamites 604

(58T) Estimating Discharge from Vertical Pipe or
Casing (General) 605

(59T) Explosives (Properties of Gelatin and Semi-
Gelatin Dynamites 605

(60T) Hammer-Drill Operational Requirements 606

(61T) Orifice Tables 607

(62T) Theoretical Discharge of Nozzles (U.S. gals/min) . . . 608

(63T) Numerical Relationship Between Permeability and
Conductivity 609

(64T) Equivalent Sieve Mesh Sizes 610

(65T) Well Screen-Selection Chart for Small-Capacity Wells . . 611

(66T) Aquifer Characteristics, Driller's Terms and
Specific Yield Estimates 612

(67T) Borehole Geophysical Logging Methods and
Their Uses in Hydrologic Studies 613

(68T) Borehole Geophysical Logging Methods and Their Uses in Hydrologic Studies (con't) . . . 614

(69T) Power Cost of Pumping Water 615

(70T) Overhead Cost of Pumping Water 616

(71T) General Water Quality Problems and Typical Solutions 617

(72T) Temperature/Pressure (General) 617

(73T) Comparison of A_e Values for Various Screen Types . . 618

(74T) Metric Conversion Tables 619

(75T) Comparison of Percent Open Area for Typical Screen Types 620

(76T) Comparison of Percent Open Area for Typical Screen Types (con't) 621

(77T) Discharge Measurements Using Small Container 621

(78T) Determining Capacity of a Pump 622

(79T) Cement Slurry Data 622

(80T) Estimating Flow from Horizontal on Inclined Pipes 623

(81T) Acid Calculations 623

(82T) Orifice Method of Measuring Water Flow 624

(83T) Water Friction in 100 Feet of Smooth Borehole 625

Table 1T

LENGTH

Unit	Equivalents of First Column						
	Centimeters	Meters	Kilometers	Inches	Feet	Yards	Miles
1 Centimeter	1	.01	.00001	.3937	.0328	.0109	.0000062
1 Meter	100	1	.001	39.37	3.2808	1.0936	.000621
1 Kilometer	100,000	1,000	1	39,370	3,280.8	1,093.6	.621
1 Inch	2.54	.0254	.0000254	1	.0833	.0278	.000016
1 Foot	30.48	.3048	.000305	12	1	.3333	.000189
1 Yard	91.44	.9144	.000914	36	3	1	.000568
1 Mile	160,935	1,609.3	1.6093	63,360	5,280	1,760	1

Table 2T

AREA

Unit	Equivalents of First Column						
	Square Centimeters	Square Meters	Square Inches	Square Feet	Square Yards	Acres	Square Miles
1 Sq. centimeter	1	.0001	.155	.00108	.00012		–
1 Sq. Meter	10,000	1	1,550	10.76	1.196	.000247	–
1 Sq. Inch	6.452	.000645	1	.00694	.000772	–	–
1 Sq. Foot	929	.0929	144	1	.111	.000023	–
1 Sq. Yard	8,361	.836	1,296	9	1	.000207	–
1 Acre	40,465,284	4,047	6,272,640	43,560	4,840	1	.00156
1 Sq. Mile	–	2,589,998	–	27,878,400	3,097,600	640	1

Table 3T

VOLUME

Unit	Equivalents of First Column						
	Cubic Centimeters	Cubic Meters	Liters	U.S. Gallons	Imperial Gallons	Cubic Inches	Cubic Feet
1 Cu. Centimeter	1	.000001	.001	.000264	.00022	.061	.0000353
1 Cu. Meter	1,000,000	1	1,000	264.17	220.083	61,023	35.314
1 Liter	1,000	.001	1	.264	.220	61.023	.0353
1 U.S. Gallon	3,785.4	.00379	3.785	1	.833	231	.134
1 Imperial Gallon	4.542.5	.00454	4.542	1.2	1	277.274	.160
1 Cu. Inch	16.39	.0000164	.0164	.00433	.00361	1	.000579
1 Cu. Foot	28,317	.0283	28.317	7.48	6.232	1,728	1

Table 4T

FLOW

Unit	Equivalents of First Column						
	Cubic Feet Per Second	Cubic Feet Per Day	U.S. Gallons Per Minute	Imp. Gallons Per Minute	U.S. Gallons Per Day	Imp. Gallons Per Day	Acre Feet Per Day
1 Cu. Foot per Sec.	1	86,400	448.83	374.03	646,323	538,860	1.983
1 Cu. Foot per Day	.0000116	1	.00519	.00433	7.48	6.233	.000023
1 U.S. Gallon per Min.	.00223	192.50	1	.833	1,440	1,200	.00442
1 Imp. Gallon per Min.	.00267	231.12	1.2	1	1,728	1,440	.0053
1 U.S. Gallon per Day	.00000155	.134	.000694	.000579	1	.833	.00000307
1 Imp. Gallon per Day	.00000186	.160	.000833	.000694	1.2	1	.00000368
1 Acre Foot per Day	.504	43,560	226.28	188.57	325,850	271,542	1

Table 5T

WEIGHT

Unit	Equivalents of First Column					
	Grams	Kilograms	Ounces (Avoir-dupois)	Pounds (Avoir-dupois)	Tons (Short)	Tons (Long)
1 Gram	1	.001	.0353	.0022	.0000011	.00000098
1 Kilogram	1000	1	35.274	2.205	.0011	.000984
1 Ounce (Avoirdupois)	28.349	.0283	1	.0625	.0000312	.0000279
1 Pound (Avoirdupois)	453.592	.454	16	1	.0005	.000446
1 Ton (Short)	907,184.8	907.185	32,000	2,000	1	.893
1 Ton (Long)	1,016,046.98	1,016.047	35,840	2,240	1.12	1

Table 6T

POWER

Unit	Equivalents of First Column				
	Watts	Kilowatts	Horsepower	Foot Pounds Per Minute	Joules Per Second
1 Watt	1	.001	.00134	44.254	1
1 Kilowatt	1000	1	1.341	44.254	1,000
1 Horsepower	746	.746	1	33,000	746
1 Foot Pound Per Minute	.0226	.0000226	.0000303	1	.0226
1 Joule Per Second	1	.001	.00134	44.254	1

Table 7T

VOLUMES AND WEIGHT EQUIVALENTS (Water at 39.2°F)

Unit	Equivalents of First Column						
	Cubic Meters	Liters	U.S. Gallons	Imp. Gallons	Cubic Inches	Cubic Feet	Pounds
1 Cu. Meter	1	1,000	264.17	220.083	61.023	35.314	2,200.83
1 Liter	.001	1	.264	.220	61.023	.0353	2.201
1 U.S. Gallon	.00379	3.785	1	.833	231	.134	8.333
1 Imp. Gallon	.00454	4.542	1.2	1	277.274	.160	10
1 Cu. Inch	.0000164	.0164	.00433	.00361	1	.000579	.0361
1 Cu. Foot	.0283	28.317	7.48	6.232	1,728	1	62.32
1 Pound	.00045	.454	.12	.1	27.72	.016	1

Table 8T

FRICTION LOSS IN SMOOTH PIPE

(approximate head loss in feet per 1000 feet of pipe)

Flow Rate in Gallons per Minute	Nominal Pipe Size in Inches					
	1¼	1½	2	2½	3	4
10	20	9	2			
15	44	20	6			
20	79	35	10	4	1	
25	123	55	16	6	2	
30	178	79	22	9	3	
40		142	40	16	5	
50		222	64	25	8	2

Table 9T

VOLUMES OF SOLIDS

Nomenclature

a, b, c, d	— Lengths of Sides
C	— Length of Chord
A	— Total Area
A_B	— Area of Base
A_L	— Area of Lateral or Convex Surfaces
A_R	— Area of Right Section
A_T	— Area of Top Section
h, h_1, h_2	— Vertical Height or Altitude
h_G	— Vertical Distance between Centers of Gravity of Areas
L, L_1, L_2	— Lateral Length or Slant Height
L_G	— Slant Height between Centers of Gravity of Areas
p	— Perimeter
p_B	— Perimeter of Base
p_R	— Perimeter of Right Section
r, r_1	— Radii
V	— Volume

18. CUBE

$A = 6a^2$

$V = a^3$

19. PARALLELOPIPED

$A = 2(ab + bc + ac)$

$V = abc$

20. GENERAL PRISM AND RIGHT REGULAR PRISM

$A_L = p_R L = p_B h$

$A = A_L + 2A_B$

$V = A_R \times L = A_B h$

21. FRUSTUM of PRISM

$V = A_B h_G$

$V = A_R L_G$

22. RIGHT REGULAR PYRAMID or CONE

$A_L = \frac{1}{2} p_B L$

$V = \frac{1}{3} A_B h$

23. GENERAL PYRAMID or CONE

$V = \frac{1}{3} A_B h$

24. FRUSTUM of RIGHT REGULAR PYRAMID or CONE

$A_L = \frac{1}{2} L (p_B + p_T)$

$A = A_L + A_B + A_T$

$V = \frac{1}{3} h (A_B + A_T + \sqrt{A_B A_T})$

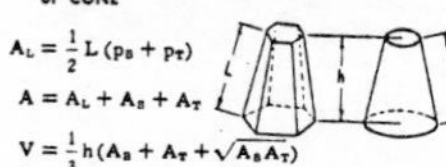

25. FRUSTUM of GENERAL PYRAMID or CONE (PARALLEL ENDS)

$V = \frac{1}{3} h (A_B + A_T + \sqrt{A_B A_T})$

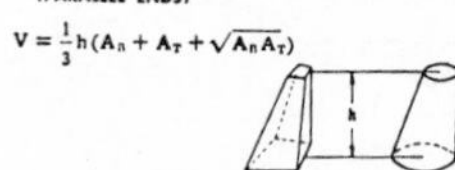

26. RIGHT CIRCULAR CYLINDER

$A_L = 2\pi rh$

$A = 2\pi r (r + h)$

$V = \pi r^2 h$

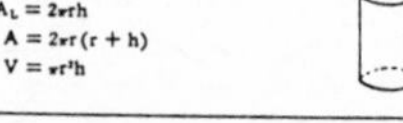

27. GENERAL CYLINDER (ANY CROSS SECTION)

$A_L = p_B h = p_R L$

$A = A_L + 2A_B$

$V = A_B h = A_R L$

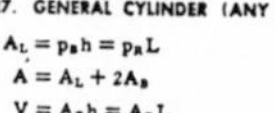

28. FRUSTUM of GENERAL CYLINDER

$V = \frac{1}{2} A_R (L_1 + L_2)$

$V = A_B h_G$

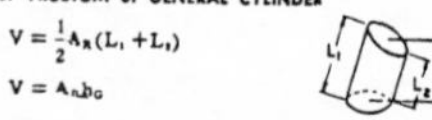

29. FRUSTUM of RIGHT CIRCULAR CYLINDER

$A_L = \pi r (h_1 + h_2)$

$A_T = \pi r \sqrt{r^2 + \left(\frac{h_1 - h_2}{2}\right)^2}$

$A_B = \pi r^2$

$A = A_L + A_T + A_B$

$V = \frac{\pi r^2}{2} (h_1 + h_2)$

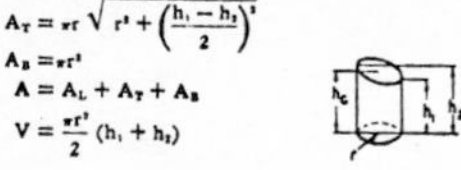

30. SPHERE

$A = 4\pi r^2 = 12.566 r^2$

$V = \frac{4}{3} \pi r^3 = 4.189 r^3$

31. SPHERICAL SECTOR

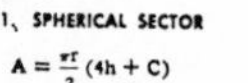

$A = \frac{\pi r}{2} (4h + C)$

$V = \frac{2}{3} \pi r^2 h = 2.0944 r^2 h$

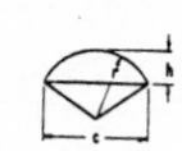

32. SPHERICAL SEGMENT

$A_T = 2\pi rh = \frac{\pi}{4} (4h^2 + C^2)$

$V = \frac{\pi}{3} h^2 (3r - h) = \frac{\pi}{24} h (3C^2 + 4h^2)$

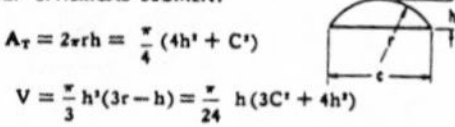

33. SPHERICAL ZONE

$A_L = 2\pi rh$

$A = \frac{\pi}{4} (8rh + a^2 + b^2)$

$V = \frac{\pi h}{24} (3C^2 + 3b^2 + 4h^2)$

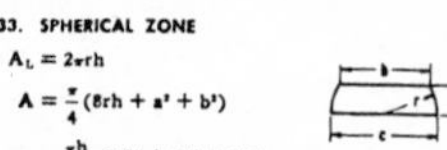

34. TORUS

$A = 4\pi^2 r r_1$

$V = 2\pi^2 r^2 r_1$

Table 10T

AREAS OF PLANE FIGURES

Nomenclature

a, b, c, d	— Lengths of Sides
A	— Area
d, d_1, d_2	— Diameters
e, f	— Lengths of Diagonals
h	— Vertical Height or Altitude
l, l_1, l_2	— Length of Arc
L	— Lateral Length or Slant Height
n	— Number of Sides
θ	— Number of Degrees of Arc
p	— Perimeter
r, r_1, r_2, R	— Radii

1. RIGHT TRIANGLE

$p = a + b + c$

$c^2 = a^2 + b^2$

$b = \sqrt{c^2 - a^2}$

$A = \frac{ab}{2}$

2. EQUILATERAL TRIANGLE

$p = 3a$

$h = \frac{a}{2}\sqrt{3} = .866\,a$

$A = a^2 \frac{\sqrt{3}}{4} = .433\,a^2$

3. GENERAL TRIANGLE

Let $s = \frac{a + b + c}{2}$

$p = a + b + c$

$h = \frac{2}{a}\sqrt{s(s-a)(s-b)(s-c)}$

$A = \frac{ah}{2}$

$A = \sqrt{s(s-a)(s-b)(s-c)}$

4. SQUARE

$a = b$

$p = 4a$

$A = a^2 = .5e^2$

$e = a\sqrt{2} = 1.414\,a$

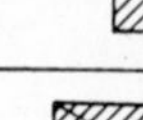

5. RECTANGLE

$p = 2(a + b)$

$e = \sqrt{a^2 + b^2}$

$b = \sqrt{e^2 - a^2}$

$A = ab$

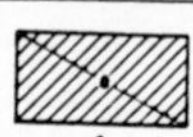

6. GENERAL PARALLELOGRAM OR RHOMBOID; AND RHOMBUS

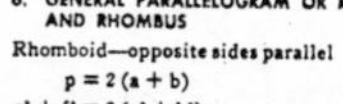

Rhomboid—opposite sides parallel

$p = 2(a + b)$

$e^2 + f^2 = 2(a^2 + b^2)$

$A = ah$

Rhombus—opposite sides parallel and all sides equal

$a = b$

$p = 4a = 4b$

$e^2 + f^2 = 4a^2$

$A = ah = \frac{ef}{2}$

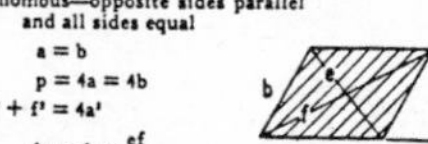

7. TRAPEZOID

$p = a + b + c + d$

$A = \frac{(a + b)}{2} h$

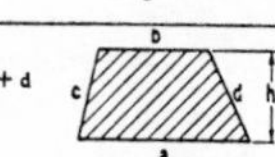

8. TRAPEZIUM

$p = a + b + c + d$

A = Sum of Areas of two major triangles

$A = \frac{(h_1 + h_2) g + f h_1 + j h_2}{2}$

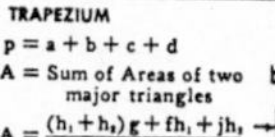

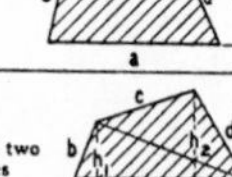

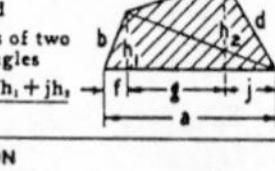

9. REGULAR POLYGON

Let n = number of sides

$p = na$

$a = 2\sqrt{R^2 - r^2}$

$A = \frac{nar}{2} = \frac{na}{2}\sqrt{R^2 - \frac{a^2}{4}}$

$= n \times$ Area of each triangle

10. CIRCLE

$p = 2\pi r = \pi d = 3.1416d$

$A = \pi r^2 = \frac{\pi d^2}{4} = .7854d^2$

$= \frac{p^2}{4\pi} = .07958p^2$

11. HOLLOW CIRCLE or ANNULUS

$A = \frac{\pi}{4} (d_2^2 - d_1^2) = .7854 (d_2^2 - d_1^2)$

$= \pi (r_2^2 - r_1^2)$

$= \pi \frac{d_1 + d_2}{2} (r_2 - r_1)$

$= \pi (r_1 + r_2)(r_2 - r_1)$

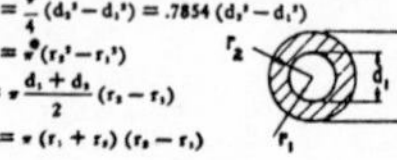

12. SECTOR of CIRCLE

$l = \frac{\pi r \theta}{180} = \frac{r\theta}{57.3} = .01745 r\theta$

$= \frac{2A}{r}$

$A = \frac{\pi \theta r^2}{360} = .0087272 \theta r^2$

$= \frac{lr}{2}$

13. SEGMENT of CIRCLE

for $\theta < 90°$

$A = \frac{r^2}{2}\left(\frac{\pi\theta}{180} - \sin\theta\right)$

for $\theta > 90°$

$A = \frac{r^2}{2}\left(\frac{\pi\theta}{180} - \sin(180 - \theta)\right)$

for chord rise, etc., see "Properties of Circle"

14. SECTOR of HOLLOW CIRCLE

$A = \frac{\pi\theta (r_2^2 - r_1^2)}{360}$

$A = \frac{r_1 - r_2}{2} (l_1 + l_2)$

15. FILLET

$A = .215 r^2$

or approximately

$A = \frac{r^2}{5}$

16. ELLIPSE

$p = \pi (a + b)$ approximately

$= \pi [1.5 (a + b) - \sqrt{ab}]$ more nearly

$A = \pi ab$

17. PARABOLA

$A = \frac{2}{3} ab$

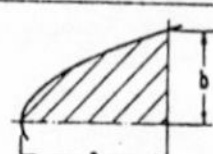

Table 11T

MISCELLANEOUS WATER DATA

A gallon of water (U.S. standard) contains 231 cubic inches and weighs approximately 8 1/3 pounds.

A cubic foot of water contains about 7½ gallons and weighs approximately 62½ pounds.

To find the pressure in pounds per square inch of a column of water, multiply the height of the column in feet by .434.

One pound of water occupies 27.7 cubic inches.

One cubic foot of salt water weighs about 64 1/3 pounds.

One standard barrel contains 31½ gallons; one barrel of oil equals 42 gallons; one average oil drum holds 55 gallons.

Friction of liquids in pipes increases as the square of the velocity.

Doubling the diameter of a pipe increases its capacity four times.

A "miner's inch" of water is approximately equal to a supply of 12 gallons per minute (9 g.p.m. in some states).

The gallons per minute which a pipe will deliver equals .0408 times the square of the diameter in inches, multiplied by the velocity of water in feet per minute.

To find the capacity of a pipe or cylinder in gallons, multiply the square of the diameter in inches by the length in inches and by .0034.

The weight of water (in pounds) in any length pipe is obtained by multiplying the length in feet by the square of the diameter in inches, and by .34.

One common water pail will hold 2.27 gallons or about 19 pounds of water.

Sharp angles or sudden bends in pipes cause an increase in friction, and consequently more power is necessary. Where change of direction is desired it should be made with long, easy curves or by using 45 degree elbows.

Barrels per day (42 gal.) x .02917 equals gal.per min.

Table 12T

CONVERSION TABLE

UNITS USED IN EXPRESSING WATER ANALYSES

Units	Parts Per Million (PPM)	Grains Per US Gallon (GPG)	Grains Per Imperial Gallon	Parts Per 100,000	Lbs. Per 1000 Gal
1 Part per Million	ONE	0.058	0.049	0.100	0.0083
1 Grain per US Gallon	17.118	ONE	0.8331	1.712	0.143
1 Grain per Imperial Gal	20.6	1.2003	ONE	2.06	0.172
1 Part per 100,000	10.00	0.583	0.486	ONE	0.0834
1 Pound per 1,000 Gal.	119.8	6.98	5.80	11.98	ONE

Table 13T

Irrigation Table

Gal min	Cu ft sec	Cu ft min	Number of Acres covered in twelve hours pumping. 1 in deep	2 in deep	3 in deep	4 in deep	6 in deep	8 in deep	10 in deep	12 in deep
20	.0446	2.675	.529	.2645	.1765	.1324	.08825	.06625	.0529	.04415
50	.1112	6.68	1.328	.664	.4425	.332	.2213	.166	.1328	.1105
100	.2225	13.37	2.96	1.325	.883	.6625	.442	.3313	.265	.221
150	.3345	20.05	3.95	1.991	1.328	.995	.664	.4975	.398	.332
225	.502	30.05	5.97	2.985	1.990	1.492	.994	.747	.597	.4975
300	.668	40.01	7.96	3.980	2.655	1.99	1.327	.995	.796	.663
400	.891	53.40	10.61	5.305	3.535	2.652	1.770	1.328	1.061	.884
700	1.560	93.50	15.58	9.28	6.18	4.64	3.095	2.32	1.858	1.548
900	2.005	120.40	23.85	11.95	7.96	5.97	3.98	2.975	2.385	1.99
1200	2.675	160.50	31.82	15.92	10.61	7.95	5.305	3.975	3.182	2.65
1600	3.565	213.50	42.35	21.20	14.15	10.61	7.075	5.305	4.235	3.535
3000	6.68	400.50	79.50	39.75	26.50	19.88	13.25	9.94	7.95	6.625
4500	10.03	602.00	119.30	59.70	39.75	29.85	19.90	14.93	11.93	9.95
6000	13.36	802.00	159.10	79.60	53.00	39.75	26.52	18.89	15.91	13.26
7000	15.61	936.00	185.70	92.80	61.90	46.45	30.95	23.20	18.57	15.47
8500	18.95	1137.00	225.50	112.80	75.20	56.35	37.60	28.19	22.55	18.79
10000	22.25	1337.00	260.00	132.50	88.30	66.25	44.20	33.15	26.50	22.10
14000	31.15	1871.00	371.00	185.50	123.70	92.75	61.80	46.35	37.10	30.95

1 Acrefoot = 1 acre covered to a depth of 1 ft = 43,560 cubic feet.

—Courtesy Ingersoll-Rand Co.

Table 14T

COMPARISON OF UNITS USED IN PETROLEUM INDUSTRY WITH UNITS USED BY GROUND WATER INDUSTRY

Ground-Water Industry Unit	Equivalent Petroleum Industry Unit
Gallon (gal.) (42 Gallons) . . . . 9,702 cu. inches 5.615 cu. feet.	1/42 Barrel (bbl.)... 1 Barrel
Q-gallons per minute (gpm)	34.29 Barrels per day (B/D)
Drawdown in feet (s) pumping level minus static water level (SWL) (s_a) - actual drawdown (s_t) - theoretical drawdown of 100% efficient well	Differential pressure 1/2.31 undisturbed formation pressure (p_o) minus flowing pressure (p_f) in pounds per square inch (psi)
Specific capacity (S). . . . . gpm per foot of drawdown	Productivity index (P.I.) 79.91 B/D per psi
Permeability:	Permeability:
meinzer - gallons per day of water at 60°F per square foot at 100% hydraulic gradient.	$\frac{1}{18.24}$ darcy - cubic centimeters per second per square centimeter at one dyne per square centimeter length and viscosity of one centipoise. 54.82 millidarcy
18.24 gallons/day/sq. foot (60°F)	1 darcy
(0.01824 gals/day/sq. foot)	1 millidarcy
Transmissibility:	Transmissibility:
gpd - ft. at prevailing temperature at 100% hydraulic gradient	$\frac{1}{20.50}$ darcy-ft. per centipoise 48.77 millidarcy-ft. per centipoise

Table 15T

CONVERSION TABLE

(Gallons per Minute--Gallons per Day--Cubic Feet per Second)

G.P.M.*	G.P.D.*	Sec.Ft*	G.P.D.*	G.P.M.*	Sec.Ft*
10	14,400	0.022	10,000	6.9	0.015
20	28,800	0.045	20,000	13.9	0.031
30	43,200	0.067	30,000	20.8	0.046
40	57,600	0.089	40,000	27.8	0.062
50	72,000	0.111	50,000	34.7	0.077
75	108,000	0.167	75,000	52.1	0.116
100	144,000	0.223	100,000	69.4	0.155
125	180,000	0.279	120,000	83.3	0.186
150	216,000	0.334	140,000	97.2	0.217
175	252,000	0.390	160,000	111.1	0.248
200	288,000	0.446	180,000	125.0	0.279
250	360,000	0.557	200,000	138.9	0.309
300	432,000	0.668	300,000	208.3	0.464
350	504,000	0.780	400,000	277.8	0.619
400	576,000	0.891	500,000	347.2	0.774
450	648,000	1.00	600,000	416.7	0.928
500	720,000	1.11	700,000	486.1	1.08
550	792,000	1.23	800,000	555.6	1.24
600	864,000	1.34	900,000	625.0	1.39
650	936,000	1.45	1,000,000	694.4	1.55
700	1,008,000	1.56	1,200,000	333.3	1.86
750	1,080,000	1.67	1,400,000	972.2	2.17
800	1,152,000	1.78	1,600,000	1111.1	2.48
850	1,224,000	1.89	1,800,000	1250.0	2.79
900	1,296,000	2.01	2,000,000	1388.9	3.09
950	1,368,000	2.12	2,500,000	1736.1	3.87
1000	1,440,000	2.23	3,000,000	2083.3	4.64
1200	1,728,000	2.67	3,500,000	2430.6	5.42
1400	2,016,000	3.12	4,000,000	2777.8	6.19
1600	2,304,000	3.57	4,500,000	3125.0	6.96
1800	2,592,000	4.01	5,000,000	3472.2	7.74
2000	2,880,000	4.46	10,000,000	6944.4	15.5

* - G.P.M.: U.S. Gallons per Minute

G.P.D.: U.S. Gallons per 24-hour Day

Sec.Ft.: Cubic Feet per Second

Table 16T

Water Well Casing—Dimensions, Weights and Test Pressures

Size: Outside Diameter, in.	Weights per Foot: Nom. Thd. & Cplg., lb.	Weights per Foot: Calculated Plain Ends, lb.	Wall Thickness, in.	Diameters: Outside, in.	Diameters: Inside, in.	Threads per Inch, No.	Couplings: Length, in.	Couplings: Outside Diameter, in.	Couplings: Calculated Weight, lb.	Test Pressures: Lap Welded Grade A, psi.
3.500	4.60	4.51	.125	3.500	3.250	14	3⅛	4.000	2.86	1100
4.000	5.65	5.53	.134	4.000	3.732	14	3⅛	4.500	3.24	1000
4.500	6.75	6.61	.142	4.500	4.216	14	3⅝	5.000	4.26	950
5.500	9.00	8.79	.154	5.500	5.192	14	4⅛	6.050	6.38	850
6.000	10.50	10.22	.164	6.000	5.672	14	4⅛	6.625	7.84	850
6.625	13.00	12.72	.185	6.625	6.255	11½	4⅝	7.390	11.88	850
8.625	17.50	16.90	.188	8.625	8.249	11½	5⅛	9.375	16.92	650

(a) The permissible variation in weight is 5 per cent above and 5 per cent below.

Table 17T

WEIGHT OF WATER AT VARIOUS TEMPERATURES

Temperature °F.	WEIGHT		Temperature °F.	WEIGHT		Temperature °F.	WEIGHT	
	Lbs./Gal.	Lbs./Cu. Ft.		Lbs./Gal.	Lbs./Cu. Ft.		Lbs./Gal.	Lbs./Cu. Ft
32	8.34	62.42	130	8.23	61.56	220	7.97	59.63
40	8.34	62.42	140	8.20	61.37	230	7.94	59.37
50	8.34	62.41	150	8.18	61.18	240	7.90	59.11
60	8.34	62.37	160	8.15	60.98	250	7.86	58.83
70	8.33	62.31	170	8.12	60.77	260	7.83	58.55
80	8.32	62.23	180	8.09	60.55	270	7.79	58.26
90	8.31	62.13	190	8.06	60.32	280	7.75	57.96
100	8.29	62.02	200	8.04	60.12	290	7.71	57.65
110	8.27	61.89	210	8.01	59.88	300	7.66	57.33
120	8.25	61.74	212	8.00	59.83			

Table 18T

FRICTION LOSS IN PIPE AND FITTINGS

The friction loss listed in this table for water flowing in pipes is calculated from the Williams and Hazen Formula using a co-efficient of 120. Friction loss in standard fittings is given in length of straight pipe in feet. The loss in fittings is based upon tests made by Crane Co.

Gallons per Minute	Nominal Size of Standard Pipe in Inches: 1	1¼	1½	2	2½	3	3½	4	5	6	8	10	12
10	8.4	2.4	1.1	.32									
15	18.0	4.8	2.2	.65	.28		Loss of Head in Feet per 100 Linear Feet of Pipe						
20	30.0	8.0	3.7	1.10	.47	.16							
30		17.0	8.0	2.30	1.00	.34	.17	.09					
40		29.0	13.5	3.9	1.7	.58	.29	.16					
50			21.0	5.8	2.6	.88	.44	.24	.08				
75				12.5	5.4	1.85	.93	.51	.17	.07			
100				21.0	9.3	3.10	1.60	.85	.28	.12			
150					19.5	6.5	3.3	1.8	.58	.25	.07		
200					33.0	11.3	5.8	3.1	1.03	.42	.11	.04	
250						17.0	8.5	4.6	1.53	.63	.17	.06	
300						24.0	12.2	6.6	2.10	.87	.24	.08	
350							16.0	8.5	2.8	1.3	.31	.10	.04
400							21.0	11.0	3.6	1.5	.39	.13	.06
500								17.0	5.3	2.3	.58	.19	.08
600								23.0	7.4	3.1	.82	.27	.11
800									13.0	5.2	1.4	.47	.19
1000									20.0	8.0	2.1	.71	.29
1200										11.5	3.0	1.00	.42
1500										17.0	4.5	1.50	.62
2000											7.7	2.6	1.05
2500											12.0	3.9	1.60
3000											16.8	5.5	2.30
4000												9.3	3.90
Standard 90° Elbow	3	4	5	6	7	9	10	12	14	16	21	27	33
Globe Valve, Open	28	38	45	57	68	85							
Angle Valve, Open	14	19	23	28	34	43							
Gate Valve Open				1.2	1.4	1.7	2.0	2.3	2.8	3.5	4.6	6.0	7.0

To allow for variation in age and condition of different pipe, multiply the values in the table by the factor selected from the following:

Ordinary new cast-iron pipe; brass or copper tubing	.86
New wrought-iron pipe; smooth and straight wood or concrete pipe; cast-iron pipe 4 to 6 years old	1.00
Cast-iron pipe 10 to 12 years old; new spiral riveted steel pipe, flow with laps	1.18
Ordinary wrought-iron or steel pipe one year old; cast-iron pipe 15 to 20 years old	1.4
Wrought-iron or steel pipe 8 to 10 years old; concrete pipe with rough texture and shoulder at joints	1.7
Corrugated pipe	3.6

—Courtesy Fairbanks, Morse & Co.

Table 19T

MAXIMUM QUANTITY OF WATER THRU PIPE

Maximum quantities of water which may be pumped through 100 feet of wrought iron pipe. At various pressures.

(in gallons per minute)

Size Pipe	½	¾	1	1¼	1½	2	2½	3	4
Pressure									
17 lbs.	3.2	9.1	18.7	33.5	51.6	106	200	290	589
30 lbs.	5	14	28	52	78	160	308	436	885
40 lbs.	6	16	33	60	90	184	350	504	1023
50 lbs.	6.5	17.5	37	70	101	206	390	564	1143
60 lbs.	7	19.5	40	76	110	226	430	617	1252
75 lbs.	7.5	22	45	85	123	253	480	690	1400
100 lbs.	9	25	52	99	142	292	558	797	1607

—Courtesy Fairbanks, Morse & Co.

Table 20T

POUNDS OF MATERIAL NEEDED
TO MAKE 100 BARRELS OF ROTARY MUD

Desired Mud Weight In			Blue or Light Colored Clays (Sp. Gr. 2.3-2.4)	Light Red Clays, Chalk (Sp. Gr. 2.4-2.6)	Dark Red Clays, Limestone (Sp. Gr. 2.6-2.8)	Baroid or Colox (Sp. Gr. 4.5-5.0)
Lbs. Per Cubic Ft.	Lbs. Per Gallon	Lbs. Per Sq.In./100 Feet				
63	8.4	43.8	310	300	290	290
64	8.6	44.4	930	920	910	850
65	8.7	45.1	1,550	1,540	1,530	1,410
66	8.8	45.8	2,170	2,160	2,150	1,970
67	8.9	46.5	2,790	2,780	2,770	2,535
68	9.1	47.2	3,410	3,400	3,390	3,100
69	9.2	47.9	4,030	4,020	4,010	3,660
70	9.4	48.6	4,650	4,640	4,630	4,225
71	9.5	49.3	5,270	5,260	5,250	4,775
72	9.6	50.0	5,890	5,880	5,870	5,340
73	9.7	50.7	6,510	6,500	6,490	5,900
74	9.8	51.4	7,130	7,120	7,110	6,465
75	10.0	52.1	7,750	7,740	7,730	7,025
76	10.1	52.8		8,360	8,350	7,590
77	10.2	53.5		8,980	8,970	8,150
78	10.4	54.2		9,600	9,590	8,710
79	10.5	54.8		10,220	10,210	9,275
80	10.6	55.5		10,840	10,830	9,840
81	10.8	56.2		11,460	11,450	10,405
82	11.0	56.9		12,080	12,070	10,975
83	11.1	57.6		12,700	12,690	11,535
84	11.2	58.3		13.320	13,310	12,100
85	11.4	59.0		13,940	13,930	12,665
86	11.5	59.7			14,550	13,230
87	11.6	60.4			15,170	13,800
88	11.8	61.1			15,790	14,360
89	11.9	61.9			16,410	14,925
90	12.0	62.5			17,030	15,485
91	12.1	63.2			17,650	16,050
92	12.2	63.9			18,270	16,615

To determine the quantity of clay necessary to increase the density of drilling mud from one weight to another, subtract the pounds necessary to make the lighter mud from the pounds required to make the fluid of the desired weight. This difference is the quantity necessary.

These figures do not account for the material that does not go into suspension which depends upon the purity of the mud.forming solids. It is usually necessary to add five or ten percent for crude pit clays.

Table 21T

Conversion Table for Hardness

Unit	Equivalent				
	Parts per Million	Grains per U.S. Gallon	Clark Degrees	French Degrees	German Degrees
One Part per million	1.0	0.058	0.07	0.10	0.056
One grain per U. S. gal.	17.1	1.00	1.20	1.71	.958
One Clark degree	14.3	.829	1.00	1.43	.80
One French degree	10.0	.583	.70	1.00	.56
One German degree	17.9	1.044	1.24	1.78	1.00

(One hydrotimetric degree = One French degree).

Parts per million and grains per US gallon as calcium carbonate.
Clark degrees are grains per Imperial gallon as calcium carbonate.
French degrees are parts per 100,000 as calcium carbonate.
German degrees are parts per 100,000 as calcium oxide.

FORMULA FOR CALCULATING TOTAL HARDNESS

Total Hardness (as ppm $CaCO_3$) = (ppm Calcium x 2.497) + (ppm Magnesium x 4.115) + (ppm Iron x 1.792) + (ppm Manganese x 1.822).

Table 22T

MUD WEIGHT CONVERSION TABLE

Lbs./Sq. In. Per 100' Depth	Lbs./Gal.	Lbs./Cu. Ft.	Specific Gravity	Lbs./Sq. In. Per 100' Depth	Lbs./Gal.	Lbs./Cu. Ft.	Specific Gravity
1 API -	*0.1925*	*1.44*	*.023108*				
40	7.7	57.6	.92	72	13.9	103.7	1.66
41	7.9	59.0	.94	73	14.1	105.1	1.69
42	8.1	60.5	.971	74	14.3	106.6	1.71
43	8.3	61.9	.993	75	14.4	108.0	1.73
*43.28	8.33	62.32	1.0000	76	14.6	109.4	1.76
**43.35	8.345	62.43	1.0018	77	14.8	110.9	1.78
44	8.5	63.4	1.02	78	15.0	112.3	1.80
45	8.7	64.8	1.04	79	15.2	113.8	1.83
46	8.9	66.2	1.06	80	15.4	115.2	1.85
47	9.1	67.7	1.09	81	15.6	116.6	1.87
48	9.2	69.1	1.11	82	15.8	118.1	1.89
49	9.4	70.6	1.13	83	16.0	119.5	1.92
50	9.6	72.0	1.16	84	16.2	121.0	1.94
51	9.8	73.4	1.18	85	16.4	122.4	1.96
52	10.0	74.9	1.20	86	16.6	123.8	1.99
53	10.2	76.3	1.22	87	16.8	125.3	2.01
54	10.4	77.8	1.25	88	16.9	126.7	2.03
55	10.6	79.2	1.28	89	17.1	128.2	2.06
56	10.8	80.6	1.29	90	17.3	129.6	2.08
57	11.0	82.1	1.32	91	17.5	131.0	2.10
58	11.2	83.5	1.34	92	17.7	132.5	2.13
59	11.4	85.0	1.36	93	17.9	133.9	2.15
60	11.6	86.4	1.39	94	18.1	135.4	2.17
61	11.7	87.8	1.41	95	18.3	136.8	2.20
62	11.9	89.3	1.43	96	18.5	138.2	2.22
63	12.1	90.7	1.46	97	18.7	139.7	2.24
64	12.3	92.2	1.48	98	18.9	141.1	2.26
65	12.5	93.6	1.50	99	19.1	142.6	2.29
66	12.7	95.0	1.53	100	19.3	144.0	2.31
67	12.9	96.5	1.55	101	19.4	145.4	2.33
68	13.1	97.9	1.57	102	19.6	146.9	2.36
69	13.3	99.4	1.59	103	19.8	148.3	2.38
70	13.5	100.8	1.62	104	20.0	149.8	2.40
71	13.7	102.2	1.64	105	20.2	151.2	2.43

* Corresponds to water at 20°C. = 68°F.

** Corresponds to 1 gm/cc; approximately to water at 4°C. = 39°F.

Table 23T

STRETCH DATA FOR DRILL PIPE, TABLE 30T TUBING AND CASING

Size of Tubings, D.P. or Casing	Length of Pipe Suspended in Well (Feet)	Stretch Per 1000 Lbs. Pull Above Weight of Pipe (Inches)	Pull Above Weight of Pipe Per In. Stretch of Pipe (Pounds)	Stretch Due To Own Weight Suspended in Water (Inches)
2" Upset Tubing 4.70 #/Ft.	500	.155	6,450	.16
	1,000	.310	3,225	.62
	2,000	.620	1,612	2.5
	3,000	.930	1,075	5.6
	4,000	1.240	806	10.
	5,000	1.550	644	16.
	10,000	3.100	322	64.
2½" Upset Tubing 6.50 #/Ft.	500	.110	9,080	.16
	1,000	.220	4,540	.62
	2,000	.440	2,270	2.5
	3,000	.660	1,513	5.6
	4,000	.880	1,135	10.
	5,000	1.100	908	16.
	10,000	2.200	454	64.
3" Upset Tubing 9.30 #/Ft.	500	.0784	12,750	.16
	1,000	.1568	6,375	.62
	2,000	.3136	3,187	2.5
	3,000	.4704	2,125	5.6
	4,000	.6262	1,593	10.
	5,000	.7840	1,274	16.
	10,000	1.568	637	64.
2⅞" Drill Pipe 10.40 #/Ft.	500	.070	14,300	.16
	1,000	.140	7,150	.62
	2,000	.280	3,575	2.5
	3,000	.420	2,383	5.6
	4,000	.560	1,787	10.
	5,000	.700	1,430	16.
	10,000	1.40	715	64.
3½" Drill Pipe 13.30 #/Ft.	500	.055	18,200	.16
	1,000	.110	9,100	.62
	2,000	.220	4,550	2.5
	3,000	.330	3,033	5.6
	4,000	.440	2,275	10.
	5,000	.550	1,820	16.
	10,000	1.10	910	64.
4½" Drill Pipe 16.60 #/Ft.	500	.0450	22,200	.16
	1,000	.0900	11,100	.62
	2,000	.180	5,550	2.5
	3,000	.270	3,700	5.6
	4,000	.360	2,775	10.
	5,000	.450	2,220	16.
	10,000	.900	1,110	64.
5½" Casing 17 #/Ft.	500	.0402	24,800	.16
	1,000	.0804	12,400	.62
	2,000	.160	6,230	2.5
	3,000	.240	4,133	5.6
	4,000	.320	3,100	10.
	5,000	.402	2,480	16.
	10,000	.804	1,240	64.
7" Casing 24#/Ft.	500	.0298	33,500	.16
	1,000	.0596	16,750	.62
	2,000	.119	8,375	2.5
	3,000	.179	5,583	5.6
	4,000	.238	4,187	10.
	5,000	.298	3,350	16.
	10,000	.596	1,675	64.

NOTE: The above figures apply only to pipe that has not been stretched, or is not being stretched beyond its elastic limit.

Table 24T

CAPACITIES OF SUCTION PITS IN CUBIC FEET PER INCH OF DEPTH

	6'0"	6'2"	6'4"	6'6"	6'8"	6'10"	7'0"	7'2"	7'4"	7'6"	7'8"	7'10"	8'0"	8'2"	8'4"	8'6"	8'8"
4'0"	2.00	2.06	2.11	2.18	2.22	2.25	2.33	2.39	2.45	2.50	2.56	2.61	2.67	2.72	2.78	2.83	2.89
4'2"	2.08	2.14	2.20	2.26	2.32	2.37	2.43	2.49	2.55	2.60	2.66	2.72	2.78	2.84	2.89	2.95	3.01
4'4"	2.17	2.23	2.29	2.35	2.41	2.47	2.53	2.59	2.65	2.71	2.77	2.83	2.89	2.95	3.01	3.07	3.13
4'6"	2.25	2.31	2.38	2.44	2.50	2.56	2.63	2.69	2.75	2.81	2.87	2.94	3.00	3.06	3.12	3.19	3.25
4'8"	2.33	2.40	2.46	2.53	2.59	2.66	2.73	2.79	2.85	2.92	2.98	3.05	3.11	3.18	3.24	3.31	3.37
4'10"	2.42	2.48	2.55	2.62	2.69	2.75	2.82	2.88	2.95	3.02	3.09	3.16	3.22	3.29	3.35	3.42	3.49
5'0"	2.50	2.57	2.64	2.71	2.78	2.85	2.92	2.98	3.06	3.12	3.19	3.27	3.33	3.41	3.47	3.54	3.61
5'2"	2.58	2.66	2.73	2.80	2.87	2.94	3.02	3.08	3.16	3.23	3.30	3.37	3.44	3.52	3.58	3.66	3.73
5'4"	2.67	2.74	2.82	2.89	2.97	3.04	3.12	3.18	3.26	3.33	3.40	3.48	3.56	3.63	3.70	3.78	3.85
5'6"	2.75	2.83	2.90	2.98	3.06	3.13	3.22	3.28	3.36	3.44	3.51	3.59	3.67	3.75	3.81	3.90	3.97
5'5"	2.83	2.91	2.99	3.07	3.15	3.23	3.31	3.38	3.46	3.54	3.62	3.70	3.78	3.86	3.93	4.01	4.09
5'10"	2.92	3.00	3.08	3.16	3.25	3.32	3.41	3.48	3.57	3.64	3.72	3.81	3.89	3.98	4.04	4.13	4.21

	8'10"	9'0"	9'2"	9'4"	9'6"	9'8"	9'10"	10'0"	10'2"	10'4"	10'6"	10'8"	10'10"	11'0"	11'2"	11'4"	11'6"
6'0"	4.42	4.50	4.58	4.67	4.75	4.83	4.92	5.00	5.08	5.17	5.25	5.33	5.42	5.50	5.58	5.67	5.75
6'2"	4.54	4.63	4.71	4.80	4.88	4.97	5.05	5.14	5.22	5.31	5.40	5.48	5.57	5.65	5.74	5.82	5.91
6'4"	4.66	4.75	4.84	4.93	5.01	5.10	5.19	5.28	5.36	5.45	5.54	5.63	5.72	5.81	5.89	5.98	6.07
6'6"	4.78	4.88	4.97	5.05	5.15	5.24	5.33	5.42	5.50	5.60	5.69	5.78	5.87	5.96	6.05	6.14	6.23
6'8"	4.91	5.00	5.10	5.18	5.28	5.37	5.46	5.56	5.65	5.74	5.83	5.93	6.02	6.11	6.20	6.30	6.39
6'10"	5.03	5.13	5.22	5.31	5.41	5.51	5.60	5.70	5.79	5.88	5.98	6.07	6.17	6.27	6.36	6.45	6.55
7'0"	5.15	5.25	5.35	5.44	5.54	5.64	5.73	5.83	5.93	6.03	6.13	6.22	6.32	6.42	6.51	6.61	6.71
7'2"	5.27	5.38	5.48	5.57	5.67	5.78	6.87	5.97	6.07	6.17	6.27	6.37	6.47	6.57	6.67	6.77	6.87
7'4"	5.39	5.50	5.61	5.70	5.81	5.91	6.01	6.11	6.21	6.31	6.42	6.52	6.62	6.72	6.82	6.92	7.03
7'6"	5.52	5.63	5.74	5.83	5.94	6.05	6.14	6.25	6.35	6.45	6.56	6.67	6.77	6.88	6.98	7.08	7.19
7'8"	5.64	5.75	5.86	5.96	6.07	6.18	6.28	6.39	6.50	6.60	6.71	6.81	6.92	7.03	7.13	7.24	7.35
7'10"	5.76	5.88	5.99	6.09	6.20	6.32	6.41	6.53	6.64	6.74	6.86	6.96	7.07	7.18	7.29	7.39	7.51

Table 25T

Lifting Power

3000 ' Velocity

Will handle chips to rate of 30' per hour;
more chips should require more air

HOLE SIZE	STEM SIZE	HOLE DEPTH AND VOLUME REQUIRED CFM						
		1000	2000	4000	6000	8000	10,000	12,000
12¼	6⅝	1700	1825	2200	2300	2500	2650	2800
	5½	1900	2100	2250	2450	2600	2800	2950
	4½	2100	2250	2400	2600	2700	2900	3050
11	6⅝	1250	1375	1600	1900	1980	2150	2300
	5½	1500	1575	1800	1950	2100	2300	2425
	4½	1600	1750	1900	2100	2200	2380	2500
9⅞	5½	1100	1220	1400	1480	1650	1820	2000
	5	1200	1290	1450	1600	1750	1900	2050
	4½	1280	1400	1530	1680	1920	1960	2090
9	5	940	1050	1200	1360	1480	1620	1730
	4½	1020	1080	1160	1400	1520	1645	1760
	3½	1150	1200	1320	1480	1600	1720	1840
8¾	5	880	960	1120	1250	1400	1520	1540
	4½	950	1040	1200	1320	1450	1590	1590
	3½	1070	1190	1290	1400	1520	1640	1760
7⅞	4½	700	820	950	1090	1200	1320	1420
	3½	860	920	1040	1160	1250	1360	1480
7⅝	3½	700	800	900	1010	1120	1210	1310
6¾	3½	575	640	760	850	960	1040	1130
6¼	3½	440	540	640	740	830	920	1000
	2⅞	540	600	680	780	880	940	1020
4¾	2⅞	280	320	400	460	560	600	660
	2⅜	300	330	430	480	590	620	680

Note: The lifting power is directly proportional to the density and to the square of the velocity. If velocities of 2000 feet per minute are desired (and this has been satisfactory, although it sacrifices penetration rate) a conversion can be made by multiplying the volume given above at desired hole and pipe size by two thirds.

The pressure requirements vary with the depth and the amount of water encountered.

Table 26T

ASCENDING VELOCITY OF MUD

(in Feet Per Minute)

Diameter (Inches)		Mud Circulation (Gallons Per Minute)									
Drill Pipe	Hole	100	200	300	400	500	600	700	800	900	1000
3 1/2	6 1/4	92	184	275							
	6 3/4	73	146	219	292						
	7 5/8	55	109	164	219	273	328				
	7 7/8	49	99	149	199	249	299	349			
	8 3/8	43	86	129	172	215	258	302	345		
	8 1/2	42	84	125	167	208	250	291	333		
	9	36	72	108	144	179	215	251	287	323	
	9 3/8	32	63	95	126	158	189	221	252	284	315
4 1/2	7 3/4	63	126	188	251	314					
	8 3/8	51	101	152	203	253	304				
	8 1/2	48	97	145	193	242	290	338			
	9	41	82	124	165	206	248	289	330		
	9 5/8	35	70	106	141	176	211	246	282	317	
	10 5/8	27	54	80	107	134	161	188	215	241	268
	11	25	50	75	100	125	149	174	199	224	249
	12 1/2	20	40	59	79	98	118	138	158	177	197
5 9/16	10 5/8	30	60	90	120	150	180	210	240	270	300
	11	28	55	83	111	139	166	194	222	249	277
	12 1/4	21	42	63	84	105	126	147	168	189	210
	13	18	36	54	72	90	108	126	144	162	180
	14 3/4	14	27	41	54	68	81	95	108	122	135
	17	10	20	30	40	50	60	70	80	90	100
	18	8	17	25	34	42	50	59	67	76	84
	20	7	13	20	26	33	39	46	52	59	65
6 5/8	11	33	65	98	130	163	195	228	260	293	325
	12 1/4	24	47	71	95	118	142	166	189	213	237
	13	20	40	60	80	100	119	139	159	179	199
	14 3/4	14	29	43	57	72	86	100	114	129	143
	17	10	20	31	41	51	61	71	82	92	102
	20	7	14	21	28	35	42	49	56	63	70
	22	6	11	17	22	28	34	39	45	50	56
	27	4	7	11	15	19	22	26	30	33	37

Table 27T

Drive Pipe—Dimensions, Weights and Test Pressures

Nom. Size, in.	Weights per Foot		Wall Thickness, in.	Diameters		Threads per Inch, No.	Couplings			Test Pressures, psi.	
	Nom. Thds. & Cplg., lb.	Calculated Plain Ends, lb.		Outside, in.	Inside, in.		Length, in.	Outside Diameter, in.	Calculated Weight, lb.	Lap Welded and Grade A	Electric Welded and Seamless Grade C
6	19.45	18.97	.280	6.625	6.065	8	5⅛	7.390	13.35	1200	2000
8	25.55	24.70	.277	8.625	8.071	8	6⅛	9.625	26.89	1200	1500
8	29.35	28.55	.322	8.625	7.981	8	6⅛	9.625	26.89	1200	1800
8	32.40	31.27	.354	8.625	7.917	8	6⅛	9.625	26.89	1200	2000
10	32.75	31.20	.279	10.750	10.192	8	6⅝	11.750	36.05	1000	1200
10	35.75	34.24	.307	10.750	10.136	8	6⅝	11.750	36.05	1000	1400
10	41.85	40.48	.365	10.750	10.020	8	6⅝	11.750	36.05	1000	1600
12	45.45	43.77	.330	12.750	12.090	8	6⅝	14.000	52.72	1000	1200
12	51.15	49.56	.375	12.750	12.000	8	6⅝	14.000	52.72	1000	1400
14 D.	57.00	54.57	.375	14.000	13.250	8	7⅛	15.000	50.22	950	1300
16 D.	65.30	62.58	.375	16.000	15.250	8	7⅛	17.000	57.17	850	1100

(a) The permissible variation in weight is 6½ per cent above and 3½ per cent below; but the carload weight shall not be more than 1¾ per cent under the nominal weight.

—Courtesy American Iron & Steel Institute

Table 28T

VOLUMETRIC REQUIREMENTS--FOR 3,000 FEET PER MINUTE ANNULAR VELOCITY (ROTARY AIR DRILLING FOR BLAST HOLES & SEISMOGRAPHIC DRILLING)

HOLE DIAM. INCHES	PIPE O.D. 2⅜" c.f.m.	PIPE O.D. 2⅞" c.f.m.	PIPE O.D. 3½" c.f.m.	PIPE O.D. 4½" c.f.m.	PIPE O.D. 5½" c.f.m.
3¾"	138	96			
3⅞"	153	111			
4"	171	129			
4⅛"	186	144			
4¼"	204	162	93		
4⅜"	219	177	108		
4½"	234	192	123		
4⅝"	249	207	138		
4¾"	267	225	156		
4⅞"	285	243	174		
5"	303	261	192		
5⅛"	321	279	210		
5¼"	342	300	231		
5⅜"	363	321	252		
5½"	387	345	276	153	
5⅝"	411	369	300	177	
5¾"	438	396	327	204	
5⅞"	468	426	357	234	
6"		456	387	264	
6⅛"		486	417	294	
6¼"			441	318	
6⅜"			468	342	
6½"			492	366	
6⅝"			519	393	
6¾"			546	420	
6⅞"			575	450	289
7"			606	480	318
7⅛"			633	507	346
7¼"			660	534	371
7⅜"			690	564	398
7½"			720	594	430
7⅝"			750	624	460
7¾"			780	654	489
7⅞"			813	687	520

Table 29T

VOLUMETRIC REQUIREMENTS--FOR 3,000 FEET PER MINUTE ANNULAR VELOCITY (ROTARY AIR DRILLING FOR GAS AND OIL WELL PRODUCTION DRILLING)

HOLE DIAM. D_2 IN.	PIPE O.D. D_1 IN.	Q Cu. Ft. per Min.	HOLE DIAM. D_2 IN.	PIPE O.D. D_1 IN.	Q Cu. Ft. per Min.
8½"	3 Inches	1353	13¾"	3	3270
	3½	1290		3½	3216
	4½	1181		4½	3084
	5½	987		5½	2919
	6⅝	764		6⅝	2730
8¾"	3	1414	15"	6⅝	3300
	3½	1361		7	3180
	4½	1231		8⅝	2748
	5½	1058	16"	6⅝	4140
	6⅝	836		7	3720
9½"	3	1682		8⅝	3582
	3½	1626	18½"	6⅝	5241
	4½	1494		7	5121
	5½	1321		8⅝	4683
	6⅝	1098	22"	6⅝	7527
9⅞"	3	1780		7	7407
	3½	1706		8⅝	6969
	4½	1674	24"	6⅝	9069
	5½	1401		7	8949
	6⅝	1177		8⅝	8511
10⅝"	3	2014	26"	6⅝	10662
	3½	1960		7	10542
	4½	1826		8⅝	10104
	5½	1652			
	6⅝	1429			
11"	3	2147			
	3½	2094			
	4½	1962			
	5½	1786			
	6⅝	1562			
12¼"	3	2606			
	3½	2573			
	4½	2440			
	5½	2260			
	6⅝	2037			

WETMOUTH FORMULA —

$$Q = K \times \frac{C2}{C1}\left(1 + (1 + C_1)^4\right)^{\frac{1}{2}}$$

K = Constant for a given annular velocity

C_1 and C_2 are functions of hole size, pipe size, sp. gravity of air or gas, annular velocity, and depth of hole.

These charts indicate only minimum air requirements. In order to allow for air line leaks, line pressure losses and for lifting moist cuttings and water when drilling through a moist area, add 30% to these figures.

In selecting compressor capacity to meet your specific requirements, select compressor or compressors having equal or greater than that indicated by the above chart.

Courtesy of Schramm, Inc.

Table 30T

TABLE 69T **STANDARD FUEL REQUIREMENTS FOR GOOD PUMPING PLANTS**

(Source: College of Agriculture, University of Nebraska)

Pumping Rate, in gpm	Head, in feet	Water Horse-power	Fuel or Energy Required: Diesel, gal per hr	Gasoline, gal per hr	Propane, gal per hr	Natural Gas, cu ft per hr	Electricity, kw-hr per hr
500	100	13	1¼	1½	2	190	14
	150	19	1¾	2¼	2¾	280	21
	200	25	2¼	3	3¾	380	29
700	100	18	1¾	2	2¾	270	20
	150	27	2½	3¼	4	400	30
	200	35	3¼	4¼	5¼	530	40
800	100	20	1¾	2½	3	300	23
	150	30	2¾	3½	4½	450	34
	200	40	3¾	4¾	6	610	46
1000	100	25	2¼	3	3¾	380	29
	150	38	3½	4½	5¾	570	43
	200	50	4½	6	7½	760	57

Table 31T

API Standard Round-Thread Tubing—
Dimensions, Weights and Test Pressures
Non-upset

Nom. Size, in.	Weights per Foot: Nom. Thd. & Cplg., lb.	Weights per Foot: Calculated: Plain Ends, lb.	Weights per Foot: Calculated: Thd. & Cplg., lb.	Wall Thickness, in.	Diameters: Outside, in.	Diameters: Inside, in.	Diameters: External Upset, in.	Couplings: Length, in.	Couplings: Outside Diameter, in.	Couplings: Cal. Weight, lb.	Test Pressures, psi. Grades: F-25	H-40	J-55	N-80
Non-Upset														
1½	2.75	2.72	2.75	.145	1.900	1.610	—	3¾	2.200	1.23	3000	3000	3000	3000
†2	4.00	3.94	4.02	.167	2.375	2.041	—	4¼	2.875	2.82	2800	3000	3000	3000
2	4.60	4.43	4.51	.190	2.375	1.995	—	4¼	2.875	2.82	3000	3000	3000	3000
2½	6.40	6.16	6.32	.217	2.875	2.441	—	5⅛	3.500	5.15	3000	3000	3000	3000
†3	7.70	7.58	7.85	.216	3.500	3.068	—	5⅝	4.250	8.17	2500	3000	3000	3000
3	9.20	8.81	9.06	.254	3.500	2.992	—	5⅝	4.250	8.17	2900	3000	3000	3000
3*	10.20	9.91	10.15	.289	3.500	2.922	—	5⅝	4.250	8.17	3000	3000	3000	3000
3½	9.50	9.11	9.42	.226	4.000	3.548	—	5¾	4.750	9.57	2300	3000	3000	3000
4	12.60	12.24	12.54	.271	4.500	3.958	—	6⅛	5.200	10.76	2400	3000	3000	3000

*** Supplied on special order only.. † Tentative.**

(a) The permissible variation in weight for any length of tubing is 6½ per cent above and 3½ per cent below; but the carload weight shall not be more than 1¾ per cent under calculated weight.

Table 32T

Square Feet of Surface per Lineal Foot of Pipe

On all lengths over 1 ft., fractions less than tenths are added to or dropped

Lgth. of pipe in ft.	Size of Pipe: ¾	1	1¼	1½	2	2½	3	4	5	6	7	8
1	.275	.346	.434	.494	.622	.753	.916	1.175	1.455	1.739	1.996	2.257
2	.5	.7	.9	1.	1.2	1.5	1.8	2.4	2.9	3.5	4.	4.5
3	.8	1.	1.3	1.5	1.9	2.3	2.7	3.5	4.4	5.2	6.	6.8
4	1.1	1.4	1.7	2.	2.5	3.	3.6	4.7	5.8	7	8.	9.
5	1.4	1.7	2.2	2.4	3.1	3.8	4.6	5.8	7.3	8.7	10.	11.3
6	1.6	2.1	2.6	2.9	3.7	4.5	5.5	7.	8.7	10.5	12.	13.5
7	1.9	2.4	3.	3.4	4.4	5.3	6.4	8.2	10.2	12.1	14.	15.8
8	2.2	2.8	3.5	3.9	5.	6.	7.3	9.4	11.6	13.9	16.	18.
9	2.5	3.1	3.9	4.4	5.6	6.8	8.2	10.6	13.1	15.7	18.	20.3
10	2.7	3.5	4.3	4.9	6.2	7.5	9.1	11.8	14.6	17.4	20.	22.6
11	3.	3.8	4.8	5.4	6.8	8.3	10.0	12.9	16.	19.1	22.	24.9
12	3.3	4.1	5.2	5.9	7.5	9	11.0	14.1	17.4	20.9	24.	27.1
13	3.6	4.5	5.6	6.4	8.1	9.8	11.9	15.3	18.9	22.6	26.	29.4
14	3.8	4.8	6.1	6.9	8.7	10.5	12.8	16.5	20.3	24.3	28.	31.6
15	4.1	5.2	6.5	7.4	9.3	11.3	13.7	17.6	21.8	26.1	30.	33.9
16	4.4	5.5	6.9	7.9	10.	12.	14.6	18.8	23.2	27.8	32.	36.1
17	4.7	5.9	7.4	8.4	10.6	12.8	15.5	20.	24.7	29.5	34.	38.4
18	5.	6.2	7.8	8.9	11.2	13.5	16.5	21.2	26.2	31.3	36.	40.6
19	5.2	6.6	8.3	9.4	11.8	14.3	17.4	22.3	27.6	33.1	38.	42.9
20	5.5	6.9	8.7	9.9	12.5	15.	18.3	23.5	29.1	34.8	40.	45.2
21	5.8	7.3	9.1	10.4	13.	15.8	19.2	24.7	30.5	36.5	42.	47.4
22	6.	7.6	9.6	10.9	13.7	16.5	20.2	25.9	32.	38.3	44.	49.7
23	6.3	8.	10.	11.3	14.3	17.3	21.1	27.	33.5	40.	46.	52.
24	6.6	8.3	10.4	11.9	14.9	18.	22.	28.2	34.9	41.7	48.	54.2
25	6.9	8.6	10.9	12.3	15.6	18.8	22.9	29.3	36.3	43.5	50.	56.4
26	7.1	9.	11.3	12.8	16.2	19.5	23.8	30.5	37.8	45.2	52.	58.6
27	7.4	9.4	11.7	13.3	16.8	20.3	24.7	31.7	39.3	47.	54.	61.
28	7.7	9.7	12.2	13.8	17.4	21.	25.6	32.9	40.7	48.7	56.	63.2
29	8.	10.	12.6	14.3	18.	21.8	26.6	34.1	42.2	50.4	58.	65.5
30	8.3	10.4	13.	14.8	18.7	22.5	27.5	35.3	43.6	52.1	60.	67.7
31	8.5	10.7	13.5	15.3	19.3	23.3	28.4	36.4	45.1	53.9	62.	70.
32	8.8	11.1	13.9	15.8	19.9	24.1	29.3	37.6	46.5	55.6	64.	72.2
33	9.1	11.4	14.3	16.3	20.5	24.8	30.2	38.8	48.	57.4	66.	74.4
34	9.4	11.7	14.7	16.8	21.2	25.6	31.1	40.	49.5	59.1	68.	76.7
35	9.6	12.1	15.2	17.3	21.8	26.3	32.	41.1	50.9	60.8	70.	79
36	9.9	12.5	15.6	17.8	22.4	27.	33.	42.3	52.4	62.6	72.	81.3
37	10.2	12.8	16.1	18.3	23.	27.8	33.9	43.5	53.8	64.3	74.	83.5
38	10.5	13.2	16.5	18.8	23.7	28.5	34.8	44.6	55.2	66.	76.	85.8
39	10.7	13.5	16.9	19.3	24.3	29.3	35.7	45.8	56.7	67.8	78.	88.
40	11.	13.8	17.4	19.8	24.9	30.1	36.6	47.	58.2	69.5	80.	90.2
41	11.3	14.2	17.8	20.3	25.2	30.8	37.6	48.2	59.6	71.3	82.	92.5
42	11.5	14.5	18.2	20.8	26.1	31.6	38.5	49.4	61.1	73.	84.	94.8
43	11.8	14.9	18.7	21.3	26.8	32.3	39.4	50.6	62.5	74.8	86.	97.
44	12.1	15.2	19.1	21.8	27.4	33.1	40.3	51.7	64.	76.5	88.	99.3
45	12.4	15.6	19.5	22.2	28.	33.8	41.2	52.9	65.5	78.2	90.	101.6
46	12.7	15.9	20.	22.7	28.6	34.6	42.2	54.	67.	80.	92.	103.8
47	12.9	16.3	20.4	23.2	29.2	35.3	43.	55.2	68.4	81.7	94.	106.
48	13.2	16.6	20.8	23.7	29.9	36.1	43.9	56.4	69.8	83.5	96.	105.4
49	13.5	17.	21.3	24.2	30.5	36.8	44.8	57.6	71.2	85.1	98.	110.5
50	13.8	17.3	21.7	24.7	31.1	37.6	45.8	58.7	72.7	87.	100.	112.8

—Courtesy Youngstown Sheet & Tube Co.

Table 33T

Pressure, Lb per Sq In. to Feet (Head) of Water
ft = 2.31 × lb/sq in.

TABLE 8T

Based on water at its greatest density (39.2°F)

Pressure Pounds Per Square Inch	Feet Head	Pressure Pounds Per Square Inch	Feet Head	Pressure Pounds Per Square Inch	Feet Head	Pressure Pounds Per Square Inch	Feet Head	Pressure Pounds Per Square Inch	Feet Head	Pressure Pounds Per Square Inch	Feet Head
1	2.31	53	122.43	105	242.55	157	362.67	209	482.79	261	602.91	365	843.15
2	4.62	54	124.74	106	244.86	158	364.98	210	485.10	262	605.22	370	854.70
3	6.93	55	127.05	107	247.17	159	367.29	211	487.41	263	607.53	375	866.25
4	9.23	56	129.36	108	249.48	160	369.60	212	489.72	264	609.84	380	877.80
5	11.55	57	131.67	109	251.79	161	371.91	213	492.03	265	612.15	385	899.35
6	13.86	58	133.98	110	254.10	162	374.22	214	494.34	266	614.45	390	900.90
7	16.17	59	136.29	111	256.41	163	376.53	215	496.65	267	616.77	395	912.45
8	18.48	60	138.60	112	258.72	164	378.84	216	498.96	268	619.08	400	924.00
9	20.79	61	140.91	113	261.03	165	381.15	217	501.27	269	621.39	405	931.55
10	23.10	62	143.22	114	263.34	166	383.46	218	503.58	270	623.70	410	947.10
11	25.41	63	145.53	115	265.65	167	385.77	219	505.89	271	626.01	415	958.65
12	27.72	64	147.84	116	267.96	168	388.08	220	508.20	272	628.32	420	970.20
13	30.03	65	150.15	117	270.27	169	390.39	221	510.51	273	630.63	425	981.75
14	32.34	66	152.46	118	272.58	170	392.70	222	512.82	274	632.94	430	993.20
15	34.65	67	154.77	119	274.89	171	395.01	223	515.13	275	635.25	435	1004.85
16	36.96	68	157.08	120	277.20	172	397.32	224	517.44	276	637.56	440	1016.40
17	39.27	69	159.39	121	279.51	173	399.63	225	519.75	277	639.87	445	1037.95
18	41.58	70	161.70	122	281.82	174	401.94	226	522.06	278	642.18	450	1039.50
19	43.89	71	164.01	123	284.13	175	404.25	227	524.37	279	644.49	455	1051.05
20	46.20	72	166.32	124	286.44	176	406.56	228	526.68	280	646.80	460	1062.60
21	48.51	73	168.63	125	288.75	177	408.87	229	528.99	281	649.11	465	1074.15
22	50.82	74	170.94	126	291.06	178	411.18	230	531.30	282	651.42	470	1085.70
23	53.13	75	173.25	127	293.37	179	413.49	231	533.61	283	653.73	475	1097.25
24	55.44	76	175.56	128	295.68	180	415.80	232	535.92	284	656.04	480	1108.80
25	57.75	77	177.87	129	297.99	181	418.11	233	538.23	285	658.35	485	1120.35
26	60.06	78	180.18	130	300.30	182	420.42	234	540.54	286	660.66	490	1131.90
27	62.37	79	182.49	131	302.61	183	422.73	235	542.85	287	662.97	495	1143.45
28	64.68	80	184.80	132	304.92	184	425.04	236	545.16	288	665.28	500	1155.00
29	66.99	81	187.11	133	307.23	185	427.35	237	547.47	289	667.59	525	1212.75
30	69.30	82	189.42	134	309.54	186	429.66	238	549.78	290	669.90	550	1270.50
31	71.71	83	191.73	135	311.85	187	431.97	239	552.09	291	672.21	575	1328.25
32	73.92	84	194.04	136	314.16	188	434.28	240	554.40	292	674.52	600	1386.00
33	76.23	85	196.35	137	316.47	189	436.59	241	556.71	293	676.83	625	1443.75
34	78.54	86	198.66	138	318.78	190	438.90	242	559.02	294	679.14	650	1501.50
35	80.85	87	200.97	139	321.09	191	441.21	243	561.33	295	681.45	675	1559.25
36	83.16	88	203.28	140	323.40	192	443.52	244	563.64	296	683.76	700	1617.00
37	85.47	89	205.59	141	325.71	193	445.83	245	565.95	297	686.07	725	1674.75
38	87.78	90	207.90	142	328.02	194	448.14	246	568.26	298	688.38	750	1732.50
39	90.09	91	210.21	143	330.33	195	450.45	247	570.57	299	690.69	775	1790.25
40	92.40	92	212.52	144	332.64	196	452.76	248	572.88	300	693.09	800	1848.00
41	94.71	93	214.83	145	334.95	197	455.07	249	575.19	305	704.55	825	1905.75
42	97.02	94	217.14	146	337.26	198	457.38	250	577.50	310	716.10	850	1963.50
43	99.33	95	219.45	147	339.57	199	459.69	251	579.81	315	727.65	875	2021.25
44	101.64	96	221.76	148	341.88	200	462.00	252	582.12	320	739.20	900	2079.00
45	103.95	97	224.07	149	344.19	201	464.31	253	584.43	325	750.75	925	2136.75
46	106.26	98	226.38	150	346.50	202	466.62	254	586.74	330	762.30	950	2194.50
47	108.57	99	228.69	151	348.81	203	468.93	255	589.05	335	773.85	975	2252.25
48	110.88	100	231.00	152	351.12	204	471.24	256	591.36	340	785.40	1000	2310.00
49	113.19	101	233.31	153	353.43	205	473.55	257	593.67	345	796.95	1500	3465.
50	115.50	102	235.62	154	355.74	206	475.86	258	595.98	350	808.50	2000	4620.
51	117.81	103	237.93	155	358.05	207	478.17	259	598.29	355	820.05	3000	6930.
52	120.12	104	240.24	156	360.36	208	480.48	260	600.60	360	831.60		

Table 34T

Pressure, Feet (head) of Water to Lb per Sq In.
lb/sq in. = .4331 × ft

Based on water at its greatest density (39.2°F.)

Feet Head	Pressure Pounds per Square Inch	Feet Head	Pressure Pounds per Square Inch	Feet Head	Pressure Pounds per Square Inch	Feet Head	Pressure Pounds per Square Inch	Feet Head	Pressure Pounds per Square Inch	Feet Head	Pressure Pounds per Square Inch
1	0.43	54	23.39	107	46.34	160	69.31	213	92.20	285	123.45
2	0.86	55	23.82	108	46.78	161	69.74	214	92.69	290	125.62
3	1.30	56	24.26	109	47.21	162	70.17	215	93.13	295	127.78
4	1.73	57	24.69	110	47.64	163	70.61	216	93.56	300	129.95
5	2.16	58	25.12	111	48.08	164	71.04	217	93.99	305	132.12
6	2.59	59	25.55	112	48.51	165	71.47	218	94.43	310	134.28
7	3.03	60	25.99	113	48.94	166	71.91	219	94.86	315	136.46
8	3.46	61	26.42	114	49.38	167	72.34	220	95.30	320	138.62
9	3.89	62	26.85	115	49.81	168	72.77	221	95.73	325	140.79
10	4.33	63	27.29	116	50.24	169	73.20	222	96.16	330	142.95
11	4.76	64	27.72	117	50.68	170	73.64	223	96.60	335	145.12
12	5.20	65	28.15	118	51.11	171	74.07	224	97.03	340	147.28
13	5.63	66	28.58	119	51.54	172	74.50	225	97.46	345	149.45
14	6.06	67	29.02	120	51.98	173	74.94	226	97.90	350	151.61
15	6.49	68	29.45	121	52.41	174	75.37	227	98.33	355	153.78
16	6.93	69	29.88	122	52.84	175	75.80	228	98.76	360	155.94
17	7.36	70	30.32	123	53.28	176	76.23	229	99.20	365	158.10
18	7.79	71	30.75	124	53.71	177	76.67	230	99.63	370	160.27
19	8.22	72	31.18	125	54.15	178	77.10	231	100.0	375	162.45
20	8.66	73	31.62	126	54.58	179	77.53	232	100.49	380	164.61
21	9.09	74	32.05	127	55.01	180	77.97	233	100.93	385	166.78
22	9.53	75	32.48	128	55.44	181	78.40	234	101.36	390	168.94
23	9.96	76	32.92	129	55.88	182	78.84	235	101.70	395	171.11
24	10.39	77	33.35	130	56.31	183	79.27	236	102.23	400	173.27
25	10.82	78	33.78	131	56.74	184	79.70	237	102.66	425	184.10
26	11.26	79	34.21	132	57.18	185	80.14	238	103.09	450	195.0
27	11.69	80	34.65	133	57.61	186	80.57	239	103.53	475	205.77
28	12.12	81	35.08	134	58.04	187	81.0	240	103.96	500	216.58
29	12.55	82	35.52	135	58.48	188	81.43	241	104.39	525	227.42
30	12.99	83	35.95	136	58.91	189	81.87	242	104.83	550	238.25
31	13.42	84	36.39	137	59.34	190	82.30	243	105.26	575	249.09
32	13.86	85	36.82	138	59.77	191	82.73	244	105.69	600	259.90
33	14.29	86	37.25	139	60.21	192	83.17	245	106.13	625	270.73
34	14.72	87	37.68	140	60.64	193	83.60	246	106.56	650	281.56
35	15.16	88	38.12	141	61.07	194	84.03	247	106.99	675	292.40
36	15.59	89	38.55	142	61.51	195	84.47	248	107.43	700	303.23
37	16.02	90	38.98	143	61.94	196	84.90	249	107.86	725	314.08
38	16.45	91	39.42	144	62.37	197	85.33	250	108.29	750	324.88
39	16.89	92	39.85	145	62.81	198	85.76	251	108.73	775	335.72
40	17.32	93	40.28	146	63.24	199	86.20	252	109.16	800	346.54
41	17.75	94	40.72	147	63.67	200	86.63	253	109.59	825	357.37
42	18.19	95	41.15	148	64.10	201	87.07	254	110.03	850	368.20
43	18.62	96	41.58	149	64.54	202	87.50	255	110.46	875	379.03
44	19.05	97	42.01	150	64.97	203	87.93	256	110.89	900	389.86
45	19.49	98	42.45	151	65.40	204	88.36	257	111.32	925	400.70
46	19.92	99	42.88	152	65.84	205	88.80	258	111.76	950	411.54
47	20.35	100	43.31	153	66.27	206	89.21	259	112.19	975	422.35
48	20.79	101	43.75	154	66.70	207	89.66	260	112.62	1000	433.18
49	21.22	102	44.18	155	67.14	208	90.10	261	113.06	1500	649.7
50	21.65	103	44.61	156	67.57	209	90.53	262	113.49	2000	866.3
51	22.09	104	45.05	157	68.0	210	90.96	270	116.96	3000	1,299.5
52	22.52	105	45.48	158	68.43	211	91.39	275	119.12		
53	22.95	106	45.91	159	68.87	212	91.83	280	121.29		

Table 35T

GENERAL METHODS OF SPECIFYING

In order to avoid confusion or misunderstanding, purchasers' inquiries and orders for tubular products should specify the following details:

1) Quantity (in linear feet, number of pieces or bundles, or weight).
2) Size.
3) Foot-weight or wall thickness.
4) Method of manufacture (welded, seamless, etc.).
5) Class of material (standard pipe, line pipe, etc.).
6) End finish (threaded and coupled, plain end, threaded only, etc.).
7) If plain end, method of joining to be used.
8) Grade of steel, where specifications provide this option.
9) Length (random, average, definite cut, or uniform).
10) Type of coating or lining, if any.
11) Applicable specifications, if any.
12) Purpose for which material is intended (flanging, bending, high temperature service, etc.).
13) Delivery date desired.
14) Inspection at mill by purchaser, if required.

—Courtesy American Iron & Steel Institute

Table 36T

CAPACITY OF PIPE, TUBING, AND DRILL PIPE

Nominal Size (Inches)	Diameters (Inches)		Gallons per Lin.Ft.	Lin. Ft. per Gallon	Sacks Cement* per Lin.Ft.	Lin.Ft. per Sk.Cement*
	O.D.	I.D.				
STANDARD PIPE						
4	4.500	4.026	.6613	1.5122	.0804	12.443
5	5.563	5.047	1.0391	.9624	.1263	7.919
6	6.625	6.065	1.5008	.6663	.1824	5.483
8	8.625	8.071	2.6577	.3763	.3230	3.096
8	8.625	7.981	2.5988	.3848	.3159	3.166
10	10.750	10.192	4.2382	.2359	.5151	1.941
10	10.750	10.136	4.1917	.2386	.5094	1.963
10	10.750	10.020	4.0963	.2441	.4978	2.009
12	12.750	12.090	5.9636	.1677	.7246	1.380
12	12.750	12.000	5.8752	.1702	.7140	1.401
14 OD	14.000	13.250	7.1629	.1396	.8705	1.149
15 OD	15.000	14.250	8.2849	.1207	1.0068	.9932
16 OD	16.000	15.250	9.4885	.1054	1.1531	.8672
17 OD	17.000	16.214	10.726	.0932	1.3034	.7672
18 OD	18.000	17.182	12.045	.0880	1.4639	.6831
20 OD	20.000	19.182	15.012	.0666	1.8244	.5481
TUBING						
1 1/4	1.660	1.380	.0777	12.870	.00944	105.93
1 1/2	1.900	1.610	.1057	9.455	.0129	77.82
2	2.375	1.995	.1623	6.158	.0197	50.68
2 1/2	2.875	2.441	.2431	4.113	.0295	33.85
3	3.500	2.992	.3652	2.737	.0444	22.52
3 1/2	4.000	3.476	.4929	2.028	.0599	16.69
4	4.500	3.958	.6391	1.564	.0777	12.87
INTERNAL UPSET DRILL PIPE**						
	2 3/8	2.000	.1632	6.1275	.0199	50.251
	2 3/8	1.815	.1344	7.4405	.0163	61.349
	2 7/8	2.323	.2202	4.5290	.0267	37.453
	2 7/8	2.151	.1888	5.2966	.0230	43.478
	3 1/2	2.900	.3431	2.9146	.0417	23.980
	3 1/2	2.764	.3117	3.2082	.0379	26.385
	4 1/2	3.958	.6392	1.5645	.0777	12.870
	4 1/2	3.826	.5972	1.6745	.0726	13.774
	5 9/16	4.975	1.0098	.9903	.1227	8.150
	5 9/16	4.859	.9633	1.0381	.1171	8.539
	5 9/16	4.733	.9140	1.0941	.1111	9.000
	6 5/8	6.065	1.5008	.6663	.1824	5.482
	6 5/8	5.965	1.4517	.6888	.1764	5.668
	6 5/8	5.761	1.3541	.7385	.1646	6.075

* Cement calculations based on the volume of an average cement mixture being 1.1 cubic feet per sack of cement.

** No allowance made for internal restrictions of upsets and tool joints.

Table 37T CAPACITY OF HOLE

Diameter of Hole (Inches)	Gallons per Lin.Ft.	Lin.Ft. per Gallon	Sacks Cement* per Lin.Ft.	Lin.Ft. per Sack Cement*
2	0.1632	6.1275	0.0199	50.2513
1/2	.2550	3.9216	.0311	32.1543
3	.3672	2.7233	.0444	22.5225
1/2	.4998	2.0008	.0607	16.4745
4	.6528	1.5319	.0791	12.6422
1/4	.7369	1.3570	.0893	11.1982
1/2	.8262	1.2104	.1006	9.9404
3/4	.9206	1.0862	.1118	8.9445
5	1.0200	.9804	.1240	8.0645
1/4	1.1246	.8892	.1367	7.3153
1/2	1.2342	.8102	.1500	6.6667
3/4	1.3489	.7413	.1639	6.1013
6	1.4688	.6808	.1785	5.6022
1/4	1.5938	.6276	.1937	5.1626
1/2	1.7238	.5801	.2095	4.7733
3/4	1.8590	.5379	.2259	4.4267
7	1.9992	.5002	.2430	4.1152
1/4	2.1445	.4663	.2606	3.8373
1/2	2.2950	.4357	.2789	3.5855
3/4	2.4505	.4081	.2978	3.3580
8	2.6112	.3830	.3173	3.1516
1/4	2.7769	.3601	.3375	2.9630
1/2	2.9478	.3392	.3583	2.7910
3/4	3.1237	.3201	.3796	2.6344
9	3.3048	.3026	.4016	2.4900
1/4	3.4910	.2865	.4243	2.3568
1/2	3.6822	.2716	.4475	2.2346
3/4	3.8785	.2578	.4714	2.1213
10	4.0800	0.2451	0.4958	2.0169
1/4	4.2865	.2333	.5209	1.9198
1/2	4.4982	.2223	.5467	1.8292
3/4	4.7150	.2121	.5730	1.7452
11	4.9368	.2026	.5999	1.6669
1/4	5.1637	.1937	.6276	1.5934
1/2	5.3958	.1853	.6557	1.5251
3/4	5.6329	.1775	.6846	1.4607
12	5.8752	.1702	.7140	1.4006
1/2	6.3750	.1569	.7748	1.2907
13	6.8952	.1450	.8380	1.1933
1/2	7.4358	.1345	.9036	1.1067
14	7.9968	.1251	.9718	1.0290
1/2	8.5782	.1166	1.0425	.9592
15	9.1800	.1089	1.1156	.8964
1/2	9.8022	.1020	1.1913	.8394
16	10.4448	.0957	1.2694	.7878
17	11.7912	.0848	1.4329	.6978
18	13.2192	.0756	1.6065	.6225
19	14.7288	.0679	1.7900	.5587
20	16.3200	.0613	1.9833	.5042
22	19.7472	.0506	2.3998	.4167
24	23.5008	.0426	2.8560	.3501
26	27.5808	.0363	3.3519	.2983
28	31.9872	.0313	3.8873	.2572
30	36.7200	.0272	4.4625	.2241
36	52.8768	.0189	6.4260	.1556

* Cement calculations based on the volume of an average cement mixture being 1.1 cubic foot per sack of cement.

—Courtesy Halliburton Oil Well Cementing Co.

Table 38T

DISPLACEMENT OF PIPE

(Displacement is amount of space taken up by metal in string of pipe.)

Size	O.D. (Ins.)	I.D. (Ins.)	Weight Per Ft. Complete (Lb.)	Displacement Per 100 Lin. Ft. Cu. Ft.	Displacement Per 100 Lin. Ft. Barrels (42 Gal.)	Sacks Cement Per 100 Lin. Ft. 15.0 Lb. Slurry	15.5 Lb. Slurry	16.0 Lb. Slurry	16.5 Lb. Slurry
TUBING Nominal (No allowance made for external upsets and couplings)									
2	2.375	1.995	4.60	0.91	0.16	0.7	0.8	0.8	0.9
2-1/2	2.875	2.441	6.40	1.26	0.22	1.0	1.1	1.1	1.2
3	3.500	2.992	9.20	1.80	0.32	1.4	1.5	1.6	1.7
DRILL PIPE O.D. (No allowance made for internal upsets and couplings)									
2-7/8	2.875	2.151	10.40	1.99	0.35	1.5	1.7	1.8	1.9
3-1/2	3.500	2.764	13.30	2.51	0.45	1.9	2.1	2.2	2.4
4-1/2	4.500	4.000	12.75	2.32	0.41	1.8	1.9	2.1	2.2
4-1/2	4.500	3.958	13.75	2.50	0.45	1.9	2.1	2.2	2.4
4-1/2	4.500	3.826	16.60	3.06	0.55	2.4	2.6	2.7	2.9
4-1/2	4.500	3.754	18.10	3.36	0.60	2.6	2.8	3.0	3.2
5-9/16	5.563	4.859	22.20	4.00	0.71	3.1	3.3	3.6	3.8
CASING O.D. (No allowance made for couplings)									
4-1/2	4.500	4.090	9.50	1.92	0.34	1.5	1.6	1.7	1.8
5	5.000	4.494	13.00	2.62	0.47	2.0	2.2	2.3	2.5
5	5.000	4.408	15.00	3.04	0.54	2.4	2.5	2.7	2.9
5-1/2	5.500	5.012	14.00	2.80	0.50	2.2	2.3	2.5	2.7
5-1/2	5.500	4.950	15.50	3.14	0.56	2.4	2.6	2.8	3.0
5-1/2	5.500	4.892	17.00	3.45	0.61	2.7	2.9	3.1	3.3
5-1/2	5.500	4.778	20.00	4.05	0.72	3.1	3.4	3.6	3.9
6	6.000	5.424	18.00	3.59	0.64	2.8	3.0	3.2	3.4
6-5/8	6.625	5.921	24.00	4.82	1.08	3.7	4.0	4.3	4.6
7	7.000	6.538	17.00	3.41	0.61	2.6	2.8	3.1	3.3
7	7.000	6.456	20.00	3.99	0.71	3.1	3.3	3.6	3.8
7	7.000	6.366	23.00	4.62	0.82	3.6	3.9	4.1	4.4
7	7.000	6.276	26.00	5.24	0.93	4.1	4.4	4.7	5.0
7-5/8	7.625	6.969	26.40	5.22	0.93	4.0	4.4	4.7	5.0
8-5/8	8.625	8.097	24.00	4.82	0.86	3.7	4.0	4.3	4.6
8-5/8	8.625	8.017	28.00	5.52	0.98	4.3	4.6	4.9	5.3
8-5/8	8.625	7.921	32.00	6.35	1.13	4.9	5.3	5.7	6.1
9-5/8	9.625	8.921	36.00	7.12	1.27	5.5	5.9	6.4	6.8
10-3/4	10.750	10.192	32.75	6.37	1.14	4.9	5.3	5.7	6.1
10-3/4	10.750	10.050	40.50	7.94	1.41	6.1	6.6	7.1	7.6
13-3/8	13.375	12.715	48.00	9.39	1.67	7.3	7.8	8.4	9.0
13-3/8	13.375	12.615	54.50	10.8	1.92	8.4	9.0	9.6	10.3
16	16.000	15.250	65.00	12.8	2.28	10.0	10.6	11.4	12.2

—Courtesy Universal Atlas Cement Co.

Table 39T

TABLE OF SUCTION LIFTS

Altitude Above Sea Level	Barometric Inches Of Mercury	Pressure Lbs. Per Sq. Inch	Equivalent Head Of Water (Ft)	Practical Suction Lift of Pump* Recip-rocal	Centri-fugal
Sea Level	30.00	14.70	33.95	22.0	15.0
500 ft.	29.43	14.40	33.26	21.5	14.5
750	29.14	14.30	32.93	21.3	14.3
1,000	28.88	14.20	32.63	21.0	14.0
1,250	28.60	14.00	32.32	20.7	13.8
1,500	28.33	13.90	32.01	20.5	13.5
1,750	28.07	13.80	31.72	20.3	13.3
2,000	27.81	13.60	31.42	20.0	13.0
3,000	26.77	13.20	30.25	19.0	12.0
4,000	25.76	12.60	29.11	18.0	11.0

* - Practical lift is equal to the vertical distance to which water is to be lifted plus friction head and any other losses.

—Courtesy Johnson National Drillers Journal

The limit of lift of a shallow well ejector or "jet" pump is similar to that shown for centrifugal pumps. Deep well jet pumps, with the nozzle or jet installed in the lower end of the pump column, are generally designed for lifts not exceeding 100 to 120 feet. The manufacturer's specifications should be consulted in any event before planning an ejector or "jet" pump installation.

Table 40T

ADVANTAGES AND DISADVANTAGES OF VARIOUS TYPES OF PUMPS

(Source: U.S. Dept. of Agriculture)

Advantages	*Disadvantages*
PLUNGER TYPE:	
Positive action (force) Wide range of speed Efficient over wide range of capacity Simple construction Suitable for hand or power operation May be used on almost any depth of well Discharge relatively constant regardless of head	Discharge pulsates Subject to vibration Deep-well type must be set directly over well Sometimes noisy
TURBINE TYPE:	
Simple design Discharge steady Suitable for direct connection to electric motor Practically vibrationless Quiet operation May be either horizontal or vertical	Must have very close clearance Subject to abrasion damage Not suitable for hand operation Speed must be relatively constant Must be set down near or in water in deep well Requires relatively large-bore well
CENTRIFUGAL TYPE:	
Simple design Quiet operation Steady discharge Efficient when pumping large volumes of water Suitable for direct connection to electric motor or for belt drive May be either horizontal or vertical	Low efficiency in low capacities Low suction-lift (6 to 8 ft.) Must be set down near or in water Requires relatively large-bore well Not suitable for hand operation Discharge decreases somewhat as discharge pressure increases
ROTARY TYPE:	
Positive action Occupies little space Wide range of speed Steady discharge	Subject to abrasion Likely to get noisy Not satisfactory for deep wells
EJECTOR TYPE:	
Simple construction Suitable either for deep or shallow wells Need not be set directly over well Quiet operation Especially suitable for use with pressure system	Jet nozzle subject to abrasion and clogging Limited to wells 120 feet or less in depth Discharge decreases somewhat as discharge pressure increases
CHAIN TYPE:	
Simple Easily installed Self-priming	Inefficient Limited to shallow wells Likely to be unsanitary
HYDRAULIC RAM:	
Simple design Low cost Uses water for power Requires little attention	Wastes water Likely to be noisy Not satisfactory for intermittent operation
SIPHON:	
Low cost Requires no mechanical or hand power except for starting	Limited to moving water to lower levels Requires absolutely airtight pipes

Table 41T

VOLUME & HEIGHT BETWEEN CASING & HOLE

Size of Casing	Diameter of Hole (inches)	Gallons per Lin.Ft.	Lin.Ft. per Gallon	Sacks Cement* per Lin.Ft.	Lin.Ft. per Sack Cement*
O.D. 4.500" or 4 1/2"	5	0.1933	5.1600	0.0236	42.3729
	6	.6426	1.5562	.0781	12.8041
	7	1.1730	.8525	.1426	7.0126
	8	1.7850	.5602	.2169	4.6104
	10	3.2538	.3073	.3954	2.5291
O.D. 5.500" or 5 1/2"	6	.2346	4.2626	.0285	35.0877
	7	.7650	1.3072	.0930	10.7527
	8	1.3770	.7262	.1673	5.9773
	9	2.0700	.4830	.2516	3.9746
	10	2.8458	.3514	.3458	2.8918
O.D. 6.625" or 6 5/8"	7	.2085	4.7962	.0253	39.5257
	8	.8204	1.2189	.0997	10.0301
	9	1.5140	.6605	.1840	5.4348
	10	2.2892	.4368	.2782	3.5945
	12	4.0845	.2448	.4964	2.0145
O.D. 8.625" or 8 5/8"	9	.2697	3.7078	.0327	30.5810
	10	1.0449	.9569	.1270	7.8740
	11	1.9017	.5258	.2311	4.3271
	12	2.8401	.3521	.3452	2.8969
	14	4.9615	.2016	.6030	1.6584
O.D. 10.750" or 10 3/4"	12	1.1603	.8618	.1410	7.0922
	13	2.1803	.4587	.2650	3.7736
	14	3.2818	.3047	.3988	2.5075
	16	5.7298	.1746	.6963	1.4362
	18	8.5043	.1176	1.0335	.9676
O.D. 13"	14	1.1016	.9078	.1339	7.4683
	15	2.2848	.4377	.2777	3.6010
	16	3.5496	.2817	.4314	2.3180
	17	4.8960	.2042	.5950	1.6807
	20	9.4248	.1061	1.1454	.8731
O.D. 14"	15	1.1832	.8452	.1438	6.9541
	16	2.4480	.4085	.2975	3.3613
	17	3.7944	.2635	.4611	2.1687
	18	5.2224	.1915	.6347	1.5755
	20	8.3232	.1201	1.0115	.9867
O.D. 16"	17	1.3464	.7427	.1636	6.1125
	18	2.7744	.3604	.3372	2.9656
	19	4.2840	.2334	.5206	1.9209
	20	5.8752	.1702	.7140	1.4006
	22	9.3024	.1075	1.1305	.8846
O.D. 20"	21	1.6728	.5978	.2033	4.9191
	22	3.4272	.2918	.4165	2.4009
	23	5.2632	.1900	.6396	1.5634
	24	7.1808	.1393	.8727	1.1459
	26	11.2608	.0888	1.3685	.7307

* Cement calculations based on the volume of an average cement mixture being 1.1 cubic foot per sack of cement.

Table 42T

Pressure in Pounds per Square Inch to Feet of Head

Pounds Pressure	Ft. of Head	Pounds Pressure	Ft. of Head
1	2.31	19	43.9
2	4.62	20	46.2
3	6.93	25	57.7
4	9.24	30	69.3
5	11.6	35	80.8
6	13.9	40	92.4
7	16.2	45	103.9
8	18.5	50	115.5
9	20.8	55	127.
10	23.1	60	138.6
11	25.4	65	150.1
12	27.7	70	161.7
13	30.	75	173.2
14	32.3	80	184.8
15	34.6	85	196.3
16	37.	90	207.9
17	39.3	95	219.4
18	41.6	100	230.9

Table 43T

AIR LIFT PUMPS

Quantity of Air Required:

$$\text{Cubic feet of free air per gallon of water} = \frac{H}{C \times \log_{10}\left(\frac{S+34}{34}\right)}$$

where H = total lift in feet; S = submergence of eductor pipe in feet when pumping; and C = a constant from the following table:

Submergence (in %)	75	70	65	60	55	50	45	40	35
C*	366	358	348	335	318	296	272	246	216
C**	330	322	306	285	262	238	214	185	162

* - Air pipe outside of eductor pipe.
** - Air pipe inside of eductor pipe.

Starting and Operating Pressures:

Starting Pressure (lbs/sq in) = 0.434 x Starting Submergence (S_s in ft)

Operating Pressure (lbs/sq in) = 0.434 x Submergence (S) in feet + friction drop (lbs/sq in) in air pipe from compressor to foot piece

Velocities:

With a straight eductor pipe the best discharge velocity of the mixture of air and water for lifts from 40 to 200 feet varies from 2000 feet per minute at 35 percent submergence to 700 feet per minute at 70 percent submergence.

With a tapered eductor pipe (smaller diameter at foot piece) the best discharge velocity is 1400 feet per minute at 35 percent submergence and 550 feet per minute at 70 percent submergence.

The best velocity for the mixture of water and air at the entrance to the bottom of the eductor pipe is 800 feet per minute at 35 percent submergence and 450 feet per minute at 70 percent submergence.

The proper size of eductor pipe can be computed from these velocities, using the formula Q = A x V, where Q is the combined volume of water and air (with allowance for pressure in calculating the air volume), A is the cross-sectional area of the eductor pipe, and V is the velocity of the air-water mixture.

Ordinary practice places velocities of the compressed air in the air pipe at from 1800 to 2400 feet per minute. The diameter of the air pipe can be computed by the formula:

$$d = 13.54\sqrt{\frac{Q}{V}}$$

where d is the diameter of the pipe in inches, Q is the volume of air in cubic feet per minute passing through the pipe, and V the velocity of flow of the air in feet per minute.

—Courtesy Compressed Air Magazine

Table 44T

TABLE 14T PIPE, CYLINDER OR HOLE CAPACITY

Diameter (Inches)	Gallons Per Foot
1½	0.09
2	0.16
2½	0.25
3	0.37
4	0.67
6	1.47
8	2.61
10	4.08
12	5.86
16	10.45
18	13.20
20	16.35
24	23.42

Table 45T

AIR LIFT PUMPS

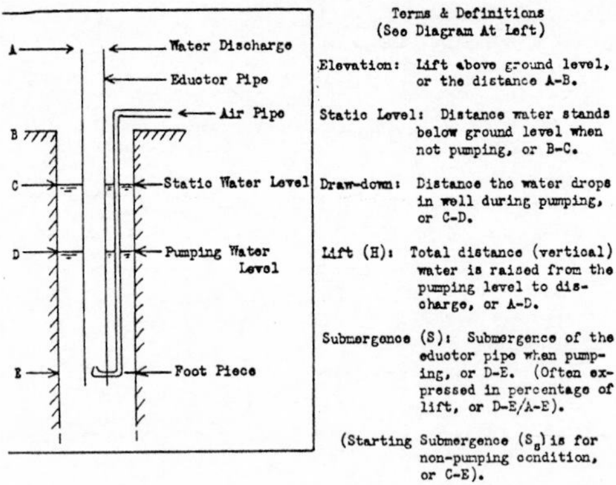

CUSTOMARY ALLOWABLE AND BEST SUBMERGENCES (S)

Lift in Feet (H)	Customary Allowable Percentage Submergence	Best Percentage Submergence	Type Compressor
20	55 to 70 %	65 to 70 %	Single Stage
30	55 to 70	65 to 70	"
40	50 to 70	65 to 70	"
50	50 to 70	65 to 70	"
60	50 to 70	65 to 70	"
80	50 to 70	65 to 70	"
100	45 to 70	65 to 70	"
125	45 to 65	65	"
150	40 to 65	60 to 65	"
175	40 to 60	55 to 60	"
200	40 to 60	55 to 60	Compound
250	40 to 60	55 to 60	"
300	37 to 55	50 to 55	"
350	37 to 55	50 to 55	"
400	37 to 50	45 to 50	"
450	35 to 45	40 to 45	"
500	35 to 45	40 to 45	"
550	35 to 45	40 to 45	"
600	35 to 45	40 to 45	"
650	35 to 45	40 to 45	"
700	35 to 40	40	"

—Courtesy Compressed Air Magazine

Table 46T

PUMPING PLANT PERFORMANCE STANDARDS

(Source: College of Agriculture, University of Nebraska)

Type of Power Unit	Standard Consumption of Fuel or Energy per Water Horsepower*
Diesel engine	0.091 gal per hr
Gasoline engine	0.116 gal per hr
Propane engine	0.145 gal per hr
Natural gas	160 cu ft per hr
Electric motor	0.885 kw-hr per hr

*Based on pump efficiency of 75 percent.

Table 47T

DEEP WELL TURBINE PUMPS

Deep well turbine pumps consist essentially of (1) a power unit (usually a hollow-shaft electric motor), (2) a suitable shaft and bearings, (3) a series of impellers, mounted in the "bowl assembly" at the lower end of the column pipe, which force the water upwards, and (4) a discharge column pipe which houses the shaft and bearings and which carries the water to the surface. The bearings along the shaft may be either water- or oil-lubricated.

Turbine pumps can be specially designed for almost any pumping situation by adding or subtracting to the number of impellers, varying the size of the impellers, and varying the size of the shaft and power unit.

The general practice, however, is to drill a well of sufficient diameter to accomodate the most efficient pump. The diameter of the well should also be great enough to permit installation of an airline alongside the pump column so that water level measurements can be made easily.

The manufacturer's representative should be consulted in advance for detailed specifications or information concerning any turbine pump installation.

ORDERING PUMPS

In ordering or inquiring about any type of pump, the following information should be specified whenever known:

(1) The exact inside diameter of the smallest part of the casing or hole in which the pump is to be installed. (Do not guess at this measurement).

(2) The static water level in the well, in feet below land surface or other reference point.

(3) The desired yield, in gallons per minute.

(4) The draw-down in feet or the lowest water level expected when pumping at the desired yield. (This should be either measured during a test of the well, or carefully estimated from the best available information on similar wells nearby).

(5) The desired water pressure at ground level (i.e. the operating pressure of any pressure tank, the height of any elevated tank or standpipe, etc.).

(6) The type of power available. If electricity, specify fully--such as 220 volts, 3 phase, alternating current.

(7) The total depth of the well.

(8) Be sure that well is straight at least as deep as the pump is to be set.

Table 48T

WINDMILL PUMPING CAPACITY

(Source: Water Systems Council, Manual of Water Supply Equipment)

[Based on wind velocity of 20 mph]

Representative Cylinder Size, inches	6½ foot mill		8 foot mill		10 foot mill	
	Depth to Water, ft.	Gals. per Hour	Depth to Water, ft.	Gals. per Hour	Depth to Water, ft.	Gals. per Hour
1-11/16	144	100	224	130	384	165
2	100	160	156	200	243	240
2-1/2	70	230	108	260	169	300
3	48	330	75	400	117	475
3-1/2	30	475	52	550	81	625
4	–	–	40	750	63	850
4-1/2	–	–	–	–	42	1000

Table 49T

DETERMINING COST OF PUMPING WATER

Frequently it is desirable to determine the cost of pumping water within reasonably close limits. The following formulas will give the approximate cost of operation per hour for any given condition, based upon the cost per kilowatt hour, per gallon of fuel or cubic foot of gas.

(1) **Electric Power.**—

$$\text{Cost per hour of operation} = \frac{\text{GPM} \times \text{Total Dynamic Head in Feet} \times .746 \times \text{rate per KWHr.}}{3960 \times \text{Overall Pump Efficiency} \times \text{Motor Efficiency}}$$

For example assume 1000 GPM, 80-foot total dynamic head, 70% Overall Pump Efficiency, 90% Motor Efficiency and cost of 2c per KWHr.

$$\frac{1000 \times 80 \times .746 \times .02}{3960 \times .70 \times .90} = 47.84\text{c per hour}$$

(2) **Diesel Engines** (based upon an average fuel requirement of .065 gallons of Diesel fuel per HP Hour).

$$\text{Cost per hour of operation} = \frac{\text{GPM} \times \text{Total Dynamic Head in Feet} \times .065 \times \text{Cost of fuel per gal.}}{3960 \times \text{Overall Pump Efficiency}}$$

For example assume 1000 GPM, 80-foot total dynamic head, 70% Overall Pump Efficiency and cost of fuel at 8c per gallon.

$$\frac{1000 \times 80 \times .065 \times .08}{3960 \times .70} = 15\text{c per hour}$$

(3) **Gasoline Engines** (based upon an average fuel requirement of .110 gallons of gasoline per horsepower hour).

$$\text{Cost per hour of operation} = \frac{\text{GPM} \times \text{Total Dynamic Head in Feet} \times .110 \times \text{Cost of fuel per gal.}}{3960 \times \text{Overall Pump Efficiency}}$$

For example assume 1000 GPM, 80-foot total Dynamic Head, 70% Overall Pump Efficiency and cost of gasoline at 20c per gallon.

$$\frac{1000 \times 80 \times .110 \times .20}{3960 \times .70} = 63\text{c per hour}$$

The above formulas give the approximate operating costs per hour without taking into consideration the initial cost of the plant, maintenance or man-hour cost of operating the plant.

—Courtesy Layne & Bowler, Inc.

Table 50T

PUMP EFFICIENCY

Symbols used:

- T.H.P. = Theoretical horsepower or water horsepower (actual work performed).
- B.H.P. = Brake horsepower, or power applied to pump shaft.
- O.A.E. = Overall efficiency or "wire-to-water" efficiency.
- KW = Kilowatts
- HP = Horsepower
- P.E. = Pump efficiency (including transmission efficiencies of any belts, chains, etc.) expressed as a decimal.
- M.E. = Motor efficiency or ratio of output of power to input of current.
- G.P.M. = U.S. Gallons per minute
- H = Total head in feet (including friction losses, discharge pressure, etc.).

$$\text{T.H.P.} = \frac{\text{G.P.M.} \times \text{H}}{3960}$$

$$\text{B.H.P.} = \frac{\text{T.H.P.}}{\text{P.E.}} = \frac{\text{G.P.M.} \times \text{H}}{3960 \times \text{P.E.}}$$

$$\text{P.E.} = \frac{\text{T.H.P.}}{\text{B.H.P.}}$$

$$\text{KW Input to Motor} = \frac{\text{B.H.P.} \times 0.746}{\text{M.E.}}$$

$$\text{O.A.E.} = \frac{\text{T.H.P.}}{\text{HP Input to Motor}} = \frac{\text{T.H.P.} \times 0.746}{\text{KW Input to Motor}} = \text{PE} \times \text{ME}$$

B.H.P. of DC motor = Volts x Amperes x 1.34 x Efficiency

B.H.P. of AC motor = Volts x Amperes x cos θ x K x 1.34 x Eff.

where cos θ = power factor of motor (generally about 0.82) and K = 1 (for single phase), K = 2 (for two-phase), and K = 1.732 (for three-phase) current.

Table 51T

Permissible Velocities in Open Channels Erosion and Deposition of Silt

Material	Bottom Velocities, in Feet Per Second at Which		
	Transportation Begins	Material in Equilibrium	Deposition Begins
Coarse Sand	1.07	0.71	0.62
Gravel (size of average pea)	0.71	0.62	
Gravel (size of small bean)	1.56	1.07	0.71
Shingle, rounded, 1" or more in Diameter	3.2	2.14	1.56
Flints, size of hen's egg	4.0	3.2	2.14

Layne & Bowler, Inc.

Table 52T

Table Showing KW/Hr. Values for 100 Feet of Field Head With Various Overall Efficiencies

Wire to Water Efficiency In %	KWHrs. per 1000 Gallons	Wire to Water Efficiency In %	KWHrs. per 1000 Gallons
20	1.57	51	.61568
21	1.4952	52	.6038
22	1.4272	53	.5924
23	1.3652	54	.5814
24	1.3083	55	.5709
25	1.256	56	.5607
26	1.2076	57	.5508
27	1.1629	58	.5413
28	1.1214	59	.5322
29	1.0827	60	.5233
30	1.0466	61	.51475
31	1.0129	62	.5064
32	.98125	63	.4984
33	.9515	64	.4906
34	.9235	65	.4830
35	.8971	66	.4757
36	.8722	67	.4686
37	.8486	68	.4617
38	.8263	69	.4550
39	.8051	70	.4485
40	.7875	71	.4422
41	.7658	72	.4361
42	.7476	73	.4301
43	.7203	74	.4243
44	.7136	75	.41866
45	.6977	76	.4131
46	.6826	77	.4077
47	.6680	78	.4025
48	.6541	79	.3974
49	.6408	80	.3925
50	.628		

To Arrive at KWHr. Value for Heads Other Than 100 Feet Proceed as Follows:

1st Example: — 200 foot head 60 Wire to Water Eff.
KWHr. for 100 Ft. @ .60 eff. = .5233

For 200 Ft. KWHrs. $= \frac{200}{100} \times .5233 = 1.0466$

2nd Example: — 35 foot head .50 Wire to Water Eff.
KWHr. for 100 Ft. @ .50 eff. = .628

For 35 Ft. KWHrs. $= \frac{35}{100} \times .628 = .2198$

—Courtesy Layne & Bowler, Inc.

Table 53T

Velocities of Water Which Will Cause Sand to Rise

Diameter of Grain, Inches	Diam. of Grain Millimeters	Retained on size Screen Number (Tyler Std. Scale)	Velocity of Water, Feet Per Second
Up to 0.0097	Up to 0.25	60	0.0 to 0.10
0.0097 to 0.0194	0.25 to 0.50	32	0.12 to 0.22
0.0194 to 0.0368	0.50 to 1.00	16	0.25 to 0.33
0.0368 to 0.0780	1.00 to 2.00	9	0.37 to 0.56
0.0780 to 0.1590	2.00 to 4.00	5	0.60 to 2.60

W. S. Tyler Co.

Table 54T

ESTIMATING FLOW FROM VERTICAL PIPE OR CASING

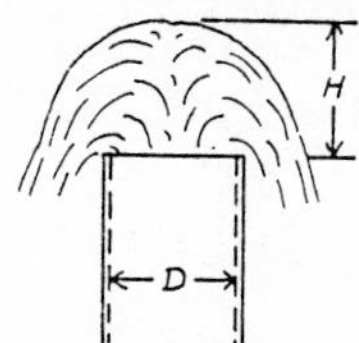

The approximate flow from vertical pipes or casings can be determined by measuring the maximum height (H) in inches to which the water jet rises above the pipe, and the inside diameter of the pipe (D) in inches.

The flow in gallons per minute is given in the following table for different sizes of Standard Pipe and for different heights of the water jets.

Height (H) in Inches	Nominal Diameter of Standard Pipe (Inches)						
	2	3	4	5	6	7	8	10
3	35	77	135	217	311	425	569	950
3 ½	38	85	149	238	341	465	626	1065
4	41	92	161	252	369	503	687	1115
4 ½	44	98	172	270	396	540	733	1200
5	47	104	182	286	420	575	779	1280
5 ½	49	109	192	301	444	606	825	1350
6	52	115	202	316	469	638	872	1415
6 ½	54	121	211	331	490	667	913	1475
7	57	126	219	345	509	700	949	1530
8	61	135	236	370	548	751	1025	1640
9	65	144	251	396	585	802	1095	1740
10	69	153	265	418	621	850	1155	1840
12	78	169	294	463	685	933	1275	2010
14	83	184	319	502	740	1020	1330	2170
16	89	197	342	540	796	1090	1480	2320
18	95	209	364	575	845	1160	1560	2460
20	101	221	386	607	890	1225	1645	2600
25	113	249	433	680	998	1375	1840	2900
30	124	273	476	746	1095	1505	2010	3180
35	135	298	516	810	1175	1630	2160	3420
40	145	318	561	865	1270	1745	2320	3680

For other pipe sizes and heights of jet, use the formula:

$$GPM = 5.68 \times C \times D^2 \times \sqrt{H}$$

where GPM = gallons per minute; D = inside diameter of pipe in inches; H = height of jet in inches; and C = a constant varying from 0.87 to 0.97 for pipes of 2 to 6 inches in diameter and heights of from 6 to 24 inches.

Table 55T

STANDARD PIPE—MILL PRACTICES

(1) Dimensions and weights shown in the tables are basic and are subject to standard mill tolerances.

(2) Pipe weights are figured on the basis of one cubic inch of steel weighing 0.2833 lb.

(3) The standard weight per foot of pipe with threads and couplings is based on a length of twenty feet over-all when the coupling is pulled tight.

(4) The outside diameter of a given size of pipe is the same regardless of the weight per foot. Variations in weight or wall thickness affect the inside diameter only.

(5) Standard pipe requiring threads is threaded to the A.S.A. Standard B2 of latest issue. Any pipe ordered to a special specification will be threaded to that specification, such as A.P.I.

(6) For plain-end pipe ordered beveled for welding, it is standard practice to bevel to an angle of 30° to 35° on the outside, with an average width of flat at the end of the pipe of $\frac{1}{16}$ inch, plus or minus $\frac{1}{32}$ inch. The angle is measured from a line perpendicular to the axis of the pipe. Pipe with bevels other than above cannot be supplied from stock and if required, is furnished on mill shipments only.

(7) Standard pipe sizes ⅛ in. to 12 in. are known by nominal inside diameter, and weight per foot should be specified. Pipe sizes 14 in. and over (line pipe) are known by the outside diameter, and when ordered, the desired wall thickness or weight per foot must be specified.

(8) Standard pipe will be cut to a definite length when so ordered.

Table 56T

ESTIMATING DISCHARGE FROM A HORIZONTAL PIPE FLOWING FULL

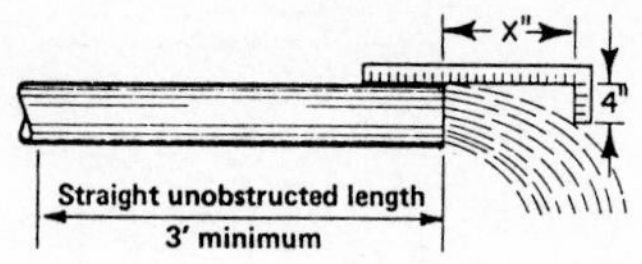

DISCHARGE RATE (Gallons Per Minute)

Horizontal Distance, x (Inches)	Nominal Pipe Diameter (Inches)				
	1	1¼	1½	2	2½
4	5.7	9.8	13.3	22.0	31.3
5	7.1	12.2	16.6	27.5	39.0
6	8.5	14.7	20.0	33.0	47.0
7	10.0	17.1	23.2	38.5	55.0
8	11.3	19.6	26.5	44.0	62.5
9	12.8	22.0	29.8	49.5	70.0
10	14.2	24.5	33.2	55.5	78.2
11	15.6	27.0	36.5	60.5	86.0
12	17.0	29.0	40.0	66.0	94.0
13	18.5	31.5	43.0	71.5	102.0
14	20.0	34.0	46.5	77.0	109.0
15	21.3	36.3	50.0	82.5	117.0
16	22.7	39.0	53.0	88.0	125.0

Table 57T
Explosives

TABLE 60T

Properties of Straight and Ammonia Dynamites

Type	Grade	Bulk Strength	Density: Ctgs per 50 lb 1¼"x 8" (1)	Velocity ft per sec	Water Resistance	Fumes (2)
Du Pont Straight	15–35% 40–50% 60%	15–35% 40–50% 60%	102 102–104 106	8,200–12,800 13,800–16,100 18,200	Poor Good Excellent	Fair Very Poor Very Poor
"Red Cross" Extra	15–35% 40–50% 60%	11–29% 35–43% 55%	110 110 110	7,400– 9,600 10,400–11,600 12,800	Fair Good Good	Fair Fair Fair
Du Pont Extra	A–B C–D E–F G–H	55–50% 45–40% 35–30% 25–20%	115–120 128–135 142–152 162–172	10,800–10,500 9,900– 9,500 9,300– 9,000 8,900– 8,800	Fair Fair Poor Poor	Good (3) Good (3) Good (3) Fair
Du Pont Extra	A-1 B-1 C-1 D-1 E-1 F-1 G-1	55–50% 45–40% 35–30% 25%	115–120 128–135 142–152 162	8,700– 8,500 7,700– 7,500 7,200– 6,900 6,500	Fair Poor Poor Poor	Good (3) Good (3) Good (3) Fair
"Red Cross" Blasting FR	25–30% 40–65%	14–16% 21–33%	125–128 (4) 133–143 (4)	3,800– 4,000 4,400– 5,100	Very Poor Very Poor	Very Poor Very Poor

NOTES: (1) Subject to a variation of plus or minus 3% from standard.
(2) Grades rated with "Very Poor" fumes are not recommended for underground use.
(3) This fume rating for the DuPont "Extra" dynamites applies only to special wrappers; in standard wrappers the fume rating is "Fair."
(4) Not made in 1¼"x 8" size cartridge counts for comparison only; these explosives are usually bag-packed.

Table 58T

ESTIMATING DISCHARGE FROM VERTICAL PIPE OR CASING

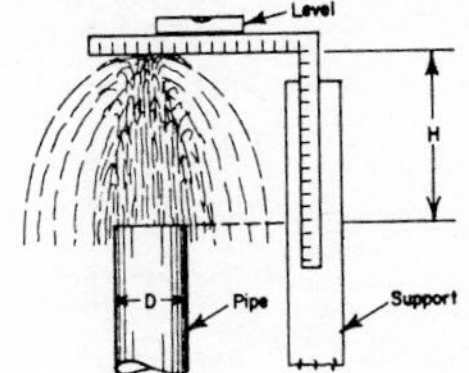

DISCHARGE RATE (Gallons Per Minute)

Height, H (Inches)	Nominal Pipe Diameter, D (Inches)		
	2	3	4
1½	22	43	68
2	26	55	93
3	33	74	130
4	38	88	155
5	44	99	175
6	48	110	190
8	56	125	225
10	62	140	255
12	69	160	280
15	78	175	315
18	85	195	350
21	93	210	380
24	100	230	400

Table 59T

Properties of Gelatin and Semi-Gelatin Dynamites

Type	Grade	Bulk Strength	Density Ctgs per 50 lb 1¼"x 8" (1)	Velocity ft per sec	Water Resistance	Fumes (2)
Du Pont Gelatin	20-60% 75-90%	30-59% 67-79%	85-96 101-107	10,500-19,700 (3) 20,600-22,300 (3)	Excellent Excellent	Excellent Very Poor
"Hi-Velocity" Gelatin	40-50% 60% 70-80%	30-38% 50% 58-68%	96-100 107 113-120	16,700-18,000 19,700 20,300-21,600	Excellent Excellent Excellent	Very Good Very Good Very Poor
Special Gelatin	30-60% 75-80% 90%	35-57% 66-70% 79%	89-98 102-107 100	13,800-15,400 (3) 16,400-17,100 (3) 19,700 (3)	Excellent Excellent Excellent	Excellent Very Good Poor
"Gelex"	No. 1 No. 2 No. 3 No. 4 No. 5	60% 45% 40% 35% 30%	110 122 130 140 150	13,100 12,600 12,300 11,800 11,300	Very Good Very Good Very Good Good Good	Excellent Excellent Excellent Excellent Excellent

NOTES: (1) Subject to a variation of plus or minus 3% from standard.

(2) Grades rated with "Very Poor" fumes are not recommended for underground use.

(3) The velocities shown for these gelatin dynamites are the high values developed when detonated unconfined with a straight dynamite primer.

—Courtesy E. I. DuPont de Nemours & Co.

Table 60T

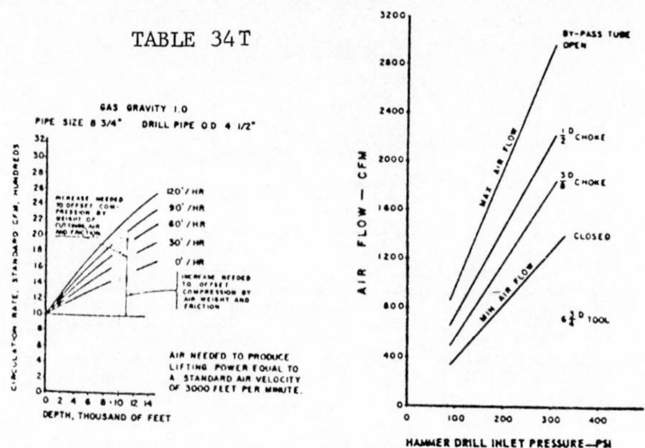

Courtesy of Mission Mfg. Co.

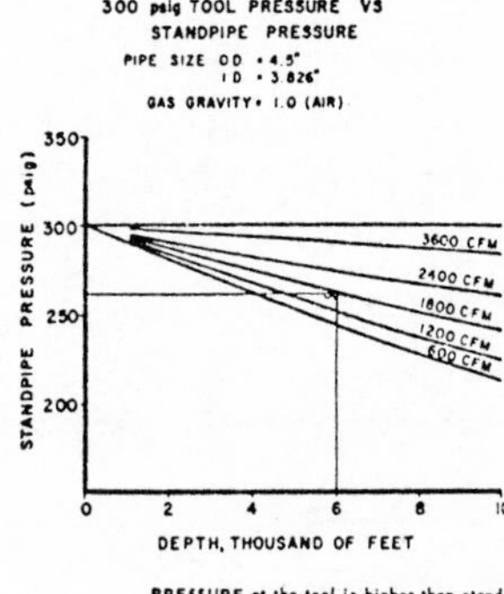

PRESSURE at the tool is higher than standpipe pressure when drill pipe friction is low; surface pressure should be cut back.

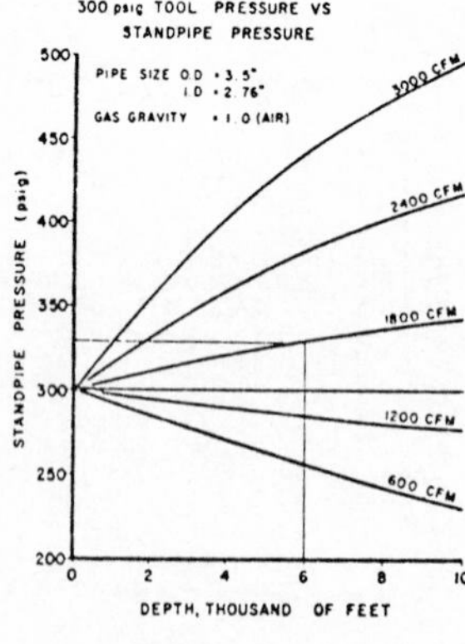

HIGHER SURFACE PRESSURE is necessary at large air rates when a small drill pipe causes high frictional losses.

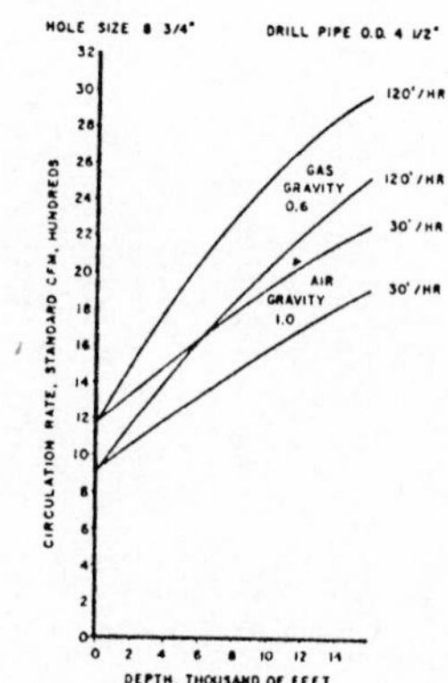

FLOW RATE MUST be higher for low density gas to obtain equal lifting power.

Table 61T

ORIFICE TABLES

For measurement of water through pipe orifices with free discharge.

The following tables have been compiled by the Engineering Department of Layne and Bowler, Incorporated, from original calibrations by Purdue University.

Head in Inches	3" Orifice		4" Orifice		5" Orifice		6" Orifice		7" Orifice	8" Orifice	Head in Inches
	4 in. Pipe	6 in. Pipe	6 in. Pipe	8 in. Pipe	6 in. Pipe	8 in. Pipe	8 in. Pipe	10 in. Pipe	10 in. Pipe	10 in. Pipe	
5	100	76	145	140	280	220	380	320			5
5.5	104	79	153	145	293	230	394	333			5.5
6	108	82	160	150	305	240	408	345			6
6.5	111	85	167	155	316	250	421	358			6.5
7	115	88	172	160	328	260	433	370			7
7.5	119	91	179	165	339	270	446	383			7.5
8	122	94	185	170	350	280	458	395	600	935	8
8.5	125	96	190	175	361	289	471	408	617	963	8.5
9	128	99	195	180	372	298	483	420	633	992	9
9.5	130	102	200	185	383	307	495	433	650	1016	9.5
10	133	104	205	190	393	316	508	445	666	1040	10
10.5	137	107	210	195	402	324	521	458	682	1060	10.5
11	140	109	215	200	412	330	533	470	698	1080	11
11.5	143	111	220	204	421	338	545	480	713	1100	11.5
12	146	114	225	208	430	346	556	490	728	1120	12
12.5	149	116	230	212	439	354	567	500	743	1139	12.5
13	151	118	234	216	448	362	578	510	757	1158	13
13.5	154	121	239	219	457	369	589	520	771	1176	13.5
14	157	123	243	224	465	376	599	530	785	1194	14
14.5	159	126	247	227	473	383	609	540	799	1212	14.5
15	162	128	250	231	480	390	618	550	812	1230	15
15.5	164	130	254	234	488	396	627	559	825	1248	15.5
16	167	132	257	238	495	402	636	568	838	1266	16
16.5	170	134	261	241	503	408	645	577	851	1284	16.5
17	172	136	264	245	510	414	654	586	863	1302	17
17.5	175	138	268	249	517	420	663	595	875	1319	17.5
18	178	140	271	252	524	426	672	604	887	1336	18
18.5	180	142	275	256	530	432	681	612	899	1353	18.5
19	183	144	278	259	536	438	690	620	910	1370	19
19.5	185	146	282	263	542	444	699	628	922	1387	19.5
20	187	148	285	266	548	449	708	636	933	1404	20
20.5	190	150	289	270	554	455	717	643	945	1421	20.5
21	192	152	292	273	560	460	726	650	956	1438	21
21.5	195	154	295	275	566	465	735	657	968	1455	21.5
22	197	156	299	279	572	470	744	664	979	1471	22
22.5	199	158	302	282	578	475	752	671	990	1486	22.5
23	201	160	305	285	584	479	760	678	1001	1500	23
23.5	203	162	307	288	590	484	768	685	1012	1515	23.5
24	205	164	310	291	596	488	776	692	1022	1529	24
24.5	207	165	314	294	602	492	784	699	1033	1543	24.5
25	210	167	317	297	608	496	791	706	1043	1557	25
25.5	212	169	320	300	614	500	798	713	1059	1571	25.5
26	214	171	323	301	620	504	805	720	1064	1585	26
26.5	216	173	326	305	626	508	812	727	1074	1599	26.5
27	219	174	329	308	632	512	818	734	1084	1613	27
27.5	221	176	332	311	638	516	825	741	1094	1627	27.5
28	222	177	335	314	644	520	831	747	1104	1641	28
28.5	224	179	337	317	650	524	838	754	1114	1655	28.5
29	226	180	340	320	656	528	844	760	1124	1669	29
29.5	228	182	343	323	662	532	851	767	1134	1683	29.5
30	230	183	346	325	668	536	857	773	1143	1697	30
30.5	232	185	348	328	674	540	863	780	1153	1711	30.5
31	235	186	351	330	680	544	869	786	1162	1725	31
31.5	236	188	354	333	686	548	876	793	1172	1739	31.5
32	239	189	357	335	692	552	882	799	1181	1753	32
32.5	240	191	360	338	697	556	889	806	1191	1767	32.5
33	242	192	363	340	703	560	895	812	1200	1791	33
33.5	244	194	366	342	709	564	901	818	1209	1795	33.5
34	246	195	369	345	715	568	907	824	1218	1809	34
34.5	248	196	372	247	720	572	913	830	1227	1823	34.5
35	250	197	375	349	726	576	919	836	1235	1837	35
35.5	252	198	377	351	732	580	925	842	1243	1851	35.5
36	254	200	380	354	737	584	931	847	1251	1865	36
36.5	256	201	383	356	743	588	937	852	1259	1879	36.5
37	257	203	385	358	748	592	943	857	1266	1893	37
37.5	259	204	388	360	754	596	949	862	1274		37.5
38	260	205	390	363	759	600	955	867	1281		38
38.5	262	206	393	365	765	604	961	872	1289		38.5
39	263	208	396	367	770	608	967	877	1295		39
39.5	265	209	398	369	776	612	974	882	1304		39.5
40	266	210	401	371	781	616	979	887	1311		40
40.5	267	211	403	373	786	620	985	891	1319		40.5
41	269	212	406	375	790	624	990	896	1326		41
41.5	271	213	408	378	795	628	996	901	1334		41.5
42	272	214	411	380	800	631	1001	906	1341		42
42.5	274	216	413	382	805	635	1007	910	1349		42.5
43	275	217	415	384	810	638	1012	915	1356		43
43.5	277	218	418	386	815	642	1018	920	1364		43.5
44	278	219	420	388	820	645	1023	925	1371		44
44.5	280	220	422	390	824	649	1029	929	1379		44.5
45	281	222	425	392	828	652	1034	934	1387		45
45.5	283	223	427	394	832	656	1040	939	1394		45.5
46	284	224	429	396	837	659	1045	944	1401		46
46.5	285	225	432	399	842	663	1051	948	1409		46.5
47	287	227	434	401	847	666	1056	953	1416		47
47.5	289	228	437	403	851	669	1062	958	1424		47.5
48	290	229	440	405	855	672	1067	963	1431		48
48.5	292	230	442	407	859	676	1073	967	1439		48.5
49	293	231	444	409	863	679	1078	972	1446		49
49.5	294	232	446	411	868	683	1084	977	1454		49.5
50	296	234	448	413	872	686	1089	982	1461		50
50.5	298	235	450	415	876	690	1095	986	1469		50.5
51	300	236	453	417	880	693	1100	991	1476		51
51.5	301	237	455	419	884	697	1105	996	1484		51.5
52	307	238	457	421	888	700	1110	1000	1491		52
52.5	303	239	459	423	892	704	1115	1005	1499		52.5
53	304	240	461	425	896	707	1120	1009	1506		53
53.5	305	241	463	427	900	711	1125	1014	1513		53.5
54	307	243	465	429	904	714	1130	1018	1520		54
54.5	309	244	467	431	908	718	1135	1023	1527		54.5
55	310	246	469	433	912	721	1140	1027	1534		55
55.5	311	247	471	435	915	725	1145	1032	1541		55.5
56	313	248	472	437	919	727	1150	1036	1548		56
56.5	314	249	474	439	923	730	1155	1040	1554		56.5
57	315	250	476	441	927	733	1160	1044	1560		57
57.5	316	251	478	443	930	736	1165	1046	1567		57.5
58	317	252	480	445	934	739	1170	1052	1574		58
58.5	319	253	482	447	938	742	1175	1056	1580		58.5
59	320	254	485	449	942	745	1180	1060	1586		59
59.5	321	256	487	451	945	748	1185	1064	1592		59.5
60	323	257	489	453	948	751	1190	1068	1598		60
60.5	324	258	491	455	951	754	1195	1072			60.5
61	325	259	492	457	955	757	1200	1076			61
61.5	326	261	494	459	958	760	1205	1080			61.5
62	328	262	496	461	961	763	1209	1084			62
62.5	329	263	498	463	964	766	1214	1088			62.5
63	330	264	500	465	968	769	1218	1092			63
63.5	331	265	502	467	971	772	1223	1096			63.5
64	333	266	504	469	974	775	1227	1099			64
64.5	334	267	507	471	977	778	1232	1103			64.5
65	335	268	509	472	981	781	1236	1106			65
65.5	336	269	511	474	984	784	1241	1110			65.5
66	338	271	513	475	988	787	1245	1113			66
66.5	339	272	515	477	991	790	1250	1117			66.5
67	340	273	517	479	995	793	1254	1120			67
67.5	341	274	518	481	998	796	1259	1124			67.5
68	343	275	520	483	1002	799	1263	1127			68

—Courtesy Layne & Bowler, Inc.

Table 62T

THEORETICAL DISCHARGE OF NOZZLES IN U. S. GALLONS PER MINUTE

Head		Velocity of Discharge	Diameter of Nozzle in Inches									
Pounds	Feet	Feet per Sec.	1/16	1/8	3/16	1/4	3/8	1/2	5/8	3/4	7/8	1
10	23.1	38.6	0.37	1.48	3.32	5.91	13.3	23.6	36.9	53.1	72.4	94.5
15	34.6	47.25	0.45	1.81	4.06	7.24	16.3	28.9	45.2	65.0	88.5	116.
20	46.2	54.55	0.52	2.09	4.69	8.35	18.8	33.4	52.2	75.1	102.	134.
25	57.7	61.0	0.58	2.34	5.25	9.34	21.0	37.3	58.3	84.0	114.	149.
30	69.3	66.85	0.64	2.56	5.75	10.2	23.0	40.9	63.9	92.0	125.	164.
35	80.8	72.2	0.69	2.77	6.21	11.1	24.8	44.2	69.0	99.5	135.	177.
40	92.4	77.2	0.74	2.96	6.64	11.8	26.6	47.3	73.6	106.	145.	189.
45	103.9	81.8	0.78	3.13	7.03	12.5	28.2	50.1	78.2	113.	153.	200.
50	115.5	86.25	0.83	3.30	7.41	13.2	29.7	52.8	82.5	119.	162.	211.
55	127.0	90.4	0.87	3.46	7.77	13.8	31.1	55.3	86.4	125.	169.	221.
60	138.6	94.5	0.90	3.62	8.12	14.5	32.5	57.8	90.4	130.	177.	231.
65	150.1	98.3	0.94	3.77	8.45	15.1	33.8	60.2	94.0	136.	184.	241.
70	161.7	102.1	0.98	3.91	8.78	15.7	35.2	62.5	97.7	141.	191.	250.
75	173.2	105.7	1.01	4.05	9.08	16.2	36.4	64.7	101.	146.	198.	259.
80	184.8	109.1	1.05	4.18	9.39	16.7	37.6	66.8	104.	150.	205.	267.
85	196.3	112.5	1.08	4.31	9.67	17.3	38.8	68.9	108.	155.	211.	276.
90	207.9	115.8	1.11	4.43	9.95	17.7	39.9	70.8	111.	160.	217.	284.
95	219.4	119.0	1.14	4.56	10.2	18.2	41.0	72.8	114.	164.	223.	292.
100	230.9	122.0	1.17	4.67	10.05	18.7	42.1	74.7	117.	168.	229.	299.

NOTE:—The actual quantities will vary from these figures, the amount of variation depending upon the shape of nozzle and size of pipe at the point where the pressure is determined. With smooth taper nozzles the actual discharge is about 94 per cent of the figures given in the tables.

—Courtesy Fairbanks, Morse & Co.

Table 63T
Numerical Relationship Between Permeability and Conductivity

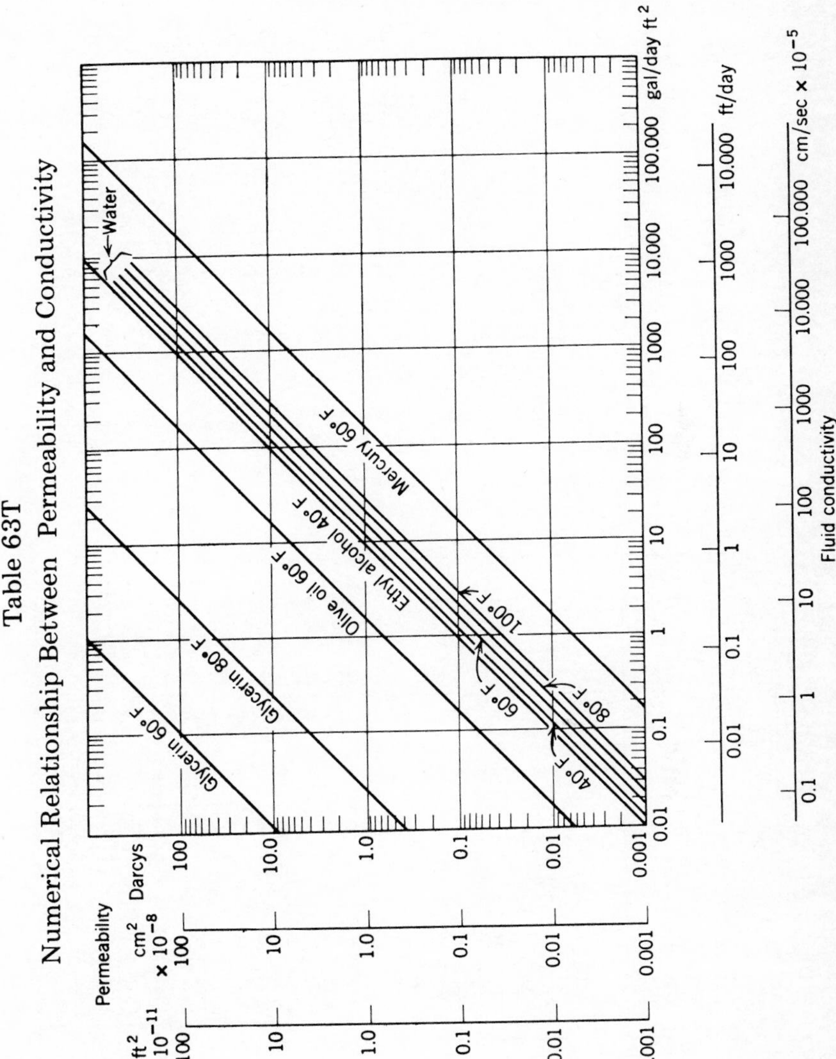

Table 64T

equivalent sieve mesh sizes

British Standard 410-1969		U.S. Standard (1924) and ASTM (E11-61) designation			U.S. Tyler (1910)		German Standard (DIN 1171-1926)		
Mesh No.	Sieve Aperture mm.	Mesh No.	ASTM designation microns	Sieve Aperture mm.	Mesh No.	Sieve Aperture mm.	Mesh per cm	Mesh per cm²	Sieve Aperture mm.
					2·5	7·925			
					3	6·680			
		3·5	5,660	5·66	3·5	5·613			
		4	4,760	4·76	4	4·699			
		5	4,000	4·00	5	3·962			
5	3·35	6	3,360	3·36	6	3·327			
6	2·80	7	2,830	2·83	7	2·794			
7	2·36	8	2,380	2·38	8	2·362			
8	2·00	10	2,000	2·00	9	1·981			
10	1·70	12	1,680	1·68	10	1·651			
							4	16	1·50
12	1·40	14	1,410	1·41	12	1·397			
14	1·18	16	1,190	1·19	14	1·168	5	25	1·20
							6	36	1·04
16	1·00	18	1,000	1·00	16	0·991			
18	0·850	20	841	0·841	20	0·833			
							8	64	0·75
22	0·710	25	707	0·707	24	0·701			
25	0·600	30	595	0·595	28	0·589	10	100	0·60
							11	121	0·54
30	0·500	35	500	0·500	32	0·495	12	144	0·50
36	0·425	40	420	0·420	35	0·417	14	196	0·43
							16	256	0·38
44	0·355	45	354	0·354	42	0·351			
52	0·300	50	297	0·297	48	0·295	20	400	0·30
60	0·250	60	250	0·250	60	0·246	24	576	0·25
72	0·212	70	210	0·210	65	0·208			
							30	900	0·20
85	0·180	80	177	0·177	80	0·175			
100	0·150	100	149	0·149	100	0·147	40	1,600	0·15
120	0·125	120	125	0·125	115	0·124			
							50	2,500	0·12
150	0·106	140	105	0·105	150	0·104			
							60	3,600	0·102
170	0·090	170	88	0·088	170	0·089	70	4,900	0·088
200	0·075	200	74	0·074	200	0·074	80	6,400	0·075
240	0·063	230	63	0·063	250	0·061	100	10,000	0·060
300	0·053	270	53	0·053	270	0·053			
350	0·045	325	44	0·044	325	0·043			
400	0·038	400	37	0·037	400	0·038			

Opening Inches		Mesh	Diameter of Wire, Decimal of an Inch	U. S. Series Equivalents (Fine Series)	
Tyler Standard Screen Scale $\sqrt{2}$	For Closer Sizing Ratio $\sqrt[4]{2}$			MICRON Designation	Number
3	------	------	.207	------	----
2	------	------	.192	------	----
1.5	------	------	.162	------	----
1.050	1.050	------	.148	------	----
------	.883	------	.135	------	----
.742	.742	------	.135	------	----
------	.624	------	.120	------	----
.525	.525	------	.105	------	----
------	.441	------	.105	------	----
.371	.371	------	.092	------	----
------	.312	2½	.088	------	----
.263	.263	3	.070	------	----
------	.221	3½	.065	5660	3½
.185	.185	4	.065	4760	4
------	.156	5	.044	4000	5
.131	.131	6	.036	3360	6
------	.110	7	.0328	2830	7
.093	.093	8	.032	2380	8
------	.078	9	.033	2000	10
.065	.065	10	.035	1680	12
------	.055	12	.028	1410	14
.046	.046	14	.025	1190	16
------	.0390	16	.0235	1000	18
.0328	.0328	20	.0172	840	20
------	.0276	24	.0141	710	25
.0232	.0232	28	.0125	590	30
------	.0195	32	.0118	500	35
.0164	.0164	35	.0122	420	40
------	.0138	42	.0100	350	45
.0116	.0116	48	.0092	297	50
------	.0097	60	.0070	250	60
.0082	.0082	65	.0072	210	70
------	.0069	80	.0056	177	80
.0058	.0058	100	.0042	149	100
------	.0049	115	.0038	125	120
.0041	.0041	150	.0026	105	140
------	.0035	170	.0024	88	170
.0029	.0029	200	.0021	74	200
------	.0024	250	.0016	62	230
.0021	.0021	270	.0016	53	270
------	.0017	325	.0014	44	325
.0015	.0015	400	.001	37	400

—Courtesy W. S. Tyler Co.

Table 65T

WELL SCREEN SELECTION CHART FOR SMALL-CAPACITY WELLS

(Source: Edward E. Johnson, Inc.)

GRADATION OF SAND	AVERAGE SLOT SIZE	MINIMUM SUGGESTED LENGTH for corresponding screen diameter* and desired well yield					
	Thousandths of an inch	1¼" SCREEN	2" SCREEN	3" SCREEN	4" SCREEN	5" SCREEN	6" SCREEN
VERY FINE SAND 6 - 7 - 8 Slot About the finest material that can be utilized for a water supply. A line composed of 12 grains would measure about 1/16".	7	300 gph – 5 ft. 450 gph – 8 ft. 600 gph – 12 ft.	450 gph – 6 ft. 600 gph – 9 ft. 900 gph – 13 ft.	600 gph – 8 ft. 900 gph – 11 ft. 1200 gph – 14 ft.	600 gph – 6 ft. 1200 gph – 10 ft. 1800 gph – 14 ft.	900 gph – 6 ft. 1200 gph – 9 ft. 1800 gph – 11 ft.	1200 gph – 8 ft. 2000 gph – 12 ft. 2400 gph – 15 ft.
FINE SAND 9 - 10 - 12 Slot Often called "sugar sand". Line of 6 or 7 average grains measures 1/16".	10	300 gph – 4 ft. 450 gph – 6 ft. 600 gph – 9 ft.	450 gph – 4 ft. 600 gph – 6 ft. 900 gph – 9 ft.	600 gph – 6 ft. 900 gph – 8 ft. 1200 gph – 10 ft.	600 gph – 4 ft. 1200 gph – 7 ft. 1800 gph – 10 ft.	900 gph – 5 ft. 1200 gph – 7 ft. 1800 gph – 8 ft.	1200 gph – 6 ft. 2000 gph – 9 ft. 2400 gph – 11 ft.
MEDIUM SAND 16 - 18 - 20 Slot Average grain size is about 4 grains to 1/16".	18	300 gph – 4 ft. 450 gph – 5 ft. 600 gph – 7 ft.	600 gph – 5 ft. 900 gph – 7 ft. 1200 gph – 9 ft.	600 gph – 4 ft. 1200 gph – 9 ft. 1800 gph – 13 ft.	600 gph – 3 ft. 1200 gph – 6 ft. 1800 gph – 9 ft.	900 gph – 4 ft. 1200 gph – 6 ft. 1800 gph – 7 ft.	1200 gph – 5 ft. 2000 gph – 8 ft. 2400 gph – 10 ft.
MEDIUM & COARSE SAND MIXED Average grain size a little less than 1/32", or between 2 and 3 grains to 1/16".	25	300 gph – 3 ft. 450 gph – 5 ft. 600 gph – 6 ft.	600 gph – 5 ft. 900 gph – 6 ft. 1200 gph – 8 ft.	600 gph – 4 ft. 1200 gph – 7 ft. 1800 gph – 11 ft.	600 gph – 3 ft. 1200 gph – 5 ft. 1800 gph – 8 ft.	900 gph – 4 ft. 1200 gph – 5 ft. 1800 gph – 6 ft.	1200 gph – 5 ft. 2000 gph – 7 ft. 2400 gph – 9 ft.
COARSE SAND Average grain size a little over 1/32" (2 grains to 1/16").	35	450 gph – 4 ft. 600 gph – 5 ft. 900 gph – 7 ft.	600 gph – 4 ft. 900 gph – 5 ft. 1200 gph – 7 ft.	900 gph – 4 ft. 1200 gph – 6 ft. 1800 gph – 10 ft.	900 gph – 3 ft. 1200 gph – 4 ft. 1800 gph – 7 ft.	1200 gph – 4 ft. 1800 gph – 6 ft. 2000 gph – 8 ft.	1200 gph – 4 ft. 2000 gph – 7 ft. 2400 gph – 8 ft.
COARSE SAND AND FINE GRAVEL MIXED Average grain size about 1/16". In coarser gravels, No. 80 and No. 100 slot are often used.	50	450 gph – 4 ft. 600 gph – 5 ft. 900 gph – 7 ft.	600 gph – 4 ft. 900 gph – 5 ft. 1200 gph – 6 ft.	900 gph – 4 ft. 1200 gph – 6 ft. 1800 gph – 10 ft.	900 gph – 3 ft. 1200 gph – 4 ft. 1800 gph – 7 ft.	1200 gph – 4 ft. 1800 gph – 6 ft. 2000 gph – 7 ft.	1200 gph – 4 ft. 2000 gph – 6 ft. 2400 gph – 8 ft.

*Nominal size of screen.

Table 66T

AQUIFER CHARACTERISTICS

DRILLERS' TERMS USED IN ESTIMATING SPECIFIC YIELD

(Source: U.S. Geological Survey)

Crystalline Bedrock (fresh)
Specific yield zero

- Granite
- Hard boulders
- Hard granite

- Hard rock
- Granite and rocks
- Rock (if in area of known crystalline rocks)

Clay and Related Materials
Specific yield 3 percent

- Adobe
- Brittle clay
- Caving clay
- Cement
- Cement ledge
- Choppy clay
- Clay
- Clay, occasional rock
- Crumbly clay
- Cube clay
- Decomposed granite
- Dirt
- Good clay
- Gumbo clay
- Hard clay
- Hardpan (H.P.)
- Hardpan shale
- Hard shale
- Hard shell
- Joint clay

- Lava
- Loose shale
- Muck
- Mud
- Packed clay
- Poor clay
- Shale
- Shell
- Slush
- Soapstone
- Soapstone float
- Soft clay
- Squeeze clay
- Sticky
- Sticky clay
- Tiger clay
- Tight clay
- Tule mud
- Variable clay
- Volcanic rock

Clay and Gravel, Sandy Clay, and Similar Materials
Specific yield 5 percent

- Cemented gravel (cobbles)
- Cemented gravel and clay
- Cemented gravel, hard
- Cement and rocks (cobbles)
- Clay and gravel (rock)
- Clay and boulders (cobbles)
- Clay, pack sand, and gravel
- Cobbles in clay
- Conglomerate
- Dry gravel (below water table)
- Gravel and clay
- Gravel (cement)
- Gravel and sandy clay
- Gravel and tough shale
- Gravelly clay
- Rocks in clay
- Rotten cement
- Rotten concrete mixture
- Sandstone and float rock
- Silt and gravel
- Soil and boulders

- Cemented sand
- Cemented sand and clay
- Clay sand
- Dry hard packed sand
- Dry sand (below water table)
- Dry sand and dirt
- Fine muddy sand
- Fine sand, streaks of clay
- Fine tight muddy sand
- Hard packed sand, streaks of clay
- Hard sand and clay
- Hard set sand and clay
- Muddy sand and clay
- Packed sand and clay
- Packed sand and shale
- Sand and clay mix
- Sand and tough shale
- Sand rock
- Sandstone
- Sandstone and lava
- Set sand and clay
- Set sand, streaks of clay
- Cemented sandy clay
- Hard sandy clay (tight)
- Sandy clay
- Sandy clay with small sand streaks, very fine
- Sandy shale
- Set sandy clay
- Silty clay
- Soft sandy clay
- Clay and fine sand
- Clay and pumice streaks

- Clay and sandy clay
- Clay and silt
- Clay, cemented sand
- Clay, compact loam and sand
- Clay to coarse sand
- Clay, streaks of hard packed sand
- Clay, streaks of sandy clay
- Clay, water
- Clay with sandy pocket
- Clay with small streaks of sand
- Clay with some sand
- Clay with streaks of fine sand
- Clay with thin streaks of sand
- Porphyry clay
- Quicksandy clay
- Sand–clay
- Sand shell
- Shale and sand
- Solid clay with strata of cemented sand
- Sticky sand and clay
- Tight muddy sand
- Very fine tight muddy sand

- Dry sandy silt
- Fine sandy loam
- Fine sandy silt
- Ground surface
- Loam
- Loam and clay
- Sandy clay loam
- Sediment
- Silt
- Silt and clay
- Silty clay loam
- Silty loam
- Soft loam
- Soil
- Soil and clay
- Soil and mud
- Soil and sandy shale
- Surface formation
- Top hardpan soil
- Topsoil
- Topsoil and sandy silt
- Topsoil–silt

- Decomposed hardpan
- Hardpan and sandstone
- Hardpan and sandy clay
- Hardpan and sandy shale
- Hardpan and sandy stratas
- Hard rock (alluvial)
- Sandy hardpan
- Semi-hardpan
- Washboard

Clay and Gravel, Sandy Clay, and Similar Materials
(continued)

- Ash
- Caliche
- Chalk
- Hard lava formation

- Hard pumice
- Porphyry
- Seepage soft clay
- Volcanic ash

Fine Sand, Tight Sand, Tight Gravel, and Similar Materials
Specific yield 10 percent

- Sand and clay
- Sand and clay strata (traces)
- Sand and dirt
- Sand and hardpan
- Sand and hard sand
- Sand and lava
- Sand and pack sand
- Sand and sandy clay
- Sand and soapstone
- Sand and soil
- Sand and some clay
- Sand, clay, and water
- Sand crust
- Sand-little water
- Sand, mud, and water
- Sand (some water)
- Sand streaks, balance clay
- Sand, streaks of clay
- Sand with cemented streaks
- Sand with thin streaks of clay

- Coarse, and sandy
- Loose sandy clay
- Medium sandy
- Sandy
- Sandy and sandy clay
- Sandy clay, sand, and clay
- Sandy clay –water bearing
- Sandy clay with streaks of sand
- Sandy formation
- Sandy muck
- Sandy sediment
- Very sandy clay

- Boulders, cemented sand
- Cement, gravel, sand, and rocks
- Clay and gravel, water bearing
- Clay & rock, some loose rock
- Clay, sand and gravel
- Clay, silt, sand, and gravel
- Conglomerate, gravel, and boulders
- Conglomerate, sticky clay, sand and gravel
- Dirty gravel
- Fine gravel, hard
- Gravel and hardpan strata
- Gravel, cemented sand
- Gravel with streaks of clay
- Hard gravel
- Hard sand and gravel
- Packed gravel
- Packed sand and gravel
- Quicksand and cobbles
- Rock sand and clay
- Sand and gravel, cemented streaks
- Sand and silt, many gravel
- Sand, clay, streaks of gravel
- Sandy clay and gravel
- Set gravel
- Silty sand and gravel (cobbles)
- Tight gravel

- Sandy loam
- Sandy loam, sand, and clay
- Sandy silt
- Sandy soil
- Surface and fine sand

- Cloggy sand
- Coarse pack sand
- Compacted sand and silt
- Dead sand
- Dirty sand
- Fine pack sand
- Fine quicksand with alkali streak
- Fine sand
- Fine sand, loose
- Hard pack sand
- Hard sand
- Hard sand and streaks of sandy clay

- Hard sand rock and some water sand
- Hard sand, soft streaks
- Loamy fine sand
- Medium muddy sand
- Milk sand
- More or less sand
- Muddy sand
- Pack sand
- Poor water sand
- Powder sand
- Pumice sand
- Quicksand
- Sand, mucky or dirty
- Set sand
- Silty sand
- Sloppy sand
- Sticky sand
- Streaks fine and coarse sand
- Surface sand and clay
- Tight sand

- Brittle clay and sand
- Clay and sand
- Clay, sand, and water
- Clay with sand
- Clay with sand streaks
- More or less clay, hard sand and boulders
- Mud and sand
- Mud, sand, and water
- Sand and mud with chunks of clay
- Silt and fine sand
- Silt and sand
- Soil, sand, and clay
- Topsoil and light sand
- Water sand sprinkled with clay

- Float rock (stone)
- Laminated
- Pumice
- Seep water
- Soft sandstone
- Strong seepage

Gravel, Sand, Sand and Gravel, and Similar Materials
Specific yield 25 per cent

- Boulders
- Coarse gravel
- Coarse sand
- Cobbles
- Cobble stones
- Dry gravel (if above water table
- Float rocks
- Free sand
- Gravel
- Loose gravel
- Loose sand
- Rocks

- Gravel and sand
- Gravel and sandrock
- Medium sand
- Rock and gravel
- Running sand
- Sand
- Sand, water
- Sand and boulders
- Sand and cobbles
- Sand and fine gravel
- Sand and gravel
- Sandy gravel
- Water gravel

Table 67T

BOREHOLE GEOPHYSICAL LOGGING METHODS AND THEIR USES IN HYDROLOGIC STUDIES

(Source: U.S. Geological Survey, 1968)

Method	Uses	Recommended conditions
Electric logging:		
Single-electrode resistance.	Determining depth and thickness of thin beds. Identification of rocks, provided general lithologic information is available, and correlation of formations. Determining casing depths.	Fluid-filled hole. Fresh mud required. Hole diameter less than 8 to 10 inches. Log only in uncased holes.
Short normal (electrode spacing of 16 inches).	Picking tops of resistive beds. Determining resistivity of the invaded zone. Estimating porosity of formations (deeply invaded and thick interval). Correlation and identification, provided general lithologic information is available.	Fluid-filled hole. Fresh mud. Ratio of mud resistivity to formation-water resistivity should be 0.2 to 4. Log only in uncased part of hole.
Long normal (electrode spacing of 64 inches).	Determining true resistivity in thick beds where mud invasion is not too deep. Obtaining data for calculation of formation-water resistivity.	Fluid-filled hole. Ratio of mud resistivity to formation-water resistivity should be 0.2 to 4. Log only in uncased part of hole.
Deep lateral (electrode spacing approximately 19 feet).	Determining true resistivity where mud invasion is relatively deep. Locating thin beds.	Fluid-filled uncased hole. Fresh mud. Formations should be of thickness different from electrode spacing and should be free of thin limestone beds.
Limestone sonde (electrode spacing of 32 inches).	Detecting permeable zones and determining porosity in hard rock. Determining formation factor in situ.	Fluid-filled uncased hole. May be salty mud. Uniform hole size. Beds thicker than 5 feet.
Laterolog	Investigating true resistivity of thin beds. Used in hard formations drilled with very salty muds. Correlation of formations, especially in hard-rock regions.	Fluid-filled uncased hole. Salty mud satisfactory. Mud invasion not too deep.
Microlog	Determining permeable beds in hard or well-consolidated formations. Detailing beds in moderately consolidated formations. Correlation in hard-rock country. Determining formation factor in situ in soft or moderately consolidated formations. Detailing very thin beds.	Fluid required in hole. Log only in uncased part of hole. Bit-size hole (caved sections may be logged, provided hole enlargements are not too great).
Microlaterolog	Determining detailed resistivity of flushed formation at wall of hole when mudcake thickness is less than three-eighths inch in all formations. Determining formation factor and porosity. Correlation of very thin beds.	Fluid-filled uncased hole. Thin mud cake. Salty mud permitted.
Spontaneous potential.	Helps delineate boundaries of many formations and the nature of these formations. Indicating approximate chemical quality of water. Indicate zones of water entry in borehole. Locating cased interval. Detecting and correlating permeable beds.	Fluid-filled uncased hole. Fresh mud.
Radiation logging:		
Gamma ray	Differentiating shale, clay, and marl from other formations. Correlations of formations. Measurement of inherent radioactivity in formations. Checking formation depths and thicknesses with reference to casing collars before perforating casing. For shale differentiation when holes contain very salty	Fluid-filled or dry cased or uncased hole. Should have appreciable contrast in radioactivity between adjacent formations.

Table 68T

BOREHOLE GEOPHYSICAL LOGGING METHODS AND THEIR USES IN HYDROLOGIC STUDIES (continued)

Method	Uses	Recommended conditions
Radiation logging: (continued)		
Gamma ray (continued)	mud. Radioactive tracer studies. Logging dry or cased holes. Locating cemented and cased intervals. Logging in oil-base muds. Locating radioactive ores. In combination with electric logs for locating coal or lignite beds.	
Neutron	Delineating formations and correlation in dry or cased holes. Qualitative determination of shales, tight formations, and porous sections in cased wells. Determining porosity and water content of formations, especially those of low porosity. Distinguishing between water- or oil-filled and gas-filled reservoirs. Combining with gamma-ray log for better identification of lithology and correlation of formations. Indicating cased intervals. Logging in oil-base muds.	Fluid-filled or dry cased or uncased hole. Formations relatively free from shaly material. Diameter less than 6 inches for dry holes. Hole diameter similar throughout.
Induction logging	Determining true resistivity, particularly for thin beds (down to about 2 feet thick) in wells drilled with comparatively fresh mud. Determining resistivity of formations in dry holes. Logging in oil-base muds. Defining lithology and bed boundaries in hard formations. Detection of water-bearing beds.	Fluid-filled or dry uncased hole. Fluid should not be too salty.
Sonic logging	Logging acoustic velocity for seismic interpretation. Correlation and identification of lithology. Reliable indication of porosity in moderate to hard formations; in soft formations of high porosity it is more responsive to the nature rather than quantity of fluids contained in pores.	Not affected materially by type of fluid, hole size, or mud invasion.
Temperature logging	Locating approximate position of cement behind casing. Determining thermal gradient. Locating depth of lost circulation. Locating active gas flow. Used in checking depths and thickness of aquifers. Locating fissures and solution openings in open holes and leaks or perforated sections in cased holes. Reciprocal-gradient temperature log may be more useful in correlation work.	Cased or uncased hole. Can be used in empty hole if logged at very slow speed, but fluid preferred in hole. Fluid should be undisturbed (no circulation) for 6 to 12 hours minimum before logging; possibly several days may be required to reach thermal equilibrium.
Fluid-conductivity logging	Locating point of entry of different quality water through leaks or perforations in casing or opening in rock hole. (Usually fluid resistivity is determined and must be converted to conductivity.) Determining quality of fluid in hole for improved interpretation of electric logs. Determining fresh-water-salt-water interface.	Fluid required in cased or uncased hole. Temperature log required for quantitative information.
Fluid-velocity logging	Locating zones of water entry into hole. Determining relative quantities of water flow into or out of these zones. Determine direction of flow up or down in sections of hole. Locating leaks in casing. Determine approximate permeability of lithologic sections penetrated by hole, or perforated section of casing.	Fluid-filled cased or uncased hole. Injection, pumping, flowing, or static (at surface) conditions. Flange or packer units required in large diameter holes. Caliper (section gage) logs required for quantitative interpretation.
Casing-collar locator	Locating position of casing collars and shoes for depth control during perforating. Determining accurate depth references for use with other types of logs.	Cased hole.

Table 69T

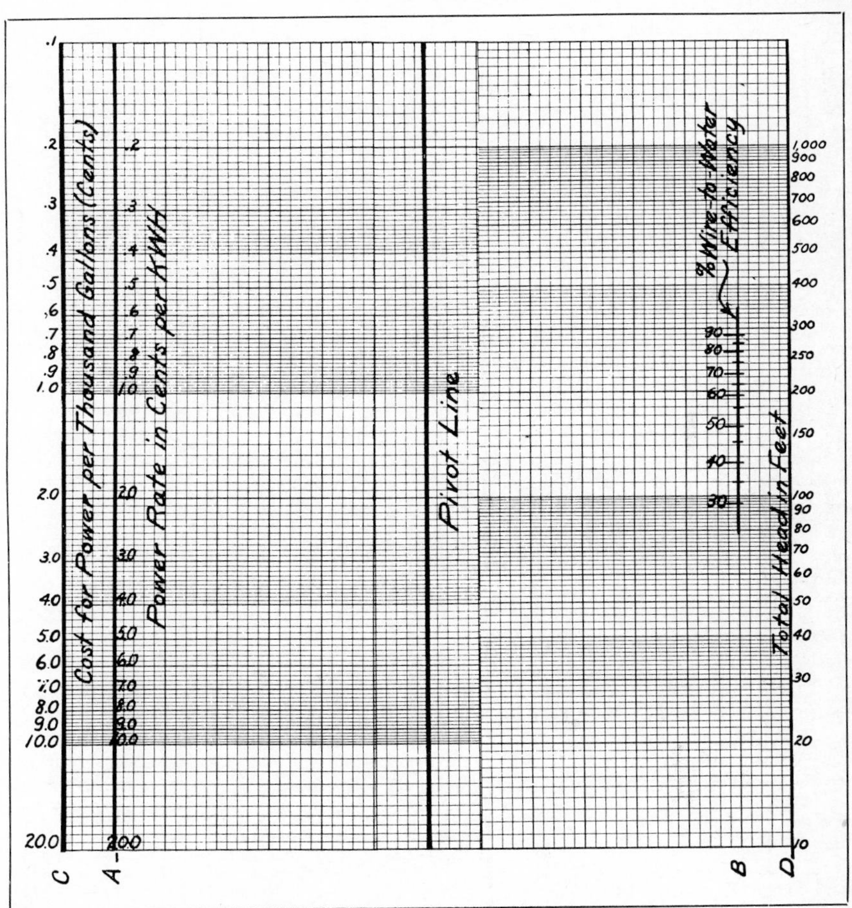

This chart is used for determining the power cost of pumping water per thousand gallons when the total head, wire-to-water efficiency and power rate are known. In using this chart a ruler is placed from line A (Power Rate in Cents per Kilowatt Hour) to line B (% Wire-to-Water Pump Efficiency). A pin or pencil point is placed against the edge of the ruler on the Pivot Line and the ruler is then pivoted to the proper point on line D (Total Head in Feet). The Cost for Power per Thousand Gallons can then be read directly from line C at the left of the chart.

Table 70T

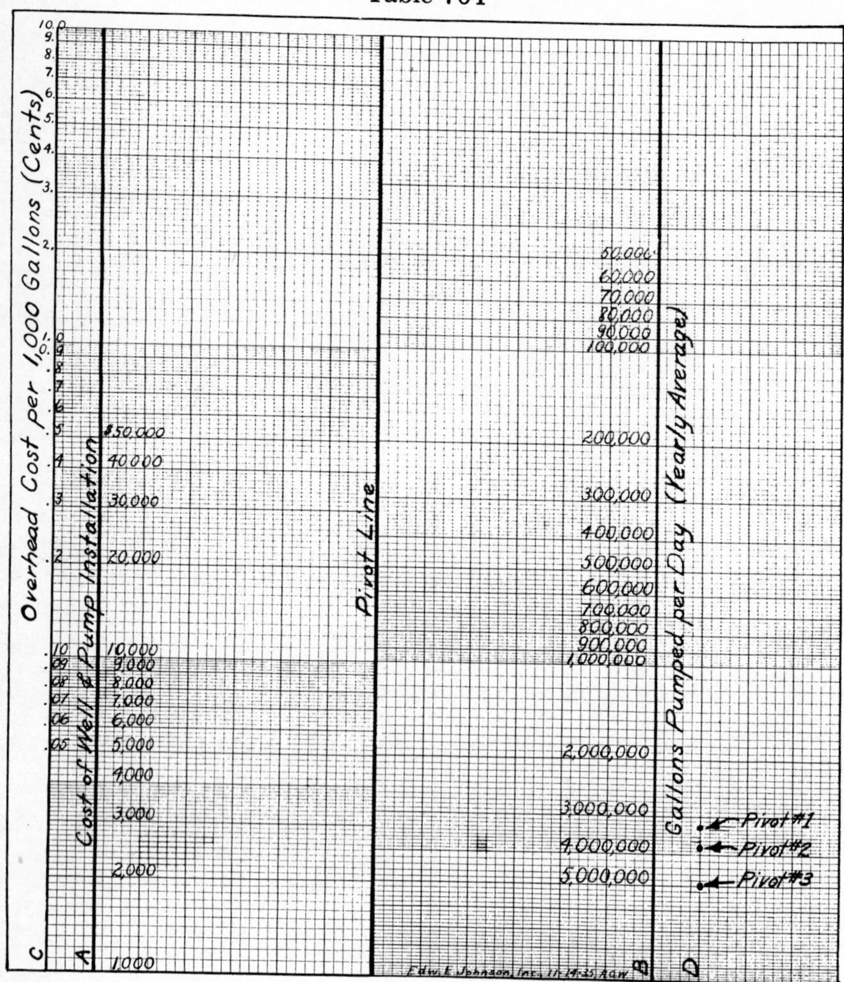

This chart is used (for determining the overhead cost of pumping water per thousand gallons) when the following factors are known: (1) The number of gallons pumped per day on a yearly basis; (2) the total cost of the installation; and (3) a general knowledge of the normal lifetime of a good installation. In using this chart a ruler is placed from line A (Cost of Well and Pump Installation) to line B (gallons pumped per day). A pin or pencil point is placed against the edge of the ruler on the Pivot Line and the ruler is then pivoted to the proper Pivot number. Pivot No. 1 is used only where wells have exceptionally long lives, and No. 3 is used where wells last only about 10 years. No. 2 is an average. The Overhead Cost per thousand gallons can then be read from the line C at the left of the chart.

Table 71T

Water quality problems and typical solutions.

Problem	Possible Cause	Test for	Typical Corrective Procedures (one or several procedures may be needed)
Diseases such as typhoid fever, dysentery, diarrhea, hepatitis, and others	Disease-producing bacteria, virus, protozoa, etc.	Coliform bacteria	Repair or reconstruct the complete water system to keep out pollutants; then disinfect, retest, and if the test results are satisfactory from 3 or more samples taken 2 weeks apart, use the water. If the results are unsatisfactory, either • Develop a new water source; disinfect; test, and if satisfactory, use; or • Install and properly operate continuous disinfection equipment.
Rust or black colored water. Stained clothes and fixtures. Reduced water flow. Discolored, unpalatable food and beverages.	Iron or Manganese	Iron and Manganese	• Aerate the water in a storage tank (pressurized or non-pressurized) and then run water through a sediment filter. • Continuously chlorinate the water and run it through a sediment filter or carbon filter. • Install a sediment filter and a softener • Install a softener, but only if the water is clear. • Install a potassium permanganate regenerated oxidizing filter and a softener. If water pH is less than 7.0, the acid must be neutralized ahead of the filter.
Rust colored slime in toilet tank and pipes. Fuzzy particles in water. Staining. Reduced water flow.	Iron bacteria	Iron bacteria	• Shock chlorinate the well and system, blow out all pipes, and shock chlorinate again. Do this when needed. • Install a continuous chlorination system and carbon filter.
Clothes becoming gray even with good washing. Scum on wash and bath water after using soap. Water heater takes long to heat water. Reduced water flow.	Hard water (dissolved minerals)	Calcium and Magnesium	• Correctly size, install, and operate a water softener. • Use small amount of packaged water softener in baths, washer, etc. • Clogged pipes, water heaters, and other equipment must be cleaned or replaced.
Smelly water -- rotten egg odor. Rapid tarnishing of silverware. Staining, black color	Hydrogen sulfide	Hydrogen sulfide	• Install a continuous chlorination system and carbon filter. • Aerate the water. Install a sediment and/or carbon filter if needed. • Install a potassium permanganate regenerated oxidizing filter. • Clogged and corroded pipes and equipment must be cleaned or replaced.
Corroded metal parts of pump and water system. Staining, green color	Low pH (Acid water)	pH	• Install a continuous soda ash solution feeder. • Install a neutralizer tank containing limestone chips.
Cloudy water	Particles of suspended or colored matter	Turbidity and color	• Install a sediment filter, and possibly a coagulant feeder ahead of the filter.
Gassy well and water	Decay of organic matter	Methane	• Install gas release valve on pressure tank and vent to outdoors. • Aerate water in a non-pressurized storage tank.

Table 72T

TEMPERATURE	PRESSURE
Degrees F = 9/5 C + 32	1 Atmosphere = 760 millimeters of mercury at 32°F. 29.921 inches of mercury at 32°F. 14.7 pounds per square inch. 2,116 pounds per square foot. 1.033 kilograms per square centimeter. 33.947 feet of water at 62°F.
Degrees C = 5/9 x (F − 32)	

Table 73T

Comparison of A_e Values for Various Screen Types
(Based on 1/2 Actual Screen Open Area in Square Feet Per Feet)

Screen Diameter inches, O.D.	Slot Size 10	20	30	60	120	190	250
Type II - Typical Wire-Wrapped Pipe Base Screen							
2	.032	.059	.082	---(n.a.)	---	---	---
4	.057	.105	.145	---	---	---	---
8	.082	.153	.215	.319 (a)	---	---	---
16	.148	.278	.390	.581 (a)	---	---	---
30 (b)	.184	.345	.485	.708 (a)	---	---	---
Type III - Typical Wire-Wrapped Rod Base Screen							
4	.059	.108	.146	.201 (a)	---	---	---
8	.097	.177	.246	.350 (a)	---	---	---
16	.121	.239	.340	.503 (a)	---	---	---
Type IV - Typical Louver Screen							
8	---	---	---	.034	.068	.101	.118
16	---	---	---	.080	.160	.260	.355
Type V - Typical Twin Louver Screen							
8	---	---	.029	.059	.122	.188	.257
16	---	---	.059	.121	.249	.384	.527
30	---	---	.113	.231	.475	.734	1.008
Type VI - Typical Shutter Screen							
4	---	---	.013	.023 (c)	.054 (d)	---	---
8	---	---	.031	.057 (c)	.136 (d)	---	---
16	---	---	.071	.130 (c)	.310 (d)	---	---
30	---	---	.129	.240 (c)	.569 (d)	---	---
Type VII - Typical Slotted Plastic Screen							
2	.010	.020	.030	.050 (a)	.100 (e)	---	---
4	.017	.033	.050	.083 (a)	.167 (e)	---	---
8	.030	.060	.090	.150 (a)	.300 (e)	---	---
16	.060	.120	.180	.300 (a)	.599 (e)	---	---
Type VIII - Typical Mill Slotted Casing: Vertical Slots Without Vertical Spacing							
2	---	.004	.006	.013 (f)	.026 (g)	---	---
4	---	.007	.011	.025 (f)	.050 (g)	---	---
8	---	---	---	.054 (f)	.118 (g)	---	---
16	---	---	---	.094 (f)	.191 (g)	---	---
Type IX - Typical Slotted Casing: 4-inch Vertical Slots Two Inches Apart With Vertical 2-inch Spacing							
4	---	---	.055	---	---	---	---
8 (h)	---	---	.100	---	---	---	---
16	---	---	.210	---	---	---	---
Type X - Typical Concrete Well Screens: Horizontal Slots Ranging from 2.5 to 6.5 Inches, Increasing With Diameters as Below							
4 (i)	---	---	---	.018	---	---	---
8	---	---	---	---	.060	---	---
16 (j)	---	---	---	---	---	.155	---
30 (k)	---	---	---	---	---	.220	---

n.a. - dashes indicate no data available.

(a) 50 used for approximate comparison.
(b) 20 used for approximate comparison.
(c) 55 used for approximate comparison.
(d) 130 used for approximate comparison.
(e) 100 used for approximate comparison.
(f) 62 used for approximate comparison.
(g) 125 used for approximate comparison.
(h) 7-5/8 used for approximate comparison.
(i) 5 used for approximate comparison.
(j) 17 used for approximate comparison.
(k) 32 used for approximate comparison.

Source: Courtesy of NWWA Research Facility.

Table 74T
Metric Conversion Tables

Recommended Units

Description	Unit	Symbol	Comments	English Equivalents
Length	meter	m	*basic SI unit*	39.37 in. = 3.28 ft = 1.09 yd
	kilometer	Km		0.62 mi
	millimeter	mm		0.03937 in.
	centimeter	cm		0.3937 in.
	micrometer	μm		3.937 $^{-5}$in. = 10^4A
Area	square meter	sq m		10.764 sq ft = 1.196 sqyd
	square kilometer	sq km		2.59 sq mi = 247 acres
	square centimeter	sq cm		0.155 sq in.
	square millimeter	sq mm		0.00155 sq in.
	hectare	ha	The hectare (10,000 sq m) is a recognized multiple unit and will remain in international use.	2.471 acres
Volume	cubic meter	cu m		35.314 cu ft = 1.3079 cu yd
	cubic centimeter	cu cm		0.061 cu in.
	liter	liter	The liter is now recognized as the special name for the cubic decimeter.	1.057 qt = 0.264 gal = 0.81 x 10^{-6} acre-ft
Mass	kilogram	kg	*basic SI unit*	2.205 lb
	gram	g		0.035 oz = 15.43 gr
	milligram	mg		0.01543 gr
	tonne	t	1 tonne = 1,000 kg	0.984 ton (long) = 1.1023 ton (short)
Time	second	sec	*Basic SI unit*	
	day	day	Neither the day nor the year is an SI unit but both are important.	
	year	yr or a		
Force	newton	N	The newton is that force that produces an acceleration of 1 m/sec^2 in a mass of 1 kg.	0.22481 lb (weight) = 7.5 poundals

Recommended Units

Description	Unit	Symbol	Comments	English Equivalents
Velocity linear	meter per second	m/sec		3.28 fps
	millimeter per second	mm/sec		0.00328 fps
	kilometers per second	km/sec		0.0103 mph
angular	radians per second	rad/sec		
Flow (volumetric)	cubic meter per second	cu m/sec	Commonly called the cumec	15.850gpm = 2.120 cfm ft/min
	liter per second	liter/sec		15.85 gpm
Viscosity	poise	poise		1.45 x 10^{-5} lb (weight) sec/sq in.
Pressure	newton per square meter	N/sq m	The newton is not yet well-known as the unit of force and kgf/sq cm will clearly be used for some time. In this field the hydraulic head expressed in meters is an acceptable alternative.	0.00014 psi
	kilonewton per square meter	kN/sq m		0.145 psi
	kilogram (force) per square centimeter	kgf/sq cm		14.223 psi
Temperature	degree Kelvin	K	*Basic SI unit*	$\frac{5F}{9}$ − 17.77
	degree Celsius	C	The Kelvin and Celsius degrees are identical. The use of the Celsius scale is recommended as it is the former centigrade scale.	
Work, energy, quantity of heat	joule	J	1 joule = 1 N-m	2.778 x 10^{-7} kw hr = 3.725 x 10^{-7} hp-hr = 0.73756 ft-lb = 9.48 x 10^{-4} Btu
	kilojoule	kJ		2.778 x 10^{-4} kw hr
Power	watt	w	1 watt = 1 J/sec	
	kilowatt	kw		
	joule per second	J/sec		

Table 75T

Comparison of Percent Open Area For Typical Screen Types

O.D. (inches)	Gauge .0075	.015	.030	.060	.120	.190	.250
Type I - Typical Continuous Wire-Wrapped Screen							
2	15.9	26.5	45.1	66.3	82.3	86.3	---(n.a.)
4	14.6	23.2	38.5	57.1	71.7	76.3	---
8	9.3	15.6	27.5	43.1	58.1	64.0	---
16	6.6	12.1	21.9	34.7	50.1	60.0	---
30	---	---	---	---	---	---	---
Type II - Typical Wire-Wrapped Pipe Base Screen							
2	10.0	17.6	31.2	---	---	---	---
4	8.9	16.0	27.8	---	---	---	---
8	6.3	11.3	20.6	34.5	---	---	---
16	5.8	10.3	18.6	31.8	---	---	---
30	---	---	---	---	---	---	---
Type III - Typical Wire-Wrapped Rod Base Screen							
2	13.5	23.9	41.1	42.7	---	---	---
4	9.3	16.3	27.9	41.8	---	---	---
8	7.6	13.4	23.6	37.5	---	---	---
16	4.6	8.6	16.2	27.2	---	---	---
30	---	---	---	---	---	---	---
Type IV - Typical Louver Screen							
2	---	---	---	---	---	---	---
4	---	---	---	---	---	---	---
8	---	---	---	3.2	6.5	9.7	11.3
16	---	---	---	3.8	7.6	12.4	17.0
30	---	---	---	---	---	---	---
Type V - Typical Twin Louver Screen							
2	---	---	---	---	---	---	---
4	---	---	---	---	---	---	---
8	---	---	2.8	5.6	11.6	18.0	24.6
16	---	---	2.8	5.8	11.9	18.3	25.2
30	---	---	2.9	5.9	12.1	18.7	25.7
Type VI - Typical Shutter Screen							
2	---	---	---	---	---	---	---
4	---	---	2.4	4.8	9.6	15.1	18.9
8	---	---	3.0	6.0	12.0	19.0	24.9
16	---	---	3.4	6.8	13.7	21.5	28.2
30	---	---	3.3	6.7	13.4	21.1	27.8

n.a. - dashes indicate no data available.

Table 76T

O.D. (inches)	Gauge .0075	.015	.030	.060	.120	.190	.250
Type VII - Typical Slotted Plastic Screens							
2	2.9	5.7	8.6	17.2	34.4	---	---
4	2.4	5.3	9.6	19.0	38.2	---	---
8	2.1	4.3	8.6	17.2	34.4	---	---
16	2.1	4.3	8.6	17.2	34.4	---	---
30	---	---	---	---	---	---	---
Type VIII - Typical Mill Slotted Casing: Vertical Slots Staggered Without Vertical Spacing							
2	---	1.5	2.6	6.0	12.3	---	---
4	---	1.0	2.3	5.2	10.6	---	---
8	---	---	2.3	5.0	10.4	---	---
16	---	---	2.3	4.5	9.1	---	---
30	---	---	---	---	---	---	---
Type IX - Typical Slotted Casing: 4-Inch Vertical Slots Two Inches Apart with Vertical 2-Inch Spacing							
2	---	---	---	---	---	---	---
4	---	---	1.0	---	---	---	---
8	---	---	1.0	---	---	---	---
16	---	---	1.0	---	---	---	---
30	---	---	1.0	---	---	---	---
Type X - Typical Concrete Well Screens: Horizontal Slots Ranging from 2.5 to 6.5 Inches, Increasing with Diameters as Below							
2	---	---	---	---	---	---	---
4	---	---	---	5.7	---	---	---
8	---	---	---	---	5.7	---	---
16	---	---	---	---	---	7.0	---
30	---	---	---	---	---	5.3	---

Note: (1) Percent open area increases proportionately to slot width.
(2) Placing slots only one inch apart doubles percent open area.

Source: Courtesy of NWWA Research Facility.

Table 77T

DISCHARGE MEASUREMENT USING SMALL CONTAINER

(Oil Drums, Stock Tanks, etc.)

$$\text{Discharge (Gallons per minute)} = \frac{\text{Volume of container (Gallons) x 60}}{\text{Time (Seconds) to fill container}}$$

Table 78T

DETERMINING THE CAPACITY OF A PUMP

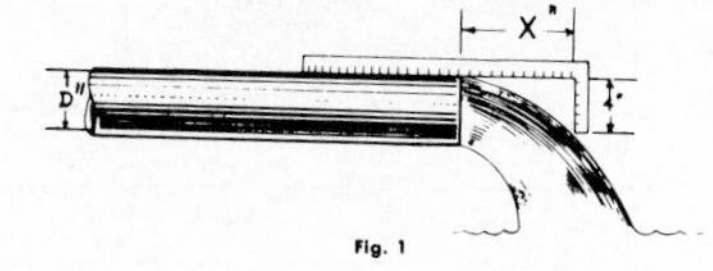

Fig. 1

BY HORIZONTAL OPEN DISCHARGE METHOD

To estimate the pumping capacity of any given unit construct an L shaped measuring instrument similar to that shown in the accompanying sketch. The shorter side should be 4" long. The longer side may be any convenient length marked in inches. With the water flowing from the horizontal open discharge, place the long side of the L along the top of the discharge pipe, allowing the shorter side to hang downward as shown in the drawing. Slide the L along the pipe until the 4" length barely touches the flow of water. Note the distance ("X") traveled by the flow of water before it drops 4". Presume, for example, that the distance is 15" and the inside diameter of the pipe (indicated by "D") is 3". Consulting the table below, find 15" in the column at extreme left headed "Horizontal Dist. X (Inches)." Then move horizontally to the right to the column showing the pipe diameter being used (3"). The discharge rate is found to be 183 gallons per minute.

Horiz. Dist. X (Inches)	DISCHARGE RATE (Gallons per minute) Nominal Pipe Diameter 1	1¼	1½	2	2½	3	4	5	6	8	10	12	Average Velocity
4	5.7	9.8	13.3	22.0	31.3	48.5	83.5						2.1
5	7.1	12.2	16.6	27.5	39.0	61.0	104	163					2.6
6	8.5	14.7	20.0	33.0	47.0	73.0	125	195	285				3.1
7	10.0	17.1	23.2	38.5	55.0	85.0	146	228	334	580			3.7
8	11.3	19.6	26.5	44.0	67.5	97.5	166	260	380	665	1060		4.2
9	12.8	22.0	29.8	49.5	70.0	110	187	293	430	750	1190	1660	4.7
10	14.2	24.5	33.2	55.5	78.2	122	208	326	476	830	1330	1850	5.3
11	15.6	27.0	36.5	60.5	86.0	134	229	360	525	915	1460	2200	5.8
12	17.0	29.0	40.0	66.0	94.0	146	250	390	570	1000	1600	2220	6.2
13	18.5	31.5	43.0	71.5	102	158	270	425	620	1080	1730	2400	6.9
14	20.0	34.0	46.5	77.0	109	170	292	456	670	1160	1860	2590	7.4
15	21.3	36.3	50.0	82.5	117	183	312	490	710	1250	2000	2780	7.9
16	22.7	39.0	53.0	88.0	125	196	334	520	760	1330	2120	2960	8.4
17		41.5	56.5	93.0	133	207	355	550	810	1410	2260	3140	9.1
18			60.0	99.0	144	220	375	590	860	1500	2390	3330	9.7
19				110	148	232	395	620	910	1580	2520	3500	10.4
20					156	244	415	650	950	1660	2660	3700	10.6
21						256	435	685	1000	1750	2800		11.4
22							460	720	1050	1830	2920		11.8
23								750	1100	1910	3060		12.4
24									1140	2000	3200		13.0

For other than standard diameter pipes the flow may be determined by using the following formula:
Q gpm = X 1.28D² where D = Inside pipe diameter
X = Horizontal open flow for drop of 4".

PROCEDURE IN DETERMINING DISTANCE TO WATER LEVEL

Install sufficient ⅛" or ¼" pipe (copper tubing may also be used) in the well so that end of pipe extends 10 to 20 feet below lowest possible pumping level. Be sure that all joints are absolutely air-tight by using white-lead or pipe compound. THE EXACT LENGTH OF PIPE OR TUBING IN THE WELL MUST BE KNOWN AND THIS INFORMATION SHOULD BE RECORDED.

Attach upper end of pipe or tubing securely at top of well. Connect a tire valve to the air line at the top of the well and also a pressure gauge. Next, connect a tire pump or other air supply to the air line and pump air into the line until the pressure gauge reaches a maximum reading. This reading is the point at which further supply of air will not increase the reading to any higher value. Record the gauge reading.

Let X = Depth to water (in feet) unknown

Y = Known length of air line (in feet)

Z = Water pressure on air line, obtained from pressure gauge reading. Altitude type gauge reads directly in feet of water. If gauge reads in pounds convert to feet by multiplying by 2.31.

X = Y − Z

Distance to water = length of air line minus gauge reading (feet).

EXAMPLE: Assume that the air pipe is 100 ft. long from center of gauge to bottom end of pipe and that the highest reading of the gauge needle is 15 lbs. = 15 × 2.31 = 34.6 feet.
Distance to water = 100 − 34.6 = 65.4 feet.

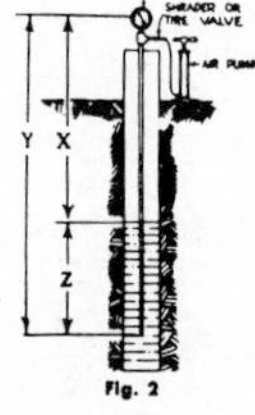

Fig. 2

Courtesy of Jacuzzi

Table 79T

Cement Slurry Data

(Based on sp. gr. of cement of 3.17)

Density of Slurry			Water Required			Volume of Slurry per Sack Cement	
Lb. per Gallon	Lb. per Cu. Ft.	Specific Gravity	% by Weight of Cement	Gal. per Sack Cement	Bbl. (42 gal.) per Sack Cement	Cu. Ft.	Bbl. (42 Gal.)
13.5	101.0	1.62	79.0	8.92	0.212	1.67	0.297
14.0	104.7	1.68	69.3	7.83	0.186	1.52	0.271
14.5	108.5	1.74	61.0	6.89	0.164	1.39	0.248
15.0	112.2	1.80	54.1	6.11	0.145	1.29	0.230
15.5	115.9	1.86	48.2	5.44	0.130	1.20	0.214
16.0	119.7	1.92	43.0	4.86	0.116	1.12	0.199
16.5	123.4	1.98	38.4	4.34	0.103	1.05	0.187
17.0	127.2	2.04	34.3	3.88	0.092	0.99	0.176
17.5	130.9	2.10	30.7	3.47	0.083	0.94	0.167

Table 80T

ESTIMATING FLOW FROM HORIZONTAL OR INCLINED PIPES

(FULL PIPES)

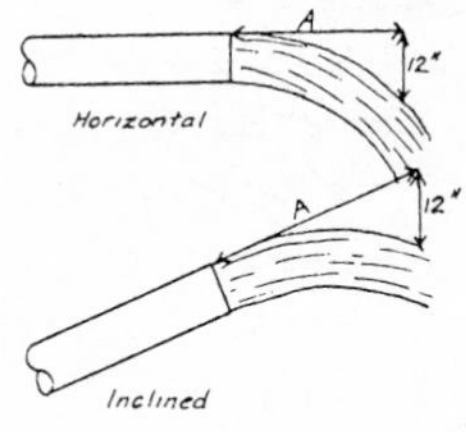

A fairly close determination of the flow from full open pipes may be made by measuring the distance the stream of water travels parallel to the pipe in falling 12 inches vertically.

Measure the inside diameter of the pipe accurately (in inches) and the distance (A) the stream travels in inches parallel to the pipe for a 12-inch vertical drop. (See diagrams)

The flow, in gallons per minute, equals the distance (A) in inches multiplied by a constant K obtained from the following table:

I.D. Pipe	K	I.D. Pipe	K	I.D. Pipe	K	I.D. Pipe	K	I.D. Pipe	K	I.D. Pipe	K
2	3.3	4	13.1	6	29.4	8	52.3	10	81.7	12	118.
1/4	4.1	1/4	14.7	1/4	31.9	1/4	55.6	1/4	85.9	1/2	128.
1/2	5.1	1/2	16.5	1/2	34.5	1/2	59.0	1/2	90.1	13	138.
3/4	6.2	3/4	18.4	3/4	37.2	3/4	62.5	3/4	94.4	1/2	149.
3	7.3	5	20.4	7	40.0	9	66.2	11	98.9	14	160.
1/4	8.5	1/4	22.5	1/4	42.9	1/4	69.9	1/4	103.	1/2	172.
1/2	10.0	1/2	24.7	1/2	45.9	1/2	73.7	1/2	108.	15	184.
3/4	11.5	3/4	27.0	3/4	49.0	3/4	77.7	3/4	113.	16	209.

(PARTIALLY FILLED PIPES)

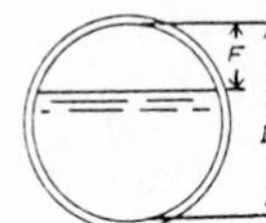

For partially filled pipes, measure the freeboard (F) and the inside diameter (D) and calculate the ratio of F/D (in percent). Measure the stream as explained above for full pipes and calculate the discharge. The actual discharge will be approximately the value for a full pipe of the same diameter multiplied by the correction factor from the following table:

F/D Percent	Factor	F/D Percent	Factor	F/D Percent	Factor	F/D Percent	Factor
5	0.981	30	0.747	55	0.436	80	0.142
10	.948	35	.688	60	.375	85	.095
15	.905	40	.627	65	.312	90	.062
20	.858	45	.564	70	.253	95	.019
25	.805	50	.500	75	.195	100	.000

Table 81T

ACID CALCULATIONS

2,000 lbs. of 20° Be. acid = 207 gal.

1 gal. of 20° Be. acid = 3.04 lbs. HCl.

207 gal. of 20° Be. acid = 629 lbs. HCl.

Gal.	Be.°	Sp. Gr.	% HCI	Wt./Gal.	Lbs. HCI/Gal.
1	20.	1.16	31.45	9.66	3.04
1	10.12	1.075	15.0	8.95	1.345
1	6.9	1.05	10.0	8.75	.875
1	5.0	1.0357	7.15	8.63	.617
1	3.5	1.032	4.95	8.596	.426
1	2.2	1.0154	3.08	8.458	.26

—Courtesy Halliburton Oil Well Cementing Co.

Table 82T

Orifice Method of Measuring Water Flow

The orifice method is a simple way to measure flow of water from a pipe discharging horizontally into the open air. The sketch shows the general arrangement.

A plate or cap is affixed to the end of the pipe with the circular orifice in the exact center of the pipe. The size of orifice should be from one-half to three-quarters of the size of the pipe, but must be of such size that it will run full of water.

A hole for 1/8″ pipe should be drilled and tapped 24″ back from the orifice and a short 1/8″ nipple screwed in until the inner end is exactly flush with the inner wall of the pipe. One end of a piece of rubber hose is slipped over the end of the nipple and the other end over a glass tube, which should be supported in a vertical position.

Ratio R is the diameter of the orifice divided by the inside diameter of the pipe, and the proper value of K is found from the curve. In the formula A is figured as the area of the orifice in square inches, G=32.2, and H is read as the height in inches of the water in the glass tube above the center of the pipe. Gallons per minute (g) then can be figured.

For convenience when determining capacities during pumping tests, tables for various orifices, covering readings at one-half inch intervals, are given.

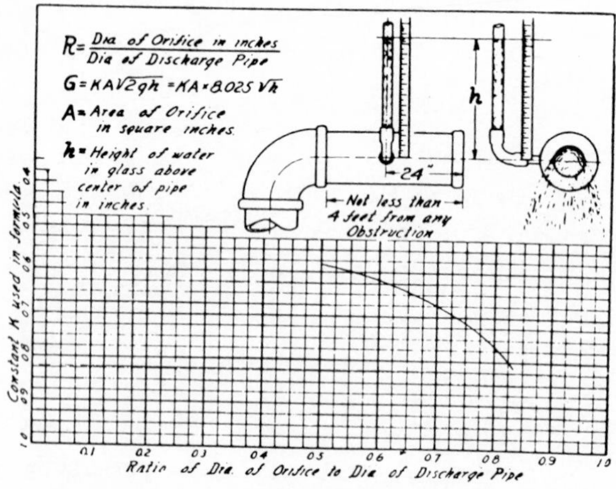

Table 83T

WATER FRICTION IN 100 FEET OF SMOOTH Bore Hole

For Various Flows and Hose Sizes, Table Gives Velocity of Water and Feet of Head Lost in Friction in 100 Feet of Smooth Bore Hose.

Sizes of Hose Shown are Actual Inside Diameters.

Flow in U. S. Gals. Per Min.	Velocity in Feet Per Sec.	Friction Head in Feet	Velocity in Feet Per Sec.	Friction Head in Feet	Velocity in Feet Per Sec.	Friction Head in Feet	Velocity in Feet Per Sec.	Friction Head in Feet	Velocity in Feet Per Sec.	Friction Head in Feet	Velocity in Feet Per Sec.	Friction Head in Feet	Flow in U. S. Gals. Per Min.
	⅝"		¾"		1"		1¼"		1½"		2"		
1.5	1.6	2.3	1.1	.97									1.5
2.5	2.6	6.0	1.8	2.5									2.5
5	5.2	21.4	3.6	8.9	2.0	2.2	1.3	.74	.9	.3			5
10	10.5	76.8	7.3	31.8	4.1	7.8	2.6	2.64	1.8	1.0	1.0	.2	10
15	2½"		10.9	68.5	6.1	16.8	3.9	5.7	2.7	2.3	1.5	.5	15
20	1.3	.32			8.2	28.7	5.2	9.6	3.6	3.9	2.0	.9	20
25	1.6	.51	3"		10.2	43.2	6.5	14.7	4.5	6.0	2.5	1.4	25
30	2.0	.70	1.4	.3	12.2	61.2	7.8	20.7	5.4	8.5	3.1	2.0	30
35	2.3	.93	1.6	.4	14.3	80.5	9.1	27.6	6.4	11.2	[illegible]	2.7	35
40	2.6	1.2	1.8	.5			10.4	35.0	7.3	14.3	4.1	3.5	40
45	2.9	1.5	2.0	.6			11.7	43.0	8.2	17.7	4.6	4.3	45
50	3.3	1.8	2.3	.7			13.1	52.7	9.1	21.8	5.1	5.2	50
60	3.9	2.5	2.7	1.0			15.7	73.5	10.9	30.2	[illegible]	7.3	60
70	4.6	3.3	3.2	1.3					12.7	40.4	[illegible]	9.8	70
80	5.2	4.3	3.6	1.7	4"				14.5	52.0	[illegible]	12.6	80
90	5.9	5.3	4.1	2.1	2.3	.5			16.3	64.2	9.2	15.7	90
100	6.5	6.5	4.5	2.6	2.5	.6			18.1	77.4	10.2	18.9	100
125	8.2	9.8	5.7	4.0	3.2	.9					12.8	28.6	125
150	9.8	13.8	6.8	5.6	3.8	1.3					15.3	40.7	150
175	11.4	18.1	7.9	7.4	4.5	1.8	5"		6"		17.9	53.4	175
200	13.1	23.4	9.1	9.6	5.1	2.3	3.3	.8	2.3	.32	20.4	68.5	200
225	14.7	29.0	10.2	11.9	5.7	2.9	3.7	1.0	2.6	.40			225
250	16.3	35.0	11.3	14.8	6.4	3.5	4.1	1.2	2.8	.49			250
275	18.0	42.0	12.5	17.2	7.0	4.2	4.5	1.4	3.1	.58			275
300	19.6	40.0	13.6	20.3	7.7	4.9	4.9	1.7	3.3	.69			300
325			14.7	23.5	8.3	5.7	5.3	2.0	3.7	.80			325
350			15.9	27.0	8.9	6.6	5.7	2.3	4.0	.90			350
375			17.0	30.7	9.6	7.4	6.1	2.6	4.3	1.0	8"		375
400					10.7	8.4	6.5	2.9	4.5	1.1	2.6	.28	400
450					11.5	10.5	7.4	3.6	5.1	1.4	[illegible]	.35	450
500					12.8	12.7	8.2	4.3	5.7	1.7	3.2	.43	500
600					15.3	17.5	9.8	6.1	6.8	2.4	3.8	.60	600
700					17.9	23.7	11.4	8.1	7.9	3.3	4.5	.80	700
800							13.1	10.3	9.1	4.2	5.1	1.1	800
900							14.7	12.8	10.2	5.2	[illegible]	1.3	900
1000							16.3	15.6	11.4	6.4	6.4	1.6	1000
1100							17.9	18.5	12.5	7.6	7.0	1.9	1100
1200									13.6	9.2	7.7	2.3	1200
1300									14.7	10.0	[illegible]	2.6	1300
1400									15.9	11.9	[illegible]	3.0	1400
1500									17.0	13.6	9.6	3.3	1500
1600											[illegible]	3.7	1600
1800											11.5	4.7	1800
2000											[illegible]	5.7	2000
2500											[illegible]	8.6	2500
3000											19.1	12.2	3000

Courtesy of Jacuzzi

18
Glossary of Inter-Industry Technology

This glossary attempts to bring together most of the common terms of the petroleum, mining and, ground water industries. The glossary, as it exists here, is comprehensive only to the extent that general terms have been chosen for preliminary definition or comment. In order to more fully assist in the transfer of inter-industry technology, a fully comprehensive glossary of terms is urgently needed. Technology can be assimilated only when terms have been fully defined and widely accepted by field and laboratory personnel.

Aa (Hawaiian). Block-lava consisting of a rough tumultuous assemblage of clinker-like scoriaceous masses.

Absorption. The penetration of molecules or ions of one substance into the interior of a solid or liquid.

Acetic Acid. Widely used as an aid in overcoming many problems encountered in well completion, stimulation, and reconditioning. Factors such as economics, handling, and the lack of technical data have limited the use over the past few years.

Acidity. The acidity of a water is usually caused by uncombined carbon dioxide, mineral acids, and salts of strong acids and weak bases such as iron and aluminum salts from mine waters or industrial wastes.

Acidizing. The process of introducing acid into the pore space of an acid-soluble formation for the purpose of enlarging the pore space by dissolving the surrounding formation. Acidizing also refers to the removal of encrustation from well screens and gravel pack, and dissolving cementitious materials.

Additive. An auxiliary agent added to the treating solution.

Adsorption. Attachment on the surfaces of solids of gases, liquids or dissolved substances with which they are in contact.

Aeration, Zone of. The zone above the water table in which interstices are partly filled with ground air except in the saturated portion of the capillary fringe.

Aerobic. Living or active only in the presence of oxygen.

Agglomerate. A chaotic assemblage of coarse angular pyroclastic materials.

Aggregate. Sand, gravel, or crushed stone to be bound by cement in order to make concrete.

Air (Drilling). Too much air causes excessive friction in pipe lines. Too little air reduces yield and causes a surging intermittent discharge.

Air Line. The smaller vertical air pipe usually submerged to within a few feet of the bottom of the eductor pipe. The length of the air line below the static water level is used in calculating the air pressure required to start the air-lift.

Alkalinity. Bicarbonate, carbonate and hydroxides in the natural or treated water usually impart alkalinity.

Alluvial Cone. A body of alluvial material deposited by a stream debouching from the region undergoing erosion above the apex of the cone.

Alteration. Change in the mineralogical composition of a rock typically brought about by the action of hydrothermal solutions. Sometimes classed as a phase of metamorphism but usually distinguished from it because milder and more localized.

Aluminum. Because it is the third most abundant element and is found in minerals, rocks and clays, aluminum is present in practically all natural waters as a soluble salt, colloid, or insoluble compound. It may appear as a residual in treated waters.

Amortize. To clear off, liquidate, or otherwise extinguish, as a debt, usually by a sinking fund.

Ampere. The rate of transfer of electricity, comparable to the fluid delivery of a pipeline.

Anaerobic. Living or active only in the absence of free oxygen.

Angle of Repose. The angle of the slope or incline at which a material comes to rest when poured or dumped into a pile or on a slope.

Angstrom unit. Measure of the wave-length of light and of other very short distances. 10^{-8}. (cm.)

Anion. A negatively charged ion or radical.

Annular Space. The space between the casing referred to and the well bore or casing surrounding it.

Annular Velocity. Determined by the area of the annulus and the pump output.

Anticlinorium. Major anticline composed of many smaller folds.

Antifoam. An agent added to acid to prevent or retard foaming during the acid reaction.

Apparent Specific Gravity (Volume-Weight). The ratio of the weight of a unit volume of oven-dry soil to that of an equal volume of water under standard conditions. This term may be applied to natural undisturbed field samples or to laboratory samples.

Aquiclude. A formation which, although porous and capable of absorbing water slowly, will not transmit it fast enough to furnish an appreciable supply for a well or spring.

Aquifer. An *aquifer* is a formation, group of formations, or part of a formation that contains sufficient saturated permeable material to yield significant quantities of water to wells and springs.

Aquifuge. A rock without inter-connected openings which can neither absorb or transmit water.

Area of Influence. The area beneath which ground-water or pressure-surface contours, modified by pumping.

Area of Pumping Depression. The area overlying the cone of pumping depression, or cone of water-table depression.

Arkose. Sedimentary rock (grain-size as in sandstone) derived from the disintegration of acid igneous rocks of granular texture. There is usually little sorting of the materials. Formed under conditions permitting little mineral decomposition, it contains a considerable proportion of feldspar.

Arroyo (Spanish). Small stream; gutter. Usage varies and in some Latin-American countries arroyo includes gorges of major proportions.

Arsenic. A toxic element that can be determined in trace amounts.

Artesian. Artesian is synonymous with *confined. Artesian water* and *artesian water body* are equivalent respectively to *confined* ground water and *confined* water body. An *artesian well* is a well deriving its water from an *artesian* or *confined* water body. The water level in an artesian well stands above the top of the artesian water body it taps.

Artesian Well. A well tapping a confined or artesian aquifer in which the static water level stands above the water table. The term is sometimes used to include all wells tapping confined water, in which case those wells with water level above the water table are said to have *positive artesian head* (pressure) and those with water level below the water table, *negative artesian head.*

Attapulgite. Has a fibrous structure and suspension properties depending on the mechanical dispersion of needle-like units which entangle to resist flow. Used as a drilling fluid component in salt source water.

Attitude. Direction and degree of dip of a structural plane (bed, fault, etc.). Attitude may be expressed in terms of Dip *and* Strike (*q.v.*)

Average Velocity. The volume of the ground water which passes through a unit cross-sectional area of the stratum divided by its porosity.

Bacteria. The nuisance organisms developing and multiplying in wells, which do not cause disease, commonly known as Clonothrix, Crenothrix, Gallionella, Leptothrix, Siderocapsa, and Sphaerotilus of the aerobic type; Sulfur Bacteria and Sulfate-Reducing Bacteria of the anaerobic type. Also Iron Bacteria.

Baking Soda (Sodium Bicarbonate). Used to counteract cement contamination of drilling mud.

Bank Storage. Storage of water in alluvial deposits adjacent to a stream during high-water stage of stream flow. The water is gradually discharged at low-water stage.

Barite. Barium sulfate; the commercial product contains small amounts of iron oxide, silica, and other minerals. It is used to make muds heavier.

Barium Sulfate. A compound also termed Barite or Barytes.

Barrel. Equals 42 U.S. gallons (9702 cubic inches or 5.615 cubic feet).

Barytes. Natural barium sulfate used as a basic material for weighting agents.

Basalt. A fine-grained basic rock usually occurring in volcanic flows, dikes, and sills.

Base Exchange. The capacity of silicates and some other substances to hold cations, such as hydrogen, sodium, calcium, magnesium, etc. in a loose union within and on the silicate particle.

Batholith. A huge intrusive body of igneous rock supposedly enlarging downward into the earth's crust. Minimum size to qualify as a batholith: 40 square miles.

Baume' Gravity. The specific gravity of a liquid measured by a graduated float hydrometer. Heavy Baume' is for liquids heavier than water; zero deg. is equivalent to one or the specific gravity of water at 4 deg. C.; 66 deg. is a specific gravity of 1.842.

$$\text{Specific Gravity} = \frac{145}{145 - \text{Baume' Gravity}}$$

Bed. Layer in body of sedimentary rock. Stratum.

Bedding, Bedding Plane. Plane of stratification. The surface marking the boundary between a bed and the bed above or below it.

Beneficiate. To reduce, as ores (Standard). As used in the Lake Superior district, beneficiation has the same meaning as concentration, but usually with the added implication that undesirable constituents (such as silica) are eliminated.

Bentonite. A highly plastic, colloidal clay composed largely of mineral montmorillonite.

Blowout. An oil or gas well commencing to flow suddenly and out of control.

Boron. In excess of 2.0 mg./1. in irrigation waters, boron is deleterious to certain plants. Some are affected by 1.0 mg./1. or less.

Bradenhead. A pack-off or seal between two casings.

Breccia. Rock consisting of fragments, more or less angular in a matrix of finer-grained material or of cementing material. May form by faulting or crushing (tectonic breccia), by erosion (clastic breccia), by collapse, by replacement bordering fractures or by volcanism (volcanic breccia).

Brecciated. Converted into a breccia.

Bridge. An obstruction to circulation (Drilling.)

Bromide. In ground water supplies probably from connate water, but on the coastal areas as a result of sea water intrusion.

Btu. A British thermal unit is the heat required to raise the temperature of one pound of water one degree fahrenheit.

Cake Thickness. The thickness of the filter cake deposited against porous media by the drilling fluid.

Calcite. $CaCO_3$. Rhombohedral. G. 2.72. H. 3. Perfect rhombohedral cleavage. Usually colorless or white, but exceptionally red, yellow, or blue.

Calcium. The most frequent cause of hardness and affects the scale-forming and corrosive properties of the water.

Capillarity. The property of tubes with minute openings which, when immersed in a fluid, raise or depress the fluid in the tubes above or below the surface of the fluid in which they are immersed.

Capillary Fringe. The zone immediately above the water table in which all or some of the interstices are filled with water that is under less than atmospheric pressure and that is continuous with the water below the water table. The water is held above the water table by interfacial forces (surface tension).

Capillary Head. The difference between capillary lift and the position of the meniscus in a capillary opening.

Capillary Interstice. An opening small enough to produce appreciable capillary rise.

Capture. Water withdrawn artificially from an aquifer is derived from a decrease in storage in the aquifer, a reduction in the previous discharge from the aquifer, an increase in the recharge, or a combination of these changes. The decrease in discharge plus the increase in recharge is termed *capture.*

Capture may occur in the form of decreases in the ground-water discharge into streams, lakes, and the ocean, or from decreases in that component of evapotranspiration derived from the saturated zone. After a new artificial withdrawal from the aquifer has begun, the head in the aquifer will continue to decline until the new withdrawal is balanced by capture.

Carbon Dioxide. Found in well waters and surface supplies, carbon dioxide may contribute to some types of corrosion. It is derived from decay of organic matter, and solution from underground sources. The gas ionizes in water forming carbonic acid.

Casing Shoe. A heavy-walled steel coupling or band at the lower extremity of the casing.

Cast Iron. Molten iron which has hardened to its proper shape in a mold and is composed of iron and 2% to 3.5% carbon.

Catalyst. A material capable of aiding a reaction by its presence but not entering into the reaction.

Cathodic Protection. An artificial electrical system constructed to redirect stray electrical currents that cause a specific type of corrosion in casing, transmission piping and drill pipe.

Cation. A positively charged ion or radical.

Caustic Soda (Flake Caustic, Lye). Used to make tannins, lignite, and chrome-lignosulfonates soluble and effective as thinners; to raise the pH of acidic makeup water; and to reduce corrosion of iron.

Cavern Flow. Sub-surface turbelent flow partly filling caverns or large open conduits.

Cement. The powdered dry cement prior to the addition of mixing water.

Cement Plug. The hardened cement slurry left in the lower portion of the casing and later drilled out after the cement has set.

Cement Slurry. A pumpable mixture of cement and water.

Cementing. The process of placing the cement slurry to provide a seal against subsurface waters.

Centipoise. A unit of viscosity. At 20 deg. C, the viscosity of water is 1.005 centipoises.

Channeling. Cement slurry by-passing a part of the rotary mud in the annular space instead of moving it all ahead of the slurry.

Chert. Cryptocrystalline variety of quartz. SiO_2 G. *2.65.* H. 7.

Chloride. The chlorides of calcium, magnesium, sodium, iron, etc. normally found in water are extremely soluble. In natural waters high chloride may indicate animal pollution, but usually is from passing through a salt formation in the earth or sea water intrusion. A salty or brackish taste may be imparted to the water.

Chlorination. The process of introducing chlorine solution into a well, pump, and formation for sterilizing and the destruction of nuisance bacteria.

Chlorine (Demand). The difference between the amount of chlorine applied and the amount of free, combined, or total available chlorine remaining at the end of the contact period. The demand varies with the amount of chlorine applied, contact time, pH, and temperature.

Chlorine (Residual). Chlorine in water may be as free available chlorine (hypochlorous acid and/or hypochlorite ion), or as combined available chlorine (chloramines and other chlorine derivatives). Both types may be present at the same time.

Chromium. Chromium salts are used as corrosion inhibitors and in industrial waters and may diffuse in natural waters from industrial wastes.

Circulation (Drilling). The drilling fluid movement from the mud pit, through the pump, standpipe, hose, drill pipe, annular space in the hole, and circulating ditch back to the mud pit.

Clay. A soft, plastic variously colored earth composed largely of hydrous silicate of alumina, formed by the decomposition of feldspar and other aluminum silicates.

Clay Minerals. A family of minerals, most of them hydrous aluminum silicates and all either finely crystalline or amorphous. All those which crystallize are monoclinic. Species are indistinguishable except by laboratory methods.

Cleavage. Mineral cleavage: The property of crystals of certain minerals by virtue of which the crystal can be broken or split along smooth planes which correspond to specific crystallographic directions. *Rock cleavage:* The ability of rocks to break along parallel surfaces of secondary origin. Schistosity is the cleavage in rocks that are sufficiently recrystallized to be called schist of gneiss. Bedding fissility is the tendency to part parallel to the stratification.

Climatic Cycle. Periodic fluctuation of climate, including a series of dry years and a proceeding or following series of years with heavy rainfall.

Climatic Year. A period used in meteorological measurements, usually beginning after the end of the rainy season.

Coagulation. Flocculation.

Coefficient of Storage. The volume of water released from or taken into storage per unit surface area of an aquifer, per unit change in the component of head normal to that surface. (S) is a dimensionless index and ranges from 3.0×10^{-1} to 1.0×10^{-5}.

Coefficient of Transmissibility. The field coefficient of permeability multiplied by the aquifer thickness in feet.

Coefficient of Viscosity. The force required to maintain a unit difference in velocity between two layers of water a unit distance apart.

Cohesion. The attractive force that holds the molecules of a substance together.

Colloid. Substances composed of extremely small particles about 0.5 to 0.005 microns in diameter. In drilling fluids, most of the colloids are dispersions.

Color. True color is the color of the water after the turbidity has been removed. Measurement of color of a water does not provide any information concerning the substance causing the color.

Compaction, Water of. Water furnished by destruction of pore space owing to compaction of sediments.

Complex Phosphates. Such as sodium tetraphosphate and sodium acid pyrophosphate, are very effective degelling agents when added to a fresh-water suspension of bentonite. It reduces the apparent viscosity and allows sand to settle.

Conductance (Specific). A measure of the ability of the water to conduct an electric current. It is related to the total concentration of ionizable solids in the water. It is inversely proportional to electrical resistance.

Conduction. The transfer of heat by contact.

Conductivity, Effective Hydraulic. The rate of flow of water through a porous medium that contains more than one fluid, such as water and air in the unsaturated zone, and should be specified in terms of both the fluid type and content and the existing pressure.

Conductivity, Hydraulic. Replaces the term "field coefficient of permeability," P_f, embodies the inconsistent units gallon, foot, and mile. If a porous medium is isotropic and the fluid is homogeneous, the *hydraulic conductivity* of the medium is the volume of water at the existing kinematic viscosity that will move in unit time under a unit hydraulic gradient through a unit area measured at right angles to the direction of flow.

Cone of Pressure Relief. An imaginary conical surface of the water level indicating pressure relief in a confined aquifer due to pumping.

Cone of Water-Table Depression. The conical surface of the water level created in an unconfined aquifer due to pumping.

Confined Ground Water. A body of ground water overlain by material sufficiently impervious to sever free hydraulic connection with overlying ground water except at the intake. Confined water moves in conduits under the pressure due to difference in head between intake and discharge areas of the confined water body.

Confining Bed. Is a Term which will now supplant the terms "aquiclude," "aquitard," and "aquifuge" in reports of the Geological Survey and is defined as a body of "impermeable" material stratigraphically adjacent to one or more aquifers.

Connate Water. Water entrapped in the interstices of a sedimentary rock at the time it was deposited.

Contact. Bounding surface between two rock units, especially the boundary between an intrusive and its host-rock.

Contamination. Denotes impairment of water quality by chemical or bacterial pollution to a degree that creates an actual hazard to public health.

Copper. Corrosion of copper, brass and bronze pipe and fittings, and copper salts used for controlling growths in reservoirs and distribution systems supply the element.

Cosmic Water. New water brought in from space with meteorites.

Crater. A funnel-shaped cavity formed by caving at the top of the well.

Critical Velocity. For porous media, the maximum velocity under which laminar flow can occur. In pipe flow, the velocity at which eddying commences is called the *higher critical velocity.* The velocity at which eddies in turbulence die out is the *lower critical velocity.*

Damage Ratio. The dimensionless ratio of transmissibility to the instantaneous productivity index (or specific capacity).

Darcy. A unit of permeability. One Darcy is the rate of viscous flow of one cubic centimeter per square centimeter per second a of a fluid of one centipoise viscosity with a pressure gradient of one atmosphere per centimeter. Equals 18.24 gallons per day per square foot (60°F.).

Deflocculation. A separation of the particles in a colloidal suspension caused by an attraction of the particles for the dispersing medium, or mutual repulsion by like electric charges on the particle.

Density. The density of a substance is its weight per unit volume.

Diameter. The greatest distance across a circle or through a sphere, the line of distance necessarily passing through the center of the circle or sphere.

Dielectric. An insulator or non-conductor of electricity.

Die Overshot. A long tapered die of heat-treated steel designed to fit over the top of the lost drill pipe and cut thread when rotated. The tool is fluted to permit the escape of metal cuttings and fluids.

Dilation, Water of. Water in excess of water of saturation held by sedimentary material in an inflated state (water of supersaturation).

Dilution. The use of more water in the cement slurry than is necessary to produce a pumpable slurry.

Dip. The inclination of bed, vein, fault, etc., measured from the horizontal, thus the angle between a line in the bed perpendicular to the strike and the horizontal plane.

Dispersion. Spreading apart of finely divided particles in a suspension.

Dolerite. (1) A field term for any fine-grained basic igneous rock whose exact identity has not been determined. (2) (British) A basic rock of the composition of gabbro but finer in grain; thus synonymous with diabase except that diabasic texture is not a necessary characteristic.

Dolomite. CaMg $(CO_3)_2$. Rhombohedral. G. 2.85. H. 3½-4. Perfect rhombohedral cleavage. Does not effervesce in dilute HCl (difference from calcite).

Dolomite (Rock). Rock composed essentially of the mineral dolomite.

Drag Bit. Equipped with short blades and a body fitted with water courses that direct the drilling fluid stream to keep the blades clean. Assists penetration by means of a jetting action against the bottom of the hole.

Drawdown. Lowering of water level caused by pumping. It is measured for a given quantity of water pumped during a specified period, or after the pumping level has become constant.

Drift. A horizontal underground passage following a vein. It is distinguished from a crosscut, which intersects a vein, or a level or gallery, which may either follow or intersect the vein.

Drift (Glacial). All the material in transport by glacier ice, and all the material predominantly of glacial origin deposited directly by glaciers, or indirectly in glacial streams, glacial lakes, and the sea.

Driller. The man in charge of the rig and crew during one tour, who handles the drilling controls.

Dry Hole or Duster. A well drilled which produces neither oil nor gas, nor water of significant quantity.

Echelon (en Echelon). An arrangement of faults, veins, etc., in which the individuals are staggered like the treads of a staircase.

Effective Grain Size. The theoretical grain size of a homogeneous material of one grain size that would transmit water at the same rate as the material investigated. Also in sand analyses the size of particle of which 90% of the sample is coarser.

Effective Porosity. The portion of pore space in saturated permeable material in which movement of water takes place. It is measured with satisfactory accuracy by specific yield.

Effective Size (of Grain). The grain size of a theoretical body of homogeneous material of one grain size that would transmit water at the same rate as the material under consideration. It is also defined as the diameter of a grain of such size that 10 per cent of the material (by weight) consists of smaller grains and 90 per cent of larger grains.

Effective Velocity. The actual or field velocity of ground water percolating through water-bearing material. It is measured by the volume of ground water passing through a unit cross-sectional area divided by effective porosity.

Effluent Seepage. Diffuse discharge of ground water to the ground surface.

Elastic Limit. That point or amount of force at which a material will not return to its original length when subjected to a straight pull. When a string of drill pipe is pulled and stretched beyond a point at which it will not return to its original length, it may be said to have been pulled beyond its elastic limit.

Electrolyte. A chemical which dissociates into positive and negative ions when dissolved in water, increasing the electrical conductivity.

Epigenetic Ore Deposit. Deposit of ore introduced into a pre-existing rock. *Examples:* Ore introduced into a vein by solutions from a magma; ore introduced by solutions and replacing limestone.

Epithermal Ore Deposit. Deposit formed by hot ascending solutions at slight depth and low temperature.

Equilization. When the hydrostatic head of the fluid in the annular space is equal to the hydrostatic head of the drilling mud and cement slurry in the casing.

Factor of Safety. The ratio of the ultimate breaking strnegth of the material to the force exerted against it. If a rope will break under a load of 6,000 lb., and it is carrying a load of 2,000 lb., its factor of safety is:

$$\frac{6{,}000}{2{,}000} = 3, \text{ or f.s. } 3.$$

Fatigue. By fatigue in referring to metals is meant the breakdown of the original crystal structure due to intermittent strain or shock which produces fractures and the likelihood of failure in service.

Faults. Displacement (on or by a fault) — A general term for the change in position of any point on one side of a fault plane relative to any corresponding point on the opposite side of the fault plane. *Slip* — Displacement as measured in the fault. Net slip is the total displacement. Dip slip is the component of slip parallel to the dip of the fault. Strike slip is the component of slip parallel to the strike of the fault. *Offset* — Horizontal separation measured perpendicular to the strike of the disrupted horizon. *Heave* — Horizontal displacement measured in a vertical section normal to the strike of a fault. *Throw* — Vertical displacement measured in a vertical plane perpendicular to the strike of the fault. *Normal fault* — One in which the hanging-wall has apparently moved downward with reference to the footwall. *Reverse fault* — One in which the hanging-wall has apparently moved upward with respect to the footwall. *Note.* By usage the terms normal and reverse refer to the apparent displacement, not the true displacement. Usage is trending toward designation by true displacement. *Dip fault* — A fault whose strike is approximately at right angles to the strike of the bedding (or the vein). *Strike fault* — A fault having the same strike as the bedding (or the vein). *Oblique fault* — Fault which strikes obliquely to the strike of the bedding (or the vein). *Dip slip fault* — Fault in which the net slip is in the direction of dip (*i.e.*, strike component is lacking). *Strike slip fault* — Fault in which the net slip is in the direction of the strike (*i.e.*, dip component is lacking). *Diagonal slip fault* — Fault in which the net slip is diagonal, *i.e.*, neither vertical nor horizontal. *Hinge fault* — A fault whose displacement is greater at one place than another and diminishes to zero at some point. *Pivotal fault* — A fault of rotary displacement in which the displacement at one place (the pivot) is zero and displacements on opposite sides of the pivot are in opposite directions. Scissors fault.

Felsite. Field term for any fine-grained acid igneous rock whose exact composition has not been determined.

Ferromagnesians. Minerals containing a high proportion of iron and magnesia, *e.g.*, pyroxene, hornblende, biotite.

Field Capacity. The capacity of the soil to hold pellicular water, measured by the soil scientist as the ration of weight of water retained by the soil to the to the weight of the dry soil. (See Specific Rentention as used by ground water invertigator.)

Filter Cake Texture. Properties of the filter cake such as toughness, slickness, etc.

Filter Loss. The amount of fluid delivered through a permeable membrane in a specified time.

Filtration Rate. Water loss.

Final Set. The cement slurry hardness when the Gilmore needle 1/24 inch in diameter, loaded to weigh 1 lb., does not make an appreciable indentation.

Fish. Any foreign material in a well which cannot be removed at will.

Fishing. The act of attempting to recover a fish.

Fishing Job. Foreign material or tools in the hole which must be removed.

Fissure. An extensive crack, break, or fracture in rock. A mere joint or crack persisting only for a few inches or a few feet is not usually termed a fissure . . . although in a strict physical sense it is one. (Ransome).

Fissure Vein. A fissure in the earth's crust filled with mineral. (Raymond). As it is now recognized that replacement as well as (or instead of) simple filling has played an important role in the formation of most veins, the term fissure vein has lost much of its meaning.

Fixed Ground Water. Water held in saturated material with interstices so small that it is permanently attached to the pore walls or moves so slowly that it is usually not available as a source of water for pumping.

Fixed Moisture. Moisture held in the soil below the hygroscopic limit.

Flash Set. Premature stiffening of a cement slurry that converts it to a non-pumpable plastic.

Flat Gel. A ten minute gel strength about equal to the initial gel strength.

Float. Loose fragments of rock, ore, or grossan found on or near the surface or in stream beds.

Flocculation. A collection of separate colloid particles into masses with a loss of colloidal properties.

Flotation. A method of concentrating ore by inducing the particles of ore to float to the surface of water or other solution (usually buoyed up by air bubbles) while the gangue particles sink to the bottom.

Flow (Steady). Occurs when at any point the magnitude and direction of the specific discharge are constant in time.

Flow (Uniform). Occurs if at every point the specific discharge has the same magnitude and direction.

Flow (Unsteady). Occurs when at any point the magnitude or direction of the specific discharge changes with time.

Foliation. (1) More or less pronounced aggregation of particular constituent minerals of a metamorphic rock into lenticles or streaks or inconstant bands, often very rich in some one mineral and contrasting with constituent lenticles or streaks rich in other minerals.
(2) (Foliate structure) — Used in a broad sense includes the textural or structural properties of certain rocks which permit them to be cleaved or parted along approximately parallel surfaces or lines. In this sense the term includes bedding fissility and schistosity.

Foot-Pound (ft.-lb.). A foot-pound is the amount of energy required to lift one pound a vertical distance of one foot.

Formation. An assemblage of rock masses grouped together into a unit that is convenient for description or mapping.

Fountain Head. The elevation of water surface in a conduit if the overlying confining stratum extends above the water table, or elevation of water table above the upper termination of the confining stratum where the latter is below the water table.

Fluid Potential. The mechanical energy per unit mass of fluid at any given point in space and time with respect to an arbitrary state and datum. Loss of fluid potential incurred as the fluid moves from a region of high potential to one of low potential represents loss of mechanical energy which is converted to heat by friction.

Fluoride. With the growth of fluoridation of water supplies as a public health measure, accurate determination has increased in importance.

Fluorite. CaF_2. Isometric. G. 3.18. H 4. Purple, red, green, and white are the colors of fluorite.

Flux (Metallurgical). A substance charged into a furnace for the purpose of combining with other substances in the ore or charge in order to form slag.

Fracture. A break. Fracture is a general term to include any kind of discontinuity in a body of rock if produced by mechanical failure, whether by shear-stress or tensile stress. Fractures include faults, shears, joints, and planes of fracture-cleavage.

Fracture System. Group of fractures (faults, joints, or veins) consisting of one or more sets, usually intersecting or interconnected. System usually implies contemporaneous age for all of the sets, but vein system is sometimes used for all veins in a given mine or district regardless of age or origin.

Free Ground Water. In the zone of saturation to the first impervious barrier, the water in the interstices which moves under the control of the water-table slope.

Free-Water Content. The amount of water in the cement slurry that is in excess of the amount needed to harden the cement.

Fuel Efficiency. The ratio of the heat produced by a fuel for doing work to the available heat of the fuel. This efficiency is determined by the non-heat-forming materials in the fuel and the nonwork-producing heat which is developed by the fuel.

Gangue. Useless minerals occurring in ore.

Gel. A colloidal suspension in which shearing stresses below a certain value fail to produce permanent deformation.

Gel Development. Associated closely with the flow properties of most waterbase muds.

Gel Strength. A measure of the effect of the forces between the particles while the mud is at rest.

Geomorphology. (1) That department of physical geography which deals with the form of the earth, the general configuration of its surface, the distribution of land and water, and the changes that take place in the evolution of land forms. (2) (Geol.). The investigation of the history of geologic changes through the interpretation of topographic forms.

Geosyncline. Basin in which many thousands of feet of sediments accumulate. *Not a synonym for synclinorium.*

Gilsonte. A selected and graded asphaltite, having a specific gravity of 1.07, which will neither accelerate nor retard the setting time of the cement. It is impervious to corrosion by waters or brines and resists the attack of acids or alkalines.

Goethite. FeO (OH). Orthorhombic. G. 4.37. H. 5-5½. Yellowish brown to dark brown. Streak yellowish brown.

Gold. Au. Isometric. G. 15.0-19.3. H. 2½ - 3.

Gossan. The oxidized equivalent of aggregated sulphide material. (Locke). (Aggregated sulphide is massive sulphide. See Chapter 10). Gossan usually consists of ferric oxide and quartz or jasper, sometimes with manganese dioxide, clay minerals, etc.

Gouge. Clay formed by comminution of rock in a fault zone.

Graded Bedding. Change of grain-size from the bottom to the top of a bed or succession of beds. Normally the gradation is from coarse at the bottom to fine at the top, with an abrupt change at the bottom of the overlying bed. In fine-grained rocks the gradation is sometimes emphasized by change of color from light to dark.

Graphite. C. Hexagonal. G. 2.3. H 1-2. Black to steel-gray, metallic. Usually foliated or in flakes.

Gravity Water. Gravity water is also used in the literature for irrigation water derived from stream flow and drainage ditches, as distinguished from. pumped water, and water furnished by a water-table well in contrast to water produced from a confined-water well.

Graywacke. (1.) Variety of sandstone (in the broader sense) derived from the disintegration of basic igneous rocks of granular texture and thus contains abundant grains of biotite, hornblende, magnetite, etc. Thus defined it is the ferromagnesian equivalent of arkose.(2) Rock whose grains are fragments of rock rather than fragments of minerals. (3) Ferromagnesian sand where cementation has advanced so far that the rock when fractured breaks across the original grains rather than around them. According to this definition graywacke would be a metamorphic rock analogous to quartzite.

Grit. A sedimentary rock composed of clastic grains coarser than sand but finer than gravel. Thus a rock intermediate in grain-size between a sandstone and a conglomerate.

Ground Water. Water in the zone of saturation.

Ground Water Cascade. Descent of ground water on a steep hydraulic gradient to a lower and flatter water-table slope. A cascade occurs below a ground-water barrier or dam which may develop effluent seepage above it, and at the contact of less permeable material with more permeable material downslope.

Ground Water (Confined). Under pressure significantly greater than atmospheric, and its upper limit is the botton of a bed of distinctly lower hydraulic conductivity than that of the material in which the confined water occurs.

Ground Water Decrement. Water abstracted from the ground water reservoir by evaporation, transpiration, spring flow, and effluent seepage, pumping wells, and outflow of ground water from underneath the area under consideration.

Ground Water Divide. The boundary of the cone of pumping depression which separates the area of influence and the area outside.

Ground Water Equation. Rainfall equals evaporation and transpiration plus stream flow plus ground water increment.

Ground Water Hydrology. The branch of the science of hydrology concerned with the occurrence, distribution, and movement of water below the surface of the earth.

Ground Water Increment. Water added to the ground water reservoir from all sources.

Ground Water Inventory. A detailed estimate of the water added to the ground water reservoir less estimates of the water removed from said reservoir.

Ground Water Mound. A mound-shaped addition to the ground water body built up by seepage, percolation, or recharge.

Ground Water (Perched). Unconfined ground water separated from an underlying body of ground water by an unsaturated zone. Its water table is a *perched water table.* It is held up by a *perching bed* whose permeability is so low that water percolating downward through it is not able to bring water in the underlying unsaturated zone above atmospheric pressure. Perched ground water may be either *permanent,* where recharge is frequent enough to maintain a saturated zone above the perching bed, or *temporary,* where intermittent recharge is not great or frequent enough to prevent the perched water from disappearing from time to time as a result of drainage over the edge of or through the perching bed.

Ground Water Province. An area in which the ground water occurs in similar formations.

Ground Water Ridge. A ridge-shaped addition to the ground water body built up by an influent stream.

Ground Water Runoff. Water runoff from spring flow and seepage.

Ground Water Trench. A trench-shaped depression of the water table caused by effluent seepage into a stream or drainage ditch or by movement of ground water to a thalweg underlying a stream.

Ground Water Turbulent Flow. Turbulent flow which occurs in large openings in the zone of saturation under high velocities.

Ground Water (Unconfined). Water in an aquifer that has a water table.

Gyp. A rock-like scale deposit in a well, usually calcium sulfate or calcium carbonate deposited from the well water.

Gypsum. $CaSO_4.2H_2O$. Monoclinic. G. 2.32. H. 2. Perfect cleavage resembles mica but has two additional cleavages. Softer than mica.

Halite (Rock Salt). NaCl. Isometric. G. 2.16. H. 2½. Colorless to white. Distinguished by taste.

Hard Formation Rock Bits. Have short teeth and are milled with a large included angle to resist breakage and battering. Requires a cutting structure designed to withstand high weights.

Hardness. Calcium and magnesium salts, present in most natural waters, cause the water to be hard. Natural waters will vary in hardness depending upon their location. When the total hardness exceeds the carbonate and bicarbonate alkalinity, the amount of hardness equivalent to the alkalinity is carbonate hardness and the remainder is non-carbonate hardness. When the carbonate and bicarbonate alkalinity equal or exceed the total hardness, all hardness is carbonate.

Head of Water Converted to Pressure. 2.307 ft. of water will exert a pressure of one pound per square inch. As generally considered, 2½ ft. of water develop a pressure of 1 lb.

Head (Static). The height above a standard datum of the surface of a column of water (or other liquid) that can be supported by the static pressure at a given point.

Head (Total). The sum of three components: (1) *elevation head,* h_e, which is equal to the elevation of the point above a datum, (2) *pressure head,* h_p, *which is the height of a column of static water that can be supported by* the static pressure at the point, and (3) *velocity head,* h_v, Which is the height the kinetic energy of the liquid is capable of lifting the liquid.

Heat of Hydration. The heat evolved during the setting and hardening of portland cement.

Hematite. Fe_2O_3. Rhombohedral. G. 5.26. H. 5½-6½. Streak red.

Homogeneity. Synonymous with uniformity. A material is homogeneous if its hydrologic properties are identical everywhere. Although no known aquifer is homogeneous in detail, models based upon the assumption of homogeneity have been shown empirically to be valuable tools for predicting the approximate relationship between discharge and potential in many aquifers.

Hornblende. $Ca_2Na(Mg,Fe'')_4(Al,Fe,''Ti)_3Si_6O_{22}(O,OH)_2$. Monoclinic. G. 3.2. H. 5-6. An amphibole.

Horsepower (hp). The rate at which work is performed, or the number of units of work performed in unit time. Any working agent is said to be developing 1 hp when it does 33,000 ft.-lb. of work in one minute; i.e., work equivalent to lifting a weight of 33,000 lb. one foot in one minute.

Horsepower (Boiler). Boiler horsepower represents the conversion of 34.5 lb. water per hour to steam at a pressure of 14.7 psi (normal atmospheric pressure at sea level), and at a temperature of 212 F.

Horsepower (Brake) (bhp). Actual power output delivered by the crankshaft of an engine. It is equal to the theoretical or indicated horsepower multiplied by the mechanical efficiency.

Horsepower (Indicated). The theoretical power output of an engine. It is determined by the mean effective pressure of the medium acting on the net area of the piston through the distance the piston travels as the engine rotates.

Hydraulic Diffusivity. The conductivity of the saturated medium when the unit volume of water moving is that involved in changing the head a unit amount in a unit volume of medium.

Hydraulic Gradient (Dimensionless). Is the change in static head per unit of distance in a given direction. If not specified, the direction generally is understood to be that of the maximum rate of decrease in head. The *gradient of the head* is a mathematical term which refers to the vector denoted by Δh or grad h, whose magnitude dh/dl is equal to the maximum rate of change in head and whose direction is that in which the maximum rate of increase occurs. The hydraulic gradient and the gradient of the head are equal but of opposite sign.

Hydrograph. A graphic plot of changes in flow of water or in elevation of water level against time.

Hydrologic Cycle. All movements of water and water vapor in the atmosphere, on the ground surface, below the surface, and return to the atmosphere by evaporation and transpiration.

Hydrologic Properties. Those properties of rocks which control the entrance of water, capacity to hold, transmit, and deliver water. They include porosity, effective porosity, specific retention, permeability, and direction of maximum and minimum permeability.

Hydrology. A science concerned with the occurrence of water in the earth, its physical and chemical reactions with the rest of the earth, and its relation to the life of the earth.

Hydrophilic. A substance which absorbs or adsorbs water.

Hydrophobic. A substance which repels water.

Hydrometer. An instrument designed to indicate the specific gravity or weight per unit volume of a fluid by the depth to which it sinks in the fluid.

Hydrothermal. Pertaining to or resulting from the activity of hot aqueous solutions originating from a magma or other source deep in the earth.

Hygroscopic Moisture. Moisture that is held in the soil in equilibrium with atmospheric water vapor at the ground surface.

Hypabyssal. A general term applied to minor intrusions, such as sills and dikes, and to rocks of which they are made, to distinguish them from volcanic rocks and formations on the one hand and "plutonic" rocks and major intrusions such as batholiths on the other.

Hypogene. Generated from depth. Refers to the effects produced by ascending (usually hydrothermal) solutions. *Cf.* Supergene.

Hypothermal Ore Deposit. Deposit formed by hot ascending solutions at great depth or at high temperature and pressure.

Igneous. (adj.) Related to or derived from molten matter that originated within the earth.

Igneous Rock. Rock made by the solidification of molten matter that originated within the earth. *Examples:* solidified lava; intrusive granite.

Impression Block. Has many forms and designs. Often used to obtain an impression of the top of the tool before attempting fishing operations. Particularly necessary in rotary-drilled, uncased holes.

Influent Seepage. Movement of gravity water in the zone of aeration from the ground surface toward the water table.

Inhibited Acid. Acid with an inhibitor added.

Inhibitor. An additive to arrest corrosive action of the acid.

Initial Set. The cement slurry hardness, when the Gilmore needle 1/12 inch in diameter, loaded to 1/4 lb., does not make an appreciable indentation.

Intensified Acid. Acid with an additive to make the acid react at a more rapid rate.

Interstream Ground-water Ridge. A ridge in the water table formed between two effluent streams, produced by percolation toward the surface streams with the development of a residual ground water ridge between them.

Intrusive. In petrology, having, while molten, penetrated into or between other rocks but solidified before reaching the surface.

Iron. In the ferric state iron is completely oxidized but in the ferrous state it is only partially oxidized. It exists in most natural waters in the ferrous state. Exposure to air or addition of chlorine oxidizes the iron to the ferric state and it may hydrolyze to form insoluble hydrated ferric oxide. The form of iron may also be changed by iron bacteria. Hence the iron may be in true solution; in a colloidal state; complexed by phosphates; as relatively coarse suspended particles; and as ferrous, ferric or both. Release of carbon dioxide will convert iron from the ferrous to the ferric state. It may be placed into solution when corrosion of iron and steel surfaces occurs.

Isohyetal Map. A map on which precipitation is plotted by connecting points of equal precipitation (isohyetal lines) and which shows rainfall distribution in the area mapped.

Isopiestic Line. A contour of the pressure surface of a confined aquifer.

Isotropic. (1) Capable of transmitting light with equal velocity in all directions. (2) Having physical properties (*e.g.*, strength characteristics) which do not vary with direction.

Isotropy. That condition in which all significant properties are independent of direction. Although no aquifers are isotropic in detail, models based upon the assumption of isotropy have been shown to be valuable tools for predicting the approximate relationship between discharge and potential in many aquifers.

Joint. A divisional plane or surface that divides a rock and along which there has been no visible movement parallel to the plane or surface.

Juvenile Water. New water of magmatic, volcanic, or cosmic origin added to the terrestrial water supply.

Kaolin. A general term for a group of clay minerals. See Clay Minerals.

Kilowatt (kw). A kilowatt is 1,000 watts or 1.341 horsepower. Watts (volume) equals volts (pressure) multiplied by amperes (flow). Thus if any two of these factors (watts, volts, amperes) are known, the other can readily be determined.

Laminar Flow. Motion of a fluid the particles of which move substantially in parallel paths. This type of flow always occurs below the lower critical velocity and may occur between the lower and higher critical velocities. Also called straight-line, streamline, or viscous flow.

Lateritic. Extreme type of weathering common in tropical climates. Iron and aluminum silicates are decomposed and silica (along with most other elements) removed by leaching. The product, laterite, is characterized by high content of alumina and/or ferric oxide.

Lead. An element not found naturally in the human body. It is cumulative and toxic. Ingestion of small quantities may cause lead poisoning.

Lime (Hydrated Lime). Sometimes added to bentonite mud to rapidly develop a stiff gel in order to "wall off" loose sands and gravels.

Limonite. A field term for hydrous iron oxides whose real identity has not been determined. May consist of lepidochrocite, goethite, or hematite, or any mixture of these, with more or less adsorbed water. Earthy, fibrous, reniform, or stalactitic.

Lithium. As a minor constituent of minerals, it may be present in fresh waters at concentrations below 10 mg./1.

Loss of Circulation. The loss of drilling fluid into formation pores or crevices.

Lubricants. Reduces frictional drag or torque.

Macroscopic, Megascopic. Visible to the unaided eye, as contrasted with microscopic.

Mafic. High in magnesia and iron and correspondingly low in silica.

Magma. Molten rock-matter together with its dissolved gas or vapor.

Magmatic Water. Water driven out of molten rock material during crystallization.

Magnesium. Magnesium and calcium salts in waters are important because of their relationship to hardness, scale-forming and corrosive properties. Determination of calcium may be made to divide the total hardness into calcium and magnesium hardness. The total hardness less the calcium hardness is the magnesium hardness.

Manganese. Most frequently found in well waters which contain iron but is usually not present in excess of 3.0 ppm. When present in very small amounts, like iron, it will deposit from solution as gray or black manganese hydroxide.

Medium-to-Hard Formation Rock Bits. Have a great number of short teeth, more closely spaced, to promote greater footage per bit.

Mesothermal Ore Deposit. Deposit formed by hot ascending solutions at intermediate depth and temperature.

Metamorphic Water. Water that is driven out of rocks by the process of metamorphism.

Metamorphism. Throughgoing change in texture or mineralogical composition of rock, usually brought about by heat, pressure, or chemically active solutions.

Meteoric Water. Water derived from the atmosphere.

Meteorology. A branch of applied physics treating the atmosphere and its phenomena, especially variations of temperature and moisture, winds, storms, etc.

Methane. A colorless, odorless, tasteless, combustible gas found in some ground waters. Escape of the gas from the water may create an explosive atmosphere. The explosive limits in air are 5-15 percent by volume. Theoretically, at sea level, a 5% concentration in air could be reached in a poorly ventilated space sprayed with hot water (68 deg. C.) having a methane concentration of only 0.7 mg./1. The vapor pressure of water at higher temperatures is so great that an explosive mixture cannot form. At lower barometric pressures, the theoretical hazardous concentration of methane in water will be reduced proportionately.

Millidarcy. 1/1000 Darcy. Equals 0.01824 gallons per day per square foot (60°F.).

Native Clay. Consists mainly of illite and kaolinite, although calcium bentonite and other clay minerals may be present. Usually, fine sand and silt are associated with the clay.

Nitrogen (Albuminoid). An approximate indication of the quantity of protein type nitrogen present in the water. Animal and plant life living in the water are largely responsible.

Nitrogen (Ammonia). A product of microbiologic activity, ammonia nitrogen may be present in variable concentrations in surface and ground waters.

Nitrogen (Nitrite). The most highly oxidized phase in the nitrogen cycle. In surface water supplies, occurrence is generally in trace quantities but some ground waters may have higher quantities.

Nitrogen(Nitrite). An intermediate stage in the oxidation or reduction process of the nitrogen cycle in water. Trace amounts in raw surface waters indicates pollution. It may also be produced by action of bacteria or other organisms on ammonia nitrogen supplied at elevated temperatures in combined residual chlorination of water.

Nitrogen (Organic). Amino acids, polypeptides, proteins and albuminoid nitrogen contribute to the organic nitrogen content of the water. A rise in the organic nitrogen content may indicate sewage or industrial waste pollution.

Oil and Grease. In natural waters, some oils may derive from decomposition of forms of aquatic life. An emulsion of oil or grease may be present from industrial wastes or similar sources.

Optimization of Operations. A method of maximizing efficiency.

Organic Polymers. Used in drilling fluids and have a strong affinity for water, (i.e., they are hydrophilic colloids). They develop highly swollen gels, are absorbed by clay particles and protect the clay from the flocculating effects of salts.

Overall Efficiency. Ratio of power output of an engine to the power input, and is the measure of the difference between indicated and brake horsepower.

Oxygen (Dissolved). Present in all surface waters and rain waters due to contact with the atmosphere, it is found in lesser quantities in well waters and may be absent in deep well supplies. Its presence creates a corrosive effect on iron and steel.

Ozone (Residual). A potent germicide also used as an oxidizing agent to destroy organic compounds producing taste and odor in water; destruction of organic coloring matter; and oxidation of reduced iron or manganese salts to insoluble oxides. A residual of 0.1 mg./1. is generally effective for disinfection.

Pellicular Front. The even front, developed only in pervious granular material, on which pellicular water depleted by evaporation, transpiration, or chemical action is regenerated by influent seepage.

Pellicular Water. Water adhering as films to the surfaces of openings and occuring as wedge-shaped bodies at junctures of interstices in the zone of aeration above the capillary fringe.

Peptization. A dispersion caused by the addition of electrolytes or other chemicals.

Peptized Bentonite. That which has been processed and chemically treated to improve the quality of standard bentonite.

Perched Ground Water. Ground water in a saturated zone which is separated from the main body of ground water by unsaturated rock.

Perlite. A volcanic material which is mined, screened and expanded by heat to form a cellular product of extremely low weight.

Permeability. The capacity of water-bearing material to transmit water, measured by the quantity of water passing through a unit cross section in a unit time under 100 percent hydraulic gradient.

Permeability Coefficient. As defined by Meinzer, the rate of flow in gallons a day through a square foot of the cross section of material, under 100 per cent hydraulic gradient, at a temperature of 60°F. In field terms it is expressed as the number of gallons of water per day at 60°F. that is conducted laterally through each mile of the water-bearing bed under investigation (measured at right angles to the direction of percolation) for each foot of thickness of the bed and for each foot per mile of hydraulic gradient.

Permeability (Intrinsic). A measure of the relative ease with which a porous medium can transmit a liquid under a potential gradient. It is a property of the medium alone and is independent of the nature of the liquid and of the force field causing movement. It is a property of the medium that is dependent upon the shape and size of the pores.

pH. A measure of the acidity of alkalinity of water.

Phenols. Trace phenol concentrations of 0.001 mg./1. can impart an objectionable taste to a water following marginal chlorination. Superchlorination, chlorine dioxide or chlorine-ammonia, ozone, and activated carbon are approaches for treatment.

Phosphate. In many natural waters, phosphate occurs in trace amounts. Raw of treated sewage, agricultural drainage, and some industrial waters contain significant concentration of phosphate. Both ortho and polyphosphates may be found in the same sample. Trace amounts may be combined with organic matter. Phosphate analysis is made primarily to control chemical dosage or to trace flow or contamination.

Phosphates. A group of glassy sodium phosphates which, when mixed with water, have a detergent effect.

Phreatophytes. Plants that habitually send their roots to the capillary fringe and feed on ground water.

Physiography. Physical geography; more specifically.

Piedmont Alluvial Deposit. A group of impinging alluvial cones built up by streams debouching from a mountain range.

Placer. Deposit of gold-bearing alluvial gravel. Also applied to similar deposits containing other metals such as tin, platinum, or tungsten.

Plutonic. Of igneous intrusive origin. Usually applied to sizable bodies of intrusive rock rather than to small dikes or sills.

Pollution. An impairment of water quality by chemicals, heat, or bacteria to a degree that does not necessarily create an actual public health hazard, but that does adversely affect such waters for normal domestic, farm, municipal, or industrial use.

Porosity. The property of a rock or soil containing interstices or voids and may for fluid transmission. It is expressed as a percentage of the total volume occupied by the interconnecting interstices. Although effective percentage. With respect to the movement of water only the system of interconnected interstices is significant.

Porosity (Effective). The amount of interconnected pore space available for fluid transmission. It is expressed as a percentage of the total volume occupied by the interconnecting interstices. Although effective porosity has been used to mean about the same thing as specific yield, such use is discouraged. It may be noted that the present definition of effective porosity differs from that of Meinzer (1923).

Potassium. Usually constitutes a major source of natural radioactivity in water supplies, due to the naturally occurring isotope K^{40}.

Potentiometric Surface. Replaces the term "piezometric surface," is a surface which represents the static head. As related to an aquifer, it is defined by the levels to which water will rise in tightly cased wells. Where the head varies appreciably with depth in the aquifer, a potentiometric surface is meaningful only if it describes the static head along a particular specified surface or stratum in that aquifer. More than one potentiometric surface is then required to describe the distribution of head. The water table is a particular potentiometric surface.

Pozzolans. Silicious materials which, in a powdered form and in the presence of water, will react chemically with lime at ordinary temperatures to form compounds that have cement-like properties.

Pressure (Static). The pressure exerted by the fluid. It is the mean normal compressive stress on the surface of a small sphere around a given point.

Pressure Relief, Cone of. An imaginary surface indicating pressure-relief conditions in a confined aquifer due to pumping.

Pressure Surface. The surface to which confined water will rise in non-pumping wells which pierce a common conduit and the water levels of which are not affected by a pumping well. It is a graphic representation of the pressure exerted by confined water on the conduit walls.

Pressure-surface Map. A map showing the contours (isopiestic lines) of the pressure surface of a confined-water system.

Primary. Of rock minerals: those originally present in the rock, not introduced or formed by alteration of metamorphism. Of ore: not enriched or oxidized by supergene processes.

Primary Openings. Those interstices that were made contemporaneously with rock formation as a result of the processes which formed the rocks that contain them.

Pumping Depression, Cone of. See Water-table Depression and Pressure Relief, Cone of.

Pyrite. FeS_2. Isometric. Commonly cubes and pyritohedrons; less commonly octahedrons. G. *5.02.* H. *6-6½.* Brassy yellow with metallic lustre.

Quartz. SiO_2. Rhombohedral. G. *2.65.* H. *7.* Colorless to white but some varieties smoky, reddish, brown or amethystine.

Raw Acid. Acid without an inhibitor.

Regolith. The mantel of loose soils, sediments, broken rock, etc. overlaying the solid rock of the earth.

Rejuvenated Water. Water returned to the terrestrial water supply by geologic processes of compaction and metamorphism. It is divided into *water of compaction* and *metamorphic water.*

Residue. The material left in the container after evaporating a sample of water and drying in an oven at a definite temperature.

Retarded Acid. Acid with an additive which causes the acid to react more slowly with limestone or dolomite formations.

Retarded Cement. A cement with a retarded initial set.

Rock Drillibility. The inverse of the resistance of a rock to be drilled. This resistance is related not only to the material composition of the rock but to its environment.

Rolling-Cutter Bits (Cone Bits). Provide a wide range of capability for drilling harder formations. May be classified according to the number of cones, the method of fluid introduction, the cutting structure, and the bearing type.

Rotary Drilling Optimization. Minimization of well cost per foot.

Roughneck. A member of a drilling crew who works on the derrick floor.

Round-Trip. Withdrawing all of the drilling shaft and bit from the bottom of the hole and returning it thereto.

Sand Content. The percentage bulk volume of sand in a drilling fluid.

Saturation, Water of. The total water that can be absorbed by water-bearing materials without dilation of the sediments.

Saturation, Zone of. The zone below the water table in which all interstices are filled with ground water.

Secondary, of Rock Minerals. Minerals introduced into the rock or formed by metamorphism or alteration. *Of ore:* Enriched by supergene processes.

Secondary Openings. Openings in rocks formed by processes affecting the rocks after they were formed.

Seconds. A.P.I. The Marsh funnel viscosity.

Sediment. Deposit of solid material (or material in transportation which may be deposited) made from any medium on the earth's surface, or in its outer crust under conditions of temperature approximating those normal to the surface.

Sedimentary Rock. Rock which originated as a sediment. The sediment may have been transported by wind, water, or ice and carried in the form of solid particles (sand, gravel, clay) or in solution (rock salt, gypsum, some calcareous sediments). Sedimentary rocks (unless still unconsolidated) have been indurated by cementation or by recrystallization.

Sedimentation. The process of deposition of sediments. In a broader sense, Sedimentation includes that portion of the metamorphic cycle from the destruction of the parent rock . . . to the consolidation of the products . . . into another rock.

Seepage. Seepage is used in this text in two distinct and different meanings: (1) The accepted usage up to the present time; *i.e.*, the appearance and disappearance of water at the ground surface. (2) As proposed in this text seepage (verb, to seep) designates the *type of movement* of water *in unsaturated material.* It is to be distinguished from percolation, which is the predominant type of movement of water in saturated material.

Selenium. A toxic element sometimes present in soil.

Shear Strength. A measure of the shear or gel properties of the drilling fluid.

Silica. In natural waters, silica is present in the soluble and colloidal forms. An abundance of silica may be found in volcanic waters. Generally silica is higher in waters of low hardness and high alkalinity. Calcium and magnesium silicate scales may be formed in recirculating cooling systems. Dense scales are formed in boilers. Silicon dioxide SiO_2. Common forms: quartz, chert, jasper, chalcedony.

Skin Effect. The reduction in permeability from damage of the bore face caused by drilling mud invasion of a producing stratum adjacent to a well or by improper completion methods.

Slip Velocity. How fast particles will fall in the annulus under varying conditions of mud flow rate and fluid properties.

Slow Setting Cement. A portland cement that hardens at a slower rate than normal.

Soda Ash. (Washing Soda, Sodium Carbonate). Used to remove hardness from water by precipitating calcium and magnesium salts; to raise the pH of acidic waters, and to treat anhydrite-contaminated mud.

Sodium. Present in most natural waters and in fairly high concentrations in waters softened by the sodium exchange process. A high sodium to total cations ratio can be detrimental to soil permeability in agriculture.

Soft Formation Rock Bits. Have widely set teeth and cut deeply so that large cuttings may be dislodged at a fast rate, prevents tracking and promotes cleaning surfaces.

Soil. The layer or mantle of mixed mineral and organic material penetrated by roots. It includes the surface soil (horizon A), the subsoil (horizon *B*), and the substratum (horizon *C*) which is the basal horizon and is limited in depth by root penetration.

Soil Moisture. Pellicular water of the soil zone. It is divided by the soil scientist into available and unavailable moisture. *Available moisture* is water easily abstracted by root action and is limited by field capacity and the wilting coefficient. *Unavailable moisture* is water held so firmly by adhesion or other forces that it cannot usually be absorbed by plants rapidly enough to produce growth. It is commonly limited by the wilting coefficient.

Soil Water. Water in the soil.

Solids, Total, Dissolved and Suspended. Suspended solids are those which are not in true solution and can be removed by filtration. They may be imparted from small particles of insoluble matter, from turbulent action of water on soil, or from domestic and industrial wastes. Dissolved solids are in true solution and cannot be removed by filtration. Their origin lies in the solvent action of the water in contact with earth minerals. Total solids represent the sum of dissolved and suspended solids.

Specific Absorption. The capacity of water-bearing material to absorb water after all gravity water has been removed. It is the ratio of the volume of water absorbed to the volume of material saturated.

Specific Capacity. The rate of discharge of water from the well divided by the drawdown of water level within the well. It varies slowly with duration of discharge which should be stated when known. If the specific capacity is constant except for the time variation, it is roughly proportional to the transmissivity of the aquifer. The relation between discharge and drawdown is affected by the construction of the well, its development, the character of the screen or casing perforation, and the velocity and length of flow up the casing. If the well losses are significant, the ratio between discharge and drawdown decreases with increasing discharge; it is generally possible to roughly separate the effects of the aquifer from those of the well by step drawdown tests. In aquifers with large tubular openings the ratio between discharge and drawdown may also decrease with increasing discharge because of a departure from laminar flow near the well, or in other words, a departure from Darcy's law.

Specific Conductance. The electrical conductivity of a water sample at 25°C (77°F), expressed in micro-ohms per centimeter.

Specific Gravity. The ratio of the mass of a body to the mass of an equal volume of water at 4 deg. C., or other specified temperatures.

Specific Heat. The ratio of amount of heat required to raise a unit weight of a material one degree to amount of heat required to raise the same unit weight of water one degree.

Specific Retention. Is the ratio of (1) the volume of water which the rock or soil, after being saturated, will retain against the pull of gravity to (2) the volume of the rock or soil.

Specific Yield. Is the ratio of (1) the volume of water which the rock or soil, after being saturated, will yield by gravity to (2) the volume of the rock of soil. The definition implies that gravity drainage is complete. In the natural environment, specific yield is generally observed as the change that occurs in the amount of water in storage per unit area of unconfined aquifer as the result of a unit change in head. Such a change in storage is produced by the draining or filling of pore space and is therefore dependent upon particle size, rate of change of the water table, time, and other variables. Hence, specific yield is only an appoximate measure of the relation between storage and head in unconfined aquifers. It is equal to porosity minus specific retention.

Spent Acid. The solution after the treating acid has reacted with the encrustation or formation and has no further dissolving power though it may retain some acidity.

Spring. Concentrated discharge of ground water issuing at the surface as a current of flowing water.

Squeeze Cementing. The use of high pump pressure for forcing a cement slurry into place.

Stabilized Acid. Acid with an additive to prevent reprecipitation of dissolved iron and aluminum compounds from the spent acid.

Standing Level. The water level in a nonpumping well. The term is used without regard to whether the well is within or outside the area of influence of pumping wells. If outside the area of influence, the term is equivalent to static level; if within the area of influence, the standing level registers one point on the cone of pumping depression.

Starting Pressure. The distance in feet from the static water level to the bottom of the air line or foot piece x 0.434 = p.s.i.

Static Level. The water level in a nonpumping well outside the area of influence of any pumping well. This level registers one point on the water table in a water-table well or one point on the pressure surface in a confined-water well.

Steel, Alloy. Iron which has been combined with small portions of carbon and small amounts of other metals.

Storage (Specific). In problems of three-dimensional transient flow in a compressible ground-water body, it is necessary to consider the amount of water released from or taken into storage per unit volume of the porous medium. The *specific storage*, is the volume of water released from or taken into storage per unit volume of the porous medium per unit change in head.

Storage Coefficient. The *storage coefficient* is the volume of water an aquifer releases from or takes into storage per unit surface area of the aquifer per unit change in head. In a confined water body the water derived from storage with decline in head comes from expansion of the water and compression of the aquifer; similarly, water added to storage with a rise in head is accommodated partly by compression of the water and partly by expansion of the aquifer. In an unconfined water body, the amount of water derived from or added to the aquifer by these processes generally is negligible compared to that involved in gravity drainage or filling or pores; hence, in an unconfined water body the storage coefficient is virtually equal to the *specific yield.*

Stratigraphy. Study of strata of sedimentary rocks, particularly with reference to correlation or determination of age. Adj. stratigraphic.

Stream (Gaining). A *gaining stream*, which replaces the term "effluent stream," is a stream or reach of a stream whose flow is being increased by inflow of ground water.

Stream (Losing). A *losing stream*, which replaces the term "influent stream," is a stream or reach of a stream that is losing water to the ground.

Stress. Stress is the force producing or tending to produce deformation or strain in a body. It is measured by the force applied per unit of area. *For example*, pounds per square inch (psi).

Strike. The bearing of a horizontal line in the plane of a bed, vein, fault, etc.

Stringer. A veinlet or small vein, usually one of a number which collectively make up a stringer lode or stockwork.

Strontium. The naturally occurring strontium is not radioactive but the radioactive strontium 90 tends to accumulate in the bone structure. Most potable supplies contain a little natural strontium and some well waters have as high as 39 mg./1.

Submergence. The length of air pipe submerged below the pumping level divided by the sum of the submerged length and the lift, the quotient being multiplied by 100 to give the result as per cent. Generally, a submergence of 60% or more is desirable.

Subsurface Water. All water below the ground surface.

Sulfate (Sulphate). Relatively abundant in hard waters, sulfate is widely distributed in nature. Its ability to combine with calcium forms calcium sulfate scale. In the presence of copper, ferrous iron can reduce sulfates in the water to sulfides. Chlorides and sulfates can cause severe corrosion under all pH conditions but under neutral, or near neutral conditions in treated water, the two anions can cause severe pitting.

Sulfide. Most commonly, sulfide is found in well waters and is a result of bacterial action on organic matter under anerobic conditions. Hydrogen sulfide imparts a "rotten egg" odor to the water, is toxic, and imparts a corrosive character to the water.

Sulfite. In general, sulfite is not found in natural waters. Its determination usually is made on boiler waters or on waters that have been treated with catalyzed sodium sulfite for corrosion prevention.

Supergene. Generated from above. Refers to the effects (usually oxidation and secondary sulphide enrichment) produced by descending ground water.

Surface Runoff. The runoff of precipitation which flows to stream channels over the surface of the ground.

Surface Tension. The free energy in a liquid surface produced by the unbalanced inward pull exerted by the underlying molecules upon the layer of molecules at the surface.

Surfactants. Concentrate in liquid interfaces, adsorb on the surfaces of solids and, thereby alter interfacial tension and wettability. When combined with detergent and water-wetting properties, increases the rate of penetration and flushing of the cuttings from the bit is enhanced.

Suspended Water. Water in the zone of aeration.

Syngenetic Ore Deposit. Deposit formed by processes similar to those which have formed the enclosing rock and in general simultaneously with it. Examples: a bed of sedimentary iron ore between quartzite and slate; a concordant band of chromite in a stratiform layer of dunite.

Tannin and Lignin. Lignin is a plant constituent often discharged as a waste during manufacture of paper pulp. Tannin may enter the water supply through vegetative degradation or wastes of the tanning industry.

Taste and Odor. Dissolved inorganic salts of iron, zinc, manganese, copper, sodium, and potassium can be detected by taste. Odors usually occur from foreign substances, usually organic.

Tectonic. Pertaining to or resulting from physical forces which have been or still are operative in the earth's crust. Tectonics is the study of such forces and their results and is more or less synonymous with structural geology.

Tensile Strength. That number of units of weight at which an object will fail or part on a dead pull.

Thinner. A substance that reduces the apparent viscosity and gel development of mud without lowering the density. The addition of thinner affects the colloidal clay fraction of mud. In terms of flow properties, the effect is a reduction in the yield point, the attractive forces between clay particles are diminished and the structural rigidity of the system is lowered.

Thixotropy. The property of some gels to repeatedly become liquid on agitation and again gelling at rest.

Tool Pusher. The foreman in charge of a drilling rig.

Torque. Torque is the effectiveness of a force to produce rotation about a center, measured by the product of the force and the perpendicular from the line of action of the force to the center about which the rotation occurs, usually measured at a one-foot radius.

Transit Time (In Sonic Logging). The time required for a sound wave to travel through a specific thickness of rock.

Transmissibility. The hydraulic conductivity multiplied by the thickness of an aquifer.

Transmissivity. The rate at which water of the prevailing kinematic viscosity is transmitted through a unit width of the aquifer under a unit hydraulic gradient. It replaces the term "coefficient of transmissibility" because by convention it is considered a property of the aquifer, which is transmissive, whereas the contained liquid is transmissible. However, though spoken of as a property of the aquifer, it embodies also the saturated thickness of the aquifer and the properties of the contained liquid. It is equal to an integration of the hydraulic conductivities across the saturated part of the aquifer perpendicular to the flow paths.

Transpiration. The discharge of water vapor by plants.

Turbidity. Suspended matter such as clay, silt, finely divided organic matter, plankton, and other microscopic organisms cause turbidity in water. The standard method for determination is the Jackson candle turbidimeter.

Turbulent Flow. The motion of a fluid which always occurs above the higher critical velocity and which may occur down to the lower critical velocity. The particles of the fluid move in sinuous paths.

Uniformity Coefficient. A ratio expressing variation in grain size of granular material. It is usually measured by the sieve aperture that passes 60 per cent of the material divided by the sieve aperture that passes 10 per cent of the material. This ratio was proposed by Hazen as a quantitative expression of degree of assortment of water-bearing sand as an indicator of porosity. The value of the coefficient for complete assortment (one grain size) is unity; for fairly even-grained sand it ranges between 2 and 3; for heterogeneous sand the coefficient may be 30.

Uraninite (Pitchblende). Complex oxide of uranium with small amounts of lead and rare elements. Isometric. G. *9.0-9.7.* H. *5½*. Black, Lustre submetallic to pitch-like. Usually massive or botryoidal.

Vadose Water. Water in excess of pellicular water seeping toward the water table; used as a synonym for gravity water.

Vanadinite. $Pb_5Cl(VO_4)_3$. Hexagonal. G. *6.7-7.1.* H. *3.* 'Ruby-red, brown, yellow. Lustre resinous to adamantine.

Velocity (Average Interstitial). Although the *specific discharge,* has the dimensions of a velocity, it expresses the average volume rate of flow rather than the particle velocity.

Very Hard Formation Bits. Have short teeth, closely spaced and intermeshed to clean the grooves.

Vibratory Explosion. A special form of shooting in which the explosive is divided into relatively small shaped charges and arranged to fire in rapid sequence, thus; providing a vibrating effect on the casing and the formation.

Viscosity, Coefficient of. The amount of force necessary to maintain a unit difference in velocity between two layers of water a unit distance apart.

Volcanic Water. Juvenile water from lava flows and volcanic centers.

Volt. The electrical pressure which, when steadily applied to a conductor whose resistance is one ohm, will cause a current of one ampere to flow. (Voltage is that which causes a flow of current comparable to pressure in a fluid being pumped.)

Wall Hook. A simple tool that can be made from steel casing, shaped with a cutting torch. A reducing sub connects the top end of the tool to the drill stem. Used to straighten the lost drill pipe in the hole in preparation for removal by the tap or overshot tools.

Water-Cement Ratio. The amount of mixing water in gallons used per sack of cement.

Water of Compaction. Water supplied by pore destruction from sediment compaction.

Water of Dilation. Water in excess of saturation held by expanded sedimentary material.

Water of Saturation. The total water that can be absorbed by water-bearing materials without dilation.

Water Spreading. Retention of water behind dams or in basins, maintenance of flow in ditches or stream channels, or feeding water down wells and shafts in order to develop influent seepage.

Water Table. That surface in an unconfined water body at which the pressure is atmospheric. It is defined by the levels at which water stands in wells that penetrate the water body just far enough to hold standing water. In wells which penetrate to greater depths, the water level will stand above or below the water table if an upward or downward component of ground water flow exists.

Water-table Depression, Cone of. A cone of depression in the water table developed around a pumping well, the periphery of which (ground water divide) delimits the ground water moving toward the well.

Water-table Map. A contour map of the upper surface of the saturated zone.

Watt. A watt is the power or work capacity of an electrical current of one ampere flowing under an electromotive force (electrical pressure of one volt). Watts equal volts multiplied by amperes, and are the units of measure of electrical energy consumed.

Well Efficiency. The actual specific capacity adjusted for well loss, divided by the theoretical specific capacity.

Wetting Agent. An additive which reduces surface tension.

Wild Well. A well flowing while out of control.

Wilting Coefficient. The ratio of the weight of water in the soil when the leaves of plants first undergo permanent reduction in their water content as the result of deficiency in the supply of soil moisture to the weight of the soil when dry.

Work Capacity. That limit of energy expended or absorbed within which a body is not unduly fatigued.

Yield Point. A measure of the forces between the particles while the mud is in motion. These forces are electrostatic in nature and involve both attraction and repulsion.

Zinc. Deterioration of galvanized iron and dezincification of brass are usually responsible for zinc in water supplies. Industrial-waste pollution may also be a cause. The zinc salts cause an unpleasant, astringent taste and an opalescence in alkaline waters.

Zone (Saturated). That part of the water-bearing material in which all voids, large and small, are ideally filled with water under pressure greater than atmospheric. The saturated zone may depart from the ideal in some respects.A rising water table may cause entrapment of air in the upper part of the zone of saturation, and the lower part may include accumulations of other natural fluids. The saturated zone has been called the phreatic zone by some.

Zone (Unsaturated). The *unsaturated zone*, which replaces the terms "zone of aeration" and "vadose zone," is the zone between the land surface and the water table. It includes the *capillary fringe.* Characteristically this zone contains liquid water under less than atmospheric pressure, and water vapor and air or other gases generally at atmospheric pressure. In parts of the zone, interstices, particularly the small ones, may be temporarily or permanently filled with water. Perched water bodies may exist within the unsaturated zone.

Zone of Aeration. The zone above the water table in which the interstices are partly filled with air.

Zone of Saturation. The zone below the water table in which all interstices are filled with ground water.

Zoning. Arrangement of minerals or mineral assemblages in zones. Also, subsurface zones of differing qualities of ground water.

19
Subject Index

The index includes key words and topical terms from all preceding chapters including the Glossary and Appendix but excluding the Annotated Bibliography. Only terms which have been covered in some detail have been indexed, although uncommon terms have been included for convenience in locating specific topic coverage.

Aa, 627
Abandonment (well), 22, 374, 415
Absorption, 627
Acetic acid (*see* Acidizing)
Acidizing, 627
  acetic acid, 368, 627
  acid-evaluation logging, 207
  acidity, 627
  advantages, 367
  calculations, 623 (81T)
  hydrochloric acid, 351, 368
  hydroxyacetic acid, 351
  incrustation control, 351
  limestone aquifer, 370
  matrix, 367
  mud acid, 367
  pressure acidizing, 369-370
  sandstone aquifers, 368
  sulfuric acid, 335
  surfactants, 62, 366
  (*See also* Stimulation, well)
Acoustic log, (*see* Cementing)
Acrylic polymers, 59
Additive, 628
Adsorption, 628
Aerated mud, (*see* Air-rotary drilling)

Aeration, zone of, 628
Aerobic, 628
Agglomerate, 628
Aggregate, 628
Agricultural chemicals, 20
Air, 628
Air compressors, 127
Air lift:
  capacity determination, 126
  pumps, 597 (43T)
  testing, 235
Air line, 628
Air-percussion rotary drilling, 132-136 (see also Air-rotary drilling)
Air-rotary drilling, 121-132, 217, 246
  aerated mud, 121
  annular velocity, 121
  button bits, 133
  downhole hammer, 132, 146
  downhole source of power, 145
  drilling rate, 121
  efficiency, 125-132
  foam drilling, 123
  gas drilling, 131
  problems, 125
  slug flow, 123
  stiff foam, 121, 123
Air velocity determination chart, 134
Air volume:
  determination, 126
  requirements, (*see* Air-rotary drilling)
Alignment (well), (*see* Plumbness)
Alkalinity, 628
Alluvial cone, 628
Alteration, 628
Aluminum, 628
American Association of Oil Well Drilling Contractors (AAODC), 71, 77
American Petroleum Institute, (*see* A.P.I.)
American Standards Association, 268
Amortize, 628
Ampere, 628
Anaerobic, 628
Angle of repose, 628
Angstrom unit, 628
Anhydrite, 199
Anion, 629
Annular space, 629
Annular velocity, 90, 121, 134, 629 (*See also* Air Rotary drilling, Hole cleaning)
Annulus air pressure, 125
Anticlinorium, 629
Antifoam, 629
A.P.I., 42, 288
  casing (*see* Casing)
  drilling fluid specifications (*see* Drilling fluids)
Apparent specific gravity, 629
Apparent viscosity, 65
Applied Science and Technology Index, 5
Aquiclude, 629
Aquifer characteristics, 173-176, 211, 612 (66T), 629
  bedrock, 16
  carbonate, 190
  closed circuit television (*see* CCTV)
  cores (*see* Coring)
  dolomite and limestone, 14
  fluid characteristics, 176
  fluid samples, 211-216
  formation evaluation, 165-222
  grain size determination, 195
  igneous and metamorphic rocks, 174
  lithology, 173
  photographic evaluation, (*see* Photographic evaluation)

pressure samples, 211-212
resolution, 174
saline aquifer 22
sampling, 211-222
specific conductance, 187
specific yield, 612 (66T)
transmissibility, 174, 226, 312,
unconfined, 178
unconsolidated, 15
water quality (*see* Water quality)
Aquifuge, 629
Areas:
of influence, 629
of pumping depression, 629
plane figures, 580 (9T)
surface/linear foot of pipe, 590 (32T)
Arkose, 199, 629
Arroyo, 629
Arsenic, 629
Artesian, 629
aquifers, 45
wells, 204, 227, 629
improper abandonment, 375-376
Asbestos, 57
Ascending velocity, (mud) 588 (26T)
Attapulgite, 54, 56, 630
Attitude, 630
Auger drilling, 139
Australia, 170, 245, 249
Average velocity, 630

B

Bacteria, 630
aerobacter aerogens, 344
aerobic, 344, 628
anaerobic, 628
bactericides, 346
coliform, 306
iron, 307, 344
leptothrix, 344
nitrate reducing, 345
proteus vulgaris, 344
pseudomonas, 343
slime-forming, 347
sulfate-reducing, 17, 307, 343
well sterilization, 303-308
Bacterial corrosion-incrustation, 345 (*see* Bacteria, Corrosion)
Bail-down method, 231, 234
Bailers, 46
dart valve, 46, 216
sand pump, 46, 216
Baking soda, 62, 630
Balling up, 135
Bank storage, 630
Barite, 61, 630
Barium sulfate, 630
Barrel, 630
Barytes, 630
Basalt, 630
Basaltic aquifers, 169
Base exchange, 630
Batholith, 630
Baume' gravity, 630
Bedding, 630
Bedrock wells, (*see* Consolidated Formations)
Beneficiate, 630
Bentonite, 290, 631
cementing, 290
drilling fluids, 53, 54
southern, 55
western, 55
Bi-metallic corrosion:
definition of, 335
(*see* Corrosion)
Bingham model, 64
Bits:
costs, 102
dressing, 45
efficiency, 37
types, 79-85
walking, 148

wear, 77
Blasting, (*see* Stimulation, well)
Blower, 127
Blow-out, 264, 631
Booting, 90
Borehole geophysics, 166-173, 613 (66T)
acoustic log, 204, 207-209
caliper log, 204, 357
cement bond log, 204, 207-209
computer analysis of logs, 209
electric logging, 166, 217
log interpretations, 170
neutron life-time log, 209
new techniques, 209
nuclear magnetic log, 209
spectral log, 209
well abandonment, 375
(*see* Aquifer characteristics)
(*see* Formation evaluation)
Borehole plug, 375
Borehole stability, 48, 110-112
Boron, 631
Bottom-hole:
pressure, 90
pneumatic tools, 134
Bradenhead, 631
Breccia, 631
Bridge, 631
Brine wells, 42
Bromide, 631
Btu, 631
Bull reel, 43
Button bit, (*see* Air rotary drilling)

C

Cable tool drilling system, 144, 231, 272
Cake thickness, 631
Calcite, 631
Calcium, 631
Calcium carbonate, (*see* Incrustation)
cable tool rig, 222
capacity, 45
casing, (*see* Casing)
costs, 47
drilling rate, 45, 47
early history of, 42
field use, 46
fuel consumption, 46
operations, 42
samples, 46
test drilling, 48
Calcium hypochlorite, (*see* Sterilization)
Caliper log, (*see* Borehole geophysics)
Capillary, 631
fringe, 631
head, 631
interstice, 631
Carbon dioxide, 632
corrosion, 331
incrustation, 347
Carbon steel, 278
Carbonates, 124
Casing (well), 24, 43, 200, 203, 206, 236, 265-281, 272, 410
API casing, 18, 266, 269, 590 (31T)
capacity of, 592 (36T)
carbon steel, 278
coatings, 341, 342
collapse pressure, 235, 271
connector, 287
corrosion of, 191, 197
cupro-nickel alloys, 278
double-walled, 272
drive pipe, 267, 588 (27T)
fishing, 281
installing of, 265
large diameter, 234
leaks, 201
life of, 269-270
line pipe, 266

local requirements, 270
longevity, 265
non-metallic, 191, 198, 279-281, 341
non-upset API casing, 269
oil string, 24
plastic casing, 279
PVC, 279
ripper, 288
reamed & drifted pipe, 266
shoes, 204, 630
single walled, 276
specifications, 583 (16T)
stainless steel, 279
standard pipe, 265, 603 (55T)
state requirements, 270
strength, 271
structural data, 57, 58
surface, 24
thick-walled, 271
transite pipe, 279
water well casing, 267
well abandonment, 374
yield strength, 276
Cast iron, 632
Catalyst, 632
Cathodic protection, 632
operation, 340
Cation, 632
Caustic soda, 57, 61, 137, 632
Cavern flow, 632
Cellulosic polymers, 57
Cementing, 201, 288-302, 632
casing, 295
casing method, 296
cement-bonding log, 204, 207-209
centering guides, 294
continuous method, 297
diatomaceous earth, 115, 290
driven casing, 300
gilsonite, 290
grouting, 289
inside-casing method, 296
local requirements, 299
materials, 622 (77T)
penetration, 292
perlite, 290
permanent-surface casing, 300
plug, 632
plunger-type receptacle method, 298
portland cement, 290, 294
pozzolans, 290
practices, 299
seals, 206, 299-301
slurry, 632
surface-apron, 301-302
techniques, 288, 292
water-cement ratio, 291
Centering guides, (*see* Cementing)
Centipoise, 632
Channeling, 632
Chemical Corrosion (*see* Corrosion)
Chert, 632
Chilled-shot core drilling, 138
Chlorine (*see* Sterilization)
effects on bacteria, 305
633
Chlorite, 54
Chloride, 632
content, 186
corrosion, 332
Chlorination, 633
Chlor-Tabs, 305
Chromelignosulfonates, 60
Chromium, 633
Chrysotile asbestos, 57
Circulating-slip overshot, 283
Circulation, 633
air, 132
rate, 261
Clay, 54, 633
minerals, 633
Cleavage, 633

Climatic:
cycle, 633
year, 633
Closed-circuit television:
formation evaluation, 329
summary of, 413
well evaluation, 329, 342
CMC (*see* Sodium carboxymethylcellulose)
Coagulation, 633
Coal, 199, 200
Coefficient of:
storage, 633
transmissibility, 633
viscosity, 633
Cohesion, 634
Coliform bacteria (*see* Bacteria)
Collar locator, 201
Collector well:
advantages of, 254
capital cost, 254
disadvantages of, 255
Colloid, 634
Color, 634
Colorado School of Mines, 5
Compaction, of water, 634
Complex phosphates, 634 (*see* Phosphates)
Compressor capacity, 133
Conductivity:
effective hydraulic, 634
electrical, 204
hydraulic, 634
log, 204-206
Cone:
bits, 80-86
pressure relief, 634
water-table, 634
Confined ground water, 634
Confining bed, 634
Conglomerate, 220
Connate water, 634
Consolidated formations, 216, 271, 300
aquifers, 239
dolomite wells, 242, 380
limestone wells, 242, 380
sandstone wells, 242, 380
well design, 242-251
Construction costs (*see* Cost analysis)
Contact, 635
Contamination, 2-10, 11-28, 374-375, 635
aquifers, 11
bacteria, 349
cement-grout seal, 19
chemical, 11
definition of, 12
drilling muds, 17, 346
gravel pack, 20
mining industry, 374
natural gas, 24
oil, 23
oil-field brines, 25
potential for, 12, 18
pump casing, 19
summary of ground water, 416
surface-derived contaminants, 15, 301
thermal, 11
urban areas, 16
well pit, 302
Contractors, European, 153
Conversion tables
ft. (head) to P.S.I., 591 (34T)
GPM - Gal./day, etc., 583 (15T)
hardness, 585 (21T)
metric, 619 (74T)
mud weight, 586 (22T)
P.S.I. to ft. (head), 591 (33T)
units in water analysis, 581 (12T)
volume to weight, 579 (7T)
Copper, 635

Cores, (*see* Aquifer characteristics) 218-222
- bits, 167
- blade-type heads, 220
- core barrels, 218
- core catcher, 218
- core recovery, 221
- diamond heads, 219, 220
- drag-type heads, 220
- drive coring, 222
- operation, 136
- recovery, 220
- sampler, 218
- wire-line, 167

Corrosion, 9, 22, 203
- acetic acid, 368
- bacterial, 343-346
- carbon dioxide, 331
- cathodic protection, 330, 340
- causes, 337
- chlorides, 332
- electrokinetic theory, 350
- field occurrence, 333-337
- galvanic, 335
- hydrogen sulphide, 24, 337
- oil well, 339
- Ph, 331, 333-335
- potential of, 331
- prevention of, 337-343
- protective coatings, 342
- rate of, 335
- saturation index, 332
- selective, 335
- stability index, 333, 334
- summary of, 412-413
- temperature, 332
- types of, 331

Corrosometer probes, 275, 342
Cosmic water, 635
Cost analysis, 377-400
- augered wells, 379
- blasting (shooting), 395
- corrosion (*see* Corrosion)
- dolomite wells, 380, 381, 395
- domestic water supply, 387
- domestic wells, 378-385
- gravel-packed wells, 386
- gravel-packing, 392
- igneous and metamorphic rock wells, 396-400
- industrial-municipal wells, 377, 385, 395
- iron removal, 386
- large capacity wells, 378
- limestone wells, 380, 395
- materials cost analysis, 392
- pump systems, 392-394, 397, 398, 599 (48T), 600 (49T, 601 (50T), 615 (69T), 616 (70T)
- sand and gravel wells, 380, 393
- sandstone wells, 380, 381, 395, 396
- sedimentary rock wells, 377-396
- softening, 385
- water treatment, 385
- well efficiency (*see* Well efficiency)

Crater, 635
Critical velocity, 635
Cross-over method, 263
Cross-roller rock bits, 83
Cupro-nickel alloys, 278
Cuttings removal, 127

D

Damage ratio, 635
- definition of, 214, 312
- drill-stem testing, 313

Data collection, 2
Definitions, 7-10
Deflocculation, 635
Density, 635
Developing countries, 46
Dezincification, (*see* Corrosion)
Diameter, 635
Diamond bits, 86

Diatomaceous earth (*see* Cement-additives)
Dielectric, 635
Die overshot, 283, 635
Dilation, water of, 635
Dilution, 635
Dip, 635
Directional drilling, 152
Disinfection (well) (*see* Sterilization)
Dispersion, 635
Dissolved solids (total) TDS
  determiniation of, 186
Dolerite, 635
Dolomite, 199, 220, 242, 636
Domestic wells:
  construction costs (*see* Cost analysis)
  well design, igneous and metamorphic rocks, 245-252
  well design, sedimentary rocks, 242-244
  (*see* Consolidated formations)
  (*see* Unconsolidated formations)
Downhole hammer: (*see* Air-rotary drilling)
Drag bit, 636
Drake oil well, 42
Drawdown, 311, 636
Drift, 636
  glacial, 636
Drillability (*see* Drilling rate)
Drill bits, 124
Drill collars, 119
Drillers' logs, 50, 169
Drilling breaks, 32
Drilling cost (*see* Drilling Optimization)
Drilling economics, 403-404
  future developments, 406
Drilling equation, 30,37,99
Drilling fluids, 51-68, 214-216
  attapulgite, 54
  bentonite, 51
  classification of, 52
  effects in drill stem testing, 212
  filtration properties, 66-67
  flow properties, 63-66
  formation damage, 312
  functions, 51
  gas-base muds, 52
  minor additives, 61-62
  mud additives, 53-61
  mud density, 63
  mud performance, 62
  mud programs, 52-53
  mud properities, 54
  native clay, 56-57
  oil-base muds, 52
  routine testing program, 68
  sand content, 67-68
  stuck pipe, 116-118
  surface system, 68
  water source, 53
Drilling, history of water well, 29
Drilling model, 37
Drilling mud (*see* Drilling fluids)
Drilling optimization, 100-104, 138, 155, 163
  bearing wear rate, 103
  bit costs, 102
  penetration rate, 103
  reduction efforts, 156
  rotary, 156
  tooth wear rate, 103
Drilling projects, 168
Drilling rate, 28, 31, 33, 147, 158, 201
  core drilling, 33
    (*See also* Air-rotary drilling: Cable tool drilling system; Rotary drilling system: Air percussion rotary drilling system; Turbine system)

equation, 35, 99
factors affecting drilling rate, 30, 69
drilling fluid properties, 93-99
formation characteristics, 70-71
hydraulic factors, 86-93
intangible factors, 100
mechanical factors, 71-86
summary of, 403
Drilling research, 7, 41, 157-162
Drilling systems (future):
downhole motor, 155
dyna-drill, 154
electric disintegration, 159
high frequency electric, 159, 161
induction, 159
nuclear, 159
terra jetter, 159
Drilling systems (others), 30
auger drilling, 139
chemical, 158
chilled-shot core drilling, 138
conventional percussive, 158
drilling fluid system, 7
electric arc, 158
electrohydraulic, 158
electron beam, 158
forced flame, 158
hydraulic-percussion drilling, 132
jet, 140
laser, 158
pellet, 158
plasma, 158
rotary shot, 138
spark, 158
spark-percussive, 158
summary of, 405
ultrasonic, 158
Drilling technology:
rotary-drilling, 39
summary of, 402
Drilling variables, 7
Drill pipe, 133, 281
structural data, 586 (23T)
Drill pipe recovery
circulating slip overshot, 283
die overshot, 283
tapered fishing tap, 283
twist off, 283
Drill-stem tests, 211
Drill-stem testing, 163
Driven casing, 300
Dust and chip catchers, 131

E

Earth-auger drilling, 139
Economic minerals, 9
Effective:
grain size, 636
porosity, 636
size (of grain), 636
velocity, 636
Efficiency (*see* Well efficiency)
Effluent seepage, 636
Elastic limit, 637
Electric logging (*see* Borehole geophysics)
Electrochemical potential, 189
Electrodrill, 153
Electrolyte, 637
Engineering Index, 5
Entrance velocity, 227
Epigenetic ore deposit, 637
Epithermal ore deposit, 637
Epoxy plastic pipe, 281
Equalization, 637
Equipment manufacturers, 27
Evaporation pits, 25
Exploratory drilling, 127
Explosive drilling system, 158
Explosives:
properties of gelatin and semi-gelatin dynamites, 605 (59T)

properties of straight and ammonia dynamites, 604 (57T)
(*See also* Stimulation, well)
Explosive gases, 309
(*See also* Well hazards)

F

Factor of safety, 637
Fatigue, 637
Faults (*see* Well design igneous and metamorphic rock), 637
Feedlots, 12
Felsite, 638
Felspathic sands, 199
Ferric hydroxide (*see* Incrustation)
Ferromagnesians, 638
Field capacity, 638
Filter cake, 56, 117, 214
texture of, 638
Filter loss, 638
Filtration rate, 638
Fishing, 281, 638
casing connector, 287
casing ripper, 288
circulating-slip overshot, 283
die overshot, 283
impression block, 283
job, 638
pipe and casing, 286
rotary drilling, 118
tapered fishing tap, 283
wall hook, 284
Flash set, 638
Flat gel, 638
Float, 638
Flocculation, 638
Flotation, 639
Flounder point, 72
Flow:
steady, 639
uniform, 639
Fluid characteristics (*see* Aquifer characteristics), 176
borehole geophysics (*see* Borehole geophysics)
water quality, 179
Fluid potential, 639
Fluid velocity, 261
log, 204-206
Fluoride, 639
Fluorite, 639
Flux, 639
Foam drilling (*see* Air-rotary drilling)
Foliation, 639
Foot-pound, 639
Formation, 639
damage, 176
logging, 46
identification, 165
resistivity, 181, 183, 185
Formation evaluations, 165-222 (*see* Aquifer characteristics)
acoustic log, 207
aquifer characteristics, 173
caliper log, 204
closed curcuit television (*see* CCTV)
conductivity log, 204
drill-stem testing, 163, 211-216
fluid velocity log, 204
gamma log, 198
novel logs, 209
photographic evaluation (*see* Photographic evaluation)
resistivity log, 191
SP log, 174, 179, 188
summary of, 407-410
temperature log, 206
Formation samples (*see* Sampling)
Fountain head, 639

Fracture, 639
  evaluation logs, 207
  system, 639
France, 143
Free-water content, 640
Friction loss, 579 (8T), 584 (18T)
Frost-heaving pressures, 302
Fuel efficiency, 640
Funnel viscosity, 64

G

Galvanic corrosion (*see* Corrosion)
Gamma log (*see* Borehole geophysics)
Gamma-gamma logs, 178
Gangue, 640
Gas drilling (*see* Air-rotary drilling)
Gas wells,
  brines, 23
  hydrogen sulphide, 24
Gel, 640
Gel cement, 56
Gel development, 64, 640
Gel-forming clay (*see* Bentonite)
Gelling, 55
Gel mud, 56
Gel strength, 64, 66, 640
Geologic marker, 198
Geologists:
  government, 10
  ground water, 9
  ground water planners, 10
  hydrologists, 9
  mineral exploration, 10
  oil, 10
  students:
    engineering, 10
    geology, 10
Geomorphology, 640
Geophysical logs, 176
Geophysical probes, 375
Geophysics (borehole) (*see* Borehole Geophysics)
Geosyncline, 640
Germany, 143
Gilsonite, 640 (*see* Cementing)
Glacial deposits, 15
Goethite, 640
Gold, 640
Gossan, 640
Gouge, 640
Graded bedding, 640
Grain size:
  determination of, 195
Granite, 174, 249
Graphite, 641
Graphitization, (*see* Dezincification)
Gravel packing, 138, 200, 225, 275
  construction costs, 394
  cross-over method, 263
  gravel pack ratio, 259
  operations, 257-265
  sterilization of, 304
Gravity water, 641
Graywacke, 641
Grit, 641
Ground water:
  cascade, 641
  contamination (*see* Contamination)
  decrement, 641
  divide, 641
  equation, 641
  hydrology, 641
  increment, 641
  inventory, 642
  mound, 642
  province, 642
  quality determination, 16
  ridge, 642

runoff, 642
sampling, 215
trench, 642
turbulent flow, 642
Grouting (*see* Cementing)
Guar gum, 57
Gypsum, 199, 642

H

Halite, 642
Hammer drills:
operational requirements, 606 (60T)
Hard formation rock bits, 83, 642
Hardness, 642
Hazards (*see* Well hazards)
Head, 643
Head of water converted to pressure, 643
Heat of hydration, 643
Hematite, 643
Hole capacity, 597 (44T), 593 (37T)
Hole cleaning, 105-110
annular velocity, 105
slip velocity, 106
velocity profile, 108-109
yield point, 106
Hole deviation, 32, 118-119, 124
well plumbness, 302-303
Holes, large diameter, 138
Hole volumes, 204
Hollow rod drilling, 139
Horsepower, 643
water, 323
HTH tablets, 305
Hydraulic:
conductivity, 174
diffusivity, 643
gradient, 643
percussion drilling, 138
Hydraulic fracturing:
definition of, 370
energy requirements, 370
permeability damage, 370
sand fracing, 370
secondary oil recovery, 371
waste disposal, 372
(*See also* Stimulation)
Hydraulics (well), 223, 235
Hydrochloric acid (*see* Acidizing)
Hydrofluoric acid (*see* Acidizing)
Hydrogen Sulphide (*see* Corrosion)
Hydrograph, 644
Hydrologic:
cycle, 644
properties, 644
Hydrology, 644
Hydrophilic, 644
colloids, 57
Hydrophobic, 644
Hydrometer, 644
Hydrothermal, 644
Hydroxyacetic acid (*see* Acidizing)
Hygroscopic moisture, 644
Hypabyssal, 644
Hypogene, 644
Hypothermal ore deposit, 644

I

Igneous and metamorphic rocks, 245-252, 644 (*see* Aquifer characteristics)
Illite, 54
Impression block, 283, 644
Incrustation:
calcium carbonate, 347
causes of, 347
control of, 348-351
definition of, 347
electrokinetic theory, 350

field occurrence, 347
research, 350
saturation index, 332
stability index, 334
summary of, 412
India, 245
Induction resistivity log, 174
Industrial growth, 39
Industrial wells, 257
radial collection well, 252
well design, igneous and metamorphic rocks, 245
well design, sedimentary rocks, 242
Infiltration galleries, 252
Influent seepage, 644
Inhibited acid, 644
Inhibitor, 645
Initial set, 645
Injectivity profiles, 178
Intensified acid, 645
Interstream ground water ridge, 645
Intrusive, 645
Iodine, 307
Iron, 645
bacteria (*see* Bacteria)
Irrigation table, 582 (13T)
Isohyetal map, 645
Isopiestic line, 645
Isotropic, 645

J

Jet drilling, 138, 140
Jetting:
hollow rod, 47
horizontal, 44
(*See also* Stimulation (well))
Jet-type roller cone bits, 84
Joint, 247, 645
Juvenile water, 645

K

Kaolin, 645
Kaolinite, 54
Kilowatt, 645

L

Laboratory drilling tests, 32
Laminar flow, 645
Langelier index, 332
Lateritic, 646
Lead, 646
swedge, 231
Legislation, 419
Lignitic material, 60, 199
Lime, 646
Limestone, 61, 174, 199, 220
wells, 242
Limonite, 646
Liquid injection, 131
Liquid-waste disposal, 11
Lithium, 646
Lithologic character (aquifer), 173
Livestock wastes, 20
Logs:
acid evaluation, 207
acoustic, 204
analysis of, 166, 209
driller's, 50, 169
electric, 217
formation, 46
fracture evaluation, 207
gamma, 170, 198-199
gamma gamma, 178
long normal, 174
radioactivity, 217
resistivity, 179, 191-198
SP, 174, 179
temperature, 206-207
Logging (*see* Borehole geophysics)
Long normal log, 174
Lost circulation, 57, 63, 112-116, 646

materials, 292
Lubricants, 62, 646

M

Macroscopic, Megascopic, 646
Mafic, 646
Magma, 646
Magnesium, 646
Maintenance (well), 351-374
(*see* Stimulation (well))
(*see* Well efficiency)
Make-up water:
characteristics of, 53
treatment of, 53
Manganese, 646
Marsh funnel, 64
Maui well, 251
Medium-to-hard formation rock bits, 83, 646
Mesothermal ore deposit, 646
Metals (*see* Corrosion)
electromotive series, 336
Metamorphic water, 646
Metamorphism, 646
Meteoric water, 647
Meteorology, 647
Methane, 647
Microlog, 176
Millidarcy, 647
Minerals:
deposits, 168
development, 168
exploration, 168
gamma log, 200
logging, 200
mineral exploration, 5, 29, 136, 168, 217, 249
phosphate, 200
radioactive, 175
resistivity log, 200
SP log, 200
stimulation of, 374
uranium, 200
Mining industry, 8, 30, 154
equipment, 127
in situ leaching, 374
open pit mining, 374
solution mining, 374
stimulation in, 374
Montmorillonite (*see* Bentonite)
Mud, 53-62
additives, 53-62
balance, 63
cake, 190
damage, 214
filtrate, 180
invasion, 193
properties, 54, 66
treatment, 54
Mud-scow drilling, 47
Municipal wells:
radial collection well, 252
well design, igneous and metamorphic, 245
well design, sedimentary rocks, 242
well field, 16

N

National Association of Corrosion Engineers, 342
Native clay, 56, 647
Natural gas:
blowout, 25
leaks in well casing, 310
(*see also* Well hazards)
Neutron-epithermal neutron logs, 178
Nitrate-reducing bacteria, 345
Nitrogen, 647
Non-metallic casing (*see* Casing)
corrosion resistance, 341
Novel drilling systems:
chemical, 156
economic potential, 163
electric arc, 155, 158

electro hydraulic, 158
electron beam, 158
erosion, 156
explosive, 158
forced flame, 155, 158
high frequency, 158
implosion, 156, 158
jet-piercing, 156
laser, 158
modified turbine, 158
nuclear, 158
pellet, 155, 158
plasma, 158
spark, 158
spark-percussive, 158
summary of, 405
ultrasonic, 158
NWWA Research Facility, 4, 5,

O

Off-shore drilling, 152
Ohio State University (The), 5
Oil and grease, 647
Oil wells, 23, 155, 156
abandoned, 25
brine disposal, 23
brines, 23, 25, 179
crude oil production, 165
drilling, 151
interpretations, 184
oil-bearing sand, 23
oil field products, 265
production practices, 23
Optimization of operations, 29, 647
Organic polymers, 53, 57, 647
Overall efficiency, 648
Oxygen, 648
Ozone, 648

P

Packers, 215
inflatable, 211, 215
seats, 204
self-sealing, 231
standard, 240-241, 231-233
Pellicular front, 648
Pellicular water, 648
Penetration rate, 133, 134
Pennsylvania State University (The), 5
Peptization, 648
Peptized bentonite, 56
Perched ground water, 648
Percussion method history, 42
Percussion reverse-circulation: drilling, 139
"Perfect cleaning," 36
Perlite, 648 (*see* Cementing)
Permeability:
coefficient of, 174, 646
damage, 237
determination of, 196
field, 312
igneous and metamorphic rocks, 248
studies, 237
Petroleum industry, 8, 136, 143, 154, 165, 235, 280, 298
exploration, 169
logging, 168
Ph, 648 (*see* Corrosion)
Phenols, 648
Phosphates, 44, 60, 199, 200, 649
exploration, 170
Photographic evaluation (wells) (*see* Aquifer characteristics)
formation evaluation, 328
stereo-photographic surveys, 328
well evaluation, 328
(*See also* Closed-circuit television)
Physiography, 649

Piedmont alluvial deposit, 649
Piezometer wells, 235
Pilot hole, 139
Pitless adapters, 302
Pitman arm, 43
Placer, 649
Plastic casing (*see* Casing)
Plastic viscosity, 64
Plumb-bob test, 302
Plumbness (well), 302-303
Plutonic, 649
Pollution (*see* Contamination)
  definition of, 12, 649
Polysaccharides, 57
Porosity, 649
Portland cement (*see* Cementing)
Potassium, 649
Potentiometric surface, 649
Pozzolans, 650 (see Cementing)
Pregelatinized starch, 57
Pressure:
  abnormal, 24
  atmospheric, 617 (72T)
  relief, 650
  surface, 650
  shut-in, 214
Pressure-surface map, 650
Primary openings, 650
Production wells, 244
Productivity index (*see* Specific capacity)
Public health, 12
Pull back method, 231
Pull down method, 134
Pump bowls, 272
Pumping depression, 650
Pumping systems:
  domestic wells, 384
  fuel requirements, 589 (30T)
  industrial-municipal, 392
  performance standards, 598 (46T)
  submersible turbine pumps, 224, 398
  vertical turbine pumps, 224, 397
  windmill, 599 (48T)
Pumps:
  advantages and disadvantages, 595 (40T)
  air lift, 597 (43T)
  application, 223
  capacity, 326, 622 (78T)
  diesel, 325
  efficiency, 314-327, 601 (50T)
  flat head, 315
  irrigation well installations, 321
  rating, 326
  steep head, 315
  suction-lift, 594 (39T)
  turbine, 599 (47T)
Pump tests, 234-237
  municipal and industrial wells, 386
Pyrite, 650

Q

Quartz, 650
Quebracho extract, 59

R

Radial-collector well, 252
  design, 252-256
Radioactive tracer, 178, 210
Radioactivity log, 217
Rate of flow:
  through pipe, 584 (19T)
  discharge measurements, 621 (77T), 604 (56T) 605 (58T), 608 (62T)
  orifice tables, 607 (61T)
Raw acid, 650
Redevelopment (well) (*see* Maintenance)
Regolith, 650

Regulatory agencies, 203, 419
Rehabilitation (well) (*see* Maintenance)
Rejuvenated water, 650
Research and development, 416-420
Residue, 650
Resistivity log, 179, 191-198
  applicators, 197
  grain-size determination, 195
  mineral exploration, 200
  permeability determination 196
Retarded acid, 650
Retarded cement, 650
Reverse-circulation rotary drilling, 136
  sampler, 138
  samples, 218
  static water levels, 137
Rice University, 5
Rig:
  cost, 102
  efficiency, 100
Rock characteristics, 7, 30-32
  consolidated formations, 135
  crystalline, 135, 245
  metamorphic, 135
  unconsolidated formations, 135
Rock compressive strength, 31, 33, 36, 70
  strength anisotropy, 32
Rock drillability, 7, 30, 35, 93, 650
  compressive strength, 30
  definition of, 39
  drilling rate equation, 99
  index, 38
  "perfect cleaning," 72
  summary of, 403
Rock mechanics, 39
Rock shear strength, 70
Rod wiper, 131
Rolling-cutter bits, 80-86, 650
Rotary drilling system, 144, 231, 275-279
  borehole stability (*see* Borehole stability)
  casing (*see* Casing)
  drilling fluid, 51-69
  drilling optimization, 651 (*see* Drilling optimization)
  drilling rate, 69
  economics, 100
  fishing (*see* Fishing)
  history, 50
  hole cleaning (*see* Hole cleaning)
  hole deviation (*see* Hole deviation)
  lost circulation (*see* Lost circulation)
  optimization (*see* Drilling optimization)
  reverse rotary drilling, 257
  samples (*see* Sampling)
  stuck pipe (*see* Stuck pipe)
  summary of, 402-403
Rotary shot drilling, 138
Roughneck, 651
Round-trip, 651
Ryzman index, 333

S

Salt, 199
Salinity content:
  determinization of, 186
Sampling:
  aquifer lithology, 216-222
  cable tool, 216
  cores (*see* Cores)
  damage ratio, 313
  drill-stem testing, 212

drive, 217
fluid, 211
lag time, 50, 217
pressure, 212
pump tests, 234
rotary drilling methods of, 50, 217
sample catcher, 218
side-wall, 222
split-drive sampler, 138
test drive, 48
water, 50

Sand:
analysis, 199
content, 651
flaps, 139

Sand and gravel:
deposits, 13, 239
wells, 242

Sand-pump:
pumping wells, 259
type bailier, 216

Sandstone, 199
deep wells, 242

Sanitary protection, 20

Saturation:
index, 332
water of, 651
zone of, 651

Scale removal:
acetic acid, 368
acidizing (*see* Acidizing)
incrustation, 351

Schedule numbers, 268

Schist, 249

Screens:
acidizing, 361
backwashing method, 44
cage-type, wire wound, 227
collapse, 235
corrosion of, 335
diameter of, 227
incrustation of, 347
inlet velocity, 601 (51T), 602 (53T)
installation, 231-234
jetting, 353
louvered, 227
open area, 225, 618 (73T) 620 (75T)
"pull back" method, 44
selection, 230, 611 (65T)
size, 228
slotted, 227, 230
yield strength, 276

Secondary:
minerals, 651
openings, 651
recovery, 237

Sediment, 651

Sedimentary basins, 168

Sedimentary formations, 242

Sedimentary rock, 651

Sedimentation, 651

Seepage, 651

Selected Water Resources Abstracts, 3

Selective corrosion (*see* Corrosion)
definition of, 335

Selenium, 651

Septic tank, 12

Shale, 54
line, 174
marker, 217

Shallow aquifers, 11, 29, 35
drilling, 41, 162
drilling economics, 39
drilling industry, 39

Shear strength, 651

Shock chlorination, 307 (*See also* Sterilization)

Shooting, 356-361
characteristics, 356
detonation velocity, 358
efficiency, 356-358

electrical, 360
primacord, 357
research, 359-361
sandstone wells, 356
vibratory explosion, 356
(*See also* Stimulation, well)
Short-normal resistivity log, 174
curve, 192
Sieve analysis, 46
mesh sizes, 610 (64T)
Silica, 651
Siltstone, 220
Skin effect, 312, 652
Slime-forming bacteria (*see* Bacteria)
Slip velocity, 652
Slow setting cement, 652
Soda ash, 53, 62, 652
Sodium, 652
Sodium bicarbonate, 62
Sodium carbonate (*see* Soda ash)
Sodium carboxymethylcellulose, 57
Sodium hypochlorite (*see* Sterilization)
Sodium silicate, 137
Soft formation rock bits, 83, 652
Soil, 652
Soil moisture, 652
Soil water, 652
Solids:
dissolved, 186, 652
suspended, 652
total, 652
Sonic logs:
logs, 174
velocity, 37
Soviet Union, 143
Specific:
absorption, 652
capacity, 227, 235, 243, 310, 372, 350 652
conductance, 187, 653
gravity, 653
heat, 653
retention, 653
power, 143
yield, 178, 653
Spectral logging, 209
Specifications (well), 418
Spindletop, 50
Splash zone, 275
Split-drive sampler, 138
SP log, 174, 179
applications, 188-191
formation resistivity, 185
mineral exploration, 200
salinity content, 186, 191
total dissolved solids, 186
water quality, 186
Spring, 653
Spudding beam, 43
Spudding gear, 43
Squeeze cementing, 292, 653
Stability index, 333
Stabilized acid, 653
Standing level, 653
Starting pressure, 653
State-of-the-art reviews, 6
State regulations, 3
Static level, 654
Static water level, 48, 50
Steel, alloy, 654
Sterilization (well), 303-307
calcium hypochlorite, 305
sod
(*See also* Bacteria)
Stiff foam (*see* Air-rotary drilling)
Stimulation (well)
acidizing, 361-370
blasting, 356-361
costs, 367
effectiveness, 372-374

hydraulic fracturing, 370-372
jetting, 353-356
limitations, 372
mining industry, 374
pay zone, 366
pressure acidizing, 369
shooting, 356-361
silicate-controlling agents, 366
summary of, 414
surging, 353, 366
Storage, 654
coefficient, 226, 312, 352
Stratigraphic correlation, 173
Stratigraphy, 654
Stream, 654
Streaming potential:
effects of incrustation, 350
(*See also* SP log)
Stress, 654
domestic, 12
farm, 12
Stringer, 654
Strontium, 654
Stuck pipe, 56, 116-117
caving, 116
drill collars, 117
Submergence, 654
Subsurface water, 655
Suction pits, 587 (24T)
Sulfamic acid (*see* Acidizing)
Sulfate, 655
Sulfate-reducing bacteria (*see* Bacteria)
Sulfide, 655
Sulfuric acid (*see* Acidizing)
Supergene, 655
Surface:
geophysical methods, 249
runoff, 655
tension, 655
Surfactants, 62, 366, 655
Surging, (*see* Stimulation)
Suspended water, 655
Sweden, 245
Swedging, 231
Syngenetic ore deposit, 655

T

Tannin and Lignin, 60, 655
Tannins, 60
Taste and odor, 655
Tattle-tell screen, 263
Technical personnel, 3
Technical supervision, 6
Tectonic, 655
Tensile strength,. 655
TDS:
corrosion, 197
determination of, 186
Temperature log, 206-207
acid-evaluation, 207
fracture evaluation, 207
location of casing gas leaks, 310
water-injection, 207
Test pumping, 48
Theis non-equilibrium formula, 312
Thinners, 59, 655
Thixotropy, 656
Timken-lubricant tester, 62
Tool pusher, 656
Torque, 656
Total dissolved solids (*see* TDS)
Toxic fluids, 17
Transit time, 37, 656
Transite pipe, 279
Transmissibility, 226, 312
definition of, 174, 656
Transmissivity, 656
Transpiration, 656
Treatment (water):
domestic wells, 385
Tremie method, 261, 295

Tubular products, 268
  structural data, 586 (23T)
Tungsten-carbide buttons, 133
Turbidity, 656
Turbine drilling system, 143-153
  compressors, 133
  drilling rate, 142
  economic potential, 152
  field operations, 147-153
  modified, 158
  operating characteristics, 147-151
Twist off, 283
Two-plug system, 297

U

Uncased holes, 197
Unconfined aquifer, 254
Unconsolidated formations, 138, 169, 221, 265, 271
  cuttings, 46
  large diameter wells, 252
  samples, 46, 217
  sands, 192
  well design, 242, 252
Uniformity coefficient, 656
Union of South Africa, 245
Units, ground water and petroleum, 582 (14T)
University of Minnesota, 5
University of Tulsa's *Petroleum Abstracts*, 4
Unsaturated zone of aeration, 177
Up-hole air velocity, 128
Uraninite, 656
Uranium, 125, 198, 199
  commercial, 200
  exploration, 168, 170, 200, 201
  ore bodies, 200
  potential of prospective aquifers, 201
U.S. Atomic Energy Commission, 131, 211
U.S. Bureau of Mines, 32
U.S. Bureau of Reclamation, 225
U.S. Department of Interior's office of Water Resources Research, 3
U.S. Geological Survey, 168
U.S. Environmental Protection Agency, 27, 179
U.S. Public Health Service, 179, 302

V

Vadose water, 656
Vanadinite, 657
Velocity, 657
  profile (*see* Hole cleaning)
Vertical turbine pumps, 223
Very hard formation bits, 83, 657
Viscometer, 64
Viscosity, 54, 63
Vocational schools, 420
Volcanic water, 657
Volt, 657
Volumes:
  of solids, 580 (9T)
  between casing and hole wall, 596 (41T)
  of casing, 594 (38T)
Volumetric requirements, air drilling:
  blast, water, and seismographic drilling, 589 (28T)
  gas and oil well drilling, 589 (29T)

W

Wall hook, 284, 657
Wall packing, 90
Wall thickness, 271, 273, 277
  minimum, 276

Wash-down method, 231
Washing soda (*see* Soda ash)
Waste disposal, 11, 168
  deep, 210
  salt water, 25
  sites, 15
Water-cement ratio, 657
Water flooding, 25, 190
  oil recovery, 166, 311
Water injection, 135, 237
Water loss, 136
Water of compaction, 657
Water of dilation, 657
Water of saturation, 657
Water quality, 197, 205
  problems and solutions, 617 (71T)
  resistivity, 179
  salinity content, 179-180
  total dissolved solids, 186
Water resistivity equivalent, 182
Water Resources Scientific Information Center, 3
Water samples, 12
Water spreading, 657
Water supplies:
  domestic, 12
  farm, 12
Water-table depression, 657
Water-table map, 658
Water-table wells, 139, 657
Weight of water, 584 (17T)
Well blowouts, 23
Well completion/development technology, 7
  summary of, 406
Well design (igneous rocks):
  faults, 247
Well design (novel):
  radial-collector wells, 252
Well design (metamorphic rocks):
  faults, 247
Well design (sedimentary rocks), 239
  dolomite, 242
  industrial, 239
  limestone, 242
  sand and gravel, 242
  sandstone, 242
Well development, 235
Well drilling market, 4
Well drilling technology, 7
Well efficiency, 225, 310-314, 658
  closed circuit television (*see* Closed circuit television)
  definition, 310
  hydraulics, 224
  oil wells, 311
  photographic evaluation (*see* Photographic evaluation)
  simple test, 313
  stimulation (*see* Stimulation, (well)
  summary of, 411
Well evaluation (*see* Photographs) (*see* Closed circuit television)
Well loss, 225-226, 313
Well maintenance technology, 7, 309-376
Well performance, 9
Well rehabilitation, 9
Well sealing, 44
Well spacing, 243-244
Well sterilization (*see* Sterilization)
  (*See also* Bacteria)
Well yield, 235
  safe yield, 206, 243
  stimulation (*see* Stimulation (well)
  (*see* Specific yield)
  (*see* Well design, igneous and

metamorphic rocks)
Workovers, 309, 341

Y

Yield point, 64, 66, 658
(*See also* Hole cleaning)
Yield test, 54
Young's modulus, 37
Yo-Yo method, 42

Z

Zone, 658
Zone of aeration, 658
Zone of saturation, 658
Zone of weathering, 248
Zoning, 658